MARINE INVERTEBRATES
OF THE PACIFIC NORTHWEST

Marine Invertebrates
of the Pacific Northwest

EUGENE N. KOZLOFF

with the collaboration of
Linda H. Price
and contributions by other specialists

UNIVERSITY OF WASHINGTON PRESS
Seattle and London

Copyright © 1987, 1996 by the University of Washington Press
First paperback edition, with additions and corrections, 1996
Printed in the United States of America

All rights reserved. No part of this publication may be reproduced or transmitted in any form or by any means, electronic or mechanical, including photocopy, recording, or any information storage or retrieval system, without permission in writing from the publisher.

Marine Invertebrates of the Pacific Northwest is a revised and greatly expanded version of *Keys to the Marine Invertebrates of Puget Sound, the San Juan Archipelago, and Adjacent Regions* (copyright © 1974 by the University of Washington Press).

Library of Congress Cataloging-in-Publication Data
Kozloff, Eugene N.
    Marine invertebrates of the Pacific Northwest / Eugene N. Kozloff ; with the collaboration of Linda H. Price, and contributions by other specialists. — 1st pbk. ed., with additions and corrections.
        p.   cm.
    Rev. and expanded ed. of: Keys to the marine invertebrates of Puget Sound, the San Juan Archipelago, and adjacent regions. [1974].
    ISBN 0-295-97562-8 (alk. paper)
    1. Marine invertebrates—Northwest, Pacific—Identification.
2. Marine invertebrates—Northwest, Pacific—Classification.
I. Price, Linda H. II. Kozloff, Eugene N. Keys to the marine invertebrates of Puget Sound, the San Juan Archipelago, and adjacent regions. III. Title.
QL365.4.N67K69   1996   96–18885
592.09632—dc20   CIP

The paper used in this publication meets the minimum requirements of American National Standard for Information Sciences—Permanence of Paper for Printed Library Materials, ANSI Z39.48-1984. ∞

CONTENTS

Preface to the 1996 Printing vii

Preface to the 1987 Printing ix

1. **Introduction** 3
2. **Phylum Porifera** 6
3. **Phylum Cnidaria** 32
4. **Phyla Ctenophora, Orthonectida, Dicyemida** 79
5. **Phyla Platyhelminthes, Gnathostomulida** 84
6. **Phylum Nemertea** 94
7. **Phyla Nematoda, Gastrotricha, Rotifera, Kinorhyncha, Priapulida** 102
8. **Phylum Annelida:** Class Polychaeta 109
9. **Phylum Annelida:** Classes Oligochaeta, Hirudinea 170
10. **Phyla Echiura, Sipuncula, Pogonophora, Vestimentifera** 180
11. **Phylum Mollusca:** Classes Aplacophora, Polyplacophora 184
12. **Phylum Mollusca:** Class Gastropoda 193
13. **Phylum Mollusca:** Class Bivalvia 259
14. **Phylum Mollusca:** Classes Scaphopoda, Cephalopoda 289
15. **Phylum Tardigrada and Phylum Arthropoda: Subphyla Chelicerata, Uniramia** 296
16. **Phylum Arthropoda: Subphylum Crustacea:** Classes Branchiopoda, Copepoda, Ostracoda, Cirripedia 303
17. **Phylum Arthropoda: Subphylum Crustacea:** Class Malacostraca: Subclasses Phyllocarida, Peracarida (in part) 320
18. **Phylum Arthropoda: Subphylum Crustacea:** Class Malacostraca: Subclass Peracarida: Order Amphipoda 346
19. **Phylum Arthropoda: Subphylum Crustacea:** Class Malacostraca: Subclass Eucarida 392
20. **Phyla Phoronida, Brachiopoda, Entoprocta** 418

21. Phylum Bryozoa (Ectoprocta)   423

22. Phylum Echinodermata   447

23. Phyla Urochordata, Hemichordata, Chaetognatha   467

Glossary   480

Additions and Corrections   487

Index   513

PREFACE TO THE 1996 PRINTING

Since 1987, when this book was first published, knowledge of the invertebrate fauna of the Pacific Northwest has increased substantially. New species have been described, and the presence of some not previously known to occur in the region has been confirmed. The scrutiny to which certain groups have been subjected has led to changes in names and to reinterpretation of the identity of numerous species. The activity of systematists has been particularly intense in connection with crustaceans of the Order Amphipoda and annelids of the Class Polychaeta, but there have been important revisions in other groups, too.

It is probably safe to say that most of the additions and corrections for this printing concern species that are not likely to be encountered except by specialists or by persons involved in very detailed faunal surveys. Many of the changes are just shifts of species from one genus to another. It is nevertheless advisable, during work with a particular group, to consult pertinent portions of the supplementary pages.

Megan Dethier, Ronald Shimek, David Egloff, Gretchen and Charles Lambert, Megumi Strathmann, George Shinn, Bill Austin, Jeff Goddard, Janice Voltzow, Steve Hulsman, Les Watling, James Carlton, and the late Chris Reed are among those who discovered errors and ambiguities, added species, called my attention to names that had been superseded, or explained their views on the systematics of certain taxa. Kerstin Wasson, Greg Jensen, and Claudia Mills provided new keys or portions of keys for several groups of invertebrates. A compilation, by José Orensanz, of recently described, renamed, and reclassified polychaete annelids was of great help to me in dealing with this taxonomically volatile assemblage. Kathy Carr and Dianne Wilkinson cheerfully saw to it that library materials I needed reached me as quickly as possible. Richard Strathmann, Bruno Pernet, Rachel Collin, Alan Kohn, and Will Jaeckle solved various little problems for me. To all of these associates, and to many others who helped in one way or another, I offer my sincere thanks.

Finally, it gives me special pleasure to acknowledge the generous collegiality of Craig Staude. He not only pointed out shortcomings on various pages of the 1987 printing and pulled together the long list of changes in the taxonomy of amphipod crustaceans, but also offered his skill with computer programs to enable me to produce the supplementary pages and enlarged index. Craig's contribution to this new printing is at least as important as my own.

Eugene N. Kozloff
Friday Harbor, Washington

PREFACE TO THE 1987 PRINTING

The origins of this book can be traced back to early efforts to catalogue the marine invertebrates of the San Juan Archipelago, Puget Sound, and adjacent areas. Much of this work was conducted at Friday Harbor Laboratories of the University of Washington, where it led to the development of a set of keys that were valuable aids in the instructional and research programs. In 1974, expanded and less tentative versions of these materials were published in book form. Having been made accessible to all who needed them, the keys were widely used in Washington and British Columbia, and to some extent in Oregon. The many errors, omissions, ambiguities, and other shortcomings that were called to my attention have been helpful to those concerned with the preparation of this revised and enlarged edition.

Contributors to the present volume are acknowledged in connection with the taxa for which they constructed keys and compiled species lists and bibliographies. I thank them all at once, without mentioning them individually. It must be said, however, that nearly all of these colleagues helped in ways that were unrelated to their primary assignments.

Three of my close associates played especially important roles in the development of this book. Paul Illg, under whose direction many of the original keys were prepared, allowed me to study his annotated copies. He also let me use his personal library and freely gave advice based on his extensive knowledge of invertebrates of our region. The late Robert Fernald, during his tenure as Director of Friday Harbor Laboratories, encouraged systematic studies by specialists and supported the formation of a reference collection. Linda Price carefully studied all parts of the manuscript for the new edition and patiently tested some of the more complex keys. Her efforts led to many improvements in content and format.

Much of the earlier work on the fauna was promoted by Dixy Lee Ray and Emery Swan. Their contributions, and those of their students and colleagues, are acknowledged with thanks. Carl Nyblade, in his capacity as supervisor of baseline studies for the National Oceanic and Atmospheric Administration and for the Department of Ecology of the State of Washington, was able to provide much valuable information. The section on gammaridean amphipods is based to a considerable extent on collections processed by his laboratory, and in constructing keys to anomuran crustaceans I relied primarily on one that he had written.

With respect to gammaridean amphipods, Edward Bousfield served as an advisor over a period of several years, and Kathleen Conlan, Norma Jarrett, John Dickinson, and Jerry Barnard also gave much assistance. Gayle Heron and Z. Kabata kindly improved the section on copepods. The keys to isopods are based primarily on materials assembled by Jarl-Ove Strömberg and Robert George. Others who helped with the development of keys, species lists, and bibliographies for crustacean groups are Charlotte Holmquist, Krispi Staude, Anamaria Escofet, Richard Albright, Richard Brusca, Jeffery Cordell, Kenneth Chew, Kevin Li, Placidus Reischman, Tom Suchanek, and David Wethey.

In preparing the section on polychaete annelids, I relied primarily on the two comprehensive treatises published by Karl Banse and the late Katharine Hobson. It is a pleasure to acknowledge their valuable contributions. José Orensanz studied my keys to polychaetes and suggested many improvements. Sally Woodin, Fred Nichols, and Colin Hermans also helped with this group.

Duane Hope supplied the bibliography for nematodes, and William Hummon read the section on gastrotrichs. Mary Arai, Anita Brinckmann-Voss, Deanna Lickey, Catherine McFadden, Ann Bucklin, Peter Corbin, Ronald Larson, and Wim Verwoort helped in connection with cnidarians, and information on ctenophores was given by Wulf Greve and Larry Madin. The authors of the section on sponges are indebted to Myriam Preker, Verena Tunnicliffe, Gerald Bakus, and Neil McDaniel, and also to various members of the staff at Bamfield Marine Station. The key to nemerteans is based to a large extent on work at Friday Harbor by Fumio Iwata, but helpful suggestions were received from Pamela Roe. Mary Rice gave advice on sipunculans.

The section on entoprocts benefited from comments by Claus Nielsen. Charles Galt, Todd Newberry, the late Donald Abbott, Richard Snyder, and Craig Young provided assistance with respect to urochordates. My compilation of echinoderms was carefully scrutinized and improved by Philip

Lambert. Scott Smiley, Arthur Fontaine, Scott McEuen, Norman Engstrom, and Lane Cameron contributed some additional information concerning this group.

For studies on molluscs and other invertebrates at the California Academy of Sciences, which I visited several times and where I took many of the photographic illustrations for this book, I received much assistance from Welton Lee, Peter Rodda, Dustin Chivers, Barry Roth, Terrence Gosliner, and Robert Van Syoc. At the Los Angeles County Museum of Natural History, James McLean generously shared information he had assembled on Pacific coast gastropods, and he and Gale Sphon made it possible for me to study and photograph specimens efficiently. David Lindberg, at the Museum of Paleontology, University of California, reviewed my treatment of limpets and provided copies of his unpublished papers on the taxonomy of this group. Ellie Dorsey, Diarmaid O'Foighil, Bob Lemon, Harry Fritchman, Alan Kohn, and the late Antonio Ferreira also helped with molluscs.

The key to siphonophores is an enlarged version of one written by George Mackie for the 1974 edition. His frequent presence at Friday Harbor and his support of systematic work on cnidarians at the University of Victoria have made him a valuable ally in our efforts to produce a better book.

In preparing the 1974 edition, I was given much help by Ralph Smith and James Carlton, editors of *Light's Manual,* which covers invertebrates of the coast of central California. They permitted me to study the manuscript and thus to become informed about many nomenclatorial changes and sytematic studies that I would otherwise have missed. In constructing keys to marine invertebrates of the Pacific Northwest, I have continued to used *Light's Manual* as a basic reference.

Much of the work on this edition was accomplished during a professional leave from the University of Washington. This support from the University is gratefully acknowledged. I express my appreciation to Dennis Willows, the present Director of Friday Harbor Laboratories, to Richard Strathmann, Associate Director, and to Kathryn Hahn, Joan Short, and Estelle Johnson, members of the office staff, for their aid. Kathy Carr, Tom Moritz, and other members of the library staff promptly supplied me with references I always seemed to need in a hurry. Megumi Strathmann urged me to reconsider some of my taxonomic decisions and she was often conveniently close when I encountered problems with the word processor I was using. Claudia Mills and Craig Staude also knew how to soothe the machine, and Rikk Kvitek more than once rescued texts that I thought had been chewed up and digested. Carol Noyes prepared numerous line drawings, and her cheerful efforts to make the illustrations accurate and attractive are warmly appreciated. My wife, Anne, was a careful proofreader, and my daughter, Rae, typed substantial portions of early drafts of the manuscript.

The list of persons to whom I and other contributors to this book are indebted could go on indefinitely. I thank everyone whose suggestions have been of assistance. Having tried so hard to eliminate errors and ambiguities, I do not look forward to discovering the ones I missed. But please tell me about them. I will appreciate all corrections and queries that are sent to me.

Eugene N. Kozloff
Friday Harbor, Washington

SOURCES OF ILLUSTRATIONS

The photographs and some of the line drawings are original or were published previously by authors who contributed to this volume. Many illustrations, however, have been taken without modification from other works. These sources are gratefully acknowledged.

Porifera: G. J. Bakus, in Journal of Zoology; M. W. de Laubenfels, in Proceedings of the United States National Museum; V. M. Koltun, in Opredeliteli po Faune SSSR

Cnidaria: C. McL. Fraser, Hydroids of the Pacific Coast of Canada and the United States; D. V. Naumov, in Opredeliteli po Faune SSSR; P. L. Kramp, in Dana Report; G. O. Mackie, in Proceedings of the Royal Society, London, B; W. Verwoort, in Fauna van Nederland; H. Boschma, in Annales de l'Institut Oceanographique

Platyhelminthes: L. H. Hyman, in Bulletin of the American Museum of Natural History

Kinorhyncha: R. P. Higgins, in Smithsonian Contributions to Zoology

Annelida: K. Banse, in Journal of the Fisheries Research Board of Canada and other serial publications; E. Berkeley & C. Berkeley, in Canadian Pacific Fauna; P. V. Ushakov, in Opredeliteli po Faune SSSR and in Fauna SSSR; M. H. Pettibone, Some Scale-bearing Polychaetes of Puget Sound and Adjacent Waters, and in Bulletin of the United States National Museum and other serial publications; O. Hartman, in Allan Hancock Foundation Pacific Expeditions and other serial publications; G. G. Martin, in Transactions of the American Microscopical Society

Mollusca: S. S. Berry, in Bulletin of the United States Bureau of Fisheries; W. H. Dall, in Bulletin of the United States National Museum; A. J. Ferreira, in The Veliger; T. Habe, in Publications of the Akkeshi Marine Biological Station; Z. A. Filatova & V. I. Zatsepin, in Opredelitel' po Faune i Flore Severnykh Morei SSSR (editor: N. S. Gaevskaia)

Arthropoda: H. V. M. Hall, in University of California Publications in Zoology; N. Ii, in Japanese Journal of Zoology; H. Richardson, in Bulletin of the United States National Museum; R. J. Menzies, in Wasmann Journal of Biology, Proceedings of the United States National Museum, and other serial publications; B. E. Stafford, in First Annual Report of the Laguna Marine Laboratory; A. L. Alderman, in University of California Publications in Zoology; E. L. Bousfield, Shallow-water Gammaridean Amphipoda of New England and in Bulletin of the Biological Society of Washington; J. L. Barnard, in Pacific Naturalist and other serial publications; D. E. Hurley, in Occasional Papers of the Allan Hancock Foundation; J. C. McCain, in Bulletin of the United States National Museum; D. R. Laubitz, in National Museum of Canada, Publications in Biological Oceanography; E. L. Mills, in National Museum of Canada, Natural History Papers; W. L. Schmitt, in University of California Publications in Zoology; J. F. L. Hart, in Proceedings of the United States National Museum

Entoprocta: C. Nielsen, in Ophelia

Bryozoa: R. C. Osburn, and R. C. Osburn & J. D. Soule, in Allan Hancock Foundation Pacific Expeditions

Echinodermata: A. M. D'iakonov, in Opredeliteli po Faune SSSR

Urochordata: W. G. Van Name, in Bulletin of the American Museum of Natural History

MARINE INVERTEBRATES
OF THE PACIFIC NORTHWEST

1
INTRODUCTION

This manual has been prepared with the hope that it will be both convenient and comprehensive. Whenever possible, the keys for identifying marine invertebrates are based on characters that are readily apparent. The organization of the material under each phylum will enable the user to associate each species with the class, order, and family to which it belongs.

Taxonomic Groups Covered

Nearly all invertebrate groups other than protozoans, flukes, tapeworms, acanthocephalans, and parasitic nematodes are dealt with. The treatment of a few taxa, however, is minimal, because the identification of species requires considerable expertise. It seems best to refer biologists concerned with groups such as insects, copepod crustaceans, and free-living nematodes to specialized works.

Geographic Region Covered

The keys, taxonomic lists, and bibliographies are relevant to invertebrates of intertidal and shallow subtidal habitats between southern Oregon and the Queen Charlotte Islands of British Columbia. For some taxa, the manual will be useful beyond the limits of the region for which it has been written. Species from subtidal habitats deeper than about 300 meters are generally excluded from the keys, although they may be entered in the taxonomic lists.

Illustrations and Glossary

It is assumed that anyone using this manual will have access to a textbook of zoology or invertebrate zoology. The characteristics of phyla, classes, and other higher taxa are therefore omitted. Although the terminology used in most keys is comparatively simple, the user will encounter terms that are not likely to be defined in texts or standard dictionaries. These terms are explained in the Glossary or illustrated in connection with the keys to which they apply.

Bibliographies

The references provided in connection with the keys and taxonomic lists for specific groups of invertebrates are, as a rule, restricted to monographs and other publications that are concerned with invertebrates of our region. Works that have been superseded or whose content has been assimilated into recent studies may be omitted, unless they are of special value because of illustrations, locality records, or other attributes. The bibliographic treatment accorded one group may be different from that given to another. Much depends on the extent to which the groups represented in our fauna have been studied. In some cases, references that deal with other geographic areas are cited because they may be helpful in recognition of genera or species that may eventually be found here, even if their presence has not yet been recorded.

A Few General References

The following books provide broad coverage of the marine invertebrates of certain portions of the Pacific coast of North America.

Austin, W. C. 1985. *An Annotated Checklist of Marine Invertebrates of the Cold Temperate Northeast Pacific.* Cowichan Bay, British Columbia: Khoyatan Marine Laboratory. xiv + 682 (in 3 vols.).

This lists nearly all marine invertebrates known from our region, as well as from California, northern British Columbia, and Alaska. It records the general habitat (littoral, shallow subtidal, and deep subtidal) and the geographic range of each species. It does not contain descriptions, keys, or illustrations, but it does give synonyms for many species names.

Kozloff, E. N. 1983. *Seashore Life of the Northern Pacific Coast. An Illustrated Guide to Northern California, Oregon, Washington, and British Columbia.* Seattle & London: University of Washington Press; Vancouver: Douglas & McIntyre. vi + 370 pp.

Seashore Life is an illustrated, nontechnical account of the common invertebrates, fishes, seaweeds, and other plants that are associated with various marine habitats, including floating docks, in the Pacific Northwest. It enables one to identify, at least to genus (and usually to species), about four hundred invertebrates.

Morris, R. H., D. P. Abbott, & E. C. Haderlie. 1980. *Intertidal Invertebrates of California.* Stanford, California: Stanford University Press. xi + 690 pp.

This book illustrates a large number of selected California invertebrates, many of which occur in our region, and provides brief accounts of their biology. Some major taxa are covered more thoroughly than others. The bibliographic citations pertaining to each species are comprehensive.

Ricketts, E. F., J. Calvin, & J. W. Hedgpeth. 1985. *Between Pacific Tides.* 5th edition, revised by D. W. Phillips. Stanford, California: Stanford University Press. xxvi + 652 pp.

Between Pacific Tides deals in considerable detail with the natural history and ecology of invertebrates of central California, and to some extent with those of areas farther south and farther north. It has a substantial bibliography.

Smith, R. I., & J. T. Carlton (editors). 1975. *Light's Manual: Intertidal Invertebrates of the Central California Coast.* Berkeley & Los Angeles: University of California Press. xvii + 717 pp.

Although *Light's Manual* is devoted primarily to the marine invertebrates of central California, its bibliographies include most of the important contributions dealing with systematics of invertebrates of the west coast of the United States and Canada. The manual also cites many papers concerned with ecology, behavior, physiology, and other aspects of the biology of Pacific coast invertebrates, and is a good guide to encyclopedic works concerned with the morphology and classification of various groups.

Arrangement of Material Pertaining to Major Taxa

As a rule, the treatment of each major taxon includes a bibliography, key to species, and taxonomic list. There are, however, exceptions. Some groups are so poorly known that keys to the few described species will almost certainly lead to incorrect identification of undescribed species. In such cases, only a bibliography and taxonomic list are provided. The authors of specific sections are indicated under the heading for the appropriate taxonomic category. When authorship is not given, the text has been prepared by the editor.

Nearly all of the keys are strictly dichotomous. Begin with the first couplet--1a and 1b--and decide which choice is the better one. Go on to the couplet to which this choice leads and continue the process until you arrive at the determination of the species. Couplets are usually organized in such a way that each character in choice a is compared against a contrasting character in choice b. Frequently, one or more noncontrasting supplementary features are added to one or both choices. When this is done, the extra material, which may help to confirm that a particular choice is the correct one, is enclosed by parentheses. In most keys, the order of the species has nothing to do with presumed affinities. In the taxonomic list that follows, however, the species are arranged according to families and higher taxa, and the authors and dates of publication of the species names are given.

In certain large assemblages, including the polychaete and oligochaete annelids, prosobranch gastropod molluscs, and bivalve molluscs, there are keys to families, then secondary keys that lead to species. Since the species are already grouped in families, the authors and dates are given in the key. In a few groups, other arrangements are dictated by practical concerns.

Species omitted from the keys but included in the taxonomic lists are indicated by an asterisk. Most of these invertebrates are not likely to be encountered within the range for which the manual has been written, or are restricted to deep water.

Nomenclature and Taxonomic Conventions

In general, the nomenclature used in this manual is that believed to be best, even if it differs from that proposed in some recent revisions. When the author and date of publication of a species name are in parentheses, this indicates that the species was placed in a different genus when it was originally described. In some groups, especially larger taxa, subgeneric names are given in addition to generic names. Following the accepted practice, the subgeneric name is in parentheses between the genus and species. Synonyms are rarely given, except in the case of species that have long been known under a different name, and in the case of names that have only recently been changed.

2

PHYLUM PORIFERA

William C. Austin and Bruce Ott

Within the region covered by this book, there are at least 200 species of sponges. Many of them have not been identified, and perhaps most of these are undescribed. The keys provided here cover 130 species, primarily from intertidal and shallow subtidal habitats. The taxonomic list includes additional species, the majority of which are from deep water.

The most useful general references on sponges of the Northwest are the chapter by Hartman in *Light's Manual* (1975) and the annotated checklist of Austin (1985) (see Introduction). The former, although concerned with species found on the central California coast, includes many that occur in Oregon, Washington, and British Columbia. Austin has evaluated the specific and generic determinations in earlier publications and has listed unidentified and undescribed species. Other papers, some of which may be helpful only to specialists who must consult original descriptions, are listed under the class of sponges to which they pertain.

Most of our species have spicules. These are calcareous or siliceous structures that give sponges firmness and often give them a feltlike surface. Spicules are lacking, however, in a few species. Some of these have a fibrous skeleton that consists of a protein called spongin. On gross examination, such sponges might be confused with compound ascidians or certain bryozoans (such as species of *Alcyonidium*).

The keys are designed to facilitate at least tentative identifications from macroscopic characters. Color is a useful key character for many living sponges and has been freely employed in the keys on the assumption that users will be working on material that is fresh. Species that are variable in form or color may occur more than once in the key. Accurate identification generally requires microscopic study, including analysis of spicule types and their arrangement. Crude microscopic preparations can be made by placing small tissue fragments from the surface and interior of a sponge on a slide, adding a drop of laundry bleach (sodium hypochlorite) and, after a few minutes, a coverslip. Avoid spillage!--bleach will damage lenses and may remove enamel from the microscope. An ocular micrometer will be needed for measuring spicules. The spicule size ranges in the keys are those measured for some populations in our region. The full range given for a particular species should not be expected in every specimen; at the same time, however, one should be aware of the possibility that either or both extremes in the range may be exceeded in some specimens.

A word of caution: the long, glassy spicules of a few sponges may penetrate human flesh and lead to effects similar to those caused by fiberglass. Some persons develop dermatitis after prolonged handling of the subtidal hexactinellid "cloud sponges" and their relatives (*Aphrocallistes* and *Chonelasma*).

Key to Classes

1a   Spicules calcareous (dissolved by hydrochloric acid) and belonging to 3 or 4 main types (fig.2.1) (species occurring in our region are usually white, occasionally light tan, and generally less than 10 cm in size)   Class Calcarea

1b   Spicules (these are absent in the demosponge genera *Aplysilla*, *Hexadella*, *Dysidea*, and *Halisarca*) siliceous (not dissolved by hydrochloric acid)   2

2a   Basic spicule type 6-rayed, but often modified by reduction or loss of certain rays; in our species, the dominant large spicules are either fused together to form a firm lattice or cover the surface as a feltlike mat; typically white or yellow, unless covered by silt, and larger than 10 cm; strictly subtidal, at depths of more than 15 cm   Class Hexactinellida

2b Spicules (unless absent, as in exceptions noted in choice 1b), not basically 6-rayed; spicules not fused to form a lattice, although in some cases they are held firmly together by fibers of spongin; some species white or dull tan, but others colorful; includes most of the intertidal and subtidal species in our region   Class Demospongiae

Class Calcarea

Our calcareous sponges are poorly known. About half of the 23 species reported for the region have been identified with complete or nearly complete certainty. Most of the rest can be assigned to genera, and some may prove to belong to species described from Europe and Asia. The literature dealing directly with calcareous sponges of the Pacific coast is scanty and difficult to work with. The early papers of Lambe (1894, 1900) and the works of Urban (1902, 1905), de Laubenfels (1932), and Johnson (1978) include original or secondary descriptions of several of our species. The monograph on Japanese Calcarea by Hozawa (1929), an update by Tanita (1943b), and keys to known species of *Leucosolenia* and *Clathrina* by Tanita (1943a), may also be useful. Burton (1963) includes a summary description for most species of Calcarea described up to the early 1950's. His many synonymies, however, are not followed here.

Burton, M. 1963. A Revision of the Classification of Calcareous Sponges; with a Catalogue of the Specimens in the British Museum (Natural History). London: British Museum (Natural History). v + 693 pp.
de Laubenfels, M. W. 1932. The marine and fresh-water sponges of California. Proc. U. S. Nat. Mus., 81(4):1-140.
Hozawa, S. 1929. Studies on the calcareous sponges of Japan. J. Fac. Sci. Imp. Univ. Tokyo, Sect. 4, Zoology, 1:277-389.
Johnson, M. F. 1978. A comparative study of the external form and skeleton of the calcareous sponges *Clathrina coriacea* and *Clathrina blanca* from Santa Catalina Island, California. Can. J. Zool., 56:1669-77.
Lambe, L. M. 1894. On some sponges from the Pacific coast of Canada and Behring Sea. Trans. Royal Soc. Canada, 11, Sect. IV, 10(1893):25-43.

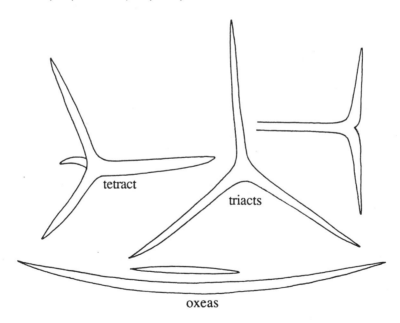

2.1 *Leucosolenia eleanor*

------. 1900. Description of a new species of calcareous sponge from Vancouver Island, B. C. Ottawa Naturalist, 13:261-3.
Sim, C.-J., & G. J. Bakus. 1986. Marine sponges of Santa Catalina Island, California. Occ. Pap. Allan Hancock Found., n. s., 5:1-23.
Tanita, S. 1943a. Key to all the described species of the genus *Leucosolenia* and their distribution. Sci. Rep., Tohoku Imp. Univ., Ser. IV, 17:71-93.
------. 1943b. Studies on the Calcarea of Japan. Sci. Rep., Tohoku Imp. Univ., Ser. IV, 17:353-490.
Urban, F. 1902. *Rhabdodermella nuttingi*, nov. gen. et nov. spec. Zeitschr. wiss. Zool., 71:268-75.
------. 1905. Kalifornische Kalkschwamme. Arch. Naturgesch., 72:33-76.

1a	Sponge tubular, saclike, vase-shaped, or nearly globular	2
1b	Sponge an encrusting or globular mass of branching tubes (a hand lens may be needed to see the tubular construction)	11
2a	Body forming a flattened sac, often wrinkled (yellow-tan)	3
2b	Body forming a sac or tube that is circular in cross-section	4
3a	Height 2-3 cm, width 1-2 cm; with a very short fringe of oxeas around the osculum; oxeas 150-500 µm; triact rays 40-150 µm; short tetract ray 50-60 µm, long tetract rays 60-90 µm *Grantia ?compressa*	
3b	Height typically 6-16 cm, width 2-5 cm; without an evident fringe of oxeas around the osculum; oxeas 400-500 µm; triact rays 120-350 µm; short tetract ray 40 µm, long tetract rays 150 µm (the only calcareous sponge of our region that has a fleshy texture) *Leucandra* similar to *levis*	
4a	With a stalk that is at least one-third of the total height (body up to 5 mm in diameter; with a short fringe of oxeas around the osculum; off-white; oxeas around osculum to 1250 µm; oxeas in wall 50-150 µm; triact rays 200-900 µm; short tetract ray 60 µm, long tetract rays 120-450 µm) *Leucilla nuttingi*	
4b	Stalk short or absent	5
5a	Globular, pear-shaped, or rarely tubular, with a thick wall, thus only slightly compressible (with long oxeas around the osculum)	6
5b	Globular or tubular, with a thin wall, readily compressible (diameter less than 2 cm)	7
6a	Appearance hispid due to many oxeas projecting from the surface; white to dirty white; size to 9 cm high by 11 cm wide; oxeas around osculum 7-15 mm; small oxeas elsewhere 50-140 µm, large oxeas elsewhere 3-10 mm; triact rays 30-210 µm; no tetracts (these are reported in California specimens of the otherwise similar *Leucandra apicalis*) *Leucandra heathi*	
	Large *Leucandra taylori* (choice 7a) may approach *L. heathi* in firmness. The microoxeas lie tangentially over the surface in *L. taylori*, but form upright, palisadelike strands in *L. heathi*.	
6b	Few if any oxeas projecting from the surface; off-white; size to 3.5 cm high by 2 cm wide; oxeas around osculum 3-7 mm: oxeas elsewhere 120 µm; triact rays 200-630 µm; short tetract ray 82 µm, long tetract rays 230 µm *Leucandra pyriformis*	
7a	Appearance hispid, due to many oxeas projecting from the body, which is round or cylindrical; with oxeas around the osculum; white to dirty white; with a leuconoid structure in section (without evident canals extending from the inner surface to near the outer surface) (size to 4.5 cm high by 2 cm wide; oxeas around osculum 1-2.5 mm; oxeas elsewhere 130-200 µm and 600-1100 µm; triact rays 45-210 µm; short tetract ray [may be difficult to detect] 25 µm, long tetract rays 45-210 µm) *Leucandra taylori*	
7b	Appearance hispid or smooth; oxeas present or absent around osculum; white to yellow-tan; with a syconoid or modified syconoid structure in section (with evident canals extending from the inner cavity to near the outer surface)	8
8a	Large (to 4-15 cm high by 1-2 cm wide); inner surface lined by bundles of oxeas (oxeas of body wall 600-1000 µm, those lining the inner surface about 350 µm; triact rays 90-190 µm; short tetract ray 25-30 µm, long tetract rays 70-120 µm) *Sycandra* close to *utriculus*	
8b	Small (1.5 cm high by less than 6 mm wide); without a lining of oxeas on the inner surface	9
9a	Surface of tube with a pattern of nearly square pores when viewed with a hand lens or dissecting microscope (0.7-5 cm high by 4-14 mm wide; oxeas present or absent around the osculum; oxeas 260-1500 µm; triact rays 60-340 µm; short tetract ray 40-50 µm, long tetract rays 150-250 µm; intertidal and on floats) *?Tenthrenodes* sp.	
	Includes some specimens referred to *Scypha* in the literature.	

9b Surface of tube with a pattern of circular pores 10
10a Surface with an obvious plush of spicules *Scypha* spp.
10b Surface usually macroscopically smooth (microscopically papillate) (height 1-3 cm or more; without oxeas around the osculum; oxeas 65 µm; triact rays 65-220 µm; long tetract ray projecting into the central cavity 280-540 µm, other rays 110 µm)
Scypha compacta
11a Some of tubes of the colony free for at least a portion of their length, and with terminal openings; oxeas present; triacts with unequal rays 12
11b None of tubes free, and openings, when visible, are along the sides of the tubes; without oxeas; triacts with equal or nearly equal rays (encrusting or stalked, white or tan)
Clathrina spp.
At least 3 species in our region.
12a Tubes smooth, typically 1 mm in diameter, but up to 2 mm in diameter in the case of some tubes that terminate in oscula, the free nonbranching portions of these latter tubes less than 3 mm long; colony forming a relatively tight hemisphere or sphere, sometimes attached by a stalk; oxeas sparse, 70-435 µm long; triact rays 80-180 µm; short tetract ray 25-40 µm, long tetract rays 80-200 µm; limited to exposed outer coast *Leucosolenia eleanor* (fig. 2.1)
12b Tubes hispid, terminating in oscula 1.5-4 mm in diameter, and with a free nonbranching portion 5-15 mm or more in length; colony a relatively loose irregular spheroid or a low encrustation; oxeas abundant, 50-1000 µm; triact rays 40-150 µm; short tetract ray 20-40 µm, long tetract rays 40-150 µm; limited to protected waters and often on floating docks
Leucosolenia nautilia
Several other species of *Leucosolenia*, with still larger tubes, occur subtidally in our area.

Species marked with an asterisk are not in the key.

Subclass Calcinea

Order Clathrinida

Family Clathrinidae

The species of *Clathrina* are not separated in the key.
Clathrina blanca (Miklucho-Maclay, 1868). Intertidal and subtidal.
Clathrina coriacea (Montagu, 1818). Intertidal and subtidal.
Clathrina sp. Intertidal and subtidal.

Subclass Calcaronea

Order Leucosoleniida

Family Leucosoleniidae

Leucosolenia eleanor Urban, 1905 (fig. 2.1). Intertidal and subtidal.
Leucosolenia nautilia de Laubenfels, 1930. Mostly in harbors, where it occurs on floats.
*Leucosolenia spp. Subtidal.

Order Sycettida

Family Sycettidae

Scypha compacta (Lambe, 1894). Subtidal.
Scypha spp., including *S. mundula* (Lambe, 1893) (subtidal) and *S. protecta* (Lambe, 1896) (intertidal).

?*Tenthrenodes* sp. Intertidal and on floats; possibly the species identified as *Scypha raphanus* by de Laubenfels, 1961.

Family Grantiidae

*Grantia comoxensis (Lambe, 1894). Subtidal; known only from the type specimen.
Grantia ?*compressa*. Intertidal and subtidal.
Leucandra heathi Urban, 1905. Intertidal and subtidal.
Leucandra similar to *levis* (Polejaef, 1884). Intertidal and subtidal.
Leucandra pyriformis Lambe, 1894. Subtidal.
Leucandra taylori Lambe, 1900. Intertidal and subtidal.
*Leucopsila stylifera (Schmidt, 1870). Subtidal.
Sycandra close to *utriculus* (Schmidt, 1870). Subtidal.

Family Amphoriscidae

Leucilla nuttingi (Urban, 1902). Intertidal and shallow subtidal, and sometimes on floating docks.

Class Hexactinellida (Hyalospongia)

Hexactinellid sponges are usually regarded as a deep-water group; in local waters, however, several species occur within depths reached by divers. These species are typically massive, often attaining sizes of 0.5-2 m. Many other species occur in deeper water off our coast; some of these have single spicules over 1 m long. References include Schulze (1887, 1899), Lambe (1893, 1894) (see references on Demospongiae), and Koltun (1967). Ijima (1903) monographed the Rossellidae (many species in our region) and later (1927) presented a useful enumeration of all the Hexactinellida known to that time.

Ijima, I. 1903. Studies on the Hexactinellida. Contribution IV. (Rossellidae). J. Coll. Sci, Imperial Univ., Tokyo, 18(7):1-307.
------. 1927. The Hexactinellida of the Siboga Expedition. Siboga-Expeditie, 6 (livr. 106). viii + 383 pp.
Koltun, V. M. 1967. [Glass, or hexactinellid, sponges of the northern and far-eastern seas of the USSR.] Opredeliteli po Faune SSSR, 94. 125 pp. (in Russian).
Schulze, F. E. 1887. Report on the Hexactinellida collected by H.M.S. Challenger during the years 1873-1876. Rep. Sci. Res. "Challenger," Zoology, 21:1-513.
------. 1899. Amerikanische Hexactinelliden nach dem Materiale der Albatross-Expedition. Jena. 126 pp.

1a   Large spicules fused together to form a firm but fragile lattice; without long spicules protruding from the surface; form various but not encrusting   2
1b   Large spicules not fused, but rather interwoven to form a feltlike surface; with or without long spicules protruding at right angles from the surface; form tubular, vaselike, or bootlike (white, or brown if covered by debris)   3
2a   Chalicelike or tubular in form; where processes extend out from the surface, these are fingerlike; inner pores 2-3 times larger than the outer pores; white or bright yellow; pinnules (200 µm) in surface layer with short teeth (acanthoxeas 500-1300 µm; scopules 200-300 µm; oxyhexasters 65-200 µm; discohexacts 50-100 µm; discohexasters 50-100 µm; hexacts to 1000 µm)   *Chonelasma calyx* (fig. 2. 4)
2b   Tubular or irregular in form; where processes extend out from the surface, these are mittenlike; inner pores approximately equal to the outer pores; white or bright yellow; pinnules of surface layer with long teeth (acanthoxeas to 2000 µm; scopules 100-200 µm; oxyhexasters 60-160 µm; discohexacts 40-60 µm; discohexasters 45-50 µm; acanthostrongylotes to 350 µm)   *Aphrocallistes vastus* (fig. 2.3)

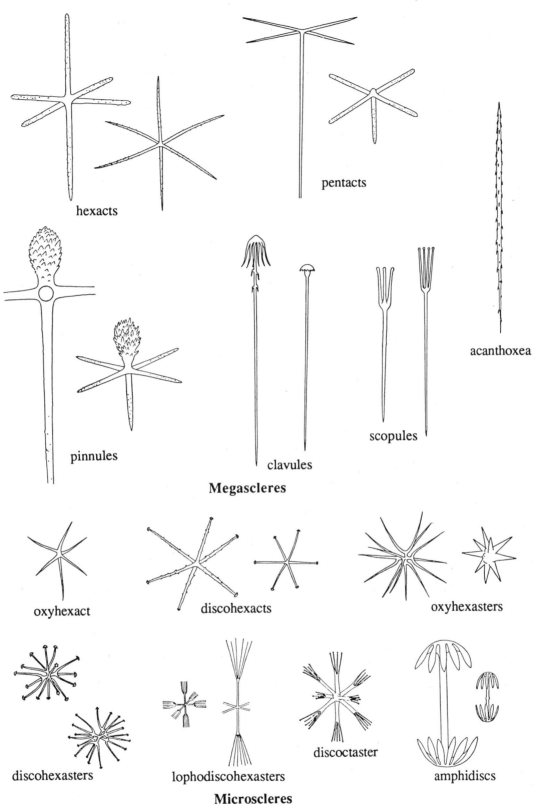

2.2 Spicules of Hexactinellida

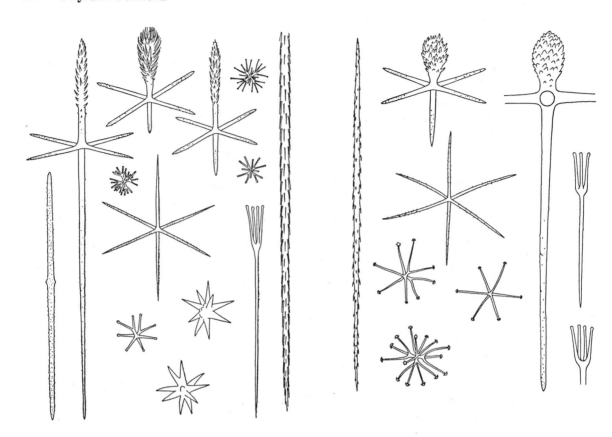

2.3 *Aphrocallistes vastus* 2.4 *Chonelasma calyx*

3a Lip of osculum thin, often with a fringe of spicules; at least some pentacts projecting above the general body surface spiny (oxeas 10-50 mm; tylostrongyles to 10 mm; projecting pentact rays 5-10 mm; pentacts within body 60-80 µm; tuberculate hexact rays 120-383 µm; oxyhexasters 60-140 µm; discohexasters 35 µm; discoctasters [may be rare] 60-100 µm; oxyhexacts 85-100 µm) *Rhabdocalyptus dawsoni*
3b Lip of osculum rounded, the spicules around it, if present, irregularly placed, not forming a distinct fringe; all pentacts projecting above the general body surface smooth (oxeas 10-60 mm; tylostrongyles 0.8-8 mm; projecting pentact rays to 8 mm; pentacts within body to 160 µm; tuberculate hexact rays 90 µm; oxyhexasters 50 µm; discohexasters 20 µm; discoctasters [may be rare] 260 µm) *Staurocalyptus dowlingi*

Species marked with an asterisk are not in the key. They are less likely than the others to be found in shallow water.

Subclass Hexasterophora

Order Lyssacinosa

Family Rossellidae

Acanthascus platei Schulze, 1899
Rhabdocalyptus dawsoni (Lambe, 1893)
Staurocalyptus dowlingi (Lambe, 1894)

Order Hexactinosa

Suborder Scopularia

Family Euretidae

Chonelasma calyx Schulze, 1887 (fig. 2.4)
**Chonelasma tenerum* Schulze, 1887

Family Aphrocallistidae

Aphrocallistes vastus Schulze, 1887 (fig. 2.3)

Class Demospongiae

Much of the literature pertaining to Demospongiae of our region has been reviewed by Bakus (1966). He also presented a detailed analysis of 23 species of the order Poecilosclerida. His work remains the most useful one for that group, even though the majority of the specimens he studied came from a relatively small area, the San Juan Archipelago. Extensive collecting, especially in British Columbia, indicates that there are over 165 species of demosponges within the range covered by this book. Some of them are primarily northern or found only in deep water. A substantial number are still unidentified or undescribed. The key provided here covers about 110 species occurring intertidally or in shallow subtidal habitats.

There are, in addition to the work of Bakus, several helpful references. *Light's Manual*, although concerned primarily with the fauna of the central California coast, deals with numerous intertidal species found in the Northwest. The extensive monographs of Koltun (1959, 1966) on demosponges of the northern and far-eastern eas of the USSR are valuable because they cover most of the genera represented in our region and many of the species that occur here. The papers of Lambe (1893, 1894, 1895, 1905), von Lendenfeld (1910), de Laubenfels (1932, 1961), and Ristau (1978) include original and secondary descriptions of species on our coast. It must be noted, however, that some of the species proposed have since been transferred to different genera, and others are now considered to be synonyms.

Bakus, G. J. 1964. Morphogenesis of *Tedania gurjanovae* Koltun (Porifera). Pacific Sci., 18:58-63.
------. 1966. Marine poecilosclerical sponges of the San Juan Archipelago, Washington. J. Zool., London, 149:415-531.
de Laubenfels, M. W. 1932. Marine and fresh-water sponges of California. Proc. U. S. Nat. Mus., 81(4):1-140.
------. 1961. Porifera of Friday Harbor and vicinity. Pacific Sci., 15:192-202.
Koltun, V. M. 1959. [Siliceous-spiculed sponges of the northern and far-eastern seas of the USSR (order Cornacuspongida).] Opredeliteli po Faune SSSR, 67. 227 pp. (in Russian).

14 Phylum Porifera

------. 1966. [Tetraxonid sponges of the northern and far-eastern seas of the USSR (order Tetraxonida).] Opredeliteli po Faune SSSR, 90. 111 pp. (in Russian).
Lambe, L. M. 1893. On some sponges from the Pacific coast of Canada and Behring Sea. Trans. Royal Soc. Canada, 10, Sect. IV (1892):67-78.
------. 1894. Sponges from the Pacific coast of Canada. Trans. Royal Soc. Canada, 11, Sect. IV (1893):113-148.
------. 1895. Sponges from the western coast of North America. Trans. Royal Soc. Canada, 12, Sect. IV (1894):113-38.
------. 1905. A new recent marine sponge (*Esperella bellabellensis*) from the Pacific coast of Canada. Ottawa Nat., 19:14-5.
Lendenfeld, R. von. 1910. The sponges. I. The Geodiidae. Mem. Mus. Comp. Zool. Harvard College, 16(1):1-259.
Ristau, D. A. 1978. Six new species of shallow-water marine demosponges from California. Proc. Biol. Soc. Wash., 91:569-89.
Sim, C.-J., & G. J. Bakus. 1986. Marine sponges of Santa Catalina Island, California. Occ. Pap. Allan Hancock Found., n. s., 5:1-23.
Simpson, T. L. 1966. A new species of clathriid sponge from the San Juan Archipelago. Postilla, 103:1-7.

1a Without spicules (other than a few from other sponges that may have been incorporated into the mass); thin encrustations, slippery to the touch unless they contain sand   2
1b With spicules; growth form various, but usually with a surface that resembles felt (but if smooth, it is rarely slippery)   6
2a Typically without spongin fibers (except in some forms of *Hexadella* sp., choice 3b)   3
2b With spongin fibers   4
3a Yellowish brown to tan   *Halisarca sacra*
3b Bright yellow in life, turning to deep purple in alcohol or on dying (subtidal, often covering the skeletons of dead specimens of the hexactinellid sponges *Aphrocallistes vastus* and *Chonelasma calyx*)   *Hexadella* sp.
    In some forms, thin spongin fibers may be visible on microscopic inspection.
4a Spongin fibers dendritic, the central axis of each fiber supporting a small cone at the surface; without sand internally or externally   5
4b Spongin fibers reticulate; sand contained within spongin fibers and sometimes so abundant as to obscure the fibers   *Dysidea fragilis* (fig. 2.16)
5a Color white to rose or red   *Aplysilla glacialis*
5b Color deep purple (the colored material exudes from the sponge when it is handled)   *Aplysilla polyraphis*
    Additional species of *Aplysilla* may occur subtidally.
6a Boring in shells of molluscs and barnacles, with exposed portions generally being small, circular patches (bright to dull yellow)   7
6b Not boring in shells   10
7a With tylostyles and microscleres   8
7b With tylostyles, but without microscleres   9
8a Tylostyles 350-550 µm (microscleres lobate spherules 20-25 µm)   *Cliona* close to *argus*
8b Tylostyles 160-190 µm (microscleres spirasters 27-32 µm)   *Cliona lobata*
9a Tylostyles 200-310 µm   *Cliona* ?*celata* subsp. *californiana*
9b Tylostyles 420-570 µm   *Cliona* similar to *warreni*
10a Unattached and containing a hermit crab (large tylostyles 250-524 µm, small tylostyles [may be absent] 90-150 µm; tylostrongyles [blunt-ended tylostyles] 135-350 µm; centrotylote microstrongyles [if present] 20-50 µm)   *Suberites* ?*suberea* forma *latus*
10b Attached to a substratum   11
11a Encrusting valves of free-living scallops   12
11b Attached to an immobile substratum   13
12a Yellow-brown to violet; styles 290-360 µm (palmate anisochelas of 2 sizes: 15-40 µm and 60-75 µm; sigmas 35-65 µm)   *Mycale adhaerens* (fig. 2.14)

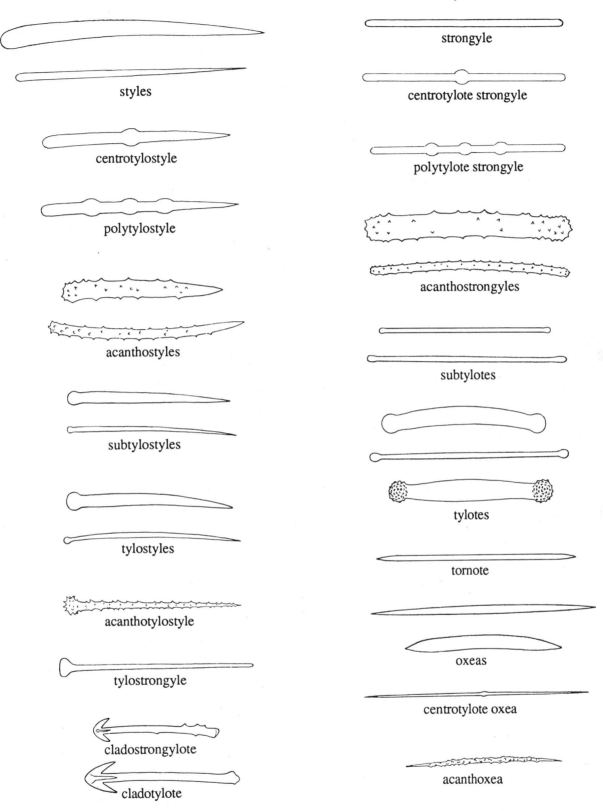

2.5 Megascleres of Demospongiae (see also Figure 2.6)

16 Phylum Porifera

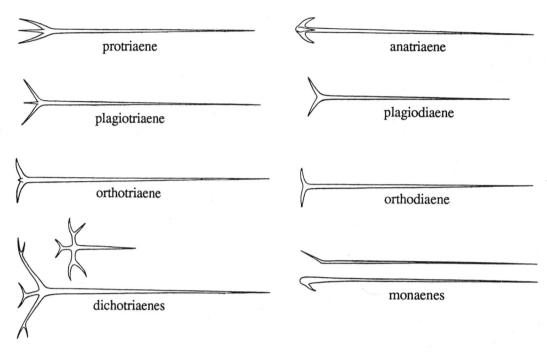

2.6 Megascleres of Demospongiae (continued)

12b Gold to gold-brown; acanthostyles 200-270 μm (tornotes 160-250 μm; tridentate isoanchors of 2 sizes: 15-20 μm and 35-70 μm; sigmas 25-53 μm) *Myxilla incrustans* (fig. 2.12)
13a Erect, the attached portion narrow 14
13b Not as described in choice 13a (if erect, the attached portion broad) 26
14a Erect, with a central stem and small filiform lateral branches similar to those on a small bottle brush (white; up to 7 cm high; styles 410-680 μm and about 1400 μm; palmate anisochelas 13 μm; forceps 32 μm) *Asbestopluma occidentalis*
14b Tubular, fingerlike, or resembling a funnel 15
15a Unbranched (occasionally bifurcated, however) 16
   Caution: early growth stages of branched species (choice 15b) may be unbranched.
15b Erect and with long, fingerlike or bladelike branches that may anastomose 20
16a Funnel-shaped or in the form of a hollow tube 17
16b Fingerlike, not hollow (there may be more than one growth form) (yellow-brown or light brown to violet; styles 275-370 μm; sigmas 37-63 μm; palmate anisochelas 15-37 μm and 59-74 μm) *Mycale adhaerens* (fig. 2.14)
17a Funnel-shaped or tubular, on a narrow stalk that is more than one-fourth the total height 18
17b Funnel-shaped, without a stalk or with only a short stalk 19
18a Surface smooth; usually funnel-shaped, sometimes tubular, occasionally bifurcated; cream to orange-brown; styles to subtylostyles 200-450 μm, shorter in the stalk *Stylissa stipitata*
18b Surface rugose; strictly funnel-shaped; cream to light brown; styles 240-500 μm, shorter in the stalk *Phakettia* close to *beringensis*
19a Styles greater than 400 μm; yellow; below diving depths but occasionally brought up by fishermen; styles 430-515 μm; palmate anisochelas 85-96 μm and 27-50 μm (to 1 m in diameter [our largest desmosponge]; sigmas [rare] 19-43 μm) *Mycale bellabellensis*
19b Styles less than 300 μm; cream to gray; below diving depths; styles 220-280 μm; palmate isochelas (may be rare or apparently absent) about 25 μm *Neoesperiosis infundibula*
20a White or cream to yellow 21
20b Dark brown, tan, red-orange, or red 22

Phylum Porifera 17

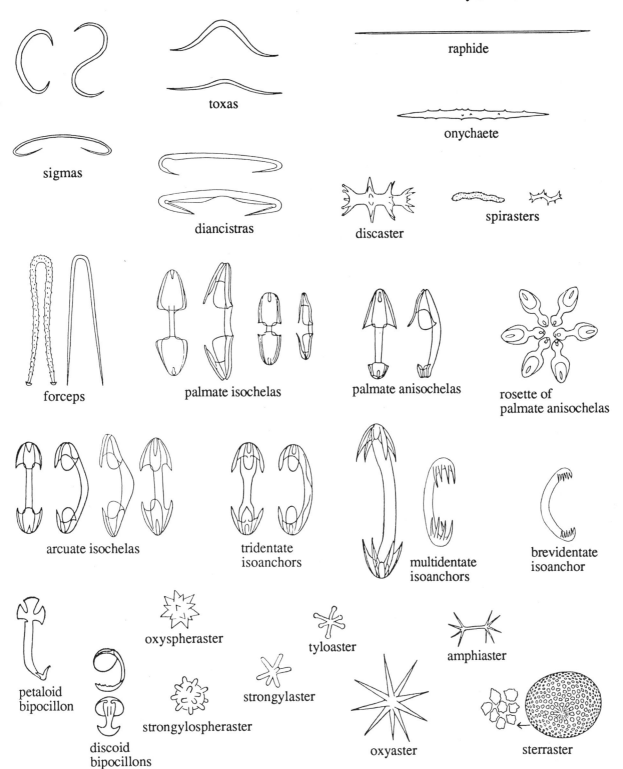

2.7 Microscleres of Demospongiae

21a Cream to yellow; surface smooth, hard (styles to subtylostyles 230-820 µm)
?Syringella amphispicula
21b White, turning black in alcohol; surface soft and compressible (acanthostyles 230-320 µm; subtylotes 235-280 µm; palmate isochelas 20-40 µm; petaloid bipocillons 14-18 µm)
Iophon chelifer var. *californiana*
22a Dark brown (styles 156-383 µm; strongyles to styles 124-228 µm; multidentate [4-5 teeth] isoanchors 29-44 µm and 65-108 µm) *Stelodoryx alaskensis*
22b Tan, red-orange, or red, or some intermediate color 23
23a Red; skeleton rigid (acanthostyles 180-320 µm; styles 430-550 µm; strongyles 350-370 µm; raphides 200-330 µm) *Hemectyon hyle*
23b Color light tan to red-orange; skeleton flexible, but tough 24
24a Palmate isochelas 19-20 µm (abundant oscula along margins of branches; styles 109-163 µm) *Neoesperiopsis vancouverensis*
24b Palmate isochelas (may be rare) 24-27 µm 25
25a With abundant oscula along margins of the branches; styles 144-216 µm
Neoesperiopsis digitata
25b With a single osculum at the apex of each branch (few, if any, along the margins); styles 170-242 µm *Neoesperiopsis rigida*
The delimitation of species in the genus *Neoesperiopsis* is being studied.
26a Sponge spherical or nearly so 27
26b Sponge forming a thin to thick encrustation, or a hemispherical mass, but not spherical 32
27a Firm, unyielding (use a probe rather than the fingers), with abundant spicules protruding from the surface 28
27b Soft, deformable, without obvious spicules protruding from the surface 30
28a Surface under the plush of spicules having the appearance of being made of paving stones (this can be seen with a strong hand lens); color white to dirty beige (sterrasters forming the surface layer 82-110 µm; small oxeas 150-600 µm; large oxeas 2-8 mm; orthotriaenes 1.5-8 mm, with rays bifurcated in some; protriaenes 2-17 mm; anatriaenes 4-23 mm; oxyasters 11-54 µm; oxyspherasters 6-32 µm; strongylasters 5-11 µm) *Geodia mesotriaena* (fig. 2.15)
28b Surface under the plush of spicules appearing smooth when viewed with a strong hand lens; color grey to dirty white 29
29a Large (diameter up to 8 cm); gray; oxeas 383-1280 µm; anatriaenes to 5 mm; protriaenes mostly 325-550 µm, but up to 11 mm; spiny spirasters 9 µm (with pointed ends when viewed with a scanning electron microscope) (intertidal and subtidal) *Craniella villosa*
29b Small (diameter up to 2 cm); white; oxeas 328-1100 µm; anatriaenes about 8 mm; protriaenes to 5.5 mm; spiny spirasters 13 µm (with rounded ends when viewed with a scanning electron microscope) (subtidal) *Craniella spinosa*
30a Color dark chocolate brown (styles 650-720 µm; discasters 55-60 µm) *Latrunculia* sp.
30b Color red-orange to yellow 31
31a Surface covered with numerous blunt projections that are reduced to flattened polygons in contracted specimens (color orange to yellow; strongyles 500-1000 µm; tylostrongyles 1300 µm; oxyspherasters 65 µm; tyloasters 30 µm) *Tethya* close to *aurantia* (fig. 2.17)
31b Surface smooth when contracted, but with abundant large pores and 1-2 oscula when expanded (polytylote styles 320-380 µm) *Stylinos* sp.
32a Forming a thick, broadly attached mass, often tending toward the shape of an irregular or oblate hemisphere (color white to yellow or yellow-brown) 33
32b Forming a thick or thin encrustation, not tending toward the shape of an irregular hemisphere 42
33a With prominent nipplelike or fingerlike protuberances, these terminating in open oscula when expanded, but appearing rounded or pointed when contracted 34
33b Without nipplelike or fingerlike protuberances (there may, however, be a few oscula on chimneys in *Merriamum oxeota*, choice 39b) 35
34a Base dark, hispid, with white to yellow protuberances (styles or subtylostyles up to more than 3 mm long) *Polymastia pachymastia*
See also couplet 50.
34b Both protuberances and base yellow-orange (soft when expanded, firm and rubbery when contracted; subtylostyles 600-700 µm) *Weberella* similar to *verrucosa*
35a Surface distinctly hispid because of projecting spicules 36

35b Surface smooth to rugose, but not hispid ... 37
36a Surface beneath the plush of spicules having the appearance of being made of paving stones when viewed with a strong hand lens; color white to dirty beige
Geodia mesotriaena (fig. 2.15) and *Geodinella robusta*
See also choices 28a and 109a.
36b Surface beneath the plush of spicules not having the appearance of being made of paving stones; dirty white (oxeas 1400-3500 μm or larger; ortho-, plagio-, dichotriaenes 1400-3500 μm; anatriaenes 110-2000 μm; oxyasters-strongylasters 9-15 μm; "grains" 3 μm; no sterrasters forming the surface layer) *Stelletta clarella*
37a Surface covered by sieve plates (circular areas about 2.5 mm in diameter) (red-orange; styles to subtylostyles 140-220 μm and 420-525 μm; subtylostrongyles 315-355 μm; arcuate isochelas 32-46 μm) *Hamigera* similar to *lundbecki*
37b Without sieve plates ... 38
38a Isoanchors present ... 39
38b Isoanchors absent (when very small spicules are present, they resemble styles, tylostyles, oxeas, strongyles, and variants of these) ... 40
39a Color yellow-orange *Myxilla lacunosa*
See also choice 56b.
39b Color rich yellow-brown (acanthostyles 300-430 μm; tornotes to oxeas 220-320 μm; brevidentate isoanchors 15-18 μm) *Merriamum oxeota*
40a Megascleres styles to subtylostyles 218-400 μm by 3-8 μm (color orange to brown; soft, porous when expanded, but firm, rubbery, smooth when contracted) *Suberites montiniger*
40b Megascleres tylostyles ... 41
41a Large tylostyles all or mostly greater than 600 μm long (range 590-1125 μm), 13-23 μm wide; small tylostyles abundant in cortex, 110-355 μm; yellowish white (less than 2 cm in diameter) *Suberites simplex*
41b Large tylostyles all or mostly less than 600 μm (range 380-590 μm), 15-25 μm wide; small tylostyles absent or rare; yellow-tan to yellow-orange *Suberites* sp.
This is considered to be different from *S.* ?*suberea* forma *latus*, which regularly contains a hermit crab and has thinner tylostyles (9-13 μm wide) and a separate class of small tylostyles.
42a Encrusting base with long (greater than 1 cm), well defined, erect, nonbranching projections other than those terminating in a single osculum ... 43
42b Encrusting and either without long projections or only with nipplelike or fingerlike protuberances that terminate in an osculum ... 48
43a Skeleton firm ... 44
43b Skeleton soft, flexible ... 46
44a Red to orange-red (projections flattened laterally; surface hispid; oscula not evident macroscopically; oxeas 500-600 μm; styles 500-900 μm) *Axinella* sp.
44b White, off-white, lavender, light to dark brown, but not red or orange-red ... 45
45a Off-white or lavender; projections in the form of columns of anastomosing branches; oxeas 260-300 μm (sigmas 40-90 μm) *Sigmadocia edaphus*
45b Off-white to gray or light brown with dark blotches; projections terminating in clusters of collared oscula; oxeas 400-2000 μm (dichotriaenes [with branching rays] 400-500 μm; bicurvate strongyles 50-210 μm; oyspherasters 9-30 μm) *Penares cortius*
46a Color red to orange-brown (branching when mature, thin and encrusting when young; thick styles to subtylostyles, microspined on their heads, 180-225 μm; thin styles to subtylostyles 155-175 μm; acanthostyles 80-90 μm; palmate isochelas 16-20 μm; toxas 15-100 μm; introduced in Willapa Bay and may be expected elsewhere) *Microciona prolifera*
46b Color white to translucent and nearly colorless ... 47
47a Projections up to 5 cm long and 2 mm wide, resembling pieces of spaghetti (do not confuse with *Leucosolenia*, which has calcareous spicules); oxeas 110-500 μm *Ciocalypta penicillus*
47b Projections up to 1 cm long by 2 mm wide; oxeas 250-1150 μm *Eumastia sitiens*
48a Nipplelike or fingerlike protuberances of a different color from the encrusting base ... 49
48b Protuberances, where present, the same color as the encrusting base ... 51
49a Smooth; dark red-brown base with orange protuberances (acanthostyles 211-316 μm; subtylotes 213-260 μm) *Podotuberculum hoffmanni*
49b Hispid; dark encrusting base with white to yellow nipplelike protuberances, many terminating in a single osculum (oscula may not be evident when the sponge is contracted) ... 50

20 Phylum Porifera

50a	Size to 2.5 cm in diameter and less than 0.7 mm high; nipplelike protuberances white, tending to be pointed and laterally flattened (when *in situ*, only the protuberances may be visible above the sediment) (styles to subtylostyles to 3 mm; tylostyles 100-200 μm and 450-700 μm) .... *Polymastia pacifica*	
50b	Size typically greater than 2.5 cm in diameter and 3 mm high; protuberances bright to dull yellow, tending to be rounded and columnar (styles to subtylostyles up to 8 mm; tylostyles up to 2.5 mm) .... *Polymastia pachymastia*	
51a	Encrustation moderately to very thick (greater than 4 mm) and with a firm to almost stony skeleton (use probe, not fingers!)	52
51b	Encrustation either thin (less than 4 mm) or, if thick, at least moderately spongy	61
52a	Contains megascleres with branches at one end	53
52b	Megascleres only of the nonbranching type	55
53a	Surface with an abundance of large, projecting spicules; color dirty white (oxeas, ortho-, plagio-, dichotriaenes 1400-3500 μm; anatriaenes 1100-2000 μm; oxyspherasters 9-15 μm) .... *Stelletta clarella*	
53b	Surface rough or smooth, but not with an abundance of large, projecting spicules; color light brown (with dark blotches), gray, or scarlet (occasionally bronze)	54
54a	Surface rough; color scarlet (occasionally bronze) (cladotylotes 135-180 μm and 356-470 μm; styles 600-790 μm; subtylotes 250-485 μm; toxas 45-645 μm and 320-340 μm; palmate isochelas 15-25 μm) .... *Acarnus erithacus* (fig. 2.11)	
54b	Surface smooth; color light brown, with dark blotches, or gray (oxeas 400-2000 μm; dichotriaenes with branching rays 400-500 μm; bicurvate strongyles 50-210 μm; oxyspherasters 9-30 μm) .... *Penares cortius*	
55a	Surface rubbery; spicules tylostyles .... *Suberites* sp. See also couplets 40 and 41.	
55b	Surface not rubbery; spicules not tylostyles	56
56a	Oxeas present, styles absent	57
56b	Oxeas absent, styles present (yellow-orange; slippery; styles to acanthostyles [these have only a few spines and are often sinuous] 170-350 μm; tornotes [often sinuous] 15-185 μm; sigmas [may be 2 size classes] 19-46 μm; tridentate isoanchors [in 2 size classes] 15-57 μm) .... *Myxilla lacunosa*	
57a	Spicules consisting entirely of oxeas	58
57b	Spicules consisting of oxeas and sigmas, or oxeas and toxas	59
58a	Color dark brown to beige-orange; thin oxeas 115-200 μm, thick oxeas 240-320 μm .... *Xestospongia trindanea*	
58b	Color white to pale yellow or tan; oxeas 150-180 μm .... *Xestospongia vanilla*	
59a	Spicules consisting of oxeas and toxas .... *Toxidocia* spp.	
59b	Spicules consisting of oxeas and sigmas	60
60a	Oxeas 270-300 μm; sigmas 40-90 μm (smaller ones C-shaped, larger ones with a sharp angular bend) (color off-white to pale lavender) .... *Sigmadocia edaphus*	
60b	Oxeas and sigmas larger or smaller than those described in choice 60a .... *Sigmadocia* spp. Several unidentified species.	
61	Note that there are 7 choices.	
61a	Color green or green with yellow areas	62
61b	Color blue or gray-blue (sometimes light orange in *Hymenamphiastra cyanocrypta*, choice 63a)	63
61c	Color lavender, violet-blue, pinkish purple, purple, or mahogany (sometimes yellow-tan in ?*Stylophus arndti*, choice 65b; sometimes buff in *Haliclona ?ecbasis*, choice 68a; sometimes dull tan in *Haliclona* close to *permollis*, choice 68b)	64
61d	Color white to ivory	69
61e	Color salmon-pink, rust-pink, red, brick-red, scarlet, red-orange, yellow-orange, or orange-brown	71
61f	Color clear yellow or sulphur-yellow (megascleres consisting of tylostyles) .... *Cliona* sp. See couplets 8 and 9 for other species of *Cliona*.	
61g	Color dull yellow, yellow-brown, buff, or brown (sometimes lavender, pinkish purple, purple, yellow, or yellow-orange in certain species that do not have microscleres, couplets 94 and 95)	89

Phylum Porifera

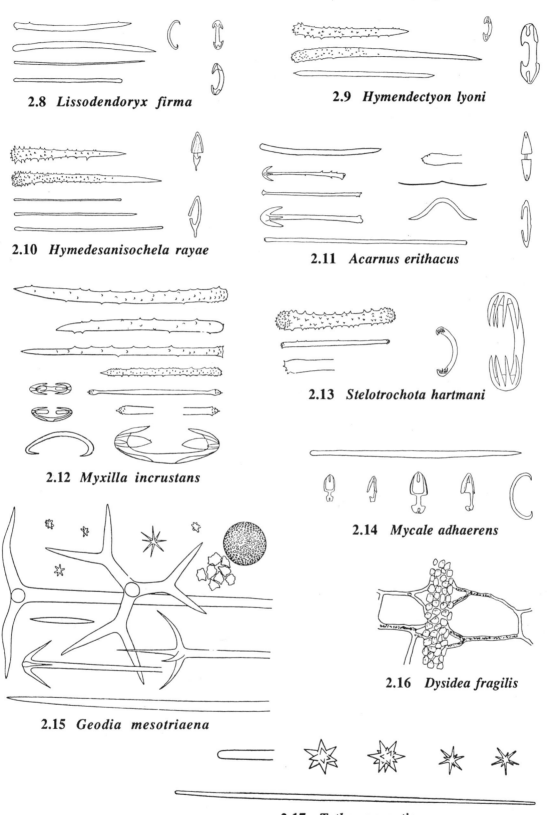

2.8 *Lissodendoryx firma*

2.9 *Hymendectyon lyoni*

2.10 *Hymedesanisochela rayae*

2.11 *Acarnus erithacus*

2.12 *Myxilla incrustans*

2.13 *Stelotrochota hartmani*

2.14 *Mycale adhaerens*

2.15 *Geodia mesotriaena*

2.16 *Dysidea fragilis*

2.17 *Tethya aurantia*

22 Phylum Porifera

62a General surface green with yellow circular areas (contracted yellow papillae) (large oxeas, with sharp central flexure, up to 1000 µm; small oxeas and acanthoxeas, with a central swelling, 110-150 µm; rare, recorded only once in our region)   ?*Higginsia* similar to *higgini*

62b Surface uniformly green or with yellow blotches, and yellow below the surface (oxeas 150 to more than 430 µm; common in lower to mid-intertidal)   *Halichondria panicea*

63a Usually cobalt-blue (but sometimes light orange) (forming thin encrustations; acanthostyles 75-300 µm; tornotes 160-175 µm; amphiasters 10-17 µm)   *Hymenamphiastra cyanocrypta*

63b Gray-blue (forming thin encrustations; oxeas 140-175 µm)   *Reniera* sp.
Species B of Hartman, 1975, in *Light's Manual*.

64a Megascleres consisting of styles or acanthosubtylostyles and tylotes to subtylotes   65

64b Megascleres limited to oxeas   66

65a Spicules consisting of styles 180-200 µm; dark purple to mahogany, buff below the surface   *Hymeniacidon ungodon*

65b Spicules consisting of acanthosubtylostyles 89-135 µm and tylotes to subtylotes (with slightly unequal ends) 127-162 µm; yellow-tan, lavender, violet-blue; forming encrustations less than 2 mm thick   ?*Stylopus arndti*

66a Spicules consisting of oxeas 200-240 µm and sigmas 34-39 µm (lavender)   *Sigmadocia* spp.

66b Spicules consisting of oxeas only   67

67a Surface layer removable in flakes (pale translucent lavender; oxeas 92-170 µm)   *Adocia gellindra*

67b Surface layer not removable in flakes   68

68a Oxeas 75-105 µm; purple, pinkish purple, to lavender, but in shaded locations becoming partly or totally buff; growth form varying from thin encrustations in which the oscula are flush with the surface to thick encrustations with oscula on chimneys (on floats in protected areas and in the intertidal region on protected to exposed coasts; may include *H. permollis* of de Laubenfels, 1961 and ?*Haliclona* sp. A of Hartman, 1975, in *Light's Manual*)   *Haliclona* ?*ecbasis*

68b Oxeas 125-175 µm; lavender or dull tan; oscula never on chimneys   *Haliclona* close to *permollis*

69a Readily torn, and surface layer removable in flakes   70

69b Moderately tough, with a spongy consistency, the surface layer not readily removable (with globular or branching projections, each terminating in an osculum; oxeas 160-230 µm, crowded within tracts that consist of spongin fibers [the tracts form a prominent network]) (growth forms various)   *Pachychalina* spp.

70a Oxeas in 2 distinct size classes: 55-100 µm and 430-1290 µm; translucent white throughout   *Topsentia disparilis*

70b Oxeas of varying size (240-295 µm), but not in 2 distinct size classes (oxeas enclosed by spongin fibers, 1-4 spicules in each tract of the network); yellowish orange below the white surface layer (oscula may be on chimneys)   ?*Adocia* sp. similar to *Reniera mollis*
May be a species of *Pellina*.

71a Spicules consisting of megascleres only   72

71b Spicules consisting of both megascleres and microscleres   74

72a Spicules consisting of styles, these in 2 size classes: 140-200 µm and 320-360 µm (color red-orange to brick-red)   *Hymeniacidon* close to *perleve*

72b Spicules consisting only of oxeas, or of oxeas and styles   73

73a Oxeas 345-530 µm; color orange to salmon-pink   ?*Adocia* sp. similar to *Reniera foraminosa*

73b Oxeas 920-1380 µm; color orange to rust-pink (styles 80-1080 µm)   *Pseudaxinella* similar to *rosacea*

74a At least some megascleres acanthose (with obvious spines when viewed with 100x magnification [megascleres with minute spines or spherules limited to rounded end(s) are not considered acanthose])   75

74b Megascleres typically with minute spines or spherules on rounded end(s), but otherwise smooth   86

75a Some acanthose megascleres with equal ends, others with unequal ends   76

75b Acanthose megascleres all with unequal ends   77

76a Megascleres including acanthostyles 177-220 µm and acanthostrongyles 120-160 µm (styles 159-275 µm; toxas 29-133 µm; palmate isochelas 19-27 µm)   *Plocamilla illgi*

76b Megascleres including acanthostyles 140-160 μm and acanthotornotes (may have only a few spines) 137-150 μm (arcuate isochelas 17-22 μm) *Jones amaknakensis*
77a Megascleres acanthosubtylostyles to acanthotylostyles 78
77b Megascleres acanthostyles 80
78a With arcuate isochelas 18 μm as well as multidentate isoanchors 15-18 μm (acanthosubtylostyles 475-600 μm; acanthotylostyles 90 μm; subtylostyles 256-305 μm; sigmas [may be growth stage of chelas] 10 μm; encrustation less than 1 mm thick) *Arndtanchora* sp.
78b With arcuate isochelas but without multidentate isoanchors 79
79a Salmon-red to yellow-orange; thin (1 mm), at least when growing on the rock scallop *Hinnites* (acanthotylostyles or acanthosubtylostyles 100-155 μm and 250-470 μm; subtylostyles 165-230 μm; arcuate isochelas 17-25 μm) *Anaata brepha*
79b Brick-red to deep red-brown; up to 5 mm thick; acanthostyles to acanthosubtylostyles 92-127 μm and 275-360 μm; styles to subtylostyles 193-253 μm; arcuate isochelas 23-28 μm and 45-50 μm *Anaata spongigartina*
80a Less than 1 mm thick 81
80b More than 2 mm thick 83
81a With amphiasters (these may be minute), but without chelas or isoanchors (light orange) *Hymenamphiastra cyanocrypta*
     See also choice 63a.
81b With isochelas 82
82a Microscleres arcuate isochelas; acanthostyles oriented vertically, with heads down and pointed ends projecting above the surface (with strongyles, subtylotes or tornotes, and arcuate isochelas; sigmas present or absent) *Hymedesmia* spp.
     Several unidentified species occur in the region. Some of them may be early growth stages of sponges of other genera.
82b Microscleres palmate isochelas; acanthostyles in plumo-reticulate tufts above the surface and in 2 size classes: 92-100 μm and 150-475 μm (subtylostyles 100-230 μm; palmate isochelas 8 μm) ?*Dictyociona asodes*
     This species, as *Leptoclathria* (or *Eurypon*) *asodes*, is reported to be yellow in California.
83a With palmate isochelas 84
83b With arcuate isochelas or isoanchors 85
     Isoanchors (=anchorate isochelas of some authors) have laterally flattened expansions on the upper and lower portions of the shaft; these are lacking in arcuate isochelas. They may be difficult to see with the light microscope; but tridentate isoanchors in our species also have a relatively straight shaft in contrast to the more arched shaft of arcuate isochelas.
84a Toxas to 325 μm, with recurved arms, microspined at tips; growths up to 1 cm thick, never branching; brilliant scarlet to red; acanthostyles 105-160 μm; thick styles, with microspined heads, 209-362 μm; thin styles to subtylostyles, with microspined heads, 119-238 μm; toxas 15-325 μm; palmate isochelas 14-17 μm *Microciona microjoanna*
84b Toxas to 100 μm, with straight, smooth arms; growths branching when mature, thin and encrusting when young; red to orange-brown; acanthostyles 80-90 μm; thick styles to subtylostyles, microspined on heads, 180-225 μm; thin styles to subtylostyles 155-175 μm; toxas 15-100 μm; palmate isochelas 16-20 μm (introduced in Willapa Bay and may be expected elsewhere) *Microciona prolifera*
85a With tridentate isoanchors, but without arcuate isochelas; brilliant yellow-orange; acanthostyles typically with few spines, 180-235 μm (subtylotes, with mucronate tips, 140-210 μm; large isoanchors 37-56 μm; small isoanchors 13-15 μm [may be a few intermediate sizes]; sigmas 15-40 μm; in our region, known only from one surge channel in British Columbia) *Myxilla behringensis*
85b With arcuate isochelas; growths up to more than 1 cm thick, yellowish brown to deep red-brown; acanthostyles 140-246 μm (styles to subtylostyles 130-241 μm; arcuate isochelas 22-30 μm) *Anthoarcuata graceae*
86a With tylotes 87
86b Without tylotes 88
87a Styles up to more than 400 μm long; not secreting obvious mucus; styles 183-687 μm; subtylostyles 84-229 μm; tylotes 84-108 μm; toxas (may not be numerous or evident) 31-72 μm; palmate isochelas 10-15 μm *Plocamilla lambei*

Phylum Porifera

87b Styles less than 300 μm long; secreting copious amounts of mucus; styles 136-258 μm; subtylostyles 132-232 μm; tylotes 187-240 μm; toxas 10-100 μm; palmate isochelas (these may be uncommon or apparently absent) 21-22 μm *Plocamia karykina*

88a With palmate isochelas; thick styles to subtylostyles 130-390 μm; thin subtylostyles 100-300 μm (palmate isochelas 13-20 μm; toxas [may be apparently absent] 110-250 μm) *Axocielita originalis*

88b Without palmate isochelas; thick subtylostyles 200-917 μm; thin subtylostyles 165-452 μm (toxas 26-165 μm) *Ophlitaspongia pennata*

89a Without microscleres ... 90
89b With one or more kinds of microscleres ... 103
90a All megascleres with equal ends (unless broken) ... 91
90b At least some megascleres with unequal ends ... 96
91a Spicules consisting of strongyles and oxeas or raphides ... 92
91b Spicules consisting only of oxeas ... 93

92a Strongyles rounded to mucronate (with a pointed nipple at their ends) 115-170 μm; oxeas 90-140 μm; color dull brown *Prianos problematicus*

92b Strongyles with multiple microspines on their ends, 200-380 μm; raphides 70-80 μm; color dull tan *?Tedanione obscurata*

93a Spicule size varying widely (by more than 100 μm) within a given specimen, and some spicules longer than 260 μm ... 94

93b Spicule size varying little within a given specimen (less than 50 μm) and no spicules longer than 260 μm ... 95

94 Note that there are 3 choices.

94a Oxeas 150 to more than 430 μm; yellow to yellow-tan throughout, or with a green surface in sunlit habitats; limited to moderately or fully exposed coasts (one of a few sponges that occur as high as mid-intertidal) *Halichondria panicea*

94b Oxeas 150-320 μm; yellow-orange to yellow-tan, never green; primarily in harbors and other protected situations *Halichondria* spp.
Includes *Halichondria bowerbanki* (common on floating docks) and a species that is of similar form but that has a different larval type.

94c Oxeas 200-672 μm; lower intertidal on exposed coast *Halichondria* sp.

95 Note that there are 4 choices.

95a Oxeas 75-105 μm; buff in shaded locations, but becoming purple or pinkish purple to lavender in sunlit habitats; growth form varying from thin encrustations to thick encrustations; with oscula on chimneys) *Haliclona ?ecbasis*
May include *H. permollis* of de Laubenfels, 1961 and *?Haliclona* sp. A of Hartman, 1975 in *Light's Manual*.

95b Oxeas 125-175 μm; dull tan or lavender; encrustations smooth, of nearly uniform thickness, the oscula never on chimneys *Haliclona* close to *permollis*

95c Oxeas 158-170 μm; dull tan; encrustations in the form of meandering ribbons *Reniera* sp.

95d Oxeas 195-262 μm; translucent white *Reniera mollis*

96a Spicules with both equal and unequal ends ... 97
96b All spicules with equal ends ... 98

97a Spicules represented by styles 296-373 μm, tornotes 200-252 μm, and onychaetes 40-110 μm and 125-265 μm; gray to cinnamon *Tedania gurjanovae*

97b Spicules represented by acanthosubtylostyles 89-135 μm and tylotes to subtylotes with slightly unequal ends 127-162 μm; yellow-tan, lavender, violet-blue (encrustations less than 2 mm thick) *?Stylopus arndti*

98a With styles only ... 99
98b With tylostyles only ... 100

99a Styles 180-200 μm; buff to dark purple or mahogany *Hymeniacidon ungodon*
99b Styles 200-450 μm; yellow *Hymeniacidon sinapium*

100a Growths more than 5 mm thick *Suberites* spp.
See couplet 41.

100b Growths less than 2 mm thick ... 101

101a Surface covered with small papillae (yellow; tylostyles up to 600 μm) *Laxosuberites* sp.
This may be a young specimen of a species of *Suberites*.

101b Surface without small papillae ... 102

102a	Tylostyles tangential to the surface	*Pseudosuberites* spp.
102b	Tylostyles extending vertically beyond the surface	*Prosuberites* spp.
103a	Microscleres including spiny and/or smooth triacts (do not confuse these with triact spicules of class Calcarea)	104
103b	Without triacts	106
104a	Rays of most large triacts and tetracts acanthose; light yellow-brown to tan (diacts 85 µm; rays of large triacts and tetracts 60 µm; rays of small triacts and tetracts 8-13 µm, lophate [with a cluster of spines] at the tips)	*Plakina* sp. similar to *Corticium bowerbanki*
104b	Rays of all triacts and tetracts smooth; light ivory	105
105a	Small triacts (rays 13 µm) lophate at the ends of 2 or 3 rays (large tetracts [1 ray typically reduced to a knob]) with rays 25 µm; diacts 62 µm)	*Plakina* close to *trilopha*
105b	Small triacts (rays 12-13 µm, lophate at their ends) with a single, short lophous structure (this may constitute, in part, a 4th ray) at the intersection of the other rays (rays of large triacts 25-28 µm; diacts 88 µm)	*Plakina* close to *brachylopha*
106a	Megacleres all with equal ends (unless broken)	107
106b	Megascleres including some spicules with unequal ends	108
107a	Megascleres oxeas 165-210 µm; dull yellow (microscleres sigmas 3-50 µm and toxas 82-86 µm)	*Orina* sp.
107a	Megascleres acanthostrongyles 199-211 µm and subtylotes 147-176 µm; light brown to gray, giving off a flaky black precipitate in alcohol (multidentate isoanchors 39-47 µm; brevidentate isoanchors 41-51 µm)	*Stelotrochota hartmani* (fig. 2.13)
108a	Without isochelas, anisochelas or isoanchors	109
108b	With either isochelas or anisochelas and/or isoanchors	111
109a	Microsleres including sterrasters (180-240 µm) that form a solid coat on the surface (the surface is hard, unyielding); color whitish tan to yellow-brown; oxeas up to 2500 µm; monaenes and plagiodiaenes to 2100 µm; strongylospherasters 7-13 µm; oxyasters 7-38 µm)	*Geodinella robusta*
109b	Microscleres include sigmas, but sterrasters absent	110
110a	Sigmas sometimes very large (range 13-414 µm); amber to light olive-brown (styles 600-1305 µm; raphides [in bundles] 90-200 µm; comma-shaped microscleres [may not be obvious] less than 10 µm)	*Biemna rhadia*
110b	Sigmas of moderate size (range 15-54 µm); yellow-bronze (tylostyles 172-253 µm; diancistras 26-41 µm)	*Zygherpe hyaloderma*
111a	Microscleres including palmate anisochelas	112
111b	Without palmate anisochelas	119
112a	Spicules include discoid bipocillons 7-10 µm; yellowish tan (slightly acanthose styles 187-300 µm; subtylotes 135-208 µm; palmate anisochelas 16-23 µm)	*Iophon piceus* var. *pacifica*
112b	Without discoid bipocillons	113
113a	With acanthostyles (growths thin [1 mm]; cinnamon to cinnamon-brown; short acanthostyles 113-152 µm; long acanthostyles 187 to more than 600 µm; subtylostyles 174-206 µm; subtylotes 150-258 µm; palmate anisochelas 12-24 µm)	*Hymedesanisochela rayae* (fig. 2.10)
113b	Without acanthostyles	114
114a	Sigmas serrated near their ends (light to dark gold; subtylostyles 312-364 µm; sigmas 34-49 µm and 119-157 µm; palmate anisochelas 14-16 µm and 27-37 µm)	*Paresperella psila*
114b	Sigmas smooth	115
115a	Toxas present (pale golden brown to olive-brown; sinuous subtylostyles 250-380 µm; toxas 32-75 µm; sigmas 34-49 µm and 60-75 µm; palmate anisochelas 13-21 µm and 30-40 µm)	*Mycale macginitiei*
115b	Toxas absent	116
116a	Large anisochelas arcuate (dull yellow; subtylostyles 525-610 µm; sigmas 23-25 µm and 78-92 µm; straight anisochelas 16-21 µm, in rosettes; arcuate anisochelas 85-92 µm)	*Mycale* close to *toporoki*
116b	All anisochelas straight or almost straight	117
117a	Sigmas 14-25 µm (yellow-rose to gold-beige; styles 196-310 µm; sigmas [many S-shaped] 14-25 µm; palmate anisochelas 29-37 µm)	*Mycale richardsoni*
117b	Sigmas larger than 30 µm	118

118a Sigmas of typical arcuate form; yellow-brown or light brown to violet; styles 260-360 µm; sigmas 34-63 µm; palmate anisochelas 15-37 µm and 55-74 µm
Mycale adhaerens (fig. 2.14)
118b Sigmas elongate, with almost a straight "back;" gold-beige to bronze; styles 192-300 µm; elongate sigmas 31-59 µm; palmate anisochelas 31-62 µm   *Mycale hispida*
119a With isoanchors, but without arcuate isochelas   120
See the note concerning tridentate isoanchors under choice 83b.
119b With arcuate isochelas, but without isoanchors   122
120a With tridentate isoanchors, but without quadridentate isoanchors; sigmas greater than 40 µm; styles-acanthostyles 150 µm   121
120b With tridentate isoanchors 45-62 µm and quadridentate isoanchors (may be difficult to count teeth with a light microscope); sigmas less than 26 µm; acanthostyles 32-65 µm and 196-314 µm (tornotes 150-209 µm)   *Ectyomyxilla parasitica*
Only the type specimen definitely fits the description. Other references to *Ectyomyxilla parasitica* are concerned with *Myxilla incrustans* (choice 21b).
121a Encrustations less than 1 mm thick; off-white; styles (acanthose on head) 600 µm (tylostyles [acanthose on head] 175 µm; tridentate isoanchors 220-290 µm)   ?*Hymenanchora* sp.
121b More than 5 mm thick; gold to light brown; styles to acanthostyles 190-320 µm (tornotes to pointed subtylotes [ends often spined] 150-252 µm; sigmas 25-60 µm; tridentate isoanchors 13-36 µm and 36-78 µm)   *Myxilla incrustans* (fig. 2.12)
122a Spicules include oxeas (yellow-orange; acanthostyles 220-400 µm; tornotes [abruptly tapering to a point] and oxeas [gradually tapering to a point] 250-320 µm; arcuate isochelas 15-21 µm)   *Merriamum oxeota*
122b Without oxeas (but there may be tornotes)   123
123a Forceps present among the microscleres (yellow; styles to acanthostyles [few spines] 170-280 µm; subtylotes 170-240 µm; arcuate isochelas 21-47 µm; sigmas 18-75 µm; forceps 10-25 µm)   *Forcepia* similar to *japonica*
123b Without forceps among the microscleres   124
124a Encrustations less than 2 mm thick   125
124b Encrustations more than 2 mm thick   126
125a Nearly all acanthostyles oriented vertically with the heads down; color various; spicules acanthostyles, tornotes, strongylotes or subtylotes, and arcuate isochelas   *Hymedesmia* spp.
125b Large acanthostyles may be oriented vertically, but small acanthostyles form a networklike pattern; persimmon to gold; spicules acanthostyles (85-136 µm and 161-370 µm), tornotes 133-166 µm) arcuate isochelas 25-35 µm)   *Hymendectyon lyoni* (fig. 2.9)
126a Sigmas absent (brown to red; acanthostyles 150-160 µm; acanthotornotes [few spines] 137-150 µm; arcuate isochelas 17-22 µm)   *Jones amaknakensis*
126b Sigmas present   127
127a With strongylotes to subtylotes (146-262 µm); gold to cinnamon-brown; styles to acanthostyles 180-366 µm; sigmas [often S-shaped] 29-56 µm; arcuate isochelas 13-52 µm
Lissodendoryx firma (fig. 2.8)
127b With anisotornotes (one end tornote, other end subtylote with a pointed nipple) 120-170 µm; color not known in live animals; acanthostyles 183-240 µm; sigmas [often S-shaped] 33-51 µm; arcuate isochelas 19-26 µm)   *Lissodendoryx* sp.

Species marked with an asterisk are not in the key.

Subclass Homoscleromorpha

Order Homoscleromorphida

Family Plakinidae

Plakina close to *brachylopha* Topsent, 1928. Subtidal.

Plakina close to *trilopha* Schulze, 1880. Subtidal.
Plakina sp. similar to *Corticium bowerbanki* Sara, 1960. Subtidal.

Subclass Tetractinomorpha

Order Choristida

Family Stellettidae

Penares cortius de Laubenfels, 1930. Intertidal and subtidal.
Stelletta clarella de Laubenfels, 1930. Intertidal and subtidal.

Family Geodiidae

Geodia mesotriaena Lendenfeld, 1910 (fig. 2.15). Subtidal.
Geodinella robusta Lendenfeld, 1910. Subtidal.

Family Pachastrellidae

Other species of this family have been collected in deep water.

?*Poecillastra rickettsi* de Laubenfels, 1930. Subtidal.

Order Spirophorida

Family Tetillidae

Craniella spinosa Lambe, 1894. Also called *Tetilla spinosa*; subtidal.
Craniella villosa Lambe, 1894. Also called *Tetilla villosa*; intertidal and subtidal.

Order Hadromerida

Family Suberitidae

Laxosuberites sp. May be a growth stage of *Suberites*; intertidal and subtidal.
Prosuberites sp. Intertidal and subtidal.
Pseudosuberites spp. Subtidal.
Suberites montiniger Carter, 1880. Subtidal.
Suberites simplex Lambe, 1894
Suberites ?*suberea* forma *latus* Lambe, 1893. Subtidal; associated with hermit crabs.
Suberites sp. Intertidal and subtidal.

Family Polymastiidae

Polymastia pacifica Lambe, 1894. Intertidal and subtidal.
Polymastia pachymastia de Laubenfels, 1932. Intertidal and subtidal.
Weberella similar to *verrucosa* Vacelet, 1960. Subtidal.

Jones amaknakensis (Lambe, 1895). Intertidal and subtidal.
Lissodendoryx firma (Lambe, 1895) (fig. 2.8). Intertidal; *L. noxiosa* de Laubenfels, 1932 is a synonym.
Lissodendoryx sp. Intertidal.
Merriamum oxeota (Koltun, 1958). Subtidal.
Myxilla behringensis Lambe, 1895. Intertidal and subtidal.
Myxilla incrustans (Esper, 1805-14) (fig. 2.12). Intertidal and subtidal.
Myxilla lacunosa Lambe, 1893. Subtidal.
Stelodoryx alaskensis (Lambe, 1895). Intertidal and subtidal.

Family Tedaniidae

*Tedania fragilis Lambe, 1895. Subtidal.
Tedania gurjanovae Koltun, 1958. Subtidal.
?*Tedanione obscurata* de Laubenfels, 1930. Intertidal.

Family Hymedesmiidae

Anaata brepha de Laubenfels, 1930. Intertidal and subtidal.
Anaata spongigartina de Laubenfels, 1930. Intertidal and subtidal.
Arndtanchora sp. Subtidal.
Hymedesanisochela rayae Bakus, 1966 (fig. 2.10). Subtidal.
Hymenamphiastra cyanocrypta de Laubenfels, 1930. Intertidal and subtidal.
Hymedesmia spp. Intertidal and subtidal.
?*Hymenanchora* sp. Subtidal.
?*Stylopus arndti* (de Laubenfels, 1930). Intertidal; ?*Astylinifer arndti* in *Light's Manual*.

Family Anchinoidae

Podotuberculum hoffmanni Bakus, 1966. Intertidal and subtidal.
Hamigera similar to *lundbecki* Hentschel, 1912. Subtidal; *Lissodendoryx* aff. *kyma* in Bakus, 1966.

Family Clathriidae

Axocielita originalis (de Laubenfels, 1930). Intertidal and subtidal; described as *Esperiopsis originalis*; *A. hartmani* Simpson, 1966 is a synonym.
?*Dictyociona asodes* (de Laubenfels, 1930). Intertidal and subtidal.
Microciona microjoanna de Laubenfels, 1930. Intertidal.
Microciona prolifera (Ellis & Solander, 1786). Intertidal and subtidal (also on floating docks) in Willapa Bay, where it has been introduced; may be expected in other bays.
**Microciona primitiva* Koltun, 1955. Subtidal.
Ophlitaspongia pennata (Lambe, 1893). Intertidal and shallow subtidal.
**Thalysias laevigata* Lambe, 1893. Subtidal; not reported since publication of the original description.

Family Plocamiidae

Anthoarcuata graceae Bakus, 1966. Subtidal.
Plocamia karykina (de Laubenfels, 1927). Intertidal.
Plocamilla illgi Bakus, 1966. Intertidal and subtidal.
Plocamilla lambei Burton, 1935. Intertidal and subtidal; *P. zimmeri* Bakus, 1966 is a synonym.
Stelotrochota hartmani Bakus, 1966 (fig. 2.13). Subtidal.

Order Haploscierida

Family Haliclonidae

?*Adocia* sp. similar to *Reniera foraminosa* Topsent, 1904. Subtidal.
?*Adocia* sp. similar to *Reniera mollis* Lambe, 1893. Intertidal and subtidal.
Adocia gellindra (de Laubenfels, 1932). Intertidal.
Haliclona ?ecbasis de Laubenfels, 1930. Intertidal.
Haliclona close to *permollis* (Bowerbank, 1866). Intertidal; *H. permollis* of various authors concerned with the fauna of the northeastern Pacific region.
Orina sp. Subtidal and intertidal.
Pachychalina spp. Several unidentified species of various growth forms; subtidal.
Reniera mollis Lambe, 1893. Intertidal and subtidal.
Reniera sp. Intertidal.
Sigmadocia edaphus (de Laubenfels, 1930). Intertidal and subtidal.
Sigmadocia spp. Intertidal and subtidal.
Toxidocia spp. Intertidal and subtidal.

Order Petrosiida

Family Petrosiidae

Xestospongia trindanea Ristau, 1978. Intertidal and subtidal.
Xestospongia vanilla (de Laubenfels, 1930). Intertidal and subtidal.

Family Dysideidae

Dysidea fragilis (Montagu, 1818) (fig. 2.16). Intertidal and subtidal.

Order Dendroceratida

Family Aplysillidae

Aplysilla glacialis (Merejkowsky, 1878). Intertidal.
Aplysilla polyraphis de Laubenfels, 1930. Intertidal.

Family Halisarcidae

Halisarca sacra de Laubenfels, 1930. Intertidal and subtidal.

Order Verongiida

Family Verongiidae

Hexadella sp. Subtidal.

3

PHYLUM CNIDARIA

Class Hydrozoa

The class Hydrozoa is covered by four keys. Two deal with the orders Siphonophora and Stylasterina. One is concerned with hydroids of the order Hydroida, and one with all hydrozoan medusae. The latter therefore includes species of the order Trachylina as well as of the order Hydroida.

Hydromedusae

Claudia E. Mills

Arai, M. N., & A. Brinckmann-Voss. 1980a. Hydromedusae of British Columbia and Puget Sound. Can. Bull. Fish. Aquat. Sci., no. 204. 194 pp.

------. 1980b. A new species of *Leuckartiara* (Pandeidae, Hydrozoa) from the east coast of Vancouver Island. Can. J. Zool., 58:1491-3.

------. 1983. A new species of *Amphinema*: *Amphinema platyhedos* n. sp. (Cnidaria, Hydrozoa, Pandeidae) from the Canadian West Coast. Can. J. Zool., 61:2179-2182.

Bigelow, H. B. 1913. Medusae and Siphonophorae collected by the U.S. Fisheries steamer "Albatross" in the northwestern Pacific, 1906. Proc. U. S. Nat. Mus., 44:1-119.

Brinckmann-Voss, A. 1985. Hydroids and medusae of *Sarsia apicula* (Murbach and Shearer, 1902) and *Sarsia princeps* (Haeckel, 1879) from British Columbia and Puget Sound with an evaluation of their systematic characters. Can. J. Zool., 63:673-81.

Fields, W. G., & G. O. Mackie. 1971. Evolution of the Chondrophora: evidence from behavioral studies on *Velella*. J. Fish. Res. Bd. Canada, 28:1595-1602.

Foerster, R. E. 1923. The hydromedusae of the west coast of North America, with special reference to those of the Vancouver Island region. Contrib. Canad. Biol., n. s., 1:219-77.

Kramp, P. L. 1959. The Hydromedusae of the Atlantic Ocean and adjacent waters. Dana Rep., 46:1-283.

------. 1961. Synopsis of the medusae of the world. J. Mar. Biol. Assoc. United Kingdom, 40:1-469.

------. 1965. The Hydromedusae of the Pacific and Indian Oceans. Dana Rep., 63:1-162.

------. 1968. The Hydromedusae of the Pacific and Indian Oceans. Sections II and III. Dana Rep., 72:1-200.

Kubota, S. 1978. The life-history of *Clytia edwardsi* (Hydrozoa; Campanulariidae) in Hokkaido, Japan. J. Fac. Sci. Hokkaido Univ., Ser. VI, Zool., 21:317-54.

------. 1981. Life-history and taxonomy of an *Obelia* species (Hydrozoa; Campanulariidae) in Hokkaido, Japan. J. Fac. Sci. Hokkaido Univ., Ser. VI, Zool., 22:379-399.

Larson, R. J. 1980. The medusa of *Velella velella* (Linnaeus, 1758) (Hydrozoa, Chondrophorae). J. Plankton Res., 2:183-6.

Mackie, G. O., & G. V. Mackie. 1963. Systematic and biological notes on living hydromedusae from Puget Sound. Nat. Mus. Canada, Bull., 199:63-84.

Mayer, A. G. 1910. Medusae of the World, vols. I, II. The Hydromedusae. Carnegie Inst. Washington Publ. 109. xv + 498 pp.

Mills, C. E. 1981. Seasonal occurrence of planktonic medusae and ctenophores in the San Juan Archipelago (NE Pacific). Wasmann J. Biol., 39:6-29.

------. 1985. A new hydrozoan, *Geomackiea zephyrolata* gen. nov., sp. nov. (Anthomedusae: Pandeidae), from inland marine waters of British Columbia and Washington State. Can. J. Zool., 63:2172-5.

Naumov, D. V. 1960. Hydroids and hydromedusae of marine, brackish and freshwater basins of the USSR. Opredeliteli po Faune SSSR, 70. 626 pp. (in Russian). (English translation, 1969. Jerusalem: Israel Program for Scientific Translation.)

Rees, J. T. 1980. The symbiotic hydrozoan *Endocrypta huntsmani*, its ascidian hosts, and its affinities with calycopsid hydromedusae. Wasmann J. Biol., 37: 48-54.

------. 1982. The hydrozoan *Cladonema* in California: a possible introduction from east Asia. Pac. Sci., 36:439-4.

Rees, J. T. & R. J. Larson. 1980. Morphological variation in the hydromedusa genus *Polyorchis* on the west coast of North America. Can. J. Zool., 58:2089-95.

Russell, F. S. 1953. The Medusae of the British Isles. Anthomedusae, Leptomedusae, Limnomedusae, Trachymedusae and Narcomedusae. Cambridge: Cambridge University Press. xiii + 530 pp.

------. 1970. The Medusae of the British Isles. II. Pelagic Scyphozoa with a Supplement to the First Volume on Hydromedusae. Cambridge: Cambridge University Press. xii + 284 pp.

1a	With one or more tentacles and a manubrium (unless the specimen is damaged); size variable, but mature specimens generally more than 2 mm high	2
1b	With neither tentacles nor a manubrium; sexually mature specimens (with a fully developed gonad along each of the 4 radial canals) less than 2 mm high	*Orthopyxis* spp.
2a	Tentacles originating at the margin of the bell	3
2b	Tentacles originating decidedly above the margin of the bell	62
3a	With not more than 4 tentacles	4
3b	With more than 4 tentacles	17
4a	With 2 tentacles, or 2 pairs of tentacles, situated on opposite sides of the bell (there may also be rudimentary tentacles elsewhere on the bell margin)	5
4b	Arrangement of tentacles not as described in choice 4a	7
5a	Bell with numerous zooxanthellae; with 2 pairs of tentacles in mature specimens, 2 tentacles in immature specimens (rare)	*Velella velella*
5b	Bell without zooxanthellae; with only 2 tentacles	6
6a	Bell usually with a conical or sharply pointed apex; with a large stomach on a broad gelatinous peduncle that hangs well below the bell margin; with 2 long tentacles and numerous rudimentary tentacles around the bell margin (common)	*Stomotoca atra* (fig 3.5)
6b	Bell with a sharply pointed apex; manubrium hanging directly from the roof of the subumbrella, never extending below the bell margin; with 2 long tentacles and numerous short but distinct tentacles around the bell margin (rare)	*Amphinema platyhedos*
7a	With 3 or 4 tentacles, each originating from a separate tentacle bulb	8
7b	With 1 to 4 tentacles originating from a single tentacle bulb (medusae are also produced asexually from this bulb) (with 3 other rudimentary tentacle bulbs, and with 5 exumbrellar nematocyst tracts running from the tentacle bulbs toward the apex)	*Hybocodon prolifer*
8a	With 3 tentacles when mature (but medusae less than 2 mm high may have only 1 tentacle)	9
8b	With 4 tentacles (many small medusae with 4 tentacles are young Anthomedusae or Leptomedusae that will have more tentacles when they mature; immature stages are usually difficult to identify and will not be dealt with in this key)	10
9a	With 3 tentacles, 1 long, 2 half as long, and 1 bulb without a tentacle (or with a rudimentary tentacle) opposite the long tentacle; gonad (on the manubrium, which may be red) usually with gametes by the time the medusa is 4 mm high	*Euphysa tentaculata* (fig.3.19)
9b	Arrangement of tentacles as described in choice 9a, but tentacle bulbs, as well as the manubrium, usually with bright red pigment; gonad not mature in medusae that are less than 6 mm high	young specimen of *Euphysa flammea* (fig. 3.16)
10a	Diameter of bell decidedly greater than the height; with 8 marginal vesicles, each with a black ocellus, spaced evenly around the bell margin	young specimen of *Tiaropsidium ?kelseyi*
10b	Diameter of bell not greater than the height; without marginal vesicles, but there may be ocelli on the tentacle bulbs	11
11a	Tentacles with swollen clusters of nematocysts along their entire length, in addition to a terminal swelling	12
11b	Tentacles with only a terminal swelling in which nematocysts are concentrated (scattered nematocysts may also occur in the tentacles, but they are not in swollen clusters)	16

34   Phylum Cnidaria

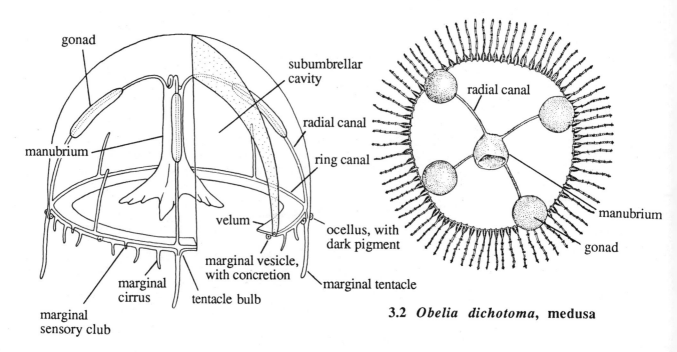

3.1 General structure of a hydrozoan medusa

3.2 *Obelia dichotoma*, medusa

12a   With a red or black ocellus on each tentacle bulb; manubrium of variable length   13
12b   Without ocelli, although tentacle bulbs may be red; manubrium not reaching the level of the
      margin of the bell and often with bright red pigment
                                                          Euphysa flammea (fig. 3.16) or *Euphysa japonica*
      Note: mature individuals of these 2 species (as in fig. 3.4) are apparently indistinguishable.
      Young stages can be differentiated because *E. japonica* has all 4 tentacles when it is released,
      whereas *E. flammea* has only 1 tentacle when it is released (the other 3 are added
      successively as the medusa grows).
13a   Bell narrow and conical, its height up to twice the diameter; height up to 4 cm, but usually
      not over 2 cm (rare)   *Sarsia princeps* (figs. 3.3 and 3.18)
13b   Height of bell only slightly greater than the diameter; height usually less than 1.5 cm   14
14a   Bell of mature medusa up to 8 mm high; tentacle bulbs and manubrium brilliant iridescent
      green (the proximal portion of the manubrium, which lacks gonads, extends to the level of
      the bell margin)   *Sarsia viridis*
14b   Bell of mature medusa more than 8 mm high; tentacle bulbs and manubrium not brilliant
      iridescent green, but may be some other color   15
15a   In relaxed specimens, proximal portion of the manubrium (the portion without gonads)
      extends at least to the level of the margin of the bell; with 4 exumbrellar nematocyst patches
      in each quadrant (2 near the margin of the bell and 2 about one-third the distance from the
      apex to the margin) (common; undescribed)   *Sarsia* sp.
15b   In relaxed specimens, proximal portion of the manubrium (the portion without gonads) less
      than one-half the height of the bell; without exumbrellar nematocyst patches, although young
      specimens may have scattered exumbrellar nematocysts that disappear by maturity (common)
                                                                                  Sarsia tubulosa (fig.3.17)
16a   Bell nearly spherical, scarcely open at the bottom; nematocysts, in loose clusters, sprinkled
      over most of the exumbrella   *Plotocnide borealis*
16b   Bell not spherical, although it is about as wide as high; nematocysts on the exumbrella
      scattered, rather than in loose clusters (rare)   *Bythotiara huntsmani*
17a   Tentacles (usually 9) branched once or twice, one branch ending in a sucker, the other 1 or 2

	branches having swollen clusters of nematocysts at the tip and elsewhere *Cladonema californicum*
17b	Tentacles not branched 18
18a	Radial canals branching, with all branches reaching the bell margin 19
18b	Radial canals not branching (they may, however, have numerous lateral diverticula) 21
19a	Bell up to 3 cm high; with 4 or more large tentacles and up to 45 small tentacles; all tentacles with a large knob of nematocysts at the tip; tentacles without distinct bulbs (rare) *Calycopsis nematophora*
19b	Bell up to 1 cm high; with up to 100 tentacles, these all the same size; tentacles with distinct bulbs, but without prominent nematocyst knobs at the tips of the tentacles 20
20a	Radial canals branching more or less dichotomously; nematocysts not commonly seen in the radial canals; exumbrella with small clusters of nematocysts between the tentacles, a short distance above the bell margin *Proboscidactyla flavicirrata* (fig. 3.12)
20b	Radial canals with fine, unsymmetrically placed branches that may contain nematocysts; exumbrella with scattered, rather than clustered, nematocysts *Trichydra pudica*
21a	With 4 or 8 radial canals 23
21b	With numerous radial canals 22
22a	Bell up to 12 cm wide; with up to 100 (or more) symmetrical radial canals; in mature specimens all radial canals reach the bell margin; gonads extending along nearly the entire length of the radial canals; up to 150 tentacles in full-grown specimens, but the tentacles are not all of the same size (common) *Aequorea victoria*
22b	Bell up to 1.2 cm wide; with fewer than 20 radial canals, these not symmetrically arranged, and some of them not reaching the bell margin; small globular gonads attached to the radial canals near the stomach; with more than 100 tentacles, all the same size (rare) *Dipleurosoma typicum*
23a	With 4 radial canals 28
23b	With 8 radial canals 24
24a	Gonads sausage-shaped, hanging freely into the subumbrellar cavity from near the proximal portions of the radial canals 25
24b	Gonads not hanging freely, but attached for most or all of their length to the radial canals 26
25a	Height of bell about twice the diameter; stomach suspended on a long peduncle *Aglantha digitale* (fig. 3.21)
25b	Height of bell about equal to the diameter; stomach attached to the roof of the subumbrella (rare) *Crossota* sp.
26a	Gonads attached to the radial canals, but free from the manubrium; with only 1 kind of tentacle, although the tentacles may be of more than 1 size 27
26b	Gonads attached by mesenteries to the radial canals and also to the manubrium; with 2 different kinds of tentacles: long tentacles having rings of nematocysts along their length, and short tentacles that lack nematocyst rings, but that have a terminal adhesive disk or sucker (rare) *Ptychogastria polaris*
27a	Mesoglea thick, especially at the apex; gonads sinuous, hanging from the radial canals; up to 90 large tentacles alternating with as many small ones; without any marginal sense organs *Melicertum octocostatum* (fig. 3.20)
27b	Mesoglea uniformly thin; gonads appressed to the subumbrella on both sides of the radial canals; up to 64 tentacles, these all alike; up to 64 club-shaped statocysts (rare) *Pantachogon* sp.
28a	Tentacles evenly distributed around the bell margin 34
28b	Most tentacles in clusters, although some single tentacles may also be present 29
29a	Manubrium with simple lips and no oral tentacles of any kind 33
29b	Manubrium with 4 oral tentacles, these either simple or much branched 30
30a	With 4 oral tentacles (these may be branched) at the corners of the mouth; with 8 clusters of tentacles on the bell margin (clusters with 3 and 5 tentacles usually alternate) *Rathkea octopunctata* (fig. 3.22)
30b	With a simple tubular mouth and 4 dichotomously branched oral tentacles inserted above the mouth opening; with 4 clusters of tentacles on the bell margin 31
31a	Oral tentacles short, branching 1 or 2 times; with up to 9 marginal tentacles originating from each bulb; ocelli on the tentacle bulbs; less than 4 mm high at maturity *Bougainvillia ramosa* (fig. 3.8)

31b Oral tentacles long, branching at least 4 times; with more than 10 marginal tentacles originating from each bulb; ocelli on the tentacle bulbs or on the bases of the tentacles; over 8 mm high at maturity 32
32a Oral tentacles branching 4-5 times; usually with 11-15 (sometimes up to 22) marginal tentacles originating from each bulb; ocelli on the bases of the tentacles; stomach suspended on a short, broad peduncle ..... *Bougainvillia superciliaris*
32b Oral tentacles branching 5-7 times; with 20-60 marginal tentacles originating from each bulb; ocelli on the tentacle bulbs; stomach attached directly to the subumbrella ..... *Bougainvillia principis* (includes *B. multitentaculata* of Foerster, 1923)
33a With a single tentacle in line with each radial canal, and with an interradial cluster of tentacles (these arising from separate bulbs) in each quadrant; tentacle bulbs with red or black ocelli ..... *Halimedusa typus*
33b With a single large tentacle in line with each radial canal and with a swollen process in each interradius, this bulb giving rise to up to 10 tentaculae; ocelli absent (rare) ..... *Geomackiea zephyrolata*
34a Tentacles with prominent rings of nematocysts along their entire length 35
34b Tentacles without prominent rings of nematocysts along their entire length 37
35a Medusa nearly disk-shaped, not deeply convex; bell of mature specimens not more than 6 mm in diameter 45
35b Medusa bell-shaped, nearly hemispherical; bell of mature specimens more than 1 cm in diameter 36
36a Tentacles with an adhesive pad located some distance proximal to the tip, and characteristically angled where the pad is located; without centripetal canals; gonads orange, red, or brown ..... *Gonionemus vertens*
36b Tentacles without an adhesive pad, but sometimes with a filamentous extension protruding at an angle from the tip; usually with numerous subumbrellar centripetal canals; gonads usually pale pink or brown ..... *Eperetmus typus* (fig. 3.15)
37a Radial canals with conspicuous lateral diverticula whose length is at least twice the diameter of the canals 38
37b Radial canals without conspicuous lateral diverticula (if the margins of the canals are irregular, they do not form distinct diverticula whose length appreciably exceeds the diameter of the canals) 40
38a Bell higher than its diameter, or height and diameter nearly equal; gonads fingerlike, in 4 groups, attached to the radial canals and hanging down into the subumbrellar cavity ..... *Polyorchis penicillatus* (fig. 3.9)
38b Diameter of bell much greater than the height; gonads closely associated with the radial canals and their diverticula, not hanging down into the subumbrellar cavity 39
39a Stomach short, quadrate; radial canals with gonads arranged in up to 30 lamelliform diverticula on both sides; with up to 500 tentacles, these without ocelli (rare) ..... *Ptychogena lactea*
39b Stomach consisting of 4 open grooves that extend along the greater part of the length of each radial canal; gonads in branched diverticula from the walls of the stomach; with up to about 4000 tentacles, each of these with an ocellus ..... *Staurophora mertensi*
40a Diameter of bell distinctly greater than the height; gonads associated with the radial canals 41
40b Height of bell greater than the diameter, or height and diameter nearly equal; gonads associated with the stomach 53
41a With ocelli on the tentacle bulbs or on the marginal vesicles 42
41b Without ocelli 44
42a With ocelli on some or all of the tentacle bulbs; without marginal vesicles, but with short marginal clubs (rare) ..... *Laodicea* sp.
42b Without tentacular ocelli; with 8 marginal vesicles, each with a black ocellus 43
43a With about 300 short tentacles, these all alike; lips of mouth much folded and crenulated ..... *Tiaropsis multicirrata*
43b With 8 large tentacles and up to 128 additional rudimentary tentacles; lips of mouth slightly frilled (rare) ..... *Tiaropsidium kelseyi*
44a With tentacles all the same size in mature specimens 45
44b With tentacles not all the same size in mature specimens 52

Phylum Cnidaria 37

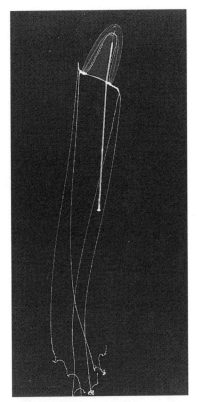

3.3 *Sarsia princeps*

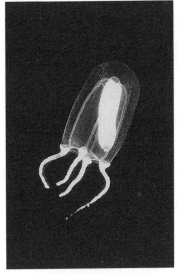

3.4 *Euphysa* sp.

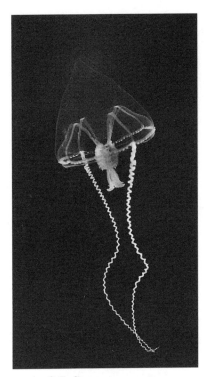

3.5 *Stomotoca atra*

3.7 *Neoturris breviconis*

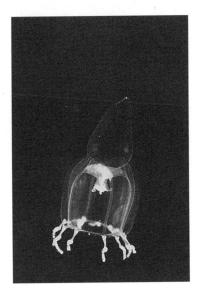

3.6 *Halitholus* sp.

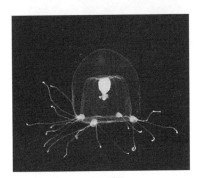

3.8 *Bougainvillia ramosa*

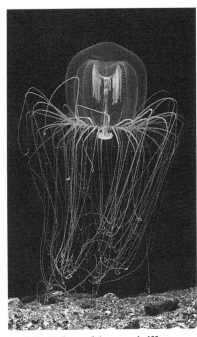

3.9 *Polyorchis penicillatus*

38   Phylum Cnidaria

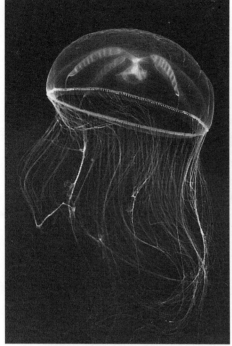

3.10 *Mitrocoma cellularia*

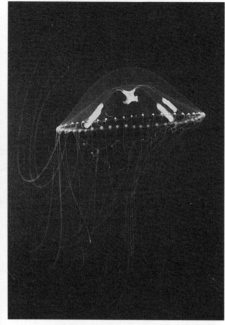

3.11 *Phialidium gregarium*

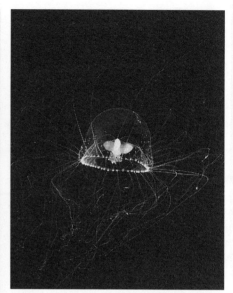

3.12 *Proboscidactyla flavicirrata*

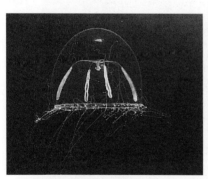

3.13 *Mitrocomella polydiademata*

3.14 *Eutonina indicans*

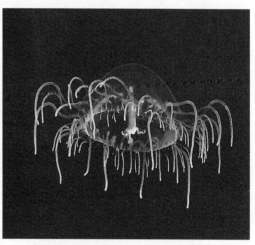

3.15 *Eperetmus typus*

45a Gonads ovoid or nearly spherical, occupying only a short section of each radial canal; tentacles more or less of fixed length (not highly extensile); diameter of bell of mature specimens not more than 6 mm 46
45b Gonads elongated, associated with the radial canals for a substantial portion of their length; tentacles highly extensile; diameter of bell of mature specimens more than 1 cm 47
46a Gonads of small individuals (up to 1 mm in diameter) near the middle of the radial canals, becoming displaced progressively farther toward the edge in specimens approaching maximum size (diameter 4 mm) .... *Obelia dichotoma* (fig. 3.2)
46b Gonads of small individuals (up to 1 mm in diameter) near the mouth, becoming displaced to near the middle of the radial canals in specimens approaching maximum size (diameter 4 to 6 mm) .... *Obelia geniculata*

47a	With stomach attached directly to the subumbrella	48
47b	With stomach suspended on a gelatinous peduncle	50
48a	With up to 80 tentacles; with 1-2 marginal vesicles between each 2 tentacles, each vesicle with one concretion; lips of mouth short	49
48b	With up to 350 tentacles; with a total of 16-24 marginal vesicles, each with numerous concretions; lips of mouth long and extended	*Mitrocoma cellularia* (fig. 3.10)
49a	Gonads not usually mature until diameter of bell is greater than 1.5 cm; gonads, when sectioned transversely, with an elliptical outline; gonads usually light-colored, but may have a stripe of dark pigment running lengthwise; bell margin may have a ring of dark pigment; up to 20 tentacles per quadrant at maturity	*Phialidium gregarium* (fig. 3.11)
49b	Gonads mature by the time the bell reaches a diameter of 1 cm; gonads, when sectioned transversely, with a circular outline; gonads usually fairly dark in color (brown, gray, or yellowish); bell margin without a ring of dark pigment; less than 10 tentacles per quadrant at maturity	*Phialidium lomae*
50a	With 8 marginal vesicles; with up to 230 marginal tentacles (colorless or with white or sepia pigment in the stomach, gonads, and tentacle bases)	*Eutonina indicans* (fig. 3.14)
50b	With more than 8 marginal vesicles; usually with less than 200 marginal tentacles	51
51a	With up to 180 marginal tentacles and with as many or more marginal vesicles; stomach, gonads, and radial canals faintly yellow, tentacle bulbs brick red	*Eirene mollis*
51b	With up to 150 marginal tentacles and about half as many marginal vesicles; stomach and radial canals (and sometimes gonads) purple	*Foersteria purpurea*
52a	With up to 64 large tentacles and 5-9 small tentacles or cirri between each two large tentacles; ring canal and gonads usually pale rose-pink; exumbrellar surface smooth and regular	*Mitrocomella polydiademata* (fig. 3.13)
52b	With up to 350 tentacles, long and short ones more or less alternating; colorless, or with the gonads and/or the exumbrella pale blue; exumbrellar surface irregularly creased	*Mitrocoma cellularia* (fig. 3.10)
53a	With more than 50 marginal tentacles	54
53b	With fewer than 50 marginal tentacles	55
54a	With a large, bulbous apical projection that is somewhat pinched off from the rest of the bell; with up to 150 tentacles, each with a black ocellus and some red or orange pigment on its bulb; bell more or less opaque	*Catablema multicirrata*
54b	Apical projection low and rounded, not at all pinched off from the rest of the bell; with up to 100 or more marginal tentacles, these without ocelli; manubrium and tentacle bases usually orangish; bell not opaque	*Neoturris breviconis* (fig. 3.7)
55a	Bell with mesoglea of nearly uniform thickness, or somewhat thicker at the apex, but the apex is not pinched off from the lower portion of the bell	56
55b	Bell with an apical projection of mesoglea that is noticeably pinched off from the lower portion of the bell	58
56a	Medusa with up to 24 large tentacles, these with large conical marginal bulbs, but without swollen nematocyst knobs at the tips; bell up to 7.5 cm high, with a deep brownish red manubrium (rare)	*Pandea rubra*
56b	Medusa with 12 or fewer tentacles, these without marginal tentacle bulbs, but with a swollen knob of nematocysts at the tip; bell less than 3 cm high	57
57a	With 8 tentacles; bell noticeably compressed from side to side so that in transverse section it is elliptical; tentacle tips without bright red pigment (rare)	*Bythotiara depressa*
57b	With 6-12 tentacles; bell not compressed from side to side, so that in transverse section it is circular; knobs at tentacle tips bright orange-red (rare)	*Heterotiara anonyma*
58a	Upper one-third or one-half of the stomach attached to the 4 radial canals by "mesenteries"	59
58b	Stomach not attached to the radial canals by mesenteries, although the top of stomach may be somewhat elongated along the proximal portions of the radial canals	60
59a	Each gonad consisting of a series of nearly parallel folds; gonads not distinctly joined into pairs by transverse bridges; tentacle bulbs without spurs running up the exumbrella	*Leuckartiara foersteri*
59b	Gonads not as above (they are usually irregularly folded and joined into pairs by transverse bridges, so that each pair is horseshoe-shaped); tentacle bulbs with or without spurs running a short distance up the exumbrella	*Leuckartiara* spp.

40 Phylum Cnidaria

3.16 *Euphysa flammea*

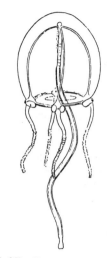

3.17 *Sarsia tubulosa*

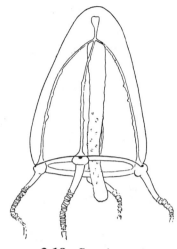

3.18 *Sarsia princeps*

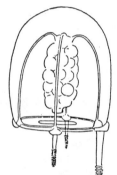

3.19 *Euphysa tentaculata*

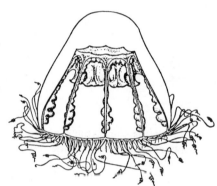

3.20 *Melicertum octocostatum*

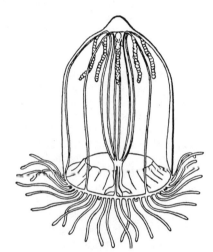

3.21 *Aglantha digitale*

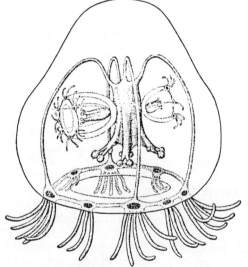

3.22 *Rathkea octopunctata*

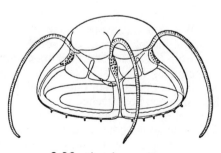

3.23 *Aegina citrea*

60a Bell up to 2.5 cm high; radial canals irregular and denticulate; apical projection nearly as large as the subumbrellar cavity *Catablema nodulosa*
60b Bell less than 1.3 cm high; radial canals smooth or slightly jagged; apical projection considerably smaller than the subumbrellar cavity 61
61a With only 8 tentacles (these with ocelli), but there may be additional rudimentary tentacle bulbs which also have ocelli; bell up to 1.2 cm high, with a round apical projection; radial canals jagged (rare) *Halitholus pauper*
61b With 8 to 16 tentacles of varying sizes (these with ocelli), and with 1-2 rudimentary marginal bulbs (with ocelli) between tentacles; bell up to 1 cm high, with a more or less conical apical projection; radial canals smooth *Halitholus* sp. (fig. 3.6)
62a With up to 100 or more tentacles, these having prominent rings of nematocysts along their entire length; with a central, pendent manubrium; gonads along the 4 radial canals *Eperetmus typus* (fig. 3.15)
62b With less than 40 tentacles, these lacking rings of nematocysts; with a broad stomach and peripheral stomach pouches, the latter having gonads on their walls; without radial canals 63
63a With 4 tentacles, and usually 8 stomach pouches *Aegina citrea* (fig. 3.23)
63b With more than 4 tentacles, and an equal number of tentacles and stomach pouches 64
64a With marginal statocysts and otoporpae (bristly tracts that run a short distance up the exumbrella from the margin) *Cunina* sp.
64b With marginal statocysts, but without otoporpae 65
65a With 8-20 (usually 16) tentacles and stomach pouches; each marginal lappet with up to 20 statocysts *Solmissus marshalli*
65b With 20-40 tentacles and stomach pouches; each marginal lappet with 2-5 statocysts (rare) *Solmissus incisa*

Order Hydroida

Suborder Athecata (Anthomedusae)

Family Corymorphidae

Plotocnide borealis Wagner, 1885

Family Euphysidae

Euphysa flammea (Linko, 1905) (fig. 3.16; see also fig. 3.4, which may be either *E. flammea* or *E. japonica*)
Euphysa japonica (Maas, 1909)
Euphysa tentaculata Linko, 1905 (fig. 3.19)

Family Tubulariidae

Hybocodon prolifer L. Agassiz, 1862

Family Velellidae

Velella velella (Linnaeus, 1758)

Family Corynidae

Sarsia sp. (undescribed)
Sarsia princeps (Haeckel, 1879) (figs. 3.3 and 3.18)

Sarsia tubulosa (M. Sars, 1835) (fig. 3.17)
Sarsia viridis Brinckmann-Voss, 1980
Sarsia spp. (at least some of these may be described species)

Family Cladonematidae

Cladonema californicum Hyman, 1947

Family Rathkeidae

Rathkea octopunctata (M. Sars, 1835) (fig. 3.22)

Family Bougainvilliidae

Bougainvillia principis (Steenstrup, 1850)
Bougainvillia ramosa (van Beneden, 1844) (fig. 3.8)
Bougainvillia superciliaris (L. Agassiz, 1849)

Family Pandeidae

Amphinema platyhedos Arai & Brinckmann-Voss, 1983
Catablema multicirrata Kishinouye, 1910
Catablema nodulosa (Bigelow, 1913)
Geomackiea zephyrolata Mills, 1985
Halitholus pauper Hartlaub, 1914
Halitholus sp. (fig. 3.6)
Leuckartiara foersteri Arai & Brinckmann-Voss, 1980
Leuckartiara spp.
Neoturris breviconis (Murbach & Shearer, 1902) (fig. 3.7)
Pandea rubra Bigelow, 1913
Stomotoca atra A. Agassiz, 1862 (fig. 3.5)

Family Halimedusidae

Halimedusa typus Bigelow, 1916

Family Calycopsidae

Bythotiara depressa Naumov, 1960
Bythotiara huntsmani (Fraser, 1911)
Calycopsis nematophora Bigelow, 1913
Heterotiara anonyma Maas, 1905

Family Trichydridae

Trichydra pudica Wright, 1858

Family Polyorchidae

Polyorchis penicillatus (Eschscholtz, 1829) (fig. 3.9)

Suborder Thecata (Leptomedusae)

Family Dipleurosomatidae

Dipleurosoma typicum Boeck, 1866

Family Melicertidae

Melicertum octocostatum (M. Sars, 1835) (fig. 3.20)

Family Laodiceidae

Laodicea sp.
Ptychogena lactea A. Agassiz, 1865
Staurophora mertensi Brandt, 1835

Family Mitrocomidae

Foersteria purpurea (Foerster, 1923)
Mitrocoma cellularia (A. Agassiz, 1865) (fig. 3.10)
Mitrocomella polydiademata (Romanes, 1876) (fig. 3.13)
Tiaropsidium kelseyi Torrey, 1909
Tiaropsis multicirrata (M. Sars, 1835)

Family Campanulariidae

Obelia dichotoma (Linnaeus, 1758) (fig. 3.2)
Obelia geniculata (Linnaeus, 1876)
Orthopyxis spp.
Phialidium gregarium (A. Agassiz, 1862) (fig. 3.11)
Phialidium lomae Torrey, 1909

Family Eirenidae

Eirene mollis Torrey, 1909
Eutonina indicans (Romanes, 1876) (fig. 3.14)

Family Aequoreidae

Aequorea victoria (Murbach & Shearer, 1902)

Suborder Limnomedusae

Family Olindiasidae

Eperetmus typus Bigelow, 1915 (fig. 3.15)
Gonionemus vertens A. Agassiz, 1862

Family Proboscidactylidae

Proboscidactyla flavicirrata Brandt, 1835 (fig. 3.12)

Order Trachylina

Suborder Trachymedusae

Family Ptychogastriidae

Ptychogastria polaris Allman, 1878

Family Rhopalonematidae

Aglantha digitale (O. F. Müller, 1776) (fig. 3.21)
Crossota sp.
Pantachogon sp.

Suborder Narcomedusae

Family Aeginidae

Aegina citrea Eschscholtz, 1829 (fig. 3.23)

Family Cuninidae

Cunina sp.
Solmissus incisa (Fewkes, 1886)
Solmissus marshalli Agassiz & Mayer, 1902

Hydroid Polyps

Claudia E. Mills and Richard L. Miller

The key is based on characters that can be seen in living polyps, and usually leads only to genus. It does not include a few genera that may be found here. The work of Fraser (1937) provides descriptions, illustrations, and distributional notes that enable one to identify many hydroids to species. Unfortunately, our knowledge of species belonging to certain genera is poor, so not all of the species dealt with by Fraser are necessarily valid. It is likely, moreover, that species and genera not listed by Fraser will be encountered. The monograph of Naumov (1960), although concerned with the fauna of the USSR, is helpful with respect to some species in our region. The revisions of Cornelius (1975a, 1975b, 1979, 1982) should be consulted in connection with several genera of thecate hydroids, but some of the species he has placed in synonymy are now thought to be valid.

Agassiz, L. 1862. Hydroidae. *In* Contributions to the Natural History of the United States of America, Second Monograph, pp. 181-380. Boston: Little, Brown, & Co.
Allman, G. J. 1871-2. A Monograph of the Gymnoblastic or Tubularian Hydroids. Parts I & II. London: Ray Society. 450 pp.
Arai, M. N., & A. Brinckmann-Voss. 1980. Hydromedusae of British Columbia and Puget Sound. Can. Bull. Fish. Aquat. Sci., 204. 192 pp.

Bouillon, J. 1985. Essai de classification des Hydropolypes-Hydroméduses (Hydrozoa-Cnidaria). Indo-Malayan Zool., 1: 29-243.

Calkins, G. N. 1899. Some hydroids of Puget Sound. Proc. Boston Soc. Nat. Hist., 28:333-47.

Cornelius, P. F. S. 1975a. The hydroid species of *Obelia* (Coelenterata, Hydrozoa: Campanulariidae), with notes on the medusa stage. Bull. British Mus. Nat. Hist. (Zool.), 28:251-93.

------. 1975b. A revision of the species of Lafoeidae and Haleciidae (Coelenterata: Hydroida) recorded from Britain and nearby seas. Bull. British Mus. Nat. Hist. (Zool.), 28:375-426.

------. 1979. A revision of the species of Sertulariidae (Coelenterata: Hydroida) recorded from Britain and nearby seas. Bull. British Mus. Nat. Hist. (Zool.), 34:243-321.

------. 1982. Hydroids and medusae of the family Campanulariidae recorded from the eastern North Atlantic, with a world synopsis of genera. Bull. British Mus. Nat. Hist. (Zool.), 42:37-148.

Fraser, C. McL. 1937. Hydroids of the Pacific Coast of Canada and the United States. Toronto: University of Toronto Press. 207 pp.

------. 1944. Hydroids of the Atlantic Coast of North America. Toronto: University of Toronto Press. 451 pp.

Hincks, T. 1868. A History of the British Hydroid Zoophytes, I. II. London. 328 pp.

Hirai, E., & M. Yamada. 1965. On a new athecate hydroid *Hataia parva* n. gen., n. sp. Bull. Mar. Biol. Sta. Asamushi, Tôhoku Univ., 12:59-62.

Naumov, D. V. 1960. [Hydroids and hydromedusae of marine, brackish and freshwater basins of the USSR.] Opredeliteli po Faune SSSR, 70. 626 pp. (in Russian). (English translation, 1969. Jerusalem: Israel Program for Scientific Translation.)

Norenburg, J. L., & M. P. Morse. 1983. Systematic implications of *Euphysa ruthae* n. sp. (Athecata: Corymorphidae), a psammophilic solitary hydroid with unusual morphogenesis. Trans. Amer. Micr. Soc., 102:1-17.

Nutting, C. C. 1900. American hydroids. Part. I. The Plumularidae. Spec. Bull. Smithsonian Inst., U. S. Nat. Mus. 142 pp.

------. 1904. American hydroids. Part. II. The Sertularidae. Spec. Bull. Smithsonian Inst., U. S. Nat. Mus. 325 pp.

------. 1915. American hydroids. Part. III. The Campanularidae and the Bonneviellidae. Spec. Bull. Smithsonian Inst., U. S. Nat. Mus. 126 pp.

Russell, F. S. 1953. The Medusae of the British Isles. Anthomedusae, Leptomedusae, Limnomedusae, Trachymedusae and Narcomedusae. Cambridge: Cambridge University Press. xiii + 530 pp.

Strong, L. H. 1925. Development of certain Puget Sound hydroids and medusae. Publ. Puget Sound Biol. Sta., 3:383-99.

Tardent, P. 1980. A giant *Tubularia* (Cnidaria, Hydrozoa) from the waters of the San Juan Islands, Washington. Syesis, 13:17-25.

Wieser, W. 1958. Occurrence of *Protohydra leuckarti* in Puget Sound. Pacific Sci., 12:106-8.

1a Benthic (solitary or colonial); polyps transparent or variously pigmented; with fixed gonophores or releasing free-swimming medusae ... 2
1b Pelagic; with an oval float and upright triangular sail; polyps deep blue or purple; releases medusae containing zooxanthellae ... *Velella velella* (fig 3.42)
2a Hydranth not enclosed by a hydrotheca or pseudohydrotheca ... 3
2b Hydranth at least partly enclosed by a hydrotheca or pseudohydrotheca ... 39
3a Hydranth with tentacles; polyp solitary or colonial ... 4
3b Hydranth without tentacles; polyp solitary (in sediment or in mats of blue-green algae) ... *Protohydra ?leuckarti* (fig. 3.35)
4a Hydranth with 1 tentacle only (colony always found subtidally on bivalve shells; with knobbed "defensive" zooids in addition to feeding zooids; releases medusae) ... *Monobrachium parasitum*
4b Hydranth with more than 1 tentacle ... 5
5a Hydranths usually with 2 or 3 tentacles when mature ... 6
5b Hydranths usually with more than 3 tentacles when mature ... 7
6a Hydranths with 2 tentacles; polyps 1-2 mm tall (found only on the rims of tubes of certain sabellid polychaetes [*Schizobranchia, Potamilla*]) ... *Proboscidactyla flavicirrata*
6b Hydranths usually with 2 or 3 tentacles (occasionally up to 5); polyps 1-2 mm tall, iridescent, with a longitudinal pink stripe near the surface ... *Rhysia* sp.

46 Phylum Cnidaria

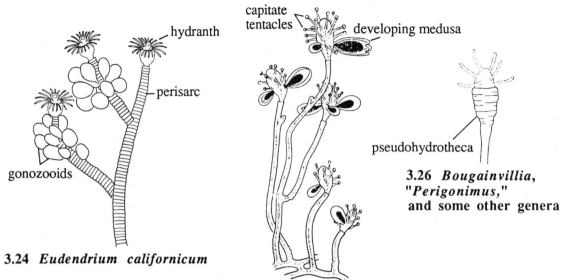

Athecate hydroids

3.24 *Eudendrium californicum*
3.25 *Sarsia tubulosa*
3.26 *Bougainvillia,* "*Perigonimus,*" and some other genera

3.27 *Obelia*
3.28 *Halecium*
3.29 *Lafoea dumosa*
3.30 Laeodiceidae
3.31 *Abietinaria*
3.32 *Thuiaria*
3.33 *Sertularella*
3.34 *Plumularia setacea*

Thecate hydroids

Terminology used in identification of hydroid polyps

Phylum Cnidaria 47

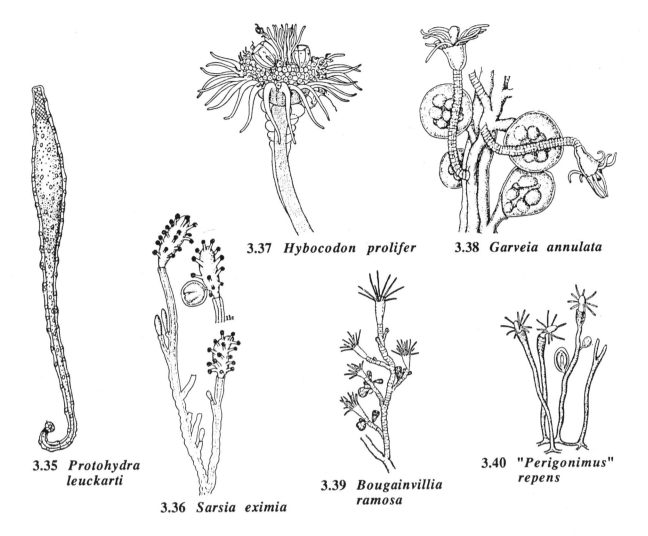

3.35 *Protohydra leuckarti*
3.36 *Sarsia eximia*
3.37 *Hybocodon prolifer*
3.38 *Garveia annulata*
3.39 *Bougainvillia ramosa*
3.40 *"Perigonimus" repens*

7a	Tentacles on the hydranths arranged in 1 or 2 whorls	8
7b	Tentacles on the hydranths scattered over the surface, not in distinct whorls	18
8a	Tentacles of hydranths limited to 1 whorl	26
8b	Tentacles of hydranths arranged in 2 whorls (1 whorl oral, the other aboral)	9
9a	With 5 or fewer tentacles in each whorl	25
9b	Usually with more than 5 tentacles in the oral whorl, and more than 10 tentacles in the aboral whorl	10
10a	Polyp solitary, not colonial; perisarc thin and flexible; stem with anchoring filaments; always found "rooted" in mud or sand, or interstitial	11
10b	Polyp either solitary or colonial; perisarc rigid; stem without anchoring filaments; always attached to a solid substratum	13
11a	Polyp usually large, up to 15 cm tall; oral and aboral tentacles filiform; epibenthic in soft sediments	*Corymorpha* sp.
11b	Polyp small, up to 8 mm tall; oral tentacles moniliform or capitate, aboral tentacles moniliform; epibenthic or interstitial	12
12a	Polyp bearing reduced gonophores (these arranged singly); perisarc inconspicuous (interstitial in gravelly sand)	*Euphysa ruthae*
12b	Polyp releasing free medusae; perisarc forming a conspicuous gelatinous tube covering the basal half of the stem (not likely to be found except in soft sediment)	*Euphysa* spp.

48   Phylum Cnidaria

13a Polyp bearing fixed medusoid gonophores that do not become free-swimming medusae   14
13b Polyp releasing free medusae that have 1-4 tentacles arising from a single marginal tentacle bulb   *Hybocodon prolifer* (fig. 3.37)
14a Polyps in dense colonies that branch near the bases of the stems or from the stolons; usually in water with a temperature above 18°C; with approximately equal numbers of oral and aboral tentacles; healthy female gonophores with 4-8 flattened bladelike tentacles at the apical end   *Tubularia crocea*
14b Polyps usually solitary or in small colonies having only a few polyps growing from the same stolon; usually in water with a temperature below 15°C; with unequal numbers of oral and aboral tentacles; healthy female gonophores with cylindrical (not flattened) tentacles, or without tentacles   15
15a Polyp stems narrow, up to 30 cm long; oral and aboral tentacle whorls with a single circle of tentacles   16
15b Polyp stems stout, up to 2 cm long; with about 60 oral tentacles in 3-4 circles, and about 50 aboral tentacles in an irregular double circle (gonophores without tentacles, radial canals, or ring canal; colony a deep claret red, living subtidally in swift current areas)   undescribed species of Tubulariidae
16a Polyp stems up to 30 cm long; with over 40 oral tentacles and about 20-30 aboral tentacles; female gonophores with 4 radial canals and a ring canal; subtidal, not usually on floats   *Tubularia indivisa*
16b Polyp stems up to 10 cm long; with about 20 oral tentacles and up to 50 aboral tentacles; female gonophores without radial canals or a ring canal; intertidal, subtidal, and on floats   17
17a Polyp stems usually 2-4 cm long; with about 18 oral tentacles and 22-26 aboral tentacles; healthy female gonophores usually with 3-4 tentacles that may be as long as the gonophores themselves (on rocks in the low intertidal and subtidal; and also on floats)   *Tubularia marina*
17b Polyp stems up to 10 cm long; with about 20 oral tentacles and 40-50 aboral tentacles; healthy female gonophores usually with 3-5 tentacles that may be more than half as long as the gonophores, but not fully as long   *Tubularia harrimani*
18a Hydranths with filiform tentacles only   19
18b Hydranths with capitate tentacles only, or with capitate and filiform tentacles   22
19a Stems of hydranths extensively branched above the stolons; gonophores located below the hydranths and on the stems; in brackish or nearly fresh water   *Cordylophora caspia*
19b Stems of hydranths not branched above the stolons; gonophores not as above; strictly marine   20
20a Polyps solitary; gonophores and new polyp buds emerging from among the tentacles   *Hataia parva*
20b Polyps colonial; gonophores or new polyp buds not emerging from among the tentacles   21
21a Gonophores located at the base of the hydrant, below the tentacles, and liberating free-swimming medusae with 4 tentacles; living within the incurrent siphon of certain ascidians, especially *Ascidia paratropa*   *Bythotiara huntsmani*
21b Gonophores arise from the stolon on separate stems; intertidal or subtidal on hard substrata   *Rhizogeton* sp.
22a Polyps with fixed gonophores that do not develop into free-swimming medusae   *Coryne* sp.
22b Polyps releasing free-swimming medusae   23
23a All polyps in the colony with capitate tentacles only   *Sarsia* spp. (includes *S. tubulosa* [fig. 3.25] and *S. eximia* [fig. 3.36])
23b At least some polyps in the colony with a single whorl of inconspicuous filiform tentacles at the base of the hydrant (below all of the capitate tentacles)   24
24a Polyps releasing medusae with exumbrellar nematocysts arranged in 8 or 16 longitudinal rows   *Sarsia japonica*
24b Polyps releasing medusae with scattered exumbrellar nematocysts   *Sarsia* sp.
25a Hydranths with tentacles arranged in 2 whorls; tentacles in both whorls capitate; polyp releasing medusae with 4 simple tentacles   *Sarsia* sp.
25b Hydranths with tentacles arranged in 2 whorls; tentacles in oral whorl capitate, tentacles in aboral whorl filiform; polyp releasing medusae with 9 branched tentacles   *Cladonema californicum*

26a	Polyps solitary, less than 2 mm tall	27
26b	Polyps colonial, usually more than 2 mm tall	29
27a	Polyp attached to the substratum; with up to 8 long, filiform or weakly capitate tentacles	28
27b	Polyp interstitial (not firmly attached to the substratum); with 4-6 short capitate tentacles	Boreohydridae (undescribed species)
28a	Hydranth emerging from a prominent perisarcal disk having a single rounded spine; with up to 8 long, slightly capitate tentacles; releasing medusae with 4 tentacles, these with scattered nematocysts	*Halimedusa typus*
28b	Polyp without a prominent perisarcal disk; with 4-6 long filiform tentacles; releasing medusae with 8 or more tentacles, these with nematocysts more or less arranged in rings	*Gonionemus vertens*
29a	Stalks of hydranths enclosed by perisarc for most of their length; colonies branched or unbranched above the stolons	37
29b	Stalks of hydranths not enclosed by perisarc for most of their length; colonies not branched above the stolons	30
30a	Stolons anastomosing to form a basal perisarcal crust from which the polyps emerge, the crust usually with spines	31
30b	Stolons not forming a basal perisarcal crust and without spines	34
31a	Gastrozooids usually with 8 tentacles	33
31b	Gastrozooids usually with 12-24 tentacles	32
32a	Spines smooth and long; adult gastrozooids usually with 12-20 tentacles; female gonophores with 1 egg	*Hydractinia milleri*
32b	Spines jagged and sometimes coalescing to form ridges; adult gastrozooids usually with 20-24 tentacles; female gonophores with several to many eggs (usually more than 6)	*Hydractinia aggregata*
33a	Spines numerous, smooth, short and slightly curved; gastrozooids uniformly pink in color (female gonophores with 1 egg)	*Hydractinia laevispina*
33b	Spines not numerous, usually smooth and long; gastrozooids with a white hypostome (female gonophores with 1 egg)	*Hydractinia* sp.
	Note: although some of the gastrozooids of this species have 8 tentacles, others have more than 8, in which case they are in 2 indistinct whorls).	
34a	Polyp with a short, smooth, collarlike perisarcal elaboration into which it can partly retract (the expanded polyp is 3 to 5 times taller than this collar); releasing medusae with 4 tentacles	*Trichydra pudica*
34b	Polyp without a short, collarlike perisarcal elaboration into which it can partly retract	35
35a	Polyps robust and conspicuous, white; 3-6 mm tall; usually with 12-20 tentacles	Hydractiniidae (undescribed species)
35b	Polyps delicate and inconspicuous; usually less than 2 mm tall; usually with fewer than 10 tentacles	36
36a	Polyps with 4-6 long filiform tentacles; polyps less than 0.5 mm tall; releasing medusae with 8 tentacle bulbs, with 1-2 tentacles on each bulb	*Rathkea octopunctata*
36b	Polyps usually with more than 5 filiform tentacles; polyps usually more than 0.5 mm tall; releasing free medusae, but these are not as described in choice 36a	various species of Pandeidae
37a	Colonies usually much branched; stems usually strongly annulated (but there may be only a few annulations); with fixed gonophores only	*Eudendrium* spp. (including *E. californicum* [fig. 3.24])
37b	Colonies branched or consisting of single polyps arising from stolons; stems usually not annulated, but they may be irregularly wrinkled; releasing free-swimming medusae	38
38a	Releasing medusae with 4 compound tentacle bulbs, each usually with several tentacles and several ocelli	*Bougainvillia* spp. (fig. 3.39)
38b	Releasing medusae with simple tentacle bulbs and various numbers of tentacles (not more than 1 per marginal bulb) (with or without ocelli)	*Leuckartiara* spp. or *Neoturris* spp.
39a	Perisarc enclosing or partly enclosing the hydranth constituting a true hydrotheca (of consistent form, and often with toothed margins, devices for closing the aperture, and other elaborations)	43

39b Perisarc enclosing or partly enclosing the hydranth constituting a pseudohydrotheca rather than a true hydrotheca (pseudohydrothecae are generally of inconsistent and somewhat irregular form, and have transverse wrinkles) 40
40a Pseudohydrotheca enclosing the lower portion of the hydranth flaring to some extent, but not forming a distinct cup; releasing free-swimming medusae 38
40b Pseudohydrotheca enclosing the lower portion of the hydranth elaborated into a distinct cup; with fixed gonophores 41
41a Perisarc around the gonophores disappearing before these are fully developed, leaving a cup-like expansion at the base; hydranths and gonophores bright orange in life 42
41b Perisarc around the gonophores persistent; hydranths and gonophores not orange .... *Bimeria* spp.
42a Stems evenly annulated or strongly wrinkled, much branched; hydranths large and conspicuous, usually with 16 tentacles .... *Garveia annulata* (fig. 3.38)
42b Stems wrinkled or weakly annulated, usually unbranched; hydranths fairly delicate, usually with 10 tentacles .... *Garveia groenlandica*
43a Hydrotheca usually wider than deep, flaring, with the margin entire, not large enough to accommodate the hydranth when it contracts .... *Halecium* spp. (figs. 3.28 and 3.46-3.48) or *Hydrodendron* spp.
43b Hydrotheca deeper than wide, large enough to accommodate the hydranth when it contracts 44
44a Hydranths generally on distinct, slender stalks (except in certain species of *Lafoea*) 54
44b Hydranths not on distinct, slender stalks, but attached directly to the main stem or its branches 45
45a Hydrotheca without an operculum; with nematophores (specialized small polyps in which nematocysts are concentrated; there is often a nearly constant number of these in relation to each hydranth) 52
45b Hydrotheca with an operculum consisting of 1 to 4 flaps; without nematophores 46
46a Hydranths arranged in a single series on one side of each branch and curved alternately to left and to right; operculum consisting of a large adcauline flap and a very small abcauline flap .... *Hydrallmania* spp.
46b Hydranths arranged in 2 series, on opposite sides of the branches, or in more than 2 series, operculum consisting of 1 or 2 flaps 47
47a Hydranths in 2 series, those on opposite sides of the branches arising at the same level and thus forming pairs 48
47b Hydranths in 2 or more series, alternating at least to some extent, not regularly forming pairs 49
48a Margin of the hydrotheca with 2 sharp, toothlike lobes; operculum consisting of 2 flaps .... *Dynamena* spp. (fig. 3.60)
  In at least 1 species (*pumila*), the contents of the gonophore emerge, but remain attached to the gonotheca.
48b Margin of the hydrotheca without toothlike lobes; operculum consisting of a single flap (distal margin of the gonotheca toothed, and the exterior of the gonotheca ribbed in some species) .... *Diphasia* spp.
49a Operculum consisting of a single flap 50
49b Operculum consisting of 2 to 4 flaps 51
50a Operculum attached on the side of the hydrotheca that is nearer the axis of the branch .... *Abietinaria* spp. (figs. 3.31 and 3.56-59)
50b Operculum attached on the side of the hydrotheca that is farther from the axis of the branch .... *Thuiaria* (including *Selaginopsis*) (figs. 3.32, 3.63, and 3.66-69)
51a Margin of the hydrotheca with 2 toothlike lobes; operculum consisting of 2 flaps .... *Sertularia* spp. (figs. 3.64 and 3.65)
51b Margin of the hydrotheca with 3 or 4 toothlike lobes; operculum consisting of 3 or 4 flaps .... *Sertularella* spp. (figs. 3.33, 3.61, and 3.62) and *Symplectoscyphus* spp.
  There are 4 lobes and opercular flaps in *Sertularella*, 3 in *Symplectoscyphus*.
52a Thecae of nematophores articulating with the branch from which they arise and thus movable to some extent .... *Plumularia* spp. (figs. 3.34, 3.44, 3.70, and 3.71)
52b Thecae of nematophores not articulating with the branch from which they arise, and thus not movable 53

Phylum Cnidaria 51

3.41 *Orthopyxis* sp.

3.42 *Velella velella*

3.43 *Obelia geniculata*

3.44 *Plumularia setacea*

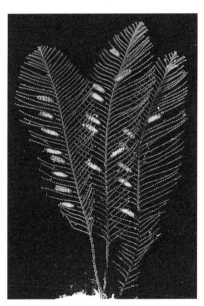

3.45 *Aglaophenia struthionides*

52 Phylum Cnidaria

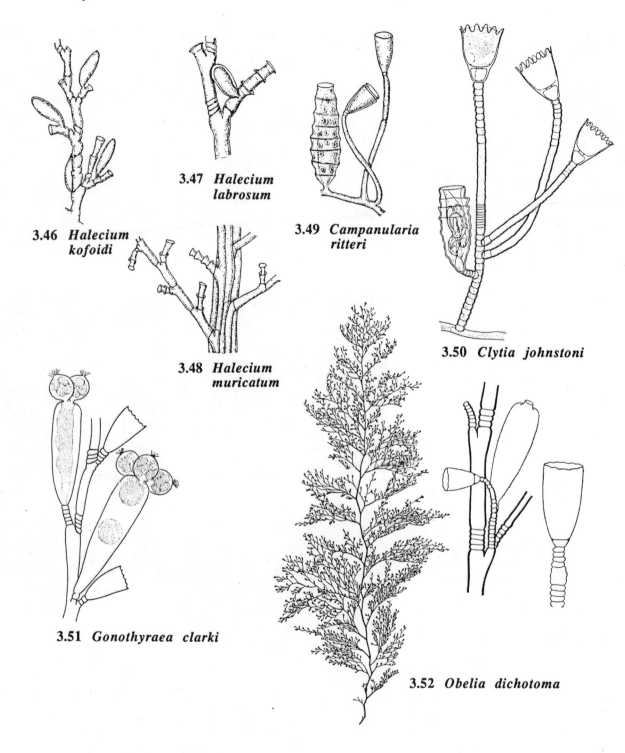

3.46 *Halecium kofoidi*
3.47 *Halecium labrosum*
3.48 *Halecium muricatum*
3.49 *Campanularia ritteri*
3.50 *Clytia johnstoni*
3.51 *Gonothyraea clarki*
3.52 *Obelia dichotoma*

Phylum Cnidaria 53

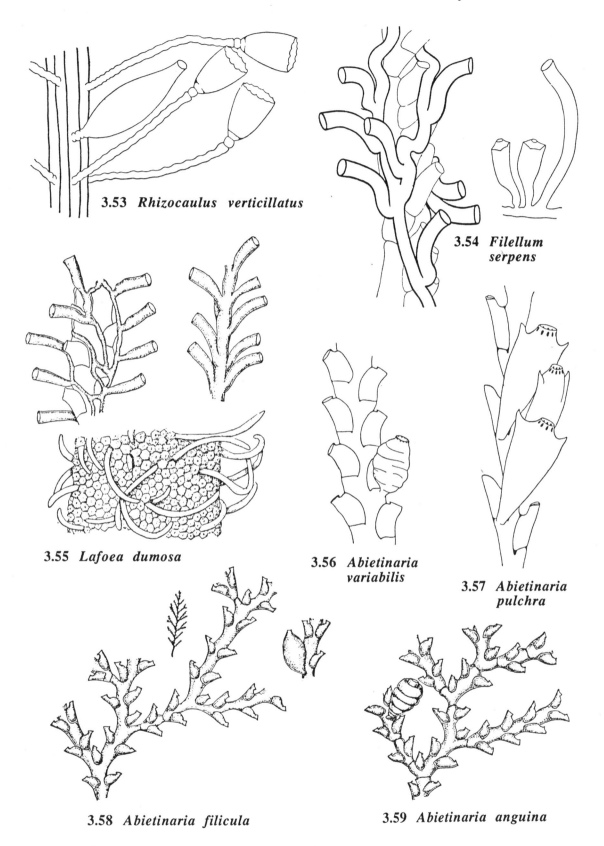

3.53 *Rhizocaulus verticillatus*

3.54 *Filellum serpens*

3.55 *Lafoea dumosa*

3.56 *Abietinaria variabilis*

3.57 *Abietinaria pulchra*

3.58 *Abietinaria filicula*

3.59 *Abietinaria anguina*

54 Phylum Cnidaria

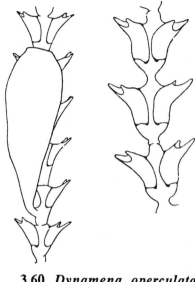

3.60 *Dynamena operculata*

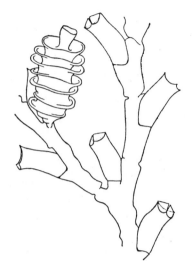

3.61 *Sertularella tricuspidata*

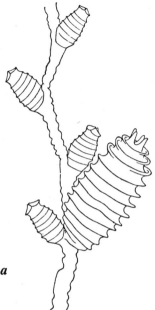

3.62 *Sertularella tenella*

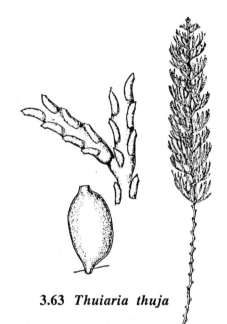

3.63 *Thuiaria thuja*

3.64 *Sertularia robusta*

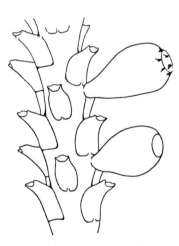

3.65 *Sertularia mirabilis*

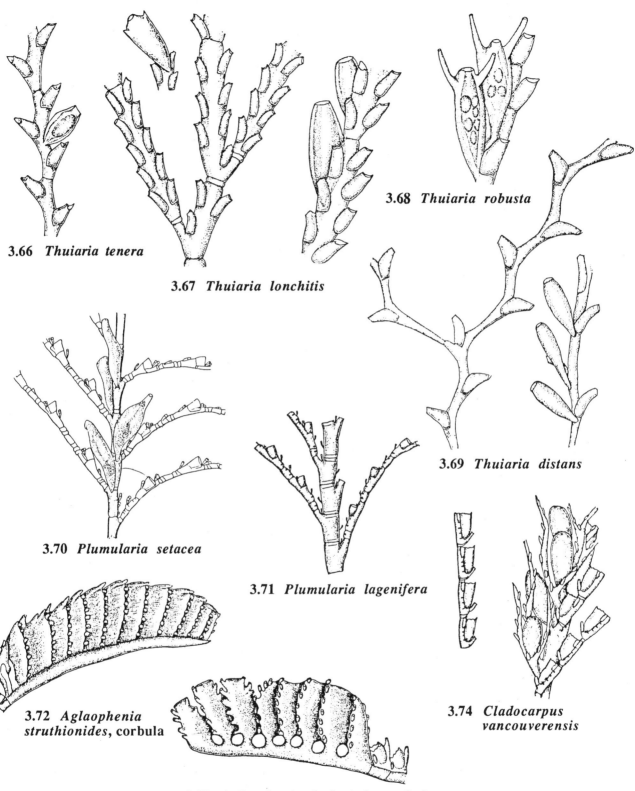

3.66 *Thuiaria tenera*
3.67 *Thuiaria lonchitis*
3.68 *Thuiaria robusta*
3.69 *Thuiaria distans*
3.70 *Plumularia setacea*
3.71 *Plumularia lagenifera*
3.72 *Aglaophenia struthionides*, corbula
3.73 *Aglaophenia latirostris*, corbula
3.74 *Cladocarpus vancouverensis*

Phylum Cnidaria

53a	Gonophores situated within corbulae, which consist of several pairs of leaflike expansions located on some of the branches *Aglaophenia* spp. (figs. 3.45, 3.72, and 3.73)	
53b	Gonophores not situated within corbulae, but associated with slender branches (phylactogonia) that lack hydranths, but that have nematophores *Cladocarpus* spp. (fig. 3.74)	
54a	Hydrotheca with an operculum of 4 or more converging flaps	55
54b	Hydrotheca without an operculum	59
55a	Opercular flaps arranged in such a way that they form 2 sloping surfaces, like the 2 halves of a roof *Stegopoma* spp.	
55b	Opercular flaps arranged in such a way that they form a conical or pyramidal configuration, or a funnel-shaped concavity	56
56a	Basal portions of opercular flaps sharply demarcated from the margin of the hydrotheca *Calycella* spp.	
56b	Opercular flaps merely extensions of the margin of the hydrotheca	57
57a	Wall of hydrotheca usually irregular and wrinkled; gonangia with up to 5 developing medusae inside; newly released medusae spherical, with 4 tentacles and scattered exumbrellar nematocysts *Eutonina indicans*	
57b	Wall of hydrotheca usually straight; gonangia and newly released medusae not as described in choice 57a	58
58a	Gonangia releasing spherical medusae having 2 tentacles and scattered exumbrellar nematocysts (on the lower half of the bell, these nematocysts form a broad ring) *Aequorea victoria*	
58b	Gonangia not as described in choice 58a various Campanulinidae and Mitrocomidae	
59a	Hydrotheca rimmed to strongly toothed, usually bell-shaped or goblet-shaped; a diaphragm always present in the lower part of the hydrotheca (except in the genus *Rhizocaulus*)	60
59b	Hydrotheca with a rim, generally tubular (but sometimes bell-shaped or goblet-shaped); diaphragm usually absent (but sometimes present)	69
60a	Colony consisting largely of free stems (these may be fascicled) that bear many hydranths	61
60b	Colony consisting largely of stolons applied to the substratum, the free branches rarely bearing more than 3 or 4 hydranths	67
61a	Colony branching sympodially, the hydranths being produced alternately on one side of a branch and then on the other; stems usually not fascicled	62
61b	Colony branching irregularly, the hydranths more or less forming whorls; stems fascicled (lacking a true hydrothecal diaphragm) *Rhizocaulus verticillatus* (fig. 3.53)	
62a	Gonangia releasing free medusae	63
62b	Gonangia with fixed gonophores, not releasing free medusae	66
63a	Released medusae disk-shaped and with 16 or more tentacles that are more or less of fixed length (not highly extensile) and have conspicuous rings of nematocysts	64
63b	Released medusae spherical or bell-shaped and with 4 or more tentacles that are highly extensile and have scattered nematocysts *Clytia* spp. (fig. 3.50)	
64a	Rim of hydrotheca always even, not sinuous or toothed; colony usually unbranched; usually on brown algae, rarely on animals or inert substrata *Obelia geniculata* (fig. 3.43)	
64b	Rim of hydrotheca even, sinuous, or toothed; colony usually branched; usually on animals or inert substrata, less often on algae	65
65a	Rim of hydrotheca with prominent teeth; perisarc not darkening as the colony ages *Obelia bidentata*	
65b	Rim of hydrotheca usually even, but may be sinuous or slightly toothed; perisarc of older colonies becoming dark brown or black *Obelia dichotoma* (fig. 3.52)	
66a	Gonangia with germ cells but never with developing medusae or medusoids *Campanularia* spp. (fig. 3.49)	
66b	Gonangia producing sessile medusoids that are extruded, but that remain attached in a cluster at the aperture *Gonothyraea* spp. (fig. 3.51)	
67a	Colony consisting of creeping stolons with 1 hydranth per upright stem; each gonangium producing a single thimble-shaped medusa that may either be retained or swim free; the medusa is less than 2 mm high and has mature gonads, but no stomach *Orthopyxis* spp.	
67b	Colony slightly branched with a few hydranths on each upright stem; gonangia and medusae not as described in choice 67a	68
68a	Gonangia producing free medusae that are spherical or bell-shaped and that have 4 or more tentacles at the time of release *Clytia* spp. (fig. 3.50)	

68b Gonangia not releasing medusae (they produce gametes, and are therefore essentially gonophores) .... *Campanularia* spp. (fig. 3.49)
69a Stalks of hydranths distinct, smooth or annulated; hydrothecae not becoming partly fused to the stems that produce them; hydranth low-domed to flat, with the mouth in a central depression .... *Bonneviella* spp.
69b Stalks of hydranths spirally twisted or lacking; hydrothecae often becoming partly fused to the stems that produce them; hydranth with a conical hypostome 70
70a Hydrotheca with a delicate diaphragm 71
70b Hydrotheca without a diaphragm 72
71a Gonangia crowded into a mufflike aggregation that surrounds the stalk, not releasing free medusae .... *Zygophylax* spp. (includes species previously assigned to *Lictorella*)
71b Gonangia not aggregated, releasing free medusae that bear gonads on the manubrium when mature .... *Hebella* spp.
72a Hydrotheca tubular or flared like a slender goblet, either on a stalk or directly attached to the stolon from which it arises; wall of the hydrotheca not adherent to the stolon .... *Lafoea* spp. (figs. 3.29 and 3.55)
72b Hydrotheca decidedly tubular, with or without a short stalk; hydrotheca either adherent to the stolon or partly or completely buried in the mass of stolons 73
73a Colony creeping (usually on other hydroids); lower portion of the hydrotheca adherent to the stolon, the upper portion bending away from the stolon .... *Filellum* spp. (fig. 3.54)
73b Colony branching free of the substratum; hydrotheca either partly adherent to the stolon, or partly or almost completely embedded in the stolon 74
74a Hydrothecae arranged irregularly or in longitudinal rows; hydrothecae buried almost completely in the stems, curving outward and away from the stolon at the distal end .... *Grammaria* spp.
74b Hydrothecae in 2 or 4 somewhat definite rows with respect to the stem from which they arise; hydrothecae bent back against the stem, so that a portion of the wall above the base is adherent to the stem (the lowermost and uppermost portions of the hydrothecae are free of the stem, however) .... *Cryptolaria* spp.

Order Hydroida

One assemblage of uncertain status, listed as "*Perigonimus*," (under family Pandeidae) is marked with an asterisk; this and other genera marked with an asterisk are not in the key.

Suborder Athecata (Anthomedusae)

Family Corymorphidae

Corymorpha sp.

Family Euphysidae

Euphysa ruthae Norenburg & Morse, 1983
Euphysa spp.

Family Tubulariidae

Hybocodon prolifer L. Agassiz, 1862 (fig. 3.37)
Tubularia crocea (L. Agassiz, 1862)
Tubularia harrimani Nutting, 1901
Tubularia indivisa Linnaeus, 1758
Tubularia marina Torrey, 1902
undescribed species

Phylum Cnidaria

Family Velellidae

Velella velella (Linnaeus, 1758) (fig. 3.42)

Family Corynidae

Coryne sp.
Sarsia japonica (Nagao, 1962)
Sarsia spp. (fig. 3.25 and 3.36)

Family Boreohydridae

undescribed species

Family Cladonematidae

Cladonema californicum Hyman, 1947

Family Rhysiidae

Rhysia sp.

Family Clavidae

Cordylophora caspia (Pallas, 1771)
Hataia parva Hirai & Yamada, 1965
Rhizogeton sp.

Family Hydractiniidae

Hydractinia aggregata Fraser, 1922
Hydractinia laevispina Fraser, 1922
Hydractinia milleri Torrey, 1902
Hydractinia sp.
undescribed species

Family Rathkeidae

Rathkea octopunctata (M. Sars, 1835)

Family Bougainvilliidae

Bimeria spp.
Bougainvillia spp. (fig. 3.39)
Garveia annulata Nutting, 1901 (fig. 3.38)
Garveia groenlandica Levinson, 1893

Family Pandeidae

Leuckartiara spp.
Neoturris spp.
*"*Perigonimus*" spp. (fig. 3.40)

Family Halimedusidae

Halimedusa typus (Bigelow 1916)

Family Calycopsidae

Bythotiara huntsmani (Fraser, 1911)

Family Trichydridae

Trichydra pudica Wright ,1858

Family Eudendriidae

Eudendrium spp. (fig. 3.24)

Family Protohydridae

Protohydra ?leuckarti Greeff, 1870 (fig. 3.35)

Suborder Thecata (Leptomedusae)

Family Tiarannidae

Stegopoma spp.

Family Laodiceidae

*Hydroid stages (fig. 3.30) of medusae of the genera *Staurophora* and *Ptychogena*

Family Mitrocomidae

*Hydroid stages of medusae of the genera *Foersteria*, *Mitrocoma*, *Mitrocomella*, *Tiaropsidium*, and *Tiaropsis*

Family Haleciidae

Halecium spp. (figs. 3.28 and 3.46-3.48)
Hydrodendron spp.

Phylum Cnidaria

Family Campanulariidae

Campanularia spp. (fig. 3.49)
Clytia spp. (fig. 3.50)
Gonothyraea spp. (fig. 3.51)
Obelia bidentata Clarke, 1875
Obelia dichotoma (Linnaeus, 1758) (fig. 3.52)
Obelia geniculata (Linnaeus, 1758) (fig. 3.43)
Orthopyxis spp. (fig. 3.41)
Rhizocaulus verticillatus (Linnaeus, 1758) (fig. 3.53)

Family Campanulinidae

Calycella spp.

Family Bonneviellidae

Bonneviella spp.

Family Lafoeidae

Cryptolaria spp.
Filellum spp. (fig. 3.54)
Grammaria spp.
Hebella spp.
Lafoea spp. (figs. 3.29 and 3.55)
Zygophylax spp.

Family Sertulariidae

Abietinaria spp. (figs. 3.31 and 3.56-3.59)
Diphasia spp.
Dynamena spp. (fig. 3.60)
Hydrallmania spp.
Sertularella spp. (figs. 3.33, 3.61, and 3.62)
Sertularia spp. (figs. 3.64 and 3.65)
Symplectoscyphus spp.
Thuiaria spp. (figs, 3.32, 3.63, and 3.66-3.69)

Family Plumulariidae

Plumularia spp. (figs. 3.34, 3.44, 3.70, and 3.71)

Family Aglaopheniidae

Aglaophenia spp. (figs. 3.45, 3.72, and 3.73)
Cladocarpus spp. (fig. 3.74)
Thecocarpus spp.

Family Eirenidae

Eutonina indicans (Romanes, 1876)

Family Aequoreidae

Aequorea victoria (Murbach & Shearer, 1902)

Suborder Limnomedusae
Family Olindiasidae

Gonionemus vertens A. Agassiz, 1862
Monobrachium parasiticum Mereschkowsky, 1877

Family Proboscidactylidae

Proboscidactyla flavicirrata Brandt, 1835

Order Stylasterina
Family Stylasteridae

Boschma, H. 1953. The Stylasterina of the Pacific. Zool. Meded., Leiden, 32:185-201.
Broch, H. 1942. Investigations of the Stylasteridae (hydrocorals). Skr. Norske Vidensk.-Akad. Oslo, matem.-naturv. Kl., 1942:1-113.
Fisher, W. K. 1931. California hydrocorals. Ann. Mag. Nat. Hist., ser. 10, 8:391-9.
------. 1938. Hydrocorals of the North Pacific Ocean. Proc. U.S. Nat. Mus., 84:493-554.
Fritchman, H. K., II. 1974. The planula of the stylasterine hydrocoral *Allopora petrograpta* Fisher: Its structure, metamorphosis and development of the primary cyclosystem. Proc. 2nd Internat. Coral Reef Symp., 2:245-58.
Ostarello, G.L. 1973. Natural history of the hydrocoral *Allopora californica* Verrill (1866). Biol. Bull., 145:548-64.

1a   Colony very rough, due to the edges of the larger dactylopores being raised to form scoop-shaped projections (colony extensively branched, sometimes more than 25 cm high; with 2-5 dactylopores in each cyclosystem; subtidal, rare)   *Errinopora pourtalesii* (Dall, 1884)
1b   Colony not especially rough, the edges of the dactylopores not raised to form scoop-shaped projections   2
2a   Cyclosystems commonly 1.25-1.5 mm in diameter and with more than 1 (generally 3-7) gastrostyles (colony encrusting, not branching, reddish purple)   *Allopora porphyra* (Fisher, 1931)
2b   Cyclosystems usually not more than 1.25 mm in diameter, and rarely with more than 1 gastrostyle   3
3a   Gastropores decidedly broader than deep, their diameter not often exceeding 0.8 mm (colony upright and branching, pink or violet; subtidal)   *Allopora venusta* Verrill, 1868
3b   Gastropores as deep or deeper than broad, their diameter usually 1-1.25 mm   4
4a   Encrusting; visible portion of the gastrostyle about one-third to one-half the diameter of the cyclosystem; reddish pink or violet   *Allopora petrograpta* Fisher, 1938
4b   Upright and branching; visible portion of the gastrostyle about one-fourth the diameter of the cyclosystem; orange-pink   *Allopora verrilli* Dall, 1884

62 Phylum Cnidaria

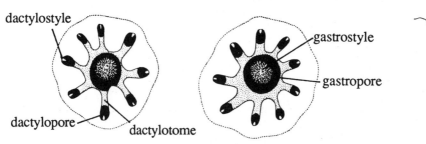
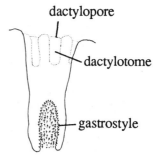

3.75 Two cyclosystems in surface view

3.76 Vertical section of a gastropore

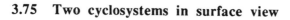

Terminology used in identifying Stylasterina

Order Siphonophora

Claudia E. Mills

Bigelow, H. B. 1913. Medusae and Siphonophorae collected by the U. S. Fisheries Steamer "Albatross" in the northwestern Pacific, 1906. Proc. U. S. Nat. Mus., 44:1-119.

Kirkpatrick, P. A. & P. R. Pugh. 1984. Siphonophores and Velellids. Synopses of the British Fauna, no. 29. Leiden: E. J. Brill/Dr. W. Backhuys. 154 pp.

Mackie, G. O. 1964. Analysis of locomotion in a siphonophore colony. Proc. Royal Soc. London, B, 159:366-91.

Purcell, J. E. 1982. Feeding and growth of the siphonophore *Muggiaea atlantica* (Cunningham, 1893). J. Exp. Mar. Biol. Ecol., 62:39-54.

Stepan'iants, S. D. 1967. [Siphonophora of seas of the USSR and of the northern part of the Pacific Ocean,] Opredeliteli po Faune SSSR, 96. 216 pp. (in Russian).

Totton, A. K. 1965a. A synopsis of the Siphonophora. London: British Museum (Natural History). vii + 230 pp.

------. 1965b. A new species of *Lensia* (Siphonophora: Diphyidae) from the coastal waters of Vancouver, B.C., and its comparison with *Lensia achilles* Totton and another new species *Lensia cordata*. Ann. Mag. Nat. Hist., ser. 13, 8:71-6.

1a	With a gas-filled float, and with numerous nectophores arranged below this, followed by a stem region bearing groups of feeding, reproductive, and buoyant zooids; terminal stem groups not becoming individually detached and living apart from the colony (suborder Physonectae)	2
1b	Without a gas-filled float, and with only 1 or 2 nectophores; terminal stem groups becoming detached and living apart from the colony as "eudoxids" (suborder Calycophorae)	4
2a	Stem elongate and trailing up to 1 m, with feeding and reproductive zooids along the entire length	3
2b	Stem reduced to the extent that the enlarged feeding and reproductive zooids form a ring immediately below the nectophores (rare)	*Physophora hydrostatica*
3a	Float, nectophores, stem, and gastrozooids with distinct chromatophores (orange when expanded, plum-colored when contracted); nectophores with ridges and truncated apical wings, and having lateral canals with pronounced sigmoid curves; colony not fragmenting in nature	*Nanomia cara* (fig. 3.77)
3b	Float, nectophores, stem, and gastrozooids colorless; nectophores not ridged, appearing heart-shaped when viewed from the ostial side, and having fairly straight lateral canals; colony fragmenting into swimming (with nectophores) and non-swimming (stem only) portions	*Cordagalma cordiformis*

Phylum Cnidaria

4a With 2 soft, gelatinous, rounded nectophores, each up to 5 cm long, arranged "back-to-back;" bracts of eudoxids with 4 major canals running through the mesoglea (rare)   5

4b With 1 or 2 firm, elongate or spherical nectophores not over 3 cm long (when there are 2, they are "in line," rather than "back-to-back"); bracts of eudoxids without canals (but see *Dimophyes arctica*, which has analogous structures)   7

5a Nectophores rounded, with a simple, curving, club-shaped somatocyst, and with 4 simple subumbrellar canals; bracts with a large, pear-shaped central vesicle (rare)
Desmophyes annectens

5b Nectophores subcylindrical and truncate at the ends, with a narrow somatocyst in the form of a branched canal, and with numerous subumbrellar canals; bracts without a large central vesicle   6

6a Somatocyst of nectophore forming a canal that divides into three branches near the apex, the right and left branches curving downward and having several lateral branches; subumbrellar canals branching dichotomously several times before they reach the margin, but not anastomosing (rare)   *Praya dubia*

6b Somatocyst of nectophore forming a canal with many short lateral branches directed away from it at right angles; subumbrellar canals anastomosing to form a reticulate network (rare)
Praya reticulata

7a Nectophore(s) elongated, approximately twice as high as wide, the hydroecium always less than half as deep as the apex of the subumbrellar cavity   8

7b Nectophore approximately spherical, the hydroecium long and curving up and over the subumbrellar cavity   *Sphaeronectes gracilis*

8a Nectophore(s) with prominent longitudinal ridges; anterior nectophore (if 2 are present) sharply pointed and about one and a half times as long as the posterior one; mouthplate of nectophore(s) divided; bract and gonophore of eudoxid with ridges   9

8b Nectophore(s) without prominent ridges; anterior nectophore rounded and about twice as long as the posterior one; mouthplate of nectophore(s) not divided; neither bract nor gonophore of eudoxid with ridges (somatocyst in bract of eudoxid long, with prominent apical and basal horns)   *Dimophyes arctica*

9a Anterior nectophore with 3 longitudinal ridges at the apex; bracteal cavity of eudoxid nearly half the height of the bract   *Chelophyes appendiculata*

9b Anterior nectophore with 5 longitudinal ridges at the apex; bracteal cavity of eudoxid substantially less than half the height of the bract   10

10a With a single nectophore up to about 5 mm long; somatocyst of the nectophore long, reaching approximately to the level of the apex of the nectosac; apex of bract of eudoxid with a flat flange, somatocyst in bract of eudoxid small and tubular   *Muggiaea atlantica* (fig. 3.78)

10b Usually with 2 nectophores, the anterior one up to 2.5 cm long; somatocyst of anterior nectophore ending some distance below the apex of the nectosac; eudoxid not as described in choice 10a   11

11a Lateral ridges of the anterior nectophore curving away from the hydroecium near the margin of the bell; base of the somatocyst above the plane of the ostium; eudoxid unknown
Lensia baryi

11b Lateral ridges of the anterior nectophore not curving, but continuing straight to the bell margin; base of the somatocyst below the plane of the ostium; eudoxid bract with conical apex and long somatocyst (rare)   *Lensia conoidea*

Suborder Physonectae

Family Agalmidae

Cordagalma cordiformis Totton, 1932
Nanomia cara A. Agassiz, 1865 (fig. 3.77)

64 Phylum Cnidaria

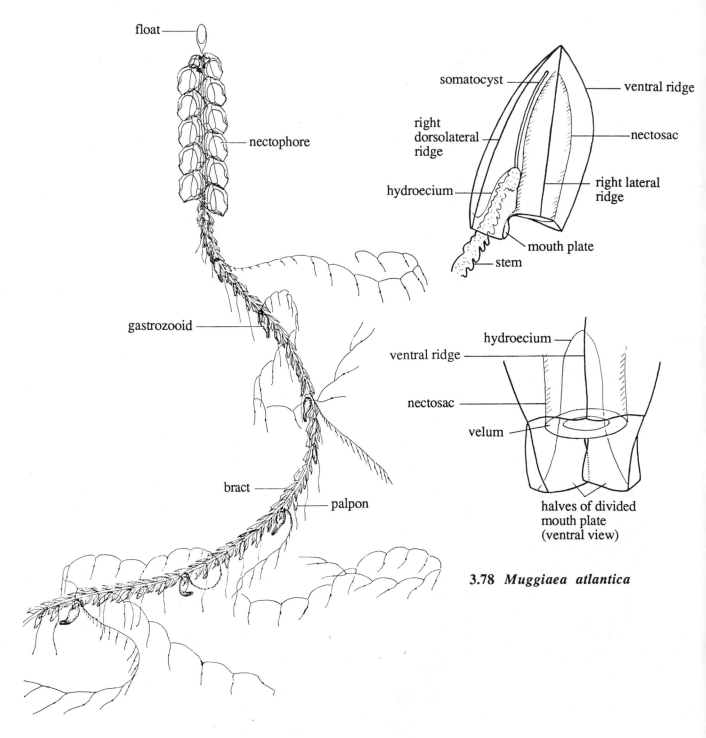

3.77 *Nanomia cara*

3.78 *Muggiaea atlantica*

Family Physophoridae

Physophora hydrostatica Forskål, 1775

Suborder Calycophorae

Family Diphyidae

Chelophyes appendiculata (Eschscholtz, 1829)
Dimophyes arctica (Chun, 1897)
Lensia baryi Totton, 1965
Lensia conoidea (Keferstein & Ehlers, 1860)
Muggiaea atlantica Cunningham, 1892 (fig. 3.78)

Family Prayidae

Desmophyes annectens Haeckel, 1888
Praya dubia (Quoy & Gaimard, 1833)
Praya reticulata (Bigelow, 1911)

Family Sphaeronectidae

Sphaeronectes gracilis (Claus, 1873)

Class Scyphozoa

Order Semaeostomeae

Claudia E. Mills

Kramp, P. L. 1961. Synopsis of the medusae of the world. J. Mar. Biol. Assoc. United Kingdom, 40:1-469.
Larson, R. J. 1976. Marine fauna and flora of the northeastern United States. Cnidaria: Scyphozoa. NOAA Tech. Rep., NMFS Circular no. 397. 18 pp.
Mayer, A. G. 1910. Medusae of the World, vol. 3. The Scyphomedusae. Carnegie Inst. Washington Publ. 109:499-735.
Naumov, D. V. 1961. [Scyphozoan medusae of seas of the USSR.] Opredeliteli po Faune SSSR, 75. 98 pp. (in Russian).
Russell, F. S. 1970. The Medusae of the British Isles. II. Pelagic Scyphozoa with a Supplement to the First Volume on Hydromedusae. Cambridge: Cambridge University Press. xii + 284 pp.
Shih, C. T. 1977. A guide to the jellyfish of Canadian Atlantic waters. Nat. Mus. Canada, Nat. Hist. Pap. 5. 90 pp.

1a Tentacles short, forming a fringe at the margin of the bell; with 8 marginal lobes; with 8 rhopalia, these with a small tentaclelike lappet on both sides   2
1b Tentacles long, arising singly from the clefts between lappets, or in groups of several to many from the underside of the bell; with 8 or 16 rhopalia, these without tentaclelike lappets   3
2a Subumbrellar canals directed toward the rhopalia not branched except proximally, and the canals midway between them not branched at all; margin not brown; gonads usually reddish, purple, or bluish   *Aurelia aurita*
2b Subumbrellar canals profusely branched, with numerous lateral diverticula that anastomose; margin dark brown (rare)   *Aurelia limbata*

66 Phylum Cnidaria

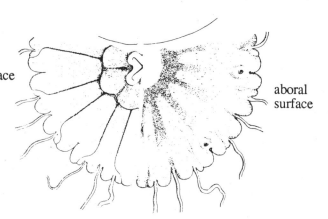

3.80 *Chrysaora melanaster*

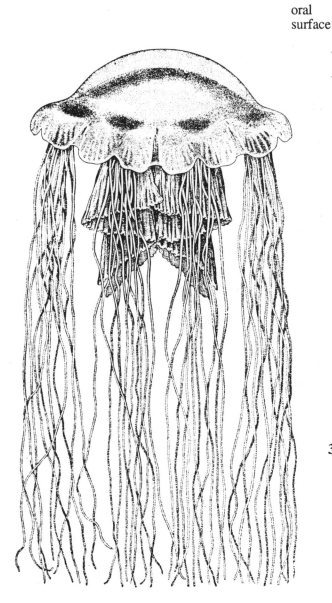

3.79 *Cyanea capillata*

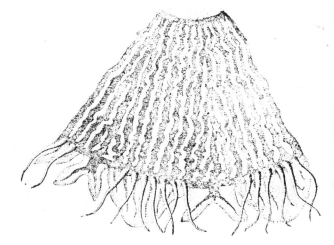

3.81 *Phacellophora camtschatica*, oral surface

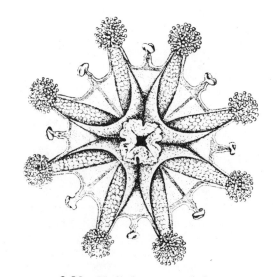

3.82 *Haliclystus salpinx*

3a Tentacles arising from the underside of the bell ... 4
3b Tentacles (usually 24) arising from the clefts between the lappets at the margin of the bell ... 5
4a With 16 nearly equal marginal lobes, these arranged in 8 pairs, with the 8 rhopalia situated between the pairs; with 8 U-shaped groups of subumbrellar tentacles (each group may have as many as 150 tentacles arranged in up to 4 rows) ... *Cyanea capillata* (fig. 3.79)
4b With 16 rounded lobes alternating with smaller lobes that are divided in such a way that they resemble fishtails (the 16 rhopalia are situated on these); with 16 linear groups of subumbrellar tentacles (up to 25 tentacles in each group) ... *Phacellophora camtschatica* (fig. 3.81)
5a Bell amber-colored, darkest near the margin an indistinct, pale star pattern radiating to the margin may or may not be present ... *Chrysaora fuscescens*
5b Bell light-colored, with a dark pattern consisting of a ring near the apex and 16 evenly-spaced streaks radiating to the margin, each streak being flanked on both sides by a dark crescent ... *Chrysaora melanaster* (fig. 3.80)

Family Pelagiidae

Chrysaora fuscescens Brandt, 1835
Chrysaora melanaster Brandt, 1835 (fig. 3.80)

Family Cyaneidae

Cyanea capillata (Linnaeus, 1758) (fig. 3.79)

Family Ulmaridae

Aurelia aurita (Linnaeus, 1758)
Aurelia limbata (Brandt, 1835)
Phacellophora camtschatica Brandt, 1835 (fig. 3.81)

Order Stauromedusae

Mayer, A. G. 1910. Medusae of the World. Carnegie Institution of Washington Publ. 109:499-735 (vol. 3).
Naumov, D. V. 1961. [Scyphozoan medusae of seas of the USSR.] Opredeliteli po Faune SSSR, 75. 98 pp. (in Russian).
Russell, F. S. 1970. Medusae of the British Isles. II. Pelagic Scyphozoa with a Supplement to the First Volume on Hydromedusae. Cambridge: Cambridge University Press. xii + 284 pp.

1a At least the outer tentacles with cushionlike swellings at their bases ... Family Cleistocarpidae: *Thaumatoscyphus hexaradiatus* (Broch, 1907)
1b None of the tentacles with cushionlike swellings at their bases ... 2
2a Marginal anchors (located between tentacle clusters) expanded into broad cups and decidedly stalked (the stalks are divided into two parts by a collarlike expansion) ... Family Haliclystidae: *Haliclystus salpinx* Clark, 1863 (fig. 3.82)
2b Marginal anchors not expanded into broad cups and not decidedly stalked ... Family Haliclystidae: *Haliclystus stejnegeri* Kishinouye, 1899

Class Anthozoa

Daphne G. Fautin, Arthur E. Siebert, and Eugene N. Kozloff

Dunn, D. F. 1982. Cnidaria. *In* Parker, S. A. (ed.), Synopsis and Classification of Living Organisms, 1:669-706. New York: McGraw-Hill Book Co.
Moore, R. C. (ed.) 1956. Treatise on Invertebrate Paleontology. Part F. Lawrence, Kansas: Geological Society of America and University of Kansas Press. 498 pp.

1a Polyps with 8 pinnately branched tentacles   Subclass Alcyonaria
1b Polyps with more than 8 tentacles, these not branched   2
2a With a ring of tentacles around the mouth, as well as on the marginal portion of the disk; inhabiting tubes that consist of discharged nematocysts embedded in a matrix of slippery mucus   Subclass Ceriantipatharia, Order Ceriantharia
2b Without a ring of tentacles around the mouth (all tentacles are on the marginal portion of the disk); not inhabiting tubes that consist of discharged nematocysts embedded in slippery mucus   Subclass Zoantharia

Subclass Alcyonaria (Octocorallia)

Bayer, F. M. 1958. Les Octocoralliares plexaurides des côtes occidentales d'Amérique. Mém. Mus. Natl. Hist. Nat. Paris, n. s., 16, A:41-56.
------. 1981. Key to the genera of Octocorallia exclusive of Pennatulacea (Coelenterata: Anthozoa), with diagnoses of new taxa. Proc. Biol. Soc. Wash., 94:902-47.
Belcik, F. P. 1977. A distribution study of the Octocorallia of Oregon. Publ. Seto Mar. Biol. Lab., 24:49-52.
Hickson, S. V. 1915. Some Alcyonaria and a *Stylaster* from the west coast of North America. Proc. Zool. Soc. London, 1915:541-57.
Kükenthal, W. 1913. Über die Alcyonarienfauna Californiens und ihre tiergeographischen Beziehungen. Zool. Jahrb., Abt. Syst., 35:219-70.
------. 1930. Pennatulacea. Das Tierreich, Lief. 43. 132 pp.
Nutting, C. C. 1909. Alcyonaria of the California coast. Proc. U. S. Nat. Mus., 35:681-727.
------. 1912. Descriptions of the Alcyonaria collected by the U. S. Fisheries Steamer "Albatross," primarily in Japanese waters. Proc. U. S. Nat. Mus., 43:1-104.
Verrill, A. E. 1870. Review of the corals and polyps of the west coast of America. Trans. Conn. Acad. Arts Sci., 1:377-567.
------. 1922. Alcyonaria and Actiniaria. Rep. Canad. Arctic Exped. 1913-18, 8, part G:1-164.

1a Colony encrusting, the polyps arising singly from creeping stolons   Order Alcyonacea, Suborder Stolonifera
1b Colony a fleshy mass in which several to many polyps are embedded, or irregularly branched or featherlike   2
2a Colony a fleshy mass in which several to many polyps are embedded   Order Alcyonacea, Suborder Alcyoniina
2b Colony irregularly branched or featherlike   3
3a Gorgonians--colony irregularly branched, attached to a hard substratum; branches with an internal core of hard material   Order Alcyonacea, Suborders Holaxonia and Scleraxonia
3b Colony featherlike, the lower portion swollen and anchoring the colony in mud or sand; rachis of the colony with a hard internal rod, but branches without comparable supporting structures   Order Pennatulacea

Order Alcyonacea

Suborder Stolonifera

Family Clavulariidae

The following species have been reported.

Clavularia moresbii Hickson, 1915. Deep subtidal.
Clavularia spp. Low intertidal, shallow subtidal.
?*Sarcodictyon* sp. Low intertidal, shallow subtidal.

Suborder Alcyoniina

Family Alcyoniidae

?*Alcyonium* sp. This species, whose whitish, pinkish, or pale orange polyp clusters are nearly flat, is common in the low intertidal and shallow subtidal, generally on shell rubble; it has often been misidentified as a *Clavularia* (suborder Stonlonifera).
?*Alcyonium* sp. Lobed, upright colonies, whitish or pale pink; subtidal; often misidentifed as *Gersemia rubiformis*.

Family Nephtheidae

Gersemia rubiformis (Ehrenberg, 1834). Deep red colonies; mostly subtidal, rarely low intertidal.

Suborder Holaxonia

This suborder includes some of the alcyonarians that are called gorgonians. Others are assigned to the suborder Scleraxonia (below).

Family Acanthogorgiidae

Calcigorgia spiculifera Broch, 1935. Shallow subtidal.

Family Plexauridae

Swiftia kofoidi (Nutting, 1909). Shallow subtidal, Oregon southward.
Swiftia simplex (Nutting, 1909). Deep Subtidal.
Swiftia spauldingi (Nutting, 1909). Shallow subtidal.
Swiftia torreyi (Nutting, 1909). Shallow subtidal.

Family Chrysogorgiidae

Radiceps sp. Deep subtidal.

Family Isididae

Acanella sp. Deep subtidal.

Family Primnoidae

Callogorgia kinoshitae Kükenthal, 1913. Deep subtidal; called *C. sertosa* Wright & Studer, 1889 by Nutting (1909).
Parastennella sp. Deep subtidal.
Primnoa willeyi Hickson, 1915 Shallow subtidal.

Suborder Scleraxonia

Family Anthothelidae

Anthothela pacifica (Kükenthal, 1913). Shallow subtidal; *Sympodium armatum* Nutting, 1909 is a synonym.

Family Paragorgiidae

Paragorgia pacifica Verrill, 1922. Shallow subtidal; the most commonly encountered gorgonian of our region; has been called *Paragorgia arborea* (Linnaeus, 1758).

Order Pennatulacea

The key includes only a few species, but these are the ones that are most likely to be found in shallow water.

1a	Rachis and branches fleshy, pale to vivid orange	*Ptilosarcus gurneyi*
1b	Rachis and branches slender, not especially fleshy, white or almost white	2
2a	Rachis smooth to the touch, without projecting calcareous spicules	*Virgularia* spp.
2b	Rachis rough to the touch, owing to projecting calcareous spicules	*Stylatula elongata*

Suborder Sessiliflorae

Family Kophobelemnidae

Kophobelemnon affine Studer, 1894. Deep subtidal.
Kophobelemnon biflorum Pasternak, 1960. Deep subtidal.
Kophobelemnon hispidum Nutting, 1912. Deep subtidal.

Family Anthoptilidae

Anthoptilum grandiflorum (Verrill, 1882). Deep subtidal.

Family Funiculinidae

Funiculina parkeri Kükenthal, 1913. Deep subtidal; called *F. armata* Verrill, 1882 by Nutting (1909).

Family Protoptilidae

Helicoptilum rigidum Nutting, 1912. Deep subtidal.

Family Scleroptilidae

Scleroptilum sp. Deep subtidal, Oregon.

Family Umbellulidae

Umbellula lindahli Kölliker, 1880. Deep subtidal; has also been called *U. loma* Nutting, 1909, *U. encrinus* (Jungersen, 1916), *U. carpenteri* Kölliker, 1880, *U. magniflora* Kölliker, 1880.

Suborder Subselliflorae
Family Virgulariidae

Balticina californica Moroff, 1902. Deep subtidal; *B. pacifica* Nutting, 1909 is a synonym.
Balticina septentrionalis (Gray, 1872). Shallow subtidal; *Verrillia blakei* Stearns, 1873 is a synonym.
Stylatula elongata (Gabb, 1863). Shallow subtidal; *S. columbiana* Verrill, 1922 is perhaps a synonym.
Virgularia spp. Shallow and deep subtidal.

Family Pennatulidae

Pennatula phosphorea (Linnaeus, 1758). Deep subtidal.
Ptilosarcus gurneyi (Gray, 1860). Shallow subtidal.

Subclass Ceriantipatharia

Order Ceriantharia

Suborder Spirularina
Family Cerianthidae

Arai, M. N. 1971. *Pachycerianthus* (Ceriantharia) from British Columbia and Washington. J. Fish. Res. Bd. Canada, 28:1677-80.

Pachycerianthus fimbriatus (McMurrich, 1910) (fig. 3.85). Shallow subtidal, common; vertical burrows in mud, sometimes more than 50 cm deep, are lined with slippery mucus in which discharged nematocysts are embedded.

Order Antipatharia

Suborder Antipathina
Family Antipathidae

Antipathes sp. Deep subtidal.

Subclass Zoantharia

1a   Colonial, the bases of the polyps connected; column with sand and other foreign material

Phylum Cnidaria

	embedded in it	*Epizoanthus scotinus,* Order Zoanthidea
1b	Not colonial (but some species form large aggregations by asexual multiplication); column without sand or other foreign material embedded in it, although such material may adhere to the outside of it, and an external calcareous skeleton may be present	2
2a	Polyps with a calcareous external skeleton	Order Scleractinia
2b	Polyps without a calcareous skeleton	3
3a	Tentacles with knobs at their tips	*Corynactis californica,* Order Corallimorpharia
3b	Tentacles without knobs at their tips	Order Actiniaria

Order Scleractinia (Madreporaria)

Durham, J. W. 1947. Corals from the Gulf of California and the North Pacific coast of America. Mem. 20, Geol. Soc. Amer. 68 pp.

------. 1952. Stony corals of the eastern Pacific collected by the Velero III and Velero IV. Allan Hancock Pacific Exped., 16:1-110.

Vaughan, T. W., & V. W. Wells. 1943. Revision of the suborders, families, and genera of the Scleractinia. Spec. Pap. 44, Geol. Soc. Amer. xv + 363 pp.

The two more commonly encountered stony corals are distinguished in the key. A few other species have been reported, however.

1a	Polyps bright orange; inner edges of calcareous septa not fused to vertical pillars; common intertidal and subtidal species	*Balanophyllia elegans*
1b	Polyps beige, brown, or pink; inner edges of some calcareous septa fused to vertical pillars that arise from the bottom of the skeletal cup; subtidal	*Caryophyllia alaskensis*

Order Caryophylliina

Family Caryophylliidae

Caryophyllia alaskensis Vaughan, 1941. Shallow subtidal.
Cyathoceras quaylei Durham, 1947. Shallow subtidal.
Desmophyllum cristagalli Milne Edwards and Haime, 1848. Shallow subtidal.
Paracyathus stearnsi Verrill, 1869. Intertidal and shallow subtidal.
Lophelia californica Durham, 1947. Shallow subtidal.
Solenosmilia variabilis Duncan, 1873. Shallow subtidal.

Suborder Dendrophylliina

Family Dendrophylliidae

Balanophyllia elegans Verrill, 1864. Intertidal and shallow subtidal.

Order Actiniaria

Carlgren, O. 1936. Some west American sea anemones. J. Wash. Acad. Sci., 26:16-23.

------. 1949. A survey of the Ptychodactiaria, Corallimorpharia and Actiniaria. Kungl. Svenska Vetenskap. Handl., ser. 4, 1:1-121.

Dunn, D. F., F. S. Chia, & R. Levine. 1980. Nomenclature of *Aulactinia* (=*Bunodactis*), with description of *Aulactinia incubans* n. sp. (Coelenterata: Actiniaria), an internally brooding sea anemone from Puget Sound. Can. J. Zool., 58:2071-80.

Fautin, D. G., & F. S. Chia. 1986. Revision of sea anemone genus *Epiactis* (Coelenterata: Actiniaria) on the Pacific coast of North America, with descriptions of two new brooding species. Can. J. Zool., 64:1663-74.
Hand, C. 1955. The sea anemones of central California. Part I. The corallimorpharian and athenarian anemones. Wasmann J. Biol., 12(1954):345-75.
------. 1955. The sea anemones of central California. Part II. The endomyarian and mesomyarian anemones. Wasmann J. Biol., 13:37-99.
------. 1956. The sea anemones of central California. Part III. The acontiarian anemones. Wasmann J. Biol., 13(1955):189-251.
Hand, C., & D. F. Dunn. 1975. Redescription and range extension of the sea anemone *Cnidopus ritteri* (Torrey) (Coelenterata: Actiniaria). Wasmann J. Biol., 32(1974):187-94.
Lawn, I. D., & D. M. Ross. 1982. The release of the pedal disk in an undescribed species of *Tealia* (Anthozoa: Actiniaria). Biol. Bull., 163:188-96.
Ross, D. M. 1979. A third species of swimming actinostolid (Anthozoa: Actiniaria) on the Pacific coast of North America? Can. J. Zool., 57:943-5.
Sebens, K. P., & G. Laakso. 1978. The genus *Tealia* (Anthozoa: Actiniaria) in the waters of the San Juan Archipelago and the Olympic Peninsula. Wasmann J. Biol., 35:152-68.
Siebert, A. E., Jr. 1973. A description of the sea anemone *Stomphia didemon* sp. nov. and its development. Pac. Sci., 27:363-76.
Siebert, A. E., Jr., & C. Hand. 1974. A description of the sea anemone *Halcampa crypta*, new species. Wasmann J. Biol. 32:327-36.
Siebert, A. E., & J. G. Spaulding. 1976. The taxonomy, development and brooding behavior of the anemone, *Cribrinopsis fernaldi* sp. nov. Biol. Bull., 150:128-38.
Stephenson, T. A. 1928. The British Sea Anemones, vol. I. London: Ray Society. xiv + 148 pp.
------. 1935. The British Sea Anemones, vol. II. London: Ray Society. xii + 426 pp.
Torrey, H. B. 1902. Papers from the Harriman Alaska Expediton, XXX. Anemones, with a discussion of variation in *Metridium*. Proc. Wash. Acad. Sci., 4:373-410.
Williams, R. B. 1975. A redescription of the brackish-water sea anemone *Nematostella vectensis* Stephenson, with an appraisal of congeneric species. J. Nat. Hist., 9:51-64.

1a Body somewhat wormlike, usually with a rounded or bulbous base (in *Halcampoides purpurea*, choice 6b, the base becomes flattened when the animal contracts), inhabiting mud or muddy sand, not attached to a firm object 2
1b Body not wormlike, with a flattened base that is attached to rock, pebble, shell, wood, or some other firm object (note: some species are partly or almost completely buried in gravel, sand, or mud, but their bases are nevertheless attached to a firm object) 7
2a With 10 tentacles *Halcampa decementaculata*
2b With 12 or more tentacles, at least in mature individuals 3
3a With 12-18 (usually 16) tentacles; gastrovascular cavity with spherical, ciliated clusters of cells circulating in it; height up to 1.5 cm; mostly in estuarine situations, especially shallow pools in salt marshes *Nematostella vectensis*
3b Usually with 12 tentacles; gastrovascular cavity without spherical, ciliated clusters of cells circulating in it; height commonly 4 cm or more; often in bays, but not typically in salt marshes or other estuarine situations 4
4a Base bulbous; part of the column covered with a brownish cuticle *Edwardsia sipunculoides*
4b Base not bulbous; column without a brownish cuticle 5
5a Column tan to brown; oral disk with distinct folds around the siphonoglyph *Peachia quinquecapitata*
5b Column translucent and colorless to opaque and beige; oral disk without folds around the siphonoglyph 6
6a Lips and pharynx beige, white, or brown; base rounded when the animal is contracted *Halcampa crypta*
6b Lips and pharynx usually red; base flattened when the animal is contracted *Halcampoides purpurea*
7a With acontia (threadlike structures that can be extruded through the mouth or through pores or breaks in the body wall when the anemone is disturbed or damaged) 8
7b Without acontia 10

74 Phylum Cnidaria

3.83 *Aulactinia incubans*

3.85 *Pachycerianthus fimbriatus*

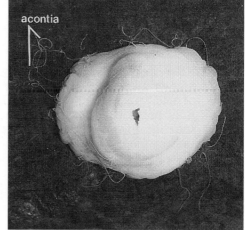

3.84 *Metridium* sp.

8a   Column green to brown, with 7-19 pale orange stripes; height not exceeding 2 cm; usually high intertidal, in cracks or shallow pools in rock   *Haliplanella lineata*
8b   Column sometimes brown, but not green, and without orange stripes; height commonly more than 2 cm; mostly low intertidal and subtidal, or on floats   9
9a   Height not more than 10 cm, and usually not more than 5 cm; column white, gray, brown, or orange, sometimes with dark spots marking the location of the pores through which acontia are extruded; generally with fewer than 100 tentacles; oral disk only slightly lobed; reproducing rapidly by pedal laceration, the pedal fragments developing into complete anemones within about 3 weeks; intertidal and subtidal   *Metridium senile*
9b   Height often more than 10 cm, and sometimes attaining 1 m; column white, brownish orange, or salmon-orange, without spots marking the location of the pores through which acontia are extruded; generally with more than 200 tentacles in larger specimens; oral disk prominently lobed, except in small individuals; not reproducing by pedal laceration; mostly subtidal   *Metridium* sp. (fig. 3.84)
10a Column with tubercles (these may, however, be small and inconspicuous), often with sand, gravel, or bits of shell adhering to them   11
10b Column without tubercles and usually without adherent material   20
11a Margin of oral disk with white or yellow spherules (these may be hidden by the tentacles)   12
11b Margin of oral disk without spherules   15
12a Tubercles restricted to the upper portion of the column (oral disk often bright pink or orange; tentacles with white bands on a gray, brown, blackish, or green background, or with the colors reversed; without symbiotic algal cells in the tissues; buried in sand or gravel, with the base attached to a rock, pebble, or shell)   *Anthopleura artemisia*
12b Tubercles present on all parts of the column   13
13a Prevailing color of column, disk, and tentacles green or olive green, due to the presence of zoochlorellae (green algal cells) and/or zooxanthellae (dinoflagellates) in the tissues (specimens growing in caves or other deeply shaded situations may lack the symbionts and therefore be nearly white)   14
13b Prevailing color of column, disk, and tentacles white, yellow, or pink (the oral disk has yellow or pink lines radiating from the mouth to the margin, and the tentacles have transverse zigzag lines) (strictly subtidal; broods internally)   *Cribrinopsis fernaldi*
14a Column green or olive green, often shading to white toward the base; tubercles usually in distinct lengthwise rows; tentacles and oral disk about the same color as the column, but the tentacles typically with pink tips and the disk often with dark radial bands; height up to about 5 cm; often in clonal masses formed by asexual multiplication, and frequently covered by sand that accumulates on rocks to which this anemone is attached; mostly mid-intertidal   *Anthopleura elegantissima*
14b Column almost uniformly green; tubercles usually not in distinct rows; tentacles and oral disk uniformly green; height commonly exceeding 5 cm; solitary, and not covered by sand; low intertidal   *Anthopleura xanthogrammica*
15a Column greenish or reddish, with lengthwise rows of tubercles that may be of the same color or lighter; oral disk with conspicuous white stripes radiating from the mouth to the margin (broods internally, releasing young through pores at the tips of the tentacles)   *Aulactinia incubans* (fig. 3.83)
15b Column greenish, olive, brownish, or red (if greenish or olive, it may have red blotches; if red, it may have greenish or olive blotches or white tubercles); oral disk without radiating white stripes   16
16a Column red, its tubercles of the same color (and not arranged in distinct rows); column accumulating sand, gravel, and bits of shell (this species is usually partly buried in coarse sand or gravel); oral disk slightly greenish   *Urticina coriacea*
16b Column red (sometimes with white tubercles), brownish, olive, or greenish (if olive or greenish, often with red blotches); column usually not accumulating sand, gravel, or bits of shell; oral disk usually not greenish   17
17a Column red (rose to maroon), with conspicuous white tubercles   18
17b Column red, brownish, olive, or greenish (if olive or greenish, often with red blotches), with inconspicuous tubercles, these not white   19
18a Tubercles smooth, arranged in lengthwise rows; column rarely more than 10 cm in diameter; intertidal and subtidal   *Urticina lofotensis*

18b Tubercles rough, arranged in circumferential rows; column frequently more than 15 cm in diameter; strictly subtidal   *Urticina columbiana*
19a Tentacles usually with transverse bands; column sometimes of a uniform color (red, brownish, olive, or greenish), sometimes with red blotches on an olive or greenish background; intertidal and subtidal, common on floating docks   *Urticina crassicornis*
19b Tentacles without transverse bands; column uniformly deep red to maroon, appearing velvety; subtidal, and usually situated atop prominences (oral disk often with alternating red and yellow radial markings)   *Urticina piscivora*
20a Column orange or white, or white with orange streaks or blotches (tentacles orange or white, sometimes with orange bands); not brooding young externally; subtidal   21
20b Column brown, orange, red, or green, sometimes with parallel light lines, but never white or with orange streaks or blotches; sometimes brooding young externally on the column; intertidal and subtidal   23
21a Column whitish, with red or orange-red streaks; tentacles with white spots at their bases; column, when extended, not more than 3 cm high, and not appreciably higher than wide; usually attached to shells of *Modiolus modiolus*   *Stomphia coccinea*
21b Column cream to pale orange (sometimes with orange or red blotches or streaks) or reddish beige; tentacles without white spots at their bases; column, when extended, often more than 5 cm high, and appreciably higher than wide; not usually attached to shells of *Modiolus modiolus*   22
22a Color cream to orange, sometimes with orange or red blotches or streaks; with at least 160 tentacles in specimens whose column height is 5 cm or more   *Stomphia didemon*
22b Color uniformly reddish beige; with not more than 130 tentacles in specimens whose column height is 5 cm or more   *Stomphia* sp.
23a Oral disk with radiating white lines; brooding young internally or externally, depending on the species   24
23b Oral disk without radiating white lines; brooding young internally (color drab green, brick-red, dark orange, or deep mustard, sometimes with lighter or darker spots; pedal disk without radiating light lines; mid- to high intertidal, mostly in caves and in surge channels shaded by logs; known only from the San Juan Archipelago)   *Epiactis fernaldi*
24a Column predominantly greenish, reddish, or brownish (sometimes orange in *Epiactis lisbethae*); radiating white lines on oral disk narrow, originating close to the mouth; pedal disk with radiating white or gray lines; column not accumulating sand grains, brooding young externally   25
24b Column generally dull red to brown, sometimes with darker spots; radiating white lines on oral disk broad, located mostly close to the tentacle bases; pedal disk with or without radiating light lines; lower part of column with adhesive areas that accumulate sand; brooding young internally (extremely flat when contracted; mostly low intertidal, often under rocks)   *Epiactis ritteri*
25a Diameter of pedal disk not more than 3.5 cm; radiating light lines on pedal disk not continuing up the column for more than part of its height; all mature individuals capable of brooding and do so throughout the year, bearing up to about 30 young of various sizes, mostly in a single circumferential row   *Epiactis prolifera*
25b Diameter of pedal disk up to about 8 cm; radiating light lines on pedal disk of mature individuals continuing up the column for its entire height (the lines are gray, blue-gray, gray-green, or white; some light colored specimens do not have distinct lines); females brooding in spring and early summer, bearing up to several hundred young of nearly uniform size in a circumferential band that is several individuals deep (females that are predominantly red, brown, or green usually have pink young; those that are orange usually have orange young)   *Epiactis lisbethae*

Species marked with an asterisk are not in the key.

Suborder Nynantheae

Family Edwardsiidae

Edwardsia sipunculoides (Stimpson, 1853)
Nematostella vectensis Stephenson, 1935

Family Halcampoididae

Halcampoides purpurea (Studer, 1878)

Family Haloclavidae

*Bicidium aequoreae McMurrich, 1913. Pelagic; juveniles sometimes found on medusae of *Aequorea victoria*.
Peachia quinquecapitata McMurrich, 1913. Adults in muddy sand, low intertidal and subtidal; juveniles on *Phialidium gregarium* and other hydromedusae.

Family Halcampidae

Halcampa crypta Siebert & Hand, 1974
Halcampa decemtentaculata Hand, 1954

Family Actiniidae

Anthopleura artemisia (Pickering, in Dana, 1848)
Anthopleura elegantissima (Brandt, 1835)
Anthopleura xanthogrammica (Brandt, 1835)
Aulactinia incubans Dunn, Chia, & Levine, 1980 (fig. 3.83)
Cribrinopsis fernaldi Siebert & Spaulding, 1976
*Cribrinopsis williamsi Carlgren, 1940. Known from Alaska, and perhaps to be expected in our region.
Epiactis fernaldi Fautin & Chia, 1986
Epiactis lisbethae Fautin & Chia, 1986
Epiactis prolifera Verrill, 1869
Epiactis ritteri Torrey, 1902
Urticina columbiana (Verrill, 1922)
Urticina coriacea (Cuvier, 1798)
Urticina crassicornis (O. F. Müller, 1776)
Urticina lofotensis (Danielssen, 1890)
Urticina piscivora (Sebens & Laakso, 1977)
*Urticina sp. Referred to as *Tealia* sp. by Lawn & Ross (1982).

Family Liponematidae

*Liponema brevicornis (McMurrich, 1893). At depths of about 100 m and more; sometimes common.

Family Actinostolidae

*Paractinostola faeculenta McMurrich, 1893. Deep subtidal, Oregon southward.

Stomphia coccinea (O. F. Müller, 1776).
Stomphia didemon Siebert, 1973
Stomphia sp. Referred to in Ross (1979).

Family Hormathiidae

*Stephanauge annularis Carlgren, 1936. Shallow subtidal, Oregon southward.

Family Metridiidae

Metridium senile (Linnaeus, 1767)
Metridium sp. (fig. 3.84)

Family Haliplanellidae

Haliplanella lineata Verrill, 1869

Order Corallimorpharia

Hand, C. 1955. The sea anemones of central California. Part I. The corallimorpharian and athenarian anemones. Wasmann J. Biol., 12(1954):345-75.

Family Corallimorphidae

Corallimorphus sp. Shallow subtidal.
Corynactis californica Carlgren, 1936. Intertidal and shallow subtidal; common.

Order Zoanthidea

Family Epizoanthidae

Epizoanthus scotinus Wood, 1958. Intertidal and shallow subtidal.

4

PHYLA
CTENOPHORA, ORTHONECTIDA, DICYEMIDA

PHYLUM CTENOPHORA

Claudia E. Mills

Agassiz, A. 1865. North American Acalephae. Mem. Mus. Comp. Zool. Harvard College, 1(2):i-xiv, 1-234.
Bigelow, H. B. 1912. Reports on the scientific results of the expedition to the eastern tropical Pacific 1904-05. XXVI. The ctenophores. Bull. Mus. Comp. Zool. Harvard College, 52:369-404.
Chun, C. 1880. Die Ctenophoren des Golfes von Neapel und der angrenzenden Meeres-Abschnitte. Fauna und Flora des Golfes von Neapel, 1. xvi + 313 pp.
Harbison, G. R., & L. P. Madin. 1982. Ctenophora. In Parker, S. P. (ed.), Taxonomy and Classification of Living Organisms, 1:707-15. New York: McGraw-Hill Book Co.
Komai, T. 1918. On ctenophores of the neighborhood of Misaki. Annot. Zool. Japon., 9:451-73.
Mayer, A. G. 1912. Ctenophores of the Atlantic coast of North America. Carnegie Inst. Washington Publ. 162. 115 pp.
Mills, C. E., & R. L. Miller. 1984. Ingestion of a medusa (*Aegina citrea*) by the nematocyst-containing ctenophore *Haeckelia rubra* (formerly *Euchlora rubra*); phylogenetic implications. Mar. Biol., 78:215-21.
Mortensen, T. 1927. Two new ctenophores. (Papers from Th. Mortensen's Pacific Expeditions 1914-16.) Vidensk. Meddel. Dansk naturh. Foren. København, 83:277-88.
Torrey, H. B. 1904. The ctenophores of the San Diego region. Univ. Calif. Publ. Zool., 2:45-51.
Vanhöffen, E. 1903. Ctenophoren. Nordisches Plankton, Zool., 6, Abt. 11:1-7.

1a Body resembling a sac, without tentacles and not divided into a pair of large lobes at the oral end ... 2
1b Body solid, not saclike, either with a pair of conspicuous tentacles arising within sheaths on opposite sides of the body or with a pair of large lobes at the oral end (in which case there are small tentacles of 2 types in the cavity enclosed by the oral lobes) ... 4
2a Meridional digestive canals with branching diverticula in a layer just below the body surface (the branches between the canals, however, do not anastomose) ... 3
2b Meridional digestive canals without diverticula just below the body surface, although there are a few diverticula directed inward toward the pharynx (with spots of golden brown pigment; an undescribed species) ... *Beroe* sp.
3a Body colorless or with pink pigmentation (this pigmentation may be on the outer surface, especially near the ctenes, or distributed throughout the mesoglea, but it is not concentrated in the lining of the pharynx); pharyngeal canals not branched ... *Beroe ?cucumis*
3b Pharynx intensely pigmented, giving the ctenophore a striking red, purple, or nearly black color; pharyngeal canals with many diverticula (rare) ... *Beroe abyssicola*
4a Body with a pair of large lobes at the oral end; tentacles small, of two types (a pair of branched tentacles in sheaths near the mouth, and numerous small tentacles in tracts leading toward the mouth; both types are in the cavity enclosed by the oral lobes) ... 5
4b Body without a pair of large lobes at the oral end; tentacles conspicuous when extended, arising from sheaths on opposite sides of the body some distance from the mouth ... 6

80 Phylum Ctenophora

5a   Body surface smooth, colorless except for rows of darkly pigmented spots in line with the comb rows, but nearer to the mouth than these; length less than 12 cm
Bolinopsis ?infundibulum
5b   Body surface covered with yellow-orange papillae; length sometimes more than 20 cm (rare; an undescribed species)
Leucothea sp.

4.1 *Pleurobrachia bachei*   4.2 *Euplokamis dunlapae*

6a   Tentacle sacs opening toward the aboral end of the body; tentacles with side branches; mouth and pharynx relatively small   7
6b   Tentacle sacs opening toward the oral end of the body; tentacles without side branches; mouth and pharynx relatively large   10
7a   Tentacle sacs angling away from the pharynx toward the body surface for their entire length   8
7b   Tentacle sacs lying close to the pharynx for most of their length   9
8a   Body nearly spherical; tentillae fairly numerous and not coiled into neat, pear-shaped bundles when contracted
Pleurobrachia bachei (fig. 4.1)
8b   Body somewhat ovoid and flattened in the tentacular plane; tentillae sparse and coiled into neat pear-shaped bundles when contracted (an undescribed species)
Euplokamis dunlapae Mills, 1987 (fig. 4.2)
9a   Body elongate, approximately cucumber-shaped, the length sometimes attaining 10 cm; tentillae sparse (rare)
Hormiphora cucumis
9b   Body ovoid, up to 2 cm long; tentillae numerous and crowded (tentacles red or pinkish purple; an undescribed species)
Mertensia sp.
10a With only 4 meridional digestive canals, these widely separated from the comb rows; tentacles with nematocysts, but without colloblasts; tentacle sheaths very long, with orange to

brick-red pigment at the tentacle bases and also near the middle of the tentacle sheaths (rare)
Haeckelia rubra
10b With 8 meridional digestive canals, these directly beneath the comb rows; tentacles without nematocysts, but with colloblasts; tentacle sheaths short and sometimes containing red pigment (additional orange-brown pigment may be present near the mouth and gonads)
Dryodora glandiformis

Order Cydippida

Family Euplokamidae

Euplokamis dunlapae Mills, 1987 (fig. 4.2)

Family Haeckeliidae

Dryodora glandiformis (Mertens, 1833)
Haeckelia rubra (Kölliker, 1853)

Family Mertensiidae

Mertensia sp.

Family Pleurobrachiidae

Hormiphora cucumis (Mertens, 1833)
Pleurobrachia bachei A. Agassiz, 1860 (fig. 4.1)

Order Lobata

Family Bolinopsidae

Bolinopsis ?infundibulum (O. F. Müller, 1776)
Leucothea sp.

Order Beroida

Family Beroidae

Beroe abyssicola Mortensen, 1927
Beroe ?cucumis Fabricius, 1780
Beroe sp.

PHYLUM ORTHONECTIDA

Only 2 named species of orthonectids, listed below, have been reported from the region. Undescribed species have been found, however, in *Mytilus edulis* (Mollusca, Bivalvia) and *Ascidia callosa*

(Urochordata, Ascidiacea), and additional species should be expected, particularly in turbellarians, nemerteans, polychaetes, and bivalves.

Kozloff, E. N. 1965. *Ciliocincta sabellariae* gen. and sp. n., an orthonectid mesozoan from the polychaete *Sabellaria cementarium* Moore. J. Parasit., 51:37-44.
------. 1969. Morphology of the orthonectid *Rhopalura ophiocomae*. J. Parasit., 55:171-95.
------. 1971. Morphology of the orthonectid *Ciliocincta sabellariae*. J. Parasit., 57:585-97.

Family Rhopaluridae

Rhopalura ophiocomae Giard, 1877. In the gonads of *Amphipholis squamata* (Echinodermata, Ophiuroidea).

Family Intoshiidae

Ciliocincta sabellariae Kozloff, 1965. In the body wall of *Sabellaria cementarium* (Polychaeta, Sabellariidae).

PHYLUM DICYEMIDA

F. G. Hochberg

Family Dicyemidae

The species described from cephalopod molluscs are arranged according to host. With the exception of *Dicyemennea brevicephaloides*, all species live in the renal coelom, where they are found attached to the renal appendages of the host.

Bogolepova-Dobrokhotova, I. I. 1962. [Dicyemidae of the far-eastern seas. II. New species of the genus *Dicyemennea*.] Zool. Zhurnal SSSR, 41:503-18 (in Russian).
Hochberg, F. G. 1982. The "kidneys" of cephalopods: a unique habitat for parasites. Malacologia, 23:121-34.
Hoffman, E. G. 1965. Mesozoa of the sepiolid, *Rossia pacifica* (Berry). J. Parasit., 51:313-20.
McConnaughey, B. 1941. Two new Mesozoa from California, *Dicyemennea californica* and *Dicyemennea brevicephala* (Dicyemidae). J. Parasit., 27:63-9.
------. 1949. Mesozoa of the family Dicyemidae from California. Univ. Calif. Publ. Zool., 55:1-34.
------. 1957. Two new mesozoans from the Pacific Northwest. J. Parasit., 43:358-64.
Nouvel, H. 1947. Les Dicyémides. 1re partie: systématique, générations vermiformes, infusorigène et sexualité. Arch. Biol., 58:59-219.

1a Axial cell sometimes branched in the region of the calotte; calotte always large and distinctly flattened; somatic cells sometimes fused to form a syncytium; vermiform embryos without an abortive second axial cell .... *Conocyema*
1b Axial cell not branched; calotte elongated or flattened; somatic cells never fused; vermiform embryos with or without an abortive second axial cell 2
2a Calotte with 4 propolar cells and 5 metapolar cells; vermiform embryos without an abortive second axial cell .... *Dicyema*
2b Calotte with 4 propolar cells and 5 metapolar cells; vermiform embryos with or without an abortive second axial cell .... *Dicyemennea*

Rossia pacifica Berry
  Dicyemennea brevicephaloides Bogolepova-Dobrokhotova, 1962. Unusual in that it lives in the branchial heart coelom. Vermiform stage: length up to 3 mm; calotte flattened; 22-24 somatic cells; verruciform cells absent. Infusoriform stage: length 25-30 µm; refringent bodies present but nonrefractive.
  Dicyemennea brevicephala McConnaughey, 1941. Vermiform stage: length not more than 1 mm; 26-28 somatic cells; verruciform cells present but not prominent. Infusoriform stage: length 35-40 µm; refringent bodies very large.
  Dicyemennea filiformis Bogolepova-Dobrokhotova, 1962 (*D. parva* Hoffman, 1965). Vermiform stage: length rarely more than 1.5 mm; calotte elongated, uniformly narrow or arrowhead-shaped; 24-26 somatic cells; uropolar cells often verruciform. Infusoriform stage: length 25 µm; refringent bodies absent.
Octopus dofleini Pickford (*O. apollyon*, as used by McConnaughey, 1957)
  Conocyema deca McConnaughey, 1957. Vermiform stage: length not more than 1 mm; calotte flattened, with 6 metapolar cells; 23 or 24 somatic cells; uropolar cells slightly verruciform. Infusoriform stage: length 35-40 µm; refringent bodies present, but very small.
  Dicyemennea abreida McConnaughey, 1957. Vermiform stage: length up to 3 mm; calotte rounded; 28-30 somatic cells; verruciform cells absent. Infusoriform stage: length 35 µm; refringent bodies present, large.
Octopus leioderma Berry
  Two species; descriptions in preparation (Hochberg).
Octopus rubescens Berry (*O. apollyon*, as used by McConnaughey, 1941, 1949 and by Nouvel, 1947)
  Dicyema apollyoni Nouvel, 1947 (*D. balamuthi* McConnaughey, 1949). Vermiform stage: length up to 3 mm; calotte elongated; 22 somatic cells; verruciform cells absent. Infusoriform stage: length 25-35 µm; refringent bodies present, small.

5
PHYLA
PLATYHELMINTHES, GNATHOSTOMULIDA

PHYLUM PLATYHELMINTHES

Class Turbellaria

Subclass Archoophora

Order Acoelida

Acoels are usually present in the top layer of sediments, as well as in detritus and growths of diatoms that coat algae and sessile invertebrates. There are probably at least 25 species in this region, but only a few have been described. Three of them are illustrated by Kozloff (1965). The monograph of Dörjes (1968), concerned with the acoel fauna of the North Sea, is valuable for recognition of families and genera.

Bush, L. F. 1981. Marine flora and fauna of the northeastern United States. Turbellaria: Acoela and Nemertodermatida. NOAA Tech. Rep. NMFS Circular 440. 70 pp.
Dörjes, J. 1968. Die Acoela (Turbellaria) der deutschen Nordseeküste und ein neues System der Ordnung. Zeitschr. Zool. Syst. Evolutionsforsch., 6:56-452.
Dörjes, J., & T. G. Karling. 1975. Species of Turbellaria Acoela in the Swedish Museum of Natural History. Zoologica Scripta, 4:175-89.
Kozloff, E. N. 1965. New species of acoel turbellarians from the Pacific coast. Biol. Bull., 129:151-66.
------. 1972. Selection of food, feeding, and physical aspects of digestion in the acoel turbellarian *Otocelis luteola*. Trans. Amer. Micr. Soc., 91:556-65.

Family Otocelididae

Otocelis luteola (Kozloff, 1965)

Family Convolutidae

Raphidophallus actuosus Kozloff, 1965
Diatomovora amoena Kozloff, 1965

Family Paratomellidae

Paratomella unichaeta Dörjes, 1966

Family Childiidae

Childia groenlandica (Levinsen, 1879). In muddy sand with eelgrass.

Order Polycladida

John J. Holleman

Assignment of polyclads to genera and families follows the scheme of Prudhoe (1985). Faubel (1983, 1984) has proposed several additional genera, mostly for species in the family Leptoplanidae, and his distribution of genera in families is slightly different from that of Prudhoe.

Bock, S. 1925. Papers from Dr. Th. Mortensen's Pacific expedition 1914-1918. XXVII. Planarians. Pt. IV. New stylochids. Vidensk. Medd. Dansk Naturh. Foren., 79:97-198.

Boone, E. 1929. Five new polyclads from the California coast. Ann. Mag. Nat. Hist., ser. 10, 3:33-46.

Ching, H. L. 1977. Redescription of *Eurylepta leoparda* Freeman, 1933 (Turbellaria: Polycladida), a predator of the ascidian *Corella willmeriana* Herdman, 1898. Can. J. Zool., 55:338-42.

------. 1978. Resdescription of a marine flatworm: *Pseudoceros canadensis* Hyman, 1953 (Polycladida: Cotylea). Can. J. Zool., 56:1372-76.

Faubel, A. 1983. The Polycladida, Turbellaria. Proposal and establishment of a new system. Part I. The Acotylea. Mitteilungen hamburg. zool. Mus. Inst., 80:17-121.

------. 1984. The Polycladida, Turbellaria. Proposal and establishment of a new system. Part 2. The Cotylea. Mitteilungen hamburg. zool. Mus. Inst., 81:189-259.

Freeman, D. 1933. The polyclads of the San Juan region of Puget Sound. Trans. Amer. Micr. Soc., 52:107-46.

Heath, H., & E. A. McGregor. 1912. New polyclads from Monterey Bay, California. Proc. Acad. Nat. Sci. Philadelphia, 64:455-87.

Holleman, J. J. 1972. Marine turbellarians of the Pacific coast I. Proc. Biol. Soc. Washington, 85:405-7.

Hyman, L. H. 1939. New species of flatworms from North, Central and South America. Proc. U. S. Nat. Mus. , 86:419-39.

------. 1939. Some polyclads of the New England coast, especially of the Woods Hole region. Biol. Bull., 76:127-52.

------. 1953. The polyclad flatworms of the Pacific coast of North America. Bull. Amer. Mus. Nat. Hist., 100:265-392.

------. 1955. The polyclad flatworms of the Pacific coast of North America: additions and corrections. Amer. Mus. Novit., no. 1704. 11 pp.

------. 1959. Some Turbellaria from the coast of California. Amer. Mus. Novit., no. 1943. 17 pp.

Prudhoe, S. 1985. A Monograph on Polyclad Turbellaria. London: British Museum (Natural History) & Oxford University Press. 259 pp.

1a	Ventral surface with a sucker (this is independent of the digestive system, and lies behind the female gonopore); tentacles, when present, situated along the margin of the body at the anterior end (suborder Cotylea)	17
1b	Ventral surface without a sucker; tentacles, when present, of the nuchal type, situated above the brain and some distance behind the anterior end (suborder Acotylea)	2
2a	A band of eyespots present along the entire margin of the body (these may not be visible except in transmitted light) (eyespots also present on the tentacles, at the bases of the tentacles, and in the region of the brain); length up to about 10 cm	*Kaburakia excelsa*
2b	Eyespots restricted to the anterior part of the body	3
3a	Eyespots scattered over much of the anterior part of the body; body slender and elongated; mouth and pharyngeal pocket located in the posterior half of the body	*Cestoplana* sp.
3b	Eyespots restricted to the tentacles and the region of the brain; body not slender and elongated; mouth and pharyngeal pocket located near the middle of the body, or slightly anterior to the middle	4
4a	With nuchal tentacles	5
4b	Without nuchal tentacles	6
5a	Chitinized penis stylet present and conspicuous; Lang's vesicle absent	*Phylloplana viridis*

86 Phylum Platyhelminthes

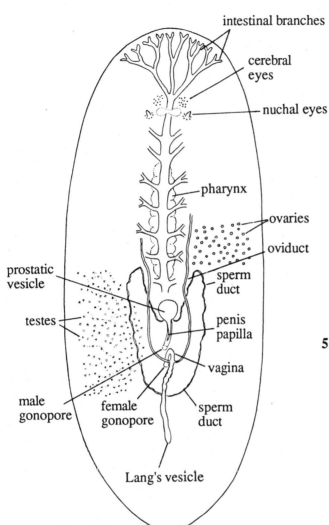

5.1 Generalized diagram of a polyclad, dorsal view

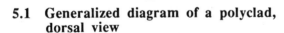

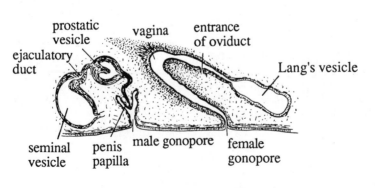

5.2 *Notoplana sanjuania*, sagittal view of copulatory organs

5.3 *Eurylepta aurantiaca*, dorsal view

5.4 *Freemania litoricola*

Phylum Platyhelminthes 87

5b Penis stylet absent; Lang's vesicle (terminal dilation of a posteriorly directed diverticulum of the vagina) present   *Pseudostylochus burchami*
6a Chitinized penis stylet present   7
6b Penis stylet absent   10
7a Penis stylet straight or arched, never coiled   8
7b Penis stylet very long and coiled   *Notoplana longastyletta*
8a Penis stylet nearly straight; gonopore single, serving both male and female reproductive systems   *Notoplana inquieta*
8b Penis stylet decidedly arched; male and female gonopores separate   9
9a Penis stylet about as long as the prostatic vesicle (a glandular bulbous structure between the short penis papilla and the seminal vesicle); free-living   *Notoplana atomata*
9b Penis stylet nearly twice as long as the prostatic vesicle; inhabiting shells occupied by hermit crabs   *Notoplana inquilina*
10a With well developed spermiducal bulbs (dilations of the sperm ducts close to the prostatic vesicle)   11
10b Without spermiducal bulbs   12
11a Spermiducal bulbs uniting medially before entering the prostatic vesicle; penis papilla large, tonguelike, often protruding in preserved specimens   *Freemania litoricola* (fig. 5.4)
11b Spermiducal bulbs separately entering an elongated seminal vesicle that leads posteriorly to the prostatic vesicle; penis papilla small, inconspicuous   *Stylochoplana chloranota*
12a Red, or the color of red wine, or marked with a large red blotch   13
12b Grayish or brownish, nearly uniform or mottled, but definitely not red or the color of red wine, and not marked with a large red blotch   14
13a Marked with a saddlelike red blotch on the central part of the dorsal surface, elsewhere gray; male and female genital pores separate   *Notoplana sanguinea*
13b Color red or wine throughout; gonopore single   *Notoplana celeris*
14a Prostatic vesicle tubular   *Leptoplana vesiculata*
14b Prostatic vesicle ovoid   15
15a Prostatic vesicle smaller than the seminal vesicle; penis papilla small, conical, not easily identified in entire specimens; not more or less obligatorily associated with oysters   16
15b Prostatic vesicle larger than the seminal vesicle; penis papilla large, bulbous, muscular, readily identified in entire specimens; regularly associated with oysters (*Ostrea lurida*), and a predator on young individuals   *Pseudostylochus ostreophagus*
16a Male and female genital pores very close together, the distance between them much shorter than the length of Lang's vesicle; capable of swimming effectively by fluttering movements of the entire body   *Notoplana natans*
16b Male and female genital pores rather widely separated, the distance between them exceeding the length of Lang's vesicle; not capable of swimming effectively   *Notoplana sanjuania* (fig. 5.2)
17a With a pair of marginal tentacles at the anterior end   19
17b Without marginal tentacles at the anterior end   18
18a Mouth located at the center of the anterior margin; male gonopore separate from the mouth   *Acerotisa alba*
18b Mouth located posterior to the brain; male antrum opening into the mouth   *Stylostomum album*
19a Background color yellowish pink or salmon, with minute white specks, or uniformly orange-red   *Eurylepta aurantiaca* (fig. 5.3)
19b Background color cream, light tan, brown, or gray, without any distinctly pink, orange, or red tones   20
20a Background color cream or light tan, marked on the dorsal surface with conspicuous reddish brown spots (marginal tentacles with orange or red bands at their base)   *Eurylepta leoparda*
20b Background color light brown to gray, mottled with darker brown or gray   *Pseudoceros canadensis*

Phylum Platyhelminthes

Species marked with an asterisk are not in the key; they have not been reported north of Oregon.

Suborder Acotylea

Family Stylochidae

Kaburakia excelsa Bock, 1925
**Stylochus atentaculatus* Hyman, 1953
**Stylochus tripartitus* Hyman, 1953

Family Leptoplanidae

Freemania litoricola (Heath & McGregor, 1912) (fig. 5.4)
Leptoplana vesiculata Hyman, 1939
Notoplana atomata (O. F. Müller, 1776)
Notoplana celeris Freeman, 1933
Notoplana inquieta (Heath & McGregor, 1912)
Notoplana inquilina Hyman, 1955. In shells occupied by hermit crabs.
Notoplana longastyletta (Freeman, 1933)
Notoplana natans Freeman, 1933
**Notoplana rupicola* (Heath & McGregor, 1912)
Notoplana sanguinea Freeman, 1933
Notoplana sanjuania Freeman, 1933 (fig. 5.2)
Phylloplana viridis (Freeman, 1933). Generally on eelgrass, *Zostera marina*.
Stylochoplana chloranota (Boone, 1929)

Family Callioplanidae

Pseudostylochus burchami (Heath & McGregor, 1912)
Pseudostylochus ostreophagus Hyman, 1955. Associated with the oyster *Ostrea lurida*, and preys on young individuals.

Family Cestoplanidae

Cestoplana sp. Introduced with oysters (Washington).

Suborder Cotylea

Family Pseudocerotidae

Pseudoceros canadensis Hyman, 1953. Generally on the compound ascidian *Distaplia occcidentalis*.

Family Euryleptidae

Stylostomum album (Freeman, 1933)
Eurylepta aurantiaca Heath & McGregor, 1912 (fig. 5.3)
Eurylepta leoparda Freeman, 1933. Generally on the solitary ascidian *Corella willmeriana*.
Stylostomum sanjuania Holleman, 1972. Generally on the compound ascidians *Distaplia occidentalis*, *Aplidium californicum*, and *Trididemnum* sp.

Subclass Eulecithophora

George L. Shinn

Order Prolecithophora

Karling (1962a, 1962b) has described several species from California. Some of them, or closely related species, probably occur in our region.

Karling, T. G. 1962a. Marine Turbellaria from the Pacific coast of North America. I. Plagiostomidae. Arkiv för Zoologi, ser. 2, 15:113-41.
------. 1962b. Marine Turbellaria from the Pacific coast of North America. II. Pseudostomidae and Cylindrostomidae. Arkiv för Zoologi, ser. 2, 15:181-209.

Order Neorhabdocoelida

Free-living neorhabdocoels are encountered in nearly all benthic situations, and there are probably at least 50 species in our fauna. The ones that have been reported from Oregon or farther north are listed below under the families Provorticidae, Koinocystididae, Polycystididae, Graffillidae, Promesostomidae, and Trigonostomidae. Some of the California species described by Karling (1962a, 1962b, 1967, 1983, 1986), Karling & Schockaert (1977), and Schockaert & Karling (1970) should also be expected.

Most of the symbiotic neorhabdocoels known from this region are in the family Umagillidae, but one belongs to the family Fecampiidae. These worms, and their hosts, are also listed below. Several species of the family Graffillidae have been encountered in molluscs, but none of these has been identified or described.

Ax, P. 1968. Turbellarien der Gattung *Promesostoma* von der nordamerikanischen Pazifikküste. Helgoländer wiss. Meeresunters., 18:116-23.
Ax, P., R. Ax, & U. Ehlers. 1979. First record of a free-living dalyellioid turbellarian from the Pacific: *Balgetia pacifica* nov. spec. Helgoländer wiss. Meeresunters., 32:259-364.
Karling, T. G. 1967. On the genus *Promesostoma* (Turbellaria), with descriptions of four new species from Scandinavia and California. Sarsia, 29:257-68.
------. 1977. Taxonomy, phylogeny, and biogeography of the genus *Austrorhynchus* Karling (Turbellaria, Polycystididae). Mikrofauna Meeresboden 61:153-65.
------. 1980. Revision of Koinocystididae (Turbellaria). Zoologica Scripta, 9:241-69.
------. 1983. Structural and systematic studies on Turbellaria Schizorhynchia (Platyhelminthes). Zoologica Scripta, 12:77-89.
------. 1986. Free-living marine Rhabdocoela (Platyhelminthes) from the N. American Pacific coast. With remarks on species from other areas. Zoologica Scripta, 15:201-9.
Karling, T. G., & E. R. Schockaert. 1977. Anatomy and systematics of some Polycystididae (Turbellaria, Kalyptorhynchia) from the Pacific and S. Atlantic. Zoologica Scripta 6:5-19.
Kozloff, E. N. 1953. *Collastoma pacifica* sp. nov., a rhabdocoel turbellarian from the gut of *Dendrostoma pyroides* Chamberlin. J. Parasit., 39:336-40.
------. 1965. *Desmote inops* sp. n. and *Fallacohospes inchoatus* gen. and sp. n., umagillid rhabdocoels from the intestine of the crinoid *Florometra serratissima* (A. H. Clark). J. Parasit., 51:305-12.
Kozloff, E. N., & G. L. Shinn. 1987. *Wahlia pulchella* sp. n., a turbellarian flatworm (Neorhabdocoela, Umagillidae) from the intestine of the sea cucumber *Stichopus californicus*. J. Parasit., 73:194-202.
Kozloff, E. N., & C. A. Westervelt, Jr. 1987. Redescription of *Syndesmis echinorum* François, 1886 (Turbellaria: Neorhabdocoela: Umagillidae), with comments on distinctions between *Syndesmis* and *Syndisyrinx*. J. Parasit., 73:184-193.
Lehman, H. E. 1946. A histological study of *Syndisyrinx franciscanus* gen. et sp. nov., an endoparasitic rhabdocoel of the sea urchin, *Strongylocentrotus franciscanus*. Biol. Bull., 91:295-311.
Shinn, G. L. 1983. *Anoplodium hymanae* sp. n., an umagillid turbellarian from the coelom

of *Stichopus californicus*, a northeast Pacific holothurian. Can. J. Zool., 61:750-60.
Shinn, G. L., & A. M. Christensen. 1985. *Kronborgia pugettensis* sp. nov. (Neorhabdocoela: Fecampiidae), an endoparasitic turbellarian infesting the shrimp *Heptacarpus kincaidi* (Rathbun), with notes on its life-history. Parasitology, 91:431-47.
Schockaert, E., & T. G. Karling. 1970. Three new anatomically remarkable Turbellaria Eukalyptorhynchia from the North American Pacific coast. Arkiv för Zoologi, 23:237-53.
Smith, N. S. 1973. A new description of *Syndesmis dendrastrorum* (Platyhelminthes, Turbellaria), an intestinal rhabdocoel inhabiting the sand dollar *Dendraster excentricus*. Biol. Bull., 145:598-606.
Stunkard, H. W., & J. O. Corliss. 1951. A new species of *Syndesmis* and a revision of the family Umagillidae Wahl, 1910 (Turbellaria: Rhabdocoela). Biol. Bull., 101:319-34.
Westervelt, C. A., Jr. 1981. *Collastoma kozloffi* sp. n., a neorhabdocoel turbellarian from the intestine of the sipunculan *Themiste dyscrita*. J. Parasit., 67:574-77.

Suborder Dalyellioida

Family Graffillidae

Breslauilla relicta Reisinger, 1929

Family Fecampiidae

Kronborgia pugetensis Shinn & Christensen, 1985. Parasitic in the hemocoel of the shrimp *Heptacarpus kincaidi*; unusual in having separate sexes (but only females have been found), and in lacking a digestive tract.

Family Provorticidae

Balgetia pacifica Ax, Ax, & Ehlers, 1979. Intertidal, in coarse sand.
Coronopharynx pusillus Luther, 1962. In pools in mudflats.

Family Umagillidae

All members of this family are symbiotic in echinoderms and sipunculans.

Anoplodium hymanae Shinn, 1983. In the perivisceral coelom of the sea cucumber *Parastichopus californicus*.
Collastoma kozloffi Westervelt, 1981. In the intestine of the sipunculan *Themiste dyscrita*.
Collastoma pacifica Kozloff, 1953. In the intestine of the sipunculan *Themiste pyroides*.
Desmote inops Kozloff, 1965. In the intestine of the crinoid *Florometra serratissima*.
Fallacohospes inchoatus Kozloff, 1965. In the intestine of the crinoid *Florometra serratissima*.
Syndesmis dendrastrorum Stunkard & Corliss, 1951. In the intestine of the sand dollar *Dendraster excentricus*.
Syndesmis spp. In the intestine of the sea urchins *Strongylocentrotus droebachiensis, S. pallidus,* and *Allocentrotus fragilis*.
Syndisyrinx franciscanus Lehman, 1946. In the intestine of the sea urchins *Strongylocentrotus droebachiensis, S. pallidus, S. franciscanus,* and *S. purpuratus*.
Wahlia pulchella Kozloff & Shinn, 1987. In the intestine of the sea cucumber *Parastichopus californicus*.

Suborder Kalyptorhynchia

Family Koinocystididae

Itaipusa bispina Karling, 1980. Intertidal, in sand around *Zostera* and among stones.
Itaipusa curvicirra Karling, 1980. Intertidal, in sand and gravel.

Family Polycystididae

Austrorhynchus californicus Karling, 1977. In sand among kelp, down to a depth of 10 m.
Austrorhynchus pacificus Karling, 1977. Intertidal, in sand and gravel among algae.
Duplacorhynchus major Schockaert & Karling, 1970. Estuarine, intertidal, at the surface of sandy mud.
Gyratrix hermaphroditus Ehrenberg, 1831. Intertidal, in sand around *Zostera*.
Gyratrix proaviformis Karling & Schockaert, 1977. Intertidal, in sand around *Phyllospadix*.
Polycystis hamata Karling, 1986. Among algae and in tide pools.
Yaquinaia microrhynchus Schockaert & Karling, 1970. Estuarine, intertidal, on the surface of sandy mud.

Suborder Typhloplanoina

Family Promesostomidae

Brinkmanniella palmata Karling, 1986. In tidepools and in holdfasts of kelp.
Promesostoma infundibulum Ax, 1968. Intertidal, in a mixture of coarse sand, gravel, and stones.
Promesostoma hymanae Ax, 1968. Intertidal, in fine sand that has a growth of diatoms.

Family Trigonostomidae

Ceratopera axi (Riedl, 1964). In shell gravel and in sand with *Zostera*.
Ceratopera pilifera Karling, 1986. Among algae.

Order Proseriata

Many species of this group occur in our fauna, and some of them have been described in papers of Ax (1966), Ax & Ax (1967, 1969), and Ax & Sopott-Ehlers (1979). Two papers of Karling (1964, 1966), on California species, and several papers of Tajika (see references below), on Japanese proseriate turbellarians, should be helpful in assigning undescribed species to family and genus.

Ax, P. 1966. Die Bedeutung der interstitiellen Sandfauna für allgemeine Probleme der Systematik, Ökologie und Biologie. Veröffentl. Inst. Meeresforsch. Bremerhaven, Sonderbd. 2:15-65.
Ax, P., & R. Ax. 1967. Turbellaria Proseriata vor der Pazifikküste der USA (Washington). I. Otoplanidae. Zeitschr. Morph. Tiere, 61:215-54.
------. 1969. Eine Chorda intestinalis bei Turbellarien (*Nematoplana nigrocapitula* Ax) als Modell für die Evolution der Chorda dorsalis. Akad. Wiss. Lit. Mainz, Abhandl. Math. Naturh. Kl., Jahrg. 1969:133-49.
Ax, P. & B. Sopott-Ehlers. 1979. Turbellaria Proseriata von der Pazifikküste der USA (Washington). Zoologica Scripta, 8:25-35.
Karling, T. G. 1964. Marine Turbellaria from the Pacific coast of North America. III. Otoplanidae. Arkiv för Zoologi, ser. 2, 16:527-41.
------. 1966. Marine Turbellaria from the Pacific coast of North America. IV. Coelogynoporidae and Monocelididae. Arkiv för Zoologi, ser. 2, 18:493-528.
Tajika, K.-I. 1979. Marine Turbellarien aus Hokkaido, Japan III. *Nematoplana* Meixner, 1938 (Proseriata, Nematoplanidae). J. Fac. Sci. Hokkaido Univ., Ser. 6, Zoology, 22:69-87.

------. 1980. Eine neue Gattung der Familie Coelogynoporidae (Turbellaria, Proseriata) aus Hokkaido, Japan. Annot. Zool. Japon., 53:18-36.

------. 1981a. Marine Turbellarien aus Hokkaido, Japan V. Coelogynoporidae (Proseriata). J. Fac. Sci., Hokkaido Univ., Ser. 6, Zoology, 22:451-73.

------. 1981b. Eine neue Art der Gattung *Archimonocelis* (Turbellaria: Proseriata: Monocelidae) aus Hokkaido, Japan. Proc. Jap. Soc. Syst. Zool., 21:1-9.

------. 1982a. Eine neue Gattung der Familie Nematoplanidae (Turbellaria, Proseriata) aus Hokkaido, Japan. Annot. Zool. Japon., 55:9-25.

------. 1982b. Marine Turbellarien aus Hokkaido, Japan IX. Monocelidae (Proseriata). Bull. Lib. Arts & Sci. Course, Sch. Med., Nihon Univ. 10:9-34.

------. 1983a. Zwei neue interstitielle Turbellarien der Gattung *Archotoplana* (Proseriata, Otoplanidae) aus Hokkaido, Japan. J. Fac. Sci., Hokkaido University, Ser. 6, Zoology, 23:179-94.

------. 1983b. Zwei neue Otoplaniden (Turbellaria, Proseriata) aus Hokkaido, Japan. Annot. Zool. Japon., 56:100-10.

------. 1983c. Zur Kenntnis der Gattung *Notocaryoplana* Steinböck, 1935 (Turbellaria, Proseriata, Otoplanidae). Bull. Nat. Sci. Mus., Tokyo, Ser. A, 9:97-104.

Family Coelogynoporidae

Coelogynopora cochleare Ax & Sopott-Ehlers, 1979. Intertidal.
Coelogynopora falcaria Ax & Sopott-Ehlers, 1979. Intertidal, in coarse sand and gravel.
Coelogynopora frondifera Ax & Sopott-Ehlers, 1979. Intertidal, in coarse sand.
Coelogynopora nodosa Ax & Sopott-Ehlers, 1979. Intertidal, in coarse sand of beaches.
Coelogynopora scalpri Ax & Sopott-Ehlers, 1979. Intertidal, in coarse sand and gravel.
Invenusta paracnida (Karling, 1966). Intertidal, in coarse sand and gravel, or in coarse sand among stones.
Vannuccia rotundouncinata Ax & Sopott-Ehlers, 1979. Intertidal, in fine sand.
Vannuccia tripapillosa Tajika, 1977 subsp. *americana* Ax & Sopott-Ehlers, 1979. Intertidal, in coarse sand and gravel.

Family Nematoplanidae

Nematoplana nigrocapitula Ax, 1966.

Family Otoplanidae

Americanaplana fernaldi Ax & Ax, 1967. Intertidal, in a mixture of coarse sand and gravel.
Pluribursaeplana illgi Ax & Ax, 1967. In coarse sediment with stones, at depths of about 50 m.
Orthoplana kohni Ax & Ax, 1967. Intertidal, in coarse sand.
Itaspiella armata Ax, 1951, subsp. *magna* Ax & Ax, 1967. Intertidal, in coarse sand and gravel.
Notocaryoplanella glandulosa Ax, 1951. Intertidal, in coarse and medium sand or in a mixture of coarse sand and gravel.
Polyrhabdoplana posttestis Ax & Ax, 1967. Intertidal, in a mixture of coarse sand and gravel.
Parotoplana pacifica Ax & Ax, 1967. In sediment, at depths of about 50 m.
Philosyrtis sanjuanensis Ax & Ax, 1967. Intertidal, in fine to medium sand.

Order Tricladida

Four species are known to occur within the range of this manual. An Alaskan species, *Nesion arcticum*, should perhaps be expected.

Ball, I. R. 1975. Contributions to a revision of the marine triclads of North America: the monotypic genera *Nexilis, Nesion, and Foviella* (Turbellaria: Tricladida). Can. J. Zool., 53:395-407.

Holleman, J. J. 1972. Marine turbellarians of the Pacific coast I. Proc. Biol. Soc. Washington, 85:405-7.
Holleman, J. J., & C. Hand. 1962. A new species, genus, and family of marine flatworms (Turbellaria: Tricladida, Maricola) commensal with molluscs. Veliger, 5:20-2.
Holmquist, D., & T. G. Karling. 1972. Two new species of interstitial marine triclads from the North American Pacific coast, with comments on evolutionary trends and systematics in Tricladida. Zoologica Scripta, 1:175-84.
Hyman, L. H. 1954. A new marine triclad from the coast of California. Amer. Mus. Novit., no. 1659. 5 pp.
------. 1956. North American triclad Turbellaria, 15. Three new species. Amer. Mus. Novit., no. 1808. 14 pp.

Suborder Maricola

Family Procerodidae

Oregoniplana opisthopora Holmquist & Karling, 1972. In the sandy substrata of *Zostera* beds, and also on rocks.
Pacifides psammophilus Holmquist & Karling, 1972. In sandy beaches.
Procerodes pacifica Hyman, 1954. Mid-intertidal, in coarse gravel.

Family Nexilidae

Nexilis epichitonius Holleman & Hand. 1962. Unusual in lacking a vagina, and in having the ovovitelline ducts entering the inflated base of the penis, just posterior to the seminal vesicle; commensal with chitons and mussels, and with gastropods of the genus *Nucella*, and sometimes found on the glass of aquaria that contain these molluscs; also in empty shells of intertidal barnacles and on the leaves of *Phyllospadix*.

Family Nesionidae

Nesion arcticum Hyman, 1956

PHYLUM GNATHOSTOMULIDA

Gnathostomulids have been found at several localities within our region, but only one species has been described. This is *Gnathostomula karlingi* Riedl, 1971 (order Bursovaginoidea, family Gnathostomulidae), which was collected at low tide levels at Yaquina Head, near Newport, Oregon. The habitat was characterized as sand, with black detritus and roots of *Zostera*.
For literature concerning this group of worms, consult the comprehensive papers of Sterrer (1977, 1982).

Riedl, R. J. 1971. On the genus *Gnathostomula* (Gnathostomulida). Int. Rev. Ges. Hydrobiol., 56:315-496.
Sterrer, W. E. 1977. Systematics and evolution within the Gnathostomulida. Syst. Zool., 21:151-73.
------. 1982. Gnathostomulida. *In* Parker, S. P. (ed.), Synopsis and Classification of Living Organisms, 1: 857-63. New York: McGraw-Hill Book Co.

6

PHYLUM NEMERTEA

Stephen A. Stricker

About 40 species of nemerteans are known to occur in the region covered by this book. The monographs of Coe (1901, 1904, 1905, 1926, 1940) are still the most useful taxonomic references. Other articles that deal with the fauna or that contain helpful illustrations are those of Coe (1944, 1954), Corrêa (1964), Gerner (1969), Gibson (1982, 1986), Stricker (1982, 1985), Stricker & Cloney (1982), Wickham (1978), and Wickham & Kuris (1985).

There are undoubtedly many species, especially in interstitial habitats, that are either undescribed or that have not yet been recorded from the region, so identifications should not be forced on specimens that do not fit clearly into the key. Although the key is based almost exclusively on external characteristics or features that can be seen in whole mounts of living specimens, much of the information used in the systematics of nemerteans is visible only in sectioned material. Thus in a few cases in which the external features of one species closely resemble those of another, histological examination may be necessary for proper identification.

Coe, W. R. 1901. Papers from the Harriman Alaska Expedition. XX. The nemerteans of the expedition. Proc. Wash. Acad. Sci., 3:1-110 (Reprinted in Harriman Alaska Expedition, 11:1-110 [1904].)
------. 1904. Nemerteans of the Pacific coast of North America. Part II. Harriman Alaska Expedition, 11:111-220.
------. 1905. Nemerteans of the west and northwest coast of America. Bull. Mus. Comp. Zool. Harvard College, 47:1-319.
------. 1926. The pelagic nemerteans. Mem. Mus. Comp. Zool. Harvard College, 49:1-244.
------. 1944. Geographical distribution of the nemerteans of the Pacific coast of North America, with descriptions of two new species. J. Wash. Acad. Sci., 34:27-32.
------. 1954. The bathypelagic nemerteans of the Pacific Ocean. Bull. Scripps Inst. Oceanogr., 6:225-86.
Corrêa, D. D. 1964. Nemerteans from California and Oregon. Proc. Calif. Acad. Sci., ser. 4, 31:515-58.
Gerner, L. 1969. Nemertinen der Gattungen *Cephalothrix* und *Ototyphlonemertes* aus dem marinen Mesopsammal. Helgoländer wiss. Meeresunters., 19:68-110.
Gibson, R. 1982. Nemertea. *In* Parker, S. A. (ed.), Synopsis and Classification of Living Organisms, 1:823-46. New York: McGraw-Hill Book Co.
------. 1986. Redescription and taxonomic reappraisal of *Nemertopsis actinophila* Bürger, 1904. (Nemertea: Hoplonemertea: Monostylifera). Bull. Mar. Sci., 39:42-60.
Gibson, R., J. Moore, & F. B. Crandall. 1982. A new semi-terrestrial nemertean from California. J. Zool., London, 196:463-74.
Griffin, B. B. 1898. Description of some nemerteans of Puget Sound and Alaska. Ann. N. Y. Acad. Sci., 11:193-218.
Humes, A. G. 1942. The morphology, taxonomy, and bionomics of the nemertean genus *Carcinonemertes*. Illinois Biol. Monogr., 18(4):1-105.
Stricker, S. A. 1982. The morphology of *Paranemertes sanjuanensis* sp. n. (Nemertea, Monostilifera) from Washington, U. S. A. Zool. Scripta, 11:107-15.
------. 1985. A new species of *Tetrastemma* (Nemertea, Monostilifera) from San Juan Island, Washington, U. S. A. Can. J. Zool., 63:682-90.
Stricker, S. A., & R. A. Cloney. 1982. Stylet formation in nemerteans. Biol. Bull., 162:387-405.

Wickham, D. E. 1978. A new species of *Carcinonemertes* (Nemertea: Carcinonemertidae) with notes on the genus from the Pacific coast. Proc. Biol. Soc. Wash., 91:197-202.

Wickham, D. E., & A. M. Kuris. 1985. The comparative ecology of nemertean egg predators. Amer. Zool., 25:127-37.

The key does not include pelagic species, but those that have been reported from the region are in the taxonomic list.

1a	Mouth separate from the proboscis pore, and both openings located on the ventral side of the head posterior to the brain; proboscis without stylets (class Anopla)	2
1b	Mouth united with the proboscis pore, the single opening located at the anterior tip of the head; proboscis armed with a stylet (except in the leechlike genus *Malacobdella*) (class Enopla)	23
2a	Head without cephalic slits (order Palaeonemertea and one species [*Baseodiscus princeps*] of the order Heteronemertea)	3
2b	Head with cephalic slits (order Heteronemertea)	10
3a	With conspicuous nematocystlike structures in the epidermis; length not exceeding 1 cm; interstitial	*Cephalothrix pacifica*
3b	Without nematocystlike structures in the epidermis; length usually much greater than 1 cm; not interstitial	4
4a	Body either with conspicuous white rings or longitudinal lines, or both	5
4b	Body color more or less homogeneous, with neither rings nor longitudinal lines	7

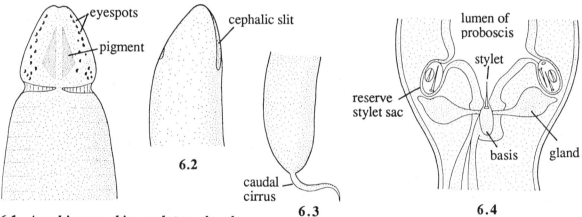

6.1 *Amphiporus bimaculatus*, head region, somewhat diagrammatic

5a	General color deep red, interrupted by white rings that are spaced at intervals about equal to the width of the body; without longitudinal lines; anterior margin of head white	*Tubulanus albocinctus*
5b	General color brownish, interrupted by white rings; with white longitudinal lines; anterior margin of head not white	6
6a	General color orange-brown to chocolate; with 5 or 6 longitudinal lines consisting of densely packed white dots (besides a mid-dorsal line, there are 2 lines on both the right and left sides of the body, and sometimes a faint mid-ventral line is present); white rings evenly spaced	*Tubulanus sexlineatus*
6b	General color dark brown throughout most of the body, although the posterior region is slightly lighter; with 3 white longitudinal lines (a continuous mid-dorsal line and 2 lateral lines that are broken into dashes anteriorly); white rings not evenly spaced (the anteriormost	

	band of white pigment lies just behind the level of the mouth and forms a V on the ventral surface)	*Tubulanus capistratus*
7a	General color cream; length not more than 50 cm	8
7b	General color yellow, ochre, orange, or reddish; length commonly exceeding 1 m, and sometimes attaining 3 m	9
8a	Generally 10-30 cm long and flattened posteriorly; head milky white; intestinal diverticula conspicuous (gravid females pinkish in the intestinal region; males that contain sperm are light orange in this part of the body)	*Carinoma mutabilis*
8b	Less than 3 cm long and very slender; head not milky white; intestinal diverticula not conspicuous (gravid females may have mid-dorsal and mid-ventral stripes between the cream-colored ovaries that are in the lateral portions of the body)	*Tubulanus pellucidus*
9a	Head without eyespots and without a pair of anteroventral grooves; color uniformly orange or reddish	*Tubulanus polymorphus*
9b	Head with numerous eyespots and with a pair of oblique grooves on the anteroventral surface (these grooves are inconspicuous); color deep ochre to brownish	*Baseodiscus princeps*
10a	Caudal cirrus present (it may be missing, however, if the posterior end has been damaged); proboscis sheath usually as long as the body	11
10b	Caudal cirrus absent; length of proboscis sheath variable (it may be much shorter than the body or nearly as long as the body)	14
11a	Body with distinct transverse or longitudinal stripes	12
11b	Body with neither transverse nor longitudinal stripes	13
12a	Color mostly deep brown to slate, sometimes with a greenish tinge; with numerous yellow rings encircling the body and with faint yellow grooves running longitudinally on the dorsal surface (anterodorsal surface of the head mostly white, and with a lemon-colored region that contains 2 oval orange spots)	*Lineus pictifrons*
12b	Color mostly deep chocolate to olive-brown; with a mid-dorsal whitish or pale yellow stripe (the stripe runs the length of the body and joins a spatulate patch of similar color on the dorsal surface of the head)	*Lineus bilineatus*
13a	Body olive-green, brown, or ochre; anterior tip of head white, with 2 transverse brown patches on the dorsal surface (often found in empty tubes of the polychaete *Sabellaria cementarium*; pharyngeal region distinctly dilated in contracted specimens)	*Lineus* sp.
13b	Body greenish brown, brown, or reddish brown; anterior tip of head not white, and with 4-8 eyespots on each side of it (during the summer, the worms usually divide into short pieces, each of which may develop into a complete worm)	*Lineus vegetus*
14a	Body relatively soft and slender, not markedly flattened or otherwise adapted for swimming; length usually less than 50 cm	15
14b	Body firm and markedly flattened posteriorly, adapted for swimming; length sometimes more than 50 cm	17
15a	Body with distinct transverse rings of a contrasting color	16
15b	Body color homogeneous (creamy tan to pale salmon) and lacking transverse rings of a contrasting color (tip of head white)	*Micrura alaskensis*
16a	Dorsal surface mostly dark brown or purple, but with regularly spaced, transverse white lines that divide the dark color into rectangles (ventral surface ivory white); dorsal surface of head with a conspicuous orange triangle	*Micrura verrilli*
16b	General color deep chestnut or some shade of brown, interrupted by irregularly spaced white rings; anterior tip and sides of head with a white border (the body fragments readily)	*Micrura wilsoni*
17a	Much of the head whitish and thus distinctly demarcated from the rest of the body	18
17b	Head not white and not distinctly different from the rest of the body (the lateral margins of the head may be lighter, however)	19
18a	General color dark brown to purplish; white area on head extending nearly to the posterior end of the cephalic slits	*Cerebratulus albifrons*
18b	General color deep red; white area on head not extending as far back as the posterior ends of the cephalic slits (length sometimes greater than 2 m)	*Cerebratulus montgomeryi*
19a	Neck region (immediately behind the head) slightly narrowed and thus separating the head from the rest of the body; cephalic slits large in proportion to the length of the body, which is usually less than 20 cm	20
19b	Neck region not obviously narrowed (head not distinct from the rest of the body); cephalic	

	slits small in proportion to the length of the body, which is usually greater than 20 cm	21
20a	Color generally some shade of dull red to grayish rose, but sometimes greenish brown (extremely fragile and likely to break up into small pieces)	*Cerebratulus californiensis*
20b	Color dark brown to purple on the dorsal surface, slightly lighter ventrally and along the margins of the head	*Cerebratulus longiceps*
21a	General color some shade of brown; length usually greater than 50 cm	22
21b	Dorsal surface chestnut-brown to reddish, ventral surface flesh-colored and with an ochre stripe along the midline; length usually 20-30 cm	*Cerebratulus occidentalis*
22a	Marginal areas of the intestinal region of the body markedly flattened, so that this part of the animal has distinct lateral flanges; general color grayish brown to dark brown, except along the lateral margins, which are lighter; length usually 50 cm to 1 m	*Cerebratulus marginatus*
22b	Body stout and not markedly flattened in the marginal areas of the intestinal region; color dark brown to reddish brown dorsally, slightly paler ventrally; sometimes more than 2 m long, 2.5 cm wide	*Cerebratulus herculeus*
23a	Body leechlike, with a muscular attachment disk at the posterior end; proboscis without a stylet; commensal in the mantle cavity of bivalve molluscs (*Siliqua patula, Macoma nasuta, M. secta,* and probably others)	*Malacobdella* spp.
23b	Body not leechlike, and without a muscular attachment disk at the posterior end; proboscis with a stylet; not commensal in the mantle cavity of bivalve molluscs	24
24a	Semi-terrestrial, generally found under rocks and driftwood at the edges of bays and saltmarshes, at a level reached only by the highest tides (dorsal surface grayish green or bluish green, with a poorly defined, darker longitudinal stripe; with 4 clusters of eyespots on the head, the anterior clusters with about 40 eyespots each, the posterior clusters with about 20 eyespots; length up to 25 cm)	*Pantinonemertes californiensis*
	See notes concerning this species in the taxonomic list.	
24b	Truly marine, not restricted to a level reached only by the highest tides	25
25a	Stylet shaft with helical grooves and ridges	26
25b	Stylet shaft smooth	29
26a	With a conspicuous statocyst dorsal to each half of the brain; length usually less than 5 mm; interstitial	*Ototyphlonemertes americana*
26b	Statocysts absent; length much greater than 5 mm; not interstitial	27
27a	Proboscis sheath less than one-fifth the length of the body; proximal portion of stylet discoidal and inconspicuous; length up to 1 m (fairly stout and conspicuously flattened; color typically tan, with light brown specks, or slightly purplish; dorsal surface darker than the ventral surface)	*Emplectonema purpuratum*
27b	Proboscis sheath more than one-fifth the length of the body; proximal piece of stylet prominent; length not often more than 30 cm	28
28a	Dorsal surface brown to purple-brown, ventral surface creamy yellow; typically with 2 or 3 sacs of reserve stylets (an average of 4 stylets in each sac)	*Paranemertes peregrina*
28b	Color uniformly flesh to pale orange; typically with 4-6 sacs of reserve stylets (about 6 stylets in each sac)	*Paranemertes sanjuanensis*
29a	Ectosymbionts on brachyuran crabs; length generally less than 1 cm; proboscis reduced and lacking reserve stylet sacs (adults orange red)	30
29b	Free-living; length greater than 1 cm (except in certain species that have 4 eyespots); proboscis with sacs of reserve stylets	31
30a	Adults inhabiting mucous tubes while on the egg mass of the host crab; epidermis without small brown spots	*Carcinonemertes epialti*
30b	Adults not inhabiting mucous tubes while on the egg mass of the host crab; epidermis with numerous small brown spots	*Carcinonemertes errans*
31a	Proboscis sheath more than half the length of the body; head with 4 eyespots that form a quadrangle; length less than 2.5 cm	32
31b	Proboscis sheath varying from less than one-third the length of the body to nearly the entire length; eyespots numerous (except in *Nemertopsis gracilis*; see choice 37a); length usually greater than 2.5 cm	35
32a	Color uniformly tan to creamy orange (no obvious markings present)	33
32b	Dorsal surface with conspicuous longitudinal stripes	34
33a	Sperm with a coiled head; found on the surfgrass, *Phyllospadix scouleri*, low intertidal (length of mature worms usually 5-10 mm)	*Tetrastemma phyllospadicola*

33b Sperm without a coiled head; not typically found on *Phyllospadix* (length of mature worms 5 mm to 2.5 cm) .... *Tetrastemma* spp., *Oerstedia* spp., *Prosorhochmus* spp.
34a Dorsal surface whitish to creamy white, and with 2 brown longitudinal stripes
.... *Tetrastemma bilineatum*
34b Dorsal surface brown, with a whitish or yellowish stripe along the midline (ventral surface whitish to yellow; tip and lateral margins of the head lighter than the rest of the dorsal surface) .... *Tetrastemma bicolor*
35a Color not uniformly light green or yellow-green on both the ventral and dorsal surface; eyespots extending posterior to the brain; basis distinctly truncate at its posterior end; stylets extremely stout .... *Zygonemertes virescens*
35b Color light green or yellow-green on both the ventral and dorsal surface; eyespots not extending posterior to the brain; basis not truncate at its posterior end; stylets not especially stout 36
36a Body long and very slender (usually 10-50 cm long, but less than 3 mm wide); proboscis sheath less than one-third the length of the body 37
36b Body not extremely slender; proboscis sheath usually more than one-third the length of the body 38
37a General color creamy tan to light brown, with 2 broad, brown longitudinal stripes on the dorsal surface; head with 4 large eyespots; stylet about the same size as the basis, which is neither curved nor swollen posteriorly .... *Nemertopsis gracilis*
37b Dorsal surface dark green to yellow-green, ventral surface yellowish; head with numerous eyespots; stylet typically more than 200 µm long and markedly curved, the basis (which is sometimes more than 2 mm long) also curved and slightly swollen posteriorly
.... *Emplectonema gracile*
38a Body relatively stout; general color pinkish, red, reddish brown, deep yellow, or purplish; head with or without conspicuous markings and sometimes lighter than the rest of the body
.... 39
38b Body not obviously stout; color uniformly whitish, gray, yellow, or pinkish orange; head (without conspicuous markings) not lighter than the rest of the body 42
39a Head demarcated dorsally from the rest of the body by a V-shaped white line or by pigmented patches, or both 40
39b Head not demarcated from the rest of the body by a white line or by conspicuous patches of pigment 41
40a Dorsal side of head with 2 triangular or trapezoidal white patches and with a V-shaped white line that separates it from the rest of the body; basis about the same length as the stylet (color purple-brown to chocolate dorsally, paler ventrally) .... *Amphiporus angulatus*
40b Head with 2 triangular dark brown patches, and not demarcated from the rest of the body by a V-shaped white line; basis about half the length of the body (color nearly uniformly reddish brown except for the pattern on the head) .... *Amphiporus bimaculatus* (fig. 6.1)
41a Length usually about 3 cm; color pinkish and slightly more pale on the ventral surface; with 6-10 eyespots on both sides of the head .... *Amphiporus rubellus*
41b Length 2.5-7.5 cm; color reddish brown; with 10-20 eyespots on both sides of the head (basis about the same size as the stylet) .... *Nipponnemertes pacificus*
42a Length usually 1-4 cm; color usually pale yellow to flesh; with a longitudinal row of 5-10 eyespots on both sides of the head; basis about the same length as the stylet and slightly more slender at its posterior end than at its anterior end .... *Amphiporus cruentatus*
42b Length usually greater than 5 cm; basis not obviously more slender at its posterior end than at its anterior end 43
43a Females usually lemon or yellow-brown (the ovaries stand out as dark olive patches along the sides of the body), males greenish yellow (testes cream); basis about twice as long as the central stylet and of uniform width .... *Amphiporus tigrinus*
43b Color not lemon, yellow-brown, or greenish yellow; basis decidedly wider at its posterior end than at its anterior end 44
44a Proboscis with 6-12 sacs of reserve stylets; color whitish to slate; length usually 5-25 cm
.... *Amphiporus formidabilis*
44b Proboscis usually with only 2 sacs of reserve stylets; color whitish to flesh, and often with a distinct orange or pinkish tinge; length usually 5-15 cm .... *Amphiporus imparispinosus*

Several species that have been reported from the region are not in the key. These are indicated by an asterisk.

Class Anopla

Order Palaeonemertea

Family Carinomidae

Carinoma mutabilis Griffin, 1898. In sand and sandy mud.

Family Cephalothricidae

Cephalothrix pacifica Gerner, 1969

Family Tubulanidae

Tubulanus albocinctus (Coe, 1904)
Tubulanus capistratus (Coe, 1901)
Tubulanus pellucidus (Coe, 1895)
Tubulanus polymorphus Renier, 1804
Tubulanus sexlineatus (Griffin, 1898)

Order Heteronemertea

Family Baseodiscidae

Baseodiscus princeps (Coe, 1901)

Family Lineidae

Cerebratulus albifrons Coe, 1901
Cerebratulus californiensis Coe, 1905
Cerebratulus herculeus Coe, 1901
Cerebratulus longiceps Coe, 1901
Cerebratulus marginatus Renier, 1804
Cerebratulus montgomeryi Coe, 1901
Cerebratulus occidentalis Coe, 1901
Lineus bilineatus (Renier, 1804)
Lineus pictifrons Coe, 1904
*Lineus ruber (O. F. Müller, 1771)
*Lineus rubescens Coe, 1904
Lineus vegetus Coe, 1931. Often abundant in intertidal mats of filamentous algae.
Lineus sp.
Micrura alaskensis Coe, 1901
Micrura verrilli Coe, 1901
Micrura wilsoni (Coe, 1904)

Class Enopla

Order Bdellonemertea

Family Malacobdellidae

Malacobdella spp. Commensal in the mantle cavity of bivalve molluscs; one undescribed species in *Siliqua patula*, another in *Macoma nasuta* and *M. secta*; perhaps other bivalves also serve as hosts. So far as is known, *Malacobdella grossa* (O. F. Müller, 1776) does not occur on the Pacific coast.

Order Hoplonemertea

Suborder Monostilifera

Family Amphiporidae

Amphiporus angulatus (Fabricius, 1774)
Amphiporus bimaculatus Coe, 1901 (fig. 6.1)
Amphiporus cruentatus Verrill, 1879
Amphiporus formidabilis Griffin, 1898. Common intertidally on rocky shores.
Amphiporus imparispinosus Griffin, 1898. Found sympatrically with *A. formidabilis*.
Amphiporus rubellus Coe, 1905
Amphiporus tigrinus Coe, 1901
Zygonemertes virescens (Verrill, 1879)
*Zygonemertes sp.

Family Carcinonemertidae

Carcinonemertes epialti Coe, 1902. On the brachyuran crabs *Pugettia producta, Hemigrapsus nudus, H. oregonensis, Pachygrapsus crassipes*, and several species of *Cancer* (but not *C. magister*).
Carcinonemertes errans Wickham, 1978. On the brachyuran crab *Cancer magister*.

Family Cratenemertidae

Nipponnemertes pacificus (Coe, 1905)
**Nipponnemertes punctatulus* (Coe, 1905)

Family Emplectonematidae

**Cryptonemertes actinophila* (Bürger, 1904). Associated with the sea anemones *Stomphia coccinea* and *S. didemon* (and elsewhere with other anemones).
Emplectonema gracile (Johnston, 1837)
Emplectonema purpuratum Coe, 1905
Nemertopsis gracilis Coe, 1904
Paranemertes peregrina Coe, 1901. Common intertidally on mudflats, and sometimes on rocky shores.
Paranemertes sanjuanensis Stricker, 1982

Family Ototyphlonemertidae

Ototyphlonemertes americana Gerner, 1969

Family Prosorhochmidae

Oerstedia spp. Possibly including *O. dorsalis* (Abildgaard, 1806).
Pantinonemertes californiensis Gibson, Moore, & Crandall, 1982. This semiterrestrial nemertean has been found at the edges of several bays from southern to northern California; it lives under rocks and wood at a level reached only by the highest tides. A grayish worm that occurs under driftwood in salt marshes of the Puget Sound region is perhaps this species.
Prosorhochmus spp. Possibly including *P. albidus* (Coe, 1905).

Family Tetrastemmatidae

Tetrastemma bicolor Coe, 1901
Tetrastemma bilineatum Coe, 1904. Common on some intertidal mudflats.
**Tetrastemma candidum* (O. F. Müller, 1774)
**Tetrastemma nigrifrons* Coe, 1904
Tetrastemma phyllospadicola Stricker, 1985. On the surfgrass, *Phyllospadix scouleri*.
Tetrastemma spp.

Suborder Polystilifera

All of our representatives of this suborder are pelagic and found in deep water.

Family Nectonemertidae

**Nectonemertes mirabilis* Verrill, 1892
**Pelagonemertes brinkmanni* Coe, 1926
**Pelagonemertes joubini* Coe, 1926

Family Protopelagonemertidae

**Plotonemertes adhaerens* Brinkmann, 1917

Family Planktonemertidae

**Crassonemertes robusta* Brinkmann, 1917

Family Dinonemertidae

**Tubonemertes wheeleri* (Coe, 1936)

7

PHYLA
NEMATODA, GASTROTRICHA, ROTIFERA, KINORHYNCHA, PRIAPULIDA

PHYLUM NEMATODA

Nematodes are abundant in sand and mud, and in the sediment that accumulates on sessile invertebrates, algal growths, and other substrata. There are also many species living as parasites or commensals in marine vertebrates and invertebrates.

The more important comprehensive works on free-living nematodes of the Pacific coast are those of Allgén (1947, 1951) and Wieser (1959a, 1959b). Although Wieser's study was confined to a few beaches in Puget Sound, he found about 80 species, distributed over 16 families. It is obvious that the fauna of nematodes in our region must be very large. Most of the publications that deal with taxonomy of free-living marine nematodes of the Pacific coast are listed below. Some particularly useful general references are also included.

Allen, M. W., & E. M. Noffsinger. 1978. A revision of the marine nematodes of the superfamily Draconematoidea Filipjev, 1918 (Nematoda: Draconematina). Univ. Calif. Publ. Zool., 109:1-133.

Allgén, C. A. 1947. West American marine nematodes. (Papers from Dr. Th. Mortensen's Pacific Expedition 1914-16, no. 75.) Vidensk. Meddel. Dansk naturh. Foren. København, 110:65-219.

------. 1951. Pacific freeliving marine nematodes. (Papers from Dr. Th. Mortensen's Pacific Expedition 1914-16, no. 76.) Vidensk. Meddel. Dansk naturh. Foren. København, 113:263-411.

Chitwood, B. G. 1960. A preliminary contribution on the marine nemas (Adenophorea) of northern California. Trans. Amer. Micr. Soc., 79:347-84.

Chitwood, B. G., & D. G. Murphy. 1964. Observations on two marine monhysterids--their classification, cultivation, and behavior. Trans. Amer. Micr. Soc., 83:311-29.

Gerlach, A., & F. Riemann. 1973. The Bremerhaven checklist of aquatic nematodes. A catalogue of Nematoda Adenophorea excluding the Dorylaimida. 1. Veröffentl. Inst. Meeresforsch. Bremerhaven, suppl., 4:1-404.

------. 1974. The Bremerhaven checklist of aquatic nematodes. A catalogue of Nematoda Adenophorea excluding the Dorylaimida. 2. Veröffentl. Inst. Meeresforsch. Bremerhaven, suppl., 4:405-736.

Heip, C., M. Vincx, N. Smol, & G. Vranken. 1982. The systematics and ecology of free-living marine nematodes. Helminth. Abstr., ser. B, Plant Nematology, 51:1-31.

Hope, W. D. 1967a. A review of the genus *Pseudocella* Filipjev, 1927 (Nematoda: Leptosomatidae) with a description of *Pseudocella triaulolaimus* n. sp. Proc. Helminth. Soc. Washington, 34:6-12.

------. 1967b. Free-living marine nematodes of the genera *Pseudocella* Filipjev, 1927, *Thoracostoma* Marion, 1870, and *Deontostoma* Filipjev, 1916 (Nematoda: Leptosomatidae) from the west coast of North America. Trans. Amer. Micr. Soc., 86:307-34.

Hope, W. D., & D. G. Murphy. 1972. A taxonomic hierarchy and checklist of the genera and higher taxa of marine nematodes. Smithsonian Contrib. Zool., 137:1-101.

Inglis, W. G. 1983. An outline classification of the phylum Nematoda. Austral. J. Zool., 31:243-55.

Jones, G. F. 1964. Redescription of *Bolbella californica* Allgén, 1951 (Enchelidiidae: Nematoda), with notes on its ecology off southern California. Pacific Sci., 18:160-5.

Maggenti, A. 1981. General Nematology. New York: Springer-Verlag. 372 pp.

Murphy, D. G. 1962. Three undescribed nematodes from the coast of Oregon. Limnol. Oceanogr., 7:386-9.

------. 1963a. A new genus and two new species of nematodes from Newport, Oregon. Proc. Helminthol. Soc. Washington, 30:73-8.

------. 1963b. Three new species of marine nematodes from the Pacific near Depoe Bay, Oregon. Proc. Helminthol. Soc. Washington, 30:249-56.

------. 1964a. *Rhynconema subsetosa*, a new species of marine nematode, with a note on the genus *Phylolaimus* Murphy, 1963. Proc. Helminthol. Soc. Washington, 31:26-8.

------. 1964b. The marine nematode genus *Pseudonchus* Cobb, 1920 with descriptions of *Cheilopseudonchus*, n. g. and *Pseudonchus kosswigi*, n. sp. Mitt. Hamburg Zool. Mus. Inst., Kosswig-Festschrift, 1964:113-8.

------. 1964c. Free-living marine nematodes, I. *Southerniella youngi, Dagda phinneyi*, and *Gammanema smithi*, new species. Proc. Helminthol. Soc. Washington, 31:190-8.

------. 1965a. Free-living marine nematodes, II. *Thoracostoma pacifica* n. sp. from the coast of Oregon. Proc. Helminthol. Soc. Washington, 32:106-9.

------. 1965b. *Thoracostoma washingtonensis*, n. sp., eine Meeresnematode aus dem pazifischen Küstenbereich vor Washington. Abhandl. Verhandl. naturw. Ver. Hamburg (n. F.), 9:211-6.

------. 1965c. *Cynura klunderi* (Leptolaimidae), a new species of marine nematode. Zool. Anz., 175:217-22.

Murphy, D. G., & H. J. Jensen. 1961. *Lauratonema obtusicaudatum* n. sp. (Nematoda: Enoploidea), a marine nematode from the coast of Oregon. Proc. Helminthol. Soc. Washington, 28:167-9.

Platt, H. M., & R. M. Warwick. 1983. Free-living Marine Nematodes. Part I. British Enoplids. Cambridge & New York: Cambridge University Press. 211 pp.

Sharma, J., B. E. Hopper, & J. M. Webster. 1979. Benthic nematodes from the Pacific coast with special reference to the cyatholaimids. Ann. Soc. Royale Zool. Belgique, 108:47-56.

Sharma, J., & M. Vincx. 1982. Cyatholaimidae (Nematoda) from the Canadian Pacific coast. Can. J. Zool., 60:271-80.

Timm, R. W. 1951. A new species of marine nematode, *Thoracostoma magnificum*, with a note on possible "pigment cell" nuclei of the ocelli. J. Wash. Acad. Sci., 41:331-3.

------. 1970. A revision of the nematode order Desmoscolecida Filipjev, 1929. Univ. Calif. Publ. Zool., 93:1-115.

Wieser, W. 1953. Free-living marine nematodes I. Enoploidea. Acta Univ. Lund. (n. f. 2), 49(6):1-155.

------. 1954. Free-living marine nematodes II. Chromadoroidea. Acta Univ. Lund. (n. f. 2), 50(16):1-148.

------. 1956. Free-living marine nematodes III. Axonolaimoidea and Monhysteroidea. Acta Univ. Lund. (n. f. 2), 52(13);1-115.

------. 1959a. Free-living nematodes and other small invertebrates of Puget Sound beaches. Univ. Wash. Publ. Biol., 19. x + 179 pp.

------. 1959b. The effect of grain size on the distribution of small invertebrates inhabiting the beaches of Puget Sound. Limnol. Oceanogr., 4:181-94.

Wright, K. A., & W. D. Hope. 1968. Elaborations of the cuticle of *Acanthonchus duplicatus* Wieser, 1959 (Nematoda: Cyatholaimidae) as revealed by light and electron microscopy. Can. J. Zool., 46:1005-11.

PHYLUM GASTROTRICHA

Only a few species of gastrotrichs have been reported from the region, and most of these were collected in Puget Sound and the San Juan Archipelago. Because a key may lead to incorrect identification of undescribed species, it seems best to provide only a checklist and bibliography of pertinent literature.

D'Hondt, J. L. 1971. Gastrotricha. Oceanogr. Mar. Biol. Ann. Rev., 9:141-92.

Hummon, W. D. 1966. Morphology, life history, and significance of the marine gastrotrich, *Chaetonotus testiculophorus* n. sp. Trans. Amer. Micr. Soc., 85:450-7.

------. 1969. *Musellifer sublitoralis*, a new genus and species of Gastrotricha from the San Juan Archipelago, Washington. Trans. Amer. Micr. Soc., 88:282-6.
------. 1976. Some taxonomic revisions and nomenclatural notes concerning marine and brackish-water Gastrotricha. Trans. Amer. Micr. Soc., 93:194-205.
------. 1982. Gastrotricha. *In* Parker, S. P. (ed.), Synopsis and Classification of Living Organisms, 1: 857-83. New York: McGraw-Hill Book Co.
Remane, A. 1925. Neue aberrante Gastrotrichen II: *Turbanella cornuta* nov. spec. und *T. hyalina* M. Schultze, 1853. Zool. Anz., 64:309-14.
------. 1936. Gastrotricha und Kinorhyncha. *In* Bronn, H. G. (ed.), Klassen und Ordnungen des Tierreichs, 4, 2:1-242.
Wieser, W. 1957. Gastrotricha Macrodasyoidea from the intertidal of Puget Sound. Trans. Amer. Micr. Soc. 76:372-81.
------. 1959. Free-living nematodes and other small invertebrates of Puget Sound beaches. Univ. Wash. Publ. Biol., 19. x + 179 pp.

Order Chaetonotida

Family Chaetonotidae

Chaetonotus testiculophorus Hummon, 1966. Intertidal, in sand.
Musellifer sublitoralis Hummon, 1969. Subtidal, in epibenthic sediment.

Order Macrodasyida

Family Macrodasyidae

Macrodasys cunctatus Wieser, 1957. Intertidal, in sand.

Family Thaumastodermatidae

Tetranchyroderma pugetensis Wieser, 1957. High intertidal, in coarse sand.

Family Turbanellidae

Paraturbanella intermedia Wieser, 1957. Low intertidal, in sand.
Turbanella cornuta Remane, 1925. Intertidal, in sand.
Turbanella mustela Wieser, 1957. High intertidal, in sand.
Turbanella sp. Reported by Wieser, 1957.

PHYLUM ROTIFERA

Almost no work has been done with marine and estuarine rotifers in our region. The list below covers most of the species that have been reported. The key of Thane-Fenchel (1968) will enable one to identify the genera known to occur in marine and brackish-water habitats in various parts of the world. It is accompanied by a useful bibliography.

Thane-Fenchel, A. 1968. A simple key to the genera of marine and brackish-water rotifers. Ophelia, 5:299-311.

Class Seisonidea
Order Seisonida
Family Seisonidae

Seison sp. On *Nebalia pugettensis* (Crustacea, Leptostraca).

Class Monogononta
Order Ploima
Family Brachionidae

Brachionus plicatilis Möbius, 1874
Keratella cochlearis subsp. *tecta* Gosse, 1851
Notholca striata (O. F. Müller, 1786)

Family Proalidae

Proales spp. On hydroids and various other invertebrates, and also on algae.

Family Trichocercidae

Trichocerca marina (Daday, 1890)

Family Synchaetidae

Synchaeta baltica Ehrenberg, 1834
Synchaeta johanseni Harring, 1921

PHYLUM KINORHYNCHA
John C. Boykin and Eugene N. Kozloff

Of the several species of kinorhynchs known to occur in the region, all but one are subtidal, living near the surface of soft mud. The only intertidal species, sometimes common in diatom-rich sediment on muddy sand, is *Echinoderes kozloffi* Higgins, 1977 (choice 4a; figs. 7.1-7.3).

Higgins, R. P. 1960. A new species of *Echinoderes* (Kinorhyncha) from Puget Sound. Trans. Amer. Micr. Soc., 79:85-91.
------. 1961. Three new homalorhage kinorhynchs from the San Juan Archipelago, Washington. J. Elisha Mitchell Sci. Soc., 77:81-8.
------. 1977. Redescription of *Echinoderes dujardinii* (Kinorhyncha) with description of closely related species. Smithsonian Contrib. Zool, no. 248.
------. 1983. The Atlantic Barrier Reef ecosystem at Carrie Bow Cay, Belize, II: Kinorhyncha. Smithsonian Contrib. Mar. Sci., no. 18. (Has keys to known species of certain genera.)

Kozloff, E. N. 1972. Some aspects of development in *Echinoderes* (Kinorhyncha). Trans. Amer. Micr. Soc., 91:119-30.

McIntyre, A. D. 1962. The class Kinorhyncha (Echinoderida) in British waters. J. Mar. Biol. Assoc. United Kingdom, 42:503-4.

1a	Cuticle of segment 3 (first trunk segment) either a closed ring or consisting of 2 half-rings; body generally circular or oval in transverse section; lateral and dorsal setae present on trunk and usually longer than one-third the diameter of the trunk (order Cyclorhagida)	2
1b	Cuticle of segment 3 with a large, strongly arched dorsal tergum, a single midventral sternum, and 2 lateroventral sterna; body triangular in transverse section; trunk setae present, but less than one-fourth the diameter of the trunk (order Homalorhagida)	6
2a	Cuticle of both segments 3 and 4 (first and second trunk segments) consisting of closed rings; median terminal spine present only in colorless juveniles that are not more than 250 µm long (family Echinoderidae)	3
2b	Cuticle of segment 4 (second trunk segment) divided into a dorsal tergum and 2 ventral sterna; median terminal spine (as well as lateral terminal spines) present in adults (family Centroderidae)	*Campyloderes* sp.

This is possibly *C. multispinosus* (McIntyre, 1962).

3a	Length less than 300 µm; cuticle colorless; smaller specimens may have a median terminal spine as well as a pair of lateral terminal spines	*Echinoderes*, juveniles
3b	Length 300 µm or more; cuticle yellow or yellow-brown; with a pair of well-developed lateral terminal spines, but without a median terminal spine (*Echinoderes*, adults)	4
4a	With a red eyespot on both sides of the head and with segmental concentrations of red pigment along the mid-dorsal line; locally abundant intertidally in diatom-rich sediment on muddy sand	*Echinoderes kozloffi* Higgins, 1977 (figs. 7.1-7.3)
4b	Without red eyespots or concentrations of red pigment along the mid-dorsal line; subtidal	5
5a	Length 310-365 µm; lateral terminal spines 50-60 µm long	*Echinoderes* sp. (undescribed)
5b	Length greater than 390 µm; lateral terminal spines 135-180 µm long	*Echinoderes pennaki* Higgins, 1960
6a	Segment 13 with a single median terminal spine or a pair of lateral terminal spines (*Pycnophyes*)	7
6b	Segment 13 with neither a median terminal spine nor a pair of lateral terminal spines (*Kinorhynchus*)	9
7a	Length less than 450 µm; with a single median terminal spine	*Pycnophyes*, younger juveniles
7b	Length greater than 450 µm; with a pair of lateral terminal spines	8
8a	Cuticle colorless	*Pycnophyes*, older juveniles
8b	Cuticle yellow or yellow-brown	*Pycnophyes sanjuanensis* Higgins, 1961
9a	Length less than 500 µm; cuticle colorless	*Kinorhynchus*, juveniles
9b	Length greater than 500 µm; cuticle yellow or yellow-brown	10
10a	Segment 3 (first trunk segment) with a pair of dorsolateral setae; segment 13 (last trunk segment) with 2 pairs of lateral setae (these setae are only 10-12 µm long, and are easily broken)	*Kinorhynchus cataphractus* (Higgins, 1961)
10b	Segment 3 without dorsolateral setae; segment 13 with only 1 pair of lateral setae	*Kinorhynchus ilyocryptus* (Higgins, 1961) (Fig. 7.4)

Order Cyclorhagida

Family Echinoderidae

Echinoderes kozloffi Higgins, 1977 (fig. 7.1-7.3)
Echinoderes pennaki Higgins, 1960
Echinoderes sp. Undescribed.

Phylum Kinorhyncha 107

7.1 *Echinoderes kozloffi*, male

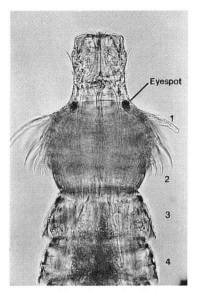

7.2 *Echinoderes kozloffi*, anterior portion of body

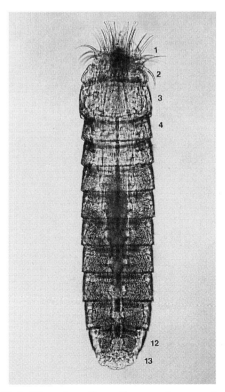

7.4 *Kinorhynchus ilyocryptus*, male

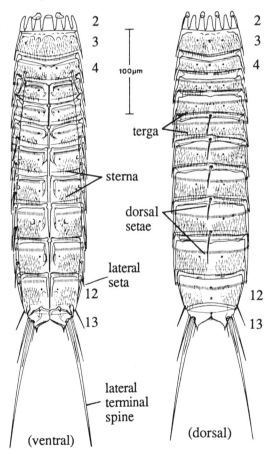

7.3 *Echinoderes kozloffi*, female

Family Centroderidae

Campyloderes sp. Possibly *C. multispinosus* (McIntyre, 1962).

Order Homalorhagida

Family Pycnophyidae

Pycnophyes sanjuanensis Higgins, 1961
Kinorhynchus cataphractus (Higgins, 1961)
Kinorhynchus ilyocryptus (Higgins, 1961) (fig. 7.4)

PHYLUM PRIAPULIDA

Order Priapulomorpha

Family Priapulidae

Priapulus caudatus Lamarck, 1816, widely distributed in the Northern Hemisphere, is the only species reported for our region. It reaches a length (exclusive of the caudal appendage) of about 5 cm, and is most easily collected by epibenthic dredging on muddy substrata. The distinctive larva, only a few mm long, is occasionally noted.

Lang, L. 1948. Contribution to the ecology of *Priapulus caudatus* Lam. Arkiv för Zoologi, 41A (5):1-12.
------. 1948. On the morphology of the larva of *Priapulus caudatus* Lam. Arkiv för Zoologi, 41A (9):1-8.

8

PHYLUM ANNELIDA: CLASS POLYCHAETA

The polychaete fauna of the Pacific Northwest is extremely rich, but many of the species are strictly subtidal. Others are so small that they are not likely to be noted in the course of field work. Careful sorting of worms from almost any habitat, especially soft sediments, will probably reveal species that would otherwise be overlooked.

For many years, biologists of this region relied on the two-part study of Berkeley & Berkeley (1948, 1952). Banse & Hobson (1974) and Hobson & Banse (1981) have provided a modernized and greatly expanded version of both volumes, and have included nearly every species that has been reported or that may be expected. These valuable works also have extensive bibliographies.

Attention is called to the two atlases of California polychaetes prepared by Hartman (1968, 1969). These are admirably illustrated and will be helpful in recognition of most genera, as well as many species, found in our region. *Light's Manual*, Banse & Hobson, and Hobson & Banse list numerous other papers by Hartman, some of which are concerned with systematics of individual families. Hartman (1951) has provided a substantially complete guide to the literature on polychaetes, on a worldwide basis, up to about 1950.

For convenience, the treatment of polychaetes in this manual is divided into two sections: a key to families, and keys to species within the families. The families, which are numbered, are presented in systematic sequence, according to the higher taxa to which they belong. The archiannelid families (55-59) and myzostomes (54) are dealt with separately, at the end of the chapter.

The illustrations are organized into three main groups. Those in the first group (8.1-8.14) are concerned primarily with terminology, although some of them show important characteristics of a few families. The figures in the second group (8.15-8.51) show one or two representatives of each of 23 families. The remaining figures illustrate additional species within these families, as well as several families not dealt with earlier.

General References on Polychaeta

Banse, K., & K. D. Hobson. 1968. Benthic polychaetes from Puget Sound, Washington, with remarks on four other species. Proc. U. S. Nat. Mus., 125:1-53.

------. 1974. Benthic errantiate polychaetes of British Columbia and Washington. Bull. 185, Fish. Res. Bd. Canada. 111 pp.

Berkeley, E., & C. Berkeley. 1948. Annelida. Polychaeta Errantia. Canadian Pacific Fauna, 9b (1). 100 pp.

------. 1952. Annelida. Polychaeta Sedentaria. Canadian Pacific Fauna, 9b (2). 139 pp.

Day, J. H. 1967a. A monograph of the Polychaeta of Southern Africa. Part I (Errantia). London: British Museum (Natural History). xxix + 258 pp.

------. 1967b. A monograph of the Polychaeta of Southern Africa. Part II (Sedentaria). London: British Museum (Natural History). xvii + 878 pp.

Fauchald, K. 1977. The polychaete worms--definitions and keys to the orders, families, and genera. Nat. Hist. Mus. Los Angeles County, Sci. Series, no. 28. 188 pp.

Fauvel, P. 1923. Polychètes Errantes. Faune de France, 5. Paris: Lechevalier. 488 pp.

------. 1927. Polychètes Sédentaires. Addenda aux Errantes, Archiannelides, Myzostomaires. Faune de France, 16. Paris: Lechevalier. 494 pp.

Hartman, O. 1951. Literature of the Polychaetous Annelids. Los Angeles (privately published). 290 pp.

------. 1968. Atlas of the Errantiate Polychaetous Annelids from California. Los Angeles: Allan Hancock Foundation, University of Southern California. 828 pp.

------. 1969. Atlas of the Sedentariate Polychaetous Annelids from California. Los Angeles: Allan Hancock Foundation, University of Southern California. 812 pp.
Hartman, O., & D. J. Reish. 1950. The marine annelids of Oregon. Oregon State Monogr., Zool., no. 6. 64 pp.
Hobson, K. D. 1971. Some polychaetes of the superfamily Eunicea from the North Pacific and North Atlantic Oceans. Proc. Biol. Soc. Washington, 83:527-44.
Hobson, K. D., & K. Banse. 1981. Sedentariate and archiannelid polychaetes of British Columbia and Washington. Can. Bull. Fish. Aquat. Sci., no. 209. viii + 144 pp.
Pettibone, M. H. 1954. Marine polychaete worms from Point Barrow, Alaska, with additional records from the North Atlantic and North Pacific. Proc. U. S. Nat. Mus., 103:203-356.
Ushakov, P. V. 1955. [Polychaete worms of the far-eastern seas of the USSR.] Opredeliteli po Faune SSSR, 56. 444 pp. (in Russian). (English translation, 1965. Jerusalem: Israel Program for Scientific Translations.)

Key to Families of Polychaeta

The Myzostomidae (54) and archiannelid families (55-59) are not in the key.

1a	Body short, grublike, with a horny plate and a tuft of filamentous gills at the posterior end	41. Sternaspidae: *Sternaspis scutata* (fig. 8.41)
1b	Body generally not grublike, and with neither a horny plate nor a tuft of filamentous gills at the posterior end	2
2a	Eyes large, bulbous, occupying about half of the dorsal surface of the prostomium (prostomium with 4 or 5 antennae; slender worms that taper gradually to the posterior end; strictly planktonic)	2. Alciopidae (figs. 8.15 and 8.57-8.60)
2b	Eyes, if present, not bulbous, and not occupying as much as half of the dorsal surface of the prostomium	3
3a	Body transparent; strictly planktonic	4
3b	Body not transparent; mostly benthic, but includes some species whose sexual individuals swim	5
4a	Prostomium extended laterally as a pair of tentaclelike structures; prostomium fused with the first and second segments, the second segment with a pair of slender tentacular cirri that are nearly or quite as long as the body (the first segment may have a pair of short tentacular cirri); without setae other than the aciculae that support the tentacular cirri of segment 2; parapodia consisting of long stalks that end in 2 broad lobes	3. Tomopteridae (fig. 8.61)
4b	Prostomium without lateral extensions, but with a fingerlike anterior prolongation; prostomium fused with the first segment, whose leaflike tentacular cirri (like those of segments 2 and 3) generally cover lateral portions of the prostomium; with aciculae and a few simple setae in segments of the posterior part of the body; setigerous lobes of the parapodia reduced, but above and below each one there is a leaflike cirrus	4. Typhloscolecidae (figs. 8.62 and 8.63)
5a	Prostomium conical, ringed, tapering to a point, and tipped with 4 small antennae (pharynx eversible, its distal end either with 4 hooklike jaws or with 2 large jaws and 2 arcs of smaller jaws)	6
5b	Prostomium, if distinct, not as described in choice 5a	7
6a	Distal end of everted pharynx with 4 hooklike jaws	5. Glyceridae (figs. 8.16 and 8.65)
6b	Distal end of everted pharyx with 2 large jaws and 2 arcs of smaller jaws (there may also be spines or chevronlike denticles on the sides of the pharynx)	6. Goniadidae (figs. 8.17 and 8.66)
7a	Most of the segments, even when contracted, longer than wide	8
7b	Few if any of the segments, when contracted, longer than wide	9
8a	First 2 or 3 setigers not distinctly demarcated from one another or from the prostomium; prostomium sometimes (*Owenia*) forming a frilly, crownlike structure	45. Oweniidae (fig. 8.45)

Phylum Annelida: Class Polychaeta

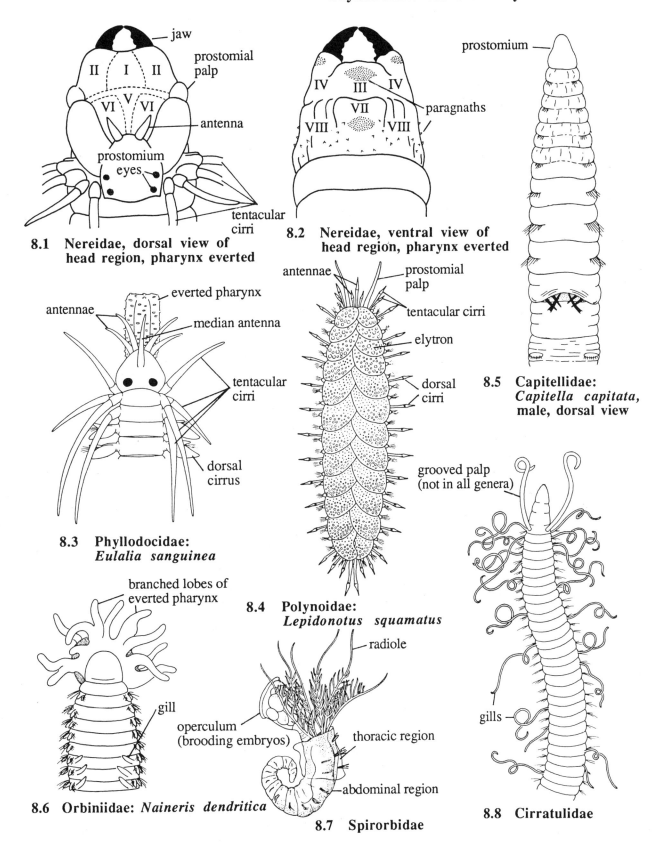

8.1 Nereidae, dorsal view of head region, pharynx everted

8.2 Nereidae, ventral view of head region, pharynx everted

8.3 Phyllodocidae: *Eulalia sanguinea*

8.4 Polynoidae: *Lepidonotus squamatus*

8.5 Capitellidae: *Capitella capitata*, male, dorsal view

8.6 Orbiniidae: *Naineris dendritica*

8.7 Spirorbidae

8.8 Cirratulidae

112 Phylum Annelida: Class Polychaeta

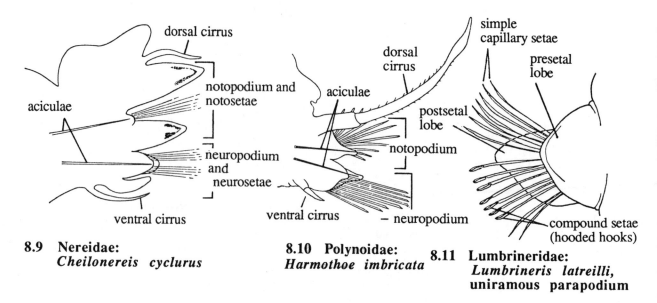

8.9 Nereidae: *Cheilonereis cyclurus*
8.10 Polynoidae: *Harmothoe imbricata*
8.11 Lumbrineridae: *Lumbrineris latreilli*, uniramous parapodium

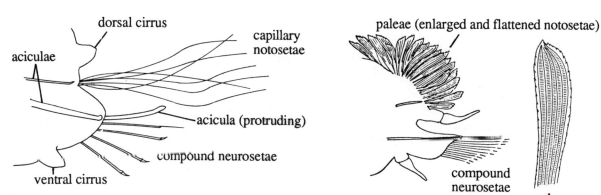

8.12 Syllidae: *Exogone lourei*
8.13 Chrysopetalidae: *Chrysopetalum occidentale*

Representative types of parapodia

8b First 2 or 3 setigers distinctly demarcated from one another, although setiger 1 may not be distinct from the prostomium; prostomium not forming a frilly, crownlike structure
 44. Maldanidae (Figs. 8.44 and 8.109-8.112)
9a All or much of the dorsal surface covered by paleae (these are flattened notosetae, arranged in transverse rows), by scalelike elytra, or by a felt that consists of hairlike, interwoven notosetae 10
9b Dorsal surface not covered by paleae, elytra, or felt 14
10a Much of the dorsal surface covered by brassy or golden paleae (length generally less than 1 cm) 17. Chrysopetalidae (fig. 8.13)
10b All or much of the dorsal surface covered by scalelike elytra or by long, hairlike notosetae that are interwoven to form a tough felt (in Aphroditidae, the felt obscures the elytra underneath it) 11
11a Eytra obvious, not covered by felt 12
11b Elytra hidden by felt that consists of long, interwoven notosetae 13. Aphroditidae (fig. 8.25)

Phylum Annelida: Class Polychaeta 113

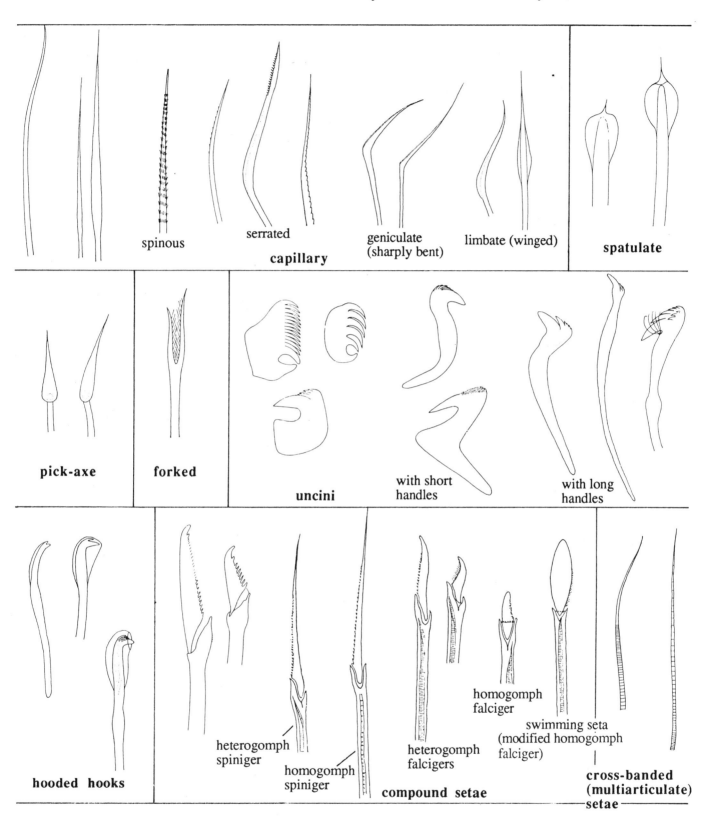

8.14 Major types of setae

114 Phylum Annelida: Class Polychaeta

12a	At least some of the neurosetae compound (divided into 2 units) (look at the neurosetae with a compound microscope!)	13
12b	All setae simple	14. Polynoidae (figs. 8.4 and 8.81-8.88)
13a	Elytra and dorsal cirri alternating regularly from setiger 5 to the posterior end	15. Polyodontidae: *Peisidice aspera*
13b	Elytra and dorsal cirri alternating only in the anterior part of the body (in the posterior part, elytra occur on all segments, and there are no dorsal cirri)	16. Sigalionidae (fig. 8.89)
14a	Slender capillary setae distinctly cross-barred, so that they appear to consist of many units (prostomium with a pair of grooved palps; body surface usually papillated)	15
14b	Capillary setae, if present, not cross-barred	16
15a	Capillary setae of setiger 1 long and forming a cage around the head region (this is not the case, however, in *Brada*); peristomium with several to many filamentous, retractable gills (living in tubes that consist of mucus, or of mud mixed with mucus)	37. Flabelligeridae (fig. 8.39)
15b	Setae of setiger 1 neither long nor forming a cage around the head region; peristomium without filamentous, retractable gills (1-4 succeeding segments, however, may have gills)	38. Acrocirridae
16a	Head region (the region consisting of the prostomium and peristomium) with at least some of the following structures: 1 or more prostomial antennae (in the Euphrosinidae, which are rare, the antennae are small and visible only from the ventral side); a pair of anteroventral palps on the prostomium; a pair of dorsal palps on the peristomium; 1 or more pairs of peristomial tentacular cirri; several to many featherlike radioles or unbranched filaments that are closely associated with the mouth and that are used in feeding (in certain families, there are also specialized large setae that are used in digging or that form part of a device for closing the tube when the animal withdraws)	17
16b	Head region without any of the structures listed in choice 15a	44
17a	With a single median prostomial antenna (no other head appendages)	31. Paraonidae (in part) (fig. 8.35)
17b	Head appendages not limited to a single median prostomial antenna	18
18a	Anterior end with a crown of several to many featherlike peristomial radioles, the complex formed by these resembling a feather-duster (in *Manayunkia*, found in fresh and brackish water, there are only a few radioles, each divided into 2 branches)	19
18b	Anterior end without a crown of featherlike radioles (there may, however, be simple filaments or other head appendages)	21
19a	Secreting a calcareous tube closed by a funnel-shaped or globular operculum (this is a modified radiole)	20
19b	Secreting a leathery, parchmentlike, or mucous tube, this not closed by an operculum	51. Sabellidae (figs. 8.48 and 8.119-8.125)
20a	Tube generally sprawling, although its oldest portion may be coiled; with more than 4 setigers composing the thoracic region	52. Serpulidae (fig. 8.51)
20b	Tube neatly coiled; with only 4 setigers composing the thoracic region	53. Spirorbidae (figs. 8.7 and 8.126-8.130)
21a	Notosetae forming transverse rows that extend nearly to the midline of the dorsal surface; gills (up to several pairs on both sides of some segments) situated behind the rows of notosetae; prostomial antennae visible only in ventral view (subtidal; rare)	19. Euphrosinidae (fig. 8.27)
21b	Notosetae not forming transverse rows that extend nearly to the midline of the dorsal surface; gills, if present, usually medial to the notosetae, rather than behind them; prostomial antennae, if present, visible in dorsal view	22
22a	Prostomium extending posteriorly on the dorsal surface as an elongated caruncle (this reaches backward for at least 2 or 3 segments and its texture is usually different from that of the segments it interrupts (with 3 antennae and a pair of palps; subtidal; rare)	18. Amphinomidae (fig. 8.26)
22b	Prostomium not extending posteriorly on the dorsal surface as an elongated caruncle	23
23a	With several to many more or less equal unbranched tentacles closely associated with the mouth (in the Ampharetidae, the tentacles can be retracted into the mouth)	24
23b	Without several to many more or less equal tentacles closely associated with the mouth (there may, however, be 1 or more prostomial antennae, a pair of prostomial palps, a pair of long,	

	grooved dorsal peristomial palps, or peristomial tentacular cirri)	28
24a	Body with a distinct caudal region that lacks setae (this may be slender and nearly as long as the rest of the body or it may be oval and dorsally concave); opening of the tube stopped by an operculum that consists in part of large, golden setae	25
24b	Body without a distinct caudal region that lacks setae; opening of the tube not stopped by a definite operculum	26
25a	Caudal region slender and nearly as long as the rest of the body, without any traces of segmentation (the caudal region is doubled back on the rest of the body, so the anus is directed toward the opening of the tube); tube sprawling, open only at one end (unless eroded), made of sand grains and bits of shell embedded in a hard, concretelike matrix, permanently affixed to rocks or shells	47. Sabellariidae (fig. 8.47)
25b	Caudal region short, oval, dorsally concave, consisting of 5 or 6 segments; tube conical, open at both ends, made of sand grains, not attached (the worms are mobile and live in sandy or muddy substrata, using their opercular setae to dig)	46. Pectinariidae (fig. 8.46)
26a	Tentacles around the mouth rather short, smooth or somewhat warty, capable of being retracted into the mouth; with 2 clusters of simple gills behind the head region (these gills are stout and may resemble tentacular cirri of Nereidae and some other polychaetes; they are also likely to break off)	48. Ampharetidae (fig. 8.49)
26b	Tentacles around the mouth long, threadlike, very extensile, not capable of being retracted into the mouth; with 1 or more pairs of simple or branched gills arising on successive segments behind the head region	27
27a	All or most thoracic segments with soft, light-colored pads on the ventral surface; neuropodial uncini of the thoracic region, like those of the abdominal region, with short handles	50. Terebellidae (figs. 8.50 and 8.115-8.118)
27b	Thoracic segments without soft pads on the ventral surface; neuropodial uncini of the thoracic region with long handles	49. Trichobranchidae (figs. 8.113 and 8.114)
28a	Without prostomial antennae or peristomial tentacular cirri; with a pair of long palps (these are usually grooved) arising from the prostomium or dorsal side of the peristomium	29
28b	With 1 or more (usually at least 2) prostomial antennae and sometimes with a pair of prostomial palps and/or 1 or more pairs of peristomial tentacular cirri	34
29a	Peristomial palps warty for all or most of their length, even when extended; head region flattened, resembling a spatula; gills absent	30. Magelonidae (fig. 8.38)
29b	Peristomial palps not warty, at least when extended; head region not flattened; gills sometimes present	30
30a	Body consisting of 2 or 3 regions, each with distinctive parapodia	31
30b	Body not consisting of 2 or 3 regions (even if the anterior portion is slightly different from the posterior portion, the parapodia of both portions are similar)	32
31a	Body consisting of 3 regions, the segments of the second region being most conspicuously modified; prostomium small, obscure; peristomium usually expanded into a liplike structure; setae of setiger 1 not especially long and not directed forward; inhabiting secreted tubes	29. Chaetopteridae (figs. 8.36 and 8.37)
31b	Body consisting of 2 regions; prostomium prominent, with a single short tentacle arising from its posterodorsal portion; peristomium not expanded into a liplike structure; setae of setiger 1 long, directed forward; not inhabiting tubes	27. Trochochaetidae: *Trochochaeta multisetosa* (fig. 8.97)
32a	Notopodia without setae other than aciculae	26. Apistobranchidae
32b	Notopodia with setae as well as aciculae	32
33a	Several to many segments with long, filamentous gills	33. Cirratulidae (in part) (figs. 8.8 and 8.101)
33b	Gills usually present (absent, however, in *Spiophanes*), arising from notopodia of most segments, but not filamentous (they are usually not more than several times as long as wide, and in some species they are pinnate)	28. Spionidae (figs. 8.34 and 8.98-8.100)
34a	Prostomium with a pair of palps (these usually originate anteroventrally, and are often divided into 2 units; when they originate more or less anterolaterally [as in *Protodorvillea*, Dorvilleidae], they are reduced to small papillae; do not confuse any of the 4 prostomial antennae of Nephtyidae [fig.8.24] with palps)	35
34b	Prostomium without palps	43

116   Phylum Annelida: Class Polychaeta

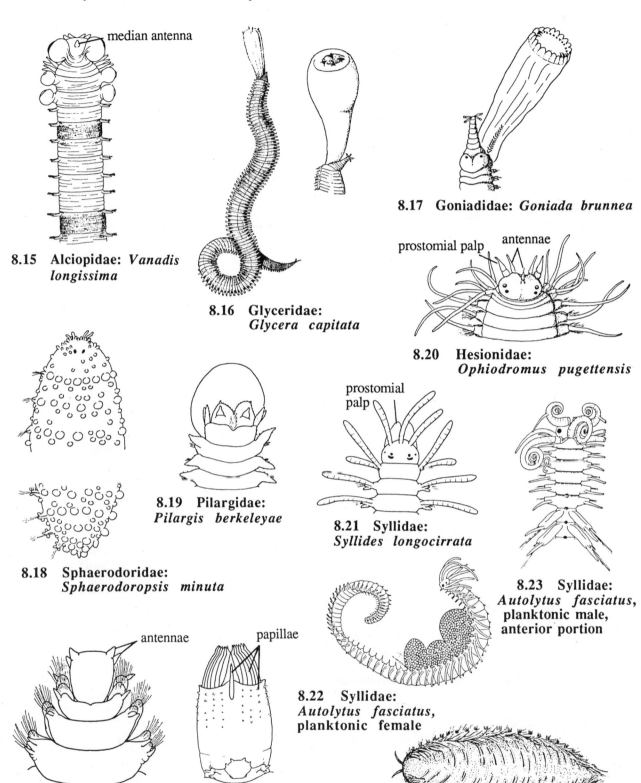

8.15 Alciopidae: *Vanadis longissima*

8.16 Glyceridae: *Glycera capitata*

8.17 Goniadidae: *Goniada brunnea*

8.18 Sphaerodoridae: *Sphaerodoropsis minuta*

8.19 Pilargidae: *Pilargis berkeleyae*

8.20 Hesionidae: *Ophiodromus pugettensis*

8.21 Syllidae: *Syllides longocirrata*

8.22 Syllidae: *Autolytus fasciatus*, planktonic female

8.23 Syllidae: *Autolytus fasciatus*, planktonic male, anterior portion

8.24 Nephtyidae

8.25 Aphroditidae

Phylum Annelida: Class Polychaeta 117

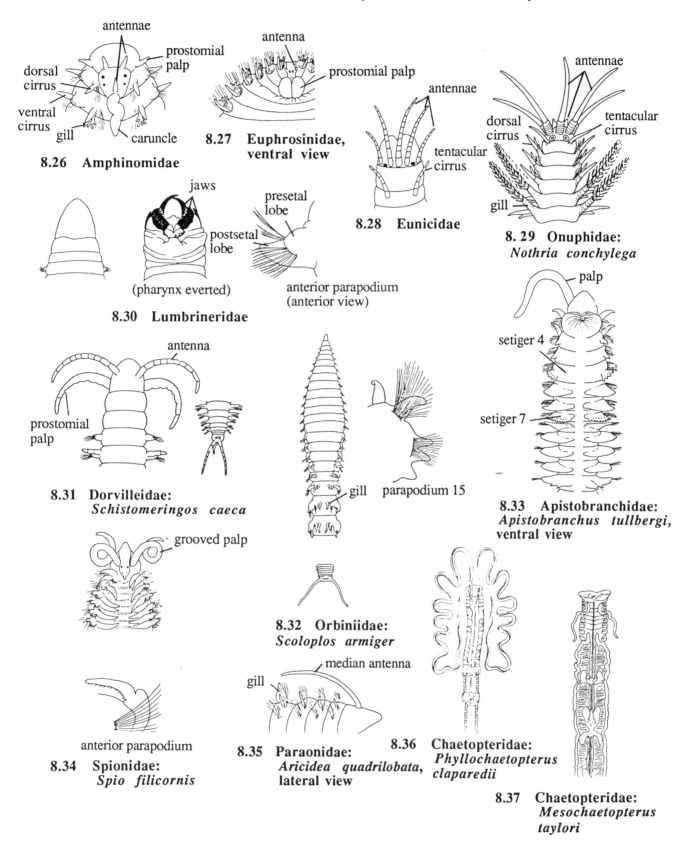

8.26 Amphinomidae
8.27 Euphrosinidae, ventral view
8.28 Eunicidae
8.29 Onuphidae: *Nothria conchylega*
8.30 Lumbrineridae
8.31 Dorvilleidae: *Schistomeringos caeca*
8.32 Orbiniidae: *Scoloplos armiger*
8.33 Apistobranchidae: *Apistobranchus tullbergi*, ventral view
8.34 Spionidae: *Spio filicornis*
8.35 Paraonidae: *Aricidea quadrilobata*, lateral view
8.36 Chaetopteridae: *Phyllochaetopterus claparedii*
8.37 Chaetopteridae: *Mesochaetopterus taylori*

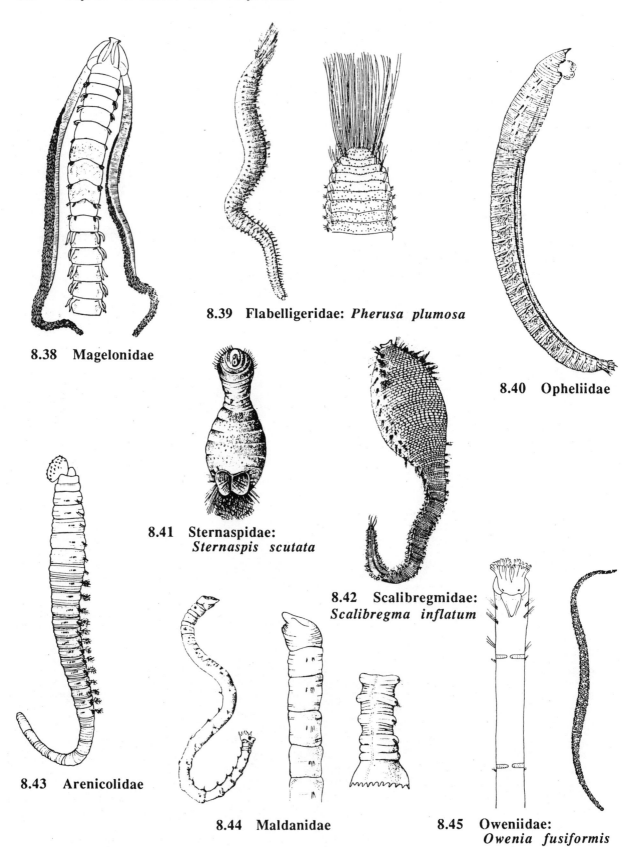

8.38 Magelonidae
8.39 Flabelligeridae: *Pherusa plumosa*
8.40 Opheliidae
8.41 Sternaspidae: *Sternaspis scutata*
8.42 Scalibregmidae: *Scalibregma inflatum*
8.43 Arenicolidae
8.44 Maldanidae
8.45 Oweniidae: *Owenia fusiformis*

Phylum Annelida: Class Polychaeta 119

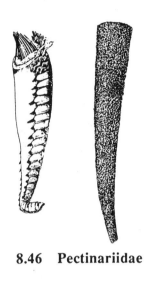

8.46 Pectinariidae

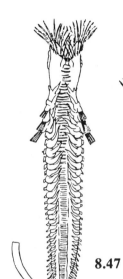

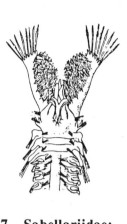

8.47 Sabellariidae: *Idanthyrsus armatus*

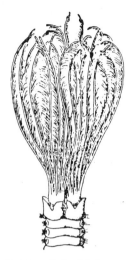

8.48 Sabellidae: *Megalomma splendida*

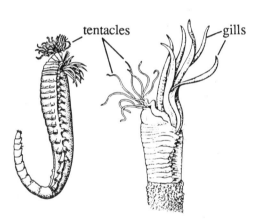

8.49 Ampharetidae

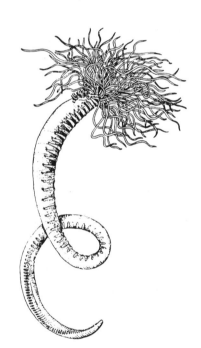

8.50 Terebellidae

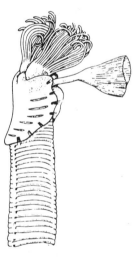

8.51 Serpulidae: *Serpula vermicularis*

35a Prostomial palps differentiated into 2 units (the distal unit may be small and nipplelike, however) 36
35b Prostomial palps not differentiated into 2 units 39
36a All setae simple (make sure by looking at them with a compound microscope!); distal unit of palps usually small and nipplelike .... 9. Pilargiidae (fig. 8.19)
36b Compound setae present; distal unit of palps usually substantial, even if smaller than the proximal unit 37
37a With 2 prostomial antennae (antennae are absent, however, in *Micronereis*); pharynx, when

everted, clearly consisting of 2 portions, with a pair of stout jaws on the distal portion and usually with conical teeth on one or more areas of both portions (to induce eversion of the pharynx, apply firm pressure just behind the head region) (proximal unit of the prostomial palps much larger than the distal unit)   11. Nereidae (figs. 8.1, 8.2, and 8.75-8.77)

37b   With 2 or 3 prostomial antennae; pharynx not divided into 2 portions, and not with the arrangement of jaws and teeth described in choice 37a (there may, however, be a pair of ventral mandibles and rows of toothed dorsal plates that form maxillae)   38

38a   With 2 or 3 prostomial antennae; peristomium with 2-8 pairs of tentacular cirri; distal unit of prostomial palps usually longer than the proximal unit; pharynx (in our species) without jaws or teeth   8. Hesionidae (in part) (fig. 8.20)

38b   With 2 prostomial antennae; peristomium without tentacular cirri; distal unit of prostomial palps much shorter than the proximal unit; pharynx with a pair of ventral mandibles and rows of toothed dorsal plates that form maxillae
    24. Dorvilleidae (in part) (figs. 8.31, 8.95, and 8.96)

39a   Peristomium without tentacular cirri; with 2 prostomial antennae (pharynx with a pair of ventral mandibles and rows of toothed dorsal plates that form maxillae; prostomial palps sometimes very small [*Ophryotrocha*, which has only a few setigers], sometimes prominent [*Protodorvillea*])   24. Dorvilleidae (in part) (figs. 8.31, 8.95 and 8.96)

39b   Peristomium with at least 1 pair of tentacular cirri; with 2 or more prostomial antennae   40

40a   Pharynx (in our species) without jaws or teeth (with 2 or 3 prostomial antennae; with 2-8 pairs of peristomial tentacular cirri)   8. Hesionidae (in part) (fig. 8.20)

40b   Pharynx with jaws or teeth   41

41a   Prostomium with 3 antennae; usually with 4 eyes (sometimes 2 or 6); pharynx with a single large tooth or a crown of small teeth, or both   10. Syllidae (figs. 8.21-8.23 and 8.69-8.74)

41b   Prostomium with 2, 5, or 7 antennae; if eyes are present, there are usually 2; pharynx with a pair of ventral mandibles and 2 or more rows of toothed dorsal pieces that form maxillae   42

42a   Prostomium with 7 antennae (the 5 arising near the posterior margin of the prostomium are usually much larger than the 2 arising near the anterior margin)
    20. Onuphidae (figs. 8. 29 and 8.90)

42b   Prostomium with 5 antennae (prostomial palps stout, globular, often indistinctly demarcated from the prostomium)   21. Eunicidae (fig. 8.28)

43a   Prostomium with 4 small antennae, those of the first pair being situated at the anterolateral corners of the prostomium, those of the second pair (which are sometimes split into 2 branches) on the lateral margins of the prostomium; without eyes; without tentacular cirri on the peristomium (with a single cirrus at the posterior end of the body)
    12. Nephtyidae (figs. 8.24 and 8.78-8.80)

43b   Prostomium with 4 or 5 antennae, these not arranged as described in choice 43a; usually with eyes; with 2-4 pairs of tentacular cirri on the peristomium
    1. Phyllodocidae (figs. 8.3 and 8.52-8.56)

44a   With not more than 15 segments when mature; most segments indistinct; length less than 1 cm   45

44b   With more than 15 segments; most segments distinct; length generally greater than 1 cm   46

45a   With both notosetae and neurosetae, these with conspicuously serrated tips (*Ctenodrilus serratus*, the only species known to occur in the region, is common in diatom-rich sediments in bays; it could be mistaken for an oligochaete)
    34. Ctenodrilidae: *Ctenodrilus serratus* (fig. 8.102)

45b   Setigers with a single bundle of setae, some of which have forked tips and some of which have bladelike distal portions   35. Parergodrilidae: *Stygocapitella subterranea*

46a   Prostomium indistinct; surface of body with prominent papillae, many of which are almost globular   7. Sphaerodoridae (figs. 8.18, 8.67, and 8.68)

46b   Prostomium distinct, even if small; if papillae are present on the surface of the body, they are not almost globular   47

47a   Notopodial gills long and threadlike; with several long, grooved filaments arising from the dorsal surface of one of the first few setigers   33. Cirratulidae (in part) (figs. 8.8 and 8.101)

47b   Notopodial gills, if present, not long and threadlike; without any grooved filaments arising from the dorsal surface of one of the first few setigers (there may, however, be a single filament growing out of setiger 2 or 3)   48

Phylum Annelida: Class Polychaeta

48a With a single threadlike filament growing out of the dorsal surface of setiger 2 or 3
　　36. Cossuridae (fig. 8.103)
48b Without a single threadlike filament growing out of the dorsal surface of setiger 2 or 3　49
49a With several pairs of large, bushy gills in the middle third of the body
　　43. Arenicolidae (fig. 8.43)
49b Without several pairs of large, bushy gills in the middle third of the body (but small, usually simple gills may be present in that region, or there may be conspicuous bushy gills near the anterior end)　50
50a Pharynx with dark, hard jaw pieces (a pair of ventral mandibles and two rows of toothed dorsal plates that form maxillae; in small worms, these may be visible through the body wall, but usually it is necessary to cause the pharynx to be everted by applying pressure behind the head region); notopodial gills absent　51
50b Pharynx without dark, hard jaw pieces; notopodial gills present or absent　52
51a Neurosetae include some hooded hooks; prostomium without eyes
　　22. Lumbrineridae (figs. 8.30, 8.91, and 8.92)
51b Neurosetae limbate capillary setae (no hooded hooks); prostomium with or without eyes
　　23. Arabellidae (figs. 8.93 and 8.94)
52a All setae of 1 type: slender capillary setae (make sure by examining them with a compound microscope)　39. Opheliidae (figs. 8.40, 8.104, and 8.105)
52b Setae of more than 1 type (there may be capillary setae, forked capillary setae, uncini, hooded hooks, etc.)　53
53a Pharynx, when everted, either in the form of a rosette around the mouth or consisting of branched lobes; capillary setae crenulated　25. Orbiniidae (figs. 8.6 and 8.32)
53b Pharynx simple, sometimes with papillae, but when everted not in the form of a rosette around the mouth and not consisting of branched lobes; capillary setae not crenulated　54
54a With simple, fingerlike notopodial gills on several segments near the anterior end, usually beginning on setiger 4 or 5　31. Paraonidae (in part) (fig. 8.35)
54b Notopodial gills, if present in the anterior part of the body, bushy (when simple gills are present, these are either limited to the posterior part of the body or are present on many segments, but not before setiger 15; in some genera, the simple gills are in clusters)　55
55a Body as a whole so slender as to be nearly threadlike when stretched out; setae consisting of capillary setae and hooded hooks, but not of forked setae or 2-toothed hooks; prostomium conical; without bushy gills near the anterior end, but sometimes with simple gills or clusters of simple gills on many segments, especially in the posterior part of the body (not before setiger 15)　42. Capitellidae (figs. 8.5 and 8.106-8.108)
55b Body not so slender as to be nearly threadlike when stretched out; setae may include slender capillary setae, forked setae, or 2-toothed hooks, but not hooded hooks; prostomium either conical, bilobed, or drawn out into 2 anterolateral horns; with either bushy gills on a few segments close to the anterior end or with simple, fingerlike gills on several segments near the posterior end　57
56a Prostomium bilobed or drawn out into 2 anterolateral horns; setae consisting of slender capillary setae and forked setae; with bushy gills on a few segments close to the anterior end
　　40. Scalibregmidae (fig. 8.42)
56b Prostomium conical; setae consisting of slender, serrated capillary setae and 2-toothed hooks; with simple, fingerlike gills on several segments near the posterior end
　　32. Questidae: *Questa caudicirra*

Order Phyllodocida

Superfamily Phyllodocidacea

1. Family Phyllodocidae

1a With 2 pairs of tentacular cirri on segment 1, none on segment 2　2
1b With 3 or 4 pairs of tentacular cirri, some of these on segment 1, the others on 1 or 2 of the succeeding segments　6

122   Phylum Annelida: Class Polychaeta

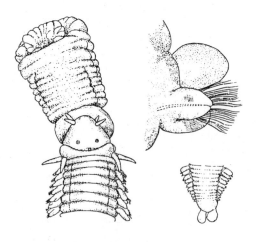

8.52 *Eteone longa*

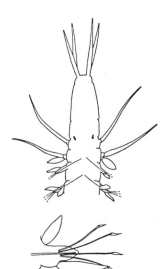

8.53 *Hesionura coineaui* subsp. *difficilis*

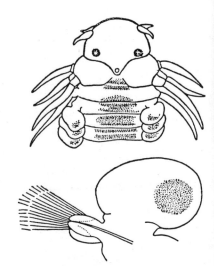

8.54 *Phyllodoce polynoides*

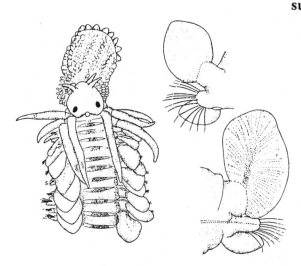

8.55 *Phyllodoce maculata*

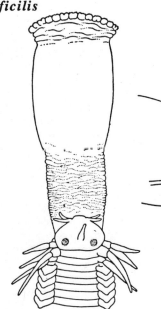

8.56 *Eulalia tubiformis*

Phyllodocidae

2a	Dorsal cirrus of parapodia broader than long, and decidedly asymmetrical	3
2b	Dorsal cirrus of parapodia usually longer than wide, and almost symmetrical	4
3a	Distal end of shaft of setae (where the blade is attached) with 2 unequal spines	*Eteone pacifica* Hartman, 1936
3b	Distal end of shaft of setae with 2 equal spines	*Eteone spetsbergensis* Malmgren, 1865
4a	Prostomium longer than wide	*Eteone tuberculata* Treadwell, 1922
4b	Prostomium about as wide as long	5

5a Cirri of pygidium about as wide as long; dorsal cirrus of parapodia near middle of body slightly longer than wide *Eteone longa* (Fabricius, 1780) (fig. 8.52)
5b Cirri of pygidium about twice as long as wide; dorsal cirrus of parapodia near middle of body wider than long *Eteone californica* Hartman, 1936
6a With 3 pairs of tentacular cirri on 2 or 3 segments 7
6b With 4 pairs of tentacular cirri on 3 segments (segments 1 and 2 may be fused, however, in which case 3 of the pairs of tentacular cirri will be on what appears to be segment 1) 8
7a Setae beginning on segment 3; dorsal and ventral cirri of parapodia oval or nearly circular, not decidedly elongated; setae of a particular parapodium with blades of unequal length (the middle setae have blades nearly twice as long as those of the upper and lower setae) *Hesionura coineaui* subsp. *difficilis* (Banse, 1963) (fig. 8.53)
7b Setae beginning on segment 2; dorsal and ventral cirri of parapodia elongated; setae of a particular parapodium with blades of about equal length *Mystides borealis* Théel, 1879
8a Parapodia biramous, the notopodium, even though small, with an acicula, and sometimes with 1 or 2 other setae 9
8b Parapodia uniramous, there being no trace of a notopodium or its acicula 10
9a Ventral tentacular cirri on segment 2 slender *Notophyllum (Notophyllum) imbricatum* Moore, 1906
9b Ventral tentacular cirri on segment 2 broad and leaflike *Notophyllum (Hesperophyllum) tectum* (Chamberlin, 1919)
10a Prostomium with 4 antennae 11
10b Prostomium with 5 antennae (1 median antenna) 20
11a With a small mid-dorsal papilla on the apparent first segment (segments 1 and 2 are fused) 12
11b Without a mid-dorsal papilla on the apparent first segment *Phyllodoce (Genetyllis) castanea* (Marenzeller, 1879)
12a Prostomium with an indentation on its posterior margin, its outline thus somewhat heart-shaped 13
12b Prostomium without an indentation on its posterior margin (its posterior portion extends backward, intruding into the fused segments 1 and 2) *Phyllodoce (Paranaitis) polynoides* (Moore, 1909) (fig. 8.54)
13a Setae beginning on segment 4 (the third apparent segment) *Phyllodoce (Anaitides) madeirensis* Langerhans, 1880
13b Setae beginning on segment 3 (the second apparent segment) 14
14a Dorsal cirrus of parapodia longer than wide, pointed or rounded distally; pharynx with a mid-dorsal row of papillae in addition to 6 lateral rows on both sides *Phyllodoce (Anaitides) medipapillata* Moore, 1909
14b Dorsal cirrus of parapodia elliptical or nearly rectangular, wider than long, not at all pointed distally; pharynx without a mid-dorsal row of papillae, and with 4, 6, or 12 rows of papillae on both sides 15
15a Pharynx with 12 irregular rows of 10-15 papillae on both sides *Phyllodoce (Anaitides) multiseriata* Rioja, 1941
15b Pharynx with 4 or 6 rows of papillae on both sides 16
16a Pharynx with 4 rows of papillae on both sides *Phyllodoce (Anaitides) citrina* Malmgren, 1865
16b Pharynx with 6 rows of papillae on both sides 17
17a Dorsal surface with 3 dark longitudinal bands; ventral cirrus of parapodia rounded 18
17b Dorsal pigmentation not concentrated in 3 longitudinal bands; ventral cirrus of parapodia pointed 19
18a Rows of papillae nearest dorsal midline of pharynx with not more than 8 papillae; ventral cirrus of parapodia extending laterally, its long axis nearly parallel to the acicula *Phyllodoce (Anaitides) maculata* (Linnaeus, 1767) (fig. 8.55)
18b Rows of papillae nearest the dorsal midline of pharynx with about 9 papillae; ventral cirrus of parapodia directed ventrolaterally, its long axis forming a 30 to 45° angle with the acicula *Phyllodoce (Anaitides) williamsi* (Hartman, 1936)
19a Rows of papillae nearest the dorsal midline of the pharynx with not more than 10 papillae *Phyllodoce (Anaitides) mucosa* Ørsted, 1843

19b Rows of papillae nearest the dorsal midline of the pharynx with not more than 12 papillae
Phyllodoce (Anaitides) groenlandica Ørsted, 1843
20a Ventral cirrus of parapodia decidedly larger than the setigerous lobe
Eulalia (Bergstroemia) nigrimaculata Moore, 1909
20b Ventral cirrus of parapodia about the same size as, or slightly smaller than, the setigerous lobe 21
21a Width of the ventral tentacular cirri on segment 2 not more than twice that of the dorsal tentacular cirri of the same segment 22
21b Ventral tentacular cirri on segment 2 leaflike, their width more than twice that of the dorsal tentacular cirri of the same segment 28
22a Segment 1 evident laterally (it bears tentacular cirri), but not on the dorsal surface 23
22b Segment 1 evident on the dorsal surface as well as laterally 25
23a Dorsal cirrus of parapodia truncate distally *Eulalia (Eumida) longicornuta* Moore, 1906
23b Dorsal cirrus of parapodia tapering distally 24
24a Dorsal cirrus of parapodia longer than wide (especially in the case of parapodia in the posterior part of the body) *Eulalia (Eumida) sanguinea* Ørsted, 1843 (fig. 8.3)
24b Dorsal cirrus of parapodia wider than long
Eulalia (Eumida) tubiformis (Moore, 1909) (fig. 8.56)
25a Dorsal cirrus of parapodia oval 26
25b Dorsal cirrus of parapodia lanceolate and pointed distally 27
26a Distal end of shaft of setae (where the blade is attached) with small spines of about equal size
Eulalia (Eulalia) bilineata (Johnston, 1840)
Perhaps an undescribed species closely related to *E. bilineata*.
26b Distal end of shaft of setae with 1 spine that is much larger than the others
Eulalia (Eulalia) levicornuta Moore, 1909
27a Setae beginning on segment 2; distal end of shaft of setae (where the blade is attached) with several spines of about equal size *Eulalia (Eulalia) viridis* (Linnaeus, 1767)
27b Setae beginning on segment 3; distal end of shaft of setae with 1 spine that is much larger than the others *Eulalia (Eulalia) quadrioculata* Moore, 1906
28a Prostomium approximately oval; blade of setae only about 6 times as long as wide
Eulalia (Pterocirrus) parvoseta Banse & Hobson, 1968
28b Prostomium approximately heart-shaped, due to an indentation on its posterior margin; blade of setae more than 10 times as long as wide *Eulalia (Pterocirrus) macroceros* (Grube, 1860)

2. Family Alciopidae

Berkeley, E., & C. Berkeley. 1957. On some pelagic Polychaeta from the northeast Pacific north of latitude 40° N. and east of longitude 175° W. Can. J. Zool., 35:573-8.

------. 1960. Some further records of pelagic Polychaeta from the northeast Pacific north of latitude 40° N. and east of longitude 175° W., together with records of Siphonophora, Mollusca, and Tunicata from the same region. Can. J. Zool., 38:787-99.

Dales, R. P. 1957. Pelagic polychaetes of the Pacific Ocean. Bull. Scripps Inst. Oceanogr., 7:99-167.

Dales, R. P., & G. Peter. 1972. A synopsis of the pelagic Polychaeta. J. Nat. Hist., 6:55-92.

Ushakov, P. V. 1972. [Polychaetes of the suborder Phyllodociformia of the Polar Basin and the northwestern part of the Pacific (families Phyllodocidae, Alciopidae, Tomopteridae, Typhloscolecidae, and Lacydoniidae).] Fauna SSSR, 102. 272 pp. (in Russian). (English translation, 1974. Jerusalem: Israel Program for Scientific Translations.)

1a All setae (other than the single large acicula that supports each parapodium) compound 2
1b With some simple setae as well as compound setae 3
2a Setigerous lobe of parapodia terminating in a single slender extension
Vanadis longissima (Levinsen, 1885) (fig. 8.15)
2b Setigerous lobe of parapodia terminating in 2 slender extensions
Alciopa reynaudii Audouin & Milne-Edwards, 1833 (fig. 8.59)

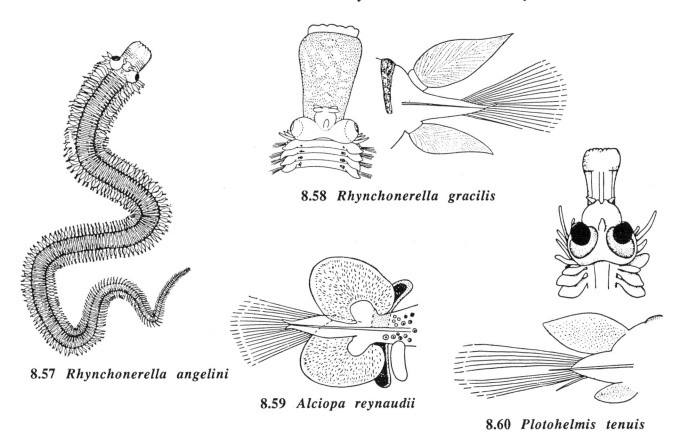

8.57 *Rhynchonerella angelini*
8.58 *Rhynchonerella gracilis*
8.59 *Alciopa reynaudii*
8.60 *Plotohelmis tenuis*

Alciopidae

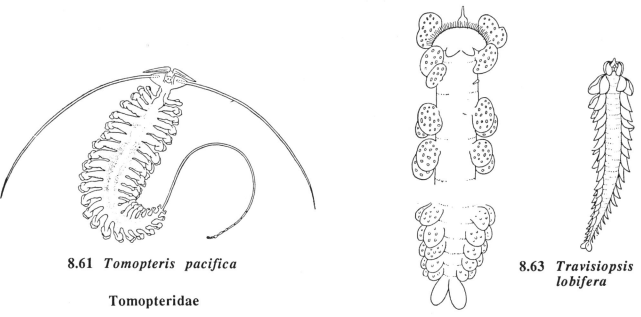

8.61 *Tomopteris pacifica*

Tomopteridae

8.62 *Typhloscolex mülleri*
8.63 *Travisiopsis lobifera*

Typhloscolecidae

3a   Setigerous lobe of parapodia tapering to a point, but not drawn out into a slender, fingerlike extension; anterior parapodia without slender, compound setae
Plotohelmis tenuis (Apstein, 1900) (fig. 8.60)
3b   Setigerous lobe of parapodia drawn out into a slender, fingerlike extension; anterior parapodia with slender compound setae   4
4a   Anterior parapodia with only 1 or 2 thick simple setae that resemble the acicula (there are also slender compound setae)   *Rhynchonerella gracilis* Costa, 1862 (fig. 8.58)
4b   Anterior parapodia with more than 2 thick simple setae that resemble the acicula (there are also slender compound setae)   *Rhynchonerella angelini* (Kinberg, 1866) (fig. 8.57)

3. Family Tomopteridae

For references, see Family 2. Alciopidae.

1a   Last few segments elongated, collectively forming a long tail; length (exclusive of tail) up to about 4 cm   *Tomopteris pacifica* Izuka, 1914 (fig. 8.61)
1b   Last few segments not elongated; length up to about 1.5 cm
Tomopteris septentrionalis Quatrefages, 1865

4. Family Typhloscolecidae

For references, see Family 2. Alciopidae.

1a   Prostomium with a dorsal and a ventral lobe that is bordered by long cilia (thus the anterior end of the body resembles that of some rotifers); pygidial cirri usually about twice as long as wide   *Typhloscolex mülleri* Busch, 1851 (fig. 8.62)
1b   Prostomium without a dorsal and a ventral lobe bordered by long cilia; pygidial cirri about as wide as long   2
2a   Prostomium with a prominent dorsal caruncle, this embraced by a pair of nuchal organs
Travisiopsis lobifera Levinsen, 1883 (fig. 8.63)
2b   Prostomium without a dorsal caruncle, but with a pair of large, nearly sigmoid nuchal organs
Sagitella kowalevskii Wagner, 1872

Superfamily Glyceracea

5. Family Glyceridae

1a   Parapodia (except for the first 2 pairs) biramous   2
1b   All parapodia uniramous   *Hemipodus borealis* Johnson, 1901 (fig. 8.64)
2a   Parapodia with 1 postsetal lobe and either 1 or 2 presetal lobes; gills absent   3
2b   Parapodia with 2 postsetal lobes and 2 presetal lobes; gills present or absent   4
3a   Parapodia in middle and posterior portions of body with 2 presetal lobes; small papillae on the pharynx without transverse ridges   *Glycera capitata* Ørsted, 1843 (fig. 8.16)
3b   Parapodia in middle and posterior portions of the body with only 1 presetal lobe; each small papilla on the pharynx with about 14 transverse ridges   *Glycera tenuis* Hartman, 1944
4a   Gills present   5
4b   Gills absent   8
5a   Gills retractile; situated on the anterior or posterior surfaces of the parapodia   6
5b   Gills not retractile, situated on the dorsal surfaces of the parapodia   7
6a   Gills branched (observe carefully, because when the gills are partly retracted, only 1 branch may be evident), located on the posterior surfaces of the parapodia
Glycera americana Leidy, 1855 (fig. 8.65)
6b   Gills globular, located on the anterior surfaces of the parapodia
Glycera gigantea Quatrefages, 1865

Phylum Annelida: Class Polychaeta

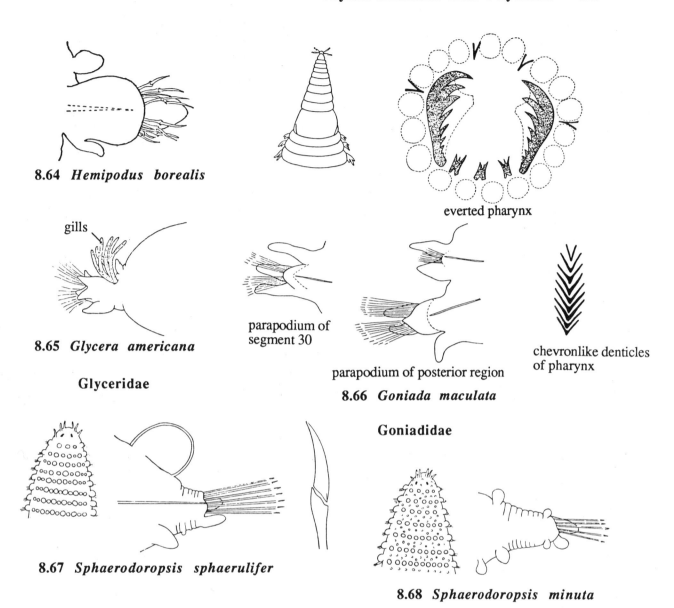

8.64 *Hemipodus borealis*

8.65 *Glycera americana*

Glyceridae

8.66 *Goniada maculata*

Goniadidae

8.67 *Sphaerodoropsis sphaerulifer*

8.68 *Sphaerodoropsis minuta*

Sphaerodoridae

7a Gills fingerlike, and longer than the parapodial lobes; postsetal lobe of neuropodia in the middle portion of the body shorter than the postsetal lobe of notopodia
Glycera convoluta Keferstein, 1862
7b Gills blisterlike, shorter than the parapodial lobes; postsetal lobes on neuropodia and notopodia in the middle portion of the body of equal length   *Glycera robusta* Ehlers, 1868
8a Postsetal lobes pointed   *Glycera siphonostoma* (delle Chiaje, 1825)
8b Postsetal lobes rounded   *Glycera tesselata* Grube, 1863

6. Family Goniadidae

1a Notosetae with a blunt outgrowth a short distance below the tip; everted pharynx without

	chevronlike denticles on both sides of its proximal portion	2
1b	Notosetae without a blunt outgrowth near the tip; everted pharynx with chevronlike denticles on both sides of the proximal portion	4
2a	Everted pharynx without small jaws on the lip ventral to the 2 large jaws *Glycinde armigera* Moore, 1911	
2b	Everted pharynx with small jaws on the lip ventral to the 2 large jaws	3
3a	Blunt outgrowth near tip of notosetae without ridges; presetal lobe of neuropodium 25 usually tapering gradually toward the tip, but sometimes with a distinctly demarcated distal portion *Glycinde picta* E. Berkeley, 1927	
3b	Blunt outgrowth near tip of notosetae with delicate ridges; presetal lobe of neuropodium 25 with a distinctly demarcated distal portion *Glycinde polygnatha* Hartman, 1950	
4a	Prostomium constricted into fewer than 6 ringlike units; everted pharynx with about 15 small jaws *Goniada annulata* Moore, 1905	
4b	Prostomium constricted into more than 6 ringlike units; everted pharynx with not more than 4 small jaws, if any	5
5a	Notopodium with a small postsetal lobe *Goniada brunnea* Treadwell, 1906 (fig. 8.17)	
5b	Notopodium without a postsetal lobe *Goniada maculata* Ørsted, 1843 (fig. 8.66)	

7. Family Sphaerodoridae

Fauchald, K. 1974. Sphaerodoridae (Polychaeta: Errantia) from world-wide areas. J. Nat. Hist., 8:257-89.

1a	Body stout, with fewer than 30 segments; dorsal surface with at least 4 globular papillae on each segment (the papillae form a transverse row, and also make longitudinal rows, the distinctness of the rows depending on the extent to which the worm is constricted or extended); all setae compound	2
1b	Body slender, with about 100 segments; dorsal surface with only 2 globular papillae on most segments; all setae simple *Sphaerodorum papillifer* Moore, 1909	
2a	Dorsal surface with 4 globular papillae on each segment; blades of setae smooth *Sphaerodoropsis biserialis* (Berkeley & Berkeley, 1944)	
2b	Dorsal surface with 8 to 14 globular papillae on each segment; blades of setae with fine serrations	3
3a	Dorsal surface with many irregularly arranged small papillae in addition to the prominent globular papillae *Sphaerodoropsis minuta* (Webster & Benedict, 1887) (figs. 8.18 and 8.68)	
3b	Dorsal surface without small papillae in addition to the prominent globular papillae *Sphaerodoropsis sphaerulifer* (Moore, 1909) (fig. 8.67)	

Superfamily Nereidacea

8. Family Hesionidae

Species of *Heteropodarke* and *Microphthalmus*, not in the key, have been reported to occur in the region.

1a	With 2 prostomial antennae	2
1b	With 3 prostomial antennae	3
2a	With 6 pairs of tentacular cirri *Micropodarke dubia* (Hessle, 1925)	
2b	With 8 pairs of tentacular cirri *Kefersteinia cirrata* (Keferstein, 1862)	
3a	With 6 pairs of tentacular cirri (a dark brown species abundant intertidally, especially on muddy sand where there are growths of *Ulva*, *Enteromorpha*, and other algae) *Ophiodromus pugettensis* (Johnson, 1901) (fig. 8.20)	
3b	With 8 pairs of tentacular cirri *Podarkeopsis brevipalpa* (Hartmann-Schröder, 1959)	

9. Family Pilargiidae (Pilargidae)

Pettibone, M. H. 1966. Revision of the Pilargidae (Annelida: Polychaeta), including descriptions of new species, and redescription of the pelagic *Podarmus ploa* Chamberlin (Polynoidae). Proc. U. S. Nat. Mus., 118:155-207.

1a	Notopodia of the middle region of the body with a stout seta whose protruding portion is bent distally *Sigambra tentaculata* (Treadwell, 1941)	
1b	Notopodia of the middle region of the body without a stout protruding seta (there is, however, an internal acicula)	2
2a	Median prostomial antenna absent; body surface papillated *Pilargis berkeleyae* Monro, 1933 (fig. 8.19)	
2b	Median prostomial antenna present; body surface smooth or wrinkled, but not papillated *Otopsis longipes* Ditlevsen, 1917)	

10. Family Syllidae

Banse, K. 1971. A new species, and additions to the descriptions of six other species of *Syllides* Örsted (Syllidae: Polychaeta). J. Fish. Res. Bd. Canada, 28:1469-81.
Gidholm, L. 1966. A revision of the Autolytinae (Syllidae, Polychaeta) with special reference to Scandinavian species, and with notes on external and internal morphology, reproduction and ecology. Arkiv för Zoologi, ser. 2, 19:157-213.

1a	Parapodia without ventral cirri	2
1b	Parapodia with ventral cirri	7
2a	Eyes large; body divided into 2 or 3 distinct regions (the second region bearing long capillary setae that function in swimming); pharynx absent (prostomial antennae bifurcated in males) planktonic sexual stages of various species of *Autolytus* (figs. 8.22 and 8.23) See Gidholm, 1966.	
2b	Eyes small; body not divided into distinct regions; pharynx present	3
3a	Upper simple setae of posterior setigers slender, with serrations along one edge of the gradually tapering tip; with segmental rings of cilia (these are visible in living specimens)	4
3b	Upper simple setae of posterior setigers stout, with serrations on a somewhat truncated portion proximal to the tapering tip; without segmental rings of cilia	5
4a	Dorsal cirrus on setiger 4 decidedly longer than that on setigers 2 and 3; trepan of pharynx with about 45 teeth of nearly equal size *Autolytus (Autolytus) magnus* E. Berkeley, 1923	
4b	Dorsal cirrus on setiger 4 not obviously longer than that on setigers 2 and 3; trepan of pharynx with 9 large teeth, these separated from one another by 2 small teeth *Autolytus (Autolytus) verrilli* Marenzeller, 1892	
5a	Trepan of pharynx with 10 large, equal teeth (with a dark mid-dorsal band and 2 lateral bands) *Autolytus (Proceraea) trilineatus* Berkeley & Berkeley, 1945	
5b	Trepan of pharynx with 9 large and 9 small teeth, these alternating	6
6a	Nuchal organs on the peristomium reaching the posterior edge of this segment; with a dark mid-dorsal band and 2 lateral bands *Autolytus (Proceraea) prismaticus* (O. F. Müller, 1776)	
6b	Nuchal organs on peristomium not reaching the posterior margin of this segment; without dark mid-dorsal or lateral bands *Autolytus (Proceraea) cornutus* Agassiz, 1862	
7a	Prostomial palps fused together for their full length, or nearly so (body length usually less than 5 mm)	8
7b	Prostomial palps separate or fused only at their bases	15
8a	Peristomium with 2 pairs of long tentacular cirri *Brania brevipharyngea* Banse, 1972 (fig. 8.73)	
8b	Peristomium with 1 pair of short tentacular cirri	9
9a	Body surface usually papillated; dorsal cirrus of parapodia usually drawn out into a slender distal portion, thus somewhat bottle-shaped	10
9b	Body surface smooth; dorsal cirrus of parapodia not drawn out into a slender distal portion	12

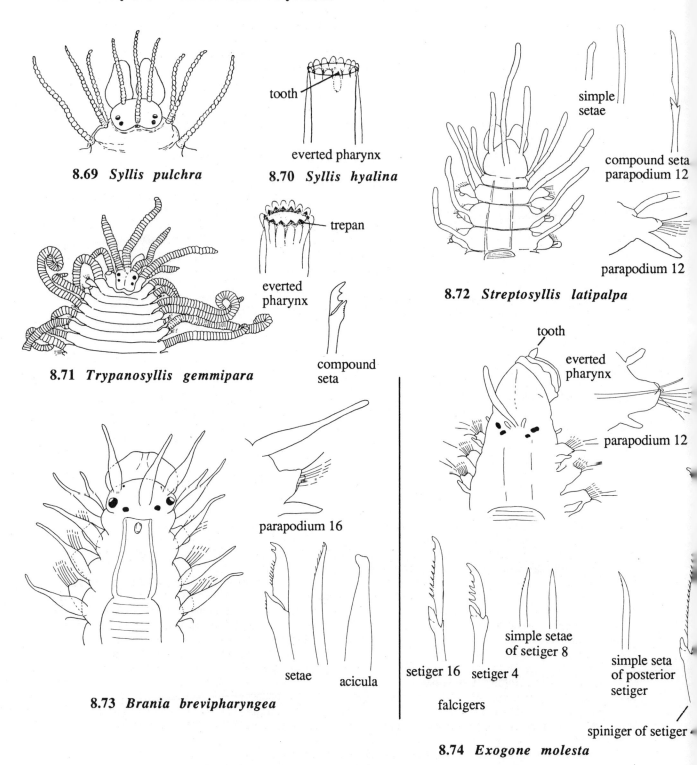

8.69 *Syllis pulchra*
8.70 *Syllis hyalina*
8.71 *Trypanosyllis gemmipara*
8.72 *Streptosyllis latipalpa*
8.73 *Brania brevipharyngea*
8.74 *Exogone molesta*

Syllidae

10a With 3 pairs of eyes *Sphaerosyllis brandhorsti* Hartmann-Schröder, 1965
10b With 2 pairs of eyes 11
11a Each parapodium (from setiger 4 posteriorly) with a capsule containing rodlike structures (the capsule is located near the proximal portions of the aciculae)
Sphaerosyllis hystrix Claparède, 1863
11b None of the parapodia with a capsule containing rodlike structures
Sphaerosyllis pirifera Claparède, 1868
12a Setigers in anterior region of the body with 1 or 2 compound setae that have short, needlelike blades; all 3 antennae of about equal length 13
12b Setigers in anterior region of the body with 1-3 compound setae that have long blades; median antenna longer than the lateral antennae 14
13a Antennae about the same length as the dorsal cirri of the parapodia
Exogone verugera (Claparède, 1868)
13b Antennae considerably longer than the dorsal cirri of the parapodia *Exogone* sp. Close to *E. gemmifera* Pagenstecher, 1862.
14a Shafts of long-bladed upper setae on setiger 2 much wider than those of other upper setae
Exogone lourei Berkeley & Berkeley, 1938 (fig. 8.12)
14b Shafts of long-bladed upper setae on setiger 2 about the same width as those of other upper setae *Exogone molesta* Banse, 1972 (fig. 8.74)
15a Antennae, peristomial tentacular cirri, and dorsal cirri of parapodia decidedly constricted into beadlike units; palps sometimes fused at the base, sometimes separate 16
15b Antennae, peristomial tentacular cirri, and dorsal cirri of parapodia only slightly constricted into units near the tip (in *Syllides*, however, the dorsal cirri are decidedly constricted into beadlike units); palps fused at their bases 32
16a Parapodia in the anterior region of the body with only compound setae, or with a mixture of compound and pseudocompound setae 17
16b Parapodia in the anterior region of the body with only simple setae
Haplosyllis spongicola subsp. *spongicola* (Grube, 1855)
17a Body dorsoventrally flattened, the width considerably greater than the thickness; pharynx with a trepan, but without a mid-dorsal tooth 18
17b Body not much wider than thick; pharynx without a trepan, but with a mid-dorsal tooth 19
18a Compound setae without a tooth near the middle of the concave side of the blade
Trypanosyllis (Trypanedenta) ingens Johnson, 1902
18b Compound setae with a tooth near the middle of the concave side of the blade
Trypanosyllis (Trypanedenta) gemmipara Johnson, 1901 (fig. 8.71)
19a With some pseudocompound setae or stout simple setae (at least in the middle region of the body) 20
19b Without pseudocompound setae or stout simple setae (but there may be slender simple setae, especially in the posterior region of the body) 22
20a Setae mostly of the pseudocompound type *Sylllis (Syllis) spongiphila* Verrill, 1885
20b Setae mostly of the compound type 21
21a Dorsal cirri of parapodia constricted into 6 or 7 units *Syllis (Syllis) gracilis* Grube, 1840
21b Dorsal cirri of parapodia constricted into 10-20 units *Syllis (Syllis) elongata* (Johnson, 1901)
22a Upper setae of parapodia with blades that are at least 4 times as long as those of the lower setae *Syllis (Ehlersia) heterochaeta* Moore, 1909
22b Upper setae of parapodia with blades that are usually considerably less than 4 times as long as those of the lower setae 23
23a Length of dorsal cirrus of each parapodium about equal to the length of the segment to which that parapodium belongs *Syllis (Typosyllis) stewarti* Berkeley & Berkeley, 1942
23b Length of the dorsal cirrus of each parapodium considerably greater than the length of the segment to which that parapodium belongs 24
24a Most dorsal cirri of parapodia with fewer than 16 units; palps separate 25
24b Most dorsal cirri of parapodia with more than 16 units; palps fused for about half their length or separate 26
25a Dorsal cirri thick and spindle-shaped, those of successive segments about the same length
Syllis (Typosyllis) armillaris (O. F. Müller, 1771)
25b Dorsal cirri of parapodia slender, those of successive segments alternating in length
Syllis (Typosyllis) hyalina Grube, 1863

132 Phylum Annelida: Class Polychaeta

26a Dorsal cirri of successive segments alternating in length, and rarely with more than 25 units; palps usually fused for about half their length 27
26b Dorsal cirri of successive segments alternating in length, and some of them with as many as 35 units; palps separate or nearly so 29
27a Dorsal surfaces of segments in middle region of the body with conspicuous circular or oval areas that lack dark pigment
Syllis (Typosyllis) adamantea subsp. *adamantea* (Treadwell, 1914)
27b Dorsal surfaces of segments in middle region of the body without circular or oval areas that lack dark pigment 28
28a Blades of upper compound setae of parapodia of the anterior region of the body from 2 to 4 times as long as those of the lower setae *Syllis (Typosyllis) alternata* Moore, 1908
28b Blades of upper compound setae of parapodia of the anterior region of the body not markedly longer than those of the lower setae (not likely to be found in our region)
Syllis (Typosyllis) fasciata Malmgren, 1867
29a Longer dorsal cirri of parapodia with about 50-70 units
Syllis (Typosyllis) pulchra Berkeley & Berkeley, 1938
29b Longer dorsal cirri of parapodia with up to about 40 units 30
30a Each parapodium with a laterally directed, tonguelike projection above the acicula
Syllis (Typosyllis) sp.
30b Parapodia without a laterally directed, tonguelike projection above the acicula 31
31a Blades of all compound setae of an individual parapodium in the middle region of the body about the same length; median antennae originating in front of the anterior pair of eyes
Syllis (Typosyllis) variegata Grube, 1860
31b Blades of the upper compound setae of an individual parapodium in the middle region of the body about 2 or 3 times as long as those of setae at lower levels; median antenna originating near the posterior edge of the prostomium
Syllis (Typosyllis) stewarti Berkeley & Berkeley, 1938
32a Length of dorsal tentacular cirri of peristomium about 5 times the width of the body; nuchal organs present (these originate on the dorsal side of the prostomium and project backward) (with only about 15 or fewer setigers) *Amblyosyllis* sp.
32b Length of dorsal tentacular cirri of peristomium not more than 3 times the width of the body; nuchal organs usually absent; generally with more than 15 setigers 33
33a Everted pharynx with several conspicuous teeth (these usually point backward) 34
33b Pharynx with or without a large mid-dorsal tooth (the anterior edge may also have small serrations), but not with several conspicuous teeth 36
34a Setigers in the posterior region of the body (starting at about setiger 40) superficially divided into 2 rings *Odontosyllis fulgurans* subsp. *japonica* Imajima, 1966
34b Setigers in the posterior region of the body not superficially divided into 2 rings 35
35a Blades of compound setae of parapodia of the anterior region of the body with hairs along one edge; dorsal surface with a dark spot on every third or fourth intersegmental furrow
Odontosyllis phosphorea Moore, 1909
35b Blades of compound setae of parapodia of the anterior region of the body without hairs; dorsal surface without dark spots on the intersegmental furrows
Odontosyllis parva E. Berkeley, 1923
36a Pharynx with a conspicuous mid-dorsal tooth 37
36b Pharynx without a conspicuous mid-dorsal tooth 42
37a Pharynx with a smooth anterior edge 38
37b Pharynx with a serrated anterior edge 39
38a Compound setae of all parapodia similar *Pionosyllis gigantea* Moore, 1908
38b Compound setae of anterior parapodia of 2 rather different types (the upper ones have long blades that are broadest in the proximal half, whereas the lower ones have shorter blades that are broadest in the distal half) *Pionosyllis uraga* Imajima, 1966
39a Posterior part of the prostomium covered by a small flap 40
39b Posterior part of the prostomium not covered by a small flap 41
40a Compound setae of the parapodia of the middle and posterior regions of the body with short blades, all of the same shape; distal ring of the everted pharynx encircled by 2 rows of soft papillae *Eusyllis blomstrandi* Malmgren, 1867
40b Compound setae of the parapodia of the middle and posterior regions of the body with blades

of 2 types, some long, some short, and these of different shapes; distal ring of the everted pharynx encircled by 1 row of soft papillae *Eusyllis japonica* Imajima & Hartman, 1964
41a Dorsal cirrus of parapodia with a prominent cirrophore (a fleshy outgrowth from which the cirrus proper arises) *Eusyllis magnifica* (Moore, 1906)
41b Dorsal cirrus of parapodia without a cirrophore *Eusyllis assimilis* Marenzeller, 1875
42a Aciculae of a few parapodia in the anterior part of the body with knoblike distal ends; dorsal cirrus of parapodia nearly smooth or only slightly constricted into units
Streptosyllis latipalpa Banse, 1968 (fig. 8.72)
42b Aciculae of parapodia in the anterior part of the body without knoblike ends; dorsal cirrus of parapodia, beginning with setiger 3, constricted into beadlike units 43
43a Upper simple setae of the first few parapodia thick, their distal ends bent, differing from the more slender, tapering simple setae of other parapodia
Syllides longocirrata Ørsted, 1845 (fig. 8.21)
43b Upper simple setae of the first few parapodia similar to those of other parapodia
Syllides japonica Imajima, 1966

11. Family Nereidae

1a Without prostomial antennae; prostomial palps small, inconspicuous
Micronereis nanaimoensis (Berkeley & Berkeley, 1953)
1b With prostomial antennae; prostomial palps conspicuous 2
2a With 3 pairs of tentacular cirri on the peristomium; notopodia without any setae other than the aciculae *Lycastopsis* sp.
Previously identified as *Namanereis quadraticeps* (Blanchard, 1849), which probably does not occur in the northern hemisphere.
2b With 4 pairs of tentacular cirri on the peristomium; notopodia with setae in addition to the aciculae 3
3a Ventral cirrus of parapodia of anterior part of the body divided near the base into 2 branches
Ceratocephale loveni Malmgren, 1867
3b Ventral cirrus of all parapodia simple 4
4a Tentacular cirri of peristomium constricted into several units
Nicon moniloceras (Hartman, 1940)
4b Tentacular cirri of peristomium not constricted into several units 5
5a Segment 2 (following peristomium) enlarged, projecting anteriorly as a collarlike structure around the peristomium (commensal with large, subtidal hermit crabs)
Cheilonereis cyclurus (Harrington, 1897) (figs. 8.9 and 8.77)
5b Segment 2 not enlarged and not forming a collarlike structure around the peristomium 6
6a Most of the paragnaths on the everted pharynx conical 7
6b At least some of the paragnaths on the everted pharynx elongated transversely, crowded into comblike configurations, or otherwise deviating from a conical form 18
7a Notopodia of posterior portion of the body with homogomph falcigerous setae 8
7b Notopodia of posterior portion of the body without homogomph falcigerous setae 15
8a Distal ring of everted pharynx without paragnaths
Nereis (Eunereis) wailesi Berkeley & Berkeley, 1954
8b Distal ring of everted pharynx with paragnaths 9
9a Upper ligule of notopodia of posterior region of the body much larger than the lower ligule 10
9b Upper ligule of notopodia of the posterior region of the body not obviously longer than the lower ligule 12
10a Upper ligule of notopodia of the posterior region of the body strap-shaped
Nereis (Nereis) vexillosa Grube, 1851 (fig. 8.75)
10b Upper ligule of notopodia of the posterior region of the body not strap-shaped 11
11a Area V of everted pharynx without paragnaths; areas VII and VIII with irregularly arranged, large paragnaths, and the anterior part of area VIII also with many small paragnaths
Nereis (Nereis) grubei (Kinberg, 1866)
11b Area V of everted pharynx with 3-5 paragnaths; areas VII and VIII with 1 or 2 anterior rows of large, often rounded paragnaths and a broad posterior band of small paragnaths
Nereis (Nereis) neoneanthes Hartman, 1948 (fig. 8.76)

134 Phylum Annelida: Class Polychaeta

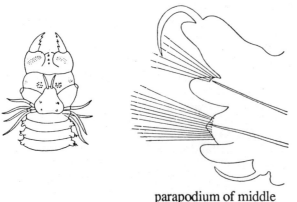
parapodium of middle region of atokous form

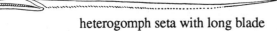
heterogomph seta with long blade

8.75 *Nereis vexillosa*

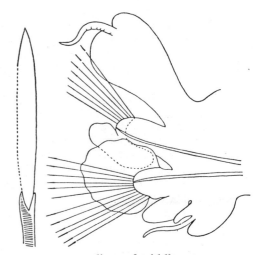
parapodium of middle region of epitokous form, and a homogomph seta

parapodium of middle region

heterogomph seta

homogomph seta

8.76 *Nereis neoneanthes*

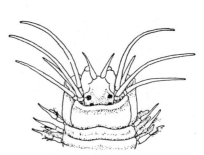
8.77 *Cheilonereis cyclurus*

Nereidae

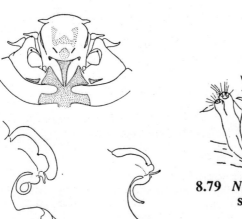
parapodium 25   posterior parapodium
8.78 *Nephtys caecoides*

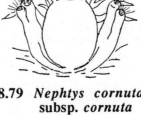
8.79 *Nephtys cornuta* subsp. *cornuta*

Nephtyidae

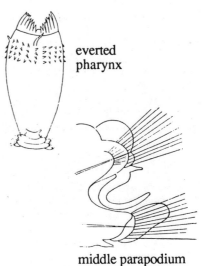
everted pharynx

middle parapodium
8.80 *Nephtys longosetosa*

12a Posteriormost dorsal tentacular cirri of peristomium much longer than the other cirri, extending at least to setiger 4   *Nereis (Nereis) procera* Ehlers, 1868
12b Posteriormost dorsal tentacular cirri of peristomium not obviously longer than the other cirri, and not extending beyond setiger 3   13
13a Paragnaths present on all areas of the everted pharynx (they are small and numerous, and form a continous band on the proximal ring)   *Nereis (Nereis) eakini* Hartman, 1936
13b Paragnaths absent on area V of the everted pharynx (areas VII and VIII have an anterior row of large paragnaths and a posterior row of small paragnaths   14
14a Upper ligule of notopodia of anterior region of the body rounded; area VI of everted pharynx with 4 or 5 paragnaths arranged in a diamond-shaped pattern; blades of notopodial homogomph falcigerous setae with a smooth margin   *Nereis (Nereis) pelagica* Linnaeus, 1761
14b Upper ligule of notopodia of anterior region of the body pointed; area VI of everted pharynx with an oval group of small paragnaths; blade of notopodial homogomph falcigerous setae serrated   *Nereis (Nereis) zonata* Malmgren, 1867
15a Proximal ring of everted pharynx usually without paragnaths, or with only a few on area VII   *Nereis (Ceratonereis) paucidentata* Moore, 1903
15b Proximal ring of everted pharynx usually with paragnaths in all areas, although they may be absent in area I or V   16
16a Dorsal ligule of parapodia not obviously larger than the ventral ligule (hermaphroditic and ovoviviparous; found in estuarine situations and in fresh water)   *Nereis (Neanthes) limnicola* Johnson, 1903
16b Dorsal ligule of parapodia leaflike and much larger than the ventral ligule   17
17a Areas VII and VIII of everted pharynx with paragnaths arranged in a band that consists of 4 or more rows   *Nereis (Neanthes) brandti* (Malmgren, 1866)
17b Areas VII and VIII of everted pharynx with paragnaths arranged in 2 or 3 irregular rows (paragnaths may be absent in area I)   *Nereis (Neanthes) virens* M. Sars, 1835
18a Area VI of everted pharynx with a large, transversely elongated paragnath   *Perinereis monterea* (Chamberlin, 1918)
18b Paragnaths on area VI of everted pharynx neither large nor transversely elongated, but compactly arranged in such a way that they form a nearly comblike configuration (common in sediment and among *Ulva* and other green algae in bays)   *Platynereis bicanaliculata* (Baird, 1863)

Superfamily Nephtyidacea

12. Family Nephtyidae

1a Interramal cirrus (a downgrowth from the notopodium) curving to the extent that it makes more than 1 complete turn, and thus appearing somewhat spiralled; everted pharynx with 14 rows of papillae   *Aglaophamus rubella* subsp. *anops* Hartman, 1950
1b Interramal cirrus not curving to the extent that it makes more than 1 complete turn, and sometimes scarcely curved at all; everted pharynx with 22 rows of papillae   2
2a Posterior prostomial antennae bifid; body not over 1.6 cm long   3
2b Posterior prostomial antennae not bifid; body generally more than 1.6 cm long   4
3a Setiger 3 with eyes; fenestrated setae present in all regions of the body   *Nephtys cornuta* subsp. *franciscana* Clark & Jones, 1955
3b Setiger 3 without eyes; fenestrated setae absent in the posterior region of the body   *Nephtys cornuta* subsp. *cornuta* Berkeley & Berkeley, 1945 (fig. 8.79)
4a Interramal cirri markedly curved, forming a C or a circle   5
4b Interramal cirri not markedly curved, and generally somewhat flattened   12
5a Interramal cirri appearing first on setiger 8 to 11   *Nephtys punctata* Hartman, 1938
5b Interramal cirri appearing first on setiger 3 to 6   6
6a Notopodium rounded, not cleft at the point where the aciculum emerges   *Nephtys longosetosa* Ørsted, 1843 (fig. 8.80)
6b Notopodium cleft at the point where the aciculum emerges   7

7a	Postsetal lobe of notopodia sufficiently prominent that it extends beyond the notopodium for a distance about equal to the length of the notopodium	8
7b	Postsetal lobe of notopodia not so prominent that it extends beyond the notopodium for a distance equal to the length of the notopodium	10
8a	Both the neuropodial and notopodial postsetal lobes large, the length of the former equal to half the width of the body; interramal cirri appearing first on setiger 4-6 *Nephtys caeca* (Fabricius, 1780)	
8b	Notopodial and neuropodial postsetal lobes relatively small, neither of them as long as one-fourth the width of the body; interramal cirri appearing first on setiger 6	9
9a	Neuropodial lobe of tenth and adjacent parapodia with a rounded projection above the point where the acicula emerges *Nephtys rickettsi* Hartman, 1938	
9b	Neuropodial lobe of tenth and adjacent parapodia without a rounded projection above the point where the acicula emerges *Nephtys assignis* Hartman, 1950	
10a	Interramal cirri appearing first on setiger 3 (with a dorsal pigment pattern consisting of segmental dark blotches, often diamond-shaped, that have narrow, laterally directed lobes) *Nephtys californiensis* Hartman, 1938	
10b	Interramal cirri appearing first on setiger 4 or 5	11
11a	Interramal cirri appearing first on setiger 4; with a dorsal pigment pattern consisting of segmental blotches, these with broad, laterally directed lobes; proximal portion of everted pharynx smooth *Nephtys caecoides* Hartman, 1938 (fig. 8.78)	
11b	Interramal cirri appearing first on setiger 4 or 5; without a distinctive dorsal pigment pattern; proximal portion of everted pharynx with small warts *Nephtys ciliata* (O. F. Müller, 1776)	
12a	Interramal cirri appearing first on setiger 5, about as broad as long, and with a small distal prolongation; without a distinctive pigment pattern *Nephtys brachycephala* Moore, 1903	
12b	Interramal cirri appearing first on setiger 3, slightly flattened, but generally elongated; with a distinctive pigment pattern (in life), consisting of several dark longitudinal lines on both sides of the midline of each segment *Nephtys ferruginea* Hartman, 1940	

Superfamily Aphroditacea

13. Family Aphroditidae

Pettibone, M. H. 1953. Some Scale-bearing Polychaetes of Puget Sound and Adjacent Waters. Seattle: University of Washington Press. iv + 89 pp.

1a	Elytra completely hidden by flexible notosetae that form a feltlike covering over the dorsal surface; protective notosetae (stiff) without barbs near the tips	2
1b	Elytra not hidden by flexible notosetae; protective notosetae with barbs near the tips *Laetmonice pellucida* Moore, 1903	
2a	Palps 8 to 11 times the length of the prostomium; tips of neurosetae hairy *Aphrodita longipalpa* Essenberg, 1917	
2b	Palps 4 to 7 times the length of the prostomium; tips of neurosetae hairy in some species, not in others	3
3a	Body with about 27 segments; ventralmost neurosetae with a little spur close to the tip; length not exceeding 2.5 cm *Aphrodita parva* Moore, 1905	
3b	Body with 38 to 43 segments; none of the neurosetae with a spur, although some of the more nearly ventral ones on the more posterior segments have several small teeth; length up to 22 cm	4
4a	Distal portion of protective notosetae with minute tubercles; median antenna club-shaped, considerably shorter than the prostomium *Aphrodita negligens* Moore, 1905	
4b	Distal portion of protective notosetae smooth; median antenna as long as, or longer, than the prostomium	5
5a	Ventralmost neurosetae coppery or gold; neurosetae sometimes with hairs or short spines *Aphrodita japonica* Marenzeller, 1879	

5b   Ventralmost neurosetae bright green; neurosetae without hairs or short spines
Aphrodita refulgida Moore, 1910

Doubtfully distinct from *A. japonica*.

14. Family Polynoidae

Pettibone, M. H. 1953. Some Scale-bearing Polychaetes of Puget Sound and Adjacent Waters. Seattle: University of Washington Press. iv + 89 pp.

1a   With 12 pairs of elytra   *Lepidonotus squamatus* (Linnaeus, 1767) (fig. 8.4)
1b   With at least 15 pairs (rarely 13 or 14 pairs) of elytra   2
2a   With 15 pairs (rarely 13 or 14 pairs) of elytra   3
2b   With at least 18 pairs of elytra   23
3a   With more than 50 segments; elytra restricted to the anterior and middle regions of the body, at least 18 of the posterior segments not covered by them   4
3b   With fewer than 50 segments; elytra covering nearly all segments   6
4a   Notosetae stouter than the neurosetae (especially common in clumps of the bivalve *Modiolus modiolus*, but also found in other habitats)   *Hermadion truncata* (Moore, 1902)
4b   Notosetae more slender than the neurosetae   5
5a   Each segment with a pair of mid-dorsal knobs; elytra colorless
Polynoe canadensis (McIntosh, 1874) (fig. 8.81)
5b   Segments without a pair of mid-dorsal knobs; elytra usually with a grayish brown patch
Polynoe gracilis (Verrill, 1874)
6a   At least some of the neurosetae bifid at the tip (the 2 teeth are not equal, however)   7
6b   None of the neurosetae bifid at the tip   12
7a   Notosetae thicker than the neurosetae   8
7b   Notosetae about the same thickness as the neurosetae   *Tenonia priops* (Hartman, 1961)
Tenonia kitsapensis Nichols, 1969 is a synonym.
8a   Anterior pair of eyes (on the dorsolateral surfaces of the prostomium) not visible from the dorsal side (they are covered by outgrowths of the prostomium) (free-living in various habitats, and also commensal with other polychaetes)
Harmothoe imbricata (Linnaeus, 1766) (figs. 8.11 and 8.82)
8b   Anterior pair of eyes visible from the dorsal side (not covered by outgrowths of the prostomium)   9
9a   Elytra nearly smooth (if microtubercles are present, they are only on the anterior portions of the elytra); elytra without globular papillae at the margins (commensal with many hosts, especially echinoderms and polychaetes)   *Harmothoe lunulata* (delle Chiaje, 1841) (fig. 8.83)
9b   Elytra with conical microtubules covering much of the surface; elytra with globular papillae at the margins   10
10a   Globular papillae at margins of elytra more slender proximally than distally (apparently free-living, but found in many habitats, including clumps of the bivalve *Modiolus modiolus*, barnacles, etc.)   *Harmothoe extenuata* (Grube, 1840)
10b   Globular papillae at margins of the elytra about as broad proximally as distally   11
11a   Elytra with some larger conical tubercles as well as microtubercles (the larger tubercles are intermediate in size between the microtubercles and the globular marginal papillae) (apparently free-living, but found in many habitats)   *Harmothoe multisetosa* (Moore, 1902)
11b   Conical tubercles on elytra all microtubercles (none intermediate in size between these and the globular marginal papillae)   *Harmothoe fragilis* (Moore, 1910)
12a   Some or all of the notosetae with capillary tips   13
12b   None of the notosetae with capillary tips   19
13a   Neurosetae of 2 types, those above the acicula with capillary tips   14
13b   Neurosetae all of 1 type, none of them with capillary tips   15
14a   Palps, antennae, and dorsal cirri smooth or nearly so, not markedly papillated (commensal with *Callianassa gigas* and *C. californiensis*)
Hesperonoe complanata (Johnson, 1901) (fig. 8.84)

138 Phylum Annelida: Class Polychaeta

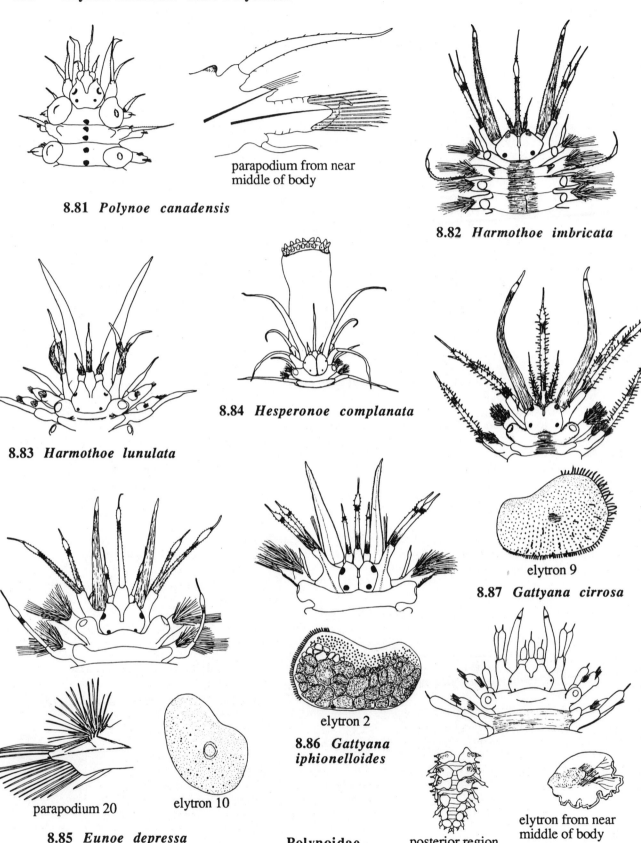

8.81 *Polynoe canadensis* — parapodium from near middle of body

8.82 *Harmothoe imbricata*

8.83 *Harmothoe lunulata*

8.84 *Hesperonoe complanata*

8.85 *Eunoe depressa* — parapodium 20, elytron 10

8.86 *Gattyana iphionelloides* — elytron 2

8.87 *Gattyana cirrosa* — elytron 9

8.88 *Arctonoe fragilis* — posterior region, elytron from near middle of body

Polynoidae

14b Palps, antennae, and dorsal cirri distinctly papillated (commensal with the echiuran *Urechis caupo*, which is not known to occur north of California)
Hesperonoe adventor (Skogsberg, 1928)
15a Notosetae thicker than the neurosetae   *Antinoella macrolepida* (Moore, 1905)
15b Neurosetae thicker than the notosetae   16
16a Lateral prostomial antennae inserted ventrally; elytra without distinct, more or less polygonal areas   17
16b Lateral prostomial antennae inserted close to the anterior edge of the prostomium; elytra with distinct, more or less polygonal areas   *Gattyana iphionelloides* (Johnson, 1901) (fig. 8.86)
17a All notosetae with fine capillary tips; elytra with a few large, conical tubercles, these with truncate, roughened summits   *Gattyana ciliata* Moore, 1902
17b Notosetae of 2 types, the upper ones stout and blunt, the lower ones slender and with capillary tips; elytra without large, conical tubercles   18
18a Microtubercles of elytra with as many as 4 prongs; bare tip of lower neurosetae not longer than the portion that has spines   *Gattyana cirrosa* (Pallas, 1766) (fig. 8.87)
18b Some microtubercles of elytra with 8 or more prongs; bare tip of lower neurosetae longer than the portion that has spines   *Gattyana treadwelli* Pettibone, 1949
19a Margins of elytra smooth, not fringed by slender papillae (elytra cream-colored; macrotubercles on elytra arising from circular, brown elevations; free-living, but also reported to be commensal with various anomuran decapod crustaceans)
Eunoe depressa Moore, 1905 (fig. 8.85)
19b Margins of elytra fringed by slender papillae   20
20a Larger tubercles on elytra unbranched, in a single row near the margin   21
20b Larger tubercles on elytra branched, not forming a distinct row   22
21a Larger tubercles on elytra nodular, roughened by small protuberances; elytrophores and tubercles on dorsal surface of segments with rounded lobes on their medial side
Eunoe nodosa (M. Sars, 1861)
21b Larger tubercles on elytra conical and smooth; elytrophores and tubercles on dorsal surface of segments without rounded lobes on their medial side
Eunoe uniseriata Banse & Hobson, 1968
22a Larger tubercles on elytra knoblike, with short branches; some notosetae with pointed tips, others with truncate tips   *Eunoe oerstedi* Malmgren, 1865
22b Larger tubercles on elytra antlerlike, with long branches; all notosetae with pointed tips
Eunoe senta (Moore, 1902)
23a With 18 pairs (rarely 19 pairs) of elytra (free-living or commensal, usually with other polychaetes)   *Halosydna brevisetosa* Kinberg, 1855
23b With more than 18 pairs of elytra   24
24a Elytra on every segment posterior to segment 38; dorsal nuchal fold prominent and covering part of the prostomium   *Hololepida magna* Moore, 1905
24b Elytra not on every segment posterior to segment 38; without a prominent nuchal fold   25
25a Lateral prostomial antennae inserted on the edge of the prostomium; notopodia abortive, with an acicula but without other setae   26
25b Lateral prostomial antennae inserted slightly ventral to the edge of the prostomium; notosetae present, though not necessarily on all notopodia, and sometimes few in number   27
26a Elytra of each pair large enough to touch or nearly touch medially; neurosetae of 3 rather distinct types: the upper ones slender and tapering to fine tips, the middle ones stouter and with bifid tips, and the lower ones short and with indistinctly bifid or simple tips
Lepidasthenia longicirrata E. Berkeley, 1923
26b Elytra of each pair so small that they are widely separated; neurosetae of 2 types: the upper ones slender and tapering to fine tips, the middle and lower ones stouter and with a row of spines that extend to the bifid tips (commensal with maldanid polychaetes)
Lepidasthenia berkeleyae Pettibone, 1948
27a Elytra on segments 2, 4, 5, then every other segment to 23, 26, 30, 32, 35; following segment 35, the arrangement of elytra is irregular, because they are not in sequence and not always in pairs; notosetae numerous, with pointed tips (commensal with the terebellid polychaetes *Thelepus crispus* and *Neoamphitrite robusta*, perhaps others)
Polyeunoa tuta (Grube, 1855)
27b Elytra on segments 2, 4, 5, then every other segment to 23, 26, 29, then every other segment

	to the end of the body; relatively few segments near the posterior end have unpaired elytra; notosetae few, with rather blunt, indistinctly bifid tips, or absent	28
28a	Margins of elytra extensively ruffled or folded (commensal with various sea stars) *Arctonoe fragilis* (Baird, 1863) (fig. 8.88)	
28b	Margins of elytra smooth or only slightly wavy	29
29a	Neurosetae (except on segment 2) of 2 types: those above the acicula slender and with blunt, bifid tips, those below the acicula stouter and with falcate, pointed tips; with a band of dark pigment extending across segments 7 and 8 (commensal with a wide variety of invertebrates, including sea stars, terebellid polychaetes, gastropods [especially *Diodora aspera*], and *Cryptochiton stelleri*) *Arctonoe vittata* (Grube, 1855)	
29b	Neurosetae essentially of 1 type: falcate and with pointed tips; without a dark band of pigment extending across segments 7 and 8 (commensal with a variety of invertebrates, including sea stars, the sea cucumber *Parastichopus californicus*, *Cryptochiton stelleri*, and terebellid polychaetes) *Arctonoe pulchra* (Johnson, 1897)	

15. Family Polyodontidae (Peisidicidae)

Peisidice aspera Johnson, 1897 is the only species known to occur in the region.

16. Family Sigalionidae

Pettibone, M. H. 1971. Descriptions of *Sthenelais fusca* Johnson, 1897 and *S. berkeleyi* n. sp. (Polychaeta: Sigalionidae) from the eastern Pacific. J. Fish. Res. Bd. Canada, 28:1393-1401.

1a	Each segment from 27 posteriorly bearing elytra; proximal portion of median prostomial antenna with a pair of lateral palplike outgrowths; ciliated cirriform gills arising from the lateral borders of the elytrophores and dorsal tubercles	2
1b	Each segment from 23 posteriorly bearing elytra; median prostomial antenna without lateral palplike outgrowths; no gills arising from the lateral borders of elytrophores or dorsal tubercles *Pholoe minuta* (Fabricius, 1780) (fig. 8.89)	
2a	Body width, exclusive of setae, up to 1 cm; ventral surface with short, closely spaced papillae between the bases of the parapodia; compound neurosetae usually falcate and with bifid tips (often among roots of *Zostera* and *Phyllospadix*, but also found in other habitats) *Sthenelais berkeleyi* Pettibone, 1971	
2b	Body width, exclusive of setae, up to 5 mm; ventral surface without papillae; compound neurosetae mostly straight and tapering to slender tips, only a few of them falcate and with bifid tips *Sthenelais tertiaglabra* Moore, 1910	

17. Family Chrysopetalidae

1a	Notosetae slender and not covering the dorsal surface *Dysponetus pygmaeus* Levinsen, 1879	
1b	Notosetae broad, flattened, and nearly covering the dorsal surface	2
2a	Only the setae arising from the lower portions of the notopodia broadened distally; dorsal surface without a golden luster; both the ventral and the dorsal cirrus of the parapodia constricted into beadlike units *Paleanotus bellis* (Johnson, 1897)	
2b	Setae arising from the lower as well as the upper portions of the notopodia broadened distally; dorsal surface with a golden luster; neither the ventral nor the dorsal cirrus of the parapodia constricted into beadlike units *Chrysopetalum occidentale* (Johnson, 1897) (fig. 8.13)	

Order Amphinomida

18. Family Amphinomidae

1a Tips of serrations on the longer lobe of the bifid notosetae of the posterior region of the body directed distally (toward the tip of the longer lobe)   *Chloeia pinnata* Moore, 1911
1b Tips of serrations on the longer lobe of the bifid notosetae of the posterior region of the body directed proximally (toward the base of the longer lobe)   *Chloeia entypa* Chamberlin, 1919

19. Family Euphrosinidae

1a Dorsal surface completely hidden by the notosetae; all notosetae deeply cleft at the tip; gills in middle region of body usually bifid, but sometimes trifid or undivided
    Euphrosine bicirrata Moore, 1905
1b Dorsal surface not quite hidden by the notosetae (a narrow mid-dorsal strip is exposed); notosetae either deeply cleft or with 1 long lobe and 1 short, spurlike lobe; gills in middle region of body (5 or more on each notopodium) usually with a least 3 branches   2
2a Notopodia of middle region of body with 5 gills   *Euphrosine arctia* Johnson, 1897
2b Notopodia of middle region of body with at least 10 gills   3
3a Notopodia of middle region of body with 10 gills; some deeply cleft notosetae smooth, others with serrations on the inner edge of 1 lobe   *Euphrosine heterobranchia* Johnson, 1901
3b Notopodia of middle region of body with 11 to 13 gills; all deeply cleft notosetae with serrations on 1 lobe   *Euphrosine hortensis* Moore, 1905

Order Eunicida

20. Family Onuphidae

1a With tentacular cirri on the peristomium   2
1b Without tentacular cirri on the peristomium   *Epidiopatra hupferiana* subsp. *monroi* Day, 1960
2a Gills with many filaments arranged in a spiral around the main axis
    Diopatra ornata Moore, 1911 (fig. 8.90)
2b Gills, if present, either unbranched filaments or branched in a comblike pattern   3
3a Gills unbranched filaments   4
3b Gills branched in a comblike pattern   *Sarsonuphis parva* (Moore, 1911)
4a Hooded hooks of more anterior parapodia bidentate at the tip (unless worn, in which case 1 of the teeth may be absent; the hood may also be worn away); tube flattened, with gravel and bits of shell embedded in it   *Nothria conchylega* (M. Sars, 1835) (fig. 8.29)
4b Most of the hooded hooks in the more anterior parapodia tridentate; tube cylindrical or absent   5
5a Gills appearing first on setiger 1   6
5b Gills appearing first on setiger 3 or more posteriorly   7
6a Gills slender on all segments that have them; hooded hooks present ventral to aciculae beginning with setiger 11-14   *Onuphis iridescens* (Johnson, 1901)
6b Gills decidedly thicker from setiger 10 to about 60 or 80 than on more anterior and more posterior segments; hooded hooks present ventral to the aciculae beginning with setiger 9 (sometimes 10 or 11)   *Onuphis elegans* (Johnson, 1901)
7a Gills appearing first on setiger 3-6, but ending about 40 segments anterior to the pygidium
    Onuphis geophiliformis (Moore, 1903)
7b Gills appearing first on setiger 19-28, and continuing nearly to the pygidium
    Mooreonuphis stigmatis (Treadwell, 1922)

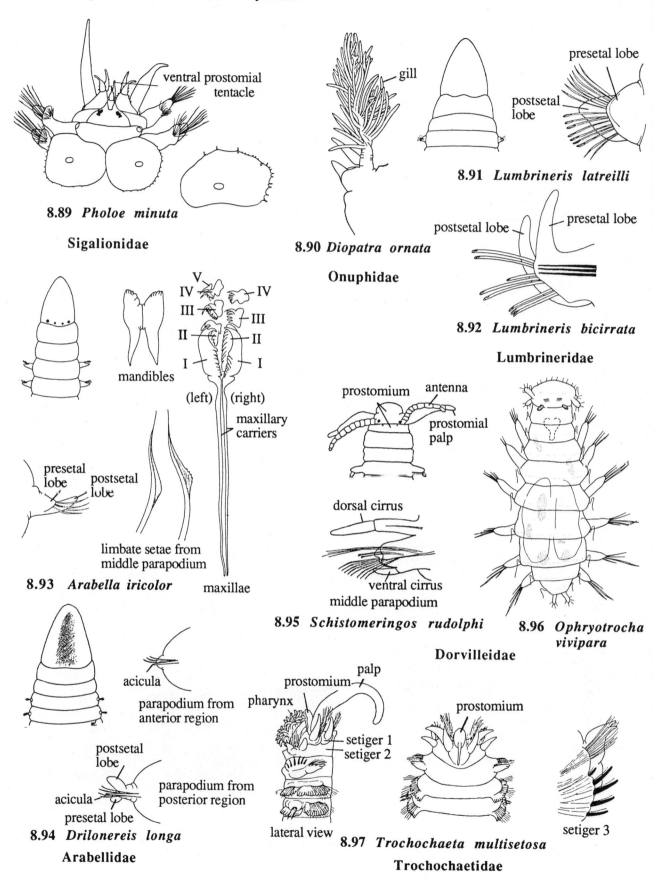

21. Family Eunicidae

Gustus, R. 1972. A species of the genus *Eunice* (Polychaeta) from the Pacific Northwest coast. Northwest Sci., 46:257-69.

1a	With a pair of tentacular cirri on the peristomium	2
1b	Without tentacular cirri on the peristomium	*Marphysa stylobranchiata* Moore, 1909
2a	Gills appearing first on setiger 3, with not more than 16 filaments; hooded hooks ventral to the aciculae yellow	*Eunice valens* (Chamberlin, 1919)
2b	Gills appearing first on setiger 5-7, with as many as 40 filaments; hooded hooks ventral to the aciculae black	*Eunice aphroditois* (Pallas, 1788)

22. Family Lumbrineridae

1a	About 20 to 40 of the more anterior setigers bearing 2 to 5 short, simple gills ventral to the postsetal lobe Similar to *N. gemmea* Moore, 1911.	*Ninoe* sp.
1b	None of the parapodia bearing gills (but each, as a rule, has a presetal and a postsetal lobe)	2
2a	All hooded hooks simple	3
2b	Compound hooded hooks present on from 7 to 30 of the more anterior setigers	7
3a	Aciculae yellow; hooded hooks appearing first on setiger 1	4
3b	Aciculae black; hooded hooks appearing first on setiger 4 or more posteriorly	6
4a	Neither the presetal nor postsetal lobe of the more posterior parapodia elongate; body often with transverse brown bands	*Lumbrineris zonata* (Johnson, 1901)
4b	The postsetal lobe (and sometimes also the presetal lobe) several times as long as wide; body without transverse brown bands	5
5a	Both the presetal and postsetal lobes of the more posterior parapodia elongated	*Lumbrineris lagunae* Fauchald, 1970
5b	Only the postsetal lobe of the more posterior parapodia elongated	*Lumbrineris luti* Berkeley & Berkeley, 1945
6a	Both the presetal and postsetal lobes of the more posterior parapodia elongated; hooded hooks appearing first on setiger 4 to 25	*Lumbrineris bicirrata* Treadwell, 1929 (fig. 8.92)
6b	Neither the presetal nor the postsetal lobe of the more posterior parapodia elongated; hooded hooks appearing first on setiger 7 to 11	*Lumbrineris similabris* Treadwell, 1926
7a	Prostomium rounded, almost globular; aciculae amber to brown (dorsal surface of first 2 segments behind the prostomium sometimes with a brown band)	*Lumbrineris inflata* Moore, 1911
7b	Prostomium conical; aciculae yellow or black	8
8a	Neither the presetal nor postsetal lobe of the more posterior parapodia elongated	9
8b	The postsetal lobe, and sometimes also the presetal lobe, of the more posterior parapodia elongated	11
9a	Aciculae yellow	*Lumbrineris latreilli* Audouin & Milne-Edwards, 1834 (fig. 8.91)
9b	Aciculae black	10
10a	Maxilla III with a single tooth	*Lumbrineris pallida* Hartman, 1944
10b	Maxilla III with 2 teeth	*Lumbrineris japonica* (Marenzeller, 1879)
11a	Only the postsetal lobe of the more posterior parapodia elongated; aciculae yellow	*Lumbrineris limicola* Hartman, 1944
11b	Both the presetal and postsetal lobes of the more posterior parapodia elongated; aciculae yellow or black	12
12a	Aciculae yellow	*Lumbrineris cruzensis* Hartman, 1944
12b	Aciculae black	*Lumbrineris californiensis* Hartman, 1944

23. Family Arabellidae

1a	Parapodia without a stout seta that resembles an acicula (only slender setae present)	2

1b Parapodia with a stout seta that resembles an acicula, as well as with slender setae 3
2a Postsetal lobes of parapodia of the posterior part of the body directed laterally, and not longer than those of parapodia of the anterior part of the body
Arabella iricolor (Montagu, 1804) (fig. 8.93)
2b Postsetal lobes of parapodia of the posterior part of the body directed slightly upward, and longer than those of the parapodia of the anterior part of the body
Arabella semimaculata (Moore, 1911)
3a Prostomium conspicuously flattened dorsoventrally; maxilla I with a sickle-shaped distal portion, and either without teeth or with teeth only on its proximal portion 4
3b Prostomium not obviously flattened; maxilla I nearly triangular, with teeth along its medial edge *Notocirrus californiensis* Hartman, 1968
4a Proximal portion of maxilla I without teeth; mandibles absent *Drilonereis nuda* Moore, 1909
4b Proximal portion of maxilla I with teeth; mandibles present or absent 5
5a Maxilla I with 2 or 3 teeth; maxilla II with 3 teeth (the most nearly proximal one is small); maxilla IV absent; mandibles present *Drilonereis falcata* subsp. *minor* Hartman, 1965
5b Maxilla 1 with 3-5 teeth; maxilla II with 6-8 teeth; maxilla IV present; mandibles present or absent *Drilonereis longa* Webster, 1879 (fig. 8.94)

24. Family Dorvilleidae

1a Palps reduced to small papillae located ventral to the antennae; adults with only 6 setigers
Ophryotrocha vivipara Banse, 1963 (fig. 8.96)
1b Palps well developed and usually rather long; with more than 6 setigers 2
2a Parapodia with 2 aciculae, the upper one supporting a slender notopodial lobe, which bears a dorsal cirrus; forked setae present or absent 3
2b Parapodia with only 1 acicula and without a distinct notopodial lobe, although there is a dorsal cirrus; forked setae present *Protodorvillea gracilis* (Hartman, 1938)
3a Forked setae absent 4
3b Forked setae present (the lobes may be unequal), and with serrations along one edge 5
4a Antennae about half the length of the palps; palps without a constriction near the distal end
Schistomeringos monilocerus (Moore, 1909)
4b Antennae longer than the palps; palps with a constriction near the distal end
Schistomeringos pseudorubrovittata (E. Berkeley, 1927)
5a Dorsal cirri present throughout the length of the body and constricted into 2 units 6
5b Dorsal cirri limited to the anterior third of the body and not constricted into 2 units
Pettiboneia pugettensis (Armstrong & Jumars, 1978)
6a Some compound setae with a long, needlelike blade that is more than 10 times as long as it is wide near the base *Schistomeringos annulata* (Moore, 1909)
6b None of the compound setae with a needlelike blade 7
7a Lobes of forked setae slender and very unequal
Schistomeringos caeca (Webster & Benedict, 1884) (fig. 8.31)
7b Lobes of forked setae stout and nearly equal 8
8a Distinctly forked setae beginning on setiger 1 *Schistomeringos japonica* (Annenkova, 1937)
8b Distinctly forked setae beginning on setiger 2-15 (usually 2-5) (setae that have 2 small, indistinct teeth at the tip may be found, however, on setiger 1-16 [usually 1-5])
Schistomeringos rudolphi (delle Chiaje, 1828) (fig. 8.95)

Order Orbiniida

25. Family Orbiniidae

1a Segment 2 (the segment succeeding the peristomium) without parapodia; body length generally less than 2 cm 2
1b Segment 2 with parapodia; body length generally exceeding 2 cm 3

2a Dorsal gills absent *Orbiniella nuda* Hobson, 1974
2b Dorsal gills present on a few setigers, beginning with 4 or 5
    Protoariciella oligobranchia Hobson, 1976
3a Prostomium rounded or truncate, not distinctly pointed 4
3b Prostomium pointed 6
4a Prostomium rounded; postsetal lobe of neuropodia in anterior portion of body with a single papilla *Naineris quadricuspida* (Fabricius, 1780)
4b Prostomium broadly truncate; postsetal lobe of neuropodia in anterior region of body either without papillae or with several papillae 5
5a Postsetal lobe of neuropodia in anterior region of body a ridge, lacking papillae; dorsal gill beginning on setiger 7 to 15 *Naineris dendritica* (Kinberg, 1867) (fig. 8.6)
5b Postsetal lobe of neuropodia in anterior region of body with several papillae; dorsal gills beginning on setiger 5 or 6 *Naineris uncinata* Hartman, 1957
6a Neuropodia in anterior region of body with stout acicular setae, these being bent near the tip (the stout setae may be short and thus not always obvious; in young specimens, moreover, they may be absent) 7
6b Neuropodia of anterior region of body without any especially stout setae 8
7a Some posterior thoracic segments and anterior abdominal segments with a small lobe beneath the neuropodium *Scoloplos armiger* (O. F. Müller, 1776) (fig. 8.32)
7b None of the segments with a lobe beneath the neuropodium
    Scoloplos acmeceps Chamberlin, 1919
8a Posterior thoracic segments with 1 or 2 prominent conical papillae (subpodial lobes) ventral and medial to the neuropodium *Leitoscoloplos panamensis* (Monro, 1933)
8b Posterior thoracic segments without papillae ventral and medial to the neuropodium
    Leitoscoloplos pugettensis (Pettibone, 1957)

Order Spionida

26. Family Apistobranchidae

1a Postsetal lobe of neuropodium of setiger 4 serrated; notopodia sometimes absent in some or all of setigers 7-11 *Apistobranchus tullbergi* (Théel, 1879) (fig. 8.33)
1b Postsetal lobe of neuropodium of setiger 4 not serrated; notopodia present on setigers 7-11
    Apistobranchus ornatus Hartman, 1965

27. Family Trochochaetidae

Pettibone, M. H. 1976. Contribution to the polychaete family Trochochaetidae Pettibone. Smithsonian Contrib. Zool., no. 230. 21 pp.

Trochochaeta multisetosa (Ørsted, 1844) (fig. 8.97) is the only species known to occur in the region.

28. Family Spionidae

Blake, J. A. 1979. Revision of some polydorids (Polychaeta: Spionidae) described and recorded from British Columbia by Edith and Cyril Berkeley. Proc. Biol. Soc. Washington, 92:606-17.
Blake, J. A., & K. H. Woodwick. 1971. A review of the genus *Boccardia* Carazzi (Polychaeta: Spionidae) with descriptions of two new species. Bull. So. Calif. Acad. Sci., 70:31-42.
Light, W. J. 1977. Spionidae (Annelida: Polychaeta) from San Francisco Bay, California: a revised list with nomenclatorial changes, new records, and comments on related species from the northeastern Pacific Ocean. Proc. Biol. Soc. Washington, 90:66-88.
------. 1981. Spionidae. Polychaeta. Annelida. (Invertebrates of the San Francisco Bay Estuary System.) Pacific Grove, California: Boxwood Press. 211 pp.

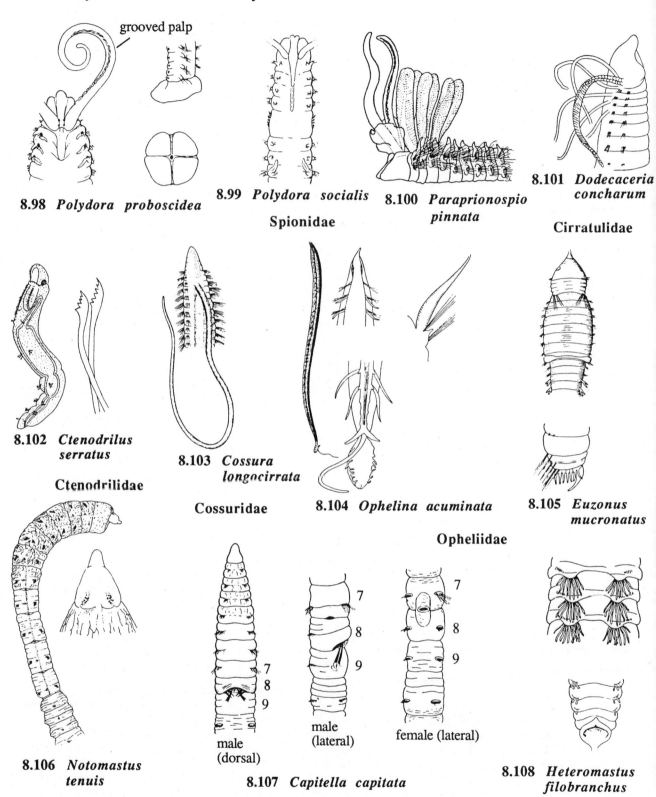

8.98 *Polydora proboscidea*
8.99 *Polydora socialis*
8.100 *Paraprionospio pinnata*
8.101 *Dodecaceria concharum*

Spionidae

Cirratulidae

8.102 *Ctenodrilus serratus*
Ctenodrilidae

8.103 *Cossura longocirrata*
Cossuridae

8.104 *Ophelina acuminata*
8.105 *Euzonus mucronatus*
Opheliidae

8.106 *Notomastus tenuis*
8.107 *Capitella capitata*
8.108 *Heteromastus filobranchus*

Capitellidae

1a Setiger 5 usually distinctly different from other setigers of the anterior region of the body (except in *Polydora kempi* subsp. *japonica*), and with stout, hooklike setae 2
1b Setiger 5 not appreciably different from other setigers of the anterior region of the body, and without stout, hooklike setae 20
2a Gills beginning on setiger 2 (but may be absent on setiger 5, and sometimes also on 4) 3
2b Gills beginning after setiger 5 7
3a All hooks of setiger 5 smooth, falcate; posterior neuropodia with thick, recurved spines
    Polydora (Boccardia) hamata Webster, 1879
3b Hooks of setiger 5 of 2 types: smooth and falcate, or clublike and hairy-tipped; posterior notopodia without thick spines 4
4a Prostomium with an anterior cleft 5
4b Prostomium without an anterior cleft 6
5a Notosetae present on setiger 1 *Polydora (Boccardia) pugettensis* Blake, 1979
5b Notosetae absent on setiger 1 *Polydora (Boccardia) polybranchia* Haswell, 1885
6a Notosetae of setiger 1 long, extending beyond the anterior end of the prostomium
    Polydora (Boccardia) columbiana E. Berkeley, 1927
6b Notosetae of setiger 1 short and scarcely noticeable
    Polydora (Boccardia) proboscidea Hartman, 1940 (fig. 8.98)
7a Setiger 5 not especially different from other anterior segments, although it has stout hooks arranged in a ʊ-shaped configuration
    Polydora (Pseudopolydora) kempi subsp. *japonica* Imajima & Hartman, 1964
7b Setiger 5 distinctly different from other anterior setigers, and its stout hooks arranged in a straight or slightly curved line 8
8a Hooks of setiger 5 with hairy tips 9
8b Hooks of setiger 5 with smooth tips 10
9a Hooks of setiger 5 falcate, the hairs just below the acute tip
    Polydora (Polydora) brachycephala Hartman, 1936
9b Hooks of setiger 5 not falcate, but with a bifid tip, the hairs between the 2 lobes
    Polydora (Polydora) quadrilobata Jacobi, 1883
10a Hooks of setiger 5 with a tooth, collar, or flange just below the tip 11
10b Hooks of setiger 5 without a tooth, collar, or flange just below the tip 19
11a Gills beginning on setiger 6; hooded hooks beginning on setiger 10 or posterior to this (commensal with hermit crabs) *Polydora (Polydora) commensalis* Andrews, 1891
11b Gills beginning on setiger 7 or posterior to this; hooded hooks beginning on setiger 7 12
12a Notosetae present on setiger 1 13
12b Notosetae absent on setiger 1 14
13a Hooks of setiger 5 with a collar near the tip; notopodia of posterior segments with a conical bundle of stout spinelike setae *Polydora (Polydora) armata* Langerhans, 1880
13b Hooks of setiger 5 with a tooth near the tip; notopodia of posterior segments with only capillary setae *Polydora (Polydora) giardi* Mesnil, 1896
14a Prostomium with an antenna *Polydora (Polydora) ligni* Webster, 1879
14b Prostomium without an antenna 15
15a Hooks of setiger 5 with a tooth near the tip 16
15b Hooks of setiger 5 without a tooth, but with a collar or flange near the tip 17
16a Palps with 4 or 5 transverse black bars; anterior end of prostomium cleft
    Polydora (Polydora) limicola Annenkova, 1934
16b Palps without black bars; anterior end of prostomium rounded
    Polydora (Polydora) pygidialis Blake & Woodwick, 1972
17a Anterior end of prostomium cleft *Polydora (Polydora) websteri* Hartman, 1943
17b Anterior end of prostomium rounded 18
18a Hooks of setiger 5 with a concavity near the tip, this concavity bordered by a flange
    Polydora (Polydora) alloporis Light, 1970
18b Hooks of setiger 5 with a collar just below the tip
    Polydora (Polydora) spongicola Berkeley & Berkeley, 1950
19a Sense organ on midline of dorsal surface of anterior segments reaching back to setiger 5 or 6; setiger 5 with 12 or more hooks *Polydora (Polydora) cardalia* E. Berkeley, 1927

19b Sense organ on midline of dorsal surface of anterior segments usually reaching back at least to setiger 7, sometimes as far back as setiger 12 (but in some specimens not reaching beyond setiger 4, 5, or 6); setiger 5 with 3-7 hooks (these protrude prominently)
  Polydora (Polydora) socialis (Schmarda, 1861) (fig. 8.99)
20a Gills absent; neuropodium of setiger 1 with a stout, curved seta 21
20b Gills present; neuropodium of setiger 1 without a stout, curved seta 23
21a Prostomium without an antenna; paired anterolateral projections of prostomium considerably longer than wide; pygidium with 2 cirri *Spiophanes bombyx* (Claparède, 1870)
21b Prostomium with an antenna (this originates close to the posterior edge of the prostomium); paired anterolateral projections of prostomium about as long as wide, bluntly rounded; pygidium with at least 6 cirri 22
22a Pygidium with 6 cirri *Spiophanes kroyeri* Grube, 1860
22b Pygidium with 8-12 cirri *Spiophanes berkeleyorum* Pettibone, 1962
23a With a single pair of gills, these on setiger 1 *Streblospio benedicti* Webster, 1879
23b With more than 1 pair of gills 24
24a Gills beginning on setiger 11-13 *Pygospio elegans* Claparède, 1863
24b Gills beginning on setiger 1 or 2 25
25a Gills present only in the anterior half of the body 26
25b Gills present over almost the entire length of the body 31
26a Hooded hooks either 2-toothed or 3-toothed at the tip 27
26b Hooded hooks with more than 3 teeth at the tip 28
27a Genital pouches first appearing on setiger 2-7 *Laonice pugettensis* Banse & Hobson, 1968
27b Genital pouches first appearing on setiger 12 or more posteriorly
  Laonice ?cirrata (M. Sars, 1851)
28a Gills (3 pairs) pinnate, beginning on setiger 1 (caution: the gills are likely to break off, but the scars may be visible; it is advisable to look at as many specimens as possible); prostomium partly enveloped by a hooklike structure
  Paraprionospio pinnata (Ehlers, 1901) (fig. 8.100)
28b Gills (4 to many pairs) simple or pinnate, beginning on setiger 2 (caution: the gills are likely to break off); prostomium not enveloped by a hoodlike structure 29
29a With 4 pairs of gills (pairs 1 and 4 pinnate, pairs 2 and 3 simple)
  Prionospio steenstrupi Malmgren, 1867
29b With at least 7 pairs of gills, all simple 30
30a With 7-10 pairs of gills; eyes reddish *Prionospio (Minuspio) multibranchiata* E. Berkeley, 1927
30b With 10-12 pairs of gills; eyes black *Prionospio (Minuspio)* sp.
31a Prostomium with a pair of anterolateral projections; notopodia in middle and posterior portions of the body without hooded hooks 32
31b Prostomium without a pair of anterolateral projections; notopodia in middle and posterior portions of the body with or without hooded hooks 33
32a Gills beginning on setiger 1; neuropodial uncini beginning on setiger 30-45, the tips of these setae with 2 teeth *Malacoceros (Malacoceros) fuliginosus* (Claparède, 1870)
32b Gills beginning on setiger 2; neuropodial uncini beginning on setiger 13-21 (usually 18 or 19), the tips of these setae with 3 teeth *Malacoceros (Rhynchospio) glutaeus* (Ehlers, 1897)
33a Prostomium usually pointed (it may be blunt in *Scololepis foliosa*); gills beginning on setiger 2; notopodia of middle and posterior portions of the body usually with hooded hooks; pygidium with an oval disk, but without cirri or lobes 34
33b Prostomium blunt and rounded or slightly bilobed; gills beginning on setiger 1-2; notopodia of middle and posterior portions of the body without hooded hooks; pygidium with cirri or 4 lobes 35
34a Hooded hooks blunt or with a single tooth at the tip; gills completely fused with the postsetal lobes of the notopodia in the anterior setigers, but only partly fused with them in the posterior setigers *Scolelepis foliosa* (Audouin & Milne-Edwards, 1833)
34b Hooded hooks mostly with 2 or 3 teeth at the tip (but sometimes with a single tooth); all gills partly fused with the postsetal lobes of the notopodia
  Scolelepis squamata (O. F. Müller, 1789)
35a Tips of hooded hooks with 3 teeth (but one of these is very small); gills on setiger 1 much smaller than those on succeeding setigers *Spio (Spio) cirrifera* (Banse & Hobson, 1968)

35b Tips of hooded hooks with 2 teeth (look carefully!); gills on setiger 1 at least half as large as those on succeeding setigers 36
36a Hooded hooks beginning on setiger 10-15; teeth at tips of hooded hooks decidedly unequal
Spio (*Spio*) sp.
Apparently undescribed; similar to *S. filicornis* (O. F. Müller, 1776) (fig. 8.34).
36b Hooded hooks beginning after setiger 19 (usually 22-24); teeth at tips of hooded hooks nearly equal *Spio* (*Spio*) *butleri* Berkeley & Berkeley, 1954

Order Chaetopterida

29. Family Chaetopteridae

1a With a small cirrus medial to the base of each palp 2
1b Without a cirrus medial to the base of each palp 3
2a Middle region of the body consisting of 2 segments; tubes up to 4 or 5 mm in diameter
Phyllochaetopterus claparedii McIntosh, 1885 (fig. 8.36)
2b Middle region of body consisting of 4-13 segments; tubes not more than 2 mm in diameter
Phyllochaetopterus prolifica Potts, 1914
3a Palps shorter than the anterior region of the body; some notopodia of the middle region fused mid-dorsally to form broad paddles (rare north of southern California)
Chaetopterus ?*variopedatus* (Renier, 1804)
C. variopedatus has been considered to be a nearly cosmopolitan species, but it is probably a complex of closely related species.
3b Palps longer than the anterior region of the body; none of the notopodia fused mid-dorsally 4
4a Notopodia of middle region of body with 2 or 3 unequal lobes; tube translucent, ringed, up to about 3 mm in diameter; setiger 4 with only 1 or 2 stout setae
Spiochaetopterus costarum (Claparède, 1870)
4b Notopodia of middle region of body with a single lobe; tube neither translucent (usually covered with sand) nor ringed, up to nearly 1 cm in diameter except at the opening, which is narrower; setiger 4 with several stout setae *Mesochaetopterus taylori* Potts, 1914

Order Magelonida

30. Family Magelonidae

Jones, M. L. 1971. *Magelona berkeleyi* n. sp. from Puget Sound (Annelida: Polychaeta), with a further redescription of *Magelona longicornis* Johnson and a consideration of recently described species of *Magelona*. J. Fish. Res. Bd. Canada, 28:1445-54.
------. 1978. Three new species of *Magelona* (Annelida, Polychaeta) and a redescription of *Magelona pitelkai* Hartman. Proc. Biol. Soc. Washington, 91:336-63.

1a Prostomium with a pair of nearly triangular lateral lobes close to its anterior end; setae of parapodium 9 similar to those of parapodia 1-8 2
1b Prostomium with or without distinct lateral lobes close to the anterior end; setae of parapodium 9 decidedly different from those of parapodia 1-8 3
2a Hooded hooks of notopodia and neuropodia of posterior part of the body with 3 teeth, and arranged in 2 series, those of 1 series oriented in such a way that they face those of the other series *Magelona berkeleyi* Jones, 1971
2b Hooded hooks of notopodia and neuropodia of posterior part of the body with 2 teeth, and arranged in 1 continuous series *Magelona longicornis* Johnson, 1901
3a Neuropodium of parapodium 9 without a small lobe medial to the seta-bearing portion; prostomium rounded at the tip *Magelona sacculata* Hartman, 1961
3b Neuropodium of parapodium 9 with a small lobe medial to the seta-bearing portion; prostomium almost truncate at the tip *Magelona hobsonae* Jones, 1978

Order Cirratulida

31. Family Paraonidae

Hobson, K. D. 1972. Two new species and two new records of the family Paraonidae (Annelida, Polychaeta) from the northeastern Pacific Ocean. Proc., Biol. Soc. Washington, 85:549-55.

1a	Notopodia in middle and posterior portions of the body with setae that have forked tips or that have a slender branch arising near the tip	2
1b	Notopodia in middle and posterior portions of the body without setae that have forked tips or that have a slender branch arising near the tip	3
2a	Notopodial setae, beginning with about setiger 5, with forked tips, the form of these gradually changing (by setiger 12 to 15) to a type in which a slender branch arises near the tip; gills beginning on setiger 5	*Cirrophorus branchiatus* Ehlers, 1908
2b	Notopodial setae, beginning with setiger 4 to 8, forked, not changing into a type in which a slender branch arises near the tip; gills beginning on setiger 4	*Cirrophorus lyra* (Southern, 1914)
3a	Prostomium with an antenna (this is branched in some species)	4
3b	Prostomium without an antenna	10
4a	Parapodia of anterior region of the body with a bifid postsetal lobe on the notopodium and an undivided postsetal lobe on the neuropodium	*Aricidea quadrilobata* Webster & Benedict, 1887 (fig. 8.35)
4b	Parapodia of anterior region of the body with an undivided postsetal lobe on the notopodium and without a postsetal lobe on the neuropodium	5
5a	Antenna branched at the tip	*Aricidea ramosa* Annenkova, 1934
5b	Antenna not branched	6
6a	Antenna constricted into 2 or more units	7
6b	Antenna not constricted into units	8
7a	Antenna constricted into 2 or 3 units; gills (9 to 14 pairs) broad, leaflike	*Aricidea minuta* Southward, 1956
7b	Antenna constricted into 3 to 6 units; gills (7 to 18 pairs) either slender or leaflike	*Aricidea wassi* Pettibone, 1965
8a	Antenna club shaped	*Aricidea neosuecica* (Hartman, 1965)
8b	Antenna tapering	9
9a	Hooklike neurosetae with hairs near the tip of the thickened portion	*Aricidea assimilis* Tebble, 1959
9b	Hooklike neurosetae without hairs near the tip of the thickened portion	*Aricidea lopezi* Berkeley & Berkeley, 1956
10a	Neuropodia in the region behind the gill-bearing segments with hooded hooks	*Tauberia gracilis* (Tauber, 1879)
10b	Neuropodia in the region behind the gill-bearing segments without hooded hooks	11
11b	Notosetae of posterior setigerous segments spinelike; prostomium almost triangular, lacking eyespots	*Paraonella spinifera* (Hobson, 1972)
11b	Notosetae of posterior setigerous segments not spinelike; prostomium slender and pointed, with 2 black eyespots (these are very small)	*Paraonella platybranchia* (Hartman, 1961)

32. Family Questidae

Questa caudicirra Hartman, 1966 is the only species recorded from the region.

33. Family Cirratulidae

1a	With a pair of long, grooved palps arising from the dorsal side of setiger 1	2
1b	With 2 groups of grooved palps arising from the dorsal side of setiger 1 or one of the next few setigers	12

Phylum Annelida: Class Polychaeta   151

2a With gills first appearing on setiger 1 (the segment that also bears a pair of grooved palps) and also on the next 2-10 setigers   3
2b With gills on many setigers, not just the first 10   4
3a With gills on the first 2-3 setigers; boring in calcareous substrata
       Dodecaceria concharum Ørsted, 1843 (fig. 8.101)
3b With gills on the first 5-10 setigers; secreting calcareous tubes
       Dodecaceria fewkesi Berkeley & Berkeley, 1954
4a All setae either of the limbate or capillary type   5
4b Setae largely of the capillary type, but stout setae present in some setigers, especially near the posterior end   8
5a Prostomium with eyes; neurosetae of posterior setigers with a serrated blade   6
5b Prostomium without eyes; neurosetae of posterior setigers with a smooth blade   7
6a Segments in middle region of body separated by deep constrictions, so that they resemble beads; posteriormost portion of body swollen; serrated neurosetae beginning on setiger 30-40   *Tharyx secundus* Banse & Hobson, 1968
6b Segments in middle region of body not separated by constrictions; posteriormost portion of body not swollen; serrated neurosetae beginning on about setiger 100
       Tharyx serratisetis Banse & Hobson, 1968
7a Length up to about 1.5 cm; mostly intertidal in estuarine situations (luminescent)
       Tharyx parvus E. Berkeley, 1929
7b Length up to 6 cm; subtidal   *Tharyx multifilis* Moore, 1909
8a Stout setae bifid or serrated at the tip   9
8b Stout setae neither bifid nor serrated at the tip   10
9a Stout setae present in neuropodia beginning with setiger 1, and in notopodia from about setiger 20 (luminescent)   *Caulleriella alata* (Southern, 1914)
9b Stout setae present in neuropodia beginning with about setiger 15 or 20, and in notopodia of posterior setigers only   *Caulleriella hamata* (Hartman, 1948)
10a Capillary setae all of the same type   11
10b Capillary setae, except in the case of the first 10-15 setigers, of 2 decidedly different lengths, the longer ones threadlike and often slightly twisted   *Chaetozone spinosa* Moore, 1903
11a Stout setae present in neuropodia beginning with setiger 18-34; in posterior segments, vertical rows of stout setae not nearly encircling the segments
       Chaetozone acuta Banse & Hobson, 1968
11b Stout setae present in neuropodia beginning with setiger 40-70; in posterior segments, vertical rows of stout setae nearly encircling each segment
       Chaetozone setosa Malmgren, 1867
12a Groups of grooved cirri on setiger 1   13
12b Groups of grooved cirri on setigers 4, 5, 6, or 7 (except in young specimens, in which they may be located more anteriorly)   *Cirriformia spirabranchia* (Moore, 1904)
13a Stout setae present in neuropodia beginning with setiger 6-12
       Cirratulus cirratus (O. F. Müller, 1776)
13b Stout setae present in neuropodia beginning with setiger 17-39
       Cirratulus spectabilis (Kinberg, 1866)

Order Ctenodrilida

34. Family Ctenodrilidae

Ctenodrilus serratus (O. Schmidt, 1857) (fig. 8.102) is abundant in sediment on muddy sand in quiet bays.

35. Family Parergodrilidae

Stygocapitella subterranea Knöllner, 1934 is interstitial in sand.

Order Cossurida

36. Family Cossuridae

1a First 2 segments without setae; only setiger 1 uniramous *Cossura soyeri* Laubier, 1961
1b First 1 or 2 segments without setae; a few setigers in addition to setiger 1 uniramous
Cossura longocirrata Webster & Benedict, 1897 (fig. 8.103)

Order Flabelligerida

37. Family Flabelligeridae

1a Secreting a mucous sheath; epidermis smooth or nearly so; notosetae of several anterior setigers forming a cage around the head *Flabelligera affinis* M. Sars, 1829
1b Not secreting a mucous sheath; epidermis usually with crowded papillae; notosetae of anterior setigers may or may not form a cage around the head 2
2a Body short and stout, resembling a maggot; setigers 4 and 5 with a pair of ventrolateral nephridial papillae; notosetae of anterior setigers fairly long, but not forming an obvious cage around the head 3
2b Body elongate; neither setiger 4 nor 5 with a pair of ventrolateral nephridial papillae; notosetae of anterior setigers forming a cage around the head region 4
3a Dorsal portion of segments in the mid-region of the body with 2 or 3 transverse rows of dome-shaped papillae, each with a hairlike tip *Brada sachalina* Annenkova, 1922
3b Dorsal portion of segments in the mid-region of the body with 3 to 12 rows of slender papillae *Brada villosa* (Rathke, 1843)
4a With 8 to many gills forming a hoodlike aggregation at the anterior end; neurosetae with simple tips 5
4b With many gills arising from a tonguelike projection at the anterior end; neurosetae with forked tips (the 2 teeth may be unequal) *Piromis eruca* (Claparède, 1870)
5a With hooded hooks beginning on setiger 4 *Pherusa negligens* (Berkeley & Berkeley, 1950)
5b Without hooded hooks beginning on setiger 4 6
6a With a single row of papillae encircling each segment *Pherusa inflata* (Treadwell, 1914)
6b With several irregular rows of papillae encircling each segment
Pherusa plumosa (O. F. Müller, 1776) (fig. 8.39)

38. Family Acrocirridae

Banse, K. 1969. Acrocirridae n. fam. (Polychaeta Sedentaria). J. Fish. Res. Bd. Canada, 26:2595-620.
------. 1979. *Acrocirrus columbianus* and *A. occipitalis*, two new polychaetes (Acrocirridae) from the northeast Pacific. Proc. Biol. Soc. Washington, 91:923-8.

1a Palps originating so close together that their bases nearly touch; usually with more than 55 segments; epidermis smooth or with small papillae; length generally greater than 4 cm 2
1b Palps separated at least to the extent of their own thickness; usually with fewer than 55 segments; epidermis obviously papillated; length generally less than 1.5 cm
Macrochaeta pege Banse, 1969
2a Neuropodia of all setigers with 1 or 2 compound setae 3
2b Neuropodia of all setigers except 11 with 1 or 2 compound setae (setiger 11 has a long hook)
Acrocirrus heterochaetus Annenkova, 1934
3a Prostomium with a posterodorsal extension that partly interrupts segment 1
Acrocirrus columbianus Banse, 1979
3b Prostomium with a posterodorsal extension that interrupts segments 1 and 2
Acrocirrus occipitalis Banse, 1979

Order Opheliida

39. Family Opheliidae

1a	Body stout, almost maggotlike, without a ventral groove (some species with a characteristic odor, resembling that of garlic or an unlighted gas jet)	2
1b	Body slender, with a ventral groove, at least in the posterior portion of the body	5
2a	Parapodial lobes in posterior portion of body not obviously longer than those at more anterior levels; up to about 8 cm long and 2 cm wide *Travisia pupa* Moore, 1906	
2b	Parapodial lobes in posterior portion of body decidedly longer than those at more anterior levels; mostly under 5 cm long (but *Travisia japonica* is sometimes longer than this)	3
3a	With 33-40 setigers and 26-35 pairs of gills; up to 7 cm long *Travisia japonica* Fujiwara, 1933	
3b	With 23-26 setigers and 18-23 pairs of gills; not more than 3 cm long	4
4a	Enlarged parapodial lobes of posterior portion of body tapering *Travisia brevis* Moore, 1923	
4b	Enlarged parapodial lobes of posterior portion of body rounded *Travisia forbesii* Johnston, 1840	
5a	Ventral groove running the entire length of the body	6
5b	Ventral groove restricted to the posterior portion of the body	8
6a	With numerous eyespots on both sides of the body *Armandia brevis* (Moore, 1906)	
6b	Without lateral eyespots	7
7a	Pygidium shaped like a spoon, the ventral depression fringed with fingerlike outgrowths and with 3 rather long cirri at its base; with 40-50 setigers *Ophelina acuminata* Ørsted, 1843 (fig. 8.104) *Ammotrypane aulogaster* Rathke, 1843 is a synonym.	
7b	Pygidium resembling a tube, its terminal opening bordered by indistinct lobes or short papillae; with about 25-32 setigers *Ophelina breviata* (Ehlers, 1913)	
8a	Gills with 2 or more branches	9
8b	Gills unbranched *Ophelia limacina* (Rathke, 1843)	
9a	Gills with 2 branches *Euzonus (Thoracophelia) mucronatus* (Treadwell, 1914) (fig. 8.105)	
9b	Gills with 2 or 3 main branches, these with irregularly arranged side branches *Euzonus (Thoracophelia) williamsi* (Hartman, 1938)	

40. Family Scalibregmidae

Kudenov, J. D., & J. A. Blake. 1978. A review of the genera and species of the Scalibregmidae (Polychaeta) with descriptions of one new genus and three new species from Australia. J. Nat. Hist., 12:427-44.

1a	Some anterior segments with branched gills *Scalibregma inflatum* Rathke, 1843 (fig. 8.42)	
1b	Without gills	2
2a	With stout setae on the first 3 setigers *Asclerocheilus beringianus* Ushakov, 1955	
2b	Without stout setae (only capillary setae) *Hyboscolex pacificus* (Moore, 1909)	

Order Sternaspida

41. Family Sternaspidae

Sternaspis scutata (Renier, 1807) (fig. 8.41) is common subtidally in soft mud.

Order Capitellida

42. Family Capitellidae

1a Peristomium with setae; males with 2 thick, spinelike copulatory setae arising from the notopodia of segments 8 and 9 (females rarely have these setae)
Capitella capitata (Fabricius, 1780) (figs. 8.5 and 8.107)
1b Peristomium without setae; neither sex with copulatory setae 2
2a Notosetae of first 4-6 setigers of the capillary type 3
2b Notosetae of first 10-11 setigers of the capillary type 8
3a With 9-10 setigers forming a thoracic region; setigers 1-4 with only capillary setae 4
3b With 11 setigers forming a thoracic region; setigers 1-5 or 1-6 with only capillary setae 6
4a Hooded hooks with more than 3 teeth above the main tooth *Mediomastus capensis* Day, 1961
4b Hooded hooks with only 3 teeth above the main tooth 5
5a Each posterior notopodium (after about segment 30) with at least 1 or 2 capillary setae
Mediomastus ambiseta (Hartman, 1947)
5b Posterior notopodia without capillary setae *Mediomastus californiensis* Hartman, 1944
6a With gills (these not conspicuous in 1 species); setigers 1-5 with capillary setae only; setigers 6-11 with hooded hooks only 7
6b Without gills; setigers 1-6 with capillary setae only; setigers 7-11 with hooded hooks only (or setigers 6-8 with a mixture of capillary setae and hooded hooks, and setigers 9-11 with hooded hooks) *Barantolla americana* Hartman, 1963
7a Gills prominent, beginning on about segment 25-40, increasing in length posteriorly
Heteromastus filobranchus Berkeley & Berkeley, 1932 (fig. 8.108)
7b Gills short, simple extensions of parapodial lobes, beginning on about segment 80
Heteromastus filiformis (Claparède, 1864)
8a With 10 setigers forming the thoracic region *Decamastus gracilis* Hartman, 1963
8b With 11 setigers forming the thoracic region 9
9a Hooks of abdominal setigers with a reduced hood *Notomastus giganteus* Moore, 1906
9b Hooks of abdominal setigers distinctly hooded 10
10a First setiger with both neurosetae and notosetae; with small gills formed by swollen edges of abdominal neuropodia *Notomastus lineatus* Claparède, 1870
10b First setiger with notosetae only; without distinct gills on the neuropodia 11
11a Prostomium with reddish eyespots arranged in 2 groups; setigers 7-10 not appreciably lighter than neighboring segments *Notomastus tenuis* Moore, 1909 (fig. 8.106)
11b Prostomium without eyespots; setigers 7-11 lighter than neighboring segments
Notomastus variegatus Berkeley & Berkeley, 1950

43. Family Arenicolidae

Healy, E. A., & G. P. Wells. 1959. Three new lugworms (Arenicolidae, Polychaeta) from the North Pacific area. Proc. Zool. Soc. London, 133:315-35.

1a Relatively large worms, length often exceeding 10 cm, width often exceeding 1 cm; not secreting tubes; posterior one-third of the body, behind the region bearing gills, without setae, and therefore resembling a tail; gills pinnate or bushy 2
1b Small worms, up to about 2.5 cm long and 0.5 mm wide; secreting membranous tubes coated with fine sand; gills and setae continuing nearly to the posterior end of the body, so that there is nothing that resembles a tail; gills consisting of 1-4 simple filaments
Branchiomaldane vincentii Langerhans, 1881
2a With several pairs of esophageal caeca, those of 1 pair being much longer than the others; neuropodia of more posterior gill-bearing segments not closely approaching the midventral line 3
2b With only 1 pair of esophageal caeca, these short; neuropodia of more posterior gill-bearing segments closely approaching the midventral line *Arenicola marina* (Linnaeus, 1758)

3a	Nephridiopores completely exposed; with 4-7 pairs of esophageal caeca; typically inhabiting muddy sand at the inner margins of bays *Abarenicola pacifica* Healy & Wells, 1959	
3b	Ventral portions of nephridiopores covered by a flap of skin; with 8 or more pairs of esophageal caeca; typically inhabiting rather clean sand at the mouths of bays or on beaches exposed to some wave action	4
4a	Usually with 8-12 pairs of esophageal caeca; usually on beaches exposed to some wave action *Abarenicola claparedi* subsp. *oceanica* Healy & Wells, 1959	
4b	Usually with 12-19 pairs of esophageal caeca; usually near the mouths of sandy bays *Abarenicola claparedi* subsp. *vagabunda* Healy & Wells, 1959	

44. Family Maldanidae

Banse, K. 1981. On some Cossuridae and Maldanidae (Polychaeta) from Washington and British Columbia. Can. J. Fish. Aquat. Sci., 38:633-7.

1a	Segments 2 and 3 with anteriorly directed, thin dermal collars, and more posterior segments (beginning with about segment 17) with posteriorly directed collars *Rhodine bitorquata* Moore, 1923	
1b	None of the segments with thin dermal collars	2
2a	Prostomium with a cephalic plate, this with a conspicuous rim that often appears leathery	3
2b	Prostomium without a cephalic plate (it may, however, have a flattened area)	22
3a	Anus dorsal to the terminal portion (which is often platelike) of the pygidium; setiger 1 without neurosetae	4
3b	Anus central, either terminal or within a funnel-like modification of the pygidium; setiger 1 usually with neurosetae	9
4a	Cephalic plate of prostomium with a slight and short median ridge, and also with a raised rim, much of which is serrated; anterior edge of setiger 1 with a collar	5
4b	Cephalic plate with a prominent median ridge that runs for nearly its entire length, and with a low, smooth rim (although this has a lateral incision); anterior edge of setiger 1 without a collar	8
5a	Lateral lobes of the cephalic plate (these lobes are demarcated from ventral and dorsal lobes by deep incisions) with a smooth, or nearly smooth, rim *Asychis similis* (Moore, 1906)	
5a	Lateral lobes of the cephalic plate distinctly serrated	6
6a	Pygidium prominently lobed, but the edges of the lobes smooth, without serrations or undulations *Asychis disparidenta* Moore, 1904	
6b	Pygidium with shallow lobes, but the edges of these serrated or undulating	7
7a	Margin of each pygidial lobe with 4 or more serrations, these rather prominent; rim of dorsal lobe of the cephalic plate with about 12 serrations *Asychis lacera* (Moore, 1923)	
7b	Margin of each pygidial lobe merely undulating or with poorly defined serrations; rim of dorsal lobe of the cephalic plate with at least 25 serrations *Asychis biceps* (M. Sars, 1861)	
8a	Posteroventral edge of pygidium serrated *Maldane glebifex* Grube, 1860	
8b	Posteroventral edge of pygidium not serrated *Maldane sarsi* subsp. *sarsi* Malmgren, 1865	
9a	Neuropodia of first 3 or 4 setigers with only a few (usually 1-3) setae, these generally spines or modified uncini; pygidium with or without a funnel	10
9b	Neuropodia of first 2 or 3 setigers with several to many uncini (in *Maldanella*, however, the first setiger does not have a neuropodium); pygidium with a funnel	19
10a	Setiger 8 with a triangular glandular area on the ventral side; pygidial funnel short or indistinct *Clymenura columbiana* (E. Berkeley, 1929)	
10b	Setiger 8 without a glandular area on the ventral side; with or without a pygidial funnel	11
11a	Pygidium with a funnel	15
11b	Pygidium without a funnel	12

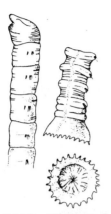

8.109 *Nicomache lumbricalis*

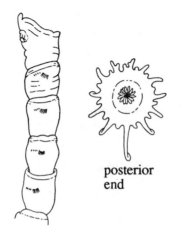

8.110 *Axiothella rubrocincta*

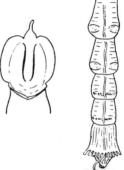

8.111 *Praxillella gracilis*

8.112 *Praxillella praetermissa*

Maldanidae

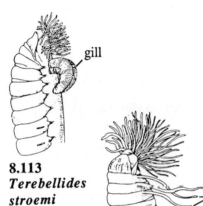

8.113 *Terebellides stroemi*

8.114 *Trichobranchus glacialis*

Trichobranchidae

8.115 *Thelepus cincinnatus*

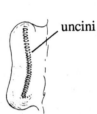

8.116 *Thelepus setosus*, thoracic segment

8.117 *Thelepus crispus*, thoracic segment

8.118 *Pista cristata*

Terebellidae

12a	Three of the segments at the posterior end without setae	13
12b	Four of the segments at the posterior end without setae	14
13a	Anteroventral portion of prostomium with a fingerlike projection	
	Praxillella gracilis (M. Sars, 1861) (fig. 8.111)	
13b	Anteroventral portion of prostomium with a short, blunt projection	
	Praxillella affinis subsp. *affinis* (M. Sars, 1872)	
14a	Setigers 1-3 with spinelike setae (the distal ends are bent, but are not toothed)	
	Praxillella affinis subsp. *pacifica* E. Berkeley, 1929	

14b Setigers 1-3 with uncini (the distal ends are toothed as well as bent)
Praxillella praetermissa (Malmgren, 1866) (fig. 8.112)
15a Setiger 5 and succeeding setigers with obvious glandular rings (these are often lighter than the other portions of these setigers); first neuropodia with a single straight spine
Isocirrus longiceps (Moore, 1923)
15b Setiger 5 and succeeding setigers with glandular rings, but these are not conspicuous; first neuropodia with spines or with uncini that are so much modified that their hooklike form is scarcely apparent 16
16a Cirri of pygidial funnel longest midventrally; with fewer than 25 setigers 17
16b Cirri of pygidial funnel all about the same length, and lacking in the midventral region; with more than 31 setigers .... *Macroclymene* sp.
17a Setiger 1 fused to the prostomial-peristomial complex; epidermis with a networklike pattern
Euclymene reticulata Moore, 1923
17b Setiger 1 distinctly separate from the prostomial-peristomial complex; epidermis smooth or ringed 18
18a Lateral notches in rim of cephalic plate distinct; setiger 1 with spines, setigers 2 and 3 with modified uncini (the bent tips of these are toothed); cirri of pygidial funnel of unequal length, the longer ones alternating with the shorter ones .... *Euclymene* sp.
This species is close to *E. zonalis*, and is perhaps within the range of variation of the latter; see choice 18b.
18b Lateral notches in rim of cephalic plate indistinct or lacking; setigers 1-3 with spines or with modified uncini; cirri of pygidial funnel all of about equal length
Euclymene zonalis (Verrill, 1874)
19a Setiger 1 without neurosetae 20
19b Setiger 1 with 5 or more neruosetae 21
20a Last segment preceding the pygidium without setae; rim of cephalic plate with a slightly scalloped edge .... *Maldanella robusta* Moore, 1906
20b Last 2 segments preceding the pygidium without setae; rim of cephalic plate with a smooth edge .... *Maldanella harai* (Izuka, 1902)
21a Anterior edge of setiger 4 with a membranous collar; cirri of pygidial funnel all about the same length .... *Clymenella torquata* (Leidy, 1855)
21b Anterior edge of setiger 4 without a membranous collar; cirri of pygidial funnel of uneven length, those of the midventral part usually longer than the others (common intertidally in muddy sand) .... *Axiothella rubrocincta* (Johnson, 1901) (fig. 8.110)
22a Capillary notosetae of middle and posterior portions of the body hairlike, twisted, sometimes longer than the width of the body; pygidium with a funnel, this sometimes scooplike 23
22b Capillary notosetae of middle and posterior portions of the body not especially long, hairlike, or twisted; pygidium without a funnel 26
23a Funnel of pygidium circular and with cirri 24
23b Funnel of pygidium elliptical and scooplike, without cirri 25
24a Last segment preceding the pygidium without setae; anterior portion of the body with conspicuous white markings .... *Nicomache personata* Johnson, 1901
24b Usually the last 2 segments preceding the pygidium without setae; anterior portion of the body without conspicuous white markings
Nicomache lumbricalis (Fabricius, 1780) (fig. 8.109)
25a With 20 setigers; edge of pygidial funnel finely scalloped
Petaloproctus tenuis subsp. *tenuis* (Théel, 1879)
25b With 21 setigers; edge of pygidial funnel smooth
Petaloproctus tenuis subsp. *borealis* Arwidsson, 1907
26a Pygidium conical; anus terminal; usually with 20 or more setigers
Praxillura maculata Moore, 1923
26b Pygidium truncate; anus dorsal; usually with 19 setigers .... *Notoproctus pacificus* (Moore, 1906)

Order Oweniida

45. Family Oweniidae

1a   Prostomium with several branched lobes (constructing a flexible, sandy tube that tapers gradually to both ends; common intertidally in sandy bays)
Owenia fusiformis delle Chiaje, 1841 (fig. 8.45)

1b   Prostomium without branched lobes   *Myriochele oculata* Zachs, 1923

Order Terebellida

46. Family Pectinariidae (Amphictenidae)

1a   Dorsal portion of the rim of the cephalic plate smooth   2
1b   Dorsal portion of the rim of the cephalic plate irregularly crenulate
Pectinaria (*Amphictene*) *moorei* Annenkova, 1929
2a   Conspicuous teeth on uncini arranged in a single row
Pectinaria (*Cistenides*) *granulata* (Linnaeus, 1767)
2b   Conspicuous teeth on uncini arranged in 2 or more rows
Pectinaria (*Pectinaria*) *californiensis* Hartman, 1941

47. Family Sabellariidae

1a   Thick setae (paleae) of the operculum arranged in 2 visible rows; without stout, hooklike setae in the dorsal cleft of the operculum   *Sabellaria cementarium* Moore, 1906
1b   Thick setae (paleae) of the operculum arranged in 3 visible rows; with stout, hooklike setae in the dorsal cleft of the operculum   2
2a   Flattened setae of the thoracic region widened near the tip, thus oarlike
Idanthyrsus armatus Kinberg, 1867 (fig. 8.47)
2b   Flattened setae of the thoracic region not widened near the tip
Idanthyrsus ornamentatus Chamberlin, 1919
Doubtfully distinct from *I. armatus*.

48. Family Ampharetidae

Banse, K. 1979. Ampharetidae (Polychaeta) from British Columbia and Washington. Can. J. Zool., 57:1543-52.

1a   With a pair of large hooks dorsally behind the gills   2
1b   Without dorsal hooks   3
2a   Tips of dorsal hooks bent at an angle of about 45°   *Melinna cristata* (M. Sars, 1851)
2b   Tips of dorsal hooks bent at an angle of about 90°   *Melinna elisabethae* McIntosh, 1922
3a   With 3 pairs of gills   *Samytha californiensis* Hartman, 1969
3b   With 4 pairs of gills   4
4a   With a bundle of setae on both sides of the body anterior to the gills   5
4b   Without a bundle of setae on both sides of the body anterior to the gills   15
5a   Uncini on 12 thoracic setigers   6
5b   Uncini on 13 or 14 thoracic setigers   11
6a   Thoracic region with 14 setigers; notopodia of thoracic region arranged in a nearly straight line   7
6b   Thoracic region with 15 setigers; notopodia of thoracic region not in a straight line (the eighth one is located farther dorsally than the others)   *Anobothrus gracilis* (Malmgren, 1866)
7a   Upper lip of prostomium with many eyespots   *Ampharete labrops* Hartman, 1961

7b	Upper lip of prostomium without eyespots	8
8a	Pygidium with many slender cirri; with 12 abdominal setigers	
	Ampharete acutifrons (Grube, 1860)	
8b	Pygidium with 2 cirri or none; with 13 or more abdominal setigers	9
9a	With 13 abdominal setigers *Ampharete finmarchica* (M. Sars, 1865)	
9b	With 16 or 17 abdominal setigers	10
10a	With 16 abdominal setigers; flattened setae (paleae) with a long filiform tip	
	Ampharete goesi subsp. *brazhnikovi* Annenkova, 1929	
10b	With 17 abdominal setigers; flattened setae rather abruptly tapered, with a short filiform tip	
	Ampharete goesi subsp. *goesi* Malmgren, 1866	
11a	Bundle of setae anterior to the gills containing some flattened setae (paleae)	12
11b	Bundle of setae anterior to the gills without any obviously flattened setae	14
12a	One pair of gills (anteromedial) flattened and with a slender, hooklike tip	
	Amphicteis scaphobranchiata Moore, 1906	
12b	None of the gills obviously flattened or with a differentiated tip	13
13a	Flattened setae in a bundle anterior to gills abruptly tapered and then extended as a long filiform tip *Amphicteis mucronata* Moore, 1923	
13b	Flattened setae in a bundle anterior to gills tapered so gradually that the filiform tip is scarcely set off from the rest *Amphicteis glabra* Moore, 1905	
14a	Setae of bundle anterior to gills short; 13 thoracic setigers with uncini; neuropodia of abdominal segments without cirri *Lysippe labiata* Malmgren, 1866	
14b	Setae of bundle anterior to gills long; 14 thoracic setigers with uncini; neuropodia of abdominal segments with slender cirri *Hobsonia florida* (Hartman, 1951)	
15a	Third from last notopodium of thoracic region enlarged, displaced dorsally, its setae arranged in such a way that they form a fan *Sosaniopsis hesslei* Banse, 1979	
15b	All notopodia of thoracic region in a nearly straight line, none especially enlarged	16
16a	With 11 thoracic setigers bearing uncini *Amage anops* (Johnson, 1901)	
16b	With 12 or more thoracic setigers bearing uncini	17
17a	Uncini present on 14 thoracic setigers *Amphisamytha bioculata* (Moore, 1906)	
17b	Uncini present on 12 thoracic setigers	18
18a	With 2 types of gills (anteriormost pair smooth, the others branched on 1 or both sides); neuropodia absent on first 3 thoracic segments *Schistocomus hiltoni* Chamberlin, 1919	
18b	All gills similar, smooth; neuropodia absent on first 2 thoracic segments	19
19a	Abdominal neuropodia with long, slender cirri (length more than 6 times the width)	
	Asabellides lineata (Berkeley & Berkeley, 1943)	
19b	Abdominal neuropodia with short cirri (length about 3 or 4 times the width)	
	Asabellides sibirica (Wirén, 1883)	

49. Family Trichobranchidae

1a	With a single mid-dorsal gill, this with 4 pectinate branches (the 4 branches are nearly fused together); all tentacles similar *Terebellides stroemii* M. Sars, 1835 (fig. 8.113)	
1b	With 2-4 pairs of gills; some tentacles much longer and more slender than the others	2
2a	With 4 pairs of gills, the more anterior ones being lanceolate, the more posterior ones being nearly globular *Novobranchus pacificus* Berkeley & Berkeley, 1954	
2b	With 2 or 3 pairs of gills, these slender filaments	
	Trichobranchus glacialis Malmgren, 1866 (fig. 8.114)	

50. Family Terebellidae

Banse, K. 1980. Terebellidae (Polychaeta) from the northeast Pacific Ocean. Can. J. Fish. Aquat. Sci., 37:20-40.

1a	Peristomium with a large, papillated, proboscislike extension *Artacama conifera* Moore, 1905	
1b	Peristomium without a proboscislike extension	2

2a	Uncini, if present in the thoracic region, in a single row (but this row may form a loop)	3
2b	Uncini in some thoracic segments (especially those in the posterior half of the thorax) in a double row, interlocking in much the same way as the teeth of a zipper	11
3a	With 2 or 3 pairs of slender, unbranched gills	4
3b	Without gills	9
4a	Capillary notosetae beginning on the first gill-bearing segment *Streblosoma bairdi* (Malmgren, 1866)	
4b	Capillary notosetae beginning on the second or third gill-bearing segment	5
5a	With 2 pairs of gills	6
5b	With 3 pairs of gills	7
6a	First 4 thoracic segments similar to succeeding thoracic segments in that they are glandular ventrally and laterally *Thelepus cincinnatus* (Fabricius, 1780) (fig. 8.115)	
6b	First 4 thoracic segments different from succeeding thoracic segments in that they are completely encircled by glandular tissue *Thelepus hamatus* Moore, 1905	
7a	Rows of uncini, beginning with segment 8, curved to form a nearly closed ellipse *Thelepus crispus* Johnson, 1901 (fig. 8.117)	
7b	Rows of uncini, on all segments that have them, nearly straight	8
8a	Notosetae absent from about the last 40 segments; gill filaments not obviously coiled *Thelepus setosus* (Quatrefages, 1865) (fig. 8.116)	
8b	Notosetae absent from only about the last 10 segments; gill filaments usually coiled *Thelepus japonicus* Marenzeller, 1884	
9a	Abdominal region without any setae *Lysilla loveni* Malmgren, 1866	
9b	Abdominal region with setae	10
10a	Abdominal region with uncini *Polycirrus californicus* Moore, 1909 There are also some undescribed species; see Hobson & Banse, 1981.	
10b	Abdominal region with straight, needlelike setae *Amaeana occidentalis* (Hartman, 1944)	
11a	Gills absent	12
11b	Gills present	15
12a	Setigers 1-6 without uncini *Laphania boecki* Malmgren, 1866	
12b	Setiger 1, and sometimes 2, without uncini, but the others with uncini	13
13a	Setiger 2 without uncini *Proclea graffii* (Langerhans, 1880)	
13b	Setiger 2 with uncini	14
14a	Segment 3 (the segment preceding the first setigerous segment) with a ridge across its dorsal part; tips of notosetae smooth *Leaena abranchiata* (M. Sars, 1865)	
14b	Segment 3 without a dorsal ridge; tips of notosetae finely toothed *Lanassa venusta* subsp. *venusta* (Malm, 1874)	
15a	With only 1 pair of gills	16
15b	With 2 or 3 pairs of gills	17
16a	Gills on segment 2, resembling a bottle brush because the branches are arranged in closely spaced whorls *Scionella estevanica* Berkeley & Berkeley, 1942	
16b	Gills on segment 4, and not resembling a bottle brush *Scionella japonica* Moore, 1903	
17a	Thoracic notosetae smooth at the tip	18
17b	Thoracic notosetae distinctly toothed at the tip	27
18a	Uncini of thoracic segments (at least the first 3 that have uncini) with long handles	19
18b	Uncini without long handles	24
19a	Last (third) pair of gills on the first segment that has uncini *Betapista dekkerae* Banse, 1980	
19b	Last (second or third) pair of gills on the segment preceding the first segment that has uncini	20
20a	All thoracic uncini with long handles; opening of tube expanded into a broad, triangular or nearly circular lobe, this with a fringe of slender outgrowths *Pista pacifica* Berkeley & Berkeley, 1942	
20b	Only the uncini of the more anterior thoracic segments with long handles; opening of tube not expanded into a broad lobe (but it may have other elaborations)	21
21a	Each gill and its crowded filaments forming a nearly club-shaped structure *Pista cristata* (O. F. Müller, 1776) (fig. 8.118)	
21b	Gills bushy or elongated and branched in such a way that they are not at all club-shaped	22
22a	Gills elongated, trailing, with many clusters of short branches *Pista moorei* Berkeley & Berkeley, 1942	

22b Gills bushy   23
23a With 2 pairs of gills, those of the first pair much shorter than those of the second pair (inhabiting mud, and constructing tubes of mud)   *Pista brevibranchiata* Moore, 1923
23b With 2 pairs of gills, those of the first pair larger than those of the second pair (inhabiting rocky habitats [often found among roots of *Phyllospadix*]; opening of tube with many filaments that form a loosely woven plug)   *Pista elongata* Moore, 1909
24a Uncini comblike, not C-shaped   *Loimia medusa* (Savigny, 1818)
24b Uncini nearly C-shaped (the teeth at one end of the C)   25
25a With 2 pairs of gills   *Nicolea zostericola* (Ørsted, 1844)
25b With 3 pairs of gills   26
26a Toothed portions of C-shaped uncini of posterior thoracic region facing each other
         Eupolymnia heterobranchia (Johnson, 1901)
26b Uncini in posterior thoracic region more nearly back to back, so that the toothed portions face away from each other   *Lanice* sp.
27a Notosetae appearing first in segment 3 (may be hidden under first pair of gills); with 2 pairs of gills   28
27b Notosetae appearing first in segment 4; with 3 pairs of gills   30
28a Uncini of abdominal region in double rows   *Neoleprea californica* (Moore, 1904)
28b Uncini of abdominal region in single rows   29
29a Notosetae on about 25 segments; body with about 15 ventral glandular shields
         Neoleprea japonica Hessle, 1917
29b Notosetae on about 35 or 40 segments; body with about 20 ventral glandular shields
         Neoleprea spiralis (Johnson, 1901)
30a Gills unbranched   *Amphitrite cirrata* O. F. Müller, 1776)
30b Gills branched   31
31a Notosetae on fewer than 25 segments; with lateral lappets in the thoracic region   32
31b Notosetae on more than 25 segments; without lateral lappets in the thoracic region
         Terebella ehrenbergi Grube, 1870
32a Notosetae on 19 segments   *Neoamphitrite groenlandica* (Malmgren, 1866)
32b Notosetae on 17 segments   33
33a Terminal branches of gills relatively long; nephridial papillae on at least 10 segments, beginning with segment 3   *Neoamphitrite robusta* (Johnson, 1901)
33b Terminal branches of gills relatively short; nephridial papillae on 9 segments, beginning with segment 3   *Neoamphitrite edwardsi* (Quatrefages, 1865)

Order Sabellida

51. Family Sabellidae

Banse, K. 1979. Sabellidae (Polychaeta) principally from the northeast Pacific Ocean. J. Fish. Res. Bd. Canada, 36:869-82.

1a Main stems of all radioles dichotomously branched 1 or more times
         Schizobranchia insignis Bush, 1904
1b Main stems of radioles not branched dichotomously (except in *Eudistylia vancouveri*, in which a few are branched)   2
2a Two of the more nearly dorsal radioles on both sides of the midline with large ocelli curled spirally around them near their tips   *Megalomma splendida* (Moore, 1905) (fig. 8.48)
2b None of the radioles with conspicuous ocelli curled around them near their tips (but there may be small ocelli scattered over the radioles)   3
3a First 5 or 6 thoracic neuropodia with avicular uncini (these have short handles or no obvious handles) and with or without pickaxe-shaped setae   4
3b First 5 or 6 thoracic neuropodia with long-handled uncini and without pickaxe-shaped setae
         15

162 Phylum Annelida: Class Polychaeta

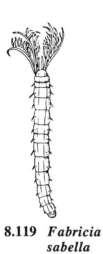
8.119 *Fabricia sabella*

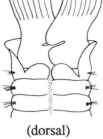
(dorsal)
8.120 *Potamilla occelata*

(dorsal)
8.121 *Jasmineira pacifica*

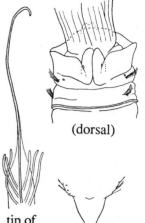
(dorsal)

tip of radiole

posterior end (dorsal)
8.122 *Chone duneri*

(dorsal)
8.123 *Oriopsis minuta*

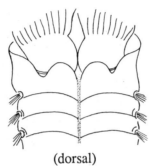
(dorsal)
8.124 *Eudistylia vancouveri*

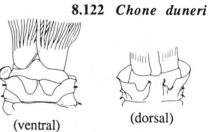
(ventral) (dorsal)
8.125 *Sabella pacifica*

Sabellidae

4a	All thoracic neuropodia with only avicular uncini	5
4b	All thoracic neuropodia with both avicular uncini and pickaxe-shaped setae	6
5a	Thoracic notopodia with both limbate and spatulate setae; uncini with broad bases	
	Laonome kröyeri Malmgren, 1866	
5b	Thoracic notopodia with only limbate or slightly spatulate setae; uncini almost S-shaped	
	Sabellastarte sp.	
6a	Lobes from which radioles originate spirally coiled	7
6b	Lobes from which radioles originate semicircular	9
7a	Notopodia of first few abdominal segments longer than the tori (reduced parapodial lobes, with uncini) of the posterior thoracic segments	8
7b	Notopodia of first few abdominal segments decidedly shorter than the tori of the posterior thoracic segments *Eudistylia catharinae* Banse, 1979	
8a	Dorsal edges of both lobes from which radioles originate with a cleft *Eudistylia polymorpha* (Johnson, 1901)	
8b	Dorsal edge of both lobes from which radioles originate without a cleft *Eudistylia vancouveri* (Kinberg, 1867) (fig. 8.124)	
9a	Thoracic notopodia with limbate and definitely spatulate setae	10
9b	Thoracic notopodia with limbate and sometimes with nearly spatulate setae or knifelike setae, but without definitely spatulate setae	13
10a	Dorsolateral portions of collar without a cleft or even a shallow incision *Potamilla neglecta* (M. Sars, 1851)	
10b	Dorsolateral portions of collar with a cleft or shallow, wide incision	11

11a Dorsal edges of lobes from which radioles originate with a cleft (common on floats, as well
    as in other habitats) .... *Potamilla occelata* Moore, 1905 (fig. 8.120)
11b Dorsal edges of lobes from which radioles originate without a cleft 12
12a Dorsolateral portions of collar with a shallow, wide incision; each radiole with more than 10
    ocelli .... *Potamilla myriops* Marenzeller, 1884
12b Dorsolateral portions of collar with a deep, narrow cleft; radioles with few or no ocelli
    .... *Potamilla intermedia* Moore, 1905
13a Collar with a distinct, even if shallow, lateral incision; all thoracic notosetae limbate
    .... *Sabella (Sabella) crassicornis* M. Sars, 1851
13b Collar without a lateral incision; short thoracic notosetae either knifelike or nearly spatulate 14
14a Free edges of collar separated dorsally by a gap that is about half as wide as the thoracic
    region; dorsomedial portions of the collar (next to the gap) not obviously expanded or folded
    into pockets .... *Sabella (Demonax) media* (Bush, 1904)
14b Free edges of collar separated dorsally by a gap that is only about one-fourth as wide as the
    thoracic region; dorsomedial portions of the collar (next to the gap) somewhat expanded and
    folded to form pockets .... *Sabella (Demonax) pacifica* (Berkeley & Berkeley, 1954) (fig. 8.125)
15a Abdominal uncini in rows that nearly encircle the segments; proximal portions of radioles
    connected by a membrane .... *Myxicola infundibulum* (Renier, 1804)
15b Abdominal uncini in short rows; radioles connected by a membrane in some species, not in
    others 16
16a With a broad furrow on the ventral side of the last few segments 17
16b Without a broad furrow on the ventral side of the last few segments 19
17a With 34-40 setigers; broad furrow on the ventral side of the posterior portion evident on 9-12
    setigers .... *Euchone analis* (Krøyer, 1856)
17b With 16 or 17 setigers; broad furrow on ventral side of posterior portion evident on only 3
    setigers 18
18a With 9 abdominal setigers .... *Euchone incolor* Hartman, 1965
18b With 8 abdominal setigers .... *Euchone hancocki* Banse, 1970
19a With only 3 abdominal setigers; length less than 1 cm 20
19b With more than 3 abdominal setigers; length generally greater than 1 cm 23
20a Anterior end with 4 radioles, each consisting of 2 pinnules, and a pair of unbranched ventral
    filaments; pygidium without eyespots (estuarine) .... *Manayunkia aestuarina* (Bourne, 1883)
    M. speciosa Leidy, 1859, with at least 10 radioles (20 pinnules) is found in fresh water.
20b Anterior end with 6 radioles, each with many pinnules, and with or without a pair of
    unbranched ventral filaments; pygidium with eyespots (visible in living specimens, but may
    not be visible in preserved specimens) 21
21a Anterior end with a pair of ventral filaments; collar more or less equally well developed all
    around the anterior end, except for a small mid-dorsal incision
    .... *Fabriciola berkeleyi* Banse, 1956
21b Anterior end without a pair of ventral filaments; collar mostly in the form of a lip on the
    ventral side, not evident dorsally 22
22a Ventral remnant of collar nearly rectangular in outline .... *Fabricia oregonica* Banse, 1979
22b Ventral remnant of collar tongue-shaped, rounded anteriorly
    .... *Fabricia sabella* (Ehrenberg, 1837) (fig. 8.119)
23a Abdominal uncini S-shaped, their middle portions longer than wide; radioles not united by a
    membrane .... *Jasmineira pacifica* Annenkova, 1937 (fig. 8.121)
23b Abdominal uncini nearly C-shaped; radioles united by a membrane in some species, not in
    others 24
24a Radioles united by a membrane; with a whitish, glandular ring behind the parapodia of
    setiger 2; length usually more than 1 cm 26
24b Radioles not united by a membrane; without a glandular ring behind the parapodia of setiger
    2; length less than 1 cm 25
25a All thoracic notosetae limbate .... *Oriopsis gracilis* Hartman, 1969
25b Some thoracic notosetae limbate, others nearly spatulate
    .... *Oriopsis minuta* (Berkeley & Berkeley, 1932) (fig. 8.123)
26a With ventral shields (aggregates of gland cells); with 12-16 radioles; length up to 1.5 cm
    .... *Chone ecaudata* (Moore, 1923)

26b Without ventral shields (the glands are distributed throughout the thoracic epidermis); with more than 16 radioles; length up to 5 cm or more ... 27
27a End portions (distal to the last pinnules) of the lateral radioles short and broad (nearly as broad as long) ... 28
27b End portions of lateral radioles much longer than broad ... 29
28a Flattened portion of spatulate setae tapering to a sharply pointed tip; tuft of setae on notopodium of setiger 1 more nearly dorsal than that of setiger 2 .... *Chone infundibuliformis* Krøyer, 1856
28b Flattened portions of spatulate setae rounded, the slender tip arising abruptly from it; tuft of setae on notopodia of setigers 1 and 2 at about the same level .... *Chone aurantiaca* (Johnson, 1901)
29a Tuft of setae on setiger 1 located in a groove that extends to the groove that marks the anterior border of setiger 2 .... *Chone mollis* (Bush, 1904)
29b Tuft of setae on setiger 1 inserted on the collar ... 30
30a Blade of spatulate setae rounded distally, although a fine hair may project from its tip .... *Chone magna* (Moore, 1923)
30b Blade of spatulate setae tapering to a sharp tip .... *Chone duneri* Malmgren, 1867 (fig. 8.122)

52. Family Serpulidae

1a Operculum well developed, at the tip of a specialized, nearly mid-dorsal radiole that has no pinnules ... 2
1b Operculum absent, or consisting of a swelling at the tip of a relatively unmodified radiole that has pinnules ... 6
2a Operculum globular, with a brown, chitinous plate at its distal end .... *Pseudochitinopoma occidentalis* (Bush, 1904)
2b Operculum funnel-shaped ... 3
3a Operculum with 2 or 3 obvious protuberances just below the funnel ... 4
3b Stalk of operculum without protuberances just below the funnel, but it has a ringlike thickening in this position .... *Serpula vermicularis* Linnaeus, 1767 (fig. 8.51)
4a Operculum with 2 protuberances just below the funnel; edge of the funnel rather irregular, due in part to the funnel being longer dorsally than ventrally and somewhat compressed laterally .... *Crucigera irregularis* Bush, 1904
4b Operculum with 3 protuberances just below the funnel; funnel opening almost perfectly circular, its edge regular .... *Crucigera zygophora* (Johnson, 1901)
5a Operculum absent ... 6
5b Operculum present, globular, transparent ... 7
6a With only 6 radioles .... *Salmacina tribranchiata* (Moore, 1923)
6b With about 120 radioles .... *Protula pacifica* Pixell, 1912
7a Abdominal setae sharply bent, with smooth blades .... *Apomatus geniculatus* (Moore & Bush, 1904)
7b Abdominal setae not sharply bent, with serrated blades .... *Apomatus timmsii* Pixell, 1912

53. Family Spirorbidae

Linda H. Price

Knight-Jones, P., E. W. Knight-Jones, & R. P. Dales. 1979. Spirorbidae (Polychaeta Sedentaria) from Alaska to Panama. J. Zool., London, 189:419-58.

1a Coiling of tube mainly dextral (the tube coils in a counterclockwise direction from the closed end of the tube when it is viewed from the exposed side, which is ventral) ... 2
1a Coiling of tube sinistral (the tube coils in a clockwise direction from the closed end of the tube when it is viewed from the exposed side) ... 7
2a Embryos attached to the inside of the tube ... 3

Phylum Annelida: Class Polychaeta 165

2b   Embryos attached to, or associated with, the outside of the tube   6
3a   Embryos in a string attached to the inside of the tube; tube not transparent (coiling of the tube sinistral in most California populations, but dextral in more northern populations; intertidal)
                                                         Spirorbis bifurcatus Knight-Jones, 1978 (fig. 8.130)
3b   All embryos attached singly to the inside of the tube; tubes may or may not be transparent   4
4a   Tube of adult animals porcelaneous or translucent, but not transparent to the extent that the animal and embryos can be seen   5
4b   Tube of adult animals transparent, revealing red embryos (juveniles have somewhat opaque tubes) (intertidal and subtidal on hard substrata; coiling primarily dextral, but some sinistral coiling has been observed)   *Paradexiospira (Spirorbides) vitrea* (Fabricius, 1780)

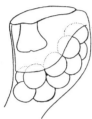
8.126 *Pileolaria similis*, operculum with brood

8.127 *Pileolaria quadrangularis*, operculum with brood

8.128 *Sinistrella abnormis*, brood chamber and 2 primary opercula

8.129 *Janua pagenstecheri*, operculum with brood

8.130 *Spirorbis bifurcatus*, tube

Spirorbidae

5a   Tube opaque and porcelaneous; intertidal and subtidal on hard substrata or on algae (coiling dextral, except in some California populations)   *Circeis armoricana* Saint-Joseph, 1894
5b   Tube translucent; subtidal and only on bryozoans and hydroids
                                                                Circeis spirillum (Linnaeus, 1758)
6a   Embyos in a inverted transparent cup within the operculum, the cup becoming detached when the brood is released (intertidal, on algae and stones)
                                       Janua (Janua) pagenstecheri (Quatrefages, 1865) (fig. 8.129)
6b   Embryos in an unattached mass behind the opercular collar
                                                   Paralaeospira malardi Caullery & Mesnil, 1897
7a   Embyos in a cup formed from an invagination of the opercular plate; surface of tube smooth or with longitudinal ridges   8
7b   Embryos in a transparent sac that is attached to the outside of the tube; surface of the tube often with transverse ridges   13
8a   Cup of opercular plate (in brooding animal) single   9
8b   Cup of opercular plate (in brooding animal) often double or multiple (sometimes single, however)   12
9a   Talon on operculum a small tubercle (color of live animal red; intertidal and subtidal on hard substrata)   *Pileolaria (Simplicaria) potswaldi* Knight-Jones, 1978

9b Talon on operculum about one-fourth to one-half the size of the operculum ... 10
10a Talon on operculum diamond-shaped; operculum domed in adult animals; surface of tube often with 2 longitudinal ridges; intertidal or subtidal, on stones and shells
Pileolaria (Jugaria) quadrangularis (Stimpson, 1854) (fig. 8.127)
10b Talon on operculum fan-shaped; operculum flattened; surface of tube smooth or with a single longitudinal ridge; subtidal ... 11
11a Setae on collar compound (a shallow incision separates the blade from the shaft); tube up to 1.5 mm in diameter, often with a rough surface but without longitudinal ridges
Pileolaria (Jugaria) similis (Bush, 1904) (fig. 8.126)
11b Setae on collar simple; tube up to 2.5 mm in diameter, smooth or with 1 longitudinal ridge
Sinistrella verruca (Fabricius, 1780)
12a Opercula usually multiple and stacked; each operculum with a talon that is prolonged into several fingerlike extensions; tube often with 3 prominent longitudinal ridges; intertidal on hard substrata and algae   *Sinistrella abnormis* (Bush, 1904) (fig. 8.128)
12b Opercula usually double (sometimes single), the more proximal operculum (the one more recently formed) with fine spines on its inner face; talons on opercula axe-shaped, without fingerlike extensions; tube smooth or with 1 inconspicuous longitudinal ridge; intertidal on hard substrata   *Sinistrella media* (Pixell, 1912)
13a Talon on operculum with a lateral spur; tube up to 5 mm in diameter, not porcelaneous; intertidal or subtidal on hard substrata   *Protolaeospira eximia* (Bush, 1904)
13b Talon on operculum without a lateral spur; tube up to 3.5 mm in diameter, porcelaneous; intertidal (rare)   *Protolaeospira capensis* (Day, 1961)

Order Myzostomida

54. Family Myzostomidae

Jägersten, G. 1940. Neue und alte *Myzostomum*-Arten aus dem Zoologischen Museum Kopenhagen. Vidensk. Meddel. Dansk naturh. Foren. København, 104:103-23.

Myzostomum pseudogigas Jägersten, 1940. On the crinoid *Florometra serratissima*.

Archiannelid Families

The polychaetes called archiannelids belong to several diverse families that do not fit into any of the recognized orders.

Ax, P. 1967. *Diurodrilus ankeli* nov. spec. (Archiannelida) von der nordamerikanischen Pazifikküste. Ein Beitrag zur Morphologie, Systematik und Verbreitung der Gattung *Diurodrilus*. Zeitschr. Morph. Ökol. Tiere, 60:5-16.
Gray, J. S. 1968. *Nerilla inopinata*, a new species of archiannelid, from the west coast of North America. Cah. Biol. Mar., 9:441-8.
------. 1969. A new species of *Saccocirrus* (Archiannelida) from the west coast of North America. Pacific Sci., 23:238-51.
Jones, E. R., & F. F. Ferguson. 1957. The genus *Dinophilus* (Archiannelida) in the United States. Amer. Midl. Nat., 57:440-9.
Martin, G. G. 1977. *Saccocirrus sonomacus* n. sp., a new archiannelid from California. Trans. Amer. Micr. Soc., 96:97-103.
Wieser, W. 1957. Archiannelids from the intertidal of Puget Sound. Trans. Amer. Micr. Soc., 76:275-85.

Phylum Annelida: Class Polychaeta

Key to Archiannelid Families

1a Head region with neither antennae nor palps ("tentacles"); without setae (in genera known to occur in our region)   56. Dinophilidae
1b Head region with antennae or palps, or both (the palps may be long and slender, in which case are often called "tentacles," or they may be short and lappetlike)   2
2a Head with 3 articulated antennae and a pair of lappetlike palps; with fewer than 15 segments; parapodia with prominent dorsal cirri and with simple or compound setae   55. Nerillidae (fig. 8.132)
2b Head without antennae, but with a pair of long palps; with more than 15 segments; parapodia, if present, without cirri, but sometimes with setae having bifid tips (bifid setae may be present even if there are no parapodia)   3
3a Palps stiff; body firm, much like that of a nematode, and without a mid-ventral ciliated band; without parapodia or setae   57. Polygordiidae

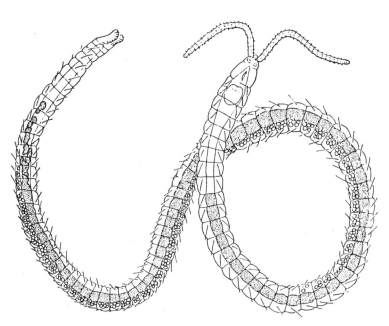

8.131 Saccocirridae: *Saccocirrus sonomacus*

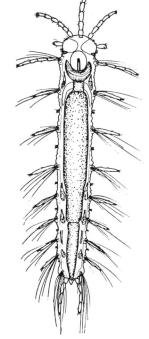

8.132 Nerillidae: *Nerilla digitata*

3b Palps slender and flexible; body flexible, and with a mid-ventral ciliated band; with or without parapodia, with or without setae   4
4a Setae present   5
4b Setae absent   58. Protodrilidae (in part)
5a Setigerous segments with cylindrical parapodia, these retractile   59. Saccocirridae (fig. 8.131)
5b Setigerous segments (each has 2 pairs of setae) without definite parapodia   58. Protodrilidae (in part)

Order Nerillida

55. Family Nerillidae

1a Setae compound   *Mesonerilla* sp.

1b Setae simple 2
2a Prostomial eyes absent; posteriormost cirri (seemingly part of the pygidium) shorter than the setae of the last setiger; cirri of first setiger only slightly longer than those of succeeding segments (in subterranean sand on beaches, above the high tide line)
Nerilla digitata Wieser, 1975 (fig. 8.132)
2b Prostomial eyes present; posteriormost cirri as long or longer than the setae of the last setiger; cirri of first setiger at least twice as long as those of succeeding segments 3
3a Cirri of setiger 2 shorter than the setae; posteriormost cirri as long as, or slightly longer than, the setae of the last setiger; most setigers with 9-19 setae on each side (intertidal, in fine sand)
Nerilla antennata O. Schmidt, 1848
3b Cirri of setiger 2 about as long as the setae; posteriormost cirri much longer than the setae of the last setiger; setigers with only about 6 setae on each side (intertidal, in shell-gravel)
Nerilla inopinata Gray, 1968

Order "Dinophilida"

56. Family Dinophilidae

This family probably belongs in the order Eunicida, even though it is commonly assigned to a separate order.

1a Prostomium, in dorsal view, somewhat trilobed, due to its differentiation into 2 small lateral protuberances and a somewhat larger anterior portion; pygidium rounded
Trilobodrilus nipponicus Uchida & Okuda, 1943
1b Prostomium, in dorsal view, simply rounded; pygidium conical or bilobed 2
2a Pygidium conical; with a pair of red eyes on the prostomium; all segments with ciliary rings (in sediment, on a substratum of mud and shell)
Dinophilus kincaidi Jones & Ferguson, 1957
2b Pygidium bilobed (the lobes unequal); without eyes; without ciliary rings (intertidal, in coarse sand-gravel)
Diurodrilus ankeli Ax, 1967

Order Polygordiida

57. Family Polygordiidae

Specimens of *Polygordius* (unidentified or undescribed) have been found on the open coast as well as in inland waters. The usual (but not exclusive) habitat is coarse sand.

Order Protodrilida

58. Family Protodrilidae

1a With 2 pairs of bifid setae on each setigerous segment; tentacles solid, generally directed anteriorly; epidermis with yellowish green inclusions (intertidal, in coarse sand)
Protodriloides chaetifer (Remane, 1926)
1b Without setae; tentacles hollow, freely orienting themselves in various directions; epidermis without yellowish green inclusions (intertidal, in coarse sand)
Protodrilus flabelliger Wieser, 1957

59. Family Saccocirridae

Saccocirrus eroticus Gray, 1969. Intertidal, in shell gravel.
Saccocirrus sonomacus Martin, 1977 (fig. 8.131). Described from northern California, but perhaps should be expected in our region.

9

PHYLUM ANNELIDA: CLASSES OLIGOCHAETA, HIRUDINOIDEA

Class Oligochaeta

The late H. R. Baker and K. A. Coates

Since the modern review of the aquatic oligochaetes by Brinkhurst & Jamieson (1971), many of the marine genera have been drastically revised. Recent studies in the northeast Pacific by Brinkhurst (1978, 1979), Baker (1981, 1982a, 1982b, 1983a, 1983b), Baker & Erséus (1982), Cook (1974), Coates & Erséus (1980), Coates (1980, 1981, 1983a, 1983b), and Tynen (1969) have established the existence of a rich marine and brackish-water oligochaete fauna in the region. The major revisionary works by Brinkhurst & Baker (1979), Baker & Brinkhurst (1981), and Coates & Ellis (1981) are the basic references.

The arrangement and shape of setae may be used to distinguish the three families of marine oligochaetes, and in the family Naididae setal characteristics are sufficient for identification of species. In the Enchytraeidae and Tubificidae, however, it is usually necessary to examine the genitalia, and the use of stains (such as borax carmine and paracarmine) may be helpful. Preliminary identifications should be confirmed by consulting the complete species descriptions which can be found in the references listed below. The general scheme of classification followed here is that of Brinkhurst (1982).

Altman, L. C. 1931. *Enchytraeus pugetensis*, a new marine enchytraeid from Puget Sound. Trans. Amer. Micr. Soc., 50:154-9.

------. 1936. Oligochaeta of Washington. Univ. Wash. Publ. Biol., 1:1-137.

Baker, H. R. 1981. *Phallodrilus tempestatis*, a new marine tubificid (Annelida: Oligochaeta) from British Columbia. Can. J. Zool., 59:1475-8.

------. 1982a. Two new phallodriline genera of marine Oligochaeta (Annelida: Tubificidae) from the Pacific Northeast. Can J. Zool., 60:2487-500.

------. 1982b. *Vadicola aprostatus* gen. nov., sp. nov., a marine oligochaete (Tubificidae; Rhyacodrilinae) from British Columbia. Can. J. Zool., 60:3232-6.

------. 1983a. New species of *Tubificoides* Lastockin (Oligochaeta; Tubificidae) from the Pacific Northeast and the Arctic. Can. J. Zool., 61:1270-83.

------. 1983b. Two new species of *Bathydrilus* Cook (Oligochaeta; Tubificidae) from British Columbia. Can. J. Zool., 61:2162-7.

Baker, H. R., & R. O. Brinkhurst. 1981. A revision of the genus *Monopylephorus* and redefinition of the subfamilies Rhyacodrilinae and Branchiurinae (Tubificidae: Oligochaeta). Can. J. Zool., 59:939-65.

Baker, H. R., & C. Erséus. 1982. A new species of *Bacescuella* Hrabĕ (Tubificidae: Oligochaeta) from the Pacific coast of Canada. Can. J. Zool., 60:1951-4.

Brinkhurst, R. O. 1978. Freshwater Oligochaeta in Canada. Can. J. Zool., 56:2166-75.

------. 1979. Distribution of aquatic Oligochaeta in some habitats of lower British Columbia. Pacific Marine Science Reports, 79-4. 15 pp.

------. 1982. Evolution in the Annelida. Can. J. Zool., 60:1043-59.

Brinkhurst, R. O., & H. R. Baker. 1979. A review of the marine Tubificidae (Oligochaeta) of North America. Can. J. Zool., 57:1553-69.

Brinkhurst, R. O., & B. G. M. Jamieson. 1971. Aquatic Oligochaeta of the World. Edinburgh: Oliver and Boyd. 860 pp.

Coates, K. A. 1980. New marine species of *Marionina* and *Enchytraeus* (Oligochaeta, Enchytraeidae) from British Columbia. Can. J. Zool., 58:1306-17.

------. 1981. New species of *Lumbricillus* (Oligochaeta, Enchytraeidae) from littoral habitats of British Columbia. Can. J. Zool., 59:1302-11.

------. 1983a. New records of marine *Marionina* (Oligochaeta, Enchytraeidae) from the Pacific Northeast, with a description of *Marionina klashkisharum* sp. nov. Can. J. Zool., 61:822-31.

------. 1983b. A contribution to the taxonomy of the Enchytraeidae (Oligochaeta). Review of *Stephensoniella*, with new species records. Proc. Biol. Soc. Wash., 96:411-9.

Coates, K. A., & D. V. Ellis. 1981. A taxonomic revision of the marine Enchytraeidae (Oligochaeta) of British Columbia. Can. J. Zool., 59:2129-50.

Coates, K. A., & C. Erséus. 1980. Two species of *Grania* (Oligochaeta: Enchytraeidae) from British Columbia. Can. J. Zool., 58:1037-41.

Cook, D. G. 1974. The systematics and distribution of marine Tubificidae (Annelida: Oligochaeta) in the Bahia de San Quintin, Baja California, with descriptions of five new species. Bull. So. Calif. Acad. Sci., 73:126-40.

Eisen, G. 1904. Enchytraeidae of the West Coast of North America. Harriman Alaska Expedition, 12:1-166.

Hiltunen, J. K., & D. J. Klemm. 1980. A guide to the Naididae (Annelida, Oligochaeta) of North America. U.S. Environmental Protection Agency, ser. EPA-600. 48 pp.

Nielsen, C.-O., & B. Christensen. 1959. Studies on Enchytraeidae. VII. Critical revision and taxonomy of European species. Nat. Jutl., 8-9:1-160.

Sperber, C. 1948. A taxonomical study of the Naididae. Zool. Bidr. Uppsala. 28:1-296.

Strehlow, D. R. 1982. *Aktedrilus locyi* Erséus, 1980 and *Aktedrilus oregonensis* n. sp. (Oligochaeta, Tubificidae) from Coos Bay, Oregon, with notes on distribution with tidal height and sediment type. Can. J. Zool., 60:593-6.

Tynen, M. J. 1969. New Enchytraeidae from the east coast of Vancouver Island. Can. J. Zool., 47:387-93.

Key to Families

1a	Setae always present, and (except for hair setae) these are sigmoid and with a nodulus, and some or all of them are bifid; hair setae sometimes present in dorsal setal bundles; male pores on segments 5 or 6 (Naididae) or 11 (Tubificidae)	2
1b	Setae, when present (they are sometimes absent), sigmoid, straight, or bent, rarely with a nodulus, and never bifid; hair setae absent; male pores on segment 12	2. Enchytraeidae
2a	Dorsal setae usually absent in segments 2 to 4 or 5, or absent in all segments (*Amphichaeta* is an exception; it has dorsal setae from segment 3 posteriorly); male pores on segment 5 or 6 (but mature specimens in which these pores can be seen may not be abundant); eyespots sometimes present	1. Naididae
2b	Dorsal setae present from segment 2; male pores on segment 11; eyespots absent	3. Tubificidae

1. Family Naididae

1a	Hair setae absent	2
1b	Hair setae present in some or all dorsal bundles	5
2a	Dorsal setae present from segment 3 posteriorly	*Amphichaeta* sp. (probably *A. leydigi* Tauber, 1879)
2b	Dorsal setae present from segment 5 or 6 posteriorly	3
3a	Dorsal setae usually present from segment 6 (some specimens lack dorsal setae), straight, thick, single	*Ophidonais serpentina* (O. F. Müller, 1773)
3b	Dorsal setae present from segment 5, sigmoid, not thick, usually more than 1 per bundle	4
4a	Body not encrusted with foreign matter; ventral setae of segment 2 in bundles of 5-7, and with the upper tooth longer than the lower tooth; in other ventral bundles, 2-3 setae, the	

upper tooth equal to or just slightly longer than the lower tooth
Paranais litoralis (O. F. Müller, 1784)

4b Body encrusted with foreign matter; ventral setae of segment 2 in bundles of 2-4, and with the upper tooth at least twice as long as the lower tooth; in other ventral bundles, 1-2 setae, the upper tooth markedly longer than the lower tooth *Paranais frici* Hrabě, 1941

5a Hair setae present from segments 5, 6, or 7, thick, stiff, markedly serrate; dorsal setae pectinate; body encrusted with foreign matter *Vejdovskyella hellei* Brinkhurst, 1971

5b Hair setae not serrate and not thick; pectinate setae absent; body not encrusted with foreign matter 6

6a Setal bundles of anterior part of body with 3-5 hair setae, plus 3-5 bifid setae; dorsal and ventral bifid setae similar, the dorsal ones slightly straighter (ventral setae of segment 2 thicker and longer than those of segments 3 and 4) *Specaria fraseri* Brinkhurst, 1978

6b Dorsal setal bundles of anterior part of body with only 1-2 hair setae and 1-2 bifid setae; dorsal bifid setae distinctly different from the ventral setae 7

7a Distal teeth of dorsal bifid setae long, parallel; upper tooth of ventral setae 1.5-2 times longer than the lower tooth from segment 6 posteriorly *Nais elinguis* O. F. Müller, 1773

7b Distal teeth of dorsal bifid setae short, often obscure; upper tooth of ventral setae from segment 6 posteriorly equal or nearly equal to the lower tooth 8

8a Teeth of dorsal bifid setae small (seen clearly only with 90x or 100x oil immersion objective), and parallel; upper tooth of ventral setae of segments 2-5 about twice as long as the lower tooth *Nais variabilis* Piguet, 1906

8b Teeth of dorsal bifid setae short (but should be clearly visible with 40x objective), divergent; upper and lower tooth of ventral setae approximately equal *Nais communis* Piguet, 1906

2. Family Enchytraeidae

1a Dorsal setae absent, at least in segment 2, and ventral setae absent in some or all segments (not more than 1-2 setae per bundle) 2

1b Dorsal and ventral setae (2 or more setae per bundle) present in all segments (fully mature specimens lack ventral setae in segment 12) 9

2a Ventral setae lacking in segments 2 and 3, and dorsal setae lacking at least in segments 1-11; where present, the setae are robust and always single; seminal vesicle and egg sac extending several segments posterior to the clitellum 3

2b Setal distribution not as described in choice 2a; seminal vesicle either small and anterior to the clitellum or absent, eggs only in clitellar segments or in the segment just behind the clitellum 4

3a Dorsal setae present from segment 13, 14, or 15 posteriorly *Grania paucispina* (Eisen, 1904)

3b Dorsal setae present from segment 17, 18, or 19 posteriorly
Grania incerta Coates & Erséus, 1980

4a Dorsal setae lacking; ventral setal bundles with 1 or 2 setae, or absent 5

4b Dorsal setae lacking only from segment 2 and, rarely, from segment 3; other dorsal and ventral bundles with 2 setae (pre-clitellar segments of living specimens intensely white or white-spotted) 8

5a Ventral setal bundles with 2 setae, and setae present in all segments from 2 posteriorly 6

5b Ventral setal bundles with only 1 seta in segments 2 to 4 or 5, or setae completely absent 7

6a With glandular epidermal pads just lateral to the setal bundles of postclitellar segments; terminal portion of spermathecal duct surrounded at the external pore by a rosette of 6 or 7 elongate gland cells *Marionina glandulifera* (Jansson, 1961) (fig. 9.2)

6b Without glandular pads; muscular wall of the terminal portion of spermathecal duct thickened just proximal to external pore, but no distinct, separate glandular cells are present
Marionina subterranea (Knöllner, 1935) (fig. 9.1)

7a Setae absent; cuticle thick, rigid (movement of living worms resembling that of nematodes)
Marionina nevisensis (Righi & Kanner, 1979)

7b Short, blunt ventral setae occurring singly in segments 2 to 4 or 5; cuticle not rigid
Marionina klaskisharum Coates, 1983

8a Sperm distributed in distinct balls in the walls of the ampullae of the spermathecae
Marionina southerni (Černosvitov, 1937)

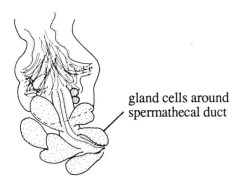

9.1 *Marionina subterranea*

9.2 *Marionina glandulifera*

50 μm

muscle layer around spermathecal duct

gland cells around spermathecal duct

8b Sperm arranged in an elongate oval bundle within the cavity of the ampullae of the spermathecae *Marionina charlottensis* Coates, 1981
9a Seminal vesicles present or absent, but if present, not divided intrasegmentally into lobes 10
9b Seminal vesicles well developed, with distinct, often pear-shaped lobes 16
10a Paired, unbranched peptonephridia present; setae straight; worms 1-15 cm long, yellowish or grayish 11
10b Peptonephridia absent; setae straight or slightly sigmoid; worms less than 1 cm long, whitish or transparent 12
11a Ampulla of each spermatheca with a single, large diverticulum; terminal portions of spermathecal ducts with small rosettes of glands at the pores and with a few scattered glands along the ducts; with 2-3 setae per bundle *Enchytraeus kincaidi* Eisen, 1904
11b Ampulla of each spermatheca with 2 large diverticula, one directed dorsally, the other ventrally; with dense glands covering the terminal halves of the spermathecal ducts, but with no distinct rosettes; with 2-3 setae per bundle *Enchytraeus multiannulatus* Altman, 1936
12a Each spermathecal ampulla with a round, subequal apical diverticulum (setae slightly sigmoid, 3 in pre-clitellar ventral bundles, 2 in other bundles; paired seminal vesicles present) *Stephensoniella trevori* (Coates, 1980)
12b Spermathecal ampullae without diverticula 13
13a Setae sigmoid, 2-6 per bundle (seminal vesicle absent; small amount of sperm in segment 11) *Marionina appendiculata* Nielsen & Christensen, 1959
13b Setae more or less straight, 2 or many in each bundle 14
14a Setae 3-6 (sometimes 7) per bundle, inner setae of bundle usually shorter than the outer setae; seminal vesicle absent (sperm morulae abundant in coelom) *Marionina vancouverensis* Coates, 1980

174 Phylum Annelida: Class Oligochaeta

14b Setal bundles with 2 setae; seminal vesicle present, dorsal 15
15a Sperm in distinct rings or balls, embedded in walls of ampullae of spermathecae
.... *Marionina sjaelandica* Nielsen & Christensen, 1959
15b Sperm in lumina of ampullae of spermathecae (ampullar walls thin)
.... *Marionina neroutsensis* Coates, 1980
16a With glandular cells covering the terminal half (or more) or each spermathecal duct, and with a distinct rosette of cells around each pore 17
16b Spermathecal ducts (which may be long or short) without glands except for a rosette around each pore 22
17a Glandular rosette of each spermathecal duct approximately equal to the length of the duct (ampullae of spermathecae elongate, onion-shaped; collars of sperm funnels wide, convoluted) .... *Lumbricillus mirabilis* Tynen, 1969
17b Height of glandular rosettes less than half the length of the spermathecal ducts 18
18a Glandular cells along each spermathecal duct sparse (ampullae onion-shaped; collars of sperm funnels not wavy, only as wide as the glandular funnel)
.... *Lumbricillus tsimpseanis* Coates, 1981
18b Glandular cells of each spermathecal duct dense, more or less completely covering the duct 19
19a Small (to 10 mm long); blood pale yellow or pink; length of each sperm funnel 1 to 2 times its width 20
19b Larger (10-20 mm long); blood usually red; length of each sperm funnel 2 or more times its width 21
20a Spermathecae attached to the gut near the middle of segment 5; lobes of pharyngeal glands extending into segment 7; vasa deferentia short, with only 1 or 2 loose loops in segment 12
.... *Lumbricillus curtus* Coates, 1981
20b Spermathecae attached to the gut at the level of the septum between segments 5 and 6; pharyngeal glands not extending into segment 7; vasa deferentia long, in a regular coil in segment 12 .... *Lumbricillus qualicumensis* Tynen, 1969
21a Diameter of each spermathecal ampulla equal or nearly equal to the width of the duct; duct 3-4 times as long as the ampulla .... *Lumbricillus annulatus* Eisen, 1904
21b Diameter of each spermathecal ampulla about 1.5 times the width of the duct; duct only 2-2.5 times as long as the ampulla .... *Lumbricillus pagenstecheri* (Ratzel, 1861)
22a Ampullae of spermathecae without distinct cavities, sperm heads embedded in the walls of the ampullae; proximal half of ampullae constricted; spermathecal ducts not more than one-fifth the length of the ampullae; length of each sperm funnel 2.5-5 times its width
.... *Lumbricillus lineatus* (O. F. Müller, 1774)
22b Ampullae of spermathecae with distinct cavities; sperm when present, in an organized bundle within the ampullae; polar regions of ampullar walls thicker than the equatorial regions; spermathecal ducts as long as the ampullae; length of each sperm funnel 2-3 times its width
.... *Lumbricillus tuba* Stephenson, 1911

3. Family Tubificidae

1a Coelomocytes large and abundant 2
1b Coelomocytes small and sparse, or absent 5
2a Dorsal setal bundles with hair setae, whose distal portions are twisted; pseudopenes with many sharp-hooked cuticular processes
.... *Monopylephorus cuticulatus* Baker & Brinkhurst, 1981
2b Setal bundles with bifid setae only (hair setae absent); pseudopenes lacking cuticular processes 3
3a With 4-7 modified penial setae per ventral bundle of segment 11; without gland cells on the atria .... *Vadicola aprostatus* Baker, 1982 (fig. 9.7)
3b Lacking penial setae; with diffuse gland cells on the atria 4
4a With only 1 spermatheca (on the left side) .... *Monopylephorus parvus* Ditlevsen, 1904
4b With 2 spermathecae .... *Monopylephorus rubroniveus* Levinsen, 1884
5a Spermatheca single, with a dorsal pore 6
5b With a pair of spermathecae, the pores ventral 8

6a   Cuticular penis sheaths present in segment 11 (sperm embedded in the distal tip of the spermatheca; each atrium with 2 stalked glands)   *Aktedrilus oregonensis* Strehlow, 1982

6b   Cuticular penis sheaths absent   7

7a   With gut diverticula in segment 9; with a large spermatheca in segment 10; without external spermatophores after mating; each atrium with 1 broadly attached gland
Limnodriloides monothecus Cook, 1974

7b   Without gut diverticula in segment 9; with small spermathecae in segment 10; with external spermatophores after mating; each atrium with 2 stalked glands
Bacescuella labeosa Baker & Erséus, 1982

8a   Body wall papillate   9

8b   Body wall not papillate   12

9a   Cuticular penis sheaths present in segment 11   10

9b   Cuticular penis sheaths absent   11

10a   Hair setae and bifid setae present; penis sheaths short and broad
Tubificoides brevicoleus Baker, 1983 (figs. 9.4, 9.5)

10b   Only bifid setae present; penis sheaths elongate, with slightly shovel-shaped tips
Tubificoides foliatus Baker, 1983

11a   Gut diverticula present in segment 9; sperm in random masses in spermathecae
Tectidrilus verrucosus (Cook, 1974)

11b   Gut diverticula absent in segment 9; sperm in organized bundles in spermathecae
Tectidrilus diversus Erséus, 1982

12a   Cuticular penis sheaths present in segment 11   13

12b   Cuticular penis sheaths absent   17

13a   Hair setae and bifid setae present   14

13b   Only bifid setae present   16

14a   Dorsal bifid setae not pectinate   *Tubificoides apectinatus* (Brinkhurst, 1965)

14b   Dorsal bifid setae pectinate   15

15a   Penis sheaths with a lateral protuberance; body with adherent foreign material; pectinate setae with diverging teeth   *Tubificoides nerthoides* (Brinkhurst, 1965)

15b   Penis sheaths without a lateral protuberance; body without adherent foreign material; pectinate setae with nearly parallel teeth   *Tubificoides kozloffi* Baker, 1983 (fig. 9.6)

16a   Spermathecae without sperm traps; posterior segments lacking glandular rings of cells about the setal bundles; segments of normal length posteriorly
Tubificoides pseudogaster (Dahl, 1960)

16b   Spermathecae with sperm traps; posterior segments with a single glandular ring of cells near the setal line; segments very elongate posteriorly
Tubificoides coatesae Brinkhurst & Baker, 1979

17a   Atria with 2 stalked glands; modified genital setae absent   *Bathydrilus torosus* Baker, 1983

17b   Atria with 1 or 2 stalked glands, or with a single, diffuse gland; modified genital setae present   18

18a   With spermathecal setae, and frequently also with penial setae   19

18b   With penial setae only   20

19a   With gut diverticula in segment 9; spermathecae in segment 10; with 1 spermathecal seta per ventral bundle of segment 10 (these setae are often missing, but the setal sacs in which they originate will still be evident); sometimes with 1 penial seta per ventral bundle of segment 11; both male systems open onto 1 large, protrusible pseudopenis
Limnodriloides victoriensis Brinkhurst & Baker, 1979

19b   Without gut diverticula in segment 9; spermathecae in segment 9; with 1 spermathecal seta per ventral bundle of segment 9; with 8-9 penial setae per ventral bundle of segment 11; both male systems open into a large median copulatory bursa
Rhizodrilus pacificus (Brinkhurst & Baker, 1979)

20a   Penial setae of 2 shapes or sizes   21

20b   Penial setae of 1 shape or size   22

21a   Penial setal bundles with 1 long, hairlike seta and 8-12 bifid setae; each atrium with a diffuse anterior gland and a stalked posterior prostate gland
Discordiprostatus longisetosus (Brinkhurst & Baker, 1979) ((fig. 9.9)

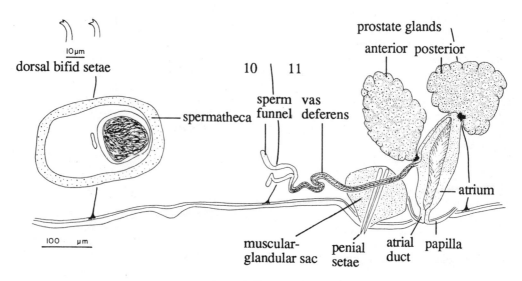

9.3 *Bathydrilus litoreus*

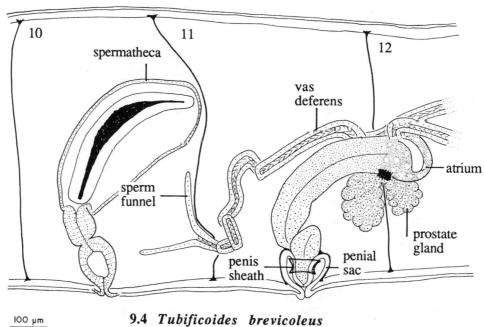

9.4 *Tubificoides brevicoleus*

9.5 *Tubificoides brevicoleus*   9.6 *Tubificoides kozloffi*

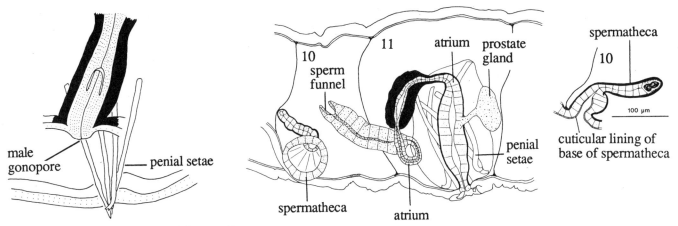

9.7 *Vadicola aprostatus* (the male gonopore is lateral, the penial setae median)

9.8 *Nootkadrilus grandisetosus*

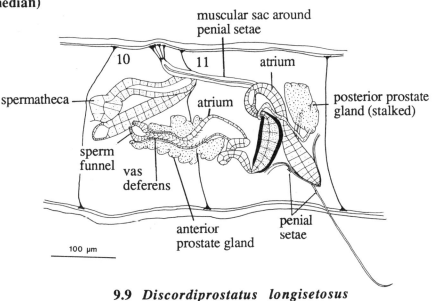

9.9 *Discordiprostatus longisetosus*

21b Penial setal bundles with 6-7 short and 4-6 longer hooked setae; part of the atrium surrounded by large, granular gland cells, and also with a stalked posterior gland
Nootkadrilus verutus Baker, 1982

22a With fewer than 8 penial setae 23
22b With more than 8 penial setae 26
23a Distal portion of penial setae flattened and hooked (with 4-6 setae in each bundle)
Nootkadrilus compressus Baker, 1982
23b Distal portions of penial setae hooked, single-pointed or bifid, but not flattened 24
24a Bundles of penial setae with 5 large, bifid setae
Nootkadrilus grandisetosus Baker, 1982 (fig. 9.8)
24b Bundles of penial setae with hooked or single-pointed setae 25
25a Bundles of penial setae with 3-5 slender, distally hooked setae at each male pore; spermathecae long and slender, the pores ventral *Phallodrilus tempestatis* Baker, 1981
25b Bundles of penial setae with 2 (sometimes 3) large, stout, single-pointed setae anterior to each male pore; spermathecae saclike, with lateral pores
Bathydrilus litoreus Baker, 1983 (fig. 9.3)

Phylum Annelida: Class Oligochaeta

26a Each bundle of penial setae with 9 (sometimes 10) large, conspicuously hooked setae
Nootkadrilus hamatus Baker, 1982
26b Each bundle of penial setae with 11-14 slender, slightly hooked setae
Nootkadrilus gracilisetosus Baker, 1982

Class Hirudinoidea

Subclass Hirudinea

Order Rhynchobdellida

Family Piscicolidae

All of the marine leeches known to occur in our region are piscicolids. They are mostly parasites of fishes, but several species have been reported from invertebrate hosts.

Burreson, E. M. 1976. *Trachelobdella oregonensis* sp. n. (Hirudinea: Piscicolidae), parasitic on the cabezon, *Scorpaenichthys marmoratus* (Ayres), in Oregon. J. Parasit., 62:793-8.
------. 1977a. *Oceanobdella pallida* n. sp. (Hirudinea: Piscicolidae) from the English sole, *Parophrys vetulus*, in Oregon. Trans. Amer. Micr. Soc., 96:526-30.
------. 1977b. Two new species of *Malmiana* (Hirudinea: Piscicolidae) from Oregon coastal waters. J. Parasit., 63:130-6.
------. 1984. A new species of marine leech (Hirudinea: Piscicolidae) from the north-eastern Pacific Ocean, parasitic on the English sole, *Parophrys vetulus* Girard. Zool. J. Linn. Soc., 80:297-301.
Epshtein, V. M. 1962. [A survey of the fish leeches (Hirudinea, Piscicolidae) from the Bering and Okhhotsk Seas and from the Sea of Japan.] Doklady Akad. Nauk SSSR, 144:1181-4. (English translation in Doklady, Biological Sciences Section, 144:648-51.)
Knight-Jones, E. W. 1962. The systematics of marine leeches. *In* Mann, K. H., Leeches (Hirudinea), Their Structure, Physiology, Ecology, and Embryology, pp. 169-86. London: Pergamon Press. x + 201 pp.
Meyer, M. C., & A. A. Barden, Jr. 1955. Leeches symbiotic on Arthropoda, especially decapod Crustacea. Wasmann J. Biol., 13:297-311.
Moore, J. P. 1952. New Piscicolidae (leeches) from the Pacific and their anatomy. Occ. Pap. Bishop Mus., 21:17-44.
Moore, J. P., & M. C. Meyer. 1951. Leeches (Hirudinea) from Alaskan and adjacent waters. Wasmann J. Biol., 9:11-77.
Soós, A. 1965. Identification key to the leech (Hirudinoidea) genera of the world, with a catalogue of the species. I. Family: Piscicolidae. Acta Zool., Budapest, 11:417-63.

Branchellion lobata Moore, 1952. On sharks and skates, California; possibly in our region.
Calliobdella knightjonesi Burreson, 1984. On *Parophrys vetulus* (Soleidae).
Johanssonia sp. On various rockfishes (Scorpaenidae) and on the anomuran crab *Chionoecetes tanneri*; egg capsules on the shrimp *Pandalus danae*.
Levinsenia rectangulata (Levinsen, 1882). On codfishes (Gadidae).
Malmiana diminuta Burreson, 1977. On various rockfishes, including *Sebastes melanops* (Scorpaenidae); on *Enophrys bison*, *Myoxocephalus polyacanthocephalus*, and *Scorpaenichthys marmoratus* (Cottidae); on *Hippoglossus stenolepis* (Pleuronectidae); on *Ophiodon elongatus* (Hexagrammidae); probably on other fishes.
Malmiana virida Burreson, 1977. On *Enophrys bison* (Cottidae).
Marsipobdella sacculata Moore, 1952. On various fishes, California; possibly in our region.
Notostomobdella cyclostoma (Johansson, 1898). A temporary parasite of various fishes, including sharks and rays; egg capsules on the anomuran crab *Paralithodes camtschatica*; not reported south of Alaska, but to be expected in our region.
Oceanobdella pallida Burreson, 1977. On *Parophrys vetulus* (Soleidae).
Ostreobdella papillata Oka, 1927. On various rockfishes (Scorpaenidae) and on *Octopus* sp.

Piscicola sp. Said to occur on various rockfishes (Scorpaenidae), but possibly not assigned to the correct genus.

Stibarobdella loricata (Harding, 1927). On various sharks, rays, and skates; widespread, and reported from California, but not as yet from our region.

Trachelobdella oregonensis Burreson, 1976. On *Scorpaenichtys marmoratus* (Cottidae).

Unidentified species. On the subtidal sea urchin *Allocentrotus pallidus*.

10

PHYLA
ECHIURA, SIPUNCULA, POGONOPHORA, VESTIMENTIFERA

PHYLUM ECHIURA

Fisher, W. K. 1946. Echiuroid worms of the North Pacific Ocean. Proc. U. S. Nat. Mus., 96:215-92.
------. 1949. Additions to the echiuroid fauna of the North Pacific Ocean. Proc. U. S. Nat. Mus., 99:479-97.
Stephen, A. C., & S. J. Edmonds. 1972. The phyla Sipuncula and Echiura. London: British Museum (Natural History). viii + 528 pp.

1a   With 2 nearly complete rings of bristlelike setae at the posterior end of the body
Echiurus echiurus subsp. *alaskanus*
1b   Without setae at the posterior end of the body   2
2a   With a pair of well developed setae near the anterior end of the trunk; proboscis (if still attached) troughlike, not branched; general coloration pinkish to reddish
Arhynchite pugettensis
When the genus *Arhynchite* was established, it was characterized by the apparent absence of a proboscis. Several species, including *A. pugettensis*, do have this structure, however. The proboscis is easily detached, and worms that have lost it may be normal in all other respects. It is possible that even the type species of *Arhynchite* will eventually be shown to have a proboscis.
2b   Without setae near the anterior end of the trunk or elsewhere; proboscis (if still attached) bifurcated; dorsal side of trunk greenish, proboscis and ventral side of trunk whitish
Nellobia eusoma

Species marked with an asterisk are not in the key.

Order Bonelloinea

Family Bonelliidae

Nellobia eusoma Fisher, 1946. Shallow subtidal.

Order Echiuroinea

Family Echiuridae

Echiurus echiurus subsp. *alaskanus* Fisher, 1946. Shallow subtidal, rarely low intertidal.
Arhynchite pugettensis Fisher, 1949. Shallow subtidal, low intertidal.

Listriolobus hexamyotus Fisher, 1949. Deep subtidal, Oregon southward.
Thalassema steinbecki Fisher, 1946. Deep subtidal, Oregon southward.

Order Xenopneusta

Family Urechidae

Urechis caupo Fisher & MacGinitie, 1928. Intertidal in muddy sand; widely distributed in California from Humboldt Bay southward, and to be expected in southern Oregon.

PHYLUM SIPUNCULA

Fisher, W. K. 1952. The sipunculid worms of California and Baja California. Proc. U. S. Nat. Mus., 102:371-450.
Rice, M. E. 1967. A comparative study of the development of *Phascolosoma agassizii*, *Golfingia pugettensis*, and *Themiste pyroides* with a discussion of developmental patterns in the Sipuncula. Ophelia, 4:143-71.
Stephen, A. C. 1964. A revision of the classification of the phylum Sipuncula. Ann. Mag. Nat. Hist., ser. 13, 7:457-62.
Stephen, A. C., & S. J. Edmonds. 1972. The phyla Sipuncula and Echiura. London: British Museum (Natural History). viii + 528 pp.

1a Tentacles inconspicuous, fingerlike, not branched 2
1b Tentacles conspicuous when extended, branching dichotomously several times and thus appearing bushy 4
2a Introvert with dark blotches and transverse streaks, and with rows of small hooks near its anterior end; tentacles arranged in a crescentic series dorsal to the mouth (body posterior to introvert sometimes with dark spots; common intertidal and shallow subtidal species, generally in gravel under rocks, in crevices, or nestling in burrows made in rock by pholadid bivalves) .... *Phascolosoma agassizii*
2b Introvert without dark spots or streaks, and without rows of small hooks near its anterior end; tentacles arranged in 1 or 2 circles around the mouth 3
3a Introvert constituting half, or slightly more than half, of the length of the body; skin nearly smooth (papillae present, but inconspicuous); whitish to dark gray; with 2 retractor muscles .... *Golfingia pugettensis*
3b Introvert constituting less than half the length of the body; skin decidedly rough; dark brown; with 4 retractor muscles .... *Golfingia vulgaris*
4a Introvert with small, black or brown spines; with 4 tentacles (these may, however, begin to branch so close to the base that there may appear to be more than 4) (common throughout the region, mostly in tight crevices or nestling in burrows made in rock by pholadid bivalves) .... *Themiste pyroides*
4b Introvert without spines; with 6 tentacles (not likely to be found north of Oregon) .... *Themiste dyscrita*

One species, marked with an asterisk, is not in the key.

Order Sipunculida

Family Golfingiidae

Golfingia pugettensis Fisher, 1952
Golfingia vulgaris (Blainville, 1827)
**Golfingia margaritacea* (M. Sars, 1851). Reported by Fisher (1952) from 2 localities in our region; needs study.
Themiste dyscrita (Fisher, 1952)
Themiste pyroides (Chamberlin, 1919)

Family Phascolosomatidae

Phascolosoma agassizii Keferstein, 1867

PHYLUM POGONOPHORA

Several species, all deep subtidal, have been reported for the region. Others should be expected.

Ivanov, A. V. 1957. Neue Pogonophora aus dem nordwestlichen Teil des Stillen Ozeans. Zool. Jahrb., Abt. Syst., Ökol. Geogr. Tiere, 85:431-500.
------. 1960 [Pogonophorans.] Fauna SSSR, n. s., no. 75. 271 pp. (in Russian). (English translation, 1963. London: Academic Press.)
------. 1961. [New pogonophorans from the eastern part of the Pacific Ocean. Part 1. *Galathealinum brachiosum* sp. n.] Zoologicheskii Zhurnal, 40:1378-84 (in Russian).
------. 1962. [New pogonophorans from the eastern part of the Pacific Ocean. Part 2. *Heptabrachia ctenophora* sp. n. and *Heptabrachia canadensis* sp. n.] Zoologicheskii Zhurnal, 41:893-900 (in Russian).
Southward, E. C. 1969a. New Pogonophora from the northeast Pacific Ocean. Can. J. Zool., 47:395-403.
------. 1969b. Growth of a pogonophore: a study of *Polybrachia canadensis* with a discussion of the development of taxonomic characters. J. Zool., London, 157:449-67.

Order Athecanephria

Family Sibloglinidae

Siboglinum fedotovi Ivanov, 1957. *S. vancouverensis* Southward, 1969 is a synonym.
Siboglinum pusillum Ivanov, 1960

Order Thecanephria

Family Polybrachiidae

Galathealinum brachiosum Ivanov, 1961
Heptabrachia ctenophora Ivanov, 1962
Polybrachia canadensis (Ivanov, 1962)

Family Lamellisabellidae

Lamellisabella coronata Southward, 1969
Lamellisabella zachsi Ushakov, 1933

PHYLUM VESTIMENTIFERA

The 3 species known to occur in our region are deep subtidal.

Jones, M. L. 1985. On the Vestimentifera, new phylum, six new species, and other taxa, from hydrothermal vents and elsewhere. Bull. Biol. Soc. Washington, no. 6:117-58.
Webb, M. 1969. *Lamellibrachia barhami*, gen. nov., sp. nov. (Pogonophora) from the northeast Pacific. Bull. Mar. Sci., 19:18-47.

Class Basibranchia

Order Lamellibrachiida

Family Lamellibrachiidae

Lamellibrachia barhami Webb, 1969. Off Oregon and southern California, at depths of about 1000-2000 m.

Order Tevniida

Family Ridgeiidae

Ridgeia piscesae Jones, 1985. Juan de Fuca Ridge, off British Columbia, at depths greater than 1500 m.
Ridgeia phaeophiale Jones, 1985. Juan de Fuca Ridge, off British Columbia, at depths greater than 1500 m.

11

PHYLUM MOLLUSCA: CLASSES APLACOPHORA, POLYPLACOPHORA

General References on Molluscs of the Pacific Northwest

Bernard, F. R. 1970. A distributional checklist of the marine molluscs of British Columbia: based on faunistic surveys since 1950. Syesis, 3:75-94.

Burch, J. Q. (ed.). 1944-46. Distributional list of the west American marine molluscs from San Diego, California, to the Polar Sea. Minutes Conchol. Club. So. Calif., nos. 45-63 (mimeographed).

Dall, W. H. 1921. Summary of the marine shell-bearing molluscs of the northwest coast of America, from San Diego, California, to the Polar Sea, mostly contained in the collection of the United States National Museum, with illustrations of hitherto unfigured species. Bull. 112, U. S. Nat. Mus. iii + 217 pp.

Hanna, G. D. 1966. Introduced mollusks of western North America. Occas. Pap. Calif. Acad. Sci., 48. 108 pp.

Keen, W. M., & E. Coan. 1974. Marine Molluscan Genera of Western North America: An Illustrated Key. 2nd ed. Stanford, California: Stanford University Press. vi + 208 pp.

MacGinitie, N. 1959. Marine Mollusca of Point Barrow, Alaska. Proc. U. S. Nat. Mus., 109:59-208.

Oldroyd, I. S. 1924. Marine shells of Puget Sound and vicinity. Publ. Puget Sound Biol. Station, 4:1-272.

------. 1925-27. The marine shells of the west coast of North America. Stanford Univ. Publ. Geol. Sci., 1:1-247; 2:1-941.

Quayle, D. B. 1964. Distribution of introduced marine Mollusca in British Columbia waters. J. Fish. Res. Bd. Canada, 21:1155-81.

Rice, T. 1971. Marine Shells of the Pacific Northwest. Edmonds, Washington: Ellison Industries. 102 pp.

Class Aplacophora

Morse, M. P. 1979. *Meiomenia swedmarki* gen. et sp. n., a new interstitial solenogaster from Washington, USA. Zool. Scripta, 8:249-53.

Scheltema, A. H. 1985. The aplacophoran family Prochaetodermatidae in the North American basin, including *Chevroderma* n. g. and *Spathoderma* n. g. (Mollusca: Chaetodermomorpha). Biol. Bull., 169:484-529.

Schwabl, M. 1963. Solenogaster mollusks from southern California. Pacific Sci., 17:261-81.

Subclass Caudofoveata

Order Chaetodermatida

Family Crystallophrissonidae

Crystallophrisson sp. Subtidal, off British Columbia; possibly one of the several species dealt with by Schwabl (1963).

Family Prochaetodermatidae

Chevroderma whitlachi Scheltema, 1985. At depths of 2800 m and greater.

Subclass Solenogastres

Order Neomeniamorpha

Family Neomeniidae

Neomenia dalyelli (Koren & Danielssen, 1877). Subtidal, off Oregon.

Family Meiomeniidae

Meiomenia swedmarki Morse, 1979. Subtidal, San Juan Archipelago.

Class Polyplacophora

Eugene N. Kozloff and Linda H. Price

Berry, S. S. 1917. Notes on west American chitons.--I. Proc. Calif. Acad. Sci., ser. 4, 7:229-48.
------. 1919. Notes on west American chitons--II. Proc. Calif. Acad. Sci., ser. 4, 9:1-36.
------. 1927. Some notes on British Columbian chitons. Proc. Malac. Soc. London, 17:159-64.
------. 1951. Notes on some British Columbian chitons--II. Proc. Malac. Soc. London, 28:213-29.
Cowan, G. I. McT., & I. McT. Cowan. 1977. A new chiton of the genus *Mopalia* from the northeast Pacific coast. Syesis, 10:45-52.
Dall, W. H. 1899. Report on the limpets and chitons of the Alaskan and Arctic regions, with descriptions of genera and species believed to be new. Proc. U. S. Nat. Mus., 1:281-344.
Eernisse, D. J. 1986. The genus *Lepidochitona* Gray, 1821 (Mollusca: Polyplacophora) in the northeastern Pacific Ocean (Oregonian and Californian Provinces). Zool. Verhandel. Leiden, no. 228. 52 pp.
Ferreira, A. J. 1978. The genus *Lepidozona* (Mollusca: Polyplacophora) in the temperate eastern Pacific, Baja California to Alaska, with the description of a new species. Veliger, 21:19-44.
------. 1982. The family Lepidopleuridae (Mollusca: Polyplacophora) in the eastern Pacific. Veliger, 22:145-65.
IAkovleva, A. M. 1952. [Shell-bearing molluscs (Loricata) of the seas of the USSR.] Opredeliteli po Faune SSSR, 45:1-127 (in Russian).
Kass, P., & R. A. Van Belle. 1985-. Monograph of the Living Chitons. Vol. 1, Order Neoloricata: Lepidopleurina, 240 pp.; vol. 2, Suborder Ischnochitonina, 198 pp.; other vols. in preparation. Leiden: E. J. Brill/Dr. W. Backhuys.
Pilsbry, H. A. 1892-4. Polyplacophora (Chitons). *In* Manual of Conchology, 14:i-xxxiv, 1-350; 15:1-133.
Smith, A. G., & I. McT. Cowan. 1966. A new deep-water chiton from the northeastern Pacific. Occ. Pap. Calif. Acad. Sci., no. 56. 15 pp.

1a All plates completely covered by the tough, reddish brown girdle; length commonly exceeding 20 cm . . . . *Cryptochiton stelleri*
1b All 8 plates visible; length not often exceeding 10 cm 2
2a Plates 2-7 divided along the midline by a strip of cartilagelike tissue (length up to 2.5 cm; rare, and not likely to be found south of British Columbia) . . . . *Schizoplax brandtii*
2b None of the plates divided along the midline 3

186   Phylum Mollusca: Class Polyplacophora

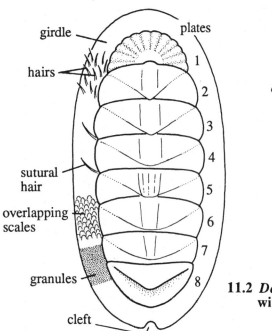

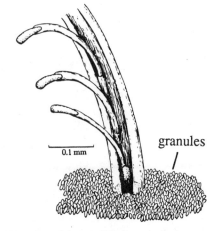

11.2 *Dendrochiton flectens*, hair with branches only on one side

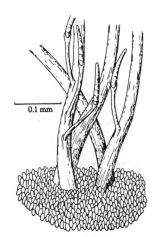

11.3 *Dendrochiton semiliratus*, hairs

11.1 Diagram of a chiton (dorsal view), showing principal features used in the key

3a   Girdle uniformly black or brownish black, covering about two-thirds of the area of each plate (visible portions of plates 2-8 not much wider than long) (girdle smooth; common mid-intertidal species)   *Katharina tunicata*

3b   Girdle not uniformly black or brownish black (except in some specimens of *Lepidochitona fernaldi*, choice 10b), and not covering as much as two-thirds of the area of each plate   4

4a   Plate 8, measured along the midline, about twice as long as plate 1   5

4b   Plate 8 not nearly twice as long as plate 1   6

5a   Dorsal surface of girdle with obvious overlapping scales, these somewhat pointed and up to about 0.3 mm long; plate 1 with 10-14 slits; plates 2-7 with 1-3 slits on both sides; foot pale; under rocks in mid-intertidal (not likely to be found north of California)   *Stenoplax heathiana*

5b   Dorsal surface of girdle with very small scales (mostly less than 0.2 mm long), these not at all pointed; plate 1 with 8-10 slits; plates 2-7 with 1 slit on both sides; foot (in life) bright orange; mostly subtidal (not likely to be found north of Oregon)   *Stenoplax fallax*

6a   Dorsal surface of girdle smooth or granular (granules, visible at a magnification of about 10x, generally have a circular outline), but without overlapping scales (scales usually have an oval outline), hairs, or spicules (there may, however, be spicules along the margin of the girdle)   7

6b   Dorsal surface of girdle with overlapping scales (clearly visible as such at a magnification of 10x), or with hairs or spicules (when hairs or spicules are present, the girdle may also be granular)   11

7a   Plates with prominent wavy or zigzag markings of white, yellow, pink, purple, or other bright colors; commonly more than 3 cm long   8

7b   Plates with various color patterns, but not predominantly with wavy or zigzag markings of bright colors; length not often exceeding 2 cm   9

8a   Anterior portions of plates 2-7 with wavy, light lines running transversely across the midline; girdle generally with subdued green blotches; mostly subtidal   *Tonicella insignis*

8b   Anterior portions of plates 2-7 without light lines running across the midline; girdle generally with conspicuous whitish, yellow, or orange blotches; common low intertidal and subtidal species   *Tonicella lineata*

9a With about 9 ctenidia on both sides, bordering the posterior third of the foot; plates without slits along their inserted margins (the anterior margin of at least 1 plate must be exposed if this feature is to be seen) (low intertidal and subtidal) *Leptochiton rugatus*
9b Usually with more than 9 ctenidia on both sides, bordering considerably more than the posterior third of the foot; plates with slits along their inserted margins 10
10a With 18-25 ctenidia on both sides, extending along at least four-fifths of the length of the foot; girdle usually with alternating light and dark bands, or with light spots; valves rarely eroded; not brooding young; eggs about 210 μm in diameter, surrounded by transparent, cone-shaped hulls; mostly low and mid-intertidal, but occurring in tidepools at higher levels
Lepidochitona dentiens
10b With 8-17 ctenidia on both sides, extending along not more than three-fourths the length of the foot; girdle usually dark, sometimes with white spots, but without alternating color bands; valves often eroded; brooding young; eggs about 260-280 μm in diameter, and nearly smooth, high intertidal *Lepidochitona fernaldi*
11a Dorsal surface of girdle without overlapping scales, but with spicules or flexible or stiff hairs
12
11b Dorsal surface of girdle with overlapping scales, but without spicules or hairs (spicules may, however, be present along the margin of the girdle) 16
12a Dorsal surface of girdle with crystalline spicules, but without hairs 13
12b Dorsal surface of girdle with hairs 23
13a With 7-14 ctenidia on both sides, bordering the posterior third of the foot; plates without slits along their inserted anterior margins (the anterior margin of at least 1 plate must be exposed if this feature is to be seen) 14
13b With at least 14 ctenidia on both sides, bordering much of the length of the foot; plates with slits along their inserted anterior margins 15
14a With 12-14 ctenidia; dorsal surface of girdle with numerous upright spicules; plates 2-7 usually without obvious sculpturing *Leptochiton nexus*
14b With 7-9 ctenidia; dorsal surface of girdle with only a few spicules; central areas of plates 2-7 with about 40-80 fine longitudinal lines *Leptochiton rugatus*
15a Exposed portions of plates 1-7 not more than twice as wide as long; with more than 50 ctenidia; color of plates and girdle (including spicules) mostly blackish brown; length sometimes exceeding 5 cm (high intertidal species, not likely to be found north of California)
Nuttallina californica
15b Exposed portions of most plates at least 3 times as wide as long; with 14-20 ctenidia; spicules glassy, but color of plates and girdle mostly reddish, orange, yellow, or olive (plate 8, however, is usually brownish or blackish); length up to 2 cm *Chaetopleura gemma*
16a Lateral areas of plates 2-7 conspicuously raised above the central area, swollen, and with pronounced tubercles; length of body almost 3 times the width (not likely to be found north of California) *Callistochiton crassicostatus*
16b Lateral areas of plates 2-7 not conspicuously raised above the central area and without pronounced tubercles (but there may be small nodules on the lateral areas); length of body not more than twice the width 17
17a Lateral areas of plates 2-7 with hemispherical or nearly globular nodules; central area usually with longitudinal ridges, or with longitudinal rows of pits separated by incipient ridges 19
17b Lateral areas of plates 2-7 without nodules; central area without longitudinal ridges or rows of pits, although there may be some delicate pitting 18
18a Lateral areas of plates 2-7 distinctly raised, usually with 2 radiating grooves that separate them into 3 flattened ribs; plate 1 with about 20 radiating grooves similar to those of lateral areas of plates 2-7; central areas of plates 2-7 with about 8 transverse rows of fine pits, although not all pits are restricted to these rows; length up to about 4 cm *Ischnochiton trifidus*
18b Lateral areas of plates 2-7 not distinctly raised, and without radiating grooves that separate them into flattened ribs; plate 1 without radiating grooves; central areas of plates 2-7 without transverse rows of fine pits (the surfaces of all plates are covered by small tubercles; length usually less than 2 cm *Ischnochiton interstinctus*
19a Central area of plates 2-7 with longitudinal rows of shallow but conspicuous pits, the longitudinal ridges between the pits often scarcely noticeable
Lepidozona retiporosa (fig. 11.7)

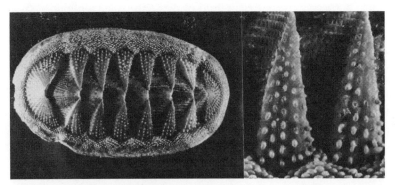
11.4 *Lepidozona mertensii*

11.5 *Lepidozona cooperi*

11.6 *Lepidozona willetti*

11.7 *Lepidozona retiporosa*

Scanning electron micrographs by Hans Bertsch, from Ferreira (1978)

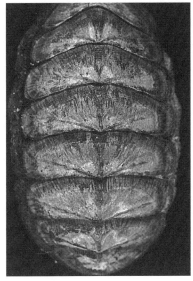
11.8 *Mopalia lignosa*

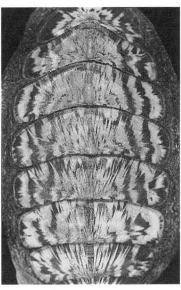
11.9 *Mopalia lignosa*

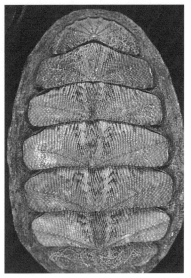
11.10 *Mopalia ciliata*

11.11 *Mopalia hindsii*

11.12 *Mopalia hindsii*

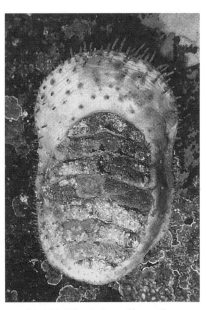
11.13 *Placiphorella velata*

19b Central area of plates 2-7 with rather distinct longitudinal ridges, even if rows of shallow pits are present between these 20

20a Longitudinal ridges near the midline of plates 2-7 diverging anteriorly to the extent that they form a wedge-shaped area; ridges prominent in all or much of the central area of plates 2-7 21

20b Longitudinal ridges near the midline of plates 2-7 not diverging anteriorly (or diverging only on plate 2); ridges in the central area of plates 2-7 indistinct (rare species) 22

21a Color olive or gray, often darker in the central portions of the plates; plates 1 and 8 with 11 slits; open coast .... *Lepidozona cooperi* (fig. 11.5)

21b Color variable, but generally reddish (sometimes greenish or purplish), occasionally speckled with cream, tan, or brown, and often with a lighter color in the central portions of the plates; plate 1 with 10 or 11 slits, plate 8 with 10 slits; inland waters as well as on the open coast .... *Lepidozona mertensii* (fig. 11.4)

22a Plate 1 with 32-50 (average 40) ribs and 10-13 (most commonly 12) slits; plate 8 with 25-40 (average 30) ribs and 11-14 (most commonly 11) slits; maximum length of girdle scales 250 µm; color uniformly orange-brown or creamy white; length up to 2.5 cm; mostly subtidal
Lepidozona scabricostata

22b Plate 1 with 20-40 (average 30) ribs and 10 or 11 (usually 11) slits; plate 8 with 15-28 ribs (average 21) ribs and 9-11 (most commonly 10) slits; maximum length of girdle scales 400 µm; color generally reddish brown; length up to 3 cm; subtidal   *Lepidozona willetti* (fig. 11.6)

23a Girdle much wider in the anterior part of the body than elsewhere
Placiphorella velata (fig. 11.13)

23b Girdle not appreciably wider in the anterior part of the body than elsewhere 24

24a Dorsal surface of girdle obviously granular, as well as with hairs 25

24b Dorsal surface of girdle not obviously granular 26

25a Central areas of plates 2-7 with a few nearly parallel longitudinal ridges on both sides of the median ridge; branching of sutural hairs (those in line with the sutures between plates) not limited to 1 side of the hairs; length not often greater than 1 cm (rare)
Dendrochiton semiliratus (fig. 11.3)

25b Central areas of plates 2-7 without parallel longitudinal ridges; sutural hairs either not branched or branched only on 1 side; length up to about 3 cm
Dendrochiton flectens (fig. 11.2)

26a Hairs stout, rubbery, and stiff (common intertidal species; length up to about 7 cm)
Mopalia muscosa

26b Hairs flexible, not thick 27

27a Length of some hairs, in addition to the sutural hairs (those in line with the sutures between plates), exceeding the width of the girdle; body length not often more than 1.2 cm; rare 28

27b Length of girdle hairs, except perhaps the sutural hairs, not exceeding the width of the girdle and generally not equaling it (in the rare *Mopalia imporcata*, some of the hairs are as long as the girdle is wide; moreover, young individuals of various species, until they have reached a length of at least 1 cm, may have proportionately long hairs) 29

28a Girdle hairs scattered, with many stout, strongly recurved branches   *Mopalia sinuata*

28b Girdle hairs branched, but the branches, which originate in a groove, are not especially recurved (besides the long sutural hairs, there are 4 hairs on the girdle in front of plate 1 and 1 hair on the girdle behind plate 8)   *Mopalia cirrata*

29a Lateral areas of plates 2-7 separated from the central area by a slight, nearly smooth ridge, or by a series of tubercles that are not considerably larger than those that form the more or less longitudinal rows on the central area and the oblique rows on the lateral areas 30

29b Lateral areas of plates 2-7 (or most of them) separated from the central area by a series of enlarged tubercles or by an obvious ridge of partly coalesced enlarged tubercles (the enlarged tubercles are located at the points where the more or less longitudinal rows of tubercles on the central area meet the oblique rows on the lateral areas) 32

30a Plates 2-7 smooth, without rows of tubercles and without pits; girdle with alternating light and dark bands, sparsely covered with branching hairs that are so delicate that the girdle appears naked; plates 2-7 generally brick-red, with zigzag turquoise markings in the lateral areas; plate 1 streaked with white; plate 8 with a triangular white patch extending posteriorly from its apex (common subtidal species)
Mopalia laevior

30b Plates 2-7 with rows of fine tubercles or with pits (the tubercles or pits are clearly visible with low magnification); girdle of a uniform color, spotted, or banded, with hairs that are easily seen, even when sparse; plates without the color patterns described in choice 30a 31

31a Plates 2-7 without obvious rows of raised tubercles, but with longitudinal rows of pits, which are especially pronounced in the central areas; lateral areas separated from the central areas by a slight, nearly smooth ridge; hairs on girdle abundant, robust and tubular; branches of hairs, if present, distributed along the entire length of the hairs; cleft in girdle at posterior end scarcely evident; interior surface of plates blue or white, chalky; ventral surface of girdle orange, even in preserved specimens (common intertidal species)
Mopalia lignosa (figs. 11.8 and 11.9)

31b Plates 2-7 with obvious rows of tubercles (clearly visible with low magnification); lateral areas separated from the central area by a series of tubercles that mark the places where the more or less longitudinal rows on the central areas meet the oblique rows on the lateral areas;

hairs on girdle sparse, very slender; branches of hairs, if present, originating only from the proximal portions of the larger hairs; cleft in girdle at posterior end pronounced; interior surface of plates white, not chalky; ventral surface of girdle not orange (common intertidal species, especially in bays) *Mopalia hindsii* (figs. 11.11 and 11.12)

32a Length of many hairs on the girdle equal to the width of the girdle; hairs abundant and scattered randomly; hairs with many branches, some of which are half the length of the hairs from which they originate; cleft in girdle at the posterior end not evident (rare)
Mopalia imporcata

32b Length of hairs on the girdle less than the width of the girdle; hairs either scattered randomly or (in *Mopalia porifera*) appearing to be in a definite pattern; branches of hairs, when present, no more than one-fourth the length of the hairs from which they originate; cleft in girdle at posterior end may or may not be evident 33

33a Hairs on girdle arranged in 3 staggered rows, the hairs of 1 row being sutural (in line with the sutures between plates 2-7); without a definite cleft in the girdle at the posterior end (branches of hairs, often calcified, originating on 1 side of each hair; rare) *Mopalia porifera*

33b Hairs on girdle scattered randomly; with a definite cleft in the girdle at the posterior end 34

34a Hairs on girdle sparse and so short (up to 0.5 mm long) that one may at first think that there are none (hairs with fine filaments coming mostly from the base, each filament tipped by a glassy spicule) *Mopalia swanii*

34b Hairs on girdle usually abundant and up to 3 or 4 mm long, unless broken off 35

35a Hairs on girdle strap-shaped, up to 3 mm long, with prominent glassy spicules along 1 side (usually the medial side); color variable, and sometimes similar to that of the next species (common intertidal and shallow subtidal species) *Mopalia ciliata* (fig. 11.10)

35b Hairs on girdle not obviously strap-shaped, up to 4 mm long, with filamentous branches tipped by glassy spicules; color almost always including bright turquoise and orange or reddish brown markings, especially on plate 2 *Mopalia spectabilis*

Species marked with an asterisk are not in the key.

Order Neoloricata

Suborder Lepidopleurina

Family Leptochitonidae

Leptochiton nexus (Carpenter, 1864)
Leptochiton rugatus (Carpenter in Pilsbry, 1892)

Family Hanleyidae

*Hanleya oldroydi (Dall, 1919). Subtidal, rare.

Suborder Chitonina

Family Ischnochitonidae

**Ischnochiton abyssicola* Smith & Cowan, 1966. Deep subtidal.
Ischnochiton interstinctus (Gould, 1846)
Ischnochiton trifidus (Carpenter, 1864). Subtidal and very low intertidal.
Stenoplax fallax (Carpenter in Pilsbry, 1892)
Stenoplax heathiana Berry, 1946
Lepidozona cooperi (Dall, 1919) (fig. 11.5)
Lepidozona mertensii (Middendorff, 1847) (fig. 11.4)
Lepidozona retiporosa (Carpenter, 1864) (fig. 11.7)

Lepidozona scabricostata (Carpenter, 1864)
Lepidozona willetti (Berry, 1917) (fig. 11.6)

Family Callistochitonidae

Callistochiton crassicostatus Pilsbry, 1892. Not likely to be found north of California.

Family Chaetopleuridae

Chaetopleura gemma Carpenter in Dall, 1879

Family Lepidochitonidae

Dendrochiton flectens (Carpenter, 1864) (fig. 11.2)
Dendrochiton semiliratus Berry, 1927 (fig. 11.3)
Lepidochitona dentiens (Gould, 1846)
Lepidochitona fernaldi Eernisse, 1986
**Lepidochitona hartwegii* (Carpenter, 1855). Not often found north of California.
Nuttallina californica (Reeve, 1847). Not likely to be found north of California.
Tonicella insignis (Reeve, 1847)
Tonicella lineata (Wood, 1815)
Schizoplax brandtii (Middendorff, 1847). Not likely to be found south of British Columbia.

Family Mopaliidae

Mopalia ciliata (Sowerby, 1840) (fig. 11.10)
Mopalia cirrata Berry, 1919
*Mopalia cithara Berry, 1951. Known only from the type specimen, collected on Vancouver Island, British Columbia.
*Mopalia egretta Berry, 1919
Mopalia hindsii (Sowerby in Reeve, 1847) (figs. 11.11 and 11.12)
Mopalia imporcata Carpenter, 1864
Mopalia laevior Pilsbry, 1918
Mopalia lignosa (Gould, 1846) (figs. 11.8 and 11.9).
Mopalia muscosa (Gould, 1846)
*Mopalia phorminx Berry, 1919
Mopalia porifera Pilsbry, 1893
Mopalia sinuata Carpenter, 1864
Mopalia spectabilis Cowan & Cowan, 1977
Mopalia swanii Carpenter, 1864
Katharina tunicata (Wood, 1815)
**Placiphorella rufa* Berry, 1917
Placiphorella velata Carpenter in Dall, 1879 (fig. 11.13). Possibly conspecific with *P. stimpsoni* (Gould, 1859).

Suborder Acanthochitonina

Family Acanthochitonidae

Cryptochiton stelleri (Middendorff, 1847)

12

PHYLUM MOLLUSCA: CLASS GASTROPODA

Key to Families of Gastropods with External Shells

Eugene N. Kozloff and Linda H. Price

The key does not include the following families: Seguenziidae (6), Cocculinidae (7), and Pseudococculinidae (8), reported only from deep water; Carinariidae (13), Atlantidae (14), and Pterotracheidae (15), restricted to oceanic plankton; Entoconchidae (40), internal parasites of sea cucumbers. In the text, these families are marked with an asterisk.

The lead to the order Patellogastropoda covers three families (Nacellidae, Acmaeidae, and Lottiidae), which can be separated by the key given under that order.

1a Shell permanently cemented to a hard substratum (usually rock), coiled loosely or irregularly and generally twisted, resembling the calcareous tube of a serpulid polychaete more than a snail shell   26. Vermetidae
1b Shell not permanently cemented to a hard substratum, either tightly coiled or tubular, conical, or cap-shaped   2
2a Shell tubular, conical, or cap-shaped, without obvious coiling   3
2b Shell obviously coiled, although the spire may be short   9
3a Shell tubular, slightly curved, the height more than twice the diameter   25. Caecidae
3b Shell conical or cap-shaped, the height rarely exceeding the greatest diameter   4
4a Shell with a dorsal opening at or near the apex, or with a slight indentation at the anterior margin (such an indentation is present only in the rare and subtidal *Arginula bella*; it marks the place where a groove on the interior of the shell, beginning at the apex, reaches the margin)   3. Fissurellidae
4b Shell with neither a dorsal opening at or near the apex nor an indentation on the anterior margin   5
5a Interior of shell with a shelf   31. Calyptraeidae
5b Interior of shell without a shelf   6
6a Apex of the shell usually at or anterior to the middle, sometimes slightly posterior to the middle (if only the shell is available, the open end of the horsehoe-shaped muscle scar faces anteriorly, except in the family Siphonariidae, in which the open end of a nearly C-shaped muscle scar faces the right side [left side, if the shell is observed in ventral view])   7
6b Apex of the shell decidedly posterior to the middle (at least as far as the beginning of the last quarter) (the foot secretes a calcareous base, and the animal remains attached to this)
                                                                                                                 30. Hipponicidae
7a Outline of the shell, in dorsal view, not symmetrical, the apex slightly to the right of the midline; muscle scar (often indistinct) nearly C shaped, its open end facing the right side (left side if the shell is observed in ventral view); with a shallow siphonal groove passing through the open end of the muscle scar   Subclass Pulmonata, Siphonariidae
7b Outline of the shell, in dorsal view, symmetrical (unless deformed by an injury), the apex on the midline; muscle scar horsehoe-shaped, its open end facing anteriorly; without a siphonal groove   8
8a Interior of shell uniformly whitish, without any color pattern; apex in the anterior third of the shell; exterior with rather conspicuous concentric lamellae (subtidal)   11. Lepetidae
8b Interior of shell usually with a color pattern of some sort (a blotch in the apical region, marginal markings, etc.); apex in the anterior or middle third of the shell (in the middle third

194 Phylum Mollusca: Class Gastropoda

in *Acmaea mitra*, the only species in which the interior is uniformly white); concentric lines not often conspicuous Order Patellogastropoda, Acmaeidae, Nacellidae, and Lottiidae

9a Abalones--shell low and earlike, with a low spire near the posterior end, and with a series of holes (some closed) near the left side; length sometimes exceeding 10 cm 2. Haliotidae

9b Shell generally not low and earlike (except in Lamellariidae and Velutinidae), and without a series of holes near the left side; only a few species larger than 10 cm, and nearly all of these have tall spires 10

10a Shell (completely or almost completely internal in Lamellariidae) with a low profile, thus resembling the shell of an abalone 11

10b Shell not resembling that of an abalone 12

11a Shell thin, translucent white, and completely or almost completely internal 34. Marseniidae

11b Shell rather firmly calcified, covered by an almost velvety periostracum, and to a large extent external 35. Velutinidae

12a Outer lip of the aperture with a deep slit (the shell, which has a short spire above the proportionately large body whorl, resembles that of a *Margarites* or *Lirularia* [Trochidae], its diameter being about equal to its height; found only at depths greater than 400 m) 1. Scissurellidae

12b Outer lip of the aperture without a deep slit (there may, however, be a notch or distinct siphonal canal at the anterior end of the aperture, and there may also be a slight notch on the outer lip near the posterior end of the aperture) (includes most intertidal and subtidal prosobranch gastropods, and some opisthobranchs and pulmonates) 13

13a Length of the aperture nearly equal to the height of the shell (if there is a spire, it is not raised) 14

13b Length of the aperture not nearly equal to the height of the shell (there is a raised spire) 15

14a Shell thick, similar in shape to that of a cowrie; with a series of teeth on the inside of the outer lip of the aperture; aperture not decidedly wider in its anterior half than elsewhere; height about 3 mm 51. Marginellidae: *Granulina margaritula* (fig. 12.58)

14b Shell thin, not resembling that of a cowrie; without teeth on the inside of the outer lip of the aperture; aperture usually widest in its anterior half; height up to nearly 2 cm, but much smaller in some species Subclass Opisthobranchia, Order Cephalaspidea

15a Much of the periostracum in the form of conspicuous hairs or bristles 16

15b Periostracum not in the form of conspicuous hairs or bristles 17

16a With a prominent siphonal canal that is about one third the total length of the aperture; height up to about 12 cm 36. Cymatiidae: *Fusitriton oregonensis*

16b Without a distinct siphonal canal, but the anterior end of the aperture is angled in such a way that it forms a small spout; height generally less than 4 cm 32. Trichotropididae

17a Anterior end of the aperture without a distinct siphonal notch, spout, or canal 18

17b Anterior end of the aperture with a distinct siphonal notch, spout, or canal 34

18a Interior of shell pearly (except in *Halistylus pupoideus*, family Trochidae); with an operculum 19

18b Interior of shell not pearly (it may, however, be colored); with or without an operculum 20

19a Operculum thin and horny, with numerous spiral lines 5. Trochidae

19b Operculum calcified and rather thick, with only a few spiral lines 4. Turbinidae

20a Diameter of the shell much greater than the height (aperture almost circular, equal to about half the total diameter, which does not exceed 4 mm; without periostracum) 21. Vitrinellidae: *Vitrinella columbiana*

20b Height of the shell equal to or greater than the diameter 21

21a Columella with 1 or more distinct folds or ridges; height less than 1 cm 22

21b Columella without any folds or ridges; height may exceed 1 cm 24

22a Columella usually with 3 folds, 1 of which is faint; shell without spiral ridges; periostracum chestnut-brown or yellowish brown; abundant among *Salicornia* and under debris in salt marshes, at levels that are not often inundated (do not confuse with species of Assimineidae or Truncatellidae, which have no folds on the columella, but which do occur in salt marshes) Subclass Pulmonata, Melampidae: *Phytia myosotis*

22b Columella with 1-3 folds; shell with or without spiral ridges; generally white or almost white, but certain species may be gray or light brown; not found in salt marshes 23

23a With spiral ridges, the furrows between these pitted; aperture about two-thirds of the height of the shell; body whorl with 2 dark brown or blackish bands, each the width of about 6

spiral ridges (there are similar bands on the spire); with 1 fold on the columella; often in eelgrass flats, but also occurring in other habitats
        Subclass Opisthobranchia, Acteonidae: *Rictaxis punctocaelatus*(fig. 12.109)

23b With or without spiral ridges, but if there are ridges, the furrows between them are not pitted; aperture not more than half the height of the shell; body whorl and spire without dark brown or blackish spiral bands; with 1-3 folds on the columella; external parasites of various invertebrates, including other molluscs Subclass Opisthobranchia, Pyramidellidae

24a Diameter about equal to the height; shell generally almost globose, consisting mostly of the body whorl (umbilicus either conspicuously open or covered by an obvious callus; height of some species attaining 10 cm)   33. Naticidae

24b Height decidedly greater than the diameter; shell not almost globose, the spire usually at least one-sixth of the total height   25

25a Height not more than twice the diameter   26

25b Height considerably more than twice the diameter   30

26a Umbilicus a slit between the columella and the body whorl (generally on eelgrass or on algae, especially kelps, mid- to low intertidal and subtidal)   16. Lacunidae

26b Umbilicus absent or indistinct (in some species, there is a narrow space between the body whorl and the edge of the inner lip of the aperture)   27

27a Height commonly exceeding 5 mm, and in some species slightly exceeding 1.5 cm; operculum horny; periostracum not uniformly tan or brown (*Algamorda subrotundata* [fig. 12.26] is brown, but usually has spiral banding) (mostly at higher tide levels on rocky shores, or on rocks, concrete and wood in bays, sometimes in salt marshes) 17. Littorinidae

27b Height not exceeding 4 mm; operculum horny or calcareous; periostracum sometimes uniformly tan or brown (*Assiminea californica*, Assimineidae, inhabits salt marshes, but members of the other families to which this couplet leads are typically found at lower levels of rocky or gravelly intertidal areas)   28

28a Exterior of shell generally pink and white, the colors often in somewhat zigzag blotches; operculum calcareous; mostly low intertidal, in gravel, under rocks, in beds of *Phyllospadix*, and on algae   4. Turbinidae: *Tricolia pulloides*

28b Exterior of shell usually tan or brown, without blotches; operculum horny or calcareous; at lower levels of rocky or gravelly intertidal areas or in salt marshes   29

29a Shell dark brown, smooth (although there may be wrinkles in the periostracum of the body whorl); spire with 4 or 5 whorls; operculum horny; common among *Salicornia* and under debris in salt marshes, mostly at a level that is not often inundated by sea water (usually occurs with *Phytia myosotis*, Melampidae)   20. Assimineidae: *Assiminea californica*

29b Shell tan or brown, smooth or with axial ribs; spire with 3 or 4 whorls; operculum horny or calcareous; on rocky and gravelly shores   18. Rissoidae

30a Shell highly polished, in some species slightly bent; sutures between whorls so slightly indented that they are barely evident (the whorls themselves, moreover, are nearly flat); parasitic on echinoderms, and strictly subtidal   39. Eulimidae

30b Shell not usually highly polished, and not bent; sutures between whorls distinct; not parasitic on echinoderms (but the Epitoniidae are parasitic on various cnidarians)   31

31a Spire decidedly tapered and generally with more than 3 whorls; not typically found at higher tide levels in salt marshes   32

31b Spire scarcely tapered, and generally with only 3 whorls (the older whorls disappear, and the uppermost surviving whorl is usually very short and smoothly rounded; typically at higher tide levels in salt marshes   22. Truncatellidae: *Cecina manchurica* (fig. 12.27)

32a Height of the spire shorter than that of the body whorl; spire usually with 4 or 5 surviving whorls; shell sculpture limited to prominent axial ribs, and when the shell is viewed with the aperture lowermost and facing the observer, the ribs are more nearly parallel to the left side of the shell than to the right side   19. Rissoinidae: *Rissoina newcombiana*

32b Height of the spire considerably greater than the height of the body whorl; spire usually with at least 6 surviving whorls; shell sculpture sometimes limited to axial ribs, but if so, the ribs are more or less parallel to both sides of the shell   33

33a With both axial ribs and spiral ridges (the intersections of these sometimes form beads), or with spiral ridges only   24. Turritellidae

33b Sculpture limited to axial ribs (except for a single spiral ridge near the base of the body whorl)   37. Epitoniidae

196 Phylum Mollusca: Class Gastropoda

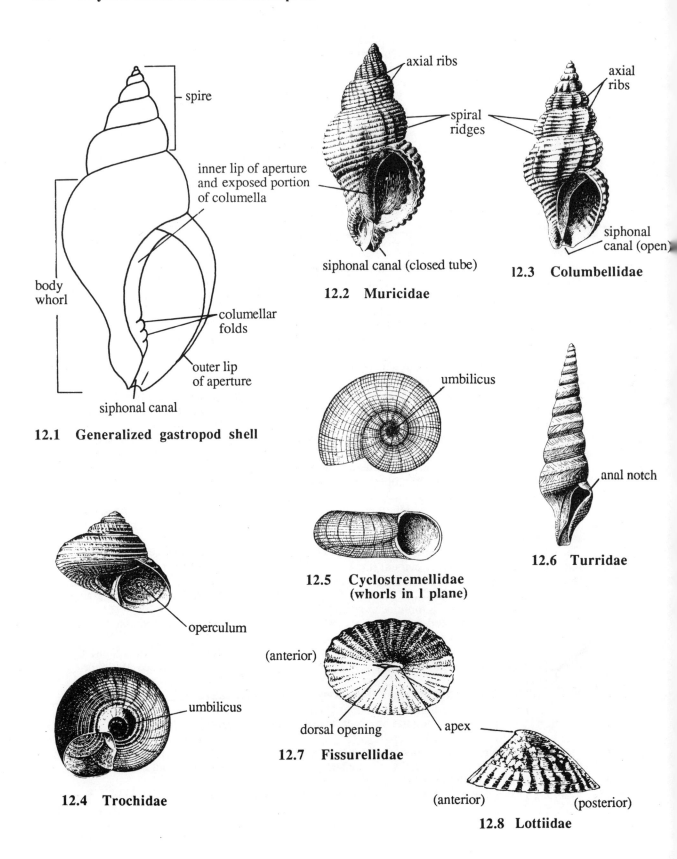

12.1 Generalized gastropod shell

12.2 Muricidae

12.3 Columbellidae

12.4 Trochidae

12.5 Cyclostremellidae (whorls in 1 plane)

12.6 Turridae

12.7 Fissurellidae

12.8 Lottiidae

34a Upper portion of the outer lip of the aperture with an anal notch
53. Turridae (figs. 12.6 and 12.92-12.94)
If the lip has been fractured, the contours of the growth lines nearest the lip are likely to indicate that an indentation had been present. In species that do not have a distinct anal notch, the following combination of characters may enable one to recognize them as members of this family: width of the aperture less than one-half, and usually less than one-third, the height of the aperture; prominent axial ribs; foot milky white. Presence of an anal notch, however, is the most reliable single feature.
In certain species of *Antiplanes,* the shell is coiled sinistrally rather than dextrally, and this is helpful in assigning them to the Turridae. Most members of the family are subtidal and they typically inhabit soft sediments. None in our region is likely to be found in a rocky intertidal habitat.
34b Upper portion of the outer lip of the aperture without an anal notch or trace of a notch 35
35a Shell highly polished, without any sculpture other than fine growth lines; widest part of the aperture (near its lower end) less than half the diameter of the shell; operculum inconspicuous; typically on sandy substrata 50. Olividae
35b Shell not polished, and usually with at least some sculpture; widest part of the aperture (generally near its middle) usually at least half the diameter of the shell; operculum usually conspicuous and large enough to close the aperture tightly; not necessarily limited to sandy substrata 36
36a Lowermost portion of the body whorl, including the siphonal canal, set off from the rest of it by a conspicuous groove that interrupts the axial ribs 46. Nassariidae (in part)
36b Lowermost portion of the body whorl not set off from the rest of it by a conspicuous groove that interrupts the axial ribs 37
37a Shell with at least 8 distinct whorls, not counting 1 or 2 that may have been worn away at the apex (in specimens that have been abruptly worn down to 4 or 5 whorls, one can estimate, on the basis of the surviving portion of the shell, that about 10 whorls have developed; height not exceeding 4 cm 38
37b Shell with not more than 7 whorls, even if none appears to have been worn away, except in specimens whose height commonly exceeds 12 cm 41
38a Siphonal canal half the total height of the aperture; axial ribs prominent, not forming beads where they intersect the faint spiral ridges 44. Neptuneidae: *Exilioidea rectirostris* (fig. 12.73)
38b Siphonal canal less than half the total height of the aperture; axial ribs and spiral ridges sometimes forming beads where they intersect 39
39a Siphonal canal short but obvious; spiral ridges distinctly (and sometimes conspicuously) beaded where intersected by the axial ribs 40
39b Siphonal canal barely evident; spiral ridges beaded in some species that have axial ribs, but not in the most common species (*Bittium eschrichtii* [fig. 12.30]); which lacks axial ribs
27. Cerithiidae
40a Siphonal canal narrow, directed toward the left at a nearly right angle to the long axis of the shell (as the shell is viewed with the aperture lowermost and facing the observer); with about 12 axial ribs; height up to 3.5 cm; usually in salt marshes
28. Potamididae: *Batillaria zonalis* (fig. 12.29)
40b Siphonal canal broad, directed to the left at about a 45º angle from the long axis of the shell; with at least 12 axial ribs (and sometimes more than 30); height not often exceeding 1 cm (some species reach 1.5 cm); often associated with sponges 29. Cerithiopsidae
41a Shell either with axial ribs as well as spiral ridges, or with only spiral ridges 42
41b Shell with only axial ribs 48. Volutidae
42a Shell limited to spiral ridges 55
42b Shell with axial ribs (these may be low and inconspicuous, however) or thin axial lamellae, as well as with spiral ridges 43
43a Axial ribs limited to the spire (the body whorl may have a few irregularly spaced grooves that are perpendicular to the spiral ridges, and young specimens of *Searlesia dira,* under 2 cm long, have axial ribs on the body whorl) 44
43b Axial ribs or lamellae extending to the body whorl, even though they may be less distinct on the body whorl than on the spire 45

44a	Columella with 2-5 folds; periostracum light brown; spiral ridges faintly evident on the inside of the outer lip of the aperture; subtidal	52. Cancellariidae: *Neoadmete modesta*
44b	Columella without folds; periostracum gray; dark lines in the furrows between spiral ridges of the body whorl visible on the inside of the aperture; intertidal and subtidal	43. Buccinidae: *Searlesia dira*
45a	Axial ribs on the body whorl restricted to the upper half of it (if continued into the lower half of the body whorl, they become much less prominent)	46
45b	Axial ribs distinct on at least much of the lower half of the body whorl, and as prominent as those on the upper half of the body whorl	49
46a	With a series of folds on the inside of the outer lip of the aperture as well as on the columella	45. Columbellidae (in part)
46b	Folds sometimes present either on the inside of the outer lip of the aperture or on the columella, but not on both	47
47a	With folds on the inside of the outer lip of the aperture (these coincide with spiral ridges on the outside of the body whorl); without folds on the columella	47. Fusinidae
47b	Without folds on the inside of the outer lip of the aperture; with or without folds on the columella	48
48a	Axial ribs rather sharp; with 2-5 (usually 2 or 3) folds on the columella	52. Cancellariidae: *Admete gracilior* (fig. 12.59)
48b	Axial ribs low and broad; without folds on the columella	44. Neptuneidae: *Plicifusus griseus*
49a	Siphonal canal (measured from the angle in the outer lip of the aperture where the canal begins) one-third to one-half the total height of the aperture	50
49b	Siphonal canal decidedly less than one-third the total height of the aperture	51
50a	Either with axial ribs and spiral ridges equally well developed and prominent, or axial ribs represented by thin lamellae and spiral ridges faint	41. Muricidae
50b	Axial ribs poorly developed in comparison with the prominent spiral ridges	44. Neptuneidae: *Neptunea tabulata* (fig. 12.79)
51a	Axial sculpture consisting of thin, frilly lamellae; spiral sculpture, which may be obscured by the lamellae, consisting of 1 or 2 prominent ridges on each whorl	42. Nucellidae: *Nucella lamellosa* (fig. 12.67)
51b	Axial sculpture consisting either of fairly broad ribs or of wrinkles (these may be irregular and discontinuous), but not of frilly lamellae; spiral sculpture consisting of closely spaced spiral ridges	52
52a	With 2 or 3 folds on the columella	52. Cancellariidae: *Cancellaria crawfordiana*
52b	Either with 1 fold or none on the columella	53
53a	Without a fold on the columella; periostracum, if persistent, straw-colored or dark brown	44. Neptuneidae (in part)
53b	With 1 fold on the columella; periostracum blackish or whitish	54
54a	Periostracum blackish, thick, cracked, but otherwise more or less continuous except where the spire is eroded; inside of aperture purplish brown; axial ribs and spiral ridges about the same size; height up to about 2.5 cm; intertidal, in mudflats	46. Nassariidae: *Ilyanassa obsoleta*
54b	Periostracum whitish or dull brown; inside of aperture white or pinkish; axial ribs much more pronounced than the spiral ridges; height up to 7 cm; subtidal	43. Buccinidae: *Buccinum plectrum* (fig. 12.72) and other species of *Buccinum*
55a	Columella with 2 or more folds	56
55b	Columella without folds	57
56a	Spiral ridges present over the entire shell; columella with about 5 folds; restricted to deep water	49. Volutomitridae
56b	Spiral ridges present only on the lowest part of the body whorl, near the siphonal canal; columella with 2 or 3 folds; intertidal to shallow subtidal	45. Columbellidae (in part)
57a	Sculpture consisting of closely spaced spiral ridges (in *Alia*, Columbellidae, these are faint and restricted to the lower half of the body whorl, and are sometimes absent)	58
57b	Sculpture consisting of only 1 or 2 prominent ridges on each whorl (the lower part of the body whorl generally has a number of smaller ridges; in occasional specimens of *Nucella emarginata*, Nucellidae, which typically has spiral ridges over most of the shell, the ridges may be obscure)	42. Nucellidae (in part)

58a Spiral ridges faint and limited to the lower half of the body whorl (they are most likely to be visible near the columella; use magnification); height up to about 1 cm
   45. Columbellidae (in part)
58b Spiral ridges (these may be faint) distributed over all of the larger whorls; height generally greater than 1 cm   59
59a With a sharp tooth on the lower half of the outer lip of the aperture
   42. Nucellidae: *Acanthina spirata* (fig. 12.66)
59b Without a sharp tooth on the lower half of the outer lip of the aperture   60
60a Siphonal canal less than one-fourth the total height of the aperture; height generally less than 4 cm; common intertidal species   42. Nucellidae (in part)
60b Siphonal canal one-fourth to one-third the total height of the aperture (the outer lip of the aperture usually has an angular indentation where the canal begins); height commonly more than 4 cm in most species; strictly subtidal   44. Neptuneidae (in part)

Subclass Prosobranchia

Order Archaeogastropoda

McLean, J. H. 1984. New species of northeast Pacific archaeogastropods. Veliger, 26:233-9.

Suborder Pleurotomariina

1. Family Scissurellidae

The only species reported for the region are small and restricted to deep subtidal situations. In both, the diameter exceeds the height, and there is a deep slit on the outer lip of the aperture.

Anatoma crispata (Fleming, 1828). Height up to 5 mm; at depths greater than 500 m.
Anatoma baxteri McLean, 1984. Height up to about 2 mm.

2. Family Haliotidae

Cox, K. W. 1962. California abalones, family Haliotidae. Bull. 118, Calif. Dept. Fish & Game. 133 pp.

1a Exterior of shell smooth, bluish black or greenish black; holes circular and flush with the surface (length up to about 15 cm; not likely to be found north of Oregon)
   Haliotis cracherodii Leach, 1814
1b Exterior of shell lumpy, wavy, or with spiral ridges crossed by raised striations; reddish (sometimes with mottling); holes oval and slightly raised   2
2a Exterior of shell with low spiral ridges crossed by closely spaced, raised striations (without a muscle scar; reddish, with pale green, blue, or nearly white mottling; length to about 15 cm, but rarely greater than 12 cm; not likely to be found north of Oregon)
   Haliotis walallensis Stearns, 1899
2b Exterior of shell lumpy or wavy   3
3a Shell thin, without an obvious muscle scar; length up to about 15 cm
   Haliotis kamtschatkana Jonas, 1845
3b Shell thick, with a prominent muscle scar; length to about 30 cm (not likely to be found north of Oregon)   *Haliotis rufescens* Swainson, 1822

3. Family Fissurellidae

Cowan, I. M. 1969. A new species of gastropod (Fissurellidae, *Fissurisepta*) from the eastern north Pacific Ocean. Veliger 12:24-6

Cowan I. M., & J. H. McLean. 1968. A new species of *Puncturella* (*Cranopsis*) from the northeastern Pacific. Veliger 11:105-8.

McLean, J. H. 1984. New species of northeast Pacific archaeogastropods. Veliger, 26:233-9.

1a   Shell with an oval or circular opening at the apex, or with a slit just anterior to the apex   2

1b   Shell without an oval or circular opening or slit (on the interior of the shell, however, there is a groove running from near the apex to the anterior margin; this takes the place of the exhalant opening typical of the family; ribs very coarse, with nodules and other roughenings; length sometimes slightly greater than 7 cm; subtidal, mostly on vertical rock cliffs)
Arginula bella (Gabb, 1865)

2a   Opening in shell oval or nearly circular, at the apex   3

2b   Opening in shell a narrow slit, just anterior to the apex   4

3a   Length of opening about one-third the length of the shell; shell not more than 2 cm long, and covered by the mantle when the animal is alive   *Fissurellidea bimaculata* Dall, 1871

3b   Length of opening about one-tenth the length of the shell; shell often 5 cm long, only a small, marginal portion of it covered by the mantle when the animal is alive
Diodora aspera (Rathke, 1833)

4a   Anterior slope of shell with a slight seam extending from the opening to the margin   5

4b   Anterior slope of shell without a seam extending from the opening to the margin   7

5a   Primary and secondary ribs of about the same size; margin nearly smooth, not obviously dentate where the ribs reach it (length up to about 2 cm)
Puncturella decorata Cowan & McLean, 1968

5b   Primary and secondary ribs of distinctly different sizes; margin obviously dentate where the primary ribs reach it   6

6a   With about 30 primary radial ribs (count near the apex), these separated, as they approach the margin, by 1-3 secondary ribs (when there are 3, the middle one may be nearly as prominent as the primary ribs, especially at the margin); marginal serrations, where the primary ribs reach the margin, not projecting more than 1 mm (length up to 3.5 cm)
Puncturella multistriata Dall, 1914

6b   With 13-23 (usually about 16) sharp, raised primary ribs, most of these separated, as they approach the margin, by 3 or 4 faint secondary ribs; marginal serrations, where the primary ribs reach the margin, projecting 2-3 mm (length up to 4 cm)
Puncturella cucullata (Gould, 1846)

7a   Height generally equal to the width, and sometimes to the length; with about 27-33 radial ribs; length up to about 1 cm   *Puncturella cooperi* Carpenter, 1864

7b   Height not quite equal to the width; with about 49-63 radial ribs; length up to 2 cm
Puncturella galeata (Gould, 1846)

Three other species have been reported from the region, but are restricted to deep water.

Fissurisepta pacifica Cowan, 1969
Puncturella expansa (Dall, 1896)
Puncturella rothi McLean, 1984

Suborder Trochina

4. Family Turbinidae

1a   Interior of shell pearly; diameter about equal to the height or greater than the height   2

1b Interior of shell not pearly; height about 1.5 times the diameter (height up to about 6 mm; generally with pinkish red, zigzag markings; usually in gravel, under rocks, in holdfasts of algae, and in beds of *Phyllospadix*) .... *Tricolia pulloides* (Carpenter, 1865) (fig. 12.11)
2a Whorls bordered by a continuous, wavy excrescence, at the base of which are 2 rows of beadlike protuberances; whorls with prominent axial (but slightly oblique) ribs; diameter sometimes exceeding 5 cm (almost completely restricted to the open coast)
.... *Astraea gibberosa* (Dillwyn, 1817) (fig. 12.10)
2b Whorls not bordered by a continuous, wavy excrescence; whorls without prominent axial ribs; diameter not exceeding 1 cm 3
3a Umbilicus open 4
3b Umbilicus closed 5
4a Diameter of umbilicus about half that of the aperture; whorls without spiral ridges; operculum not appreciably calcified; diameter up to about 2.5 mm, considerably greater than the height (caution: do not confuse with young *Margarites pupillus* [fig. 12.12] or other species of the family Trochidae) .... *Spiromoelleria quadrae* (Dall, 1897)
4b Diameter of umbilicus much less than half that of the aperture; whorls with spiral ridges; operculum well calcified; diameter up to 4 mm, only slightly greater than the height
.... *Homalopoma lacunatum* (Carpenter, 1864)
5a Shell with prominent spiral ridges (about 15 on the body whorl); diameter up to about 1 cm
.... *Homalopoma luridum* (Dall, 1885) (fig. 12.9)
5b Shell nearly smooth, appearing slightly shiny, but nevertheless with fine spiral ridges; diameter up to 3 mm (base of body whorl, near the columella, white)
.... *Homalopoma subobsoletum* (Willett, 1937)

5. Family Trochidae

1a Height more than twice the diameter; interior of shell not pearly (aperture about one-fourth the height of the shell; lower portions of whorls sometimes bulging slightly; spiral ridges numerous on each whorl, varying slightly in width and spacing; height not more than 7 mm; shallow subtidal, on sandy and gravelly substrata) .... *Halistylus pupoideus* (Carpenter, 1864)
1b Height much less than twice the diameter; interior of shell pearly 2
2a Columella with 1 or 2 small nodes 3
2b Columella without any nodes 5
3a Umbilicus covered by a callus, nearly always closed 4
3b Umbilicus open (periostracum brown) .... *Tegula pulligo* (Gmelin, 1791)
4a Shell with a purplish black periostracum, and with a scaly band below the suture; with 2 nodes on the columella (the lower one often worn away); mid-intertidal on open coast
.... *Tegula funebralis* (A. Adams, 1855)
4b Shell with a brown periostracum, and without a scaly band below the suture; with 1 node on the columella; low intertidal, and not likely to be found north of Oregon
.... *Tegula brunnea* (Philippi, 1848)
5a Umbilicus decidedly open 6
5b Umbilicus closed (caution: in young specimens it may be open) 15
6a Umbilical area bordered by a beaded or crenulate spiral ridge (or 2 such ridges) that is decidedly more conspicuous than the other spiral ridges of the basal part of the body whorl; subtidal 7
6b Umbilical area not bordered by a beaded or crenulated spiral ridge that is decidedly more conspicuous than the other spiral ridges of the basal portion of the body whorl; includes intertidal and subtidal species 9
7a With 6 or 7 whorls; height (up to 2 cm); equal to, or slightly greater than, the diameter; periostracum tan, with darker axial streaks (in addition to pronounced spiral ridges, there are slight axial ribs) .... *Solariella peramabilis* Carpenter, 1864 (fig. 12.44)
7b With 5 whorls; height (not often exceeding 1 cm) slightly less than the diameter; periostracum without darker axial streaks 8
8a With axial ribs .... *Solariella vancouverensis* (E. A. Smith, 1887)
8b Without axial ribs .... *Solariella obscura* (Couthouy, 1838) (fig. 12.45)

202 Phylum Mollusca: Class Gastropoda

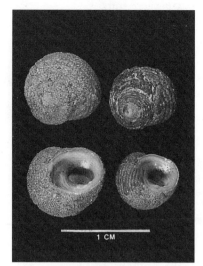

12.9 *Homalopoma luridum*

12.10 *Astraea gibberosa*

12.11 *Tricolia pulloides*

12.12 *Margarites pupillus*

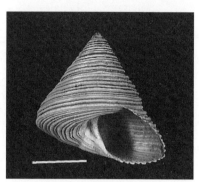

12.13 *Calliostoma canaliculatum*

12.14 *Margarites marginatus*

12.15 *Lirularia lirulata*

12.16 *Lirularia succincta*

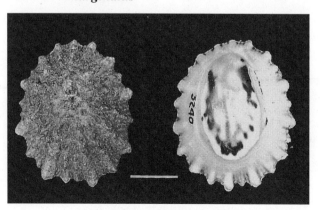

12.17 *Macclintockia scabra*

12.18 *Lottia digitalis*

9a With obvious spiral ridges on the spire and on the body whorl (the spiral ridges can usually be seen without magnification) 10
9b Without obvious spiral ridges on the spire and on the body whorl above the aperture (faint spiral ridges may be present, but may not be visible except at a magnification of at least 10x) 14
10a Base of body whorl with about 10-12 spiral ridges below the one that is at the level of the uppermost part of the aperture (in *Margarites rhodia* [fig. 12.49], however, the spiral ridges on the base are faint) 11
10b Base of body whorl with not more than 8 spiral ridges below the one that is at the level of the uppermost part of the aperture 13
11a Spiral ridges on base of body whorl prominent 12
11b Spiral ridges on base of body whorl faint, intersected by cross-striations of about the same size (height up to about 7 mm) .... *Margarites (Margarites) rhodia* Dall, 1921 (fig. 12.49)
12a Height up to about 1.7 cm, equal to or slightly greater than the diameter; spiral ridges prominent, and those on the body whorl above the aperture usually of 2 different sizes; base of body whorl, between the umbilicus and the periphery, sloping uniformly
    Margarites (Valvatella) pupillus (Gould, 1841) (figs. 12.12 and 12.48)
12b Height up to about 6 mm, equal to or slightly less than the diameter; spiral ridges not prominent, and those on the body whorl above the aperture all of about the same size; base of body whorl, between the umbilicus and the periphery, not sloping uniformly (several of the spiral ridges occupy a flattened or slightly depressed area that runs toward the aperture)
    Lirularia succincta (Carpenter, 1864) (figs. 12.16 and 12.46)
13a Spiral ridges on most whorls distinct, but not especially angular or otherwise pronounced; without axial sculpture in the form of sharp, evenly spaced lamellae (common intertidal and shallow subtidal species) .... *Lirularia lirulata* (Carpenter, 1864) (fig. 12.15)
    This is an extremely variable species, and some specimens approach those of *L. parcipicta* (fig. 12.47), choice 13b.
13b Spiral ridges on most whorls somewhat angular, pronounced; with axial sculpture in the form of sharp, evenly spaced lamellae .... *Lirularia parcipicta* (Carpenter, 1864) (fig. 12.47)
14a Spire and upper part of the body whorl with faint, low spiral ridges similar to those on the base of the body whorl; interior of aperture iridescent, but not markedly so (common intertidal and shallow subtidal species)
    Margarites (Margarites) marginatus Dall, 1919 (fig. 12.14)
    Commonly called *M. helicinus* (Phipps, 1874), a separate species that does not occur in our region.
14b Spire and upper part of the body whorl without even faint spiral ridges (there may, however, be faint spiral ridges at the base of the body whorl); interior of aperture highly iridescent (intertidal and subtidal, but uncommon, and not likely to be found as far south as Washington) .... *Margarites (Margarites) beringensis* (E. A. Smith, 1899)
15a Shell without spiral ridges, and with an almost waxy appearance (color usually white, yellowish, or pinkish); rare, not often encountered at depths of less than 100 m
    Calliostoma platinum Dall, 1890
15b Shell with spiral ridges, and without a waxy appearance; includes some intertidal as well as subtidal species 16
16a Spiral ridges beaded 17
16b Spiral ridges not beaded 20
17a Beads on spiral ridges pronounced; base of body whorl decidedly convex; sutures between whorls deeply indented; periostracum gray, brown, or olive-green; subtidal 18
17b Beads on spiral ridges small and delicate; base of body whorl nearly flat; sutures between whorls not deeply indented; periostracum not gray, brown, or olive-green; includes intertidal and subtidal species 19
18a Diameter about three-fourths of the height; periostracum gray or brown; height up to 4 cm; shallow subtidal .... *Cidarina cidaris* (Carpenter, 1864) (fig. 12.43)
18b Diameter more than three-fourths of the height; periostracum olive-green; height up to 6 cm; mostly at depths of more than 350 m .... *Bathybembix bairdi* (Dall, 1889)
19a Shell usually orange-yellow, dotted with brown and with a bright purple or violet band encircling the lower edge of each whorl; with 8 or 9 whorls; body of living animal orange, with brown dorsal spots .... *Calliostoma annulatum* (Lightfoot, 1786)

19b Shell almost uniformly yellowish, tan, or pale pink; with 6-8 whorls; body of living animal cream, with brown dorsal spots *Calliostoma variegatum* (Carpenter, 1864)
20a Whorls, as the shell is viewed in profile, almost flat; periostracum more nearly tan or brown than pinkish brown or reddish brown; not common
Calliostoma canaliculatum (Lightfoot, 1786) (fig. 12.13)
20b Whorls slightly convex; periostracum usually pinkish brown or reddish brown; common
Calliostoma ligatum (Gould, 1849)

The following have also been reported, but most of them are restricted to deep water.

Calliotropis carlotta (Dall, 1902). At depths of 1600 m or more.
Calliotropis ceratophora (Dall, 1896). At depths of 1600 m or more.
Calliostoma bernardi McLean, 1984. Shallow subtidal, but very rare.
Margarites simblus Dall, 1903. At depths of 1000 m or more.
Solariella nuda Dall, 1896. At depths of 360 m or more.

6. Family *Seguenziidae

The following species occur in deep water.

Carenzia inermis Quinn, 1983. At depths of 2800 m or more, off Oregon.
Seguenzia cervola Dall, 1919. At depths of 1500 m or more.
Seguenzia quinni McLean, 1985. At depths of 2800 m or more.
Seguenzia stephanica Dall, 1908. At depths of 750 m or more.
Seguenzia megaloconcha Rokop, 1972. At depths of 2700 m or more.

7. Family *Cocculinidae

Cocculina cowani McLean, 1986. Deep subtidal, off Queen Charlotte Islands.
Gymnocrater baxteri McLean, 1986. Deep subtidal, Alaska to Vancouver Island.

8. Family *Pseudococculinidae

Colotrachelus careyi McLean, 1986. At depths greater than 2700 m, off Oregon.

Order Patellogastropoda

Suborder Nacellina

Lindberg, D. R. 1981. Acmaeidae. Gastropoda. Mollusca. (Invertebrates of the San Francisco Bay Estuary System.) Pacific Grove, California: Boxwood Press. xii + 122 pp.

1a Interior of shell uniformly white or faintly yellowish, without any dark markings; exterior whitish, unless it has a thin, light brown periostracum or is blackened by a deposit; apex at the end of the first one-third or slightly anterior to this; with delicate, scaly radial ribs or with threadlike striations that may have scales or beads where they are intersected by concentric growth lines (the striations are sometimes obscure, but at least a few of them should be visible when the shell is examined with the aid of 10x magnification); primarily subtidal in our region 11. Lepetidae
1b Interior of shell not uniformly white (it is either uniformly dark or has dark markings, such as an apical blotch, marginal band, or series of marginal streaks or spots; the only exception

is *Acmaea mitra*, Acmaeidae, whose shell is nearly conical, with the apex close to the center); apex generally near the end of the first one-third, but sometimes considerably anterior to this, and sometimes central; radial ribs, when present, usually prominent and rather stout; mostly intertidal or shallow subtidal 2

2a Interior of shell uniformly white; exterior white (unless coated by an encrusting coralline alga, which is usually the case); shell nearly conical, about as high as wide, with the apex close to the center; without radial ribs 10. Acmaeidae: *Acmaea mitra*

2b Interior of shell not uniformly white, except in *Niveotectura funiculata*, Lottiidae (it may be almost completely dark, or have an apical blotch, marginal band, or marginal streaks or spots); exterior rarely white, unless eroded by a fungus or abraded; shell not usually conical or as high as wide (there are exceptions); radial ribs present in some species 3

3a Height generally about one-fourth the length of the shell (but sometimes higher); margins prominently scalloped (the places where the pronounced radial ribs meet the margin may project as points); animal white, with dark spots on the head and sides of the foot
 9. Nacellidae: *Macclintockia scabra* (fig. 12.17)

3b Height generally (but not always) more than one-fourth the length of the shell; margins not so prominently scalloped that the places where the radial ribs meet the margin project as points; animal sometimes white, but if so, without dark spots on the head and sides of the foot (includes most limpets of our region) 12. Lottiidae

9. Family Nacellidae

Macclintockia scabra (Gould, 1846) (fig. 12.17). This is the only representative of the family in our region. It was previously called *Collisella scabra* and assigned to the Acmaeidae. As a rule, the animals live in shallow depressions in rock, returning to these after foraging. The body is white, with dark spots on the head and foot. The margin of the shell is prominently scalloped, and the places where the radial ribs meet the margin usually project as points. This species is not likely to be found north of Oregon.

10. Family Acmaeidae

Acmaea mitra Rathke, 1833. Low intertidal and subtidal; generally encrusted by a coralline red alga; common throughout the region.

11. Family Lepetidae

McLean, J. H. 1985. Two new northeastern Pacific gastropods of the families Lepetidae and Seguenziidae. Veliger, 27:336-8.

1a Shell with broad, flat radial ribs that are visible without magnification (the ribs have imbricating scales where they are intersected by concentric growth lines, and the distance between them is about 4 or 5 times their width); shell translucent; length up to 1 cm
 Iothia lindbergi McLean, 1985

1b Shell with threadlike radial striations that may not be visible without magnification (the striations are sometimes scaly or have small pustules where they are intersected by the concentric growth lines); shell opaque; length frequently more than 1 cm 2

2a Radial striations ornamented by small pustules where they are intersected by concentric growth lines; length up to 1.5 cm *Lepeta caeca* (O. F. Müller, 1776)

2b Radial striations scaly or smooth, but not ornamented by pustules; length up to about 2 cm 3

3a Radial striations very close together, somewhat flat-topped, not roughened by scales; width of shell usually close to three-fourths the length; height usually more than one-third the width
 Cryptobranchia concentrica (Middendorff, 1847)

3b Radial striations rather widely spaced, often roughened by scales; width of shell usually

about four-fifths the length; height usually about one-third the width
Limalepeta caecoides (Carpenter, 1864)

12. Family Lottiidae

Fritchman, H. K. 1960. *Acmaea paradigitalis* sp. nov. (Acmaeidae, Gastropoda). Veliger, 2:53-7.
Lindberg, D. R. 1981. Acmaeidae. Gastropoda. Mollusca. (Invertebrates of the San Francisco Bay Estuary System.) Pacific Grove, California: Boxwood Press. xii + 122 pp.
------. 1986. Name changes in the "Acmaeidae." Veliger, 29:142-8.

Macclintockia scabra, Nacellidae, is included in the key because it could be mistaken for a member of the Lottiidae.

1a   Apex (located near the end of the first one-eighth) decidedly lower than the highest part of the shell, which is near the middle; length up to 10 cm (not often found north of California)
Lottia gigantea Sowerby, 1834
1b   Apex the highest point of the shell; length rarely more than 5 cm   2
2a   Interior of shell uniformly white, without an apical blotch or marginal markings (apex at the end of the first one-third; shell with prominent, flat radial ribs; length not exceeding 2 cm; associated with encrusting coralline algae and often coated by a bryozoan of the genus *Rhynchozoon*; subtidal, rare)   *Niveotectura funiculata* (Carpenter, 1864)
2b   Interior of shell not uniformly white (it may be uniformly dark or have markings, such as an apical blotch, marginal band, or marginal streaks or spots)   3
3a   Right and left margins of the shell nearly parallel for much of their length, the outline of the aperture thus somewhat elongated   4
3b   Right and left margins of the shell curving outward, the outline of the shell thus oval (a form of *Lottia pelta* that lives on the kelp *Egregia*, and forms of *L. pelta* and *L. ochracea* that live on stipes of laminarians, may also have nearly parallel margins; see choices 4a and 7a)   8
4a   Right and left margins lower than the anterior and posterior margins, so the shell can be rocked when it is on a flat surface (on stipes of laminarian kelps, such as *Pterygophora*, to which it clings after the fashion of a saddle; length up to about 2 cm; shell brown) (caution: forms of *Lottia pelta* and *L. ochracea* also occur on the stipes of laminarians, and the shape of their shells may resemble that of this species; as a rule, however, they are not predominantly brown)   *Lottia instabilis* (Gould, 1846)
4b   Right and left margins at essentially the same horizontal level as the anterior and posterior margins, so the shell has little tendency to rock when on a flat surface   5
5a   Length about 4 times the width; apex near the end of the first quarter (restricted to the leaves of *Phyllospadix*; shell brown; length up to about 8 mm)   *Tectura paleacea* (Gould, 1853)
5b   Length not more than 3 times the width; apex near the end of the first third or closer to the middle   6
6a   Length about 3 times the width (exterior of shell mostly whitish, but with brown radial markings and sometimes with a brown spot at the apex; on coralline red algae; length up to about 8 mm; rare)   *Lottia triangularis* (Carpenter, 1864)
6b   Length not more than twice the width   7
7a   Exterior and interior of shell dark brown; length up to 2 cm; on the kelp *Egregia*, whose stipes it erodes (height about three-fourths the width)   *Discurria insessa* (Hinds, 1842)
A form of *Lottia pelta* that lives on *Egregia* may also have nearly parallel lateral margins and be brown externally; internally, however, the shell is mostly bluish, and the height of the shell is only about half the width.
7b   Exterior of shell mostly dark brown, but with light markings (these are more or less radial or contribute to a checkerboard pattern), interior bluish, with a brown apical stain and with external markings showing through; length up to 1.2 cm; on leaves of eelgrass, *Zostera marina* (not likely to be found south of Washington)   *Lottia alveus* (Conrad, 1831)
8a   Exterior of shell mostly pink, with white and/or light brown markings (length up to about 8 mm; on coralline red algae; subtidal, rare)   *Tectura rosacea* (Carpenter, 1864)
8b   Exterior of shell not mostly pink, unless overgrown by coralline red algae   9

9a Exterior of shell mostly whitish, but with brown radial markings and a brown spot at the apex; apex nearly in the middle; on coralline red algae; length up to about 8 mm (rare)
....*Lottia triangularis* (Carpenter, 1864)
9b Exterior of shell not mostly whitish (unless encrusted by a coralline alga that becomes white), with or without markings, but not typically with a brown spot at the apex; apex usually decidedly anterior to the middle; except for *Lottia painei* and a form of *L. ochracea*, not generally on coralline red algae; length commonly more than 1 cm, except in *Lottia asmi* (includes most of the intertidal and shallow subtidal limpets) 10
10a Shell grayish or brownish black externally, uniformly black internally; length not often more than 1 cm; mostly on *Tegula funebralis*, sometimes on *Mytilus californianus*, rarely on rock
....*Lottia asmi* (Middendorff, 1847) (fig. 12.21)
10b Color not uniformly black internally, although the exterior may be mostly dark; length commonly more than 1 cm, except in *Lottia ochracea*; except for some small *L. pelta*, not likely to be found on *Tegula funebralis* (large *L. pelta*, however, occur on *Mytilus californianus*) 11
11a Shell with prominent ribs radiating from the apex to the margin 12
11b Shell without prominent ribs radiating from the apex to the margin (if there are ribs, they are delicate and closely spaced) 15
12a Apex anterior to the end of the first quarter, and sometimes nearly level with the anterior margin; radial ribs more prominent posterior to the apex than anterior to it (common high intertidal species) ....*Lottia digitalis* (Rathke, 1833) (fig. 12.18)
12b Apex at about the end of the first quarter or first third; ribs anterior to the apex about as well developed as those posterior to it 13
13a Height generally about one-fourth of the length (but sometimes more than one-fourth); margins prominently scalloped (the places where the radial ribs meet the margin may project as points) (length up to 3 cm; animal white, with dark spots on the sides of the foot; generally in depressions in rock, to which the animals return after foraging; not likely to be found north of Oregon) .... Family Nacellidae: *Macclintockia scabra* (Gould, 1846) (fig. 12.17)
13b Height generally more than one-fourth of the length; margins not prominently scalloped (the places where the radial ribs meet the margin are rounded and do not project as points) 14
14a Ribs with overlapping scales, thus rough to the touch
....*Lottia limatula* (Carpenter, 1864) (fig. 12.20)
14b Ribs without overlapping scales, not rough to the touch .... *Lottia pelta* (Rathke, 1833)
15a Length usually not more than 1.3 times the width 16
15b Length usually at least slightly more than 1.3 times the width 17
16a Height usually not more than one-third the width; exterior usually with a pattern of light and dark streaks or blotches; length sometimes exceeding 5 cm (common mid- and low intertidal and shallow subtidal species) .... *Tectura scutum* (Rathke, 1833) (fig. 12.22)
16b Height usually more than one-third the width, exterior mostly uniformly brown, olive, or gray, but sometimes with a few white spots; length not often exceeding 3 cm (mostly on smooth boulders and rocks in the mid- and low intertidal) .... *Tectura fenestrata* (Reeve, 1855)
17a Shell thin, translucent; coloration of exterior resembling a checkerboard of white and dark (usually brownish or reddish) markings (these show through to the interior) (length up to about 1 cm) .... *Lottia ochracea* (Dall, 1871)
17b Not generally thin or translucent except in young specimens; coloration of exterior sometimes with light spots and light and dark radial streaks, but not resembling a checkerboard 18
18a Interior of shell white or faintly bluish, with a continuous dark border, or with a border of dark spots, but without a blotch at the apex (exterior of shell often so badly pitted by a fungus that it is uniformly dull gray; length not often more than 1.5 cm; mostly mid-intertidal)
....*Lottia strigatella* (Carpenter, 1864) (fig. 12.19)
This species shows considerable variation in its wide geographic range, which extends to Mexico. Southern specimens are usually rather different from those of our region, but not in radular characters. *Lottia paradigitalis* (Fritchman, 1960) is considered to be a synonym.
18b Interior of shell with a blotch at the apex (in *Lottia painei*, however, the blotch is small and faint) 19
19a Apical blotch on interior of shell large, dark brown; margin of interior with a continuous dark band; shell length commonly exceeding 3 cm; not regularly associated with coralline red algae or encrusted by an alga of this type 20

19b Apical blotch on interior of shell small, faint, grayish; margin of interior with short, nearly radial streaks, or with both streaks and a narrow band; shell length rarely exceeding 2 cm; associated with coralline red algae and typically encrusted by an alga of this type
Lottia painei Lindberg, 1987
20a Apex at the end of the first one-third, often inclined slightly forward; posterior and lateral slopes of shell markedly convex; lateral margins of shell often slightly higher than the anterior and posterior margins, so that the shell can be rocked from side to side when placed on a flat surface; apical blotch usually at least half the length of the shell; length up to about 5 cm; at higher levels of the intertidal region, and mostly at the edges of the undersides of boulders, except when foraging *Tectura persona* (Rathke, 1833)
20b Apex slightly behind the end of the first one-third, not often inclined forward; posterior and lateral slopes of the shell slightly if at all convex; lateral margins of the shell usually at the same horizontal level as the anterior and posterior margins; apical blotch usually about one-third the length of the shell; length up to about 4 cm; mostly mid- and low intertidal
Lottia pelta (Rathke, 1833)

Order Mesogastropoda

Suborder Heteropoda

13. Family *Carinariidae

The following species, all pelagic and primarily oceanic, have been reported to occur in our region.

Carinaria cristata (Linnaeus, 1776)
Carinaria japonica Okutani, 1955. Perhaps just a form of *C. cristata*.
Carinaria lamarcki Peron & Lesueur, 1810
Carinaria latidens Dall, 1919

14. Family *Atlantiidae

Atlantia gaudichaudi Souleyet, 1852. May be expected in oceanic plankton; perhaps a form of *A. peronii* Lesueur, 1817.

15. Family *Pterotracheidae

The following species, all pelagic and primarily oceanic, have been reported.

Pterotrachea coronata Forskål, 1775
Pterotrachea hippocampus Philippi, 1836
Pterotrachea minuta Bonnevie, 1920
Pterotrachea scutata Gegenbaur, 1855

Suborder Taenioglossa

16. Family Lacunidae

The key is not reliable for very young specimens (shell height under 3 mm).

1a Height of aperture nearly or fully two-thirds the height of the shell 2
1b Height of aperture about half the height of the shell (in some specimens of *Lacuna variegata* (fig. 12.24) the height of the aperture is slightly more than half the height of the shell, but definitely not two-thirds the height) 3

2a Shell glossy brown, often with lighter mottling, without wrinkles, but sometimes with faint spiral lines; height up to about 6 mm (not likely to be found north of Oregon)
Lacuna (*Lacuna*) *marmorata* Dall, 1919
2b Shell dull brown, usually without mottling, but with microscopic wrinkles (in general, the wrinkles are oriented in much the same way as spiral lines or ridges, but they are more irregular) *Lacuna* (*Lacuna*) *porrecta* Carpenter, 1864
3a Shell mostly light tan, with chevron markings of brown under a thin, dull periostracum; without distinct spiral lines; larger whorls sometimes with a slight, flattened shoulder; height up to about 7 mm; common on *Zostera* in bays (not likely to be found south of Washington)
Lacuna (*Epheria*) *variegata* Carpenter, 1864 (fig. 12.24)
3b Shell tan or brown, usually with a broad, whitish spiral band under a glossy periostracum; with fine spiral lines; whorls evenly rounded; height up to 1.3 cm; on algae on rocky shores (not likely to be found south of Washington)
Lacuna (*Epheria*) *vincta* (Montagu, 1803) (fig. 12.23)
L. carinata (Gould, 1849) is a synonym.

17. Family Littorinidae

Mastro, E., V. Chow, & D. Hedgecock. 1982. *Littorina scutulata* and *Littorina plena*: sibling species status of two prosobranch gastropod species confirmed by electrophoresis. Veliger, 24:239-46.
Murray, T. 1979. Evidence for an additional *Littorina* species and a summary of the reproductive biology of *Littorina* from California. Veliger, 2:469-74.
------. 1982. Morphological characterization of the *Littorina scutulata* species complex. Veliger, 24:233-8.
Rosewater, J. 1978. A case of double primary homonymy in eastern Pacific Littorinidae. Nautilus, 92:123-5.

1a Diameter of shell almost equal to the height; often with rather prominent spiral ridges (height up to about 2 cm; common, but mostly north of Oregon; primarily on rocky shores, but also on rocks, wood, and debris in bays and salt marshes) *Littorina sitkana* Philippi, 1845
1b Height of shell decidedly greater than the diameter; without obvious spiral ridges 2
2a Spire generally consisting of 3 whorls; usually with a white band inside the lower portion of the aperture, near the columella (color, except when the shell is eroded, usually brown or nearly olive, with irregular lighter mottling; not likely to be found north of Oregon)
Littorina keenae Rosewater, 1978 (fig. 12.25)
2b Spire generally consisting of 4 whorls; without a white band inside the aperture 3
3a Shell thick (typical for *Littorina*); without an umbilicus; height to 1.5 cm; color usually dark brown, purple, or black, often with a checkerboard pattern of whitish areas; common on rocky shores and also found on rocks in bays *Littorina scutulata* Gould, 1849
Littorina plena Gould, 1849, known to occur in Oregon and California, is similar to *L. scutulata* and generally cannot be distinguished from the latter on the basis of shell characters. The penis of male *L. plena* has a distinct marginal projection near the middle of its length, and the portion distal to this projection is more slender than the proximal portion and usually curved; in *L. scutulata*, the penis tapers almost gradually to a blunt tip, there being neither a prominent projection nor a slender distal portion. Egg capsules produced by female *L. plena* resemble automobile wheels in having equal rims; in egg capsules of *L. scutulata*, one rim projects much farther than the other. For more details on these differences, as well as on differences obtained by electrophoretic studies, see the papers of Mastro, Chow, & Hedgecock (1982) and Murray (1979, 1982).
3b Shell thinner than typical for *Littorina*; with an umbilicus that is partly covered by the flangelike inner lip of the aperture; height to 8 mm; color brown (there are generally 2 or 3 wide, spiral bands of dark brown separated by narrow bands of lighter brown); restricted to salt marshes, where it is found on vegetation
Algamorda subrotundata (Carpenter, 1864) (fig. 12.26)

18. Family Rissoidae

Bartsch, P. 1911. The recent and fossil mollusks of the genus *Alvania* from the west coast of America. Proc. U. S. Nat. Mus., 41:333-62.

1a	Shell smooth (except perhaps for a faint spiral ridge on the body whorl); aperture nearly circular; operculum calcareous	2
1b	Shell with spiral ridges or axial ribs, or both; aperture more nearly oval than circular; operculum horny	4
2a	Body whorl with a faint spiral ridge (this makes the body whorl appear slightly angular); height to about 4 mm	*Barleeia acuta* (Carpenter, 1864)
2b	Body whorl without a faint spiral ridge and not at all angular; height to about 3.5 mm	3
3a	Whorls decidedly convex, often separated by deep indentations	*Barleeia subtenuis* (Carpenter, 1864)
3b	Whorls nearly flat, not separated by deep indentations	*Barleeia haliotiphila* (Carpenter, 1864)
4a	Spiral ridges rather pronounced and beaded, but axial ribs faint or lacking; aperture slightly less than one-third the height	*Alvania carpenteri* (Weinkauff, 1885)
4b	Both spiral ridges and axial ribs prominent and about equal, the sculpture thus cancellate (axial ribs may be absent on the lower part of the body whorl, however); aperture one-third to two-fifths the height	6
5a	Aperture about one-third the height	*Alvania sanjuanensis* Bartsch, 1920
5b	Aperture about two-fifths the height	*Alvania compacta* (Carpenter, 1864)

A. rosana Bartsch, 1911 is perhaps synonymous; it is, on the average, not quite so tall as *compacta* (the height of most specimens is slightly less than twice the diameter).

19. Family Rissoinidae

Bartsch, P. 1915. The recent and fossil mollusks of the genus *Rissoina* from the west coast of America. Proc. U. S. Nat. Mus., 49:33-62.

Rissoina newcombiana Dall, 1897

20. Family Assimineidae

Assiminea californica Tryon, 1865. Common in salt marshes, under *Salicornia* and under debris; usually with *Phytia myosotis* (subclass Pulmonata, order Archaeopulmonata, family Melampidae).

21. Family Vitrinellidae

Vitrinella columbiana (Bartsch, 1921). The only species likely to be found in our region.

22. Family Truncatellidae

Cecina manchurica A. Adams, 1861 (fig. 12.27). Common in some salt marshes in Washington and British Columbia; introduced from the Orient.

23. Family *Choristidae

Choristes carpenteri Dall, 1896. Deep; reported off Washington.
Choristes coani Marincovitch, 1975. Deep; reported off Oregon.

12.19 *Lottia strigatella*

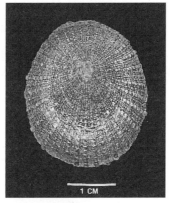
12.20 *Lottia limatula*

12.21 *Lottia asmi*

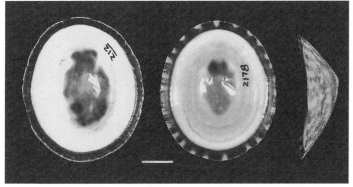
12.22 *Tectura scutum*

12.23 *Lacuna vincta*

12.24 *Lacuna variegata*

12.25 *Littorina keenae*

12.26 *Algamorda subrotundata*

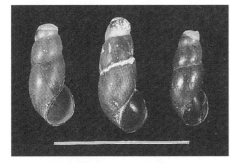
12.27 *Cecina manchurica*

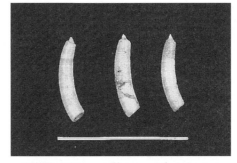
12.28 *Micranellum crebricinctum*

12.29 *Batillaria zonalis*

24. Family Turritellidae

1a Without axial ribs; spiral ridges almost flat, separated by shallow channels about half their width; height to 3 cm   *Tachyrhynchus erosus* (Couthouy, 1838)
1b With axial ribs; spiral ridges convex (even if faint); height not more than 2 cm   2
2a Spiral ridges and axial ribs intersecting, but not always forming definite beads; height up to 1 cm   *Turritellopsis acicula* (Stimpson, 1851)
2b Spiral ridges conspicuously beaded where they intersect the axial ribs; height up to about 2 cm   *Tachyrhynchus lacteolus* (Carpenter, 1864) (fig. 12.50)

25. Family Caecidae

1a Shell encircled by prominent, closely spaced circular rings; periostracum light tan or almost white; height to about 7 mm   *Micranellum crebricinctum* (Carpenter, 1864) (fig. 12.28)
1b Shell almost smooth, the circular rings being scarcely evident even if the periostracum is worn away; periostracum brown; length to about 4 mm   *Fartulum occidentale* Bartsch, 1920

26. Family Vermetidae

Keen, A. M. 1961. A proposed reclassification of the gastropod family Vermetidae. Bull. British Mus. Nat. Hist. (Zool.), 7:183-214.

1a Shell loosely coiled into a flat spiral, so that the entire shell of some individuals in the population is distinct; exterior rough, due to intersection of axial ribs with raised growth lines; diameter of aperture to about 4 mm   *Dendropoma lituella* (Mörch, 1886)
1b Coiling indistinct, the numerous members of the population forming contorted masses; exterior almost smooth; diameter of aperture to about 2.5 mm
          Petaloconchus compactus (Carpenter, 1864)

27. Family Cerithiidae

Bartsch, P. 1911. The recent and fossil mollusks of the genus *Bittium* from the west coast of America. Proc. U. S. Nat. Mus., 40:383-414.

1a Without axial ribs on the body whorl and whorl above it, so that the spiral ridges on these whorls are not noticeably beaded; height up to about 2.2 cm
          Bittium eschrichtii (Middendorff, 1849) (fig. 12.30)
1b With axial ribs on the body whorl, or at least on the whorl above it (as well as on the upper whorls), so that the spiral ridges are slightly to conspicuously beaded; height to about 1.2 cm   2
2a Spiral ridges conspicuously beaded on all whorls
          Bittium munitum (Carpenter, 1864) (fig. 12.31 and 12.51)
    B. vancouverense Dall & Bartsch, 1910, *B. fastigiatum* (Carpenter, 1864), and *B. oldroydae* Bartsch, 1911 are here considered to be synonyms.
2b Spiral ridges only faintly beaded   *Bittium attenuatum* Carpenter, 1864 (fig. 12.32)
    B. subplanatum Bartsch, 1911 is here considered to be a synonym.

28. Family Potamididae

Batillaria zonalis (Bruguière, 1792) (fig. 12.29). Common in some salt marshes, where it has been introduced from the Orient; *Batillaria attramentaria* (Sowerby, 1855) is here considered to be a synonym.

29. Family Cerithiopsidae

Bartsch, P. 1909. A new species of *Cerithiopsis* from Alaska. Proc. U. S. Nat. Mus., 37:399-400.
------. 1911. The recent and fossil mollusks of the genus *Cerithiopsis* from the west coast of North America. Proc. U. S. Nat. Mus., 40:327-67.
------. 1917. Descriptions of new west American marine mollusks and notes on previously described forms. Proc. U. S. Nat. Mus., 52:637-81.
------. 1921. New marine mollusks from the west coast of America. Proc. Biol. Soc. Washington, 34:34-40.

This group needs careful study. Some of the species names proposed by Bartsch are almost certainly synonyms, and are here treated as such.

1a   Basal portion of the body whorl, to the left of the aperture, with 3 distinct spiral ridges (these cross the columella, reaching just to the inside of the aperture); beads of the uppermost spiral ridge on each whorl usually about equal in size to those of the other 2 beaded ridges
                                                            Cerithiopsis columna Carpenter, 1864 (fig. 12.33)
       C. cosmia Bartsch, 1907 is probably a synonym.
1b   Basal portion of the body whorl nearly smooth, without 3 distinct spiral ridges; beads of the uppermost spiral ridge on each whorl usually slightly smaller than those of the other 2 beaded ridges   2
2a   Lowermost portion of the siphonal canal turned slightly to the left (as the aperture is lowermost and facing the observer)   *Cerithiopsis stejnegeri* Dall, 1884
       The following are here considered to be synonyms: *C. charlottensis* Bartsch, 1917; *C. onealensis* Bartsch, 1921; *C. paramoea* Bartsch, 1911; *C. stephensae* Bartsch, 1909.
2b   Lowermost portion of the siphonal canal straight   *Cerithiopsis signa* Bartsch, 1921

30. Family Hipponicidae

Cowan, I. M. 1974. The west American Hipponicidae and the application of *Malluvium*, *Antisabia*, and *Hipponix* as generic names. Veliger, 16:377-80.

Hipponix cranioides Carpenter, 1864. Found on rocky shores of the open coast; native.
Sabia conica (Schumacher, 1817). Widespread in the western Pacific, locally introduced in British Columbia.

31. Family Calyptraeidae

1a   Shell conical, almost circular in outline; with a spiral septum (shelf), the free edge of which is twisted   *Calyptraea fastigiata* Gould, 1846
1b   Shell usually cap-shaped (sometimes nearly conical), generally not circular in outline and often slightly irregular; with a simple septum   2
2a   Septum attached more broadly on one side (the left side, as the aperture faces the observer and the shelf is uppermost), so that part of it projects like a tongue
                                                       Crepidula dorsata (Broderip, 1834)
2b   Septum attached to about the same extent on both sides, its free margin therefore nearly straight   3
3a   Shell definitely arched (often higher than wide), the apex prominent and beaklike; periostracum dark brown; on shells of living *Tegula*, *Calliostoma*, and other gastropods
                                                       Crepidula adunca Sowerby, 1825 (fig. 12.34)

214 Phylum Mollusca: Class Gastropoda

12.30 *Bittium eschrichtii*

12.31 *Bittium munitum*

12.32 *Bittium attenuatum*

12.33 *Cerithiopsis columna*

12.34 *Crepidula adunca*

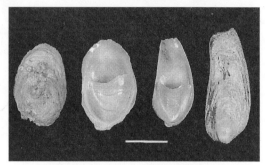
12.35 *Crepidula perforans*

12.36 *Trichotropis cancellata*, with *Odostomia columbiana* (arrow) attached

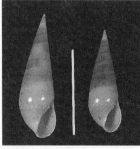
12.37 *Eulima micans*

12.38 *Eulima thersites*

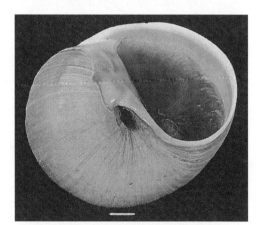
12.39 *Polinices lewisii*

12.41 *Nitidiscala indianorum*

12.42 *Nitidiscala tincta*

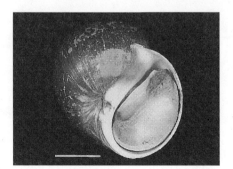
12.40 *Natica clausa*

Phylum Mollusca: Class Gastropoda   215

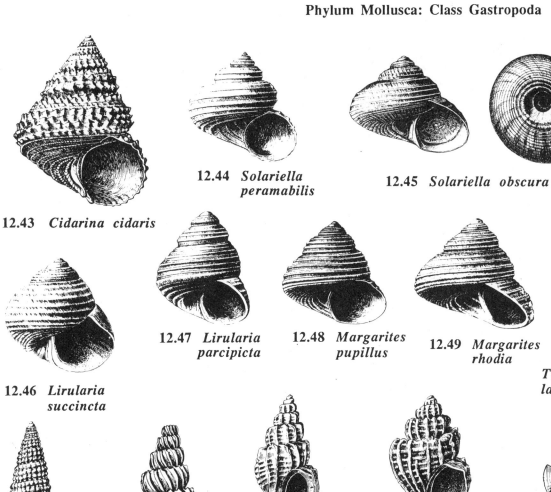

12.43 *Cidarina cidaris*
12.44 *Solariella peramabilis*
12.45 *Solariella obscura*

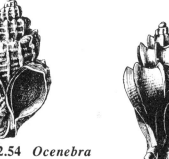

12.46 *Lirularia succincta*
12.47 *Lirularia parcipicta*
12.48 *Margarites pupillus*
12.49 *Margarites rhodia*
12.50 *Tachyrhynchus lacteolus*

12.51 *Bittium munitum*
12.52 *Nitidiscala caamanoi*
12.53 *Ocenebra orpheus*
12.54 *Ocenebra painei*
12.55 *Trophonopsis clathratus*

12.56 *Trophonopsis pacificus*
12.57 *Volutomitra alaskana*
12.58 *Granulina margaritula*
12.59 *Admete gracilior*
12.60 *Aforia circinata*

3b Shell low (never as high as wide), the apex not prominent and often not even distinct; exterior whitish or with a light periostracum, but not dark brown; frequently on empty shells of bivalves or gastropods, including those occupied by hermit crabs, but not attached to the shells of living *Tegula, Calliostoma,* or other gastropods   4
4a Apex of shell directed laterally and united with the margin (introduced with oysters from the Atlantic coast, and found in only a few areas)   *Crepidula fornicata* Linnaeus, 1758
4b Apex of shell directed posteriorly, often overhanging the posterior margin of the shell   5
5a Shell whitish, generally almost flat, with a shaggy, yellow-brown periostracum
    Crepidula nummaria Gould, 1846
5b Shell whitish, almost flat, concave, or misshapen (depending on habitat), with a thin, brown periostracum or without any obvious periostracum (often in burrows of pholadid bivalves, in empty shells, and inside shells occupied by hermit crabs, fitting the contour of the concave surface   *Crepidula perforans* (Valenciennes, 1846) (fig. 12.35)

32. Family Trichotropididae

1a Diameter about two-thirds of the height; height of aperture about half or slightly less than half the total height of the shell   2
1b Diameter more than two-thirds of the height; height of aperture about two-thirds the total height of the shell   3
2a Height of aperture usually slightly less than half the height of the shell (common shallow subtidal species, and occasionally found intertidally)
    Trichotropis cancellata Hinds, 1849 (fig. 12.36)
2b Height of aperture about half, or slightly more than half, the height of the shell (rare subtidal species)   *Trichotropis borealis* Broderip & Sowerby, 1829
3a With 2 pronounced spiral ridges on the body whorl, these not separated by smaller ridges; lowermost part of aperture rather sharply angled; height up to 5 cm (rare subtidal species, not likely to be found south of British Columbia)   *Trichotropis bicarinata* (Sowerby, 1825)
3b With 2 pronounced ridges on the body whorl, these separated by several smaller ridges; lowermost part of aperture not sharply angled (the outline of the aperture as a whole is nearly circular); height to 2.5 cm (rare subtidal species)   *Trichotropis insignis* Middendorff, 1849

33. Family Naticidae

1a Umbilicus open (at least slightly); operculum horny   2
1b Umbilicus closed by a callus; operculum calcified (periostracum brown or chestnut; subtidal)
    Natica clausa Broderip & Sowerby, 1829 (fig. 12.40)
2a Height up to slightly more than 3 cm; with 4 or 5 whorls; color usually whitish, grayish brown, or cream   *Polinices pallidus* Broderip & Sowerby, 1829
2b Height commonly more than 5 cm; with 5 or 6 whorls; color cream to pinkish, beige, or golden brown   3
3a Inner lip of aperture with an almost tongue-shaped callus that spreads onto the body whorl, covering much of the umbilicus (thus the opening is not more than about one-eighth the diameter of the shell); height up to about 14 cm; abundant intertidally on sandflats, but also subtidal   *Polinices lewisii* (Gould, 1847) (fig. 12.39)
3b Inner lip of aperture without an extensive callus that covers much of the umbilicus (thus the umbilicus, measured across its greatest diameter, is about one-third the diameter of the shell); height up to about 9 cm; subtidal   *Polinices draconis* Dall, 1903

The following species, strictly subtidal, have also been recorded.

Amauropsis purpurea Dall, 1871
Calinaticina oldroydi (Dall, 1897)
Neverita lamonae Marincovich, 1975. Deep.

34. Family Marseniidae (Lamellariidae)

Behrens, D. W. 1980. The Lamellariidae of the north eastern Pacific. Veliger, 22:323-39.
Lambert, G. 1980. Predation by the prosobranch mollusk *Lamellaria diegoensis* on *Cystodytes lobatus*, a colonial ascidian. Veliger, 22:340-4.

1a	Dorsal part of mantle with a slit through which the shell can be seen when the mantle is retracted	2
1b	Dorsal part of mantle without a slit through which the shell can be seen	3
2a	Mantle smooth, white or pinkish white, with evenly spaced, small pores; length up to 2 cm (on the compound ascidian *Trididemnum opacum*, and perhaps on other species) *Marsenina stearnsii* Dall, 1871	
2b	Mantle smooth or warty, ranging from colorless to gray, yellow, or orange, with or without darker spots or blotches, but without evenly spaced pores; length up to 4 cm (on the solitary ascidian *Ascidia paratropa* and the compound ascidians *Cystodytes lobatus* and *Aplidium* spp., and perhaps on other species) *Marsenina rhombica* (Dall, 1871)	
3a	Dorsal surface of mantle divided into 6 areas by low ridges that radiate from a central hexagon, but without any tubercles or branching projections; color white to orange, red, or brown, and with a dark blotch in the central hexagon and each lateral area (on *Botrylloides* sp., and perhaps on other compound ascidians) *Marseniopsis sharonae* (Willett, 1939)	
3b	Dorsal surface of mantle not divided into 6 areas by ridges that radiate from a central hexagon, but often with tubercles or branching projections; color white to purple or brown, with darker blotches and radial markings (on *Polyclinum planum*, *Cystodytes lobatus*, *Aplidium* spp., *Trididemnum opacum*, and perhaps on other compound ascidians) *Lamellaria diegoensis* Dall, 1885	

35. Family Velutinidae

1a	Periostracum usually brown, generally with spiral ridges that diverge as they approach the edge of the aperture (commonly associated with the solitary ascidian *Styela gibbsii*) *Velutina plicatilis* O. F. Müller, 1776	
1b	Periostracum usually yellowish brown, without spiral ridges *Velutina prolongata* Carpenter, 1864	

36. Family Cymatiidae

Fusitriton oregonensis (Redfield, 1848). Common subtidal species, and found intertidally in Washington and British Columbia.

37. Family Epitoniidae

DuShane, H. 1979. The family Epitoniidae (Mollusca: Gastropoda) in the northeastern Pacific. Veliger, 22:91-134.

The key is based on shell characters, but the monograph cited above deals also with characteristics of the radula teeth.

1a	With a spiral ridge near the base of the body whorl (the axial ribs may or may not terminate at this ridge)	2
1b	Without a spiral ridge near the base of the body whorl (the axial ribs continue to the base of this whorl)	3
2a	Usually with 7 axial ribs (intertidal and subtidal) *Opalia borealis* Keep, 1881	

Opalia chacei Strong, 1937 and *O. wroblebskii* (Mörch, 1876) are synonyms.

2b   With 8-10 axial ribs (intertidal and subtidal)   *Opalia montereyensis* (Dall, 1907)
3a   Generally with more than 20 axial ribs (range 19-30) (height to 2.2 cm; subtidal)
      Nitidiscala catalinensis (Dall, 1917)
3b   Generally with fewer than 20 axial ribs (*Nitidiscala sawinae* sometimes has 20 or 21)   4
4a   Height about twice the diameter, or only slightly more than twice the diameter, but not as much as 2.5 times the diameter (with 11-13 axial ribs; height up to 1.7 cm; subtidal)
      Nitidiscala caamanoi (Dall & Bartsch, 1910) (fig. 12.52)
4b   Height usually considerably more than twice the diameter (generally 2.5 to 3 times the diameter)   5
5a   Generally with not more than 14 axial ribs   6
5b   Generally with more than 14 axial ribs (with as few as 10, however, in *Nitidiscala indianorum* [fig. 12.41])   7
6a   With 8-14 axial ribs; upper portions of whorls usually with a brownish or purplish band just below the suture; axial ribs evenly curved, neither angular nor with projections near the sutures (length up to 1.5 cm; common intertidal and shallow subtidal species feeding on the sea anemones *Anthopleura elegantissima* and *A. xanthogrammica*)
      Nitidiscala tincta (Carpenter, 1865) (fig. 12.42)
6b   With 7-14 axial ribs; upper portions of whorls without a brownish or purplish band below the suture; axial ribs often angular or with a spine just below the suture (length up to 2.6 cm; intertidal and subtidal)   *Nitidiscala hindsii* (Carpenter, 1856)
7a   Axial ribs decidedly angular just below the sutures, and sometimes with a spine (with 12-20 axial ribs; length up to 1.8 cm; subtidal)   *Nitidiscala catalinae* (Dall, 1908)
7b   Axial ribs evenly curved, not angular just below the sutures (spines may or may not be present)   8
8a   With 14-21 axial ribs; spines generally present on the axial ribs just below the sutures; height up to 2.4 cm (subtidal)   *Nitidiscala sawinae* (Dall, 1903)
8b   With 10-17 axial ribs; spines rarely present on the axial ribs; height up to nearly 4 cm (intertidal and subtidal)   *Nitidiscala indianorum* (Carpenter, 1865) (fig. 12.41)

38. Family *Janthinidae

Janthina prolongata Blainville, 1822. Pelagic in the open ocean; rarely washed ashore in our region.

39. Family Eulimidae (Melanellidae)

The gastropods of this family are associated with sea stars and other echinoderms.

1a   Shell straight   2
1b   Shell decidedly bent   5
2a   Lip of anterior portion of aperture flaring somewhat with respect to the left margin of the body whorl (as the shell is viewed with the aperture lowermost and facing the observer), thus not continuing the nearly straight line of the left margin   3
2b   Lip of anterior portion of aperture not flaring appreciably with respect to the left margin of the body whorl, thus continuing the nearly straight line of the left margin   4
3a   Whorls distinctly convex; height up to about 8 mm   *Eulima randolphi* (Vanatta, 1899)
3b   Whorls flat; height up to about 1.3 cm   *Eulima micans* (Carpenter, 1864) (fig. 12.37)
      E. tacomaensis (Bartsch, 1917) and *E. oldroydi* (Bartsch, 1917) are probably synonyms.
4a   Diameter less than one-third of the height; translucent   *Eulima rutila* (Carpenter, 1864)
4b   Diameter about two-fifths of the height; not especially translucent
      Eulima thersites (Carpenter, 1864) (fig. 12.38)
5a   Height about 3.5 times the diameter; shell bent twice   *Balcis columbiana* (Bartsch, 1917)
5b   Height about 2.5 times the diameter; shell bent once   *Balcis montereyensis* (Carpenter, 1864)

Sabinella ptilocrinicola (Bartsch, 1907). Occurs on the deep-water crinoid *Ptilocrinus pinnatus*.

40. Family Entoconchidae

Members of the Entoconchidae are wormlike parasites of sea cucumbers. They lie in the coelom but are attached to, and covered by, the intestinal peritoneum. Two species from this region were described as new by Tikasingh (1961): *Comenteroxenos parastichopoli*, in *Parastichopus californicus*, and *Thyonicola americana*, in *Eupentacta quinquesemita* and *E. pseudoquinquesemita*. Kincaid (1964) questioned the validity of *C. parastichopoli*, indicating that it appears to be identical with *Enteroxenos oestergreni* Bonnevie, 1902, which parasitizes *Stichopus tremulus* in Europe.

Ivanov, A. V. 1945. [A new endoparasitic mollusc *Parenteroxenos dogieli* nov. gen., nov. sp.] Doklady Akad. Nauk SSSR, 48:450-2.
Kincaid, T. 1964. A gastropod parasitic in the holothurian, *Parastichopus californicus* (Stimpson). Trans. Amer. Micr. Soc., 83:373-6.
Lützen, J. 1979. Studies on the life history of *Enteroxenos* Bonnevie, a gastropod endoparasitic in aspidochirote holothurians. Ophelia, 18:1-51.
Tikasingh, E. S. 1960. Endoparasitic gastropods of some Puget Sound holothurians. J. Parasit., 46 (suppl.):13.
------. 1961. A new genus and two new species of endoparasitic gastropods from Puget Sound, Washington. J. Parasit., 47:268-72.

Enteroxenos parastichopoli (Tikasingh, 1961). In *Parastichopus californicus*; may be synonymous with *Enteroxenos oestergreni* Bonnevie, 1902.
Thyonicola americana Tikasingh, 1961. In *Eupentacta quinquesemita* and *E. pseudoquinquesemita*.
Thyonicola dogieli (Ivanov, 1945). In *Cucumaria miniata*, *C. japonica*, and *C. obunca*. Although Ivanov's material of *C. miniata* was collected in the USSR, it is likely that specimens from *C. miniata* on the northwest coast of North America belong to his species. Tikasingh (1960), however, referred parasitic entoconchids from *C. miniata* and *C. lubrica* to *T. mortenseni* Mandahl-Barth, 1941, originally described from *Thyone serrata*, an abyssal sea cucumber of the region of the Cape of Good Hope.

Order Neogastropoda

Suborder Rachiglossa

41. Family Muricidae

Radwin, G., & A. d'Attilio. 1976. Murex Shells of the World. An Illustrated Guide to the Muricidae. Stanford, California: Stanford University Press. 284 pp.

1a   Edges of the siphonal canal touching or nearly touching for at least part of the canal, so this is essentially a closed tube   2
1b   Edges of the siphonal canal not touching or nearly touching, so the canal is an open trough for its entire length   7
2a   With a large tooth on the outer lip of the aperture; with 3 winglike axial excrescences, as well as unevenly spaced spiral ridges; height up to about 8 cm (common intertidal and shallow subtidal species)   *Ceratostoma foliatum* (Gmelin, 1791)
2b   Without a large tooth on the outer lip of the aperture; without winglike axial excrescences, but with axial ribs and usually also with distinct spiral ridges; height generally less than 4 cm   3
3a   Outer lip of aperture, in mature or nearly mature specimens, thickened to the extent that it is more than half the width of the aperture (height up to about 3 cm; locally introduced with the Pacific oyster, *Crassostrea gigas*)   *Ceratostoma inornatum* (Recluz, 1851) (fig. 12.61)
3b   Outer lip of aperture not thickened to the extent that it is half the width of the aperture   4

220 Phylum Mollusca: Class Gastropoda

12.61 *Ceratostoma inornatum*

12.62 *Ocenebra interfossa*

12.63 *Ocenebra lurida*

12.64 *Urosalpinx cinereus*

12.65 *Trophonopsis lasius*

12.66 *Acanthina spirata*

12.67 *Nucella lamellosa*

12.68 *Nucella lima*

12.69 *Nucella canaliculata*

4a   Axial ribs thin lamellae, about equal to the spiral ridges, but slightly more widely spaced than these; upper portions of whorls flattened nearly at right angles to the long axis of the shell, thus very distinctly set off from one another; height up to 1.5 cm (subtidal, rare)
Ocenebra painei (Dall, 1903) (fig. 12.54)
4b   Axial ribs stout, not thin lamellae; upper portions of whorls not flattened; height often more than 1.5 cm 5
5a   Axial ribs broad, nearly contiguous, rounded, not sharply elevated (height up to about 2.5 cm; common intertidal species)   *Ocenebra lurida* (Middendorff, 1849) (fig. 12.63)
5b   Axial ribs narrower than the furrows between them, sharply elevated 6
6a   Spiral ridges mostly narrower than the furrows between them; height not often exceeding 2 cm   *Ocenebra interfossa* Carpenter, 1864 (fig. 12.62)
6b   Spiral ridges at least as wide as the furrows between them; height up to about 4 cm
Ocenebra sclera (Dall, 1919)
7a   Height decidedly more than twice the diameter   *Trophonopsis lasius* (Dall, 1919) (fig. 12.65)
7b   Height not more than twice the diameter 8
8a   Axial ribs stout, not at all like lamellae 9
8b   Axial ribs rather thin lamellae, sometimes frilly 12
9a   Height much less than twice the diameter; outer lip of aperture, in mature or nearly mature specimens, thickened to the extent that it is more than half the width of the aperture (height up to about 3 cm; locally introduced with the Pacific oyster, *Crassostrea gigas*)
Ceratostoma inornatum (Recluz, 1851)
9b   Height about twice the diameter; outer lip of aperture not thickened to the extent that it is half the width of the aperture 10
10a Interior of aperture purplish or brownish; spiral ridges more obvious between the axial ribs than where they intersect the ribs; on oysters and barnacles in bays where the Atlantic oyster, *Crassostrea virginica*, has been introduced   *Urosalpinx cinerea* Say, 1822 (fig. 12.64)
10b Interior of aperture usually light, not purplish or brownish; spiral ridges generally as prominent where they intersect the axial ribs as between the ribs; on rocky shores 11
11a Axial ribs broad, nearly contiguous, rounded, not sharply elevated; height up to about 2.5 cm (common intertidal species)   *Ocenebra lurida* (Middendorff, 1849) (fig. 12.63)
11b Axial ribs narrower than the furrows between them, sharply elevated; height not often exceeding 2 cm   *Ocenebra interfossa* Carpenter, 1864 (fig. 12.62)
12a Upper portion of each whorl with a shoulder formed by the axial ribs, which may extend upward above the level of the suture 13
12b Upper portion of each whorl without a shoulder formed by the axial ribs
Trophonopsis pacificus (Dall, 1902) (fig. 12.56)
13a With spiral ridges that intersect the axial lamellae (the ridges are more obvious, however, between the lamellae)   *Ocenebra orpheus* (Gould, 1829) (fig. 12.53)
13b Without obvious spiral ridges   *Trophonopsis clathratus* (Linnaeus, 1767) (fig. 12.55)

The following species have been reported, but most of them are northern and/or restricted to deep water.

Ocenebra triangulata (Stearns, 1873)
Trophonopsis dalli (Kobelt, 1878)
Trophonopsis disparilis Dall, 1891
Trophonopsis kamtchatkanus (Dall, 1902)
Trophonopsis macouni (Dall & Bartsch, 1910)
Trophonopsis scitulus Dall, 1891
Trophonopsis staphylinus Dall, 1919
Trophonopsis tripherus Dall, 1902

42. Family Nucellidae (Thaisidae)

1a   Outer lip of the aperture with a sharp tooth slightly below the middle (smaller specimens may, however, lack the tooth); with interrupted dark spiral bands (uncommon north of California)   *Acanthina spirata* (Blainville, 1832) (fig. 12.66)

Phylum Mollusca: Class Gastropoda

12.70 *Nucella emarginata*

12.71 *Searlesia dira*

12.72 *Buccinum plectrum*

12.73 *Exilioidea rectirostris*

12.74 *Colus halli* 12.75 *Colus adonis*

12.76 *Beringius eyerdami*

12.77 *Beringius kennicotti*

12.78 *Neptunea phoenicia*

1b Outer lip of aperture without a sharp tooth; without interrupted dark spiral bands 2
2a Axial sculpture consisting of frilly lamellae; spiral sculpture (which may be obscured by the lamellae) consisting of 1 or 2 prominent ridges on each whorl (height up to 10 cm)
Nucella lamellosa (Gmelin, 1791) (fig. 12.67)
Specimens with frilly lamellae are generally subtidal or found in intertidal situations where the water is relatively calm; see also choice 3a.
2b Axial sculpture, if noticeable, consisting either of broad ribs or low wrinkles; spiral sculpture consisting either of 1 or 2 prominent ridges on each whorl, or of numerous closely spaced ridges 3
3a Spiral sculpture consisting of 1 or 2 prominent ridges on each whorl (but the lower part of the body whorl usually has several smaller ridges), the whorls flattened between the ridges and the sutures, thus appearing angled; height up to about 10 cm
Nucella lamellosa (Gmelin, 1791) (fig. 12.67)
Specimens that fit under this choice are typical of situations where currents are strong or where there is considerable wave action.
3b Spiral sculpture consisting of closely spaced ridges (in some specimens of *Nucella emarginata*, however, the ridges are obscure); height rarely exceeding 3 cm 4
4a Spiral ridges generally of 2 sizes, the smaller ones alternating with the larger ones; whorls not separated from one another by a deep groove 5
4b Spiral ridges generally of nearly equal size; whorls usually separated from one another by a deep groove *Nucella canaliculata* (Duclos, 1832) (fig. 12.69)
5a Shell relatively thin; width of the aperture about half the diameter of the shell (not likely to be found south of the northern part of Vancouver Island)
Nucella lima (Gmelin, 1791) (fig.12.68)
5b Shell relatively thick, the thickness being especially noticeable on the outer lip of the aperture; width of aperture (the actual opening) less than half the diameter of the shell (interior of aperture often purplish or purplish brown; common)
Nucella emarginata (Deshayes, 1839) (fig. 12.70)

43. Family Buccinidae

1a Dark lines between prominent and evenly spaced spiral ridges showing through on the inside of the aperture; axial ribs nearly parallel to the long axis of the shell, generally not extending to the body whorl except in specimens less than 3 cm high (height to 5 cm; common intertidal and shallow subtidal species) *Searlesia dira* (Reeve, 1846) (fig. 12.71)
1b No dark lines on the inside of the aperture (the spiral ridges are almost microscopic, and there are no obvious dark lines between them); axial ribs extending to the body whorl, and often somewhat irregular as well as oblique to the long axis of the shell (height to about 7 cm; moderately common subtidal species) *Buccinum plectrum* Stimpson, 1865 (fig. 12.72)

Other species of *Buccinum*, some of which are similar to *B. plectrum* (fig. 12.72), others distinctive in one way or another, have been reported for the region. Most of them, however, are rare or restricted to deep water, or both.

Buccinum aleuticum Dall, 1894
Buccinum castaneum (Dall, 1877)
Buccinum diplodetum Dall, 1907. Oregon southward.
Buccinum glaciale Linnaeus, 1761
Buccinum planeticum Dall, 1919
Buccinum scalariforme Møller, 1842
Buccinum strigillosum Dall, 1891

224 Phylum Mollusca: Class Gastropoda

12.79 *Neptunea tabulata*

12.80 *Neptunea smirnia*

12.81 *Volutharpa ampullacea*

12.82 *Alia carinata*

12.83 *Alia permodesta*

12.84 *Alia tuberosa*

12.85 *Amphissa versicolor*

12.86 *Amphissa reticulata*

12.87 *Nassarius fraterculus*

12.88 *Nassarius mendicus*

44. Family Neptuneidae

Smith, A. G. 1959a. A new *Neptunea* from the Pacific Northwest. Veliger, 11:117-20.
------. 1959b. A new *Beringius* from the Pacific Northwest, with comments on certain described forms. Nautilus, 73:1-9.

1a Height about 3 times the diameter; siphonal canal about half the total length of the aperture
Exilioidea rectirostris (Carpenter, 1865) (fig. 12.73)
1b Height less than 3 times the diameter; siphonal canal usually less than half the total height of the aperture 2
2a Height of aperture about four-fifths the height of the shell
Volutharpa ampullacea (Middendorff, 1848) (fig. 12.81)
2b Height of aperture not more than two-thirds the height of the shell 3
3a Outer lip of aperture curving convexly until it reaches the tip of the siphonal canal (it may, however, be slightly sinuous, especially in species that have spiral ridges on the body whorl); oldest whorls, at the top of the spire, often conspicuously bulbous 4
3b Outer lip of aperture generally with an obvious indentation just above the siphonal prolongation 5
4a Height of aperture, including the siphonal canal, about half the height of the shell; spiral ridges pronounced, convex, some of them not much wider than the furrows between them; periostracum largely limited to the furrows between the ridges; with or without axial ribs
Beringius crebricostatus (Dall, 1877)
B. undatus Dall, 1919, which has axial ribs, is here considered to be a synonym.
4b Height of aperture slightly more than half the height of the shell; spiral ridges, if prominent, not markedly convex and much wider than the furrows between them (in some specimens, spiral ridges are evident only on the body whorl); without axial ribs (periostracum fibrous, abundant, and usually continuous)
Beringius eyerdami A. G. Smith, 1959 (fig. 12.76)
B. kennicotti (Dall, 1907) (fig. 12.77) is a rare northern species. It has faint spiral lines, except on the body whorl, and about 9 stout axial ribs.
5a Shell smooth or with closely spaced, contiguous spiral ridges, all ridges on a particular whorl being of about the same size (axial ribs may also be present); whorls without angular shoulders 6
5b Shell either with widely spaced spiral ridges (these may be restricted to the upper whorls) or with spiral ridges of decidedly different sizes; whorls usually with angular shoulders 10
6a Without axial ribs, with or without closely spaced spiral ridges (includes one fairly common subtidal species, *Colus halli* [fig. 12.74]) 7
6b With rather prominent axial ribs, as well as with closely spaced spiral ridges (rare subtidal species) 9
7a Spiral ridges prominent; periostracum usually worn away (length to 7 cm)
Colus spitzbergensis (Reeve, 1855)
7b Spiral ridges faint, but usually visible; periostracum persistent 8
8a Height nearly 3 times the diameter; periostracum tan; height up to 4.5 cm
Colus adonis Dall, 1919 (fig. 12.75)
8b Height slightly more than twice the diameter; periostracum usually chestnut-brown or tan (especially in younger specimens); height up to 6 cm (the most common species of *Colus* in our region) *Colus halli* Dall, 1873 (fig. 12.74)
C. jordani Dall, 1913 and *C. dalmasius* (Dall, 1919) are considered to be synonyms. The following are also probably synonyms: *C. errones* Dall, 1919; *C. georgianus* Dall, 1920; *C. halimeris* Dall, 1919; *C. herendeeni* Dall, 1902; *C. mordita* Dall, 1919; *C. tahwitanus* Dall, 1919. Several other species of *Colus* have also been reported for the region: *C. aphelus* Dall, 1890 (deep); *C. dimidiatus* (Dall, 1919); *C. halidonus* Dall, 1919; *C. roseus* (Dall, 1877); *C. rophius* (Dall, 1919). Not all of the names are necessarily valid, however:
9a With about 23-28 axial ribs on each whorl; height up to 3 cm *Plicifusus griseus* (Dall, 1890)
9b With 12-15 axial ribs on each whorl; height up to 1.8 cm *Mohnia frielei* (Dall, 1891)

10a Shoulders on upper portions of whorls flattened and terracelike, nearly at right angles to the long axis of the shell (spiral ridges rather close together, larger ones alternating with smaller ones) .... *Neptunea tabulata* (Baird, 1863) (fig. 12.79)
10b Shoulders on upper portions of whorls not especially flattened and not terracelike or nearly at right angles to the long axis of the shell 11
11a Spiral ridges scarcely evident (they are more likely to be present on the upper whorls rather than on the lower whorls) .... *Neptunea smirnia* (Dall, 1919) (fig. 12.80)
11b Spiral ridges obvious, and sometimes prominent 12
12a Spiral ridges rather evenly spaced and mostly of about the same size (on the body whorl, however, especially on its lower portion, smaller ridges alternate with larger ridges, but the spacing of the larger ridges is nearly the same throughout); height up to about 11 cm .... *Neptunea phoenicia* (Dall, 1919) (fig. 12.78)
12b Spiral ridges not evenly spaced, those on the upper part of the body whorl being at least twice as far apart as those on the lower part (in large specimens, the ridges are generally not as well defined as they are in specimens of medium size); height up to about 16 cm .... *Neptunea lyrata* (Gmelin, 1791)

The following species of *Neptunea* are also known to occur in the region, but all are rare: *N. amianta* (Dall, 1890) (shallow); *N. humboldtiana* A. G. Smith, 1968 (shallow); *N. ithia* (Dall, 1891) (deep); *N. pribiloffensis* (Dall, 1919) (shallow); *N. stilesi* A. G. Smith, 1968.

45. Family Columbellidae (Pyrenidae)

1a Spiral ridges and grooves limited to the lower portion of the body whorl 2
1b Spiral ridges and grooves not limited to the lower portion of the body whorl 5
2a Upper portion of body whorl with a well developed ridge (this is often of a lighter color than the rest of the shell); outer lip of aperture with a prominent indentation below the point where the ridge meets the lip; color generally yellow-brown to brown, sometimes with white or dark brown mottling (height to about 1 cm) .... *Alia carinata* (Hinds, 1844) (fig. 12.82)
2b Upper portion of the body whorl without a well developed ridge (in *A. gausapata*, there may be a slight ridge); outer lip of aperture without a prominent indentation (if there is a slight ridge on the body whorl, there may be an outwardly directed projection where the ridge meets the lip); color usually brown or tan 3
3a Outer lip of aperture evenly curved (periostracum brown; length to 1.5 cm; subtidal) .... *Alia permodesta* (Dall, 1890) (fig. 12.83)
3b Outer lip of aperture not evenly curved (a portion of it is nearly straight) 4
4a Siphonal region, when the shell is viewed with the aperture facing away from the observer, not obviously set off from the rest of the body whorl; height to 1.1 cm; intertidal and subtidal .... *Alia gausapata* (Carpenter, 1864)

A. gouldi (Carpenter, 1864) is a synonym.

4b Siphonal region, when the shell is viewed with the aperture facing away from the observer, obviously set off from the rest of the body whorl; height rarely more than 9 mm; subtidal .... *Alia tuberosa* (Carpenter, 1864) (fig. 12.84)
5a Axial ridges obviously not parallel to the long axis of the shell (they are directed slightly to the left as they extend downward); height rarely attaining 1.5 cm (often with distinctive light and dark markings, unless the shell is covered with foreign material) .... *Amphissa versicolor* Dall, 1871 (fig. 12.85)
5b Most axial ridges almost parallel to the long axis of the shell 6
6a Middle third of the outer lip of the aperture (in mature specimens, at least 2 cm high) parallel to the long axis of the shell; height to 3 cm; common intertidal and subtidal species .... *Amphissa columbiana* Dall, 1916
6b Middle third of the outer lip of the aperture (in mature specimens, at least 1 cm high) curved; height to 1.5 cm; subtidal, rare .... *Amphissa reticulata* Dall, 1916 (fig. 12.86)

46. Family Nassariidae

Demond, J. 1952. The Nassariidae of the west coast of North America between Cape San Lucas, Lower California, and Cape Flattery, Washington. Pacific Sci., 6:300-17.

1a Without a pronounced groove separating most of the body whorl from its lower portion (up to about 3 cm high; usually blackish brown; common in some muddy bays to which it has been introduced from the Atlantic coast) .... *Ilyanassa obsoleta* (Say, 1922)
1b With a pronounced groove separating most of the body whorl from its lower portion 2
2a Axial ribs prominent, but spiral ridges evident only on the body whorl (there are about 3 or 4 of them just above the groove) (up to about 1.5 cm high; introduced into a few bays in the Puget Sound region) .... *Nassarius fraterculus* (Bruguière, 1789) (fig. 12.87)
2b Spiral ridges prominent on most whorls 3
3a Inner lip of aperture with a broad, usually orange callus that spreads out over the body whorl, well above the termination of the outer lip; axial ribs (which may or may not be prominent) not reaching the groove near the base of the body whorl; height up to 4.5 cm .... *Nassarius fossatus* (Gould, 1850)
3b Inner lip of the aperture without a large orange callus (if the callus spreads out over the body whorl, it does not go higher than the termination of the outer lip); axial ribs reaching the groove near the base of the body whorl; height rarely exceeding 3 cm 4
4a Axial ribs (of which there are more than 20) about equal to the spiral ridges, the two intersecting in such a way as to form nearly evenly spaced beads 5
4b Axial ribs (of which there are about 12) much more pronounced and more widely spaced than the spiral ridges (the two do not intersect in such a way that they form nearly evenly spaced beads) .... *Nassarius mendicus* (Gould, 1850) (fig. 12.88)
5a Axial ribs generally not prominent, but the beads formed where they intersect with the spiral ridges are obvious without magnification; spiral ridges on body whorl all about the same size (there is no regular alternation of large and small ridges); height to about 3 cm (not likely to be found north of Oregon) .... *Nassarius rhinites* Berry, 1953
5b Axial ribs and spiral ridges about equally prominent, intersecting to form a cancellate pattern, but beading is not evident without magnification; spiral ridges on body whorl not all of the same size (larger ribs alternate with smaller ones); height to about 2 cm .... *Nassarius perpinguis* (Hinds, 1844)

47. Family Fusinidae

1a Axial ribs more prominent on the upper portions of the whorls than on the lower portions, forming shoulders that are about level with the sutures .... *Fusinus monksae* (Dall, 1915)
1b Axial ribs not more prominent on the upper portions of the whorls than below, and not forming shoulders that are about level with the sutures 2
2a Siphonal canal constituting about one-third of the height of the aperture; diameter slightly less than half the height; height up to about 6 cm .... *Fusinus harfordii* (Stearns, 1871) (fig. 12.89)
2b Siphonal canal constituting about half of the total height of the aperture; diameter less than one-third of the height; height sometimes attaining 13 cm (Oregon southward) .... *Fusinus barbarensis* (Trask, 1855)

48. Family Volutidae

Arctomelon stearnsii Dall, 1872. Shallow subtidal.

49. Family Volutomitridae

Volutomitra alaskana Dall, 1902 (fig. 12.57). Deep subtidal.

50. Family Olividae

Olsson, A. A. 1956. Studies on the genus *Olivella*. Proc. Acad. Nat. Sci. Philadelphia, 108:155-225.

1a   Width of columellar callus, at the base of the body whorl, about one-third of the height of the shell; height up to 3 cm   *Olivella biplicata* (Sowerby, 1825) (fig. 12.90)
1b   Width of columellar callus, at the base of the body whorl, about one-fourth the height of the shell; height not often more than 2 cm   2
2a   Diameter about half the height; often with brown, zigzag lines running lengthwise (not common north of Oregon)   *Olivella pycna* Berry, 1935
2b   Diameter about two-fifths of the height; brown lines, if present, not zigzag   *Olivella baetica* Carpenter, 1864 (fig. 12.91)

51. Family Marginellidae

Coan, E., & B. Roth. 1966. The west American Marginellidae. Veliger, 8:276-99.

Granulina margaritula (Carpenter, 1857) (fig. 12.58) is the only species in our region. It is often abundant intertidally, especially around holdfasts of kelps.

Suborder Toxoglossa

52. Family Cancellariidae

1a   Axial ribs absent or scarcely evident; upper portions of whorls with slightly flattened shoulders (height to nearly 4 cm)   *Neoadmete modesta* (Carpenter, 1865)
1b   Axial ribs rather prominent, at least on the spire and upper part of the body whorl   2
2a   Axial ribs extending to the lowest portion of the body whorl; height up to about 5 cm   *Cancellaria crawfordiana* (Dall, 1891)
2b   Axial ribs present on the upper part of the body whorl, but not extending to the lower half; height up to about 2 cm   *Admete gracilior* (Carpenter, 1866) (fig. 12.59)

The following species have also been reported from the region.

Admete californica Dall, 1908. Deep.
Neoadmete circumcincta (Dall, 1873). Shallow.

53. Family Turridae

Ronald L. Shimek

1a   Basic shell color white or tan (although the body whorl may be pink), and without colored bands   17
1b   Basic shell color various (if white, the shell has prominent colored bands; if not white, it may have white, tan, red, or violet spiral bands)   2
2a   Shoulders of whorls sharply angled   3
2b   Shoulders of whorls indistinctly angled   10
3a   Axial ribs or varices absent   4
3b   Axial ribs or varices present   6
4a   Spiral sculpture consisting of raised, cordlike ridges   5

Phylum Mollusca: Class Gastropoda 229

4b   Spiral sculpture consisting of distinctly incised lines, not of raised, cordlike ridges
Megasurcula carpenteri (Gabb, 1865)

5a   Spiral sculpture consisting of several distinct and prominent ridges; height seldom exceeding 1.5 cm   *Taranis strongi* Arnold, 1903

5b   Spiral sculpture consisting of a single prominent ridge on the equator of each whorl; height sometimes exceeding 6 cm   *Aforia circinata* (Dall, 1873) (fig. 12.60)

6a   Spiral sculpture absent   7
6b   Spiral sculpture present   8
7a   Color uniformly brown, tan, or gray   *Cytharella victoriana* (Dall, 1897)
7b   Color pattern including bands of white, brown, or violet   *Kurtziella plumbea* (Hinds, 1843)

8a   Spiral sculpture consisting of raised, beaded ridges (the beads are visible at 10x magnification) (axial sculpture consisting of sharp ribs; color brown, tan, or russet)
Kurtzia arteaga (Dall & Bartsch, 1910)

8b   Spiral sculpture not as described in choice 8a   9

9a   Spiral sculpture consisting of fine, raised, unbeaded ridges; color pattern including bands of white, brown, or violet   *Kurtziella plumbea* (Hinds, 1843)

9b   Spiral sculpture consisting of incised grooves; apical whorls typically gray, tan, or violet, body whorl often intensely violet (axial sculpture consisting of rounded ribs)
Oenopota levidensis (Carpenter, 1864)

10a   Shoulders of whorls indistinctly angled, with incised sutures   12
10b   Shoulders of whorls rounded (sutures may or may not be incised)   11

11a   Height of aperture equal to, or greater than, half the height of the shell
Mitromorpha gracilior Hemphill, 1864

11b   Height of aperture about one-third the height of the shell (height rarely more than 1 cm; found on rocks, cobble, or silt-covered rocks, seldom on sand or mud)
Clathromangelia interfossa (Carpenter, 1864)

12a   Sutures incised   *Oenopota alaskensis* (Dall, 1871)
12b   Sutures not incised   13

13a   Shell coiling sinistral (when the spire is uppermost and the aperture is facing the observer, the aperture is on the observer's left)   14
13b   Shell coiling dextral (the aperture is on the observer's right)   15

14a   Shell color consisting of red, pink, or brown spiral bands alternating with white spiral bands; height 2 to 3 times the diameter   *Antiplanes voyi* (Gabb, 1866)

14b   Shell color olive to tan, sometimes white, without colored bands, but generally with an olivaceous periostracum that flakes off when it dries; height sometimes more than 4 times the diameter and often exceeding 5 cm   *Antiplanes perversa* (Gabb, 1865) (fig. 12.93)

15a   Basic shell color white, with spiral bands of tan or some other color (spiral sculpture consisiting of fine ridges; periostracum not evident; height to 2 cm)
Kurtziella plumbea (Hinds, 1843)

15b   Basic shell color tan, gray, or violet   16

16a   Anal notch decidedly above the widest portion of the aperture
Antiplanes (Rectiplanes) thalea (Dall, 1902)

16b   Anal notch at the widest point of the aperture (shell gray or brown, with thin, black, incised lines; height up to 4 cm; sometimes found intertidally)
Ophiodermella inermis (Hinds, 1843) (fig. 12.94)

17a   Sculpture absent or faint   18
17b   Either spiral or axial sculpture, or both, prominent   20

18a   Sculpture absent or microscopic (shell appears smooth to the unaided eye) (periostracum, if present, flakes off when dried; it may be sculptured, but the underlying calcareous portion of the shell is smooth)   19

18b   Sculpture consisting of low but distinct raised axial ribs and spiral ridges that give a finely cancellate appearance to the shell (length seldom exceeding 1.3 cm; common in silty habitats)
Ophiodermella cancellata (Carpenter, 1865)

19a   Coiling sinistral (see choice 13a)   *Antiplanes perversa* (Gabb, 1865)
19b   Coiling dextral (see choice 13b)   *Antiplanes (Rectiplanes) thalea* (Dall, 1902)

20a   Shoulders of whorls angular (the angle varies, depending on the species, from a right angle to an obtuse angle)   21
20b   Shoulders of whorls rounded   36

21a	Shoulders distinctly angular, but not turriculate or tabulate	22
21b	Shoulders turriculate or tabulate	30
22a	Axial ribs curved (they may or may not be parallel to the long axis of the shell)	23
22b	Axial ribs straight and parallel to the long axis of the shell	26
23a	Height of aperture greater than half the height of the shell	24
23b	Height of aperture less than, or just equal to, half the height of the shell	25
24a	With more than 15 minute spiral ridges on the shoulder above the shoulder cord; individual radular teeth long and narrow, their length more than 5 times their width *Oenopota alitakensis* (Dall, 1919)	
24b	With fewer than 12 spiral ridges on the shoulder above the shoulder cord; individual radular teeth stubby, their length less than 5 times their width *Oenopota popovia* (Dall, 1919)	
25a	Outside of siphonal canal smooth *Oenopota elegans* (Möller, 1842)	
25b	Outside of siphonal canal with spiral grooves *Oenopota pleurotomaria* (Couthouy, 1838)	
26a	Axial sculpture faint, but consisting of more than 26 ribs on the body whorl (with pronounced spiral ridges) *Oenopota maurelli* (Dall & Bartsch, 1910)	
26b	Axial sculpture ranging from faint to pronounced, but consisting of not more than 24 ribs on the body whorl	27
27a	With more than 13 axial ribs on the body whorl	28
27b	With fewer than 13 axial ribs on the body whorl *Oenopota harpularia* (Couthouy, 1838)	
28a	Cordlike ridge on shoulder faint; outer lip of aperture generally angled	29
28b	Cordlike ridge on shoulder pronounced; outer lip of aperture with a distinct ridge at the angle (height greater than twice the diameter; individual radular teeth stubby, their length less than 5 times their width) *Oenopota turricula* (Montagu, 1803)	
29a	Height more than twice the diameter; individual radular teeth slender, their length greater than 5 times their width; with several prominent spiral, cordlike ridges, the shoulder cord being no more pronounced than the others *Oenopota viridula* (O. Fabricius, 1780)	
29b	Height less than twice the diameter; individual radular teeth stubby, their length less than 5 times their width; spiral cord at the shoulder more pronounced than the other spiral ridges (height seldom exceeding 1.5 cm) *Oenopota excurvata* (Carpenter, 1864)	
30a	Axial sculpture mostly parallel to the long axis of the shell	31
30b	Axial sculpture mostly not parallel to the long axis of the shell *Oenopota babylonia* (Dall, 1919)	
31a	Shoulders turriculate (there are large nodes where the axial ribs cross the cordlike shoulder ridge)	32
31b	Shoulders tabulate	35
32a	Spiral sculpture consisting of raised ridges, with intervening incised areas that are as wide as, or wider than, the ridges	33
32b	Spiral sculpture consisting of narrow incised grooves (height of aperture less than half the height of the shell; height up to 2 cm; common in sand, silt, and habitats in which sand and shell fragments are mixed) *Oenopota fidicula* (Gould, 1849) (fig. 12.92)	
33a	Shoulders of whorls strongly turriculate (raised nodes at the shoulders are wider than the axial ribs and are connected to one another by a prominent, cordlike spiral ridge)	34
33b	Shoulders of whorls weakly turriculate to nodose (the raised nodes are distinct, but not generally wider than the axial ribs, and they are not connected by an especially prominent cordlike ridge) (length of radular teeth greater than 5 times their width) *Oenopota viridula* (O. Fabricius, 1780)	
34a	Spiral sculpture consisting of numerous distinct ridges, none of which is especially prominent (length of radular teeth greater than 5 times their width) *Oenopota sculpturata* (Dall, 1886)	
34b	Cordlike spiral ridge at the shoulder prominent, and larger and more distinct than the other ridges (thus the shoulder nodes are larger than the other nodes formed where the spiral ridges cross the axial ribs) (most individuals less than 1.5 cm high, but some exceed 2.5 cm) *Oenopota turricula* (Montagu, 1803)	
35a	Tabulate shoulder distinctly separated from the spiral sculpture below it; spiral and axial sculpture (except for the cordlike shoulder ridge) about equal; height generally more than twice the diameter (height seldom exceeding 1.5 cm; found in sand mixed with shell fragments, or on rock; body whorl usually pink when the animal is alive) *Oenopota tabulata* (Carpenter, 1864)	

12.89 *Fusinus harfordi*

12.90 *Olivella biplicata*

12.91 *Olivella baetica*

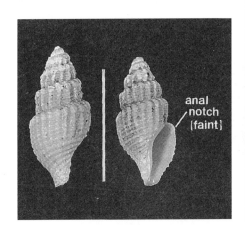

12.92 *Oenopota fidicula*

12.93 *Antiplanes perversa*

12.94 *Ophiodermella inermis*

35b   Tabulate shoulder generally merging with the spiral sculpture below it; spiral sculpture less distinct than the axial sculpture (except at the shoulder, where they are about equal); height generally less than twice the diameter (found in silt or in sand with shell fragments; body whorl never pink) .... *Oenopota excurvata* (Carpenter, 1864)

36a   Axial ribs curved, either parallel or not parallel to the long axis of the shell   37
36b   Axial ribs straight and parallel to the long axis of the shell   45
37a   With 20 or more axial ribs on the body whorl   38
37b   With fewer than 20 axial ribs on the body whorl   39

38a Spiral ridges as prominent as, or more prominent than, the axial ribs (the sculpture has a reticulate appearance); height of aperture at least two-thirds the height of the shell
Oenopota reticulata (Brown, 1827)
38b Spiral ridges not as prominent as the axial ribs, although they may cross the axial ribs; height of aperture less than two-thirds the height of the shell *Oenopota krausei* (Dall, 1886)
39a With a prominent shoulder cord *Oenopota elegans* (Möller, 1842)
39b Without a prominent shoulder cord 40
40a Spiral sculpture present; body whorl not pink 41
40b Spiral sculpture absent or very faint; body whorl pink in living animals (height to 3 cm)
Oenopota pyramidalis (Strøm, 1788)
41a Sutures between whorls incised 42
41b Sutures between whorls not incised *Oenopota elegans* (Möller, 1842)
42a Spiral grooves on outside of siphonal canal faint or absent 43
42b Spiral grooves on outside of siphonal canal deep and obvious, at least as pronounced as grooves elsewhere on the shell *Oenopota pleurotomaria* (Couthouy, 1838)
43a Spiral sculpture on body whorl consisting of ridges that are about the same width as the spaces between them 44
43b Spiral sculpture on body whorl consisting of ridges that are mostly distinctly wider than the spaces between them *Oenopota alaskensis* (Dall, 1871)
44a Operculum present; diameter of shell greater than one-third the height
Oenopota kyskana (Dall, 1919)
44b Operculum absent; diameter of shell less than one-third the height
Kurtzia variegata (Carpenter, 1864)
45a Height of aperture more than three-fifths the height of the shell 46
45b Height of aperture less than three-fifths the height of the shell 47
46a Spiral grooves on the outside of siphonal canal deep and obvious; generally with more than 20 axial ribs on the body whorl *Oenopota harpa* (Dall, 1885)
46b Spiral grooves on the outside of siphonal canal faint, not more distinct than elsewhere on the shell; generally with fewer than 16 axial ribs on the body whorl *Oenopota solida* (Dall, 1886)
47a Spiral grooves on the outside of siphonal canal pronounced, at least as deep as elsewhere on the shell (body whorl often pinkish in the living animals, particularly in northern Vancouver Island and father north) *Oenopota rosea* (M. Sars in Lovén, 1846)
47b Spiral grooves on the outside of siphonal canal faint or absent
Oenopota crebricostata (Carpenter, 1864)

Subclass Opisthobranchia

Gordon A. Robilliard and Eugene N. Kozloff

The references listed here are those that deal with opisthobranchs in general. Works concerned with individual orders or suborders will be found under the headings for these groups.

Behrens, D. W. 1980. Pacific Coast Nudibranchs: A Guide to the Opisthobranchs of the Northeastern Pacific. Los Osos, California: Sea Challengers. 112 pp.

Bergh, L. S. R. 1894. Die Opisthobranchien. *From* Reports on the dredging operations off the west coast of Mexico and in the Gulf of California in charge of Alexander Agassiz, carried on by the U. S. Fish Commission Steamer "Albatross" during 1891. Bull. Mus. Comp. Zool. Harvard College, 25:125-233.

Ghiselin, M. T. 1966. Reproductive function and phylogeny of opisthobranch gastropods. Malacologia, 3:327-78.

Goddard, J. H. R. 1984. The opisthobranchs of Cape Arago, Oregon, with notes on their biology and a summary of benthic opisthobranchs known from Oregon. Veliger, 27:143-63.

Gosliner, T. M. 1981. Origins and relationships of primitive members of the Opisthobranchia (Mollusca: Gastropoda). Biol. J. Linn. Soc., 16:197-225.

Hurst, A. 1967. The egg masses and veligers of thirty northeast Pacific opisthobranchs. Veliger, 9:255-88.

Lambert, P. 1976. Records and range extensions of some northeastern Pacific opisthobranchs (Mollusca: Gastropoda). Can. J. Zool., 54:293-300.

Lance, J. R. 1961. A distributional list of southern California opisthobranchs. Veliger, 4:64-9.

------. 1966. New distributional records of some northeastern Pacific Opisthobranchiata (Mollusca-Gastropoda) with descriptions of two new species. Veliger, 9:68-91.

McDonald, G. R., & J. W. Nybakken. 1981. Guide to the Nudibranchs of California. Melbourne, Florida: American Malacologists, Inc. 72 pp.

MacFarland, F. M. 1906. Opisthobranchiate Mollusca from Monterey Bay, California and vicinity. Bull. U. S. Bur. Fish., 25:109-57.

------. 1966. Studies of opisthobranchiate mollusks of the Pacific coast of North America. Mem. 6, Calif. Acad. Sci. xvi + 546 pp.

Marcus, E. 1961. Opisthobranch mollusks from California. Veliger, 3 (suppl.):1-84.

Roller, R. A. 1970. A list of recommended nomenclatural changes for MacFarland's "Studies of opisthobranchiate mollusks of the Pacific coast of North America." Veliger, 12:371-4.

Steinberg, J. E. 1963. Notes on the opisthobranchs of the west coast of North America. IV. A distributional list of opisthobranchs from Point Conception to Vancouver Island. Veliger, 6:68-73.

The key to orders omits consideration of any members of the pelagic order Thecosomata other than *Limacina helicina*.

1a With an external shell large enough to accommodate all (or nearly all) of the animal when it withdraws   2
1b Either without an external shell, or with a shell that is only partly visible externally and that cannot accommodate more than a small part of the animal   4
2a Shell coiled sinistrally, but otherwise shaped much like that of *Tegula*, *Margarites* and other trochid prosobranchs (pelagic; swimming with oarlike flaps of the much modified foot)
    Order Thecosomata: *Limacina helicina*
2b Shell either not coiled or coiled dextrally, and not shaped like that of a trochid prosobranch   3
3a Shell either slender and with a whorl, spire taller than the body or more or less ovoid, or (in *Cyclostremella*) planispiral and with the diameter greater than the height; aperture not more than half the height of the shell (except in *Cyclostremella*)   Order Pyramidellacea
3b Shell either nearly cylindrical and with a short spire or no visible spire, or narrowly ovoid; aperture at least slightly more than half the height of the shell, and sometimes equal to the height   Order Cephalaspidea (in part)
4a Pelagic; part of the mantle, just behind the head, elaborated into flipperlike structures used for swimming   Order Gymnosomata
4b If pelagic, only temporarily so; lateral extensions of the body, if present, forming sheetlike flaps along much of the length of the foot, and not primarily for swimming   5
5a With a single gill, this located on the right side of the body, between the overlapping mantle margin and the foot   Order Notaspidea: *Berthella californica*
5b With or without gills, but not with a single gill located on the right side of the body between the overlapping mantle margin and the foot   6
6a With a gill chamber (its length about one-fifth the length of the body) located about halfway back on the dorsum, just to the right of the midline (with black longitudinal stripes on a green or yellow-green background)   Order Anaspidea: *Phyllaplysia taylori*
6b Without a gill chamber on the dorsum (none of the species in our region has longitudinal black stripes on a green or yellow-green background)   7
7a Shell present, although it may be small and delicate and partly or completely hidden by the mantle; cephalic shield present; fleshy lateral outgrowths of the mantle turned up dorsally but overlapping one another only slightly, if at all   Order Cephalaspidea (in part)
7b Shell absent; cephalic shield absent; lateral outgrowths of the body, if present, forming sheetlike flaps along much of the length of the foot, and if turned up dorsally, overlapping one another to a considerable extent   8

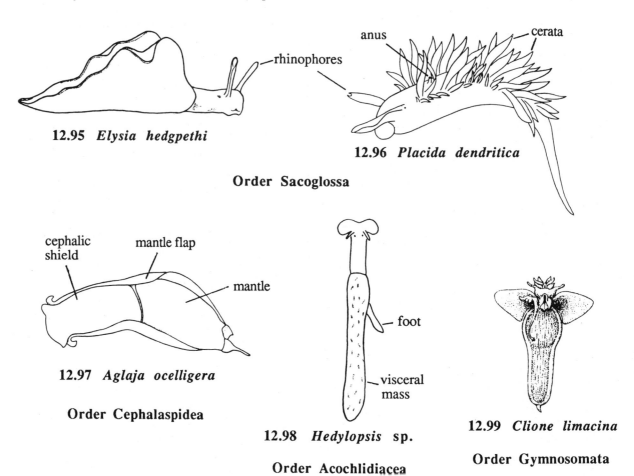

12.95 *Elysia hedgpethi*

12.96 *Placida dendritica*

Order Sacoglossa

12.97 *Aglaja ocelligera*

Order Cephalaspidea

12.98 *Hedylopsis* sp.

Order Acochlidiacea

12.99 *Clione limacina*

Order Gymnosomata

8a Small (maximum length 3 or 4 mm); living in sand; visceral mass distinct from head and foot, and extending posteriorly beyond the foot ..... Order Acochlidiacea

8b Not especially small (commonly more than 3 or 4 mm long, except when young); not living in sand; visceral mass not distinct from the head and foot, and not extending posteriorly beyond the foot 9

9a With a lung, whose opening is on the posteroventral side of the mantle immediately behind the anus; eyes at the tips of distinct stalks ..... Subclass Gymnomorpha, Order Onchidiacea: *Onchidella borealis*

9b Without a lung; eyes not at the tips of stalks 10

10a Rhinophores absent ..... Order Sacoglossa (in part)

10b Rhinophores present 11

11a Clavus of rhinophores perfoliate, or with longitudinal ridges, or with vertical pinnate plumes beside the clavus ..... Order Nudibranchia (in part)

11b Clavus of rhinophores smooth and not distinct from the stalk (in cross-section, the rhinophores may be solid or rolled into cylinders) 12

12a Rhinophores retractile into sheaths ..... Order Nudibranchia (in part)

12b Rhinophores not retractile into sheaths 13

13a Dorsum with elongate outgrowths, such as cerata 14

13b Dorsum without elongate outgrowths ..... Order Sacoglossa (in part)

14a Rhinophores rolled into cylinders ..... Order Sacoglossa (in part)

14b Rhinophores solid in cross-section 15

15a	Anus on the midline, just posterior to the rhinophores	Order Sacoglossa (in part)
15b	Anus on the right side of the body	Order Nudibranchia (in part)

Order Pyramidellacea

Pyramidellaceans, which some malacologists believe to be prosobranchs rather than opisthobranchs, are suctorial parasites of various invertebrates, especially bivalve and gastropod molluscs, echinoderms, polychaete annelids, and sea anemones. With a few exceptions, those in our region are subtidal. The last comprehensive work on Pacific coast species was that of Dall & Bartsch (1909), but numerous names have been proposed since then. Positive identification of almost any specimen is difficult because the distinctions between many of the so-called species seem to be based on minor variations that may be expected within species. Some names have already been placed in synonymy by compilers.

Two genera, *Odostomia* and *Turbonilla*, are particularly well represented in our region. Until the systematics of these genera has been studied carefully, it will not be possible to make a reliable key to the species. Thus the key provided here will be helpful only for separating *Odostomia* from *Turbonilla*, and for identifying a few species that belong to two other genera.

Bartsch, P. 1917. Descriptions of new west American marine mollusks and notes on previously described forms. Proc. U. S. Nat. Mus., 52:637-81.
------. 1920. The Caecidae and other marine mollusks from the northwest coast of America. J. Wash. Acad. Sci., 10:565-72.
------. 1927. New west American marine mollusks. Proc. U. S. Nat. Mus., 70(11):1-36.
Dall, W. H., & P. Bartsch. 1907. The pyramidellid mollusks of the Oregonian faunal area. Proc. U. S. Nat. Mus., 33:491-534.
------. 1909. A monograph of west American pyramidellid mollusks. Bull. 68, U. S. Nat. Mus. xii + 258 pp.
Jordan, E. K. 1920. Notes on a collection of shells from Trinidad, California. Proc. U. S. Nat. Mus., 58:1-5.
Robertson, R. 1973. *Cyclostremella*: a planispiral pyramidellid. Nautilus, 87:88.

1a	Coiling planispiral (i.e., nearly in one plane, much like that of a freshwater planorbid snail), the concave spire encircled by the body whorl (diameter up to about 3 mm)	Cyclostremellidae: *Cyclostremella concordia*
1b	Coiling not planispiral, the spire prominent and extending well above the body whorl	2
2a	Aperture about half the height of the shell; columella with a slight swelling, but without a distinct fold; outer lip of aperture somewhat scalloped (with prominent spiral ridges and thin, closely spaced axial lamellae, especially on the body whorl)	3
2b	Aperture less than half the height of the shell; columella with a distinct fold (in some species of *Turbonilla*, however, the fold is deep within the aperture); outer lip of aperture usually not scalloped	4
3a	Body whorl with 9 spiral ridges; color generally yellowish brown (length up to 6 mm; often associated with *Mytilus edulis*)	Pyramidellidae: *Iselica ovoidea* (fig. 12.100)
3b	Body whorl with 7 spiral ridges; color generally pinkish or rose (length up to about 6 mm; subtidal, usually on gravel)	Pyramidellidae: *Iselica obtusa*
4a	Shell usually with at least 7 whorls; aperture less than one-fourth the total height of the shell; body whorl less than half the total height (shell sometimes with only faint axial ribs, sometimes with prominent axial and spiral sculpture)	Pyramidellidae: *Turbonilla* (figs. 12.105-12.106)
4b	Shell usually with 5 or 6 whorls; aperture more than one-fourth the total height of the shell; body whorl at least half the total height (shell sometimes smooth, sometimes with prominent sculpture)	Pyramidellidae: *Odostomia* (figs. 12.36 and 12.101-12.104)

Many of the names in the species list are probably synonyms. A considerable number of names that have already been placed in synonymy, rightly or wrongly, are not mentioned here. Unless otherwise indicated, all species are subtidal.

Family Pyramidellidae

Iselica obtusa (Carpenter, 1864). Low intertidal and subtidal.
Iselica ovoidea (Gould, 1853) (fig.12.100)
Odostomia angularis Dall & Bartsch, 1907
Odostomia barkleyensis Dall & Bartsch, 1910 (fig. 12.101)
Odostomia canfieldi Dall, 1908
Odostomia cassandra Bartsch, 1912
Odostomia chinooki Bartsch, 1927
Odostomia columbiana Dall & Bartsch, 1907. Commonly associated with *Trichotropis cancellata* (Mesogastropoda: Trichotropididae); mostly subtidal, but occasionally intertidal.
Odostomia cypria Bartsch, 1912
Odostomia engbergi Bartsch, 1920
Odostomia grippiana Bartsch, 1912
Odostomia kennerlyi Dall & Bartsch, 1907. Low intertidal and subtidal.
Odostomia nuciformis Carpenter, 1864
Odostomia oregonensis Dall & Bartsch, 1907
Odostomia quadrae Dall & Bartsch, 1910 (fig. 12.103)
Odostomia satura Carpenter, 1864
Odostomia spreadboroughi Dall & Bartsch, 1910
Odostomia tacomaensis Dall & Bartsch, 1907
Odostomia tenuisculpta Carpenter, 1864
Odostomia vancouverensis Dall & Bartsch, 1910 (fig. 12.102)
Odostomia willetii Bartsch, 1917
Odostomia youngi Dall & Bartsch, 1917 (fig. 12.104)
Turbonilla alaskana Dall & Bartsch, 1909
Turbonilla aurantia (Carpenter, 1864). Low intertidal and subtidal.
Turbonilla barkleyensis Bartsch, 1917
Turbonilla engbergi Bartsch, 1920
Turbonilla lordi (E. A. Smith, 1880)
Turbonilla lyalli Dall & Bartsch, 1907. Low intertidal and subtidal.
Turbonilla pesa Dall & Bartsch, 1909 (fig. 12.105)
Turbonilla pugetensis Bartsch, 1917
Turbonilla rinella Dall & Bartsch, 1910 (fig. 12.106)
Turbonilla stylina (Carpenter, 1864)
Turbonilla taylori Dall & Bartsch, 1907
Turbonilla torquata (Gould, 1852)

Family Cyclostremellidae

Cyclostremella concordia Bartsch, 1920

Order Acochlidiacea

One species, apparently undescribed but tentatively placed in the genus *Hedylopsis* (fig. 12.98) (Hedylopsidae), has been found in sand collected at various localities in Puget Sound and the San Juan Archipelago.

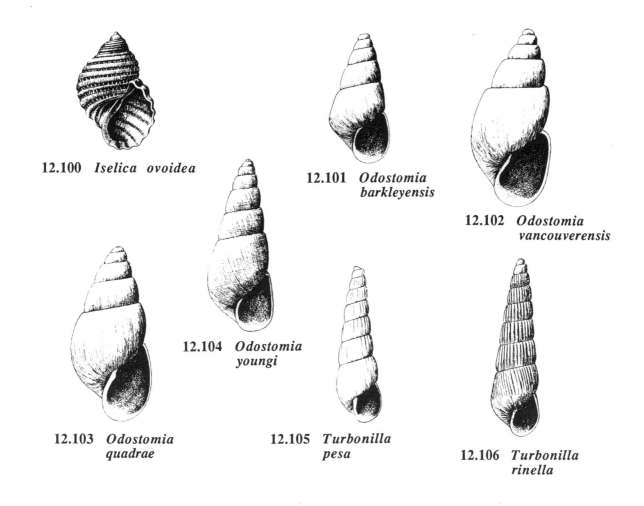

12.100 *Iselica ovoidea*
12.101 *Odostomia barkleyensis*
12.102 *Odostomia vancouverensis*
12.103 *Odostomia quadrae*
12.104 *Odostomia youngi*
12.105 *Turbonilla pesa*
12.106 *Turbonilla rinella*

Order Cephalaspidea

Gosliner, T. M. 1979. A review of the systematics of *Cylichnella* Gabb (Opisthobranchia: Scaphandridae). Nautilus, 93:85-92.

Marcus, E. duB.-R. 1974. On some Cephalaspidea (Gastropoda: Opisthobranchia) from the western and middle Atlantic. Bull. Mar. Sci., 14:300-71.

------. 1977. On the genus *Tornatina* and related forms. J. Moll. Stud., Suppl. 2:1-35.

Rudman, W. B. 1971. On the opisthobranch genus *Haminoea* Turton and Kingston. Pacific Sci., 25:549-59.

------. 1973. The genus *Philine* (Opisthobranchia: Gastropoda). Proc. Malac. Soc. London, 40:171-87.

------. 1974. A comparison of *Chelidonura*, *Navanax*, and *Aglaja* with other genera of the Aglajidae (Opisthobranchia: Gastropoda). Zool. J. Linnean Soc. London, 54:178-212.

Steinberg, J. E. 1963. Notes on the opisthobranchs of the west coast of North America. II. The order Cephalaspidea from San Diego to Vancouver Island. Veliger, 5:114-7.

1a	Shell either not visible externally or too small to accommodate the animal when it retracts	2
1b	Shell large enough to accommodate the animal when it retracts	7
2a	Shell visible externally, but partly covered by the mantle	3
2b	Shell not visible externally	4

3a   Width of the aperture, at the level where the shell is broadest, slightly more than one-third the width of the shell; height up to about 2 cm .... *Haminoea vesicula* (fig. 12.107)
3b   Width of the aperture, at the level where the shell is broadest, about two-thirds the width of the shell; height up to about 1.5 cm .... *Haminoea virescens* (fig. 12.108)
4a   Body completely dark brownish purple or blue-black .... *Melanochlamys diomedea*
4b   Body not completely dark brownish purple or blue-black 5
5a   Body light to dark reddish brown, with numerous yellow spots; with yellow lines on the edge of the cephalic shield, foot, and borders of the mantle flaps
.... *Aglaja ocelligera* (fig. 12.97)
5b   Body not reddish brown, and without yellow spots; without yellow lines on the cephalic shield, foot, and mantle flaps 6
6a   Body pale yellow, with scattered clusters of small red dots .... *Gastropteron pacificum*
6b   Body pale yellow to white, without clusters of red dots .... *Philine bakeri*
7a   Aperture slightly more than half the height of the shell; body whorl with 2 broad spiral bands, each consisting of several gray or gray-brown parallel lines (similar bands are present on the spire) .... *Rictaxis punctocaelatus* (fig. 12.109)
7b   Aperture at least three-fourths the height of the shell, and sometimes equal to the height; without broad bands that consist of several gray or gray-brown lines 8
8a   Spire apparent, even if abraded down to the upper edge of the body whorl 9
8b   Spire completely concealed by the body whorl 12
9a   Upper part of the body whorl, and sometimes the whorls of the spire, with irregular, closely spaced axial riblets; spire usually prominent, with a turreted appearance due to flattening of the uppermost portions of the whorls (including the body whorl) .... *Cylichnella harpa*
9b   Neither the body whorl nor spire with axial riblets; spire, when prominent, without a decidedly turreted appearance 10
10a Columella with a prominent fold that emerges into the aperture at the top of its lowermost third (the course of the fold is somewhat spiral and does not conform perfectly to the general downward course of the columella); spire usually prominent; body whorl (and sometimes other whorls) with irregular, reddish or brownish spiral lines 11
10b Columella without a prominent fold (there is a slight fold, and its course conforms closely to that of the columella); spire usually eroded, so that only its lower portion is evident within a depression at the top of the body whorl; whorls without reddish or brownish spiral lines
.... *Cylichna alba*
11a Posterolateral extension of the mantle curving to the left along the dorsal side of the body whorl; height of shell not often exceeding 1 cm .... *Cylichnella cerealis*
11b Posterolateral extension of the mantle curving to the right and extending downward to the ventral side of the body whorl; height of shell up to 1.7 cm
.... *Cylichnella culcitella* (fig. 12.111)
Cylichnella eximia is similar, and its systematic status is uncertain. It may be a synonym of *C. culcitella*.
12a Apex of the shell, where it is reached by the upper part of the lip of the aperture, forming a somewhat conical projection .... *Volvulella cylindrica*
12b Apex of the shell, where it is reached by the upper part of the lip of the aperture, not forming a conical projection .... *Cylichna attonsa* (fig. 12.110)

Two species, marked with an asterisk, are not in the key.

Family Acteonidae

Rictaxis punctocaelatus (Carpenter, 1864) (fig. 12.109)
**Microglyphis estuarinus* Dall, 1908

Family Atyidae

Haminoea vesicula Gould, 1855 (fig. 12.107)
Haminoea virescens (Sowerby, 1833) (fig. 12.108)

Phylum Mollusca: Class Gastropoda 239

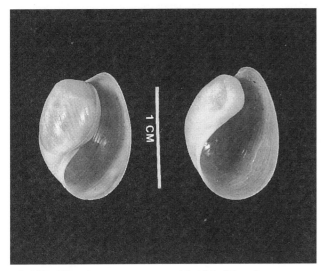

12.107 *Haminoea vesicula*

12.108 *Haminoea virescens*

12.109 *Rictaxis punctocaelatus*

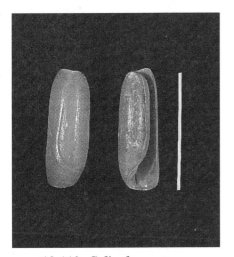

12.110 *Cylinchna attonsa*

12.111 *Cylichnella culcitella*

12.112 *Clione limacina*

Family Retusidae

Volvulella cylindrica (Carpenter, 1864)

Family Diaphanidae

Diaphana brunnea Dall, 1919

Family Philinidae

Philine bakeri Dall, 1900

Family Aglajidae

Aglaja ocelligera (Bergh, 1894) (fig. 12.97)
Melanochlamys diomedea (Bergh, 1894)

Family Gastropteridae

Gastropteron pacificum Bergh, 1894

Family Cylichnidae

Cylichna alba (Carpenter, 1864)
Cylichna attonsa (Brown, 1827) (fig. 12.110)
Cylichnella cerealis (Gould, 1852)
Cylichnella culcitella (Gould, 1852) (fig. 12.111)
Cylichnella harpa (Dall, 1871)
Cylichnella eximia (Baird, 1863). See key, choice 11b.

Order Anaspidea

Beeman, R. O. 1968. The order Anaspidea. Veliger, 3 (suppl., part 2):87-102.

Family Aplysiidae

Phyllaplysia taylori Dall, 1900. Found on leaves of eelgrass, *Zostera marina*, a background against which it is rather cryptically colored and patterned.

Order Notaspidea

Family Pleurobranchidae

Berthella californica (Dall, 1900). White all over and strikingly visible on subtidal rocky bottoms; apparently preys on ascidians, and perhaps also on sponges; uncommon.
Roya spongotheres Bertsch, 1980. Subtidal, on a sponge of the genus *Hexadella*; assignment to genus *Roya* doubtful.

Order Thecosomata

Limacina helicina (Phipps, 1774), characterized by a sinistrally coiled shell, is the most commonly observed species in the region. Several other members of this strictly pelagic group have been reported, however. In the Cavoliniiidae (suborder Euthecosomata), the shell is either elongated and conical or nearly bivalved, with or without projections. In the family Cymbuliidae (suborder Pseudothecosomata), the "shell" is an internal, cartilaginous structure called the pseudoconch.

Suborder Euthecosomata

Family Cavoliniidae

Cavolinia gibbosa (Orbigny, 1836
Cavolinia tridentata (Niebuhr, 1775)
Clio polita Pelseneer, 1888. Deep.
Clio pyramidata Linnaeus, 1767
Clio recurva (Childern, 1823)
Creseis virgula (Rang, 1828)
Diacria trispinosa (Blainville, 1821)
Styliola subula (Quoy & Gaimard, 1827)

Family Limacinidae

Limacina helicina (Phipps, 1774)

Suborder Pseudothecosomata

Family Cymbuliidae

Corolla spectabilis Dall, 1871

Order Gymnosomata

Only 3 species have been reported with certainty from the region. The key differentiates the 2 that are most commonly observed. If a few crystals of urethane are added to a small amount of water containing the specimens, the buccal appendages, important in identification, will usually be protruded.

1a With stellate chromatophores on the body; lateral gill a membranous, unpigmented band near the middle of the body, posterior gill with 4 radial crests; with 2 lateral buccal appendages, each with numerous stalked suckers; proboscis large and protrusible
Pneumodermopsis macrochira
1b Without chromatophores; without gills; with 3 pairs of buccal appendages (buccal cones), these without stalked suckers; without a proboscis   *Clione limacina* (fig. 12.99 and 12.112)

Suborder Gymnosomina

Family Clionidae

Clione limacina (Phipps, 1774) (fig. 12.99 and 12. 112)

Family Pneumodermatidae

Pneumoderma atlanticum (Oken, 1815)
Pneumodermopsis macrochira Meisenheimer, 1905

Order Sacoglossa

Crane, S. 1971. The feeding and reproductive behavior of the sacoglossan gastropod *Olea hansineensis* Agersborg 1923. Veliger, 14:57-9.
Gonor, J. J. 1961. Observations on the biology of *Hermaeina smithi*, a sacoglossan opisthobranch from the west coast of North America. Veliger, 4:85-98.
Lance, J. R. 1962. A new *Stiliger* and a new *Corambella* (Mollusca: Opisthobranchia) from the northwest Pacific. Veliger, 5:33-8.
MacFarland, F. M. 1966. Studies on opisthobranchiate mollusks of the Pacific coast of North America. Mem. 6, Calif. Acad. Sci. xvi +546 pp.
Marcus, E. 1961. Opisthobranch mollusks from California. Veliger, 3 (suppl.):1-84.
Millen, S. V. 1980. Range extensions, new distribution sites, and notes on the biology of sacoglossan opisthobranchs (Mollusca: Gastropoda) in British Columbia. Can. J. Zool., 58:1207-9.
Williams, G. C., & T. M. Gosliner. 1973. Range extensions for four sacoglossan opisthobranchs from the coasts of California and the Gulf of California. Veliger, 16:112-6.

1a Cerata present (there may also be a pair of rhinophores); without parapodia (broad lateral outgrowths of the foot) 2
1b Cerata absent; with well developed parapodia that may be turned up to cover the dorsum or spread flat against the substratum .... *Elysia hedgpethi* (fig. 12.95)
2a Cerata (about 10) limited to the posterior half of the body; body mostly black (head and tail pale gray) .... *Olea hansineensis*
2b Cerata (usually more than 10) on both the anterior and posterior halves of the body; body not black (except in *Aplysiopsis smithi*) 3
3a Without rhinophores; body slightly flattened; foot broader than the body, tapering abruptly, not extended posteriorly as a tail (head region with a pronounced dorsolateral fold; body greenish yellow, with black streaks) .... *Alderia modesta*
3b With rhinophores; body nearly cylindrical; foot relatively narrow and tapering gradually to form a pointed tail 4
4a Rhinophores solid (body translucent white, with an intricate reddish brown pattern on the cerata as well as on the dorsum) .... *Stiliger fuscovittatus*
4b Rhinophores rolled up to form a hollow cylinder, at least for much of their length (the edges may not quite overlap, however) 5
5a Basal portion of rhinophores solid (only the distal part is rolled up); body black, with white around the mouth, anterior edge of the foot, tips of the cerata, and within the rhinophores .... *Aplysiopsis smithi*
5b Rhinophores rolled up for their entire length; body pale, with a branching green or red pattern on the dorsum and within the cerata (the pattern is formed by branches of the gut and digestive gland) 6
6a With as many as 50 slender cerata (body creamy white, with a green digestive tract) .... *Placida dendritica* (fig. 12.96)
6b With only about 18-24 cerata, these plump rather than slender 7
7a Cerata of nearly uniform size and shape, and with a yellow cap at the tip; tail short; digestive tract reddish brown .... *Hermaea oliviae*
7b Cerata of varying size and shape, and without a color cap at the tip; tail extending considerably beyond the ceratal area, and may even equal it in length; digestive tract golden brown .... *Hermaea vancouverensis*

Family Stiligeridae

Alderia modesta (Lovén, 1844). On *Vaucheria*, high intertidal.
Aplysiopsis smithi Marcus, 1961. On *Urospora*, *Chaetomorpha*, *Rhizoclonium*, and *Enteromorpha*.
Hermaea oliviae MacFarland, 1966
Hermaea vancouverensis O'Donoghue, 1924

Placida dendritica Alder & Hancock, 1843. On *Codium* and *Bryopsis*, and perhaps other green algae.
Stiliger fuscovittatus Lance, 1962. On *Polysiphonia*.

Family Oleidae

Olea hansineensis Agersborg, 1923. On egg masses of *Melanochlamys*, *Haminoea*, and *Gastropteron*.

Family Elysiidae

Elysia hedgpethi Marcus, 1961. On *Codium* and *Bryopsis*, and perhaps other green algae.

Order Nudibranchia

Most of the references listed under Subclass Opisthobranchia are concerned to at least some extent with nudibranchs. The especially important ones are listed again here. Papers that deal with specific suborders will be found under the headings for these groups.

Behrens, D. W. 1980. Pacific Coast Nudibranchs: A Guide to the Opisthobranchs of the Northeastern Pacific. Los Osos, California: Sea Challengers. 112 pp.
Bergh, L. S. R. 1879. On the nudibranchiate gastropod Mollusca of the north Pacific Ocean, with special reference to those of Alaska. Part I. Proc. Acad. Nat. Sci. Philadelphia, 1879:71-132.
------. 1880. On the nudibranchiate gastropod Mollusca of the north Pacific Ocean, with special reference to those of Alaska. Part II. Proc. Acad. Nat. Sci. Philadelphia, 1880:40-127.
------. 1894. Die Opisthobranchien. *From* Reports on the dredging operations off the west coast of Mexico and in the Gulf of California in charge of Alexander Agassiz, carried on by the U. S. Fish Commission Steamer "Albatross" during 1891. Bull. Mus. Comp. Zool. Harvard, 25:125-233.
Goddard, J. H. R. 1984. The opisthobranchs of Cape Arago, Oregon, with notes on their biology and a summary of benthic opisthobranchs known from Oregon. Veliger, 27:143-63.
McDonald, G. R. 1983. A review of the nudibranchs of the California coast. Malacologia, 24:114-276.
McDonald, G. R., & J. W. Nybakken. 1981. Guide to the Nudibranchs of California. Melbourne, Florida: American Malacologists, Inc. 72 pp.
MacFarland, F. M. 1906. Opisthobranchiate Mollusca from Monterey Bay, California and vicinity. Bull. U. S. Bur. Fish., 25:109-57.
------. 1966. Studies of opisthobranchiate mollusks of the Pacific coast of North America. Mem. 6, Calif. Acad. Sci. xvi + 546 pp.
Marcus, E. 1961. Opisthobranch mollusks from California. Veliger, 3 (suppl.):1-84.
O'Donoghue, C. N. 1921. Nudibranchiate Mollusca from the Vancouver Island region. Trans. Royal Can. Inst., 13:147-209.
------. 1922. Notes on the nudibranchiate Mollusca from the Vancouver Island region. III. Records of species and distribution. Trans. Royal Can. Inst., 14:145-67.
------. 1924. Notes on the nudibranchiate Mollusca from the Vancouver Island region. IV. Additional species and records. Trans. Royal Can. Inst., 15:1-33.
------. 1926. A list of the nudibranchiate Mollusca recorded from the Pacific coast of North America with notes on their distribution. Trans. Royal Can. Inst., 15:199-247.
------. 1927. Notes on the nudibranchiate Mollusca from the Vancouver Island region. V. Two new species and one new record. Trans. Royal Can. Inst., 16:1-12.
Robilliard, G. A. 1971. Range extensions of some northeast Pacific nudibranchs (Mollusca: Gastropoda: Opisthobranchia) to Washington and British Columbia, with notes on their biology. Veliger, 14:162-5.
Russell, H. D. 1971. Index Nudibranchia, a Catalog of the Literature 1554-1965. Greenville: Delaware Museum of Natural History. 141 pp.
Sphon, G. G. 1972. Some opisthobranchs (Mollusca: Gastropoda) from Oregon. Veliger, 12:153-7.

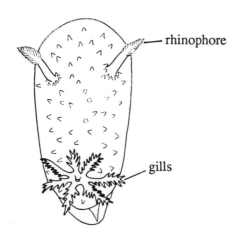

12.113 *Acanthodoris brunnea*

Suborder Doridacea

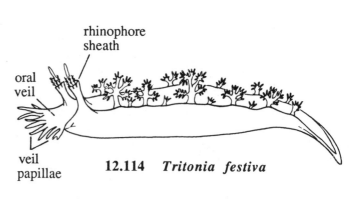

12.114 *Tritonia festiva*

12.116 *Hermissenda crassicornis*

Suborder Aeolidacea

12.115 *Dendronotus subramosus*

Suborder Dendronotacea

Steinberg, J. E. 1961. Notes on the opisthobranchs of the west coast of North America. I. Nomenclatural changes in the order Nudibranchia (southern California). Veliger, 4:57-63.

------. 1963a. Notes on the opisthobranchs of the west coast of North America. III. Further nomenclatural changes in the order Nudibranchia. Veliger, 6:63-7.

------. 1963b. Notes on the opisthobranchs of the west coast of North America. IV. A distributional list of opisthobranchs from Point Conception to Vancouver Island. Veliger, 6:68-73.

1a Anus, partly or completely encircled by gills, located on the midline of the dorsum, one-half to three-fourths of the body length from the anterior end   Suborder Doridacea (in part)

1b Anus not located on the midline of the dorsum (it is either on the right side, on the dorsum near the right side, or at the posterior end of the body; if gills are present, they are not arranged in a partial or complete circle around the anus)   2

2a Anus at the posterior end of the body, on the midline, in a deep groove in the mantle between the foot and the edge of the dorsum; dorsum without any outgrowths other than the rhinophores   Suborder Doridacea (in part)

2b Anus on the right side of the body, or on the dorsum near the right side; dorsum usually with outgrowths (simple or branched cerata, or gills) in addition to rhinophores (if dorsal outgrowths are not present, the dorsum has several white longitudinal ridges on a pinkish brown background)   3

3a   Without dorsal outgrowths other than the rhinophores; dorsum with several white longitudinal ridges on a pinkish brown background
                                Suborder Arminacea: *Armina californica* (fig. 12.118)

3b   With dorsal outgrowths (cerata or gills) in additon to the rhinophores; dorsum without white longitudinal ridges   4

4a   Clavus of the rhinophores partly or fully retractile into the sheath; dorsal outgrowths either club-shaped cerata (with rings of prominent tubercles), paddlelike or bushy cerata, or bushy gills (as a rule, there are 4 to 10 cerata or numerous gills arranged in each of 2 dorsolateral longitudinal series)   Suborder Dendronotacea

4b   Clavus of rhinophores not retractile into a sheath (each rhinophore stands free; it has no sheath); dorsal cerata cylindrical, lanceolate, or leaflike, usually numerous (more than 20) and usually arranged in transverse rows   5

5a   Anus on a conspicuous papilla near the right side of the posterior third of the dorsum, usually among the posterior third of the cerata   Suborder Arminacea (in part)

5b   Anus on an inconspicuous papilla near the right side of the anterior half of the dorsum, usually among the anterior third of the cerata   Suborder Eolidacea

Suborder Doridacea

Bloom, S. A. 1976. Morphological correlation between dorid nudibranch predators and sponge prey. Veliger, 18:289-301.

Burn, R. 1968. *Archidoris odhneri* (MacFarland, 1966) comb. nov., with some comments on the species of the genus on the Pacific coast of North America. Veliger, 11:90-2.

Cockerell, T. D. A. 1915. The nudibranch genus *Triopha* in California. J. Entom. Zool. Pomona College, 8:228-9.

Ferreira, A. J. 1977. A review of the genus *Triopha* (Mollusca: Nudibranchia). Veliger, 19:387-402.

Gosliner, T. M., & G. C. Williams. 1975. A genus of dorid nudibranch previously unrecorded from the Pacific coast of the Americas, with the description of a new species. Veliger, 17:396-405.

Lance, J. R. 1962. A new *Stiliger* and a new *Corambella* (Mollusca: Opisthobranchia) from the northwestern Pacific. Veliger, 5:33-8.

MacFarland, F. M. 1925. The Acanthodorididae of the California coast. Nautilus, 39:49-65.

------. 1926. The Acanthodorididae of the California coast. Nautilus, 39:94-105.

MacFarland, F. M., & C. H. O'Donoghue. 1929. A new species of *Corambe* from the Pacific coast of North America. Proc. Calif. Acad. Sci., ser. 4, 18:1-27.

Millen, S. V. 1982. A new species of dorid nudibranch (Opisthobranchia: Mollusca) belonging to the genus *Anisodoris*. Can. J. Zool., 60:2694-705.

------. 1985. The nudibranch genera *Onchidoris* and *Diaphorodoris* (Mollusca, Opisthobranchia) in the northeastern Pacific. Veliger, 28:80-3.

Millen, S. V., & T. M. Gosliner. 1985. Four new species of dorid nudibranchs belonging to the genus *Aldisa* (Mollusca: Opisthobranchia), with a revision of the genus. Zool. J. Linnean Soc., 84:195-233.

Rivest, B. R. 1984. Copulation by hypodermic injection in the nudibranchs *Palio zosterae* and *Palio dubia* (Gastropoda, Opisthobranchia). Biol. Bull., 167:543-54.

Robilliard, G. A. 1971. A new species of *Polycera* (Opisthobranchia: Mollusca) from the northeastern Pacific, with notes on other species. Syesis, 4:235-43.

1a   Body, including the gills and long dorsal outgrowths (which resemble cerata of eolid nudibranchs) rose-pink (length up to 2 cm)   *Hopkinsia rosacea*

1b   Body not primarily rose-pink   2

2a   Anus on the midline at the posterior end of the body, situated in a mantle groove between the foot and edge of the dorsum; gills located on both sides of the anus; dorsum without any outgrowths other than the rhinophores; pale ochre spots forming an intricate pattern on the otherwise white dorsum, the pattern closely resembling that of the bryozoan *Membranipora*,

	on which these dorids feed	3
2b	Anus on the midline of the dorsum, one-half to one-fourth of the body length from the posterior end; gills in a circle or semicircle around the anus; dorsal outgrowths include gills and rhinophores, and may also include various types of papillae or tubercles; body may be white, but an intricate pattern of ochre spots is absent, even in a few species that are found on bryozoans	4
3a	With a distinct notch on the posterior border of the dorsum; with 6-14 gills on each side of the anus	*Corambe pacifica*
3b	Without a notch on the posterior border of the dorsum; with 2-4 gills on each side of the anus	*Doridella steinbergae*
4a	Dorsum usually with some conspicuous large outgrowths in addition to the rhinophores and gills, these outgrowths generally near the rhinophores or gills or along the lateral margins of the dorsum	5
4b	Dorsum without large outgrowths other than the rhinophores and gills, although it may be almost completely covered by tubercles or slender, nearly microscopic papillae	11
5a	Body firm, due to spicules in the dorsum; with short, blunt, cylindrical papillae arranged in 4-6 irregular longitudinal rows on the dorsum; body white to cream, with brown to black spots scattered over the dorsum	*Aegires albopunctatus*
5b	Body relatively soft (if there are spicules on the tubercles of the dorsum, they do not contribute to a supporting network); papillae, if present, scattered, not arranged in irregular longitudinal rows; body without brown to black spots on the dorsum	6
6a	Body somewhat flattened dorsoventrally; with numerous (over 40) slender, club-shaped papillae all around the margin of the dorsum, and also with a few short papillae in the central area of the dorsum; body white or cream, with bright orange on the tips of the papillae and on the rhinophores, but not on the gills and not on the dorsum itself	*Laila cockerelli*
6b	Body not flattened dorsoventrally; with fewer than 30 papillae of various shapes around the margins of the dorsum, but none of these are slender and club-shaped; bright orange, if present in the color pattern, found on the gills and in spots on the dorsum itself, as well as on the papillae and rhinophores	7
7a	With a few (about 10-15) small coronate papillae along the lateral margins of the dorsum; with about 8-15 small, irregularly papillated outgrowths on the edge of the oral veil; body white to pale yellow, with deep orange on the tips of the gills, rhinophores, tail, dorsal papillae, and oral veil papillae, and also in spots on the dorsum itself	*Triopha catalinae*
7b	Without coronate papillae on the lateral margins of the dorsum; without papillated outgrowths on the oral veil; body with little or no orange anywhere	8
8a	With long, cylindrical, tapered papillae on the dorsolateral margins between the rhinophores and gills, and on the central area of the dorsum anterior to the gills; with 14 pinnate gills	*Okenia vancouverensis*
8b	If long, cylindrical, tapered papillae are present, they occur only at the edge of the oral veil, lateral to the gills, and just anterior to the rhinophores (papillae not usually present on the dorsolateral margins between the rhinophores and gills, or on the central area of the dorsum anterior to the gills; *Palio zosterae*, however, may have short, blunt papillae scattered over much of the body); with fewer than 8 pinnate gills	9
9a	Background color white or cream, with markings of yellow or of both yellow and black	10
9b	Body deep olive to black	*Palio zosterae*
10a	With only yellow pigment on the tips of the papillae, gills, and rhinophores; with 3 parallel yellow lines between the rhinophores and gills; with 2 long, simple papillae on both sides of the body just anterior to the rhinophores; with 3 or 4 erect papillae on both sides of the gills; with 2 papillae on the right and left edges of the oral veil	*Ancula pacifica*
10b	On all papillae, gills, and rhinophores, the color changes, proximally to distally, from black to yellow to white; with yellow spots scattered over the dorsum and with a yellow line on the edge of the foot; without papillae just anterior to the rhinophores; with 3-6 erect papillae on both sides of the gills; with 6-12 papillae on the right and left edges of the oral veil	*Polycera tricolor*
11a	Gills completely retractile into sheaths; rhinophores relatively stout, not usually tapering gradually from the base to the tip	12
11b	Gills not completely retractile into sheaths; rhinophores relatively slender (when fully extended, usually at least 4 times as long as wide at the base), tapering gradually from the	

	base to the tip	25
12a	Without large, rounded tubercles on the dorsum, the entire surface being either nearly smooth, minutely villous, or with very small tubercles visible only with magnification	13
12b	With both large and small, rounded tubercles scattered over the dorsum, but the surface otherwise smooth	14
13a	Body mostly white to dusky gray, with markings up to 1.5 cm in diameter on the dorsum, each marking usually consisting of a white ring that encloses a brown ring with a dark brown to black center	*Diaulula sandiegensis*
13b	Body off-white to pale yellow (seldom dusky gray), without markings that consist of concentric white and brown rings (small black spots, less than 1 mm in diameter, are present, however)	*Discodoris heathi*
14a	Body red or orange-red, sometimes with a few very small black spots on the dorsum; rhinophores with 10-12 vertically oriented leaves and terminating in slender, simple papillae	*Rostanga pulchra*
14b	Body white, yellow, orange, or red, sometimes with yellow or dark spots; rhinophores perfoliate (the leaves oriented horizontally, as typical for dorids)	15
15a	Body either yellow (ranging from pale yellow to lemon-yellow to dusky, greenish yellow), orange-yellow, orange, or red, with darker (usually black) spots on the dorsum (but some specimens of *Aldisa sanguinea*, which is red, lack spots); dorsum not edged with a line of yellow pigment	16
15b	Body white, cream, or greenish (if white or cream, then sometimes with yellow spots on the tubercles of the dorsum); without dark spots on the dorsum; dorsum sometimes edged with a line of yellow pigment	20
16a	Body orange or red, usually with black spots on the dorsum	17
16b	Body some shade of yellow, greenish yellow, or orange-yellow (but not decidedly orange or red), with at least some spots (except in unusual specimens)	18
17a	With up to 8 black spots, usually arranged in a single row on the dorsal midline, from just anterior to the rhinophores to the gills; body as a whole generally ranging from pale orange to red (but some specimens are yellow, in which case the row of black spots on the midline is the best diagnostic character)	*Aldisa cooperi*
17b	With up to 5 black spots scattered over the dorsum, not arranged in a row on the midline; body as a whole bright red (not likely to be found north of Oregon)	*Aldisa sanguinea*
18a	Length up to 1.5 cm; body mostly pale yellow, with reddish brown flecks scattered over the dorsum and with a few larger, dark brown spots near the midline, between the rhinophores and gills; distal portions of rhinophores brownish maroon; with reddish maroon spots at the bases of the gills (uncommon north of Oregon)	*Hallaxa chani*
18b	Length commonly more than 5 cm; body yellow, greenish yellow, or orange-yellow, usually with black spots (these are rarely absent) on the dorsum; rhinophores paler or duskier than the body, but not brownish maroon in their distal portions; without reddish maroon spots at the bases of the gills	19
19a	Body yellow to orange-yellow; black spots usually present only between the tubercles, rarely on them (in unusual specimens, the black spots are absent); gills and rhinophores usually paler than the body; often exuding a considerable quantity of sweet-smelling mucus when disturbed	*Anisodoris nobilis*
19b	Body lemon-yellow to greenish yellow, rarely orange-yellow; black spots present on the tubercles as well as between them; gills and rhinophores usually duskier than the body; exuding little mucus and almost odorless	*Archidoris montereyensis*
20a	Dorsum almost entirely greenish, but with a white margin	*Aldisa albomarginata*
20b	Dorsum not greenish	21
21a	Dorsum white (rarely pale custard in *Archidoris odhneri*) or translucent white with spots of white pigment, but without any clear yellow spots or lines	22
21b	Dorsum mostly white, but with yellow spots and/or a yellow line around the margin (in addition, the clavus of the rhinophores usually has some yellow to dark brown pigment)	23
22a	Dorsum low, not arched, translucent white (the internal organs are partly visible) with white spots of pigment; with a ring of tubercles around the pits from which the rhinophores and gills emerge; length not more than 3 cm	*Aldisa tara*
22b	Dorsum high, arched, opaque white (internal organs not visible); without a ring of tubercles around the pits from which the rhinophores and gills emerge; length commonly exceeding	

248   Phylum Mollusca: Class Gastropoda

        5 cm, and sometimes reaching 10 cm *Archidoris odhneri*

23a With a continuous lemon-yellow line around the margin of the dorsum and on the posterior margin of the foot; gills and most of the dorsal tubercles with yellow tips; rhinophores usually white, except for yellow at their tips *Cadlina luteomarginata*

23b Without a yellow line around the margin of the dorsum and part of the foot; gills without yellow tips, and yellow spots on the dorsum are generally large and mostly near the margin; rhinophores dusky brown to nearly black   24

24a Rhinophore leaves brown to almost black; dorsum and lip tentacles with microscopic brown spots *Cadlina flavomaculata*

24b Rhinophore leaves pale brown; dorsum, lip tentacles, and other parts of the body without microscopic brown spots *Cadlina modesta*

25a Dorsum covered with slender, flexible, close-set papillae, so that the surface appears fuzzy; papillae (which are conical or nearly so) without spicules protruding from their tips   26

25b Dorsum with stout, rounded tubercles, so that the surface is firm, rather than fuzzy; tubercles (which may be conical, nearly conical, or mushroom-shaped) often with spicules protruding from their tips   28

26a Body white or slightly off-white (no other color on the dorsum or its outgrowths) *Acanthodoris pilosa*

26b Body mostly white or some shade of gray, but with yellow, gray, or brownish red (or some combination of these colors) on the dorsum or its outgrowths   27

27a Body mostly white, without gray on the dorsum, but with a continuous lemon-yellow line around the margin of the dorsum; with yellow on the tips of the rhinophores and gills; usually with 5 gills *Acanthodoris hudsoni*

27b Body mostly white to mauve-gray (there are many intergrades; darker animals have dark gray splotches), without a yellow line around the margin of the dorsum; with yellow on the tips of the papillae and brownish red on the tips of the rhinophores and gills; usually with 9 gills *Acanthodoris nanaimoensis*

28a Body dull cream with irregular brown bands on the dorsum, or brown in general; length usually at least 1.5 cm   29

28b Body usually white, cream, or pale orange, without any brown; length usually less than 1.5 cm   30

29a Gills pinnate, almost erect, 16 or more in number, arranged in a broad horseshoe-shaped pattern around the anus (the opening of the horseshoe is directed posteriorly); dorsal tubercles mushroom-shaped, with spicules protruding from their tips; body usually dull cream, with dark brown pigment in 2 or 3 irregular longitudinal bands on the dorsum; without yellow on the gills, tubercles, rhinophores, or edge of the foot *Onchidoris bilamellata*

29b Gills bipinnate, spreading outward when expanded, usually 5-7 in number, completely encircling the anus; most dorsal tubercles conical or nearly conical, almost none of them with protruding spicules; body pale brown, with small splotches of black scattered between the tubercles; with yellow on the tips of the gills, tubercles, and rhinophores, and often around the edge of the foot *Acanthodoris brunnea* (fig. 12.113)

30a Gills enclosed by a common sheath, this not closing over the gills when they are retracted; tail with a mid-dorsal ridge *Diaphorodoris lirulatocauda*

30b Gills originating within separate pits, not enclosed by a common sheath; tail without a mid-dorsal ridge   31

31a Many or all tubercles on the dorsum mushroom-shaped, sometimes with bundles of spicules protruding from their distal ends; spicules in the dorsum tending not to radiate from the bases of the tubercles, but rather to form parallel bundles between the tubercles; usually with a large, white tubercle just posterior to the anus   32

31b Few if any of the tubercles on the dorsum mushroom-shaped (most of them are nearly conical, inflated, and pointed at the tip), but most tubercles with bundles of spicules protruding from their distal ends; large spicules in the dorsum tending to radiate from the bases of the tubercles; without a large, white tubercle just posterior to the anus (length up to about 1 cm) *Adalaria* sp.

32a Length usually more than 1.5 cm; usually found in close association with barnacles, upon which this species feeds (brown-mottled specimens, described in choice 29a, often occur in the same population) *Onchidoris bilamellata*

Phylum Mollusca: Class Gastropoda 249

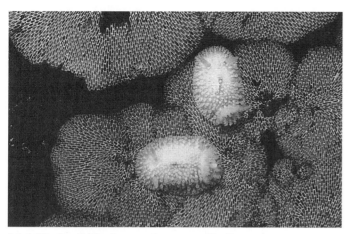

12.117 *Onchidoris muricata,* on the bryozoan *Membranipora membranacea*

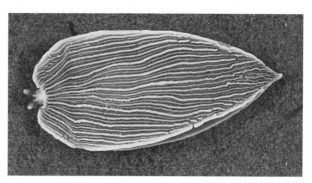

12.118 *Armina californica*

12.120 *Dirona albolineata*

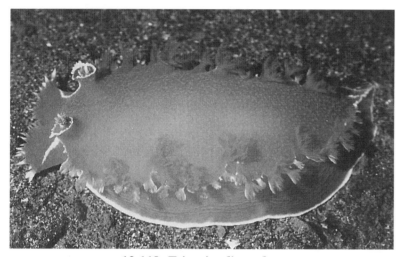

12.119 *Tritonia diomedea*

12.122 *Aeolidea papillosa*

12.121 *Tochuina tetraquetra*

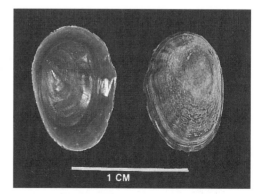

12.123 *Siphonaria thersites*

Phylum Mollusca: Class Gastropoda

32b Length usually less than 1.5 cm; usually found on *Membranipora*, *Schizoporella*, and other encrusting bryozoans, upon which it feeds .... *Onchidoris muricata* (fig. 12.117)

Family Corambidae

Corambe pacifica MacFarland & O'Donoghue, 1929. Feeds on the bryozoan *Membranipora membranacea*.
Doridella steinbergae (Lance, 1962). Feeds on the bryozoan *Membranipora membranacea*.

Family Goniodorididae

Ancula pacifica MacFarland, 1905. Feeds on the entoproct *Barentsia ramosa*.
Cargoa vancouverensis (O'Donoghue, 1921)
Hopkinsia rosacea MacFarland, 1905. Feeds on the bryozoan *Eurystomella bilabiata*.

Family Onchidorididae

**Acanthodoris armata* O'Donoghue, 1927. Feeds on bryozoans.
**Acanthodoris atrogriseata* O'Donoghue, 1927
Acanthodoris brunnea MacFarland, 1905 (fig. 12.113). Feeds on bryozoans.
Acanthodoris hudsoni MacFarland, 1905
Acanthodoris nanaimoensis O'Donoghue, 1921. Feeds on compound ascidians.
Acanthodoris pilosa (O. F. Müller, 1776). Feeds on bryozoans.
Adalaria sp. Similar to *A. proxima* (Alder & Hancock, 1854), originally described from England. The status of *A. pacifica* Bergh, 1880, described from Alaska, remains uncertain.
Diaphorodoris lirulatocauda Millen, 1985
Onchidoris bilamellata (Linnaeus, 1767). Feeds on barnacles; juveniles feed on bryozoans.
Onchidoris muricata (O. F. Müller, 1776) (fig. 12.117). Feeds on bryozoans.

Family Polyceridae

Aegires albopunctatus MacFarland, 1905. Feeds on the sponge *Leucilla nuttingi*.
*Crimora coneja Marcus, 1961
Laila cockerelli MacFarland, 1905. Feeds on bryozoans.
Palio zosterae O'Donoghue, 1924. Feeds on bryozoans.
Polycera tricolor Robilliard, 1971. Feeds on bryozoans.
Triopha catalinae (Cooper, 1863). Feeds on bryozoans.

Family Chromodoridae

Cadlina flavomaculata MacFarland, 1905. Feeds on the sponge *Aplysilla glacialis*.
Cadlina luteomarginata MacFarland, 1966. Feeds on sponges.
Cadlina modesta MacFarland, 1966. Feeds on the sponge *Aplysilla glacialis*.

Family Actinocyclidae

Hallaxa chani Gosliner & Williams, 1975. Feeds on the compound ascidian *Didemnum carnulentum*.

Family Aldisidae

Aldisa albomarginata Millen, 1985

Aldisa cooperi Robilliard & Baba, 1972
Aldisa sanguinea (Cooper, 1863). Feeds on red sponges.
Aldisa tara Millen, 1985. Feeds on red sponges.

Family Rostangidae

Rostanga pulchra MacFarland, 1905. Feeds on red sponges.

Family Dorididae

*Doris odonoghuei Steinberg, 1963

Family Archidorididae

Archidoris montereyensis (Cooper, 1862). Feeds on sponges.
Archidoris odhneri (MacFarland, 1966). Feeds on sponges.

Family Discodorididae

*Anisodoris lentiginosa Millen, 1982
Anisodoris nobilis (MacFarland, 1905). Feeds on sponges.
Diaulula sandiegensis (Cooper, 1862). Feeds on sponges.
Discodoris heathi MacFarland, 1905. Feeds on sponges.

Suborder Dendronotacea

Robilliard, G. A. 1970. The systematics and some aspects of the ecology of the genus *Dendronotus* (Gastropoda: Nudibranchia). Veliger, 12:433-79.
Thompson, T. E. 1971. Tritoniidae from the North American Pacific coast. Veliger, 13:333-8.

1a Body stout, much more massive than most slugs; dorsal outgrowths represented by numerous small gill tufts concentrated along a definite dorsolateral margin on both sides of the body ... 2
1b Body sluglike; dorsal outgrowths represented by cerata that are either bushy, club-shaped (with rings of tubercles), or paddlelike, these cerata being arranged in a longitudinal row on both sides of the body (but there are no definite dorsolateral margins) ... 5
2a With white gill tufts that form unbroken rows extending from the rhinophores to the tail region; anterior border of the oral veil white, crenulate, but without papillae; dorsum apricot-orange, with patches of white pigment, the sides lighter ... *Tochuina tetraquetra* (fig. 12.121)
2b Gill tufts of each row separated by obvious spaces; anterior border of the oral veil with simple papillae; body color mostly pale pink to deep salmon or rose ... 3
3a Dorsum pale pink, usually with an intricate pattern of white lines; edge of rhinophore sheath smooth; oral veil with about 10 long, slender papillae ... *Tritonia festiva* (fig. 12.114)
3b Dorsum salmon or rose, without a pattern of white lines, the sides lighter; edge of rhinophore sheath crenulate; oral veil with 10-30 short papillae ... 4
4   Although external characters are helpful in separating the next two species, dissection may be necessary to confirm the identification.
4a Body about as wide as high; length up to 8 cm; oral veil with up to 25 (usually 10-16) papillae; dissection reveals a white oral canal, a short stout penis, and denticles on the masticatory border ... *Tritonia exsulans*
4b Body decidedly wider than high; length up to 12 cm; oral veil usually with 18 papillae;

dissection reveals a reddish brown or orange oral tube, an elongate, clubbed penis, and absence of denticles on the masticatory border   *Tritonia diomedea* (fig. 12.119)

5a With a large oral hood that has numerous filiform tentacles around its margin; usually with 4-6 pairs of paddle-shaped cerata; oral hood with small, earlike rhinophores (body translucent yellowish gray, with opaque brown hepatic diverticula)   *Melibe leonina*

5b Without an oral hood, although an oral veil may be present; cerata bushy or club-shaped; rhinophores large and prominent   6

6a Cerata club-shaped, with 3-6 whorls of rounded tubercles; clavus of rhinophores smooth; rhinophore sheath may be crenulate, but not drawn out into prominent papillae that give it a crownlike appearance   7

6b Cerata bushy; clavus of rhinophores perfoliate (with encircling flanges) rhinophore sheath drawn out into 4-10 simple or branched papillae that give it a crownlike appearance   9

7a Most or all of the tubercles on the cerata whitish or yellowish gray, encircled by black rings (dorsum and sides of the body dark brown; length up to 1.2 cm)   *Doto columbiana*

7b Tubercles on cerata not yellowish gray and not encircled by black rings   8

8a Cerata orange or orange-pink, the tubercles with yellow or white tips; dorsum yellowish, without dark pigment (length up to 1.2 cm)   *Doto amyra*

8b Cerata whitish; dorsum with blackish brown blotches (length up to 1 cm)   *Doto kya*

9a Oral veil large, wide, overhanging the mouth region, and usually with 3 pairs of long, branched papillae on its anterolateral third; with a white line around the edge of the foot, or yellowish white spots liberally scattered over the whole body, including the cerata, rhinophores, and veil papillae (background color pale red-brown, with darker reddish brown patches)   *Dendronotus albopunctatus*

9b Oral veil not large and wide, and only slightly overhanging the mouth region, and usually with more than 3 pairs of papillae on its anterolateral third (if only 3 pairs of papillae are present on the veil, these are seldom long and branched); usually without a white line around the edge of the foot, but if such a line is present, there are few or no white spots scattered over the body   10

10a With a vertical row of 3-6 small, bushy outgrowths on the posterior border of the rhinophore stalk; with an opaque white line around the edge of the foot (body color gray or gray-brown to deep orange-red; with 4-7 pairs of tall, bushy cerata tipped with dark purple, orange, yellow, or white, or some combination of these colors; length up to 30 cm, but usually 6-18 cm)   *Dendronotus iris*

10b Without outgrowths on the posterior border of the rhinophore stalk; without an opaque white line around the foot   11

11a Without lateral papillae on the rhinophore stalks; cerata tall and stout, branching in a rosette pattern about halfway up; body usually pale to dark reddish brown, mottled, and with 4 parallel reddish brown longitudianal lines on the dorsum, 2 on each side, joining the bases of the rhinophores and each successive pair of cerata (occasionally, the body is white, yellow, or orange, but at least portions of the 4 reddish brown lines will be evident) (length up to 5 cm)   *Dendronotus subramosus* (fig. 12.115)

11b Usually with lateral papillae on the rhinophore stalks (these papillae may, however, be reduced or absent in *Dendronotus albus* and *D. diversicolor*); cerata branched in a fanlike pattern, with the fan oriented at a right angle with respect to the surface of the body; dorsum without 4 parallel reddish brown longitudinal lines   12

12a Length usually more than 4 cm, and sometimes up to 30 cm; with 6-8 pairs of very bushy cerata; with 4 or 5 pairs of large, extensively branched papillae on the oral veil; with 20 or more simple or branched lip papillae   13

12b Length usually 1 to 4 cm, but sometimes up to 6 cm; with 4-8 pairs of sparsely branched cerata; with 3 or 4 pairs of papillae on the oral veil; with up to 8 simple lip papillae   14

13a With a red line around the foot; body usually grayish white, with magenta pigment on the distal portion of the cerata, rhinophores, oral veil, lip, and lateral papillae (sometimes mottled brownish red and pale pink all over); with 6-8 pairs of tall, extensively branched cerata (the branches long); with numerous small accessory cerata between the principal cerata; anterior lip papillae large and branched, but the posterior and ventral lip papillae usually simple; oral veil with 5 pairs of extensively branched papillae; rhinophores with 5 pairs of extensively branched crown papillae; length up to 28 cm, but usually 8-15 cm   *Dendronotus rufus*

13b Without a red line around the foot; body translucent white to pink, with opaque white pigment on the distal third of the cerata and rhinophores; with 6 or 7 pairs of short, extensively branched cerata (the branches relatively short); without accessory cerata between the principal cerata; lip papillae (there are often 30 or more of these) small, simple, fingerlike; oral veil with 4 or 5 pairs of short, moderately branched papillae; rhinophores with 4-12 (usually 5) moderately branched crown papillae (but there may be more of these on one rhinophore than on the other); length up to 14 cm, usually 4-8 cm *Dendronotus dalli*

14a Body white or lilac, with white and/or orange on the tips of the cerata; with a dorsomedial white line extending from the tip of the tail to at least the posteriormost pair of cerata 15

14b Body pale brown to dark reddish brown, sometimes whitish; without a dorsomedial white line between the tip of the tail and the posteriormost pair of cerata (length up to 5 cm, usually 1.5 to 3 cm; with 5-7 pairs of moderately branched cerata; 4 pairs of simply branched papillae on the oral veil; up to 6 simple lip papillae; body delicate, sluglike) *Dendronotus frondosus*
The following color morphs prevail: 1) body pale brown, with yellow caps on the tubercles, and with small yellow or white flecks on the outgrowths of the dorsum; 2) body whitish, with a few small pinkish brown spots on the dorsum; 3) body dark reddish brown, often with large patches of white pigment between the cerata and/or on the sides, ventral to the cerata, and with white or yellow spots on the dorsum and its outgrowths.

15a With 5-8 pairs of moderately branched cerata; often with a few nearly conical papillae on the dorsum; body translucent white; with opaque white pigment on the distal portions of the cerata, rhinophore stalk and crown papillae, veil papillae, and lateral papillae; with a dorsomedial white line extending from the tip of the tail to the level of the 4th to 2nd pairs of cerata (in some individuals, the core of the cerata shows a gradation, proximally to distally, from tan to golden yellow to copper to white) (length up to 4.5 cm, usually 1-3 cm)
Dendronotus albus

15b Usually with 4 pairs (rarely 5) of tall, slender, sparsely branched cerata; dorsum without papillae; body translucent white or lilac, with intergrades occasionally seen; with opaque white and/or metallic orange pigment on the distal portions of the cerata and on the rhinophores and veil papillae; with a dorsomedial white line extending from the tip of the tail to the posteriormost pair of cerata; length up to 7 cm, usually 3-6 cm
Dendronotus diversicolor

Family Tritoniidae

Tochuina tetraquetra (Pallas, 1788) (fig. 12.121). Feeds on sea pens and other alcyonaceans.
Tritonia diomedea Bergh, 1894 (fig. 12.119). Feeds on sea pens.
Tritonia exsulans Bergh, 1894
Tritonia festiva (Stearns, 1873) (fig. 12.114). Feeds on sea pens and other alcyonaceans.

Family Dendronotidae

Dendronotus albopunctatus Robilliard, 1972. Feeds on sponges.
Dendronotus albus MacFarland, 1966. Feeds on hydroids.
Dendronotus dalli Bergh, 1879. Feeds on hydroids of the genus *Abietinaria*.
Dendronotus diversicolor Robilliard, 1970. Feeds on hydroids.
Dendronotus frondosus (Ascanius, 1774). Feeds on hydroids.
Dendronotus iris Cooper, 1863. Feeds on the ceriantharian *Pachycerianthus fimbriatus*.
Dendronotus rufus O'Donoghue, 1921. Feeds on scyphistomae of scyphozoans.
Dendronotus subramosus MacFarland, 1966 (fig. 12.115). Feeds on hydroids of the genus *Aglaophenia*.

Family Tethydidae

Melibe leonina (Gould, 1853). Feeds on small crustaceans.

Family Dotoidae

Doto amyra Marcus, 1961. Feeds on hydroids.
Doto columbiana O'Donoghue, 1921. Feeds on hydroids.
Doto kya Marcus, 1961. Feeds on hydroids.

Suborder Arminacea

Gosliner, T. M. 1982. The genus *Janolus* (Nudibranchia: Arminacea) from the Pacific coast of North America, with reinstatement of *Janolus fuscus* O'Donoghue, 1924. Veliger, 24:155-8.
Hurst, A. 1966. Description of a new species of *Dirona* from the north-east Pacific. Veliger, 9:9-15.
MacFarland, F. M. 1912. The nudibranch family Dironidae. Zool. Jahrb., suppl., 15:515-36.

1a	Without cerata; dorsum with several white longitudinal ridges on a pinkish brown background (these ridges converge anteriorly toward the rhinophores) (on sandy substrata)	*Armina californica* (fig. 12.118)
1b	With cerata; dorsum without white longitudinal ridges	2
2a	Cerata slender, circular or only slightly oval in cross-section, and approximately the same diameter over most of their length; without a prominent oral veil	3
2b	Cerata leaflike (broad in the middle and tapering toward both ends), distinctly flattened in cross-section; with a prominent oral veil on the head	4
3a	Body whitish, translucent, without blue spots on the dorsum; with an irregular dorsomedial red band (this may be interrupted) running from in front of the rhinophores to near the last pair of cerata; crest between the rhinophores distinctly red; terminal portions of cerata with a proximal orange band followed by a transparent colorless band and finally by an opaque white tip	*Janolus fuscus*
3b	Body whitish, translucent, with small (visible with magnification) blue spots scattered over the dorsum; without a dorsomedial red line; crest between the rhinophores sometimes pale orange but not red; terminal portions of cerata with a proximal orange band succeeded by white, or with an orange band succeeded by white bands and finally by blue bands	*Janolus barbarensis*
4a	Body orange, with white splotches on the dorsum (with white lines on the edges of the cerata, but not around the foot and oral veil, or between the rhinophores)	*Dirona aurantia*
4b	Body not orange, and without white splotches on the dorsum	5
5a	Cerata smooth, each bordered by a white line; body usually whitish, translucent (some specimens are mauve, and occasionally the cerata are pinkish); with white lines around the edge of the foot and oral veil, as well as between the rhinophores	*Dirona albolineata* (fig. 12.120)
5b	Cerata usually roughened, each with a pale red spot, but not bordered by a white line; body light brown to greenish gray, with small yellowish, greenish, and pink dots; without white lines around the edge of the foot and oral veil, or between the rhinophores (not likely to be found north of Oregon)	*Dirona picta*

Family Arminidae

Armina californica (Cooper, 1863) (fig. 12.118). Feeds on sea pens.

Family Dironidae

Dirona albolineata MacFarland in Cockerell & Eliot, 1905 (fig. 12.120). Feeds on small snails.
Dirona aurantia Hurst, 1966. Feeds on bryozoans.
Dirona picta MacFarland in Cockerell & Eliot, 1905. Feeds on hydroids, bryozoans.

Family Zephyrinidae

Janolus barbarensis (Cooper, 1863). Feeds on hydroids, bryozoans.
Janolus fuscus O'Donoghue, 1924

Suborder Aeolidacea

Behrens, D. W. 1971. *Eubranchus misakiensis* Baba, 1960 (Nudibranchia: Eolidacea) in San Francisco Bay. Veliger, 14:214-5.
Edmunds, M., & A. Kress. 1969. On the European species of *Eubranchus* (Mollusca: Opisthobranchia). J. Mar. Biol. Assoc. United Kingdom, 49:879-912.
Gosliner, T. M., & S. V. Millen. 1984. Records of *Cuthona pustulata* (Alder & Hancock, 1854) from the Canadian Pacific. Veliger, 26:183-7.
Roller, R. A. 1969. Nomenclatural changes for the new species assigned to *Cratena* by MacFarland, 1966. Veliger, 11:421-3.
Williams, G. C., & T. M. Gosliner. 1979. Two new species of nudibranchiate molluscs from the west coast of North America, with a revision of the family Cuthonidae. Zool. J. Linnean Soc., 67:203-23.

1a Each ceras with a sail-like ridge on its posterior side (the ridge begins near the base and runs for much of the length of the ceras) (length up to 2 cm)   *Fiona pinnata*
1b Cerata without a sail-like ridge on the posterior side   2
2a At least a few rows of cerata located anterior to the bases of the rhinophores   3
2b None of the rows of cerata located anterior to the bases of the rhinophores (some may be at the same level as the rhinophores, however)   7
3a Longest cerata more than half the length of the body; cerata usually oriented approximately at right angles to the anterior-posterior axis of the body; foot proportionately broad (usually less than 3 times as long as wide), tapering to a point at the tail   *Cumanotus beaumonti*
3b Longest cerata less than one-third the length of the body; cerata usually oriented approximately parallel to the anterior-posterior axis of the body; foot relatively narrow (usually more than 4 times as long as wide), bluntly rounded at the tail   4
4a Anterolateral corners of foot projecting as acute pedal tentacles; body covered with gray to brown spots, these being most dense on the cerata (some specimens, especially from floats, lack pigment); usually with a prominent triangular white patch just anterior to the rhinophores; hepatic diverticula not clearly visible within the cerata of pigmented specimens; usually associated with anemones, especially *Metridium* and *Anthopleura*; length often attaining 5 cm   *Aeolidea papillosa* (fig. 12.122)
4b Anterolateral corners of foot rounded or somewhat triangular, but not projecting as acute pedal tentacles; body not covered with gray to brown spots; without a prominent triangular white patch anterior to the rhinophores; hepatic diverticula clearly visible within the cerata; usually associated with hydroids; length generally less than 2 cm   5
5a With 10-12 rows of cerata; external surfaces of the cerata (except the tips, which are white) metallic silver-blue; cores of cerata purple-brown to rust   *Cuthona concinna*
5b With more than 12 rows of cerata; external surfaces of cerata not metallic silver-blue (the tips, however, are white, as in *Cuthona concinna*); cores of cerata not purple-brown to rust, although they may be some shade of brown or orange   6
6a Body translucent white or pink; cores of cerata rose to burnt umber; with 3-8 rows of cerata anterior to the rhinophores   *Cuthona divae*
6b Body translucent white; cores of cerata rose-orange to salmon; with 7-11 rows of cerata anterior to the rhinophores   *Cuthona nana*
7a Anterolateral corners of foot elongated, forming rather prominent pedal tentacles   8
7b Anterolateral corners of foot broadly rounded or somewhat triangular, but not forming prominent tentacles   13
8a Basic body color purple; rhinophores and cerata bright orange for their entire length   *Flabellina iodinea*

256   Phylum Mollusca: Class Gastropoda

8b Basic body color not purple, usually translucent white or grayish; rhinophores and cerata not orange for their entire length (they may, however, be partly orange) 9

9a With an orange band beginning just anterior to the rhinophores and passing between them, reaching the first group of cerata (a similar orange band may be present on the cardiac region); orange areas usually bordered by broad, opaque white or luminous blue dorsomedial lines that begin on the oral tentacles and continue to the tip of the tail; tips of cerata usually orange or chrome yellow, but with some white at the extremity .... *Hermissenda crassicornis* (fig. 12.116)

9b Without orange bands on the dorsum; white lines on the dorsum, if present, narrow and often interrupted; tips of cerata may be opaque white, but rarely orange 10

10a With a dorsomedial white line (this may be interrupted or restricted to the tail region) or with 3 white lines, 1 dorsomedial and 2 lateral 11

10b Without longitudinal white lines (rhinophores pale yellowish green distally; oral tentacles with opaque white dots) .... *Flabellina pricei*

11a With 3 narrow white lines: 1 dorsomedial, running from the tip of the tail to a level just anterior to the rhinophores, where it divides, 1 branch going to the tip of each oral tentacle; 1 on each side of the body, just ventral to the cerata, beginning at the level of the first group of cerata and joining the dorsomedial line posterior to the last group of cerata .... *Flabellina trilineata*

11b With a dorsomedial white line, this interrupted and often restricted to the tail region, but without lateral white lines 12

12a Oral veil, between the oral tentacles, usually projecting dorsally as a blunt, triangular, median papilla; tail short, blunt, extending just beyond the last group of cerata; pedal tentacles long; hepatic diverticula slender (less than one-third the diameter of the cerata); white dorsomedial line extending from the tip of the tail to the most posterior group of cerata only; distal one-fourth of each ceras opaque white .... *Flabellina fusca*

12b Oral veil, between the oral tentacles, rounded, not projecting dorsally; tail long, slender, acute, extending one-third to one-half the total body length beyond the last group of cerata; pedal tentacles short; hepatic diverticula at least three-fourths the diameter of the cerata; white dorsomedial line extending from the tip of the tail to the rhinophores, although often interrupted; only the tips of the cerata opaque white .... *Flabellina verrucosa*

13a Surface of cerata usually smooth; cerata may be held down on the body; cerata usually not inflated (or only slightly inflated), although in some species they may be rather flat, tapering from a wide central portion to a bluntly rounded tip; cerata not readily shed; hepatic diverticula, before entering the cerata, forming a neat symmetrical pattern, usually in transverse rows running across the body 14

13b Surface of cerata slightly bumpy; cerata usually held erect; cerata cylindrical, and slightly inflated in the central portion; cerata often shed when the animal is disturbed; hepatic diverticula, before entering the cerata, forming an irregular network 17

14a Each rhinophore and oral tentacle encircled by a purple band near its middle (sometimes there is a white band above and/or below the purple band, and there may also be some purple at the bases of the rhinophores and oral tentacles); proximal third of the hepatic diverticulum in each ceras dark red, the central third olive-green to brown, the distal third pale yellow, and the tip colorless (3 opaque white bands, which may be solid or broken, accentuate the color changes of the hepatic diverticula) .... *Cuthona abronia*

14b Rhinophores and oral tentacles not encircled by purple bands; hepatic diverticula usually of the same color throughout 15

15a With a more or less solid accumulation of opaque white pigment on the rhinophores, on the dorsal side of the head anterior to the rhinophores, on the dorsum from the rhinophores to the last pair of cerata and down the sides to the lateral cerata, on the anterolateral face of the proximal portion of each ceras, and on the entire surface of the distal portion (except for the tip, which is clear); hepatic diverticula may be reddish brown or green, but no other pigmentation prominent .... *Cuthona albocrusta*

15b White pigment, if present, not so extensive as described in choice 15a, usually limited to patches on the cerata or rhinophores, and/or the dorsum; other pigmentation usually present 16

16a With small bright orange spots densely scattered on the proximal portions of the oral tentacles and on the distal portions of the rhinophores; opaque white pigment usually present on the tips of the rhinophores, on the dorsal side of the oral tentacles (overlain by orange spots), on the anterior faces and tips of the cerata, and sometimes on the dorsal side of the head anterior to the rhinophores; hepatic diverticula usually bright red (they may, however, be yellowish; their color depends on the prey that has been eaten)   *Catriona columbiana*

16b   Without small bright orange spots on the proximal portions of the oral tentacles and distal portion of the rhinophores; opaque white pigment present on the dorsal side of the oral tentacles, on the tips of the rhinophores, and on the tips of the cerata, but not on the anterior faces of the cerata; hepatic diverticula chocolate brown   *Cuthona cocoachroma*

17a   Body pale translucent yellowish green, with small specks of brown, black, and white; hepatic diverticula (including those in the cerata) greenish; usually with fewer than 15 cerata; associated with hydroids of the genus *Obelia*   *Eubranchus olivaceus*

17b   Body translucent white, sometimes with rust-brown splotches on the dorsum and cerata, and with small white spots on the cerata, rhinophores, and oral tentacles; hepatic diverticula not greenish (usually reddish or yellowish brown); usually with more than 15 cerata; not associated with hydroids of the genus *Obelia*   18

18a   With opaque white pigment on the tips of the cerata and often on the dorsum anterior to the region of the heart; hepatic diverticula red or reddish tan, giving the animal as a whole a reddish color; usually associated with the hydroid *Sertularella tricuspidata*   *Eubranchus sanjuanensis*

18b   With splotches of rust-brown pigment, as well as with opaque white pigment on the tips of the rhinophores and cerata, and in a ring near the middle of each ceras (distal to the white ring is a subterminal ring of rust-colored spots; there is also a ring of rust-colored spots near the middle or beginning of the distal quarter of each oral tentacle and rhinophore); hepatic diverticula brown, sometimes with yellow spots; associated with the hydroid *Plumularia lagenifera*   *Eubranchus rustyus*

A few species, one of which is not known to occur in our region, are not in the key. They are marked with an asterisk.

Family Flabellinidae

Flabellina fusca O'Donoghue, 1921
Flabellina iodinea (Cooper, 1862). Feeds on hydroids, compound ascidians.
Flabellina pricei (MacFarland, 1966)
Flabellina trilineata (O'Donoghue, 1921). Feeds on hydroids.
Flabellina verrucosa (M. Sars, 1829)

Family Cumanotidae

Cumanotus beaumonti (Eliot, 1906). Feeds on hydroids of the genus *Tubularia*.

Family Eubranchidae

Eubranchus misakiensis Baba, 1960. Introduced into San Fransisco Bay from Japan, and possibly has been introduced into some localities of our region; feeds on hydroids.
Eubranchus olivaceus (O'Donoghue, 1922). Feeds on hydroids, especially of the genus *Obelia*.
Eubranchus rustyus (Marcus, 1961). Feeds on hydroids.
Eubranchus sanjuanensis Roller, 1972. Feeds on hydroids of the genus *Sertularia*.

Family Tergipedidae

Catriona columbiana (O'Donoghue, 1922). Feeds on hydroids.
Cuthona abronia (MacFarland, 1966)
Cuthona albocrusta (MacFarland, 1966)

Cuthona cocoachroma Williams & Gosliner, 1979
Cuthona concinna (Alder & Hancock, 1843)
Cuthona divae (Marcus, 1961). Feeds on athecate hydroids.
Cuthona nana (Alder & Hancock, 1843)
 *Cuthona pustulata (Alder & Hancock, 1854). Feeds on hydroids of the genus *Halecium*.

Family Fionidae

Fiona pinnata (Eschscholtz, 1831). Feeds on barnacles of the genus *Lepas*, and also on the floating cnidarian *Velella velella*.

Family Facelinidae

Hermissenda crassicornis (Eschscholtz, 1831) (fig. 12.116). Feeds on hydroids, sea pens, ascidians, and various other organisms.

Family Aeolididae

Aeolidea papillosa (Linnaeus, 1761) (fig. 12.122). Feeds on many species of anemones.

Subclass Gymnomorpha

Order Onchidiacea

Onchidella borealis Dall, 1871. This species is abundant throughout the region. It is found in the mid-intertidal and also at higher levels; its usual habitats are crevices and holdfasts of large seaweeds.
Onchidella carpenteri (Binney, 1860). The systematic status and distribution of this species are not clear.

Subclass Pulmonata

Order Archaeopulmonata

Family Melampidae

Phytia myosotis (Draparnaud, 1801). Abundant in salt marshes, especially under debris and growths of *Salicornia* at levels that are flooded only at the highest tides; usually accompanied by *Assiminea californica*, a mesogastropod of the family Assimineidae.

Order Basommatophora

Family Siphonariidae

Siphonaria thersites Carpenter, 1864 (fig. 12.123). Found in the mid-intertidal of rocky shores; distribution spotty, however.

13

PHYLUM MOLLUSCA: CLASS BIVALVIA

Bernard, F. R. 1983. Catalogue of the living Bivalvia of the eastern Pacific Ocean. Canad. Spec. Publ. Fish. Aquat. Sci., 61. vii + 102 pp.
Fitch, J. E. 1953. Common marine bivalves of California. Calif. Dept. Fish & Game, Fish Bull. 90. 102 pp.
Knudsen, J. 1970. The systematics and biology of abyssal and hadal Bivalvia. Galathea Rep., 11. 241 pp.
Quayle, D. B. 1969. Intertidal Bivalves of British Columbia. Handbook 17, British Columbia Provincial Museum, Victoria. 2nd ed. 104 pp.
Skarlato, O. A. 1960. [Bivalve molluscs of the far-eastern seas of the USSR. Order Dysodonta.] Tabl. Anal. Faune, Akademiĭa Nauk, SSSR, 71. 127 pp. (in Russian).
------. 1981. [Bivalve molluscs of temperate latitudes of the western part of the Pacific Ocean.] Opredeliteli po Faune SSSR, 126. 461 pp. (in Russian).

Key to Families

The following families are omitted: Nucinellidae (2), Arcidae (8), Limopsidae (10), Galeommatidae (22), Kelliellidae (34), Vesicomyidae (35), Laternulidae (41), Verticordiidae (49), Poromyidae (51). With the exception of Galeommatidae, represented by a species that lives attached to a sea cucumber, *Leptosynapta clarki*, these families are largely restricted to deep water.

1a Shipworms--body wormlike, the shell greatly reduced and covering only a small part of the animal; burrowing in wood, forming tunnels lined by a whitish, calcareous deposit; siphonal tube with a pair of calcareous pallets (paddlelike or featherlike structures) for closing the opening of the burrow 46. Teredinidae
1b Body not wormlike, the shell usually covering a substantial part of the animal; not burrowing in wood (exception: *Xylophaga*), although some species may nestle in holes made by other animals; siphonal tube without calcareous pallets 2
2a Valves of shell decidedly unequal in form as well as size; 1 valve almost always permanently cemented to a hard substratum, such as rock, concrete, shell, or wood (some oysters, however, lie on a muddy substratum) 3
2b Valves of shell essentially similar to one another, although 1 valve may be slightly smaller or less convex than the other; neither valve permanently cemented to the substratum (but the animal may be attached by a byssus of organic threads) 6
3a Attached valve thin, flat or conforming to the shape of the substratum, with a prominent hole near the hinge (the byssal material that cements the animal to the substratum emerges through this hole) 16. Anomiidae: *Pododesmus* (*Monia*) *cepio*
3b Attached valve not thin and without a hole near the hinge 4
4a Anterior and posterior adductor muscles separate, their scars distinct on both valves 25. Chamidae
4b Anterior and posterior adductor muscles united into a single large muscle 5
5a Oysters--valves generally higher than long; without regularly spaced ribs radiating from the umbones (if riblike elevations are present, they follow no particular pattern); valves without winglike extensions on both sides of the umbones; animal without eyes along the edge of the mantle 14. Ostreidae
5b Rock scallops--valves very thick, almost circular in outline; with rather regularly spaced, rough ribs radiating from the umbones; with winglike extensions on both sides of the

umbones (these persist to a varying extent, and are not always prominent); with a row of eyes along the edge of the mantle (interior of both valves with a purple blotch near the hinge)
15. Pectinidae: *Hinnites giganteus*

6a	Scallops and limas--both valves with winglike extensions on both sides of the umbo (the extension on one side may be small, however)	7
6b	Neither valve with winglike extensions on both sides of the umbo	8
7a	Height about twice the length	13. Limidae
7b	Height not much, if at all, greater than the length	15. Pectinidae and Propeamussidae
8a	Anterior two-thirds of shell nearly globular and radially ribbed, the posterior third much narrower, almost cylindrical, comparatively smooth, and truncate; length rarely exceeding 1.5 cm (without definite ctenidia; strictly subtidal)	50. Cuspidariidae
8b	Shell not divided into a nearly globular anterior portion and a much narrower, almost cylindrical posterior portion; length frequently greater than 1.5 cm	9
9a	Hinge plate of both valves with taxodont dentition (several to many nearly identical small teeth on both sides of the umbo); almost strictly subtidal	10
9b	Hinge plate without taxodont dentition	11
10a	Hinge ligament external; shell about as high as long; valves with prominent radial ribs	11. Glycymerididae
10b	Hinge ligament internal or external; shell decidedly longer than high; valves without radial ribs (in *Acila castrensis*, however, there is radial sculpture that forms chevronlike patterns)	Order Nuculoida
11a	Exterior of both valves divided into 2 or 3 regions of distinctly different texture, the anterior region being rasplike or filelike for burrowing into hard clay, rock, shell, or wood (caution: *Petricola pholadiformis*, a member of the Petricolidae, burrows into clay, and its rasplike radial sculpture is especially prominent on the anterior half of the shell; this bivalve does not, however, have the other features of the family to which this choice leads); usually with a myophore (a projection for attachment of the foot muscle, not to be confused with a chondrophore, which is a socketlike, spoonlike, or shelflike structure to which the internal ligament of some bivalves is attached); shell gaping anteriorly, where the foot emerges, unless the gape is closed by a callus (typical of adults of some genera)	12
11b	Exterior of valves not usually divided into 2 or 3 regions of distinctly different texture (in some the valves are divided into 2 sectors by a groove that originates near the umbo, but the sectors have a similar texture) and generally without an anterior portion that is rasplike or filelike for burrowing into clay, rock, or wood (there are a few species that burrow into clay or rock, but with the exception of *Petricola pholadiformis* these do not have an anterior rasplike or filelike portion that is distinctly different from the rest of the shell); without a myophore; shell not often gaping anteriorly, even though it may gape posteriorly	13
12a	Shell almost globular, but with a notch in the ventral side of the anterior portion of both valves; length less than 1 cm; boring in waterlogged wood (subtidal and rare; do not confuse with *Bankia* or *Teredo*, family Teredinidae)	45. Xylophagaidae: *Xylophaga washingtona*
12b	Shell decidedly elongated, without a notch on the ventral side of the valves near the anterior end (the anterior end may, however, be abruptly truncate); burrowing in firm mud, clay, or shale	44. Pholadidae
13a	With a shiny brown periostracum that extends well beyond the edges of the valves; shell so thin that it usually cracks when the periostracum dries (interior of valves with a slight radial rib just anterior to the scar of the posterior adductor muscle; dorsal and ventral margins almost perfectly parallel for much of their length; hinge without teeth; subtidal)	1. Solemyidae
13b	Periostracum, whether shiny or not, not extending well beyond the edges of the valves; shell not usually so thin that it cracks when the periostracum dries (except in some species of Lyonsiidae)	14
14a	Shell as a whole very flat, the width not more than about one-sixth of the height; right valve flat, fitting into the slightly convex left valve; without a true hinge plate, the right valve having 2 or 3 ridges that articulate with grooves in the left valve	47. Pandoridae
14b	Shell as a whole at least one-fourth as wide as high; neither valve flat; with a true hinge plate, this sometimes provided with teeth or a chondrophore, or both	15
15a	Both valves with a single deep groove that runs from near the umbo to the posterior margin, this groove segregating a small posterodorsal sector; hinge without teeth; strictly subtidal	18. Thyasiridae

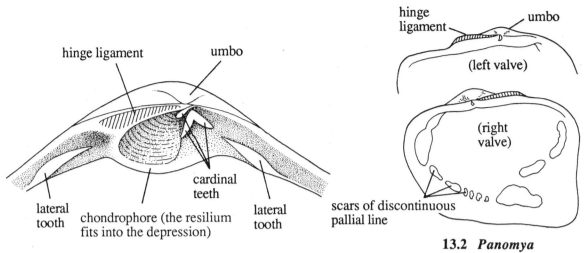

13.1 Diagram of hinge region

13.2 *Panomya*

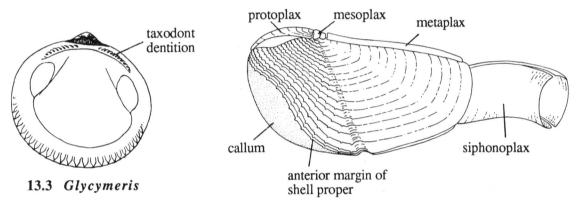

13.3 *Glycymeris*

13.4 Pholadidae

15b Neither valve with a deep groove that segregates a small posterodorsal sector; hinge with or without teeth; includes intertidal and subtidal species 16

16a With a single adductor muscle scar not far from the center of each valve; height up to about 5 mm, slightly greater than the length (valves with prominent beaks, covered externally with a thick periostracum, pearly inside; attached by a byssus, usually to coralline algae)
.... 9. Philobryidae

16b With 2 adductor muscle scars in each valve, although the anterior scar may be small and easily overlooked; height generally greater than 5 mm, and usually equaled or exceeded by the length 17

17a Mussels--scars of the anterior and posterior adductor muscles decidedly unequal (the anterior scar is the smaller one, and may be scarcely noticeable); umbones generally, but not always, much nearer the anterior end than the middle, hinge plate without teeth and without a chondrophore; usually attached to a hard substratum by a byssus of organic threads (includes some species that bore in rock or hard clay) .... 12. Mytilidae

17b Scars of anterior and posterior adductor muscles of almost equal size (they may, however, be of different shape); umbones not usually nearer to the anterior end than to the middle (but there are exceptions); hinge plates usually with teeth or with a chondrophore, sometimes both; usually not attached by a byssus of organic threads (but there are exceptions) 18

Phylum Mollusca: Class Bivalvia

18a   Hinge plate of both valves without teeth and without a chondrophore; shell without radial ribs   19
18b   Hinge plate of 1 or both valves either with teeth or with a chondrophore, or both; shell with or without radial ribs   21
19a   Umbones so close together that they touch, the left one usually wearing a depression or hole in the right one; strictly subtidal   40. Thraciidae
19b   Umbones not so close together that they touch; intertidal and subtidal   20
20a   Interior of valves at least slightly pearly; with a continous pallial line and pallial sinus (in *Mytilimeria*, which is typically embedded in compound ascidians, the pallial line and pallial siphon may not be distinct); tips of siphons not bright red   48. Lyonsiidae
20b   Interior of valves chalky, not at all pearly; pallial line discontinuous, pallial sinus absent; tips of siphons bright red   43. Hiatellidae (in part)
21a   Razor clams and jackknife clams--shell more than twice as long as high; periostracum glossy, usually yellowish brown, yellowish olive, or greenish   22
21b   Shell not more than twice as long as high (except in *Petricola pholadiformis*); periostracum sometimes glossy and sometimes greenish, but not in species that are nearly or more than twice as long as high   23
22a   About 4 times as long as high; dorsal margin slightly concave for its entire length   29. Solenidae: *Solen (Solen) sicarius*
22b   Less than 3 times as long as high; dorsal margin not concave   30. Cultellidae
23a   With radial ribs distributed over most of the shell (in *Lucina tenuisculpta*, Lucinidae, the radial ribs may not be evident except where the periostracum has been worn away; in *Serripes groenlandicus*, Cardiidae, and *Semele rubropicta*, Scrobiculariidae, the ribs are often faint, and they may be apparent only on certain portions of the shell)   24
23b   Without radial ribs (except in *Cryptomya californica*, Myidae, which has faint radial ribs in the posterior portion, but which is distinctive in having a shelflike chondrophore projecting from the hinge plate of the left valve)   32
24a   Length of the shell not appreciably greater than the height (not as much as 1 and one-sixth times the height), and sometimes slightly less than the height   25
24b   Length of the shell definitely greater than the height, at least 1 and one-sixth times the height   28
25a   Posterior quarter of both valves with rather prominent concentric ridges as well as low radial ribs, so that the sculpture on this part of the valve is decidedly different from that over the rest of the surface, where only the closely spaced radial ribs are distinct   27. Cardiidae: *Nemocardium centifilosum* (fig. 13.32)
25b   Posterior quarter of valves not with concentric and radial sculpture that is decidedly different from the sculpture characteristic of the rest of the surface   26
26a   Radial ribs often faint, and may not be visible where covered by periostracum, but distinct on portions of the shell from which the periostracum has been worn away (shell chalky white, except where covered by a brownish or greenish periostracum; umbones near the middle; length up to about 1.5 cm; subtidal)   17. Lucinidae: *Lucina (Parvilucina) tenuisculpta* (fig. 13.21)
26b   Radial ribs distinct even on portions of the shell where there is abundant periostracum   27
27a   Hinge with both cardinal teeth and lateral teeth; with more than 30 radial ribs   27. Cardiidae (in part)
27b   Hinge with only cardinal teeth; with fewer than 30 radial ribs   24. Carditidae (in part)
28a   Umbones anterior to the end of the first one-fifth of the shell; anterior and ventral margins forming nearly a right angle (the corner of the angle, however, is rounded); with about 14 or 15 radial ribs; height up to about 1 cm   24. Carditidae: *Glans carpenteri*
28b   Umbones not as far forward as the end of the first one-fifth of the shell; anterior and ventral margins not forming nearly a right angle; with more than 15 radial ribs, even if some or all are faint; height usually more than 1 cm   29
29a   Shell about 2.5 times as long as wide; radial ribs especially prominent in the anterior half of the shell (burrowing in clay, firm mud, or soft shale)   38. Petricolidae: *Petricola pholadiformis*
29b   Shell not more than twice as long as wide; radial ribs not obviously more prominent in the anterior half of the shell than elsewhere   30

Phylum Mollusca: Class Bivalvia   263

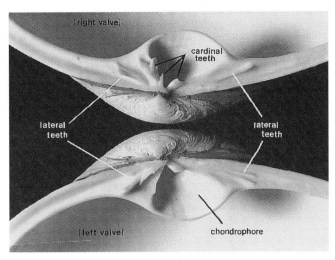

13.5 *Tresus capax*

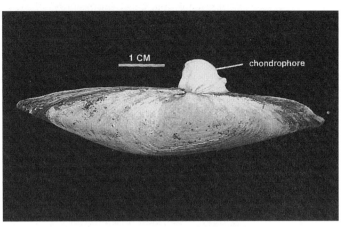

13.6 *Mya arenaria*, left valve

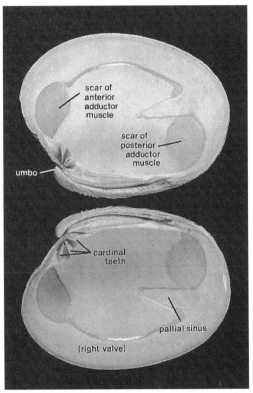

13.7 *Protothaca tenerrima*

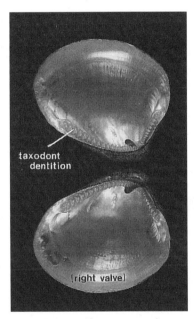

13.8 *Acila castrensis*

13.9 *Chlamys hastata*

264   Phylum Mollusca: Class Bivalvia

30a Without a pallial sinus (umbones near the end of the second one-fifth of the shell; radial ribs low and broad, often obscure; periostracum tan or faintly greenish, generally somewhat glossy and tightly adherent; length up to 7 cm)   27. Cardiidae: *Serripes groenlandicus*
30b With a pallial sinus   31
31a Radial ribs low, often less prominent than the concentric growth lines; exterior of valves usually with pink radial rays; interior of valves sometimes with a purple stain at both ends of the hinge plate (length up to 3 cm; intertidal and subtidal)
         32. Scrobiculariidae: *Semele rubropicta* (fig. 13.35)
31b Radial ribs almost always more prominent than the concentric growth lines; exterior of valves without pink radial rays; interior of valves without a purple stain at both ends of the hinge plate (but there may be a purple stain near the posterior end of each valve)
         36. Veneridae (in part)
32a With a shelflike or spoonlike chondrophore projecting from the hinge plate of the left valve (there is no chondrophore of this type on the right valve); hinge plate of both valves without teeth; hinge ligament largely or completely internal   42. Myidae
32b Without a shelflike or spoonlike chondrophore projecting from the hinge plate of the left valve, but sometimes with a socketlike chondrophore in both valves; hinge plate of both valves with teeth; hinge ligament generally external but sometimes almost completely internal
         33
33a Valves without a pallial sinus   34
33b Valves with a pallial sinus   43
34a Valves with pronounced concentric ridges, and not nearly circular in outline   26. Astartidae
34b Valves with faint concentric growth lines, but without pronounced concentric ridges (except in *Lucinoma annulata*, which is distinctive in having nearly circular valves)   35
35a Height approximately equal to the length, the outline of the valves nearly circular   36
35b Length decidedly greater than the height, the outline of the valves not nearly circular   38
36a With some conspicuous concentric ridges, the shell thus noticeably roughened; length up to 5 cm   17. Lucinidae: *Lucinoma annulata* (fig. 13.22)
36b Without conspicuous concentric ridges (only fine growth lines); length not exceeding 3.5 cm
         37
37a Exterior of valves pale green, glistening when fresh; length up to about 7 mm
         18. Thyasiridae: *Axinopsida serricata* (fig. 13.23)
37b Exterior of valves whitish, not glistening; length up to 3.5 cm, but usually smaller
         19. Ungulinidae
38a Ventral margin of valves nearly straight or slightly concave; up to about 2 cm long; attached by a byssus to the ventral surface of the abdomen of the mud shrimp *Upogebia pugettensis* and the ventral surface of polychaetes of the genus *Aphrodita*
         23. Montacutidae: *Pseudopythina rugifera*
38b Ventral margin of valves convex; up to 2.5 cm long (*Kellia suborbicularis*), but generally less than 1 cm; sometimes nestling in holes made by other invertebrates, sometimes closely associated with other invertebrates, but not attached to *Upogebia* or *Aphrodita*   39
39a Periostracum thin, greenish or yellowish; mantle may be extended to cover nearly the whole shell (inhalant siphon directed ventrally or anteriorly, exhalant siphon directed posteriorly)
         20. Kelliidae
39b Periostracum not greenish or yellowish; mantle cannot be extended to cover a large part of the shell   40
40a Umbones closer to the posterior (siphonal) end than to the middle, and inclined posteriorly (unusual for a bivalve); with abundant brown periostracum; interior of valves slightly pearly (length up to 5 mm)   23. Montacutidae: *Mysella tumida* (fig. 13.27)
40b Umbones close to the middle or about halfway between the anterior end and the middle; without abundant brown periostracum; interior of valves not pearly   41
41a Exterior of valves purplish near the umbones, or entirely reddish; umbones about halfway between the anterior end and the middle; dorsal margin of right valve not serrated; length not more than 5 mm   42
41b Exterior of valves not purplish near the umbones and not reddish; umbones near the middle; dorsal margin of right valve serrated, especially posterior to the umbo; length up to about 1.5 cm   23. Montacutidae: *Pseudopythina compressa*
42a Exterior of valves purplish near the umbones, interior white   37. Turtoniidae

42b Exterior and interior of valves reddish 21. Lasaeidae
43a Pallial line consisting of a series of discontinuous scars (with 1 small tooth on the hinge plate of each valve; shell gaping at both ends; posterior end truncate; length up to 7 cm)
43. Hiatellidae: *Panomya beringiana*
43b Pallial line continuous 44
44a With a single tooth on the hinge plate of each valve; shell large, the length sometimes exceeding 15 cm, but not large enough to accommodate the immense siphon, and gaping widely except in the region of the hinge 43. Hiatellidae: *Panope abrupta*
44b With more than 1 tooth on the hinge plate of each valve; shell generally less than 10 cm long (except in *Tresus*), large enough to accommodate the siphon when it is withdrawn, gaping only at the posterior end, if at all 45
45a Shell gaping broadly at the posterior end (the width of the gape is more than one-fourth the width of the shell); length commonly exceeding 12 cm 28. Mactridae: *Tresus*
45b Shell closing tightly or gaping only slightly at the posterior end; length rarely exceeding 10 cm 46
46a Hinge ligament almost completely internal 28. Mactridae (in part)
46b Hinge ligament largely external 47
47a With 3 cardinal teeth (and sometimes a pair of lateral teeth) in both valves
36. Veneridae (in part)
47b With fewer than 3 cardinal teeth in at least 1 valve 48
48a With 3 cardinal teeth in 1 valve, 2 in the other 49
48b With 2 cardinal teeth in both valves 50
49a Valves oval, rather delicate, polished; siphons not tipped with bright purple
39. Cooperellidae: *Cooperella subdiaphana*
49b Valves somewhat elongated, and usually narrower posteriorly than anteriorly, rather thick, chalky (often distorted as a result of conforming to the shape of pholad burrows in which this species nestles); siphons tipped with bright purple
38. Petricolidae: *Petricola* (*Rupellaria*) *carditoides* (fig. 13.37)
50a Outline of valves almost elliptical (about twice as long as high); hinge plate of both valves, behind the umbo, with a thin dorsal extension to which the ligament is attached; exterior of valves usually with pinkish or purplish radial rays; siphons short, united; length up to about 7 cm 33. Psammobiidae: *Gari* (*Gobraeus*) *californica*
50b Outline of valves not almost elliptical (the 2 ends are different, one being less evenly rounded than the other); hinge plate of valves without a thin dorsal extension; exterior of valves without pinkish or purplish radial rays (various tints, however, may be present internally or externally); siphons long and separate; maximum length ranging from 1 to 10 cm
31. Tellinidae

Families marked with an asterisk are not in the key to families. Representatives of these families are, as a rule, restricted to very deep water.

Subclass Cryptodonta

Order Solemyoida

1. Family Solemyidae

Bernard, F. R. 1980. A new *Solemya* s. str. from the northeastern Pacific (Bivalvia: Cryptodonta). Venus (Jap. J. Malacol.), 39:17-23.

Solemya (*Solemya*) *reidi* Bernard, 1980. To 6 cm long; periostracum extending far beyond the edge of the shell; shell thin, cracking as it dries; outline of valves elliptical, but the posterior end not quite so rounded as the anterior end; shallow subtidal.
Acharax johnsoni (Dall, 1891). Deep water.

2. Family *Nucinellidae

Huxleyia munita (Dall, 1878). Shallow subtidal, Oregon southward.

Subclass Palaeotaxodonta

Order Nuculoida

Key to Families

1a   Ligament external; interior of valves porcelaneous, not pearly; with a pallial sinus (animal with siphons)   2
1b   Ligament internal; interior of valves pearly, especially when fresh; without a pallial sinus (animal without siphons)   3. Nuculidae
2a   With a chondrophore on the lower part of the hinge plate   3
2b   Without a chondrophore   4
3a   Shell closing tightly; pallial sinus shallow   4. Nuculanidae
3b   Shell gaping posteriorly; pallial sinus deep   5. Yoldiidae
4a   Series of hinge teeth posterior to the umbo nearly twice as long as the series anterior to the umbo; posterior ends of valves rounded   6. Malletiidae
4b   Series of hinge teeth posterior to the umbo not much longer than the series anterior to the umbo; posterior ends of valves somewhat pointed, due to the way the curved ventral margin meets the nearly straight dorsal margin   7. Tindariidae

3. Family Nuculidae

1a   Valves with a distinct radial sculpturing that forms chevronlike patterns
     Acila (Truncacila) castrensis (Hinds, 1843) (fig. 13.8)
1b   Valves without radial sculpturing (only concentric growth lines are present)
     Nucula (Leionucula) tenuis (Montagu, 1808)

N. (Leionucula) quirica Dall, 1916 is similar.

The following species have been collected in deep water.

Nucula (Nucula) chrysocoma Dall, 1908
Nucula (Lamellinucula) carlottensis Dall, 1897
Nucula (Lamellinucula) darela Dall, 1916
Nucula (Leionucula) cardara Dall, 1916
Nucula (Leionucula) linki Dall, 1916

4. Family Nuculanidae

1a   Dorsal margin, posterior to the umbones, markedly concave (the depth of the concavity equal to about one-fourth of its length)
     Nuculana (Thestyleda) hamata (Carpenter, 1864) (fig. 13.10)
1b   Dorsal margin, posterior to the umbones, almost straight   2
2a   Hinge plate with 12 teeth in the anterior series, 14 teeth in the posterior series
     Nuculana (Nuculana) minuta (Fabricius, 1776)
2b   Hinge plate with 22 teeth in the anterior series, 16 teeth in the posterior series
     Nuculana (Saccella) cellulita (Dall, 1896)

Phylum Mollusca: Class Bivalvia 267

13.10 *Nuculana hamata*

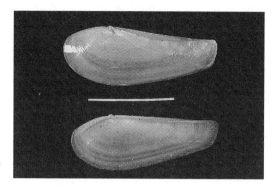
13.11 *Nuculana conceptionis*

13.12 *Nuculana fossa*

13.13 *Nuculana taphria*

13.14 *Yoldia thraciaeformis*

13.16 *Yoldia amygdalea*

13.15 *Yoldia scissurata*

13.17 *Philobrya setosa*

Phylum Mollusca: Class Bivalvia

The following species, not all necessarily valid, have been reported for the region, but most of them are restricted to deep water.

Nuculana (Nuculana) amiata (Dall, 1916)
Nuculana (Nuculana) conceptionis (Dall, 1896) (fig. 13.11)
Nuculana (Nuculana) gomphoidea (Dall, 1916)
Nuculana (Nuculana) lomaensis (Dall, 1919)
Nuculana (Nuculana) pernula (O. F. Müller, 1779)
Nuculana (Nuculana) tenuisculpta (Couthouy, 1838)
Nuculana (Saccella) fossa (Baird, 1863) (fig. 13.12). Shallow.
Nuculana (Saccella) liogona (Dall, 1916)
Nuculana (Saccella) penderi (Dall & Bartsch, 1910). Shallow.
Nuculana (Saccella) taphria (Dall, 1897) (fig. 13.13)
Nuculana (Thestyleda) extenuata (Dall, 1897)
Nuculana (Thestyleda) leonina (Dall, 1896)
Nuculana (Thestyleda) spargana (Dall, 1916). Shallow.

5. Family Yoldiidae

Ockelmann, K. W. 1954. On the interrelationship and zoogeography of northern species of *Yoldia* Møller, s. str. (Mollusca, Fam. Ledidae), with a new subspecies. Medd. Gronl. Kom. Videnskab. Undersog., 107:1-32.

1a Shell swollen; posterior end broadly truncate
   Yoldia (Megayoldia) thraciaeformis (Storer, 1838) (fig. 13.14)
1b Shell rather flat, not swollen; posterior end not truncate 2
2a Concentric sculpture not coinciding with growth lines (there are slight concentric ridges that cross the growth lines)   *Yoldia (Cnesterium) scissurata* Dall, 1897 (fig. 13.15)
2b Without any concentric sculpture other than growth lines 3
3a Umbones slightly posterior to the middle; length twice the height
   Yoldia (Yoldia) amygdalea (Valenciennes, 1846) (fig. 13.16)
3b Umbones at the middle; length slightly less than twice the height
   Yoldia (Yoldia) myalis (Couthouy, 1838)

The following species have been reported for the region, but most of them are restricted to deep water.

Katadesmia gibbsii (Dall, 1897)
Portlandia (Portlandia) dalli Krause, 1885
Yoldia (Megayoldia) beringiana (Dall, 1916)
Yoldia (Megayoldia) martyria Dall, 1897. Shallow.
Yoldia (Megayoldia) montereyensis Dall, 1893
Yoldiella (Yoldiella) capsa (Dall, 1916)
Yoldiella (Yoldiella) dicella (Dall, 1908)
Yoldiella (Yoldiella) orcia (Dall, 1916)
Yoldiella (Yoldiella) sanesia (Dall, 1916)

6. Family Malletiidae

The following species, among others (not all names are necessarily valid) have been reported, mostly from depths greater than 300 m.

Malletia (Malletia) faba Dall, 1897
Malletia (Malletia) pacifica Dall, 1897

Malletia (Malletia) talama Dall, 1916
Malletia (Malletia) truncata Dall, 1908

7. Family Tindariidae

The following species, among others, have been reported; none is likely to be found at depths of less than 300 m.

Saturnia (Saturnia) brunnea (Dall, 1916)
Saturnia (Saturnia) cervola (Dall, 1916)
Saturnia (Saturnia) kennerlyi (Dall, 1897)
Tindaria compressa Dall, 1908
Tindaria dicofania Dall, 1916
Tindaria panamensis Dall, 1908

Subclass Pteriomorphia

Order Arcoida

8. Family *Arcidae

Bathyarca nucleator Dall, 1908. At depths greater than about 2000 m.

9. Family Philobryidae

Philobrya (Philobrya) setosa (Carpenter, 1864) (fig. 13.17). Intertidal and shallow subtidal; attached by a byssus, usually to coralline red algae.

10. Family *Limopsidae

Empleconia vaginata (Dall, 1891). At depths of at least 400 m.
Limopsis (Limopsis) akutanica Dall, 1916. At depths of at least 130 m.
Limopsis (Limopsis) dalli Lamy, 1912. At depths of at least 3000 m.

11. Family Glycymerididae

Willett, G. 1944. Northwest American species of *Glycimeris*. Bull. South. Calif. Acad. Sci., 42:107-14.

1a Radial ribs obvious, even if rather faint (they are least likely to be visible in small specimens); hinge with about 15 or 16 teeth, the anterior and posterior sets decidedly separated; periostracum thick, dark brown; common subtidal species
Glycymeris (Axinola) subobsoleta (Carpenter, 1864)
1b Radial ribs so faint that they are not likely to be visible without magnification; hinge with about 20 teeth, the anterior and posterior sets close together; periostracum thin, tan; rare
Glycymeris (Axinola) corteziana Dall, 1916

Order Mytiloida

12. Family Mytilidae

Bernard, F. R. 1978. New bivalve molluscs, subclass Pteriomorphia, from the northeastern Pacific. Venus (Jap. J. Malacol.), 37:61-75.

Glynn, P. W. 1964. *Musculus pygmaeus* spec. nov., a minute mytilid of the high intertidal zone at Monterey Bay, California (Mollusca: Pelecypoda). Veliger, 7:121-8.

Soot-Ryen, T. 1955. A report on the family Mytilidae. Allan Hancock Pac. Exped., 20. 154 pp.

1a	Shell decidedly longer than high; umbones usually at or close to the anterior end (in *Adula*, near the end of the first quarter)	2
1b	Shell about as high or higher than long; umbones near the middle of the dorsal margin	13
2a	Umbones essentially terminal; anterior ends of valves acute	3
2b	Umbones not quite terminal; anterior ends of valves rounded	4
3a	Valves with coarse radial ribs as well as irregular growth lines; periostracum of larger specimens often partly or completely worn off; length frequently exceeding 10 cm; in relatively exposed situations, mostly on the open coast *Mytilus* (*Mytilus*) *californianus* Conrad, 1837	
3b	Valves without coarse radial ribs, and with rather regular growth lines; periostracum usually shiny black (brown in small specimens) and generally persistent; length rarely exceeding 7 cm; common in protected situations, but also found on the open coast *Mytilus* (*Mytilus*) *edulis* Linnaeus, 1758	
4a	Greatest height of shell usually only slightly greater than the height at the level of the umbones (except in *Adula diegensis*); width sometimes greater than the height; mostly burrowing in shale (except *Adula diegensis*, which is attached to rocks, pilings, etc.)	5
4b	Greatest height of shell usually about 1.5 times the height at the level of the umbones; width not greater than the height; attached to firm substrata, but not boring	8
5a	Posterodorsal slopes of valves hairy, and generally with an accumulation of clay particles; boring in shale or attached to rocks, pilings, etc.	6
5b	Posterodorsal slopes of valves with rough, chalky encrustations and transverse wrinkles, but not hairy; boring in limestone, sometimes other rocks (not likely to be found north of California) *Lithophaga* (*Diberus*) *plumula* (Hanley, 1844)	
6a	Valves with filelike vertical striations; length sometimes exceeding 6 cm; boring in shale (not likely to be found north of Oregon) *Adula falcata* (Gould, 1851)	
6b	Valves smooth, or with fine radiating striations on their anterior portions; length not attaining 5 cm	7
7a	Valves generally tapering posteriorly (the height at the level of the umbones is not usually exceeded behind the middle, but there are exceptions); generally boring in shale, but sometimes attached to rocks, other mussels, etc. *Adula californiensis* (Philippi, 1847)	
7b	Valves generally higher posteriorly than near the middle (the greatest height is usually about 1.5 times the height at the level of the umbones); attached to rocks or pilings *Adula diegensis* (Dall, 1911)	
8a	Length from about 2 to about 3 times the height (if slightly less than 2 times the height, then relatively large--up to more than 10 cm long)	9
8b	Length about 1.5 to 1.75 times the height	11
9a	Maximum length about 3 cm; periostracum greenish, often with wavy, brownish markings; periostracum not elaborated into soft bristles (introduced, along with oysters, into estuarine situations) *Musculista senhousia* (Benson, 1842)	
9b	Length commonly greater than 5 cm; periostracum usually brown or blackish brown; periostracum of young specimens elaborated into soft bristles	10
10a	Length about twice (or slightly less than twice) the height and width *Modiolus* (*Modiolus*) *modiolus* (Linnaeus, 1758)	
10b	Length about 3 times the height and width (not likely to be found north of California) *Modiolus* (*Modiolus*) *rectus* (Conrad, 1837)	

Modiolus flabellatus (Gould, 1850) is here considered to be a synonym.

Phylum Mollusca: Class Bivalvia 271

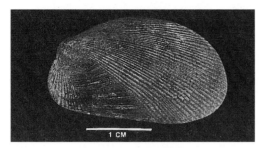
13.18 *Musculus discors*

13.19 *Limatula subauriculata*

13.20 *Parvamussium alaskensis*

13.21 *Lucina tenuisculpta*

13.22 *Lucinoma annulata*

13.23 *Axinopsida serricata*

13.24 *Conchocele bisecta*

13.25 *Diplodonta orbella*

11a With numerous delicate, radiating ribs, obvious to the unaided eye, extending across much of the shell; length up to about 8 cm *Musculus (Musculus) niger* (Gray, 1824)
11b Without radiating ribs obvious to the unaided eye (except for some in the anterior part of the shell); length not often exceeding 4 cm 12
12a With a line running posteroventrally from near the umbo of each valve, the valve being slightly raised anterior to this line; length up to about 4 cm *Musculus (Musculus) discors* (Linnaeus, 1767) (fig. 13.18)
12b Without a line running posteroventrally from the umbo; length about 5 mm *Musculus (Musculus) taylori* (Dall, 1897)
13a Shell with radial ribs (these may be delicate and inconspicuous); subtidal, but not necessarily in deep water 14
13b Shell smooth, without radial ribs; restricted to deep water *Dacrydium (Dacrydium) pacificum* Dall, 1916
14a Radial ribs delicate, inconspicuous *Megacrenella columbiana* (Dall, 1879)
14b Radial ribs prominent *Crenella decussata* (Montagu, 1808)

The following have been collected in deep water.

Dacrydium (Dacrydium) rostriferum Bernard, 1978. At depths of more than 2000 m.
Habepegris washingtonia Bernard, 1978. At depths of more than 2500 m.

Order Limoida

13. Family Limidae

Limatula subauriculata (Montagu, 1808) (fig. 13.19) is the most commonly encountered species in our region. It occurs mostly at depths between 50 and 350 m.

The following species have also been reported.

Limatula attenuata Dall, 1916. Shallow.
Limatula saturna Bernard, 1978. Shallow.
Limatula vancouverensis Bernard, 1978. Deep.

Order Ostreoida

Suborder Ostreina

14. Family Ostreidae

1a Valves generally much roughened and conspicuously frilly; height up to about 25 cm (introduced, but now widespread) *Crassostrea gigas* (Thunberg, 1793)
1b Valves not much roughened and rarely showing any frills (the sculpturing is usually limited to irregular concentric lines); height not exceeding 15 cm 2
2a Height up to nearly 15 cm (but generally much less than this); scar of adductor muscle deep blue-black (introduced in a few localities) *Crassostrea virginica* (Gmelin, 1791)
2b Height up to about 5 cm; scar of adductor muscle not much darker than the rest of the interior of the valve *Ostrea (Ostrea) lurida* Carpenter, 1864

Suborder Pectinina

15. Families Pectinidae and Propeamussidae

Bernard, F. R. 1978. New bivalve molluscs, subclass Pteriomorphia, from the northeastern Pacific. Venus (Jap. J. Malacol.), 47:61-75.
Grau, G. 1959. Pectinidae of the eastern Pacific. Allan Hancock Found. Pac. Exped., 23. viii + 308 pp.

Except for one species indicated as belonging to the family Propeamussidae, the key is concerned with members of the family Pectinidae.

1a Only the right valve decidedly convex (the left valve is nearly flat); height commonly exceeding 10 cm *Patinopecten caurinus* (Gould, 1850)
1b Both valves decidedly convex; height less than 7 cm (except in the case of *Hinnites giganteus*, which reaches a height of more than 10 cm; this species, after reaching a height of about 1 cm, becomes permanently cemented to rock) 2
2a Interior of both valves with purple blotch; free-living until about 1 cm high, then becoming cemented to rock by the right valve; height often exceeding 10 cm
 Hinnites giganteus (Gray, 1825)
2b Interior of valves without a purple blotch near the hinge; not permanently cemented to rock (but it may be temporarily attached by a byssus); height not more than 6 cm 3
3a With true external ribs (there may also be ruffles or spines, but if these are worn away, the ribs will still be evident); height commonly greater than 3 cm 4
3b Either without ribs or only with pseudoribs, which consist of raised ruffles of the concentric ridges (if the ruffles, which may overlap one another, are scraped away, no rib remains) 6
4a Radial ribs of both valves with very prominent spines or ruffles, rough to the touch
 Chlamys (Chlamys) hastata (Sowerby, 1843) (fig. 13.9)
4b Valves nearly or quite smooth to the touch (if the ribs have spines or knoblike excrescences, these are so slight as to be invisible without magnification; the excrescences, moreover, are generally restricted to the right valve) 5
5a Both valves with about 20-30 prominent ribs (ribs of the right valve with ruffles, but those of the left valve smooth; common shallow subtidal species)
 Chlamys (Chlamys) rubida (Hinds, 1845)
5b Both valves with about 10 especially prominent ribs, between which are less prominent ribs (rare subtidal species in our region) *Chlamys (Chlamys) behringiana* (Middendorff, 1849)
6a With raised internal ribs (height up to 2.5 cm)
 Family Propeamussidae: *Parvamussium alaskensis* (Dall, 1871) (fig. 13.20)
6b Without raised internal ribs 7
7a Either without pseudoribs or with pseudoribs that are about the same caliber as the concentric ridges (height up to 3 cm) *Delectopecten randolphi* (Dall, 1897)
 Includes subspecies *tillamookensis* (Arnold, 1906), characterized by pseudoribs, which are lacking in typical *randolphi*.
7b With pseudoribs that are much more pronounced than the concentric ridges
 Delectopecten vancouverensis (Whiteaves, 1893)

The following species and subspecies, not in the key, have been reported, mostly from deep water.

Family Pectinidae

Chlamys (Chlamys) jordani (Arnold, 1903). Shallow.
Cyclopecten argenteus Bernard, 1978
Cyclopecten bistriatus Dall, 1916
Cyclopecten carlottensis Bernard, 1968
Cyclopecten knudseni Bernard, 1978

Cyclopecten squamiformis Bernard, 1978
Hyalopecten neoceanus (Dall, 1908)

Family Propeamussidae

Propeamussium malpelonium (Dall, 1908). At depths of about 3000 m.

16. Family Anomiidae

Pododesmus (*Monia*) *cepio* (Gray, 1850) is common intertidally and subtidally. It is attached by a byssus that emerges through the large hole in the right valve, which is tightly applied to the substratum (rock, concrete, shell, wood, plastic).

Subclass Heterodonta

Order Veneroida

17. Family Lucinidae

1a   Radial ribs so low that they are overshadowed by the concentric ridges (with a grayish or greenish periostracum; up to about 1 cm long; subtidal)
                          Lucina (*Parvilucina*) *tenuisculpta* (Carpenter, 1864) (fig. 13.21)
1b   Radial ribs prominent, not overshadowed by the concentric ridges (some of the latter are prominent, however, and the valves are noticeably roughened) (length up to 5 cm; subtidal)
                          Lucinoma annulata (Reeve, 1850) (fig. 13.22)

18. Family Thyasiridae

1a   Outline of valves not nearly circular; both valves with a deep groove that runs from near the umbo to the posterior margin, thus segregating a small posterodorsal sector from the rest of the valve   2
1b   Outline of valves not nearly circular; neither valve with a deep groove that runs from the umbo to the posterior margin (length about 5 mm; intertidal and subtidal)
                          Axinopsida serricata (Carpenter, 1864) (fig. 13.23)
      A. viridis (Dall, 1901) (subtidal) is similar.
2a   Valves with prominent concentric growth lines; umbones decidedly anterior, in line with the nearly straight, vertical anterior margin   3
2b   Valves almost smooth, the growth lines being very fine; umbones near the middle   4
3a   Ventral margin evenly curved (length to about 8 cm; common subtidal species)
                          Conchocele bisecta (Conrad, 1849) (fig. 13.24)
      C. disjuncta (Dall, 1901) is similar.
3b   Ventral margin somewhat angular, due to slight projections being produced where 2 grooves on the interior of each valve meet the margin   *Thyasira cygnus* Dall, 1916
4a   Umbones prominent; interior of valves glossy, with radiating lines (length less than 1 cm)
                          Thyasira gouldii (Philippi, 1847)
4b   Umbones not prominent; interior of valves not glossy (length less than 1 cm)
                          Thyasira barbarensis Dall, 1901

The following species may also be expected.

Conchocele excavata (Dall, 1901). At depths of 800 m or more.
Axinulus redondoensis T. Burch, 1941. Shallow subtidal; not yet reported north of Oregon.

19. Family Ungulinidae

1a   Shell rather polished   *Diplodonta orbella* (Gould, 1851) (fig. 13.25)
1b   Shell dull   *Diplodonta impolita* Berry, 1953

20. Family Kelliidae

1a   Periostracum of live specimens faintly greenish; length up to slightly more than 2 cm (common, intertidal and subtidal)   *Kellia suborbicularis* (Montagu, 1803)
1b   Periostracum, if evident, not greenish; length rarely more than 1 cm   2
2a   Anterior and posterior halves of shell nearly the same size and shape; umbones not obviously inclined toward the anterior end (rare, intertidal and subtidal, and mostly south of our region)   *Rhamphidonta retifera* (Dall, 1899)
2b   Anterior and posterior halves of shell rather distinctly different (anterior half more broadly rounded); umbones obviously inclined toward the anterior end (subtidal)   *Odontogena borealis* (Cowan, 1964)

21. Family Lasaeidae

Lasaea subviridis Dall, 1899 (fig. 13.26). Common in mussel beds and in growths of tufted algae. There is electrophoretic evidence that there are two sympatric species in the region; these are difficult to distinguish on the basis of shell characters.

22. Family *Galeommatidae

O'Foighil, D. 1984. The morphology, reproduction, and ecology of *Scintillona bellerophon* spec. nov. (Galeommatacea). Veliger, 27:72-80.

Scintillona bellerophon O'Foighil, 1984. Lives attached to the upper side of the apodan sea cucumber *Leptosynapta clarki*.

23. Montacutidae

O'Foighil, D. 1985. Form, function, and origin of temporary dwarf males in *Pseudopythina rugifera* (Carpenter, 1864) (Bivalvia: Galeommatacea). Veliger, 27:245-52.
Rosewater, J. 1984. A new species of leptonacean bivalve from off northwestern Peru (Heterodonta: Veneroida: Lasaeidae). Veliger, 27:81-99.

1a   Umbones close to the posterior end of the shell and inclined posteriorly (unusual for a bivalve); length not more than 5 mm (free-living in sediment or associated with burrowing crustaceans and polychaetes)   *Mysella (Rochefortia) tumida* (Carpenter, 1864) (fig. 13.27)
       Mysella aleutica (Dall, 1899) has been reported from the region. It is subtidal, and its shell, which has a polished periostracum, usually has 3 or 4 concentric dark zones.
1b   Umbones near the middle of the shell and inclined slightly anteriorly; length up to nearly 2 cm   2
2a   Dorsal margin of right valve as smooth as that of the left valve; thickness more than one-third of the length; ventral margin generally either straight or indented (attached by a byssus to the ventral side of the abdomen of the decapod crustacean *Upogebia pugettensis* and to the ventral surface of polychaetes of the genus *Aphrodita*)   *Pseudopythina rugifera* (Carpenter, 1864)
2b   Dorsal margin of right valve serrated, especially posterior to the umbo; thickness about one-

third of the length; ventral margin convex to nearly straight, but not indented
Pseudopythina compressa (Dall, 1899)

24. Family Carditidae

1a Length of shell about 1.5 times the height; interior of valves usually purplish
Glans carpenteri (Lamy, 1922)
1b Length of shell not more than 1.25 times the height (in some species it is only equal to, or less than, the height); interior of valves generally not purplish 2
2a With fewer than 15 radial ribs 3
2b With more than 15 radial ribs 4
3a Height about equal to the length; with 12-14 ribs
Crassicardia crassidens (Broderip & Sowerby, 1829)
3b Height generally slightly greater than the length; with 10 or 11 ribs
Miontodiscus prolongatus (Carpenter, 1864)
4a With about 18-20 ribs *Cyclocardia* (*Cyclocardia*) *ventricosa* (Gould, 1850)
4b With about 25 ribs *Cyclocardia* (*Cyclocardia*) *crebricostata* (Krause, 1885) (fig. 13.28)

25. Family Chamidae

Bernard, F. R. 1976. Living Chamidae of the eastern Pacific (Bivalvia: Heterodonta). Nat. Hist. Mus. Los Angeles County, Contr. Sci., no. 278. 43 pp.

1a Shell attached to the substratum by its right valve; shell white, sometimes with red or green markings; concentric lamellae often raised up into spinelike projections (not likely to be found north of Oregon) *Pseudochama exogyra* (Conrad, 1837)
1b Shell attached by its left valve; exterior waxy-translucent, sometimes with pink or red streaks; concentric lamellae close-set, generally foliaceous *Chama* (*Chama*) *arcana* Bernard, 1976

26. Family Astartidae

Dall, W. H. 1903. Synopsis of the family Astartidae, with a review of the American species. Proc. U. S. Nat. Mus., 26:933-51.

1a Concentric ridges irregular (some are incomplete, discontinuous, or divided into 2 ridges) 2
1b Concentric ridges rather regular 3
2a Concentric ridges mostly well separated, many of them irregular; length up to about 1.5 cm (common subtidal species) *Astarte* (*Rictocyma*) *esquimalti* (Baird, 1863) (fig. 13.29)
2b Concentric ridges crowded, only a few of them irregular; length up to about 1.2 cm
Astarte (*Astarte*) *compacta* Carpenter, 1864
  A. (*A.*) *willetti* Dall, 1917, which attains a length of 2 cm, is similar.
3a Outline of the dorsal margin of the shell, from the umbones to the anteriormost tip, with a slight but distinct concavity (length to 2.5 cm; common subtidal species)
Tridonta alaskensis Dall, 1903 (fig. 13.30)
3b Outline of the dorsal margin of the shell, from the umbones to the anteriormost tip, nearly straight *Astarte* (*Astarte*) *undata* Gould, 1841

27. Family Cardiidae

Keen, A. M. 1980. The pelecypod family Cardiidae. A taxonomic summary. Tulane Stud. Geol. Paleontol. 16:1-44.

Phylum Mollusca: Class Bivalvia 277

13.26 *Lasaea subviridis*

13.27 *Mysella tumida*

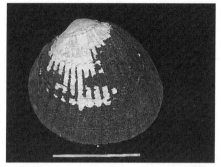
13.28 *Cyclocardia crebricostata*

13.29 *Astarte esquimalti*

13.30 *Tridonta alaskensis*

13.31 *Serripes groenlandicus*

13.32 *Nemocardium centifilosum*

13.33 *Clinocardium blandum*

1a Length about 1.25 times the height; radial ribs faint, though rather broad, and the surface of the shell nearly smooth (length up to 8 cm, but mostly less than 5 cm; subtidal)
Serripes groenlandicus (Bruguière, 1789) (fig. 13.31)
1b Length decidedly less than 1.25 times the height (sometimes just equal to or less than the height); radial ribs pronounced 2
2a Posterior quarter of valves with rather prominent concentric ridges and low radial ribs, so that the sculpture on this part of the shell is very different from that on the rest of the surface, where only the closely spaced radial ribs are distinct
Nemocardium (*Keenaea*) *centifilosum* (Carpenter, 1864) (fig. 13.32)
2b Valves without a substantial area in which the sculpture differs from that on the rest of the surface 3
3a Shell as high as or slightly higher than long; height up to about 8 cm; common intertidal and shallow subtidal species *Clinocardium* (*Clinocardium*) *nuttallii* (Conrad, 1837)
3b Length about equal to or slightly greater than the height; height of most species not exceeding 4 cm (but in *C. californiense* it may nearly reach 7 cm); subtidal 4
4a Length about equal to the height (height up to about 3 cm; with over 50 radiating ribs, these rather delicate) *Clinocardium* (*Clinocardium*) *blandum* (Gould, 1850) (fig. 13.33)
4b Length slightly greater than the height (ribs rather coarse) 5
5a With 32-38 radial ribs; height usually under 4 cm
Clinocardium (*Ciliatocardium*) *ciliatum* (Fabricius, 1780)
5b With 40 or more radiating ribs; height may reach or exceed 4 cm 6
6a Height up to 7 cm; with about 40-45 ribs
Clinocardium (*Clinocardium*) *californiense* (Deshayes, 1839)
6b Height up to slightly more than 4 cm; with about 45-50 ribs
Clinocardium (*Clinocardium*) *fucanum* (Dall, 1907)

28. Family Mactridae

Pearce, J. B. 1965. On the distribution of *Tresus capax* and *Tresus nuttallii* in the waters of Puget Sound and the San Juan Archipelago (Pelecypoda: Mactridae). Veliger, 7:166-71.

1a Width of gape at posterior end more than one-fourth the width of the shell; length sometimes attaining 20 cm; burrowing in mud and clay, intertidal and subtidal 2
1b Width of gape at posterior end less than one-fourth the width of the shell; length rarely exceeding 10 cm; mostly in sand, and primarily subtidal S
Spisula (*Symmorphomactra*) *falcata* (Gould, 1850)
2a Shell generally about 1.5 times as long as high; umbones near the end of the anterior third; common and widely distributed in our region *Tresus capax* (Gould, 1850) (fig. 13.5)
2b Shell more than 1.5 times as long as high; umbones near the end of the anterior quarter; uncommon in our region *Tresus nuttallii* (Conrad, 1837)

29. Family Solenidae

Solen (*Solen*) *sicarius* Gould, 1850. More common subtidally than intertidally; inhabits mud.

30. Family Cultellidae

1a Length sometimes exceeding 18 cm; color mostly uniformly olive-green, even in small specimens (some, however, have radiating purplish bands near the umbones); anterior edge of internal supporting rib diverging at an angle of about 30° from the dorsoventral axis (intertidal on exposed sandy beaches, and also subtidal)
Siliqua (*Siliqua*) *patula* (Dixon, 1788)

1b Length up to about 5 cm; color not primarily olive-green; anterior edge of internal supporting rib diverging not more than 15-20° from the dorsoventral axis   2
2a Valves with concentric brown bands alternating with light cream or tan bands; subtidal
  Siliqua (Siliqua) sloati Hertlein, 1961
2b Valves with radial bands of a purplish color on a whitish tan background; low intertidal and subtidal (not likely to be found north of California)   *Siliqua (Siliqua) lucida* (Conrad, 1837)

31. Family Tellinidae

Coan, E. V. 1971. The northwest American Tellinidae. Veliger, 14 (suppl.):1-63.
Dunnill, R. M., & D. V. Ellis. 1969. Recent species of the genus *Macoma* (Pelecypoda) in British Columbia. Nat. Mus. Canada Natur. Hist. Pap., no. 45. 34 pp.

1a Hinge plate with lateral teeth (these may be small) in addition to the cardinal teeth   2
1b Hinge plate without lateral teeth   5
2a Shell about twice as long as high; length commonly more than 1 cm   3
2b Shell much less than twice as long as high; length up to 1 cm or slightly more   4
3a Interior of both valves with a ridge just behind the scar of the anterior adductor muscle; length up to about 2 cm   *Tellina (Angulus) modesta* (Carpenter, 1864)
3b Neither valve with a ridge behind the scar of the anterior adductor muscle; length up to about 5 cm   *Tellina (Moerella) bodegensis* Hinds, 1845 (fig. 13.34)
4a Shell slightly more than 1.5 times as long as high; valves becoming narrowed toward the posteroventral corner; exterior white to pink   *Tellina (Angulus) carpenteri* Dall, 1900
4b Shell about 1.3 times as long as high; valves rounded posteriorly; exterior whitish or yellow, but interior often pink   *Tellina (Cadella) nucleoides* (Reeve, 1854)
  The genus *Macoma* is represented in Oregon, Washington, and British Columbia by 12 species. Most of these are primarily or strictly subtidal, and the distinctions between some of them cannot be expressed concisely in a key. For species not in the key (see list below), consult the works of Coan (1971) and Dunnill & Ellis (1969).
5a Length about twice the height; rare   *Macoma (Psammacoma) yoldiformis* Carpenter, 1864
5b Length much less than twice the height; common intertidal and subtidal species   6
6a Posterior portions of both valves distinctly bent to the right
  Macoma (Heteromacoma) nasuta (Conrad, 1837)
6b Posterior portions of valves not bent to the right   7
7a Periostracum polished, usually transparent; hinge ligament not more than one-fourth the length of the slope from the umbones to the posterior end; length commonly exceeding 5 cm, and sometimes attaining 10 cm   *Macoma (Rexithaerus) secta* (Conrad, 1837)
7b Periostracum not polished or transparent; hinge ligament at least one-third the length of the slope from the umbones to the posterior end; length not often exceeding 5 cm   8
8a Outline of valves almost perfectly oval; the posterior end nearly as rounded as the anterior end; valves (especially the interior) often yellow, pink, or bluish; length to about 2 cm
  Macoma (Macoma) balthica (Linnaeus, 1758)
8b Outline of valves not almost perfectly oval, the posterior end being narrowed and less rounded than the anterior end; valves whitish, except for the brown periostracum; length up to about 5 cm   *Macoma (Heteromacoma) inquinata* (Deshayes, 1855)

The following species are not in the key.

Macoma (Macoma) brota Dall, 1916
Macoma (Macoma) calcarea (Gmelin, 1791)
Macoma (Macoma) eliminata Dunnill & Coan, 1968
Macoma (Macoma) moesta (Deshayes, 1855)
Macoma (Macoma) obliqua (Sowerby, 1817)

Macoma (Psammacoma) carlottensis Whiteaves, 1880
Macoma (Rexithaerus) expansa Carpenter, 1864

32. Family Scrobiculariidae

1a   Outline of valves approximately elliptical, both ends being rounded to about the same extent (exterior of valves usually with pink radial rays and with some greenish periostracum; interior with a purple stain at both ends of the hinge plate; subtidal)
Semele rubropicta Dall, 1871 (fig. 13.35)

1b   Outline of valves not approximately elliptical, the anterior end rounded, the posterior end narrowed to a nearly pointed tip (not likely to be found north of California)
Cumingia californica Conrad, 1837

Abra (Abra) profundorum (E. A. Smith, 1885) has been found at depths of about 3000 m.

33. Family Psammobiidae

Gari (Gobraeus) californica (Conrad, 1837) is our only representative of the family. Its length may attain 7 cm. The valves, about twice as long as high, are nearly elliptical in outline; the exterior usually has pinkish or purplish radial rays.

34. Family *Kelliellidae

Kelliella galatheae Knudsen, 1970 has been collected at depths of about 3000 m and more.

35. Family *Vesicomyidae

The following species, all from depths of more than 500 m, have been reported.

Calyptogena (Calyptogena) kilmeri Bernard, 1974
Calyptogena (Calyptogena) pacifica Dall, 1891
Calyptogena (Archivesica) gigas (Dall, 1896)
Vesicomya (Vesicomya) lepta (Dall, 1896)
Vesicomya (Vesicomya) ovalis (Dall, 1896)
Vesicomya (Vesicomya) stearnsii (Dall, 1895)

36. Family Veneridae

Eugene N. Kozloff and Alan R. Kabat

Bernard, F. R. 1982. *Nutricola* n. gen. for *Transennella tantilla* (Gould) from the northeastern Pacific (Bivalvia: Veneridae). Venus (Jap. J. Malacol.), 41:146-9.

Gray, S. 1982. Morphology and taxonomy of two species of the genus *Transennella* (Bivalvia: Veneridae) from western North America and a description of *T. confusa* sp. nov. Malacological Rev., 15:107-17.

1a   With distinct radial ribs (these may, however, be delicate or faint)   2
1b   Without radial ribs   4
2a   Radial ribs typically more prominent than the concentric ridges (but this is not always the case); hinge ligament about one-third the length of the shell; length not more than 6 cm   3
2b   Radial ribs less prominent than the concentric ridges; hinge ligament nearly half the length of the shell; length sometimes attaining 8 cm
Protothaca (Callithaca) tenerrima (Carpenter, 1857) (fig. 13.7)

13.34 *Tellina bodegensis*

13.35 *Semele rubropicta*

13.36 *Humilaria kennerlyi*

13.37 *Petricola carditoides*

13.38 *Platyodon cancellatus*

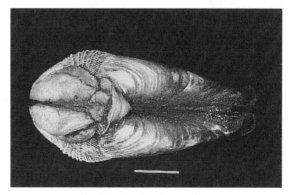
13.39 *Penitella penita*

Phylum Mollusca: Class Bivalvia

3a With a row of small teeth on the inside of the valves, close to their ventral margins; interior of valves whitish, without a purple stain near the posterior end
Protothaca (Protothaca) staminea (Conrad, 1857)

3b Without a row of teeth on the inside of the valves near their ventral margins; interior of valves yellowish, with a purple stain near the posterior end
Tapes (Ruditapes) philippinarum (Adams & Reeve, 1850)

4a Length less than 8 mm; height usually more than three-fourths the length; angle formed by the dorsal margin at the umbones not more than 110°; 3 cardinal teeth on the hinge plate of both valves; embryos brooded internally 5

4b Length generally greater than 2 cm; height not often more than three-fourths the length; angle formed by the dorsal margin at the umbones more than 110° 8

5a Middle cardinal tooth considerably larger than the other 2 teeth, but not bifid; lateral teeth absent 6

5b Middle cardinal tooth bifid; left valve with an anterior lateral tooth, right valve with a corresponding socket 7

6a Umbones almost in the middle; outline of shell resembling that of an isosceles triangle whose corners have been rounded; intertidal and shallow subtidal   *Psephidia lordi* (Baird, 1863)

6b Umbones at the end of the second one-fifth of the shell; outline of shell oval; subtidal
Psephidia ovalis Dall, 1902

7a Siphons fused for about half their length; tentacles on siphons (9-12 on the exhalant siphon; 10-14 on the inhalant siphon) long and flexible; posterior end of shell usually brown or purplish; umbones generally eroded (widespread in sandy and muddy habitats throughout our region)   *Transennella tantilla* (Gould, 1852)
Bernard (1982) has proposed that this species be transferred to a new genus, *Nutricola*.

7b Siphons fused for about four-fifths of their length; tentacles on siphons (10-14 on the exhalant siphon; 11-16 on the inhalant siphon) short and stiff; posterior end of shell not pigmented, but there may be 1 or more brownish spots anterior to the umbones; umbones not often eroded (not reported north of Coos Bay, Oregon)   *Transennella confusa* Gray, 1982

8a Concentric ridges conspicuously raised; posteriormost cardinal tooth on right valve not bifid 9

8b Concentric lines of growth not conspicuously raised, the exterior of the shell thus nearly smooth; posteriormost cardinal tooth on right valve bifid (length up to about 6 cm)
Compsomyax subdiaphana (Carpenter, 1864)

9a Umbones at the end of the second one-fifth of the shell; periostracum yellowish, polished; length up to 2.5 cm; subtidal   *Liocyma fluctuosa* (Gould, 1841)

9b Umbones at the end of the first one-fifth of the shell; periostracum not yellowish and not polished; length generally exceeding 3 cm 10

10a Shell closing tightly at the posterior end; concentric ridges bent in the direction of the umbones; length up to 10 cm; subtidal   *Humilaria kennerlyi* (Reeve, 1863) (fig. 13.36)

10b Shell gaping slightly at the posterior end; concentric ridges, even if sharp, not bent in the direction of the umbones; length often exceeding 10 cm; common intertidally
Saxidomus giganteus Deshayes, 1839

37. Family Turtoniidae

Ockelmann, K. W. 1964. *Turtonia minuta* (Fabricius), a neotenous veneracean bivalve. Ophelia, 1:121-46.

Turtonia minuta (Fabricius, 1780) is the only species in our region. It occurs intertidally, attached by a byssus to algae, sessile animals, or rocks. It reaches a length of 2 mm, and is characteristically glossy brown, with a purplish or rose suffusion at the umbones.

38. Family Petricolidae

1a   Valves with prominent radial ribs, especially in the anterior one-third; length about 2.5 times the height; burrowing in clay and firm mud (introduced from the Atlantic)
Petricola (Petricola) pholadiformis Lamarck, 1818
1b   Radial sculpture barely apparent, consisting of closely spaced fine lines (these may be completely worn away); length less than twice the height; nestling in crevices and in burrows originally made by rock-boring bivalves (the shell is generally misshapen, conforming to the limited space in which the animal is living)
Petricola (Rupellaria) carditoides (Conrad, 1837) (fig. 13.37)

39. Family Cooperellidae

Cooperella subdiaphana (Carpenter, 1864)

40. Family Thraciidae

1a   Anterior and posterior ends of valves almost identically rounded; umbones almost exactly in the middle   *Thracia (Crassithracia) beringi* Dall, 1915
1b   Anterior and posterior ends of valves dissimilar; umbones slightly nearer to the anterior end than to the posterior end   2
2a   Umbones about halfway between the end of the first one-third and the middle; length up to 5 cm   3
2b   Umbones near the end of the first one-third; length not often more than 3 cm
Thracia (Ixartia) curta Conrad, 1837
3a   Posterior end of shell nearly truncate   *Thracia (Thracia) trapezoides* Conrad, 1849
3b   Posterior end of shell broadly rounded, not truncate
Thracia (Crassithracia) challisiana Dall, 1915

41. Family *Laternulidae

Laternula (Laternulina) limicola (Reeve, 1864) was introduced to Coos Bay from Japan, but there are no recent records of its occurrence in our region.

Order Myoida

Suborder Myina

42. Family Myidae

1a   Pallial sinus at least as deep as high (except in *Sphenia ovoidea*, in which it is not quite as deep as high); umbones at least slightly anterior to the middle; anterior and posterior halves of shell dissimilar   2
1b   Pallial sinus much less deep than high; umbones almost exactly in the middle; anterior and posterior halves of shell similar (siphons opening into burrows of *Upogebia* and *Callianassa*, rather than at the surface)   *Cryptomya (Cryptomya) californica* (Conrad, 1837)
2a   Chondrophore projecting only slightly, and elongated along the longitudinal axis of the shell; umbones decidedly anterior to the middle (anterior and posterior portions of valves very dissimilar, the posterior portion being slightly narrower and almost truncate; in burrows of other animals and in algal holdfasts)   *Sphenia ovoidea* Carpenter, 1864
2b   Chondrophore projecting nearly or quite to the extent of its width; umbones only slightly anterior to the middle   3

3a Posterior (siphonal) end of shell somewhat less smoothly rounded than the anterior end, but not truncate; length sometimes exceeding 10 cm; especially common in estuaries and other situations where the salinity is reduced by influx of fresh water
Mya (Arenomya) arenaria Linnaeus, 1758 (fig. 13.6)
3b Posterior end of shell almost truncate; length rarely exceeding 7 cm; not usually found in situations where the salinity is reduced 4
4a Shell about 1.5 times as long as high; periostracum thick, brown, wrinkled; interior of valves yellowish, not pearly
Mya (Mya) truncata Linnaeus, 1758
4b Shell almost twice as long as high; periostracum thin, yellowish; interior of valves pearly
Platyodon (Platyodon) cancellatus (Conrad, 1837) (fig. 13.38)

43. Family Hiatellidae

1a Pallial line discontinuous, consisting of a series of separate scars 2
1b Pallial line continuous (length sometimes exceeding 15 cm; the mantle cavity is not large enough to accommodate the immense siphon; the shell gapes widely except in the region of the hinge; deep burrower in mud, intertidal and shallow subtidal)
Panope abrupta (Conrad, 1849)
Panope generosa (Gould, 1851), by which name this species has long been known, is a synonym.
2a Shell gaping widely at both ends; hinge plate of both valves with a single pointed tooth; scars forming the discontinuous pallial line conspicuous; length up to about 7 cm; tips of siphons not bright red (shallow subtidal)
Panomya chrysis Dall, 1909
2b Shell gaping only slightly; hinge plate of valves of adults without teeth (in young specimens, there are 2 teeth on the hinge plate of both valves); scars forming the discontinuous pallial line faint; length generally less than 5 cm; tips of siphons bright red (nestling in algal holdfasts as well as in burrows made by rock-boring bivalves, and in other protected situations; capable also of burrowing into shell and soft rock)
Hiatella arctica (Linnaeus, 1767)
Hiatella pholadis (Linnaeus, 1767) is here considered to be a synonym.

Suborder Pholadina

44. Family Pholadidae

Evans, J. W., & D. Fisher. 1966. A new species of *Penitella* (family Pholadidae) from Coos Bay, Oregon. Veliger, 8:222-4.
Kennedy, G. L. 1974. West American Cenozoic Pholadidae (Mollusca: Bivalvia). Mem. 8, San Diego Soc. Nat. Hist. 127 pp.
Turner, R. D. 1955. The family Pholadidae in the western Atlantic and the eastern Pacific. Part 2. Martesiinae, Jouannetiinae, and Xylophaginae. Johnsonia, 3:65-160.

1a Rasping anterior portion of valves abruptly separated by a groove from the non-rasping portion 2
1b Rasping anterior portion of valves not abruptly separated from the non-rasping portion (burrowing in firm mud or clay; length to about 5 cm; not likely to be found north of southern Washington)
Barnea (Anchomasa) subtruncata (Sowerby, 1834)
2a Without a myophore (apophysis) in either valve; posterior portion of shell, in mature specimens, tapering to a point, thus resembling a bird's beak; anterior portion of shell gaping as if abruptly cut off; siphonoplax (consisting of a pair of flaps that enclose the siphons) calcified
Netastoma rostrata (Valenciennes, 1845)
Netastoma japonica (Yokoyama, 1920) has also been reported.
2b With a myophore in both valves; posterior portion of shell not tapering sufficiently to resemble a bird's beak (it may taper, however, especially in *Pholadidea conradi*); anterior

	portion of shell gaping (except in full-grown specimens, in which the gape is closed by a callum), but not appearing to have been abruptly cut off; siphonoplax not calcified	3
3a	Rasping anterior portion of each valve occupying about half the area of the valve; valves without a shieldlike plate (protoplax) dorsal to the rasping portion; anteroventral gape not becoming closed, when the animal is full-grown, by a callum; length up to 15 cm; burrowing in firm mud and clay *Zirfaea pilsbryii* Lowe, 1931	
3b	Rasping portion of each valve occupying less than half the area of the valve; valves with a thick, shieldlike plate dorsal to the rasping portion; anteroventral gape becoming closed, in full-grown specimens, by a callum; length not exceeding 8 cm; generally boring in shale, sometimes in hard clay	4
4a	Posterior (siphonal) ends of valves of mature specimens with a siphonoplax (this consists of a pair of flaps that enclose the siphons); siphons smooth	5
4b	Posterior ends of valves of mature specimens without a siphonoplax; siphons smooth or slightly warty	6
5a	Siphonoplax consisting of flexible flaps, not lined by calcareous granules; length to about 8 cm; generally boring in shale *Penitella penita* (Conrad, 1837) (fig. 13.39)	
5b	Siphonoplax not flexible, lined with calcareous granules; length to about 2.5 cm; generally boring in shells of abalones (*Haliotis*), *Mytilus californianus*, and other molluscs *Penitella conradi* Valenciennes, 1846	
6a	Siphons smooth; umbonal reflections (calcified plates between the protoplax and the rasping portion of each valve) tightly applied to the shell *Penitella turnerae* Evans & Fisher, 1966	
6b	Siphons warty; umbonal reflections free of the shell for about half their length *Penitella gabbii* (Tryon, 1863)	

45. Family Xylophagaidae

Knudsen, J. 1961. The bathyal and abyssal *Xylophaga* (Pholadidae, Bivalvia). Galathea Rep., 5:163-209.
Turner, R. D. 1955. The family Pholadidae in the western Atlantic and the eastern Pacific. Part 2. Martesiinae, Jouanetiinae, and Xylophaginae. Johnsonia, 3:65-160.

The only species in our region is *Xylophaga washingtona* Bartsch, 1921, which bores in waterlogged wood. It is subtidal and rare.

46. Family Teredinidae

Bartsch, P. 1922. A monograph of the American shipworms. Bull. 122, U. S. Nat. Mus. 51 pp.
Turner, R. D. 1966. A Survey and Illustrated Catalogue of the Teredinidae (Mollusca: Bivalvia). Cambridge, Massachusetts: Museum of Comparative Zoology, Harvard University. vii + 265 pp.

1a	Pallets consisting of a series of cones set one inside the other, but on the whole somewhat featherlike *Bankia (Bankia) setacea* (Tryon, 1865)
1b	Pallets approximately paddlelike (uncommon north of Oregon) *Teredo (Teredo) navalis* Linnaeus, 1758

Phylum Mollusca: Class Bivalvia

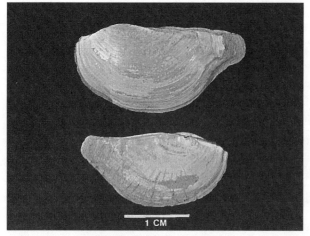

13.40 *Pandora filosa*

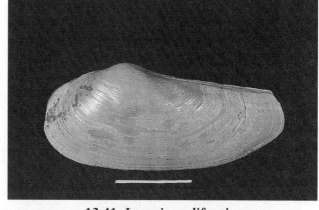

13.41 *Lyonsia californica*

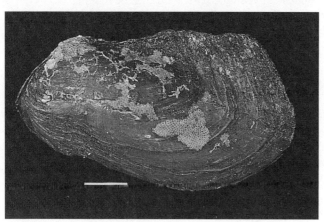

13.42 *Agriodesma saxicola*

13.43 *Cardiomya californica*

13.44 *Cardiomya planetica*

13.45 *Cardiomya oldroydi*

Subclass Anomalodesmata

Order Pholadomyoida

47. Family Pandoridae

1a Interior of right valve with 3 articulating ridges; dorsal edge of shell, posterior to the umbones, markedly concave   *Pandora (Pandorella) punctata* Conrad, 1837
1b Interior of right valve with 2 articulating ridges; dorsal edge of shell, posterior to the umbones, nearly straight or slightly convex   2
2a Posterior portion of shell drawn out into a truncate, beaklike prolongation
  Pandora (Pandorella) filosa (Carpenter, 1864) (fig. 13.40)
2b Posterior portion of shell not drawn out into a truncate, beaklike prolongation   3
3a Left valve with a groove that extends from the umbo to the ventral margin, thus dividing the surface into a small anterior portion and a large posterior portion   4
3b Left valve without a groove that divides it into 2 portions, but both valves (especially the left valve) have 2 ridges extending from near the umbones to the posterior end
  Pandora (Pandorella) bilirata Conrad, 1855
4a Groove on left valve distinct; dorsal margin of the shell, behind the umbones, slightly convex; length up to 5 cm   *Pandora (Pandorella) wardiana* A. Adams, 1859
  Pandora grandis Dall, 1877 is a synonym.
4b Groove on left valve indistinct; dorsal margin of the shell, behind the umbones, straight; length up to about 2 cm (rare)   *Pandora (Pandorella) glacialis* Leach, 1819

48. Family Lyonsiidae

1a Valves generally nearly circular in outline, but not consistently the same shape, and often obviously deformed; embedded in colonies of compound ascidians, with the siphons exposed
  Mytilimeria nuttalli Conrad, 1837
1b Valves decidedly longer than high, not nearly circular in outline (often deformed, however); not embedded in colonies of compound ascidians, but may be embedded in sponges or nestling in burrows made by other animals   2
2a Shell not often appearing to be deformed; valves with delicate radial ribs (these are not obvious on portions where the periostracum has persisted); pearly internally and externally, except where there is periostracum   *Lyonsia (Lyonsia) californica* Conrad, 1837 (fig. 13.41)
  Lyonsia bracteata (Gould, 1850) and *L. nesiotes* Dall, 1915 have also been reported, but their taxonomic status is not clear. One or both may be variants of *L. californica*.
2b Shell usually at least slightly deformed; valves without radial ribs; chalky rather than pearly externally, and usually with considerable periostracum (valves tending to crack when dried)   3
3a Valves gaping posteriorly or ventrally; length sometimes exceeding 10 cm
  Agriodesma saxicola (Baird, 1863) (fig. 13.42)
3b Valves not gaping; length up to about 2.5 cm (often in sponge masses) (not likely to be found north of California)   *Entodesma pictum* (Sowerby, 1834)

49. Family *Verticordiidae

The following species have been reported from the region.

Halicardia perplicata (Dall, 1890). Mostly at depths greater than 1000 m.
Lyonsiella parva Okutani, 1962. At depths greater than 300 m.
Policordia alaskana Dall, 1895. At depths greater than 800 m.

Order Septibranchida

Bernard, F. R. 1974. Septibranchs of the eastern Pacific (Bivalvia: Anomalodesmata). Allan Hancock Found. Monogr. Mar. Biol., no. 8. 279 pp.

------. 1979. New species of *Cuspidaria* from the northeastern Pacific (Bivalvia: Anomalodesmata), with a proposed classification of septibranchs. Venus (Jap. J. Malacol.), 38:14-24.

Okutani, T., & S. Sakurai. 1964. Genus *Cardiomya* (Mollusca: Lamellibranchiata) from Japanese waters. Bull. Nat. Sci. Mus. Tokyo, 7:17-32.

1a   Shell becoming abruptly narrowed posteriorly into a truncate beak (the narrowing may be more obvious in dorsal view than in lateral view)   50. Cuspidariidae
1b   Shell not becoming abruptly narrowed into a beak   51. Poromyidae

50. Family Cuspidariidae

All of the species in the key are subtidal, but may be expected at depths of less than 200 m.

1a   Without radial ribs, but external surfaces of valves covered with small tubercles (length up to about 2.5 cm)   *Plectodon scaber* Carpenter, 1864
1b   With radial ribs   2
2a   With 30-40 radial ribs; length sometimes exceeding 2.5 cm   *Cardiomya planetica* (Dall, 1908) (fig. 13.44)
2b   With not more than 20 radial ribs; length not often greater than 1.5 cm   3
3a   Length twice (or slightly more than twice) the height; with 16-20 (usually 18 or 19) radial ribs (some may be faint)   *Cardiomya californica* (Dall, 1886) (fig. 13.43)
3b   Length less than twice the height (except in some specimens of *Cardiomya pectinata*, in which it is just twice the height); with 12-16 radial ribs (some may be faint)   4
4a   Length of beak, as measured along its dorsal midline from the posteriormost radial rib, nearly or quite half the length of the shell; height of the shell more than three-fifths of the length, greatest slightly anterior to the middle   *Cardiomya pectinata* (Carpenter, 1865)
4b   Length of beak, as measured along its dorsal midline from the posteriormost radial rib, about two-fifths the length of the shell; height of shell about three-fifths of the length, greatest near the middle   *Cardiomya oldroydi* Dall in Oldroyd, 1924 (fig. 13.45)

The following species have been reported for the region, but are not likely to be taken at depths of less than 500 m.

Cardiomya curta (Jeffreys, 1881)
Cuspidaria (Cuspidaria) apodema Dall, 1916
Cuspidaria (Cuspidaria) cowani Bernard, 1967
Cuspidaria (Cuspidaria) filatovae Bernard, 1979
Cuspidaria (Cuspidaria) glacialis Dall, 1913
Cuspidaria (Cuspidaria) variola Bernard, 1979
Myonera tillamookensis Dall, 1916

51. Family *Poromyidae

Poromya (Dermatomya) canadensis Bernard, 1969. Deep subtidal.
Poromya (Dermatomya) beringiana (Dall, 1916). Deep subtidal.
Poromya (Dermatomya) leonina (Dall, 1916). Deep subtidal.
Poromya (Dermatomya) tenuiconcha Dall, 1913. From depths of less than 300 m.
Poromya (Cetoconcha) malespinae Ridewood, 1903. Deep subtidal.

14

PHYLUM MOLLUSCA: CLASSES SCAPHOPODA, CEPHALOPODA

Class Scaphopoda

Elsie Marshall and Ronald L. Shimek

Clark, R. B. 1963. The economics of *Dentalium*. Veliger, 6:9-19.
Emerson, W. K. 1962. A classification of the scaphopod mollusks. J. Paleontol., 36:461-82.
------. 1978. Two new eastern Pacific species of *Cadulus*, with remarks on the classification of the scaphopod mollusks. Nautilus, 92:117-23.
Marshall, E. 1980. *Pulsellum salishorum* spec. nov., a new scaphopod from the Pacific Northwest. Veliger, 23:149-52.
Pilsbry, H. A., & B. Sharp. 1897-98. Class Scaphopoda. *In* Manual of Conchology, 17:i-xxxii, 1-280.

1a	Median tooth of radula twice as wide as long; foot conical; shell length commonly exceeding 2 cm	2
1b	Median tooth of radula nearly as long as wide; foot wormlike, but can be expanded at the tip into a disk; shell length generally less than 2 cm	4
2a	Shell robust, with a narrow slit originating at the posterior aperture (this portion of the shell may be broken off, however) (inhabiting gravel that consists to a large extent of shell fragments)	*Dentalium pretiosum* (fig. 14.1)
2b	Shell fragile, without a narrow slit originating at the posterior aperture even if this portion of the shell has not been broken off	3
3a	Surface of shell smooth (ovary of female yellow, visible through shell; inhabiting silt or sand)	*Dentalium rectius* (fig. 14.2)
3b	Surface of shell with fine ribs or lines	*Dentalium agassizi*
4a	Diameter of shell greatest at the anterior aperture; shell not polished; length seldom exceeding 1 cm (ovary of female pink)	*Pulsellum salishorum*
4b	Diameter of shell greatest posterior to the anterior aperture (but the widest part may be close to the anterior aperture); shell highly polished in adults; length commonly more than 1 cm	5
5a	Diameter of shell greatest just posterior to the anterior aperture, and only slightly greater than the diameter of the aperture (ovary of female brown)	*Cadulus aberrans*
5b	Diameter of shell decidedly greater in the middle third than just posterior to the anterior aperture	6
6a	Edge of posterior aperture smooth; shell robust, opaque, white	*Cadulus californicus*
6b	Edge of posterior aperture with 5 delicate, regularly spaced lobes; shell relatively thin and hyaline, some of the internal organs visible through the shell (ovary of female brown)	*Cadulus tolmiei* (fig. 14.3)

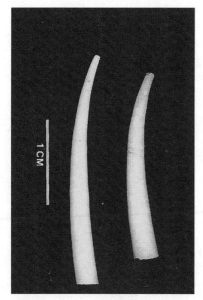

14.1 *Dentalium pretiosum*

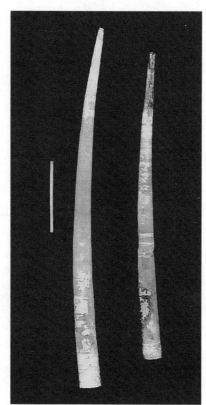

14.2 *Dentalium rectius*

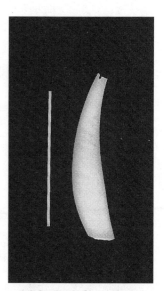

14.3 *Cadulus tolmiei*

Order Dentaliida

Family Dentaliidae

Dentalium agassizi Pilsbry & Sharp, 1897
Dentalium pretiosum Sowerby, 1860 (fig. 14.1)
Dentalium rectius Carpenter, 1864 (fig. 14. 2). *Dentalium dalli* Pilsbry & Sharp, 1897 is a synonym.

Order Gadilida

Family Gadilidae

Cadulus aberrans Whiteaves, 1887. *Cadulus hepburni* Dall, 1897 is a synonym.
Cadulus californicus Pilsbry & Sharp, 1898
Cadulus tolmiei Dall, 1897 (fig. 14.3)

Family Pulsellidae

Pulsellum salishorum Marshall, 1980

Class Cephalopoda

F. G. Hochberg

Berry, S. S. 1911. Preliminary notices of some new Pacific cephalopods. Proc. U. S. Nat. Mus., 37:407-19.

------. 1912. A review of the cephalopods of western North America. Bull. U. S. Bur. Fisheries, 30(1910):267-336.

------. 1953. Preliminary diagnoses of six West American species of *Octopus*. *In* Berry, S. S., Leaflets in Malacology, 1(10):51-8.

Fields, W. G. 1965. The structure, development, food relations, reproduction, and life history of the squid *Loligo opalescens* Berry. Calif. Dept. Fish & Game, Fish Bull. no. 131. 108 pp.

Fields, W. G., & V. A. Gauley. 1971. Preliminary descriptions of an unusual gonatid squid (Cephalopoda: Oegopsida) from the north Pacific. J. Fish. Res. Bd. Canada, 28:1796-1801.

Jefferts, K. 1985. *Gonatus ursabrunae* and *Gonatus oregonensis*, two new species of squids from the northeastern Pacific Ocean (Cephalopoda: Oegopsida: Gonatidae). Veliger, 28:159-74.

Kubodera, T., & K. Jefferts. 1984. Distribution and abundance of the early life stages of squid, primarily Gonatidae (Cephalopoda, Oegopsida) in the northern North Pacific (part 2). Bull. Nat. Sci. Mus. Tokyo, Ser. A (Zoology), 10:165-93.

Nesis, K. N. 1982. [Brief Guide to the Cephalopod Molluscs of the World Oceans.] Moskva: Izdatel'stvo Legkaîa i Pischevaîa Promyshlennost'. 365 pp. (in Russian).

Pattie, B. H. 1968. Notes on giant squid *Moroteuthis robusta* (Dall) Verrill trawled off the southwest coast of Vancouver Island, Canada. Wash. Dept. Fish. Res. Rep., 3:47-50.

Pearcy, W. G. 1965. Species composition and distribution of pelagic cephalopods from the Pacific Ocean off Oregon. Pacific Sci., 19:261-6.

Pearcy, W. G., & Voss, G. L. 1963. A new species of gonatid squid from the northeastern Pacific. Proc. Biol. Soc. Wash., 76:106-12.

Pickford, G. E. 1964. *Octopus dofleini* (Wülker), the giant octopus of the north Pacific. Bull. Bingham Oceanogr. Coll., 19(1):1-68.

Recksiek, C. W., & H. W. Frey (eds.). 1978. Biological, oceanographic, and acoustic aspects of the market squid, *Loligo opalescens* Berry. Calif. Dept. Fish & Game, Fish Bull., 169:1-185.

Young, R. E. 1972. The systematic and areal distribution of pelagic cephalopods from the seas off southern California. Smithsonian Contr. Zool., no. 97. iii + 159 pp.

Some information for the portion of the key dealing with species of *Octopus* was contributed by Ellie Dorsey.

1a	With 8 arms; arms with sessile suckers; body ovoid and without fins; without an internal shell or "pen"	2
1b	With 2 tentacles in addition to 8 arms; arms with stalked suckers and/or hooks; body rounded to elongate, with fins; with an internal shell or "pen"	4
2a	Skin of body smooth except for a ridge that borders the mantle on both sides; ability to change color limited to lighter and darker shades of brick-orange; white spots not present on the dorsal mantle or on the web in front of the eyes; mantle length (measured from the body apex to the midpoint between the eyes on the dorsal mantle) less than 5 cm; arms 2-3 times the body length; males without special enlarged suckers on the arms; without a planktonic larval stage (hectocotylus large, about one-fifth the length of the third right arm; with 11-12 lamellae on the outer demibranch of each gill)	*Octopus leioderma* (fig. 14.4)
2b	Skin of body not smooth, but with extensible folds and/or papillae; capable of changing color freely; color red to red-brown, often mottled with white; with conspicuous white spots on the dorsal mantle and on the web in front of the eyes; mantle length usually greater than 5 cm; arms 3-5 times the body length; males with 1 or 2 enlarged suckers on the arms; with a planktonic larval stage	3

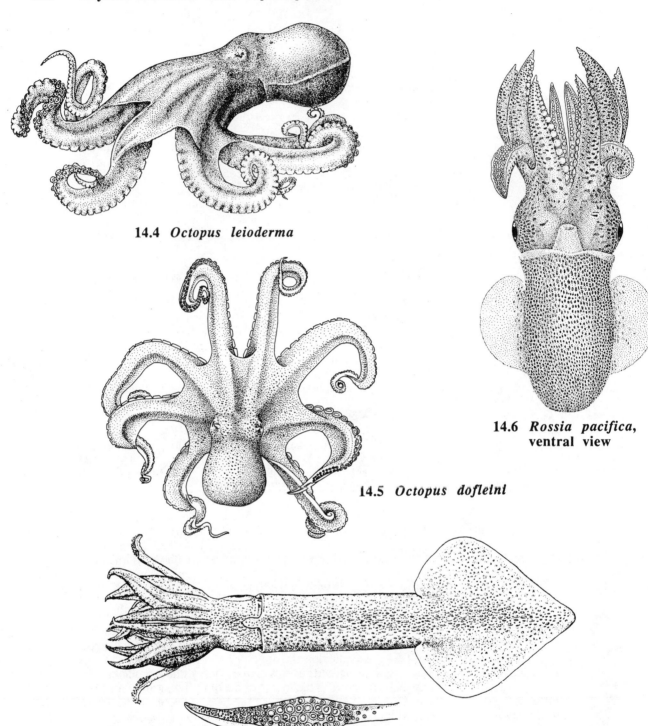

14.4 *Octopus leioderma*

14.5 *Octopus dofleini*

14.6 *Rossia pacifica*, ventral view

14.7 *Loligo opalescens*, dorsal view, and a tentacular club

3a Mantle length frequently greater than 20 cm, weight sometimes exceeding 50 kg; body with skin folds and large truncate papillae; hectocotylus large, about one-fifth the length of the third right arm; with 12-15 lamellae on the outer demibranch of each gill; eggs 6-9 mm long; planktonic larva with a single row of chromatophores on each arm
Octopus dofleini (fig. 14.5)

3b Mantle length less than 10 cm, weight less than 200 g; body with small, pointed papillae; hectocotylus small, about one-tenth the length of the third right arm; with 11-13 lamellae on the outer demibranch of each gill; eggs 3-4 mm long; planktonic larva with a double row of chromatophores on each arm
Octopus rubescens

4a Body rounded; mantle length (measured from the body apex to the free edge of the dorsal mantle) less than 5 cm; mantle free all around; fins semicircular, almost as long as the mantle, with a broad free lobe; arms short, circular in transverse section; tentacles retractile into special pits; dorsal arms of male modified for spermatophore transfer
Rossia pacifica (fig. 14.6)

4b Body elongate and tapering; mantle length usually greater than 5 cm; mantle attached; fins triangular, generally about half as long as the mantle; arms long, angular in transverse section; tentacles not retractile into special pits; dorsal arms of male not modified for spermatophore transfer   5

5a Tentacle clubs narrow, with 2 rows of hooks and with a distinct cluster of modified suckers that form a "fixing apparatus;" length of fins greater than their width, and more than half the mantle length; mantle covered with fleshy longitudinal ridges; mantle length commonly exceeding 150 cm
Moroteuthis robusta

5b Tentacle clubs without hooks and without a "fixing apparatus;" length of fins equal to or less than their width, and generally less than half the mantle length; mantle smooth, without ridges; mantle length less than 30 cm   6

6a Tentacle clubs narrow, the suckers large and in 4 rows; arms with 2 rows of suckers, without hooks; arms of unequal length, the ventral pair long and broad, and the left ventral arm of the male modified for spermatophore transfer; eyes covered by a continuous membrane (cornea); length of fins equal to their width; mantle length not exceeding 15 cm
Loligo opalescens (fig. 14.7)

6b Tentacle clubs expanded, with numerous minute suckers, 15-20 in each row; arms with 4 rows of armature, the outer rows with suckers, the inner rows with hooks on all but the ventral arm pair; arms nearly equal in length, and the left ventral arm of the male not modified for spermatophore transfer; membrane over the eyes perforated; width of fins greater than their length; mantle length up to 30 cm (specimens collected at floating docks are usually small, with a mantle length of less than 10 cm)
Berryteuthis magister

Species marked with an asterisk are not in the key. Most of them will not be encountered near shore, and some are restricted to deep water.

Subclass Coleoidea

Order Sepioidea

Family Sepiolidae

Rossia pacifica Berry, 1911 (fig. 14.6)

Order Teuthoidea

Suborder Myopsida

Family Loliginidae

Loligo opalescens Berry, 1911 (fig. 14.7)

Suborder Oegopsida

Family Enoploteuthidae

Abraliopsis felis McGowan & Okutani, 1968

Family Octopoteuthidae

Octopoteuthis deletron Young, 1972

Family Onychoteuthidae

Moroteuthis robusta (Verrill, 1876)
Onychoteuthis borealijaponica Okuda, 1927

Family Gonatidae

Berryteuthis anonychus (Pearcy & Voss, 1963)
Berryteuthis magister (Berry, 1913)
Gonatopsis borealis Sasaki, 1923
Gonatus berryi Naef, 1923
Gonatus californiensis Young, 1972
Gonatus madokai Kubodera & Okatani, 1977
Gonatus onyx Young, 1972
Gonatus oregonensis Jefferts, 1985
Gonatus pyros Young, 1972
Gonatus ursabrunae Jefferts, 1985

Family Architeuthidae

Architeuthis japonica Pfeffer, 1912

Family Histioteuthidae

Histioteuthis dofleini (Pfeffer, 1912)

Family Ommastrephidae

Ommastrephes bartrami (Lesueur, 1821)
Todarodes pacificus Steenstrup, 1880

Family Chiroteuthidae

Chiroteuthis calyx Young, 1972
Valbyteuthis oligobessa Young, 1972

Family Cranchiidae

Cranchia scabra Leach, 1817
Galiteuthis phyllura Berry, 1911
Leachia dislocata Young, 1972
Taonius pavo (Lesueur, 1821)

Order Vampyromorpha

Family Vampyroteuthidae

Vampyroteuthis infernalis Chun, 1903

Order Octopoda

Suborder Cirrata

Family Cirroteuthidae

Cirroteuthis muelleri Eschricht, 1836

Family Opisthoteuthidae

Opisthoteuthis californiana Berry, 1949

Suborder Incirrata

Family Bolitaenidae

Japatella diaphana Hoyle, 1885

Family Octopodidae

Benthoctopus abruptus (Sasaki, 1920)
Benthoctopus profundum Robson, 1932
Octopus dofleini (Wülker, 1910) (fig. 14.5)
Octopus leioderma (Berry, 1911)
Octopus rubescens Berry, 1953 (fig. 14.4)

15

PHYLUM TARDIGRADA AND PHYLUM ARTHROPODA: SUBPHYLA CHELICERATA, UNIRAMIA

PHYLUM TARDIGRADA

Crisp, D. J., & J. Hobart. 1954. A note on the habitat of the marine tardigrade *Echiniscoides sigismundi* (Schultze). Ann. Mag. Nat. Hist., ser. 12, 7:554-60.

Marcus, E. 1929. Tardigrada. *In* Bronn, H. G., Klassen und Ordnungen des Tierreichs, V: IV: 3. vii + 608 pp.

Renaud-Mornant, J., & L. W. Pollock. 1971. A review of the systematics and ecology of marine Tardigrada. Smithsonian Contr. Zool., no. 76:106-17.

Schulz, E. 1955. Studien an marinen Tardigraden. Kieler Meeresforsch., 11:73-9.

Schuster, R. O., & A. A. Grigarick. 1965. Tardigrada from western North America, with emphasis on the fauna of California. Univ. Calif. Publ. Zool., 76:1-67.

Echiniscoides sigismundi (M. Schultze, 1865) (family Echiniscoididae), a widely distributed species, is found on filamentous green algae that grow on rocks, barnacles, and mussels, and it is also associated with the red alga *Endocladia muricata*. In general, its occurrence is restricted to upper levels of the intertidal region. *Batillipes mirus* Richters, 1909 (family Batillipedidae) has been reported from intertidal sands in British Columbia.

In terrestrial habitats, tardigrades are abundant in growths of mosses and some lichens. The monograph of Schuster & Grigarick (1965) covers the species that are likely to be encountered.

PHYLUM ARTHROPODA

Subphylum Chelicerata

Class Pycnogonida

Joel W. Hedgpeth

The key is based on characters of adult pycnogonids. Differences between the sexes are avoided as much as possible. It must be remembered, however, that these animals are sexually dimorphic, and that the males carry the eggs. Mature specimens without eggs may be females, but not necessarily so, unless they also lack ovigerous legs. Because pycnogonids do not have the characteristic complement of segments and appendages until they are fully adult, it may be impossible to assign juveniles to genus or even to family. One should try to see as many specimens as possible from a particular situation, with the hope that some of them will be mature. Young specimens are often found in hydroids, molluscs, or other soft-bodied invertebrates.

Cole, L. J. 1904. Pycnogonida of the west coast of North America. Harriman Alaska Expedition, 10:249-98.
Exline, H. I. 1936. Pycnogonida from Puget Sound. Proc. U. S. Nat. Mus., 83:413-22.
Hedgpeth, J. W. 1941. A key to the Pycnogonida of the Pacific coast of North America. Trans. San Diego Soc. Nat. Hist., 9:253-64.
------. 1951. Pycnogonids from Dillon Beach and vicinity, California, with descriptions of two new species. Wasmann J. Biol., 9:105-17.
Scott, F. M. 1913. On a species of *Nymphon* from the north Pacific. Ann. Mag. Nat. Hist., ser. 8, 10:206-9.

1a   Legs thick, shorter than (or at least not conspicuously longer than) the combined length of the proboscis and trunk; outline of the animal as a whole oval or elliptical; without chelicerae or pedipalps; without conspicuous spiny projections   2
1b   Legs usually much longer than the combined length of the proboscis and trunk (pycnogonids of this type are usually awkward and gangly, but sometimes are fairly compact); outline of the animal as a whole approximately circular; with either chelicerae or pedipalps, or both; often spiny or setose   3
2a   Color uniform (ivory to pink); height of dorsal tubercles, when these are present, not greater than their basal diameter   *Pycnogonum stearnsi* (fig. 15.4)
2b   Color not uniform (generally light brown with dark lines, or with clear areas between dark patches); height of dorsal tubercles greater than their basal diameter   *Pycnogonum rickettsi*
3a   With both chelicerae and pedipalps (the pedipalps are conspicuous and have several articles); chelicerae either with functional chelae or with chelae reduced to knobs (juveniles of some species, however, have chelae that are lost as the animals mature); ovigerous legs usually present in both sexes and conspicuous   4
3b   With chelicerae, but without pedipalps; chelicerae with chelae and conspicuous; ovigerous legs reduced or entirely absent in females   17
4a   Chelicerae with 2 basal articles; small chelae sometimes present (the chelae may, however, be reduced to nonfunctional knobs)   5
4b   Chelicerae with a single basal article (scape); with a chela or a knoblike second article (or the entire appendage may be reduced to a knob or to an elongate, tuberclelike structure)   7
5a   Very spiny in appearance, with prominent double rows of tall, spiny projections on the longer articles of the legs and a single median row of spiny projections on the dorsum of the trunk   *Nymphopsis spinosissima* (fig. 15.3)
5b   Smooth or spiny, but if spiny, without tall projections arranged as described in choice 5a   6
6a   Trunk compact, circular; legs without long spines; with 2 or 3 tubercles along the median part of the dorsum of the trunk   *Ammothella tuberculata*
6b   Trunk not so compact as to be circular; legs with long spines, especially on the longer articles; without tubercles along the median part of the dorsum of the trunk   *Ammothella* sp.
7a   Chelicerae reduced to knobs or tubercles   8
7b   Chelicerae consisting of a basal article (scape) and either a well developed terminal chela that extends beyond the proboscis or with a knoblike article (if functional chelae are present in sub-adults, these do not extend beyond the proboscis)   11
8a   Each chelicera consisting of an easily recognized, papillate process; a gangly species with rather long and slender legs, these often with brownish purple bands; pedipalps with 9 articles   *Ammothea hilgendorfi*
8b   Chelicerae small and difficult to see; animal rather compact; pedipalps with 6 or fewer articles   9
9a   Body covered with short setae, thus appearing hairy; associated with the alcyonacean *Gersemia rubiformis*   *Tanystylum anthomasti* (fig. 15.6)
9b   Body not covered with short setae (it may, however, have sparse spines); not normally associated with *Gersemia*   10
10a Proboscis about twice as long as its diameter at the base, rounded at the tip; trunk very compact, the lateral projections indistinct   *Tanystylum occidentalis*

298 Phylum Arthropoda: Class Pycnogonida

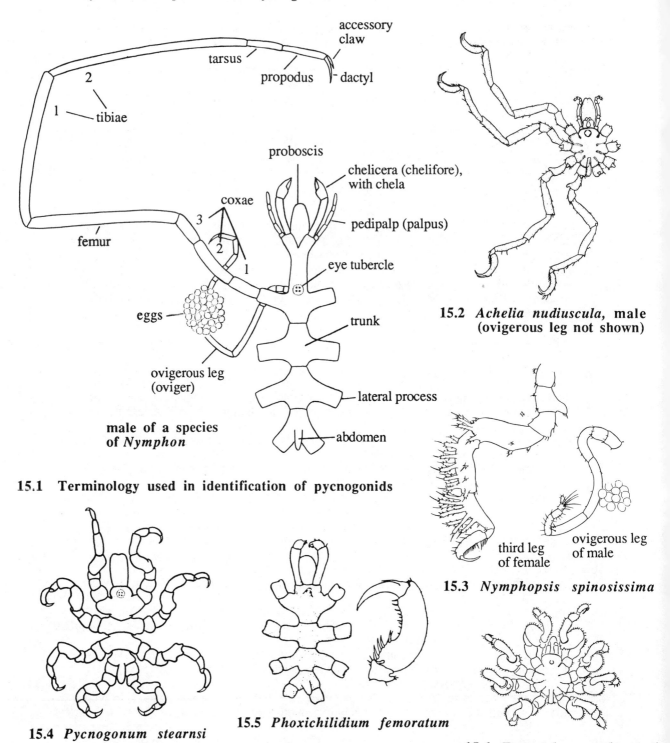

15.1 Terminology used in identification of pycnogonids

15.2 *Achelia nudiuscula*, male (ovigerous leg not shown)

15.3 *Nymphopsis spinosissima*

15.4 *Pycnogonum stearnsi*

15.5 *Phoxichilidium femoratum*

15.6 *Tanystylum anthomasti*, female

10b Proboscis at least 3 times as long as its diameter at the base, tapering to a blunt, subconical tip; trunk not so compact as in *T. occidentalis*, the lateral projections touching, but distinct
Tanystylum sp.
11a Chelicerae conspicuous, chelate, extending beyond the proboscis; pedipalps slender, with 5 articles; total diameter, when the legs are outstretched, sometimes exceeding 5 cm; usually subtidal 12
11b Chelicerae usually reduced to knobs in the adult, but if chelae persist, the chelicerae are shorter than the proboscis; pedipalps with 8 articles; total diameter less than 2 cm; includes some intertidal species 13
12a Chelae slender, the dactyl and fixed finger about as long as the palm, and gracefully curved; tarsus of legs as long as, or longer than, the propodus; legs slender *Nymphon pixellae*
12b Chelae stout, the dactyl and fixed finger less than half as long as the palm; tarsus of legs usually shorter than the propodus; legs stout, sparsely spinose (a highly variable species)
Nymphon grossipes
13a Adults not chelate 14
13b Adults chelate, the dactyl and fixed finger arcuate (a seasonal parasite of mussels in central California; found at least as far north as Boiler Bay, Oregon, but the habitat on the Oregon coast is not known) *Achelia chelata*
14a Dorsodistal edges of first coxal articles with fingerlike projections or short outgrowths 15
14b Dorsodistal edges of first coxal articles without fingerlike projections or short outgrowths (there may, however, be low knobs or wartlike eminences on the lateral projections of the trunk) *Achelia latifrons*
15a Projections on first coxal articles less than half as long as the diameter of the articles; proboscis broadly ellipsoidal (usually in bays) *Achelia nudiuscula* (fig. 15.2)
15b Projections on first coxal articles at least half as long as the diameter of the articles; proboscis narrowly ellipsoidal 16
16a Projections on first coxal articles at least three-fourths as long as the diameter of the articles; eye tubercle erect *Achelia gracilipes*
16b Projections on first coxal articles about half as long as the diameter of the articles; eye tubercle low and broad *Achelia alaskensis*
17a Lateral projections very close together; auxiliary claws minute and inconspicuous; legs slender (intestine bright green, its branches clearly visible in the legs; eye tubercle at least twice as high as broad) *Anoplodactylus viridintestinalis*
17b Lateral projections separated by at least half their diameter; auxiliary claws small, but well developed; legs stout *Phoxichilidium femoratum* (fig. 15.5)

A few species, marked with an asterisk, are not in the key. Some of them are restricted to deep water.

Family Ammotheidae

Achelia alaskensis (Cole, 1904)
Achelia chelata (Hilton, 1939)
**Achelia discoidea* (Exline, 1936)
*Achelia echinata Hodge, 1864
Achelia gracilipes (Cole, 1904)
Achelia latifrons (Cole, 1904)
Achelia nudiuscula (Hall, 1913) (fig. 15.2)
Ammothea hilgendorfi (Böhm, 1879)
Ammothella tuberculata Cole, 1904
Ammothella sp.
Nymphopsis spinosissima (Hall, 1912) (fig. 15.3)

Family Colossendeidae

**Colossendeis angusta* Sars, 1877
**Colossendeis colossea* Wilson, 1881

Colossendeis tenera Hilton, 1943
Hedgpethia californica (Hedgpeth, 1939)

Family Nymphonidae

Nymphon grossipes (Fabricius, 1780)
**Nymphon longitarse* Krøyer, 1844
Nymphon pixellae Scott, 1913
**Nymphon rubrum* Hodge, 1865

Family Phoxichilidiidae

**Anoplodactylus erectus* Cole, 1904
Anoplodactylus viridintestinalis (Cole, 1904)
Phoxichilidium femoratum (Rathke, 1799) (fig. 15.5)

Family Pycnogonidae

Pycnogonum rickettsi Schmitt, 1934. On the anemones *Anthopleura xanthogrammica* and *Metridium senile*, on the hydroid *Aglaophenia* sp., and on the ascidian *Clavelina huntsmani*.
Pycnogonum stearnsi Ives, 1892 (fig. 15.4). On the anemones *Anthopleura xanthogrammica* and *Metridium senile* and on the hydroid *Aglaophenia* sp.

Family Tanystylidae

Tanystylum anthomasti Hedgpeth, 1949 (fig. 15.6). On the alcyonacean *Gersemia rubiformis*.
Tanystylum occidentalis (Cole, 1904)
Tanystylum sp.

Class Arachnida

Order Pseudoscorpionida (Chelonethida)

The only distinctly maritime species reported from this area is *Halobisium occidentale* Beier, 1931. Its maximum length is about 4 mm. It is most likely to be found at the edges of *Salicornia* marshes, under rocks or debris just above the high tide line.

Chamberlin, J. C. 1931. The arachnid order Chelonethida. Stanford Univ. Publ. Biol., Sci., 7:1-284.

Order Acarida

Most of the genera of truly marine mites belong to the family Halacaridae of the suborder Trombidiformes. There are, however, marine representatives of 2 other suborders, Parasitiformes and Sarcoptiformes.
    The Halacaridae probably do not constitute a monophyletic assemblage; they are believed to have evolved from several groups of marine mites. About 15 genera assigned to the family have been reported from the northeastern Pacific region, and most of these are in the key prepared by Newell for *Light's Manual*. This key does not deal with species, of which there must be many, mostly still undescribed. Anyone who contemplates working with halacarids will find the various papers of Newell indispensable; those considered to be especially important for this region are listed below.

Two distinctive mites found in the upper reaches of the intertidal zone will be singled out for mention here. One is *Neomolgus littoralis* (Linnaeus, 1758) (suborder Trombidiformes, family Bdellidae), found on rocky shores, often where lichens or the green alga *Prasiola* are growing. It is bright red, about 3 mm long, and feeds by sucking juice from small flies.

The other species is *Gammaridacarus orchestoideae* (Hall, 1912) (order Parasitiformes, family Laelaptidae), which is parasitic on at least 2 amphipods characteristic of the drift zone of sandy beaches: *Megalorchestia californiana* and *Traskorchestia traskiana*. This mite was originally described by Hall (1912) as *Seius orchestoideae*. Canaris (1962), being unaware of Hall's work, described it as *Gammaridacarus brevisternalis*.

Canaris, A. G. 1962. A new genus and species of mite (Laelaptidae) from *Orchestoidea californiana* (Gammaridea). J. Parasit., 48:467-9.
Hall, H. V. M. 1912. Some marine and terrestrial Acarina of Laguna Beach. Laguna Mar. Lab., First Annual Rep., 177-86.
Newell, I. M. 1947. A systematic and ecological study of the Halacaridae of eastern North America. Bull. Bingham Oceanogr. Coll., 10(3):1-232.
------. 1949. New genera and species of Halacaridae (Acari). Amer. Mus. Novit., no. 1411. 22p.
------. 1950. New species of *Copidognathus* from the Aleutians. Amer. Mus. Novit., no. 1476. 19 pp.
------. 1951. New species of *Agaue* and *Thalassarachna* from the Aleutians. Amer. Mus. Novit., no. 1489. 19 pp.
------. 1952a. *Copidognathus curtus* Hall, 1912, and other species of *Copidognathus* from western North America. Amer. Mus. Novit., no. 1499. 27 pp.
------. 1952b. Further studies on Alaskan Halacaridae. Amer. Mus. Novit., no. 1536. 56 pp.
------. 1953. The natural classification of the Rhombognathinae (Acari, Halacaridae). Syst. Zool., 2:119-35.
------. 1971. Halacaridae (Acari) collected during Cruise 17 of the R/V Anton Bruun in the southeastern Pacific Ocean. Anton Bruun Report Number 8 (Scientific Results of the Southeast Pacific Expedition):1-58.

Subphylum Uniramia

Class Insecta

Relatively few insects are found in habitats that are truly marine. Nevertheless, marine representatives of several families of Diptera, Hemiptera, and Coleoptera do occur in our region, either in rocky intertidal areas or in pools of high salinity. At the fringes of the sea--under driftwood and decaying seaweeds high on sandy beaches or in salt marshes--there is an interesting variety of insects. *Light's Manual* has a good account of the insect fauna found in marine and maritime situations in central California, and an extensive bibliography for the Pacific coast. The references cited below should be especially helpful in our region.

General References

Cheng, L. 1976. Marine Insects. Amsterdam: North Holland Publishing Co.; New York: American Elsevier Publishing Co. 581 pp.
Essig, E. O. 1958. Insects and Mites of Western North America. 2nd ed. New York: Macmillan Co. 1050 pp.
Merritt, R. W. & K. W. Cummins. 1978. An Introduction to the Aquatic Insects of North America. Dubuque, Iowa: Kendall/Hunt Publishing Co. 441 pp.
Saunders, L. G. 1928. Some marine insects of the Pacific coast of Canada. Ann. Entom. Soc. Amer., 21:521-45.
Usinger, R. L. (ed.) 1956. Aquatic Insects of California. Berkeley & Los Angeles: University of California Press. ix + 508 pp.

Order Coleoptera

Arnett, R. H. 1960. The Beetles of the United States. A Manual for Identification. Washington, D. C.: Catholic University of America Press. xi + 1112 pp.

Blackwelder, R. E. 1931. The genus *Endeodes* Le Conte (Coleoptera, Melampyridae). Pan-Pacific Entomol., 8:128-36.

Hatch, M. H. 1957. The Beetles of the Pacific Northwest. Part II. Staphyliniformia. Seattle: University of Washington Press. ix + 384 pp.

------. 1965. The Beetles of the Pacific Northwest. Part IV. Macrodactyles, Palpicornes, and Heteromera. Seattle: University of Washington Press. viii + 268 pp.

Moore, I. 1954. Notes on *Endeodes* Le Conte with a description of a new species from Baja California. Pan-Pacific Entomol., 30:195-8.

------. 1965a. A revision of the Pacific coast Phytosi with a review of the foreign genera (Coleoptera: Staphylinidae). Trans. San Diego Soc. Nat. Hist., 12:103-52.

------. 1965b. Notes on some intertidal Coleoptera with descriptions of the early stages (Carabidae, Staphylinidae, Malachiidae). Trans. San Diego Soc. Nat. Hist. 12:207-30.

Spilman, T. J. 1967. The heteromerous intertidal beetles. Pacific Insects, 9:1-21.

Order Collembola

Dexter, R. W. 1943. *Anurida maritima*: an important sea-shore scavenger. J. Econ. Entomol., 36:797.

Order Diptera

Bryce, D. & A. Hobart. 1972. The biology and identification of the larvae of the Chironomidae (Diptera). Ent. Gaz. 23:175-217.

Cole, F. R. 1969. The Flies of Western North America (with collaboration of E. I. Schlinger). Berkeley & Los Angeles: University of California Press. 693 pp.

Johannsen, O. A. 1938. Aquatic Diptera, Part IV. Chironomidae subfamily Chironominae. Mem. 210, Cornell Univ. Agric. Exp. Sta. 56 pp.

Morley, R. L. & R. A. Ring. 1972a. The intertidal Chironomidae (Diptera) of British Columbia I. Keys to their life stages. Canad. Entomol., 104:1093-8.

------. 1972b. The intertidal Chironomidae of British Columbia II. Life history and population dynamics. Canad. Entomol., 104:1099-121.

Saunders, L. G. 1928. Some marine insects of the Pacific coast of Canada. Ann. Entomol. Soc. Amer., 21:521-45.

Order Hemiptera

Herring, J. L. 1961. The genus *Halobates* (Hemiptera: Gerridae). Pacific Insects, 3:223-305.

Hutchinson, G. E. 1931. On the occurrence of *Trichocorixa* Kirkaldy (Corixidae, Hemiptera-Heteroptera) in salt marshes and its zoogeographical significance. Amer. Nat., 65:573-4.

Order Thysanura

Benedetti, R. 1973. Notes on the biology of *Nemachilis halophila* in a California sandy beach. Pan-Pacific Entomol., 49:246-9.

16

PHYLUM ARTHROPODA: SUBPHYLUM CRUSTACEA: CLASSES BRANCHIOPODA, COPEPODA, OSTRACODA, CIRRIPEDIA

Class Branchiopoda

Subclass Diplostraca

Order Cladocera

Suborder Eucladocera

Family Podonidae

Apstein, C. 1901. Cladocera (Daphnidae), Wasserflöhe. Nordisches Plankton, Zool., 4, Abt. VII:11-15.
------. 1911. Die Cladoceren (Daphniden). Nachtrag. Nordisches Plankton, Zool., 4, Abt. VII:17-20.
Baker, H. M. 1938. Studies on the Cladocera of Monterey Bay. Proc. Calif. Acad. Sci., ser. 4, 23:311-65.
Mordukhaĭ-Boltovskoĭ, F. D. 1978. [A contribution to the taxonomy of marine Podonidae (Cladocera).] Zoologicheskii Zhurnal, 57:523-9 (in Russian).
Rammner, W. 1930. Phyllopoda. *In* Grimpe, G., & E. Wagler (eds.), Die Tierwelt der Nord- und Ostsee, 10a. 32 pp.
------. 1939. Cladocera. Fiches d'Identification du Zooplancton. Conseil Permanent International pour l'Exploration de la Mer, no. 3.

1a   Head separated from the rest of the body by a deep groove; dorsal pouch bulbous, nearly spherical   2
1b   Head not separated from the rest of the body by a deep groove; dorsal pouch tapering, so that in side view the outline of the body as a whole is nearly triangular   3
2a   Exopodite of legs 1-3 with 1 seta   *Podon leuckarti* G. O. Sars, 1862
2b   Exopodite of legs 1-3 with 3 setae   *Pleopis polyphemoides* (Leuckart, 1859)
3a   Exopodite of legs 1 and 2 with 2 setae; exopodite of leg 3 with 1 seta; exopodite of leg 4 with 1 seta   *Evadne nordmanni* Lovén, 1835
3b   Exopodite of legs 1 and 2 with 3 setae; exopodite of leg 3 with 3 setae; exopodite of leg 4 with 2 setae   *Pseudevadne tergestina* (Claus, 1877)

Class Copepoda

Order Calanoida

The most important modern reference on marine calanoid copepods of the Pacific Northwest is the work of Gardner & Szabo (1982). It provides keys and good illustrations for all species known to occur in British Columbia, and its bibliography is substantially complete. Another reference that may be useful is that of Fulton (1972). The older monograph of Davis (1949) is considered to be obsolete. A few other contributions that deal with species that have been reported from our region or that may be expected to occur are also listed below.

Brodskiĭ (Brodsky), K. A. 1950. [Calanoida of the far eastern seas of the USSR and of the Polar basin.] Opredeliteli po faune SSSR, 35. 442 pp. (in Russian). (English translation, 1967, by Israel Program for Scientific Translations. U. S. Department of Commerce, TT67-51200).
Campbell, M. H. 1929. Some free-swimming copepods of the Vancouver Island region. Trans. Royal Soc. Canada, Sect. V, ser. 3, 23:303-32.
------. 1930. Some free-swimming copepods of the Vancouver Island region. II. Trans. Royal Soc. Canada, Sect. V, ser. 3, 24:177-84.
Damkaer, D. M. 1971. *Parastephos occatum*, a new species of hyperbenthic copepod (Calanoida: Stephidae) from the inland marine waters of Washington State. Proc. Biol. Soc. Washington, 83:505-14.
Davis, C. C. 1949. The pelagic Copepoda of the northeastern Pacific Ocean. Univ. Wash. Publ. Biol., 14. iv + 117 pp.
Esterly, C. O. 1924. The free-swimming Copepoda of San Francisco Bay. Univ. Calif. Publ. Zool., 26:81-129.
Frost, B. W. 1971. Taxonomic status of *Calanus finmarchicus* and *C. glacialis* (Copepoda), with special reference to adult males. J. Fish. Res. Bd. Canada, 28:23-30.
------. 1974. *Calanus marshallae*, a new species of calanoid copepod closely allied to the sibling species *C. finmarchicus* and *C. glacialis*. Mar. Biol., 26:77-99.
Frost, B., & A. Fleminger. 1968. A revision of the genus *Clausocalanus* (Copepoda: Calanoida) with remarks on distributional patterns in diagnostic characters. Bull. Scripps Inst. Oceanogr., 12:1-99.
Fulton, J. D. 1972. Keys and references to the marine Copepoda of British Columbia. Fish. Res. Bd. Can. Tech. Rep., no. 313. 63 pp.
Gardner, G. A., & I. Szabo. 1982. British Columbia pelagic marine Copepoda: an identification manual and annotated bibliography. Can. Spec. Publ. Fish. Aquat. Sci., no. 62. 536 pp.
Heron, G. A. 1964. Seven species of *Eurytemora* (Copepoda) from northwestern North America. Crustaceana, 7:199-211.
Johnson, M. W. 1932. Seasonal distribution of plankton at Friday Harbor, Washington. Univ. Wash. Publ. Oceanogr., 1:1-38.
Park, T. S. 1966. A new species of *Bradyidius* (Copepoda: Calanoida) from the Pacific coast of North America. J. Fish. Res. Bd. Canada, 23:805-11.
------. 1967. Two new species of calanoid copepods from the Strait of Georgia, British Columbia, Canada. J. Fish. Res. Bd. Canada, 24:231-42.
von Vaupel-Klein, J. C. 1970. Notes on a small collection of calanoid copepods from the northeastern Pacific, including the description of a new species of *Andinella* (Fam. Tharybidae). Zool. Verhandel. Leiden, 110:1-43.
Wilson, C. B. 1932. The copepods of the Woods Hole region, Massachusetts. Bull. 158, U. S. Nat. Mus. xix + 635 pp.

Order Harpacticoida

The most important single work on marine harpacticoids of the Pacific coast is that of Lang (1965). Although Lang was concerned primarily with species found in central California, he attempted to evaluate the lists of harpacticoids reported for other areas of the Pacific coast. Of the nearly 100 species he recorded, about 80 were new. Chappuis (1958), however, listed 38 species for Puget Sound (many of them referred to also by Wieser, 1959), only 3 of which were new. Lang suggested that Chappuis had given existing names to some species that had not been described. Nevertheless, of the 75 species recorded by Kask, Sibert, & Windecker (1982) from the Nanaimo estuary, on the east side of Vancouver Island, a substantial number had previously been found in the Atlantic, Mediterranean, and other regions.

The papers listed in the bibliography are the more comprehensive ones dealing with Pacific coast harpacticoids. They will lead to most of the minor contributions that the specialist may need to consult.

Attention is also called to the study of Fahrenbach (1962) which deals principally with *Diarthrodes cystoecus*, found in close association with certain red algae; his work also briefly mentions *Thalestris rhodymeniae*, which inhabits the bladders of the red alga *Halosaccion glandiforme* (in California, this species harbors *D. cystoecus*).

Campbell, M. H. 1929. Some free-swimming copepods of the Vancouver Island region. Trans. Royal Soc. Canada, Sect. V, ser. 3, 23:303-32.
------. 1930. Some free-swimming copepods of the Vancouver Island region. II. Trans. Royal Soc. Canada, Sect. V, ser. 3, 24:177-84.
Chappuis, P. A. 1958. Harpacticoïdes psammiques marins des environs de Seattle (Washington, U. S. A.). Vie et Milieu, 8 (1957):409-22.
Fahrenbach, W. H. 1962. The biology of a harpacticoid copepod. La Cellule, 62:301-76.
Kask, B. A., J. R. Sibert, & B. Windecker. 1982. A check list of marine and brackish water harpacticoid copepods from the Nanaimo estuary, southwestern British Columbia. Syesis, 15:35-8.
Lang, K. 1948. Monographie der Harpacticiden. 2 vols. Lund: Haken Ohlssons Boktryckeri (reprinted by Otto Koeltz Science Publishers, Koenigstein). 1683 pp.
------. 1965. Copepoda Harpacticoidea from the Californian Pacific coast. Kungl. Svenska Vetenskapsakad. Handl., ser. 4, 10, no. 2. 554 pp.
Monk, C. R. 1941. Marine harpacticoid copepods from California. Trans. Amer. Micr. Soc., 60:75-99.
Wells, J. B. J. 1976. Keys to Aid in the Identification of Marine Harpacticoid Copepods. Aberdeen: University of Aberdeen Press. 215 pp.
Wieser, W. 1959. Free-living nematodes and other small invertebrates of Puget Sound beaches. Univ. Wash. Publ. Biol., 19. xi + 179 pp.

Order Cyclopoida

Most of the marine cyclopoid copepods found along the Pacific coast are parasites of invertebrates and fishes. The only free-living marine species in the region belong to the genus *Oithona* (family Oithonidae). Many genera formerly included in the Cyclopoida are now assigned to the order Poecilostomatoida.

Family Oithonidae

The species of *Oithona* are dealt with by Gardner & Szabo (1982), Fulton (1972), and Davis (1949) (see bibliography for order Calanoida). The following reference is also pertinent.

Nishida, S., O. Tanaka, & M. Omori. 1977. Cyclopoid copepods of the family Oithonidae in Suruga Bay and adjacent waters. Bull. Plankton Soc. Japan, 24:119-58.

Families Ascidicolidae and Notodelphyidae

There are numerous species of these families in ascidians. Most of the genera previously placed in the Notodelphyidae are now assigned to the Ascidicolidae. The following monographs are comprehensive references for our fauna.

Dudley, P. L. 1966. Development and systematics of some Pacific marine symbiotic copepods. A study of the biology of the Notodelphyidae, associates of ascidians. Univ. Wash. Publ. Biol., 21. 282 pp.
Illg, P. L. 1958. North American copepods of the family Notodelphyidae. Proc. U. S. Nat. Mus., 107:463-649.
Illg, P. L., & P. L. Dudley. 1980. The family Ascidicolidae and its subfamilies (Copepoda, Cyclopoida), with descriptions of new species. Mém. Mus. Nat. Hist. Nat. Paris, Zool., 117:1-192.

Ooishi, S. 1980. the larval development of some copepods of the family Ascidicolidae, subfamily Haplostominae, symbionts of compound asdicians. Publ. Seto Mar. Biol. Lab., 25:253-92.

Ooishi, S., & P. L. Illg. 1977. Haplostominae (Copepoda, Cyclopoida) associated with compound ascidians from the San Juan Archipelago and vicinity. Special Publ. Seto Mar. Biol. Lab., ser. V. 154 pp.

Order Poecilostomatoida

Families Oncaeidae and Corycaeidae

Several genera of free-living Poecilostomatoida occur in the plankton of this region. The family Oncaeidae is represented by *Conaea*, *Oncaea*, *Lubbockia*, and *Pseudolubbockia*; the family Corycaeidae is represented by *Corycaeus* and *Farranula*. These are dealt with by Gardner & Szabo (1982), Fulton (1972), and Davis (1949) (see bibliography for order Calanoida). The following references should also be consulted.

Heron, G. A., & D. M. Damkaer. 1978. Seven *Lubbockia* species (Copepoda: Cyclopoida) from the plankton of the northeast Pacific, with a revision of the genus. Smithsonian Contr. Zool., no. 267. 36 pp.

Heron, G. A., T. S. English, & D. M. Damkaer. 1984. Arctic Ocean Copepoda of the genera *Lubbockia*, *Oncaea*, and *Epicalymma* (Poecilostomatoida: Oncaeidae), with remarks on distributions. J. Crust. Biol., 4:448-90.

Family Clausidiidae

Gooding, R. V. 1960. North and South American copepods of the genus *Hemicyclops* (Cyclopoida : Clausidiidae). Proc. U. S. Nat. Mus., 112:159-95.

Light, S. F., & O. Hartman. 1937. A review of the genera *Clausidium* Kossmann and *Hemicyclops* Boeck (Copepoda, Cyclopoida), with the description of a new species from the northeast Pacific. Univ. Calif. Publ. Zool., 41:173-87.

Clausidium vancouverense (Haddon, 1912). A reddish species regularly found in the branchial chamber of the ghost shrimps, *Callianassa californiensis* and *C. gigas*, and the mud shrimp, *Upogebia pugettensis* (see Light & Hartman, 1937).

Hemicyclops thysanotus Wilson, 1935 (*H. pugettensis* Light & Hartman, 1937). Occurring with *Clausidium vancouverense* in the branchial chamber of *Callianassa gigas* (farther south, it has been taken also on *C. californiensis*, *Upogebia pugettensis*, and the nudibranch *Hermissenda crassicornis*).

Hemicyclops subadhaerens Gooding, 1960. In burrows of *Callianassa californiensis*.

Family Myicolidae

Illg, P. L. 1960. Marine copepods of the genus *Anthessius* from the northeastern Pacific Ocean. Pacific Sci., 14:337-72.

Anthessius nortoni Illg, 1960. Associated with the keyhole limpet, *Diodora aspera*.

Family Lichomolgidae

Humes, A. G. 1982. A review of Copepoda associated with sea anemones and anemone-like forms (Cnidaria, Anthozoa). Trans. Amer. Phil. Soc., 72:1-120.

Humes, A. G., & J. H. Stock. 1973. A revision of the family Lichomolgidae Kossmann, 1877, cyclopoid copepods mainly associated with marine invertebrates. Smithsonian Contr. Zool., no. 127. v + 368 pp.

Doridicola ptilosarci Humes & Stock, 1973. Between the flaps and on the surface of the stem of the sea pen, *Ptilosarcus gurneyi*.

Family Sabelliphilidae

Several species of *Hermannella* parasitize the ctenidia of various bivalve molluscs; most of them were described by Illg (1949) as species of *Paranthessius* (see Humes & Stock, 1973, under family Lichomolgidae, above). An undescribed sabelliphilid occurs on the sabellid polychaete *Myxicola infundibulum*.

Illg, P. L. 1949. A review of the copepod genus *Paranthessius* Claus. Proc. U. S. Nat. Mus., 99:391-428.

Family Pseudanthessiidae

Illg, P. L. 1950. A new copepod, *Pseudanthessius latus* (Cylopoida: Lichomolgidae), commensal with a marine flatworm. J. Wash. Acad. Sci., 40:129-33.

Pseudanthessius latus Illg, 1950. Commensal with a polyclad turbellarian, *Kaburakia excelsa* (*Cryptophallus magnus*).

Family Splanchnotrophidae

Ho, J.-S. 1981. *Ismaila occulata*, a new species of poecilostomatoid copepod parasitic in a dendronotid nudibranch from California. J. Crust. Biol., 1:130-6.

Family Gastrodelphyidae

Several species of *Gastrodelphys* and *Sabellacheres* associated with sabellid polychaete annelids are referred to by Dudley (1964).

Dudley, P. L. 1964. Some gastrodelphyid copepods from the Pacific coast of North America. Amer. Mus. Novit., no. 2194. 51 pp.

Family Philichthyidae

Sekerak, A. D. 1970. Parasitic copepods of *Sebastes alutus* including *Chondracanthus triventricosus* and *Colobomatus kyphosus* sp. nov. J. Fish. Res. Bd. Canada, 27:1943-60.

Colobomatus kyphosus Sekerak, 1970. On fishes.

Families Bomolochidae and Ergasilidae

Members of these families are parasites of fishes. See also references cited under the order Siphonostomatoida.

Kabata, Z. 1971. Four Bomolochidae (Copepoda) from fishes of British Columbia. J. Fish. Res. Bd. Canada, 28:1563-72.
Wilson, C. B. 1911. North American parasitic copepods belonging to the family Ergasilidae. Proc. U. S. Nat. Mus., 39:263-400.

Family Chondracanthidae

This family has many species which are parasitic on fishes. In addition to the papers cited below, see the comprehensive references listed under the order Siphonostomatoida.

Fraser, C. McL. 1920. Copepods parasitic on fish from the Vancouver Island Region. Trans. Royal Soc. Canada, sect. V, ser. 3, 13:45-67.
Ho, J.-S. 1970. Revision of the genera of Chondracanthidae, a copepod family parasitic on marine fishes. Beaufortia, 17:105-218.
Kabata, Z. 1968. Some Chondracanthidae (Copepoda) from fishes of British Columbia. J. Fish. Res. Bd. Canada, 25:321-45.
------. 1969. *Chondracanthus narium* sp. n. (Copepoda: Chondracanthidae), a parasite of nasal cavities of *Ophiodon elongatus* (Pisces: Teleostei) in British Columbia. J. Fish. Res. Bd. Canada, 26:3043-7.
------. 1984. A contribution to the knowledge of Chondracanthidae (Copepoda: Poecilostomatoida) parasitic on fishes of British Columbia. Can. J. Zool., 62:1703-13.
Sekerak, A. D. 1970. Parasitic copepods of *Sebastes alutus* including *Chondracanthus triventricosus* and *Colobomatus kyphosus* sp. nov. J. Fish. Res. Bd. Canada, 27:1943-60.
Wilson, C. B. 1912. Parasitic copepods from Nanaimo, British Columbia, including eight species new to science. Contrib. Canada Biol. Fish., 1906-1910:85-101.

Family Mytilicolidae

Humes, A. G. 1954. *Mytilicola porrecta* n. sp. (Copepoda: Cyclopoida) from the intestine of marine pelecypods. J. Parasit., 40:186-94.

Mytilicola orientalis Mori, 1935. In the intestine of *Mytilius edulis*, *Crassostrea gigas*, and other bivalves.

Order Siphonostomatoida

The order Siphonostomatoida includes the copepods that have in the past been assigned to the orders Caligoida and Lernaeopodida. Most of them parasitize fishes, but some are associated with invertebrates or marine mammals, and a few are free-living. Only six families will be mentioned here. The following comprehensive references deal with all of these, as well as with other families that are represented by fish parasites. Important works concerned with individual families are also cited below.

Kabata, Z. 1979. Parasitic Copepoda of British Fishes. Ray Society (London), Monograph 152. 468 pp.
Love, M. S., & M. Moser. 1983. A checklist of parasites of California, Oregon and Washington marine and estuarine fishes. NOAA Tech. Rep. NMFS SSRF-777. 576 pp. (Provides global coverage; many of the species listed do not occur on the Pacific coast.)
Margolis, L., & J. R. Arthur. 1979. Synopsis of the parasites of fishes of Canada. Bull. Fish. Res. Bd. Canada, no. 199. vi + 269 pp.
Wilson, C. B. 1908. North American parasitic copepods. A list of those found upon the fishes of the Pacific coast, with descriptions of new genera and species. Proc. U. S. Nat. Mus., 35:431-81.

------. 1932. The copepods of the Woods Hole Region, Massachusetts. Bull. 158, U. S. Nat. Mus. xix + 635 pp.
------. 1935. Parasitic Copepoda from the Pacific coast. Amer. Midl. Nat., 16:776-97.
------. 1944. Parasitic copepods in the United States National Museum. Proc. U. S. Nat. Mus., 94:529-82.
Yamaguti, S. 1963. Parasitic Copepoda and Branchiura of Fishes. New York & London: Interscience Publishers. vii + 1104 pp.

Family Caligidae

Kabata, Z. 1973. The species of *Lepeophtheirus* (Copepoda: Caligidae) from fishes of British Columbia. J. Fish. Res. Bd. Canada, 30:729-59.
------. 1974. *Lepeophtheirus cuneifer* sp. nov. (Copepoda: Caligidae), a parasite of fishes from the Pacific coast of North America. J. Fish. Res. Bd. Canada, 31:43-7.
Margolis, L., Z. Kabata, & R. R. Parker. 1975. Catalogue and synopsis of *Caligus*, a genus of Copepoda (Crustacea) parasitic on fishes. Bull. Fish. Res. Bd. Canada, no. 192. 117 pp.
Wilson, C. B. 1905. North American parasitic copepods belonging to the family Caligidae. Part 1. Caliginae. Proc. U. S. Nat. Mus., 28:479-672.
------. 1907a. North American parasitic copepods belonging to the family Caligidae. Part 2. The Trebinae and Euryphorinae. Proc. U. S. Nat. Mus., 31:669-720.
------. 1907b. North American parasitic copepods belonging to the family Caligidae. Parts 3 and 4. A revision of the Pandarinae and the Cecropinae. Proc. U. S. Nat. Mus., 33:323-490.

Family Lernaeopodidae

Kabata, Z. 1970. Some Lernaeopodidae (Copepoda) from fishes of British Columbia. J. Fish. Res. Bd. Canada, 27:865-85.
Kabata, Z., & A. V. Gusev. 1966. Parasitic Copepoda of fishes from the collection of the Zoological Institute in Leningrad. J. Linn. Soc. (Zool.), 46:155-207.
Wilson, C. B. 1915. North American parasitic copepods belonging to the Lernaeopodidae, with a revision of the entire family. Proc. U. S. Nat. Mus., 47:565-79.

Family Pennellidae

Kabata, Z. 1967. The genus *Haemobaphes* (Copepoda: Lernaeoceridae) in the waters of British Columbia. Can. J. Zool., 45:853-75.

Family Pandaridae

Cressey, R. 1967. Revision of the family Pandaridae (Copepoda: Caligoida). Proc. U. S. Nat. Mus., 121:1-133.

Family Herpyllobiidae

Members of this family are parasites of polychaete annelids.

Lützen, J. 1966. The anatomy of the family Herpyllobiidae (parasitic copepods). Ophelia, 3:45-64.

Family Nicothoidae

Heron, G. A., & D. M. Damkaer. 1986. A new nicothoid copepod parasitic on mysids from northwestern North America. J. Crust. Biol., 6:652-65.

Hansenulus trebax Heron & Damkaer, 1986. On the mysids *Neomysis mercedis, Alienacanthomysis macropsis, Proneomysis wailesi*; also on *Xenacanthomysis pseudomacropsis* (Bering Sea).

Order Monstrilloida

Family Monstrillidae

The members of this group are parasitic (largely in polychaete annelids) during most of their life history. When they reach maturity, they emerge and become free-swimming. The adults lack mouthparts and cannot feed, so must subsist on reserves accumulated during the parasitic phase; they live only for a short time.

Five species have been reported from this region: *Monstrilla longiremis* Giesbrecht, 1892 (see Wailes, 1929); *M. canadensis* McMurrich, 1917; *M. wandelii* Stephensen, 1913; *M. helgolandica* Claus, 1863; *M. spinosa* Park, 1967. The last 3 were carefully redescribed by Park (1967). Davis (1949) reviewed the Monstrilloida in general.

Davis, C. C. 1949. A preliminary revision of the Monstrilloida, with descriptions of two new species. Trans. Amer. Micr. Soc., 68:245-55.
Park, T. S. 1967. Two unreported species and one new species of *Monstrilla* (Copepoda: Monstrilloida) from the Strait of Georgia. Trans. Amer. Micr. Soc., 86:144-52.
Wailes, G. H. 1929. The marine zoo-plankton of British Columbia. Art, Hist. Sci. Assoc., Vancouver, B. C., Mus. Art Notes, 4:159-69.

Subclass Branchiura

Order Arguloida

Family Argulidae

These are parasites of fishes. Only two marine species have been reported in the region: *Argulus borealis* Wilson, 1912 and *A. pugettensis* Dana, 1853 (*A. niger* Wilson, 1902).

Cressey, R. F. 1972. The genus *Argulus* (Crustacea: Branchiura) of the United States. Identification Manual No. 2, Biota of Freshwater Ecosystems, U. S. Environmental Protection Agency. viii + 14 pp.
Meehean, O. L. 1940. A review of the parasitic Crustacea of the genus *Argulus* in the collections of the United States National Museum. Proc. U. S. Nat. Mus., 88:459-522.
Yamaguti, S. 1963. Parasitic Copepoda and Branchiura of Fishes. New York & London: Interscience Publishers. vii + 1104 pp.

Class Ostracoda

Jack Q. Word

The ostracodes of our region are still poorly known. Lucas (1931) and Smith (1952) reported 35 species, mostly from near Nanaimo, and Wieser (1959) found 7 species in Puget Sound.

The list of species has been compiled partly from publications dealing with the ostracode fauna of Puget Sound, the Strait of Juan de Fuca, and Strait of Georgia. It has been augmented by records obtained during recent sampling in the same areas. In the case of most of the species listed, the information on habitats is not definitive, because some of the species have been collected only a few times.

The bibliography includes publications pertinent to this region, as well as some comprehensive papers that are likely to be helpful in recognition of genera, if not species.

Baker, J. H. 1978. Life history patterns of the myodocopid ostracod *Euphilomedes producta*, Poulsen, 1962. *In* Löffler, H. & D. Danielpol (eds.), Aspects of Ecology and Zoogeography of Recent and Fossil Ostracoda, p. 245. The Hague: Dr. W. Junk.

Cohen, A. C. 1982. Ostracoda. *In* Parker, S. P. (ed.), Synopsis and Classification of Living Organisms, 2:181-202. New York: McGraw-Hill Book Co.

Cohen, A. C., & L. S. Kornicker. 1975. Taxonomic indexes to Ostracoda (suborder Myodocopina) in Skogsberg (1920) and Poulsen (1962, 1965). Smithsonian Contr. Zool., no. 204. 29 pp.

Elofson, O. 1941. Zur Kenntnis der marinen Ostracoden Schwedens. Zool. Bidr. Uppsala, 19:215-534.

Hart, C. W., Jr. 1971. A new species of parasitic ostracod of the genus *Acetabulosoma* (Paradoxostomatidae, Paradoxostominae) with a discussion of the copulatory appendage homologies. Notulae Naturae Acad. Nat. Sci. Philadelphia, no. 442. 11 pp.

Hulings, N. C. 1971. Summary and current status of the taxonomy and ecology of benthic Ostracoda including interstitial forms. Smithsonian Contr. Zool., no. 76:91-7.

Juday, C. 1907. Ostracoda of the San Diego region. II. Littoral forms. Univ. Calif. Publ. Zool., 3:135-56.

Klie, W. 1929. Ostracoda. *In* Grimpe, G., and E. Wagler, Die Tierwelt der Nord- und Ostsee. Teil 10b. 56 pp.

Kornicker, L. S. 1978. *Harbansus*, a new genus of marine Ostracoda, and a revision of the Philomedidae (Myodocopina). Smithsonian Contr. Zool., no. 260. iii + 75 pp.

------. 1981. Revision, distribution, ecology, and ontogeny of the ostracode subfamily Cyclasteropinae (Myodocopina: Cylindroleberididae). Smithsonian Contr. Zool., no. 317. 548 pp.

Lie, U. 1968. A quantitative study of benthic infauna in Puget Sound, Washington, USA, in 1963-1964. Fiskerdir. Skr., Ser. Havunders., 14:229-556.

Lucas, V. Z. 1931. Ostracoda of the Vancouver Island region. Contrib. Can. Biol. Fish., n. s., 6:397-416.

McHardy, R. A. 1964. Marine ostracods from the plankton of Indian Arm, British Columbia, including a diminutive subspecies resembling *Conchoecia alata major* Rudjakov. J. Fish. Res. Bd. Canada, 21:555-76.

Neale, J. W. (ed.) 1969. The Taxonomy, Morphology, and Ecology of Recent Ostracoda. Edinburgh: Oliver & Boyd. ix + 553 pp.

Poulsen, E. M. 1962. Ostracoda-Myodocopa. Part I. Cypridiniformes-Cypridinidae. Dana Report, no. 57. 414 pp.

------. 1965. Ostracoda-Myodocopa. Part II. Cypridiniformes-Rutidermatidae, Sarsiellidae and Asteropidae. Dana Report, no. 65. 484 pp.

Sars, G. O. 1922-28. Ostracoda. *In* An Account of the Crustacea of Norway, vol. 9. 277 pp.

Skogsberg, T. 1920. Studies on marine ostracods, Part I. Cypridinids, Halocyprids, Polycopids. Zool. Bidr. Uppsala, Suppl., 1:1-784.

------. 1928. Studies on marine ostracods, Part II. External morphology of the genus *Cythereis* with descriptions of twenty-one new species. Occ. Pap. Calif. Acad. Sci., no. 15. 155 pp.

Smith, V. Z. 1952. Further Ostracoda of the Vancouver Island region. J. Fish. Res. Bd. Canada, 9:16-41.

Watling, L. 1970. Two new species of Cytherinae from central California. Crustaceana, 19:251-63.

------. 1972. A new species of *Acetabulastoma* Skornikov from central California with a review of the genus. Proc. Biol. Soc. Wash., 85:481-8.

Wieser, W. 1959. Free-living nematodes and other small invertebrates of Puget Sound beaches. Univ. Wash. Publ. Biol., 19. xi + 179 pp.

Subclass Myodocopa

Order Myodocopida

Suborder Myodocopina

Family Cypridinidae

Gigantocypris agassizii Müller, 1895. Pelagic.

Family Philomedidae

Euphilomedes carcharodonta (V. Z. Smith, 1952). Subtidal, in fine sediments.
Euphilomedes longiseta (Juday, 1907). Subtidal, in fine sediments.
Euphilomedes producta Poulsen, 1962. Subtidal, in fine sediments.
Harbansus mayeri Kornicker, 1978. Shallow subtidal, in fine sediments.
Philomedes dentata Poulsen, 1962. Subtidal, 45-70 m.
Scleroconcha trituberculatum (Lucas, 1931). Subtidal, in fine sediments.

Family Rutidermatidae

?*Rutiderma rostratum* Juday, 1907. Subtidal sediments between rocks.

Family Cylindroleberididae

Specimens that have previously been identified as *Cylindroleberis mariae* (Baird, 1850) may belong to one or more of the first four genera listed below.

Asteropella sp.
Bathyleberis sp.
Diasterope sp.
Parasterope sp.
Sarsiella sp. In bays, mostly among algae.
Vargula sp.
Leuroleberis sharpei Kornicker, 1981. Subtidal, in sand (not reported from our region, but to be expected).

Order Halocyprida

Suborder Halocypridina

Family Halocyprididae

All three species of *Conchoecia* are found mostly in midwater in Puget Sound and other deep inland waters.

Conchoecia alata subsp. *minor* McHardy, 1964
Conchoecia elegans G. O. Sars, 1928
Conchoecia spinirostris Claus, 1891

Subclass Podocopa

Order Podocopida

Suborder Podocopina

Family Cytherideidae

Jonesia rostrata (Lucas, 1930). Subtidal, 36-90 m.
Sahnia sp.
Tontocythere sp.

Family Cytheridae

The species of *Cythere* occur among algae in tide pools, and all three of them may be found in the same pool.

Cythere lutea O. F. Müller, 1785
Cythere alveolivalva V. Z. Smith, 1952
Cythere unifalcata V. Z. Smith, 1952

Family Cytheruridae

Cytherura clathrata M. Sars, 1865. Littoral.

Family Trachyleberididae

Some species that have been assigned to *Cythereis* may not belong to this genus, and they may even belong to either the family Cytheridae or the family Hemicytheridae.

Cythereis aurita Skogsberg, 1928. Littoral, among algae
Cythereis pacifica Skogsberg, 1928. Littoral, among algae.
Cythereis dunelmensis Norman, 1865. In muddy sediments.
Cythereis glauca Skogsberg, 1928. In holdfasts of *Laminaria*.
Cythereis montereyensis Skogsberg, 1928. In holdfasts of *Laminaria*.
Cythereis longiductus Skogsberg, 1928. Littoral to about 50 m, on gray clay with scattered algae.
Cythereis polita Skogsberg, 1928. Littoral.
Cythereis arachis Lucas, 1931. At moderate depths, 20-40 m.
Cythereis filoplumosa Lucas, 1931. At moderate depths, 20-40 m.
Cythereis serridentata V. Z. Smith, 1952. At moderate depths, 20-40 m.

Family Xestoleberididae

Xestoleberis depressa M. Sars, 1865. At depths of about 35 m.
Xestoleberis dispar Müller, 1894. Littoral, with algae (including coralline red algae).

Family Hemicytheridae

Hemicythere bicarina V. Z. Smith, 1952. On holdfasts of *Laminaria*.
Hemicythere obesa (Lucas, 1931). Littoral, among algae.

Family Loxoconchidae

Loxoconcha dentiarticula V. Z. Smith, 1952. On algae.
Loxoconcha fragilis G. O. Sars, 1928. On hydroids of the genus *Obelia*, including those on floats.
Loxoconcha tenuiungula V. Z. Smith, 1952. Littoral, among algae.

Family Paradoxostomatidae

Acetabulosoma kozloffi Hart, 1971. Symbiotic with the amphipod *Ampithoe humeralis*.
Cytherois pusilla G. O. Sars, 1928. At moderate depths (20-40 m).
Paradoxostoma cuneata Lucas, 1931. At depths of less than 100 m.
Paradoxostoma fraseri V. Z. Smith, 1952. At depths of less than 100 m.
Paradoxostoma striungulum V. Z. Smith, 1952. On hydroids of the genus *Obelia*, including those on floats.

Suborder Metacopina

Family Pontocyprididae

Argilloecia cylindrica G. O. Sars, 1928. At moderate depths (35-145 m).
Argilloecia conoidea G. O. Sars, 1928. At moderate depths (35-145 m).
Pontocypris clemensi V. Z. Smith, 1952. In tidal pools.

Class Cirripedia

Order Thoracica

Cornwall, I. E. 1969. The barnacles of British Columbia. 2nd ed. Handbook No. 7, British Columbia Provincial Museum, Victoria. 69 pp.
Haven, S. B. 1973. Occurrence and identification of *Balanus balanoides* (Crustacea: Cirripedia) in British Columbia. Syesis, 6:97-9.
Henry, D. P. 1940a. The Cirripedia of Puget Sound with a key to the species. Univ. Wash. Publ. Oceanogr., 4:1-48.
------. 1940b. Notes on some pedunculate barnacles from the North Pacific. Proc. U. S. Nat. Mus., 88:225-36.
------. 1942. Studies on the sessile Cirripedia of the Pacific coast of North America. Univ. Wash. Publ. Oceanogr., 4:95-134.
Newman, W. A., & A. Ross. 1976. Revision of the balanomorph barnacles; including a catalog of the species. Mem. 9, San Diego Society of Natural History. 108 pp.
Pilsbry, H. A. 1907. Cirripedia from the Pacific coast of North America. Bull. U. S. Bur. Fish., 26:193-204.
------. 1916. The sessile barnacles (Cirripedia) contained in the collections of the U. S. National Museum; including a monograph of the American species. Bull. 93, U. S. Nat. Mus. xi + 366 pp.
------. 1921. Barnacles of the San Juan Islands, Washington. Proc. U. S. Nat. Mus., 59:111-5.
Zullo, V. A. 1979. Marine Flora and Fauna of the Northeastern United States. Arthropoda: Cirripedia. U. S. Dept. Commerce, NOAA Tech. Rep., NMFS Circular 425.

Key to Suborders

1a	Attached to the substratum by a somewhat flexible stalk	Suborder Lepadomorpha
1b	Not attached by a stalk	Suborder Balanomorpha

Phylum Arthropoda: Class Cirripedia   315

Suborder Lepadomorpha

The key does not include species found on whales. These are described by Cornwall (1969).

1a   Capitulum with more than 10 plates; attached to rocks   2
1b   Capitulum with either 5 plates or only 2 small plates; attached to floating timbers, glass floats, or jellyfishes, or secreting a firm, gelatinous float   3
2a   Capitulum with numerous small plates in addition to 5 larger plates; carina without spines arising from ridges; common mid-intertidal species in areas where there is considerable wave action, and generally associated with *Mytilus californianus* (also subtidal in situations where there is a swift current)   *Mitella polymerus*

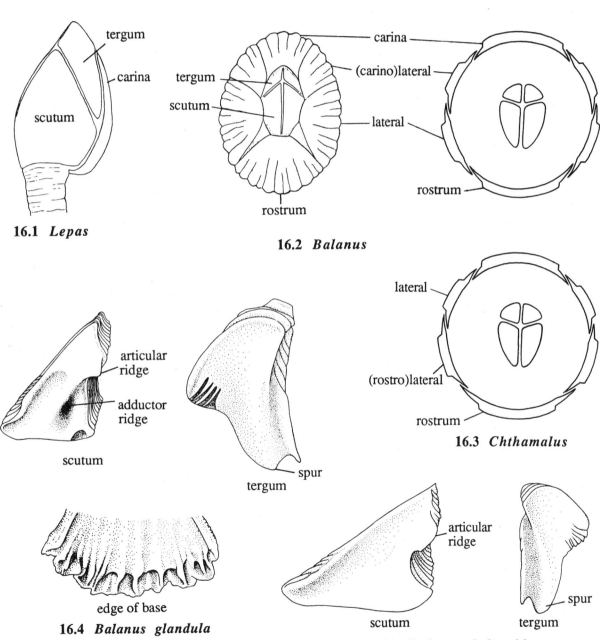

16.1 *Lepas*

16.2 *Balanus*

16.3 *Chthamalus*

16.4 *Balanus glandula*

16.5 *Semibalanus balanoides*

2b	Capitulum with 13 or 14 plates; carina with distinct spines arising from ridges; subtidal, rare	*Scalpellum columbianum*
3a	Capitulum with only 2 small, Y-shaped plates; attached to scyphozoan jellyfishes	*Alepas pacifica*
3b	Capitulum with 5 plates; attached to floating timbers, glass, floats, and similar objects, or secreting a firm, gelatinous float	4
4a	Capitulum flattened; attached to floating timbers and similar objects	5
4b	Capitulum nearly globular; secreting a firm, gelatinous float (as a rule, several to many individuals contribute to a common float)	*Lepas fascicularis*
5a	Tergum with a rather distinct notch on the side that borders the scutum	*Lepas pacifica*
5b	Tergum without a notch on the side that borders the scutum	6
6a	Plates with fine striations; with not more than 2 filamentous outgrowths originating at the base of each of the first cirri	*Lepas anatifera*
6b	Plates smooth; with 3 or more filamentous outgrowths originating at the base of each of the first cirri	*Lepas hilli*

Family Scalpellidae

Mitella polymerus Sowerby, 1833
Scalpellum columbianum Pilsbry, 1909

Family Lepadidae

Alepas pacifica Pilsbry, 1907
Conchoderma auritum (Linnaeus, 1758). Attached to shells of *Coronula* (order Balanomorpha) on the humpback whale, *Megaptera nodosa*, and sometimes to ship bottoms and floating objects.
Conchoderma virgatum (Spengler, 1790). Attached to shells of *Coronula* on the humpback whale, and sometimes to ship bottoms and floating objects; also occasionally attached to large copepods parasitic on whales and fishes.
Lepas anatifera Linnaeus, 1758
Lepas fascicularis Ellis & Solander, 1758
Lepas hilli (Leach, 1818)
Lepas pacifica Henry, 1940

Suborder Balanomorpha

The key does not include species of the family Coronulidae, found only on whales. These are described by Cornwall (1969).

1a	Rostrum overlapped by the wall plates on both sides of it; cover plates, when closed, fitting together in such a way that the lines between them intersect to form a cross; diameter, at base, not more than 1 cm (color generally gray or brownish gray; high intertidal)	*Chthamalus dalli*
1b	Rostrum overlapping the wall plates on both sides of it; cover plates, when closed, not fitting together in such a way that the lines between them intersect to form a cross; diameter, at base, often more than 1 cm, and in some species attaining 5 cm or more	2
2a	Tips of terga drawn out and curved in such a way that they form a beak	3
2b	Tips of terga not drawn out and curved in such a way that they form a beak	6
3a	Ridges on outside of wall narrow, closely spaced, and giving rise (especially in the lower half) to downward-pointing, fingerlike projections (this characteristic may be obscure on crowded or eroded specimens); base of shell not extensively calcified, and when the barnacle is knocked off the substratum, the base and some soft tissue generally remain on the substratum (common in mid- and low intertidal, and also in shallow subtidal; sometimes attaining a diameter of 4 cm or more)	*Semibalanus cariosus*

3b Ridges of wall, if present, not giving rise to fingerlike projections; base of shell extensively calcified, and when the barnacle is knocked off a hard substratum, a calcareous scar remains on the substratum, but the rest of the barnacle is intact   4
4a Diameter of shell commonly exceeding 5 cm, and sometimes exceeding 8 cm; exterior generally much eroded, so that the low ridges characteristic of the wall of young specimens become obscured (mostly subtidal, but also found on floating docks and in low intertidal on the open coast)   *Balanus nubilus*
4b Diameter of shell rarely attaining 5 cm, and usually much smaller than this; exterior generally not much eroded, so that low ridges, when present, persist   5
5a Tubes within wall plates with transverse septa, at least in the upper part of the wall (to see this feature, file the plates to expose the tubes, then blow out the powder); overlapping portions of plates not especially glossy; diameter rarely attaining 3 cm (subtidal)   *Balanus balanus*
5b Tubes within wall plates lacking transverse septa; overlapping portions of plates usually glossy; diameter not often more than 3 cm, but sometimes reaching 5 cm (mostly subtidal [sometimes encountered on floating docks and on rocks in tidal streams])   *Balanus rostratus*
6a Orifice small, generally only about one-fourth the diameter of the base; wall with ridges beginning at the edge of the orifice; subtidal, and almost always associated with hydrocorals of the genus *Allopora*   *Solidobalanus engbergi*
6b Orifice from one-third to more than one-half the diameter of the base; when there are ridges on the wall, these do not necessarily begin at the edge of the orifice; includes intertidal and subtidal species   7
7a Lines of contact between the terga and scuta approximately ∫∖-shaped   8
7b Lines of contact between the terga and scuta nearly straight or slightly sinuous, not ∫∖-shaped   9
8a Interior of base of shell with numerous centripetal ridges; interior surface of scutum with a prominent adductor ridge (this has a deep depression alongside it below the place where it meets the articular ridge); spur on tergum not as long as wide (mostly mid-and high intertidal; also on floating docks, where generally concentrated at or close to the water line; highly variable in general shape and in the extent to which the wall is ridged; crowded specimens tend to be tall and narrow and to lack ridges, whereas widely spaced specimens tend to be low and broad and to have prominent ridges)   *Balanus glandula* (fig. 16.4)
8b Interior of base of shell without centripetal ridges; interior surface of scutum without an obvious adductor ridge; spur on the tergum at least as long as wide (intertidal, with a wide vertical range, on rocks, wood, and sometimes on seaweed; not known to occur south of Washington)   *Semibalanus balanoides* (fig. 16.5)
9a Wall plates with internal tubes (these are exposed by filing the wall; they have transverse septa, at least in the upper part of the wall)   10
9b Wall plates without internal tubes (subtidal)   *Solidobalanus hesperius*
10a Exterior surface of scutum generally somewhat concave; wall with or without ridges (ridges are usually prominent only in young specimens); interior surface of scutum without an adductor ridge; spur of tergum wider than long (mostly subtidal; often found on ship bottoms and on floating docks; also low intertidal, and sometimes mid-intertidal, especially on relatively smooth boulders and cobble; not in brackish water)   *Balanus crenatus*
10b Exterior surface of scutum generally flat; wall usually without ridges; interior surface of scutum with an adductor ridge; spur of tergum longer than wide (restricted to brackish water; uncommon north of Oregon)   *Balanus improvisus*

Family Chthamalidae

Chthamalus dalli Pilsbry, 1916

Family Coronulidae

Coronula diadema (Linnaeus, 1767). On *Megaptera nodosa* (humpback whale).

Cryptolepas rachianecti Dall, 1872. On *Eschrichtius glaucus* (California gray whale); almost wholly embedded in the skin.
Xenobalanus globicipitis Steenstrup, 1851. On *Balaenoptera borealis* (sei whale), *Balaenoptera physalis* (common finback whale), and *Globicephala ventricosa* (blackfish, or pilot whale).

Family Archaeobalanidae

Semibalanus balanoides (Linnaeus, 1767) (fig. 16.5)
Semibalanus cariosus (Pallas, 1788)
Solidobalanus engbergi (Pilsbry, 1921)
Solidobalanus hesperius (Pilsbry, 1916)

Family Balanidae

Balanus balanus (Linnaeus, 1758)
Balanus crenatus Bruguière, 1789
Balanus glandula Darwin, 1854 (fig. 16.4)
Balanus improvisus Darwin, 1854
Balanus nubilus Darwin, 1854
Balanus rostratus Hoek, 1883

Order Rhizocephala

Boschma, H. 1962. Rhizocephala. Discovery Rep., 33:55-92.
------. 1970. Notes on Rhizocephala of the genus *Briarosaccus*, with the description of a new species. Proc. Nederl. Akad. Wetensch., C, 73:233-42.
Boschma, H., & E. Haynes. 1969. Occurrence of the rhizocephalan *Briarosaccus callosus* Boschma in the king crab *Paralithodes camtschatica* (Tilesius) in the northwest Pacific Ocean. Crustaceana, 16:97-8.
McMullen J. C., & H. T. Yoshihara. 1970. An incidence of parasitism of deepwater king crab, *Lithodes aequispina*, by the barnacle *Briarosaccus callosus*. J. Fish. Res. Bd. Canada, 27:818-21.
Reinhard, E. G. 1944. Rhizocephalan parasites of hermit crabs from the northwest Pacific. J. Wash. Acad. Sci., 34:49-58.
Reischman, P. G. 1959. Rhizocephala of the genus *Peltogaster* from the coast of the state of Washington and the Bering Sea. Proc. Nederl. Akad. Wetensch., C, 62:409-35.
Veillet, A. 1962. Sur la sexualité de *Sylon hippolytes* M. Sars, cirripède parasite de crevettes. C. r. Acad. Sci. Paris, 254:176-7.

Suborder Kentrogonida

The following species have either been reported from our region or may be expected to occur.

Family Clistosaccidae

Angulosaccus tenuis Reinhard, 1944. On the hermit crab *Parapagurus pilosimanus* subsp. *benedicti* ("*Pagurus armatus*"), collected in very deep water off the coast of Washington.
Clistosaccus paguri Lilljeborg, 1861. On the hermit crabs *Pagurus capillatus*, *P. dalli*, and *Labidochirus splendescens* (Bering Sea, Alaska).

Family Peltogastridae

Briarosaccus callosus Boschma, 1962. On the lithodid crabs *Lithodes aequispina*, *L. couesi*, and *Paralithodes camtschatica*.
Peltogaster boschmae Reinhard, 1944. On the hermit crab *Discorsopagurus schmitti*.
Peltogaster paguri Rathke, 1842. On hermit crabs of the genus *Pagurus*.
Peltogasterella gracilis (Boschma, 1927). On hermit crabs: *Discorsopagurus schmitti* and species of *Pagurus* and *Elassochirus*; *Peltogasterella socialis* Krüger, 1912 and *P. subterminalis* Reinhard, 1944 are synonyms.

Family Sacculinidae

Loxothylacus panopaei (Gissler, 1884). On the brachyuran crab *Lophopanopeus bellus* subsp. *bellus*.

Family Sylonidae

Sylon hippolytes M. Sars, 1870. On shrimps of the genera *Hippolyte*, *Heptacarpus*, *Spirontocaris*, *Sclerocrangon*, and *Pandalus*.

Suborder Akentrogonida

Family Akentrogonidae

Thompsonia sp. On the hermit crab *Discorsopagurus schmitti*.

Order Acrothoracica

The cirripedes of this order burrow into shells of other barnacles, and also into shells of molluscs and calcareous skeletons of corals. One species, *Trypetesa lateralis* Tomlinson, 1953, is abundant in central California, its habitat being gastropod shells that are occupied by hermit crabs. It should be expected in our region.

Tomlinson, J. T. 1959. A burrowing barnacle of the genus *Trypetesa* (order Acrothoracica). J. Wash. Acad. Sci., 43:373-81.
------. 1969. The burrowing barnacles (Cirripedia: order Acrothoracica). Bull. 296, U. S. Nat. Mus. v + 162. pp.

Order Ascothoracica

An undescribed species of *Dendrogaster* parasitizes the sea star *Mediaster aequalis*; another occurs in *Solaster stimpsoni*, *Dermasterias imbricata*, and *Crossaster papposus*. The paper of Grygier (1982), which deals with 3 species in sea stars dredged off the coast of southern California, has an extensive bibliography concerned with ascothoracicans.

Grygier, M. J. 1982. *Dendrogaster* (Crustacea: Ascothoracica) from California: sea-star parasites collected by the *Albatross*. Proc. Calif. Acad. Sci, 42:443-54.

17

PHYLUM ARTHROPODA: SUBPHYLUM CRUSTACEA: CLASS MALACOSTRACA: SUBCLASSES PHYLLOCARIDA, PERACARIDA (in part)

Subclass Phyllocarida

Order Leptostraca

Family Nebaliidae

In our area, *Nebalia pugettensis* (Clark, 1932) is abundant in the lower intertidal, and also subtidally. It prefers situations where algae and other organic detritus are decomposing. Clark originally described it as *Epinebalia pugettensis*, on the basis of the fact that the antenna 2 of the male is sickle-shaped instead of straight. The importance of this feature for separating the two genera is questionable.

Another species of *Nebalia*, in which the antenna 2 of the male is straight, also occurs in this region; it is evidently undescribed. The females of the 2 species are almost indistinguishable.

Cannon, H. G. 1960. Leptostraca. *In* Bronn, H. G., Klassen und Ordnungen des Tierreichs, 5, Abt. 1, Buch 4, Teil 1. 81 pp.
Clark, A. E. 1932. *Nebaliella caboti* n. sp., with observations on other Nebaliacea. Trans. Royal Soc. Canada, Sect. V, ser. 3, 26:217-35.

Subclass Peracarida

Order Mysidacea

Kendra L. Daly

The basic references for mysids of this region are those by Banner (1948a, 1948b, 1950), Tattersall (1932, 1933, 1951), Holmquist (1958, 1973, 1975, 1979, 1980, 1981a, 1981b, 1982) and Daly & Holmquist (1986). It will be advisable to consult descriptions and illustrations in order to confirm identifications made by using this key.

Banner, A. H. 1948a. A taxonomic study of the Mysidacea and Euphausiacea (Crustacea) of the northeastern Pacific. Part I. Mysidacea, from Family Lophogastridae through Tribe Erythropini. Trans. Royal Canad. Inst., 26:345-99.
------. 1948b. A taxonomic study of the Mysidacea and Euphausiacea (Crustacea) of the northeastern Pacific. Part II. Mysidacea, from Tribe Mysini through Subfamily Mysidellinae. Trans. Royal Canad. Inst., 27:65-125.
------. 1950. A taxonomic study of the Mysidacea and Euphausiacea (Crustacea) of the northeastern Pacific. Part III. Euphausiacea. Trans. Royal Canad. Inst., 28:1-63. (Includes keys and bibliography for both Mysidacea and Euphausiacea).
------. 1954a. New records of Mysidacea and Euphausiacea from the northeastern Pacific and adjacent areas. Pacific Sci., 8:125-39.
------. 1954b. A supplement to W. M. Tattersall's review of the Mysidacea of the United States National Museum. Proc. U. S. Nat. Mus., 103:575-83.

Daly, K. L., & C. Holmquist. 1986. A key to the Mysidacea of the Pacific Northwest. Can. J. Zool., 64:1201-10.
Gleye, L. G. 1981. *Acanthomysis nephrophthalma* and *Mysidella americana* (Mysidacea, Mysidae) along the coast of southern California. Crustaceana, 40:220-1.
Gordon, J. 1957. A bibliography of the order Mysidacea. Bull. Amer. Mus. Nat. Hist., 112:283-393.
Holmquist, D. 1958. On a new species of the genus *Mysis*, with some notes on *Mysis oculata* (O. Fabricius). Meddelelser om Grønland, 159:1-17.
------. 1973. Taxonomy, distribution and ecology of the three species *Neomysis intermedia* (Czerniavsky), *N. awatschensis* (Brandt) and *N. mercedis* Holmes (Crustacea, Mysidacea). Zool. Jahrb., Abt. Syst. Ökol. Geogr. Tiere, 100: 197-222.
------. 1975. A revision of the species *Archaeomysis grebnitzkii* Czerniavsky and *A. maculata* (Holmes) (Crustacea, Mysidacea). Zool. Jahrb., Abt. Syst. Ökol. Geogr. Tiere, 102:51-71.
------. 1979. *Mysis costata* Holmes, 1900, and its relations (Crustacea, Mysidacea). Zool. Jahrb., Abt. Syst. Ökol. Geogr. Tiere, 106:471-99. (Fig. 7, E should be labelled *Holmesimysis nudensis*).
------. 1980. *Xenacanthomysis*--a new genus for the species known as *Acanthomysis pseudomacropsis* (W. M. Tattersall, 1933) (Crustacea, Mysidacea). Zool. Jahrb., Abt. Syst. Ökol. Geogr. Tiere, 107:501-10.
------. 1981a. *Exacanthomysis* gen. nov., another detachment from the genus *Acanthomysis* Czerniavsky (Crustacea, Mysidacea). Zool. Jahrb., Abt. Syst. Ökol. Geogr. Tiere, 108:247-63.
------. 1981b. The genus *Acanthomysis* Czerniavsky, 1882 (Crustacea, Mysidacea). Zool. Jahrb., Abt. Syst. Ökol. Geogr. Tiere, 108:386-415.
------. 1982. Mysidacea (Crustacea) secured during investigations along the west coast of North America by the National Museum of Canada, 1955-1956, and some inferences drawn from the results. Zool. Jahrb., Abt. Syst. Ökol. Geogr. Tiere, 109:469-510.
Ii, N. 1964. Mysidae (Crustacea). *In* Fauna Japonica. Tokyo: Biogeographical Society of Japan, National Science Museum. x + 610 pp.
Mauchline, J. 1980. The biology of mysids and euphausiids. Adv. Mar. Biol., 18:1-681.
Mauchline, J., & M. Murano. 1977. World list of the Mysidacea, Crustacea. J. Tokyo Univ. Fish., 64:39-88.
Tattersall, W. M. 1932. Contributions to a knowledge of the Mysidacea of California. Univ. Calif. Publ. Zool., 37:301-47.
------. 1933. Euphausiacea and Mysidacea from western Canada. Contrib. Canad. Biol. Fish., 8:183-205.
------. 1951. A review of the Mysidacea of the United States National Museum. Bull. 201, U. S. Nat. Mus. x + 292 pp.
Tattersall, W. M., & O. S. Tattersall. 1951. The British Mysidacea. London: Ray Society. viii + 460 pp.

1a	Telson cleft	2
1b	Telson not cleft	6
2a	Outer margin of scale of antenna 2 without setae	3
2b	Outer margin of scale of antenna 2 with setae	4
3a	Outer margin of scale of antenna 2 without teeth, the scale terminating in a stout spine that projects beyond the truncated apex	*Archaeomysis grebnitzkii*
3b	Outer margin of scale of antenna 2 armed with 6 large teeth, the most distal tooth not beyond the apex of the scale	*Inusitatomysis insolita*
4a	Scale of antenna 2 about as long as the peduncle of antenna 2	5
4b	Scale of antenna 2 longer than the peduncle of antenna 2 (telson with spines along the entire length of its lateral margins)	*Mysis litoralis*
5a	Telson with spines along the entire length of the lateral margins; endopodite of thoracopod 3 thickened and stout, its carpopropodus with strong spines	*Heteromysis odontops*
5b	Telson with spines only on about the distal half of the lateral margins; endopodite of thoracopod 3 neither thickened nor with spines on the carpopropodus	*Mysidella americana*
6a	Carapace and abdomen with numerous long spines; cornea of each eye divided into 2 distinct portions; body somewhat expanded laterally	*Caesaromysis hispida*
6b	Carapace and abdomen without numerous long spines; cornea of eye not divided into 2 portions; body not expanded laterally	7

322 Phylum Arthropoda: Order Mysidacea

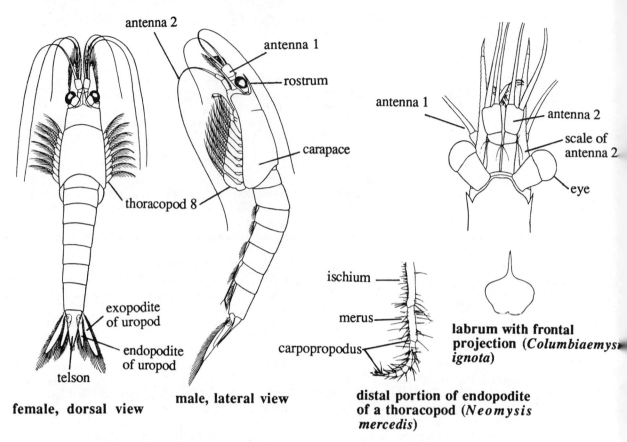

17.1 Terminology used in identification of mysids

7a Scale of antenna 2 terminating in a stout tooth or spine, and the outer margin of the scale not setose ... 8
7b Scale of antenna 2 without a terminal tooth or spine, and both margins of the scale setose ... 11
8a Telson without spines on the lateral margins (the truncated apex, however, has pairs of spines and a median pair of plumose setae) ... *Meterythrops robusta*
8b Telson with spines on the distal half to two-thirds of the lateral margins ... 9
9a Eyes well developed; carapace with a short, broad rostrum (apex of the telson truncated, with 2 pairs of spines and a median pair of plumose setae) ... *Holmesiella anomala*
9b Eyes rudimentary; carapace without a rostrum (its anterior margin is slightly curved, but there is no projection) ... 10
10a Eyes united to form a single large plate across the front of the carapace, the plate with a median notch and serrated anterolateral margins; lateral margins of the distal half of the telson with a few short spines; apex of the telson broad, with several stout spines (the outermost ones are the shortest) and a median pair of plumose setae ... *Pseudomma berkeleyi*
10b Eyes separate, each with a small anteromedian pointed extension; lateral margins of the distal two-thirds of the telson with many closely spaced spines; apex of the telson with closely spaced spines ... *Amblyops abbreviata*
11a Scale of antenna 2 with a pointed apex ... 12
11b Scale of antenna 2 with a rounded apex ... 14
12a Telson with 12-15 widely spaced spines on both lateral margins ... *Neomysis mercedis*
12b Telson with more than 15 spines on both lateral margins ... 13
13a Marginal spines of the telson shorter than the distance between their bases, thus appearing to be widely spaced ... *Neomysis rayii*

13b Marginal spines on the distal half of the telson longer than the distance between their bases, thus appearing to be closely spaced *Neomysis kadiakensis*
14a Posterior margin of the last abdominal segment with an acute mid-dorsal projection; telson with 4 stout spines at its apex 15
14b Posterior margin of the last abdominal segment almost straight, without a mid-dorsal projection; telson either with 2 stout spines at its apex, or without such spines 19
15a Only abdominal segment 6 with a mid-dorsal posterior projection 16
15b At least some abdominal segments other than 6 with a mid-dorsal posterior projection 17
16a All abdominal segments with at least 2 transverse dorsal folds *Holmesimysis costata*
16b Abdominal segments usually smooth, rarely with a single transverse dorsal fold on one or more of the last 3 segments or with 2 folds on segments 3 or 4 *Holmesimysis nuda*
17a Abdominal segments 4-6 with a mid-dorsal posterior projection; segments 5 and 6 also with a projection on both sides lateral to the midline (abdominal segment 3 sometimes with a mid-dorsal projection, and all abdominal segments with at least 2 transverse dorsal folds) *Holmesimysis sculpta*
17b Only abdominal segments 5 and 6 with a mid-dorsal posterior projection; segment 5 with very small projections on both sides lateral to the midline 18
18a Abdominal segments with at least 2 transverse dorsal folds; uropods with 3-5 spines on the inner margin of the statocyst *Holmesimysis sculptoides*
18b Abdominal segments smooth, rarely with 1 transverse dorsal fold on 1 or more of the last 3 segments; uropods with 4-8 spines on the inner margin of the statocyst *Holmesimysis nudensis*
19a Abdominal segments with transverse dorsal folds, segment 6 always with 3 pairs; lateral margins of the telson with several small spines between the larger spines; apex of telson with 4 spines (2 large, 2 small) 20
19b Abdominal segments smooth, without transverse dorsal folds; telson armature variable 21
20a Telson with a narrow distal portion; larger spines along margins of telson conspicuously increasing in size up to the narrowed distal portion; 12 small spines on the narrow distal portion of the telson *Exacanthomysis davisi*
20b Telson without a narrow distal portion; large spines along the lateral margins of the telson nearly equal; about 3 small spines on the distal portion of the telson *Exacanthomysis alaskensis*
21a Carapace with anterolateral corners rounded 22
21b Carapace with anterolateral corners projecting and acute or nearly acute 23
22a Spines along the lateral margins of the telson arranged in such a way that each 2 larger spines are separated by several smaller spines *Pacifacanthomysis nephrophthalma*
22b Spines along the lateral margins of the telson of nearly uniform size *Stilomysis grandis*
23a Anterior margin of carapace with a supraocular spine on each side *"Acanthomysis" columbiae*
23b Anterior margin of the carapace without supraocular spines 24
24a Telson with spines on only the distal half of its lateral margins (these spines short and stout) (some specimens have, in addition, a single spine on the proximal portion of the telson) *Proneomysis wailesi*
24b Telson with spines along the entire length of the lateral margins 25
25a Apex of telson broadly rounded, with many closely spaced spines of about equal size; scale of antenna 2 about as long as the peduncle of antenna 2 26
25b Apex of telson rounded or truncated, with 2 larger spines and a smaller median pair; scale of antenna 2 much longer than the peduncle of antenna 2 27
26a Telson about 3 times as long as its width at the base, its lateral margins with equal, widely spaced spines on the proximal portion, and with slightly unequal, more closely spaced spines on the middle portion; labrum with a small, acute frontal projection; inner flagellum of antenna 2 of male with small striated structures *Xenacanthomysis pseudomacropsis*
26b Telson about 2.5 times as long as its width at the base, its lateral margins with widely spaced spines on the proximal portion and a few closely spaced small spines between the larger spines on the distal portion; labrum without a frontal projection; inner flagellum of antenna 2 of male without striated structures *Alienacanthomysis macropsis*
27a Apex of telson with spines of about equal length; lateral margins of telson with closely spaced small spines of unequal length *Disacanthomysis dybowskii*

27b   Apex of telson with a pair of small spines between 2 much larger spines; lateral margins of telson with closely spaced spines of equal length   *Columbiaemysis ignota*

Suborder Mysina

Family Mysidae

"*Acanthomysis*" *columbiae* (W. M. Tattersall, 1933). Shallow water (5-7 m) over sandy bottoms.
Alienacanthomysis macropsis (W. M. Tattersall, 1932). In shallow water of bays and estuaries, among eelgrass or algae.
Amblyops abbreviata (M. Sars, 1868). Midwater plankton to epibenthos (150-1000 m).
Archaeomysis grebnitzkii Czerniavsky, 1882. Common on sandy beaches, in surf at low tide; may burrow in sand during the day.
Caesaromysis hispida Ortmann, 1893. Midwater plankton, coastal (50 to more than 2000 m); *C. vanclevei* Banner, 1948 is a synonym.
Columbiaemysis ignota Holmquist, 1982. Intertidal, on muddy and rocky bottoms with algae.
Disacanthomysis dybowskii (Derzhavin, 1913). Shallow water (0-21 m), among eelgrass or algae.
Exacanthomysis alaskensis (Banner, 1948). Shallow water (9-25 m) over clay or rocky bottoms that have growths of algae.
Exacanthomysis davisi (Banner, 1948). Littoral, in surface water and in swarms near shore (associated with algae and eelgrass).
Heteromysis odontops Walker, 1898. Midwater plankton.
Holmesiella anomala Ortmann, 1908. Plankton, coastal (0-900 m).
Holmesimysis costata (Holmes, 1900). Common intertidally among eelgrass and algae on sandy or rocky beaches; found near river mouths and in bays and lagoons, but generally restricted to higher salinities.
Holmesimysis nuda (Banner, 1948). May occur in moderate surf, in surface plankton, or in swarms near shores of clay, sand, or rock; associated with algae and eelgrass.
Holmesimysis nudensis Holmquist, 1979. Coastal; open water or near shore, including surf of beaches.
Holmesimysis sculpta (W. M. Tattersall, 1933). Littoral, shallow coastal waters; known only from a few localities in British Columbia.
Holmesimysis sculptoides Holmquist, 1979. Near shores with muddy, sandy, or rocky bottoms; associated with dense algal growths and eelgrass.
Inusitatomysis insolita Ii, 1940. Midwater plankton (150-200 m).
Meterythrops robusta S. I. Smith, 1879. Midwater plankton to epibenthos (50 to more than 200 m).
Mysidella americana Banner, 1948. Bottoms, nearshore to deep (30-600 m).
Mysis litoralis (Banner, 1948). Shallow (in eelgrass) to deep.
Neomysis kadiakensis Ortmann, 1908. In bays and inlets (0-100 m), but primarily coastal and avoiding water of low salinities.
Neomysis mercedis Holmes, 1897. Euryhaline; fresh water (Lake Washington) to brackish water; common near shore in areas of freshwater influence; will swarm in shallow water at head of bays and estuaries.
Neomysis rayii (Murdoch, 1884). Plankton (0-100 m); avoids lower salinites.
Pacifacanthomysis nephrophthalma (Banner, 1948). Coastal, from surface plankton to epibenthos (0-300 m).
Proneomysis wailesi W. M. Tattersall, 1933. Plankton, shallow to midwater (0-50 m).
Pseudomma berkeleyi W. M. Tattersall, 1933. Epibenthos (about 120 m).
Stilomysis grandis (Goës, 1868). Midwater plankton to epibenthos (about 220 m).
Xenacanthomysis pseudomacropsis (W. M. Tattersall, 1933). Plankton (0-175 m) of open waters of coastal areas.

Order Cumacea

Josephine F. L. Hart

The Cumacea of our region have been little studied. When Calman (1912) reported on the collections in the United States National Museum, there was not much material from Oregon, Washington, and British Columbia. Hart (1930) recorded 17 species from various localities around Vancouver Island. Zimmer (1936, 1943) reported 4 species from Puget Sound, and some that he described from other areas have since been found in the region covered by this manual. Wieser (1956, 1959) reported 3 species, and Lie (1968, 1969, 1971) recorded 19 species. Altogether, these papers account for about 30 species in the area between 47° and 50° north latitude. Other species have been collected but not yet dealt with in publications.

Due to the present state of our knowledge, keys to the genera would seem to be the most useful. Many species are widely distributed, and for this reason the key provided includes some genera which have not yet been recorded here but which may be expected because they have been found either to the north or to the south of our region. For those wishing more detail, Jones (1963) has an excellent general account of the characteristics of the Order Cumacea; papers by Given (1961, 1964) and Lomakina (1958) should also be consulted.

Two keys are required because of sexual dimorphism. As the males mature, various structural differences develop that distinguish them from females and immature individuals (fig. 17.2 and 17.3). The antenna 2 of males becomes very long, the tip of its flagellum reaching nearly to the posterior end of the body. Males may also have pleopods and larger, or additional, exopodites on the thoracic appendages. Females and immature individuals have a rudimentary antenna 2, no pleopods, and fewer and less well developed exopodites. Males of the genera *Lamprops*, *Mesolamprops*, and *Hemilamprops* can be segregated on the basis of the absence of pleopods or the number of pleopods, but females and immature specimens show no such consistent differences.

Barnard, J. L., & R. R. Given. 1961. Morphology and ecology of some sublittoral cumacean Crustacea of southern California. Pacific Nat., 2:153-65.
Calman, W. T. 1912. The Crustacea of the order Cumacea in the collections of the United States National Museum. Proc. U. S. Nat. Mus., 41:603-76.
Given, R. R. 1961. The cumacean fauna of the southern California continental shelf. No. 1. Family Leuconidae. Bull. So. Calif. Acad. Sci., 60:130-46.
------. 1964. The cumacean fauna of the southern California continental shelf. No. 2. The new family Mesolampropidae. Crustaceana, 7:284-92.
Hart, J. F. L. 1930. Some Cumacea of the Vancouver Island region. Contrib. Canad. Biol. Fish., n. s., 6:25-40.
Jones, N. S. 1963. The marine fauna of New Zealand: Crustaceans of the order Cumacea. Bull. New Zealand Dept. Sci. Industr. Res., no. 152. 80 pp.
------. 1969. The systematics and distribution of Cumacea from depths exceeding 200 meters. Galathea Rep., 10:99-180.
------. 1976. British Cumaceans. (Synopses of the British Fauna, new series, no. 7.) London: Linnean Society and Academic Press. 62 pp.
Lie, U. 1968. A quantitative study of benthic infauna in Puget Sound, Washington, U. S. A., in 1963-1964. Fiskerdir. Skr., Ser. Havunders., 14:229-556.
------. 1969. Cumacea from Puget Sound and off the northwestern coast of Washington, with descriptions of 2 new species. Crustaceana, 17:19-30.
------. 1971. Additional Cumacea from Washington, U. S. A., with description of a new species. Crustaceana, 21:23-6.
Lomakina, N. B. 1958. [Cumacean crustaceans (Cumacea) of the seas of the USSR.] Opredeliteli po Faune SSSR, 66. 301 pp. (in Russian).
Sars, G. O. 1899-1900. Cumacea. *In* An Account of the Crustacea of Norway, with Short Descriptions and Figures of All the Species, vol. 3. 115 pp.
Stebbing, T. R. R. 1913. Cumacea (Sympoda). Das Tierreich, 39. 210 pp.
Wieser, W. 1956. Factors influencing the choice of substratum in *Cumella vulgaris* Hart (Crustacea, Cumacea). Limnol. Oceanogr., 1:274-85.

------. 1959. Free-living nematodes and other small invertebrates of Puget Sound beaches. Univ. Wash. Publ. Biol., 19. x + 179 pp.
Zimmer, C. 1936. California Crustacea of the order Cumacea. Proc. U. S. Nat. Mus., 83:423-39.
------. 1943. Cumaceen des Stillen Ozeans. Arch. Naturgesch., n. F., 12:130-74.

Males

1a	With a distinct telson	2
1b	Without a telson	9
2a	Telson with 3 or more terminal spines	3
2b	Telson with 2 or no terminal spines	5
3a	Pleopods present	4
3b	Pleopods absent	*Lamprops*
4a	With 2 pairs of pleopods	*Mesolamprops*
4b	With 3 pairs of pleopods	*Hemilamprops*
5a	Telson with 2 terminal spines	6
5b	Telson without terminal spines	8
6a	Free pereonite 4, in dorsal view, longer than wide	*Diastylopsis*
6b	Free pereonite 4, in dorsal view, not longer than wide	7
7a	Telson short, bulbous, with not more than 1 pair of lateral spines	*Leptostylis*
7b	Telson long and tapered distally, usually with numerous lateral spines	*Diastylis*
8a	Telson short, rounded, and not reaching beyond the anal valves	*Colurostylis*
8b	Telson long, reaching beyond the anal valves and tapering to an acute and upturned tip	*Oxyurostylis*
9a	Pleopods present	10
9b	Pleopods absent	16
10a	With 2 pairs of pleopods	11
10b	With 3 or 5 pairs of pleopods	13
11a	Carapace acute anteriorly	*Leucon*
11b	Carapace truncate anteriorly	12
12a	Exopopodite of uropod shorter than the endopodite	*Eudorella*
12b	Exopodite of uropod longer than the endopodite	*Eudorellopsis*
13a	With 3 pairs of pleopods	*Leptocuma*
13b	With 5 pairs of pleopods	14
14a	Exopodite on pereopod 1 only	*Cyclaspis*
14b	Exopodites on pereopods 1-4	15
15a	Eye pigmented	*Vaunthompsonia*
15b	Eye not pigmented	*Bathycuma*
16a	Carapace bulbous and extending backwards over the free pereonites; anterolateral angle of the carapace usually rounded	*Campylaspis*
16b	Carapace not bulbous and not extending backwards over the free pereonites; anterolateral angle of the carapace acute	*Cumella*

Females and Immature Males

1a	With a distinct telson	2
1b	Without a telson	7
2a	Telson with 3 or more terminal spines	*Lamprops, Hemilamprops, Mesolamprops*
2b	Telson with 2 or no terminal spines	3
3a	Telson with 2 terminal spines	4
3b	Telson without terminal spines	6
4a	Free pereonite 4, in dorsal view, longer than wide	*Diastylopsis*
4b	Free pereonite 4, in dorsal view, not longer than wide	5
5a	Telson short, somewhat bulbous, with not more than 1 pair of lateral spines	*Leptostylis*
5b	Telson long, tapered distally, with several lateral spines	*Diastylis*
6a	Telson short, not reaching beyond the anal valves	*Colurostylis*

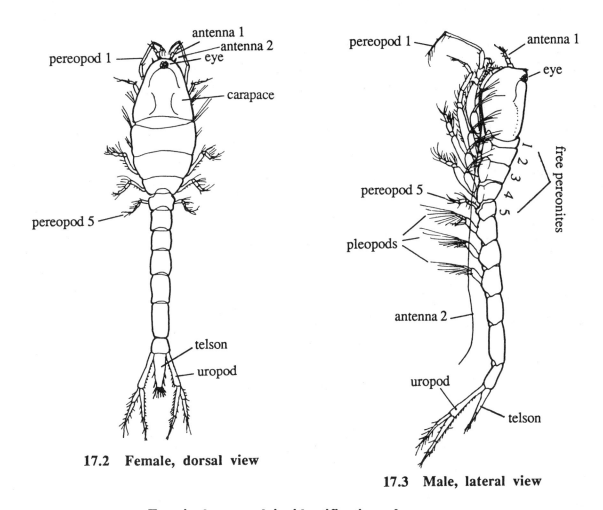

17.2 Female, dorsal view

17.3 Male, lateral view

Terminology used in identification of cumaceans

6b	Telson long, reaching beyond the anal valves and tapering to an acute and upturned tip	*Oxyurostylis*
7a	With a double row of spines or spinules on the dorsal crest of the carapace	8
7b	Without a double row of spines or spinules on the dorsal crest of the carapace	9
8a	Eye pigmented	*Vaunthompsonia*
8b	Eye not pigmented	*Bathycuma*
9a	Exopodites on pereopod 1 only	*Cyclaspis*
9b	Exopodites on pereopods 1 and 2	10
10a	Exopodites on at least pereopods 1 and 2	11
10b	Exopodites on pereopods 1-3	12
11a	Carapace bulbous, extending backwards over the free pereonites; anterolateral angle of the carapace usually rounded	*Campylaspis*
11b	Carapace not bulbous and not extending backwards over the free pereonites; anterolateral angle of the carapace acute	*Cumella*
12a	Carapace acute anteriorly	13
12b	Carapace truncate anteriorly	14
13a	Eye present	*Leptocuma*
13b	Eye absent	*Leucon*

14a Exopodite of uropod shorter than the endopodite ... *Eudorella*
14b Exopodite of uropod longer than the endopodite ... *Eudorellopsis*

Family Lampropidae

Hemilamprops gracilis Hart, 1930. Shallow subtidal.
Lamprops beringi Calman, 1912. Intertidal.
Lamprops carinata Hart, 1930. Shallow subtidal.
Lamprops fuscata G. O. Sars, 1865
Lamprops krasheninnikovi Derzhavin, 1926. Intertidal.
Lamprops serrata Hart, 1930. Subtidal.
Mesolamprops californiensis Zimmer, 1936. Shallow subtidal.

Family Diastylidae

Diastylis alaskensis Calman, 1912. Shallow subtidal.
Diastylis aspera Calman, 1912. Subtidal.
Diastylis bidentata Calman, 1912. Shallow subtidal.
Diastylis dalli Calman, 1912. Shallow subtidal.
Diastylis dawsoni S. I. Smith, 1880. Intertidal.
Diastylis nucella Calman, 1912. Shallow subtidal.
Diastylis paraspinulosa Zimmer, 1926. Subtidal.
Diastylis pellucida Hart, 1930. Subtidal.
Diastylis rathkei (Krøyer, 1841). Shallow subtidal.
Diastylis umatillensis Lie, 1971. Subtidal.
Leptostylis villosa G. O. Sars, 1869. Subtidal.

Family Leuconiidae

Eudorellopsis integra S. I. Smith, 1880. Subtidal.
Eudorella pacifica Hart, 1930. Subtidal.
Eudorella tridentata Hart, 1930. Shallow subtidal.
Leucon fulvus G. O. Sars, 1864. Shallow subtidal.
Leucon nasica Krøyer, 1841. Subtidal.

Family Nannastacidae

Campylaspis hartae Lie, 1969. Subtidal.
Campylaspis rubromaculata Lie, 1969. Subtidal.
Campylaspis rufa Hart, 1930. Subtidal.
Cumella vulgaris Hart, 1930. Intertidal; common in muddy sand in bays.

Family Bodotriidae

Vaunthompsonia pacifica Zimmer, 1943. Intertidal.

Family Colurostylidae

Colurostylis occidentalis Calman, 1912

Order Tanaidacea

Fee, A. R. 1927. The Isopoda of Departure Bay and vicinity, with descriptions of new species, variations, and colour notes. Contrib. Canad. Biol. Fish., n. s., 3:13-47.
Hatch, M. H. 1947. The Chelifera and Isopoda of Washington and adjacent regions. Univ. Wash. Publ. Biol., 10:155-274.
Lang, K. F. 1957. Tanaidacea from Canada and Alaska. Contr. Départ. Pêch. Québec, 52:1-54.
------. 1961. Further notes on *Pancolus californiensis* Richardson. Arkiv för Zoologi, ser. 2, 13:573-7.
------. 1973. Taxonomische und phylogenetische Untersuchungen über die Tanaidaceen (Crustacea). 8. Die Gattungen *Leptochelia* Dana, *Paratanais* Dana, *Heterotanais* G. O. Sars und *Nototanais* Richardson. Dazu einige Bemerkungen über die Monokonophora und ein Nachtrag. Zool. Scripta, 2:197-229.
Menzies, R. J. 1953. The apseudid Chelifera of the eastern tropical and north temperate Pacific Ocean. Bull. Mus. Comp. Zool. Harvard, 107:442-96.
Richardson, H. 1905a. A monograph on the isopods of North America. Bull. 54, U. S. Nat. Mus. lii + 727 pp.
------. 1905b. Description of a new genus of Isopoda belonging to the family Tanaidae and of a new species of *Tanais*, both from Monterey Bay, California. Proc. U. S. Nat. Mus., 28:367-70.
Sieg, J. 1980. Taxonomische Monographie der Tanaidae Dana, 1849. Abhandl. senckenb. naturforsch. Ges., 537:7-267.

1a Pleon with 5 free pleonites, plus the pleotelson 2
1b Pleon with 3 free pleonites, plus the pleotelson *Pancolus californiensis*
2a With 3 pairs of pleopods; uropods uniramous *Zeuxo normani*
2b With 5 pairs of pleopods; uropods biramous (but the exopodite may be very small) 3
3a Length of pereonite 1 less than one-third the length of pereonite 4; movable and stationary claws of the chelae slender, about equal, and lacking teeth *Pseudotanais oculatus*
3b Length of pereonite 1 at least half the length of pereonite 4; movable and stationary claws of the chelae either decidedly unequal or toothed 4
4a Endopodite of uropod with 5 articles; eyes present, pigmented; common intertidally on muddy substrata, especially where there are growths of *Ulva*, *Enteromorpha*, and other green algae, as well as diatoms *Leptochelia savignyi* (fig. 17.4)
4b Endopodite of uropod with 2 articles; eyes absent; subtidal, at depths of about 10-20 m *Leptognathia gracilis*

Suborder Tanaidomorpha

Family Paratanaidae

Leptochelia savignyi (Krøyer, 1842) (fig. 17.4)
Leptognathia gracilis (Krøyer, 1842)

Family Pseudotanaidae

Pseudotanais oculatus Hansen, 1913

Family Tanaidae

Pancolus californiensis Richardson, 1905
Zeuxo normani (Richardson, 1905)

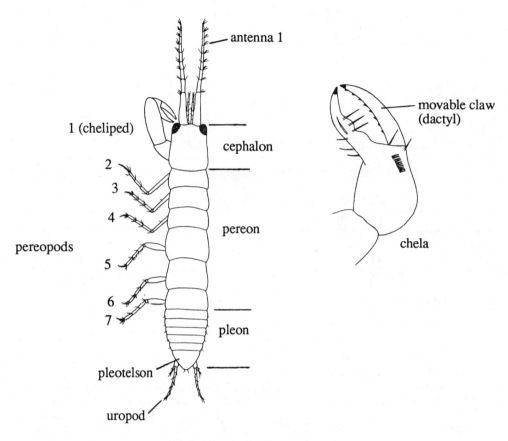

17.4 *Leptochelia savignyi*

Order Isopoda

Fee, A. R. 1926. The Isopoda of Departure Bay and vicinity, with descriptions of new species, and color notes. Contr. Can. Biol. Fish. (n. s.), 3:13-46.
Hatch, M. H. 1947. The Chelifera and Isopoda of Washington and adjacent regions. Univ. Wash. Publ. Biol., 10:155-274.
Miller, M. A. 1968. Isopoda and Tanaidacea from buoys in coastal waters of the continental United States, Hawaii, and the Bahamas (Crustacea). Proc. U. S. Nat. Mus., 125:1-53.
Richardson, H. 1905. A monograph on the isopods of North America. Bull. 54, U. S. Nat. Mus. liii + 727 pp.
Schultz, G. A. 1969. How to Know the Marine Isopod Crustaceans. Dubuque, Iowa: W. C. Brown Co. vii + 359 pp.
Van Name, W. G. 1936. American land and fresh-water isopod Crustacea. Bull. Amer. Mus. Nat. Hist., 71:1-535.

1a Adults parasitic on other Crustacea (Cirripedia, Mysidacea, Decapoda); females generally grotesque, often asymmetrical, with most appendages reduced or absent; males small, symmetrical (a male is usually associated with each female)   Suborder Epicaridea
1b Adults free-living, parasitic on fishes, or commensal with other isopods; females and males recognizable as normal isopods, even if specialized for clinging to fishes   2
2a Uropods ventral to the pleotelson or decidedly lateral to the pleotelson   3
2b Uropods terminal, close to the midline   5
3a Uropods ventral, forming a valvelike covering over the pleopods   Suborder Valvifera

3b Uropods lateral, flattened, and, together with the pleotelson, forming a caudal fan (figs. 17.5 and 17.6) 4
4a Body at least 7 times as long as wide, nearly cylindrical; exopodites of uropods arching over the pleotelson; pereopod 1 larger than the others and subchelate (pereopods 2 and 3 may also be subchelate) .... Suborder Anthuridea
4b Body usually not more than 5 times as long as wide; exopodites of uropods not arching over the pleotelson; pereopod 1 neither especially enlarged nor subchelate (except in the genus *Tecticeps*, in which it is subchelate) .... Suborder Flabellifera
5a Body slender, at least 7 times as long as wide; pleon with 2 free pleonites plus the pleotelson, these 3 units being nearly equal; pereopod 1 subchelate; generally under 2 mm long; living between sand grains .... Suborder Microcerberidea
5b Body usually robust, much less than 7 times as long as wide; pleon not consisting of 3 nearly equal units (if there are 3, the pleotelson is much larger than the free pleonites); pereopod 1 generally not subchelate (but sometimes subchelate); mostly more than 2 mm long (but some species may be smaller); in various habitats 6
6a Aquatic; pleon generally with not more than 2 free pleonites (sometimes 3) plus the pleotelson; both antenna 1 and antenna 2 well developed .... Suborder Asellota
6b Essentially terrestrial, although certain species are restricted to moist situations near the high tide line; pleon generally with 5 free pleonites plus the pleotelson; antenna 1 rudimentary, with only 2 or 3 articles .... Suborder Oniscoidea

Suborder Microcerberidea

Family Microcerberidae

The members of this group are small and occur typically in substrata of sand and fine gravel. Only 1 species, *Microcerberus abbotti* Lang (1961), has been described from the Pacific coast. Originally found in central California, it has since been reported from southern California. A subspecies found on San Juan Island, Washington, was named *M. abbotti* subsp. *juani* by Coineau & Delamare Debouteville (1967).

Coineau, N., & C. Delamare Debouteville. 1967. Étude des microcerbérides (Crustacés, Isopoda) de la côte Pacifique des États-Unis. Ire partie:systématique. Bull. Mus. Nat. Hist. Nat., Sér. 2, 39:955-64.

Lang, K. 1961. Contributions to the knowledge of the genus *Microcerberus* Karaman (Crustacea: Isopoda) with a description of a new species from the central Californian coast. Arkiv för Zoologi, ser. 2, 13:493-510.

Suborder Anthuridea

Family Anthuridae

Two species, *Cyathura carinata* (Krøyer, 1849) and *Haliophasma geminata* Menzies & Barnard, 1959, have been reported for the region. Some other species of the families Anthuridae and Paranthuridae, described by Menzies (1951) and Menzies & Barnard (1959) from California, should be expected. These small isopods, generally under 3 mm long, often occur in deposits of sand and other gritty material that accumulates on holdfasts of kelps, and in similar situations.

Menzies, R. J. 1951. New marine isopods, chiefly from northern California, with notes on related forms. Proc. U. S. Nat. Mus., 101:105-56.

Menzies, R. J., & J. L. Barnard. 1959. Marine Isopoda on coastal shelf bottoms of southern California: systematics and ecology. Pacific Nat., 1:3-35.

Suborder Flabellifera

Brusca, R. C. 1981. A monograph on the Isopoda Cymothoidae (Crustacea) of the eastern Pacific. Zool. J. Linnean Soc., 73:117-79.

Hoestlandt, H. 1973. Étude systématique de trois espèces Pacifiques nordaméricaines du genre *Gnorimosphaeroma* Menzies (Isopodes Flabellifères). I. Considérations générales et systématique. Arch. Zool. Expér. Gén., 114:349-95.

Kussakin, O. G. 1979. [Marine and brackish-water Isopoda of the cold and temperate waters of the Northern Hemisphere. Suborder Flabellifera.] Opredeliteli po Faune SSSR, 122:1-470 (in Russian).

Menzies, R. J. 1954. A review of the systematics and ecology of the genus "*Exosphaeroma*," with the description of a new genus, a new species, and a new subspecies (Crustacea, Isopoda, Sphaeromidae). Amer. Mus. Novit., no. 1683. 24 pp.

------. 1957. The marine borer family Limnoriidae (Crustacea, Isopoda). Bull. Mar. Sci. Gulf Carib., 7:101-200.

Riegel, J. A. 1959. A revision in the sphaeromid genus *Gnorimosphaeroma* Menzies (Crustacea: Isopoda) on the basis of morphological, physiological and ecological studies on its "subspecies." Biol. Bull., 117:154-62.

1a	Pleon with 4 or 5 visible free pleonites, plus the pleotelson	2
1b	Pleon with not more than 3 visible free or partly free pleonites, plus the pleotelson (pleonite 1 is hidden beneath pereonite 7)	15
2a	Uropod with both rami well developed, these usually lamellar in form; not capable of rolling up into a ball; not boring in wood or kelps	5
2b	Uropod with a rudimentary and more or less clawlike exopodite; capable of rolling up into a ball; boring in wood or kelps	3
3a	Boring in holdfasts of kelps; incisor process of mandibles with neither a rasplike nor a filelike specialization	*Limnoria (Phycolimnoria) algarum* (fig. 17.8)
3b	Boring in wood; incisor process of right mandible with filelike ridges, that of the left mandible with scales that form a rasp	4
4a	Dorsal surface of pleotelson without tubercles, but with a slight median ridge that bifurcates posteriorly	*Limnoria (Limnoria) lignorum* (fig. 17.9)
4b	Dorsal surface of pleotelson with 3 small tubercles	*Limnoria (Limnoria) tripunctata* (fig. 17.10)
5a	None of the pereopods with large, hooklike dactyls for attachment to fishes	6
5b	Pereopods 1-3, or all pereopods, with large, hooklike dactyls for attachment to fishes	9
6a	Posterior border of telson fringed by about 26 short, simple setae; anterior edge of cephalon with an inconspicuous rostrum; on rocky shores, among mussels, etc.	*Cirolana harfordi* (fig. 17.6)
6b	Posterior border of telson fringed by long, plumose setae; anterior edge of cephalon with an obvious spatulate rostrum; on sandy shores	7
7a	Telson broadly rounded or almost truncate posteriorly	*Excirolana linguifrons*
7b	Telson somewhat angular posteriorly	8
8a	Tip of angular portion of telson rather sharply pointed	*Excirolana kincaidi*
8b	Tip of angular portion of telson rounded	*Excirolana vancouverensis*
9a	Pereopods 1-3 prehensile, with hooklike dactyls for attachment to fishes (but some species are free-living part of the time)	10
9b	All pereopods prehensile, with hooklike dactyls for attachment to fishes	14
10a	Flagellum of antenna 1 with numerous articles; propodus of prehensile pereopods not expanded	*Aega symmetrica*
10b	Flagellum of antenna 1 with only 4 to 6 articles; propodus of prehensile pereopods expanded along the inner margin	11
11a	Cephalon drawn out into a spatulate rostrum, this flanked on both sides by a small projection	*Rocinela tridens*
11b	Cephalon drawn out into a more or less triangular rostrum, but the tip of this may be truncated	12

Phylum Arthropoda: Order Isopoda 333

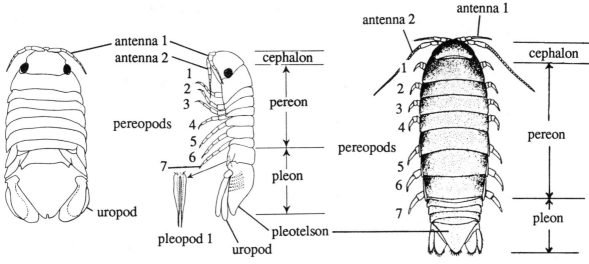

17.5 General anatomy of an isopod of the Suborder Flabellifera

17.6 *Cirolana harfordi* (Suborder Flabellifera)

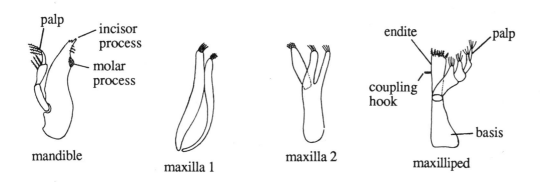

17.7 Mouthparts of an isopod

12a	Tip of rostrum pointed	*Rocinela belliceps*
12b	Tip of rostrum truncated	13
13a	Propodus of prehensile pereopods about as long as the dactyl	*Rocinela propodialis*
13b	Propodus of prehensile pereopods about half as long as the dactyl	*Rocinela angustata*
14a	Distance between the eyes equal to about 3 times the diameter of each eye; antenna 2 with 10 or 11 articles; anterior margin of cephalon broadly rounded or nearly truncate; pleotelson of adult female almost twice as wide as long	*Lironeca vulgaris*
14b	Distance between the eyes equal to about 1.5 times the diameter of each eye; antenna 2 with 8 or 9 articles; anterior margin of cephalon with a blunt projection; pleotelson of adult female about as wide as long	*Lironeca californica*
15a	Not capable of rolling up into a ball; pereopod 1 subchelate in both sexes, pereopod 2 subchelate in the male	*Tecticeps pugettensis*
15b	Capable of rolling up into a ball; none of the pereopods subchelate in either sex	16
16a	Pleotelson either trilobed or deeply notched at the apex	17
16b	Pleotelson neither trilobed nor deeply notched at the apex	20
17a	Pleotelson trilobed at the apex	*Cymodoce japonica*
17b	Pleotelson deeply notched at the apex	18

18a Dorsal surface of pleotelson smooth, without ridges or rows of fused tubercles
　　　　　　　　　　　　　　　　　　　　　　　　　　　　　　　　Dynamenella glabra
18b Dorsal surface of pleotelson with 4 ridges, or 4 rows of fused tubercles　　19
19a Dorsal surface of pleotelson with 4 ridges　　*Dynamenella benedicti*
19b Dorsal surface of pleotelson with 4 rows of fused tubercles　　*Dynamenella sheareri*
20a Two or 3 of the visible pleonites (these are free laterally, but not along the midline) reaching the lateral margins; pleopods without respiratory folds　　21
20b Only 1 pleonite reaching the lateral margins; pleopods 4 and 5 with respiratory folds　　23
21a Three pleonites reaching the lateral margins　　22
21b Two pleonites reaching the lateral margins (the third visible pleonite is "crowded out") (in decidedly brackish or nearly fresh water)　　*Gnorimosphaeroma insulare* (fig. 17.11)
22a Basal articles of antennae 1 touching each other on the midline　　*Gnorimosphaeroma noblei*
22b Basal articles of antennae 1 distinctly separated from one another by a short rostrum (common intertidal species; usually under rocks, but also found on floats and pilings and in various other situations)　　*Gnorimosphaeroma oregonense* (fig. 17.12)
23a Pleotelson triangular, its length decidedly more than one-third the total length of the body; dorsal surface with prominent tubercles　　*Exosphaeroma amplicauda*
23b Pleotelson less than one-third the total length of the body, its posterior portion rounded, slightly truncate, or expanded into a nearly rhomboidal projection　　24
24a Posterior portion of the pleotelson expanded into a nearly rhomboidal projection
　　　　　　　　　　　　　　　　　　　　　　　　　　　　　　　　Exosphaeroma rhomburum
24b Posterior tip of pleotelson rounded or slightly truncated　　25
25a Posterior tip of pleotelson rounded　　*Exosphaeroma crenulatum*
25b Posterior tip of pleotelson slightly truncated　　*Exosphaeroma inornata*

Family Aegidae

Aega symmetrica Richardson, 1905
Rocinela angustata Richardson, 1898. On halibut and *Raja binoculata*.
Rocinela belliceps (Stimpson, 1864). On *Hydrolagus colliei*, species of *Raja*, cod, rockfish, and halibut.
Rocinela propodialis Richardson, 1905. On *Raja binoculata*, halibut, rockfish, and other fishes.
Rocinela tridens Hatch, 1947. On various fishes.

Family Cirolanidae

Cirolana harfordi Lockington, 1877 (fig. 17.6)
Excirolana kincaidi (Hatch, 1947)
Excirolana linguifrons (Richardson, 1899)
Excirolana vancouverensis (Fee, 1926)

Family Cymothoidae

Cymodoce japonica Richardson, 1906
Lironeca californica Schiödte & Meinert, 1883-84
Lironeca vulgaris Stimpson, 1857

Family Limnoriidae

Limnoria (Phycolimnoria) algarum Menzies, 1956 (fig. 17.8)
Limnoria (Limnoria) lignorum Rathke, 1799 (fig. 17.9)
Limnoria (Limnoria) tripunctata Menzies, 1951 (fig. 17.10)

Phylum Arthropoda: Order Isopoda 335

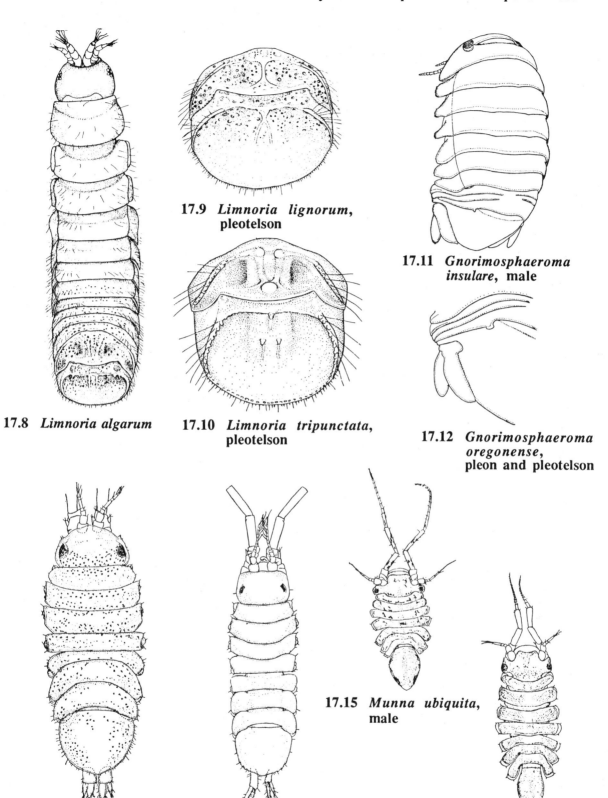

17.8 *Limnoria algarum*
17.9 *Limnoria lignorum*, pleotelson
17.10 *Limnoria tripunctata*, pleotelson
17.11 *Gnorimosphaeroma insulare*, male
17.12 *Gnorimosphaeroma oregonense*, pleon and pleotelson
17.13 *Ianiropsis magnocula*, female
17.14 *Ianiropsis tridens*, male
17.15 *Munna ubiquita*, male
17.16 *Munna chromatocephala*, male

Family Sphaeromatidae

Dynamenella benedicti (Richardson, 1899)
Dynamenella glabra (Richardson, 1899)
Dynamenella sheareri Hatch, 1947
Exosphaeroma amplicauda (Stimpson, 1857)
Exosphaeroma crenulatum Richardson, 1902
Exosphaeroma inornata Dow, 1958. *E. media* George & Strömberg, 1968 is a synonym.
Exosphaeroma rhomburum (Richardson, 1899)
Gnorimosphaeroma insulare (Van Name, 1940) (fig. 17.11). *G. luteum* Menzies, 1954 is a synonym.
Gnorimosphaeroma noblei Menzies, 1954
Gnorimosphaeroma oregonense Dana, 1854-55 (fig. 17.12)
Tecticeps pugettensis Hatch, 1947

Suborder Asellota

George, R. Y., & J.-O. Strömberg. 1968. Some new species and new records of marine isopods from San Juan Archipelago, Washington, U. S. A. Crustaceana, 14:225-54.

Menzies, R. J. 1952. Some marine asellote isopods from northern California, with descriptions of nine new species. Proc. U. S. Nat. Mus., 102:117-59.

------. 1951. New marine isopods, chiefly from northern California, with notes on related forms. Proc. U. S. Nat. Mus., 101:105-56.

Wilson, G. D. 1982. Two new natatory asellote isopods (Crustacea) from the San Juan Archipelago, *Baeonectes improvisus* n. gen., n. sp. and *Acanthamunnopsis milleri* n. sp., with a revised description of *A. hystrix* Schultz. Can. J. Zool., 60:3332-43.

1a	Pereonites 5-7 not fused together; pereopods 5-7 ambulatory, not obviously broadened, and without long, plumose setae	2
1b	Pereonites 5-7 fused together dorsally (only their lateral margins are distinct); pereopods 5-7 modified for swimming, the carpus and propodus broadened, paddlelike, and with rows of long, plumose setae	13
2a	Uropods minute, without a peduncle, but almost always with both an endopodite and an exopodite	9
2b	Uropods generally either with a peduncle or uropods absent (if the uropods consist of a single article, the body as a whole is either narrow and with nearly parallel margins, or there are posterolateral processes on the pleotelson)	3
3a	Articles 1 to 5 of palp of maxilliped all narrow and of about equal width, less than half as wide as the endite	*Jaeropsis setosa*
3b	Articles 1 to 3 of palp of maxilliped conspicuously wider than articles 4 to 5, at least as wide as the endite	4
4a	Posterior margin of propodus of pereopod 1 toothed for about one-third of its length	5
4b	Posterior margin of propodus of pereopod 1 not toothed	6
5a	Rostrum of cephalon elongated and pointed; associated with the sea star *Solaster stimpsoni*	*Janiralata solasteri*
5b	Rostrum of cephalon short, triangular; apparently free-living	*Janiralata occidentalis*
6a	Pleotelson with 3 spinelike serrations on each side	*Ianiropsis tridens* (fig. 17.14)
6b	Pleotelson with more than 3 spinelike serrations on each side	7
7a	Pleotelson as long as, or longer than, wide	8
7b	Pleotelson much wider than long	*Ianiropsis analoga*
8a	Length of uropods exceeding half the length of the pleotelson	*Ianiropsis kincaidi*
8b	Length of uropods not exceeding half the length of the pleotelson	*Ianiropsis magnocula* (fig. 17.13)
9a	Cephalon wide, not enclosed by lateral portions of pereonite 1; eyes well developed, with numerous ommatidia	10
9b	Cephalon small and narrow, enclosed by lateral portions of pereonite 1; eyes usually absent	

(but if present, separate ommatidia are not recognizable)   12
10a   Uropods small and leaflike; pleotelson with lateral serrations
                                                    Munna (*Uromunna*) *ubiquita* (fig. 17.15)
10b   Uropods not leaflike (rounded in transverse section); pleotelson without lateral serrations   11
11a   Uropods with at least 1 large, medially recurved apical spine; maxilliped with 3 coupling hooks; ventral surface of operculum of female with 2 longitudinal rows of setae
                                                      Munna (*Neomunna*) *chromatocephala* (fig. 17.16)
11b   Uropods without recurved apical spines; maxilliped with 2 coupling hooks; ventral surface of operculum of female with numerous setae, these not in 2 rows   *Munna* (*Munna*) *fernaldi*
12a   Eyes absent; epimera pointed and clearly visible in dorsal view   *Pleurogonium rubicundum*
12b   Eyes present; epimera rounded and hardly visible in dorsal view   *Munnogonium tillerae*
13a   Planktonic; body translucent, elongate, and with spines on the dorsal surface; anterior pereopods much longer than the body   *Acanthamunnopsis milleri*
13b   Benthic; body opaque, more or less oval in outline, with some pigment on the edges of the segments but without spines on the dorsal surface; anterior pereopods only slightly longer than the body   *Baeonectes improvisus*

Family Eurycopidae

Acanthamunnopsis milleri Wilson, 1982

Family Jaeropsididae

Jaeropsis setosa George & Strömberg, 1968
Janiralata occidentalis (Walker, 1898)
Janiralata solasteri (Hatch, 1947)

Family Janiridae

Ianiropsis analoga Menzies, 1952
Ianiropsis kincaidi Richardson, 1904
Ianiropsis magnocula Menzies, 1952 (fig. 17.13)
Ianiropsis tridens Menzies, 1952 (fig. 17.14)

Family Munnidae

Munna (*Munna*) *fernaldi* George & Strömberg, 1968
Munna (*Neomunna*) *chromatocephala* Menzies, 1952 (fig. 17.16)
Munna (*Uromunna*) *ubiquita* Menzies, 1952 (fig. 17.15)
Munnogonium tillerae (Menzies & Barnard, 1959). *M. waldronense* George & Strömberg, 1968 is a synonym.

Family Munnopsidae

Baeonectes improvisus Wilson, 1982

Family Pleurogoniidae

Pleurogonium rubicundum G. O. Sars, 1863

Suborder Valvifera

Menzies, R. J. 1950. The taxonomy, ecology, and distribution of northern California isopods of the genus *Idothea* with the description of a new species. Wasmann J. Biol., 8:155-95.
Menzies, R. J., & M. A. Miller. 1972. Systematics and zoogeography of the genus *Synidotea* (Crustacea: Isopoda). Smithsonian Contr. Zool., no. 102. 33 pp.
Rafi, F. 1972. *Idotea obscura*, a new species of Idoteidae (Isopoda, Valvifera) from the North American Pacific coast. Can. J. Zool, 50:781-6.

1a Body nearly cylindrical; pereon with only 6 free pereonites (pereonite 1 is fused to the cephalon); pereopods 1-4 smaller and much more setose than pereopods 5-7, and not used for locomotion .... *Idarcturus hedgpethi*
1b Body rather definitely flattened dorsoventrally; with 7 free pereonites; all pereopods similar 2
2a Lateral margins of cephalon with a deep incision; eyes decidedly dorsal; in brackish water and fresh water .... *Saduria entomon*
2b Lateral margins of cephalon without a deep incision; eyes lateral; mostly marine (a few species, however, occur in situations where the salinity is slightly reduced) 3
3a Pleon consisting entirely of a pleotelson, although a pair of lateral incisions indicate the posterior limits of 1 partly free pleonite 4
3b Pleon consisting of 2 free pleonites plus the pleotelson, which has a pair of lateral incisions indicating the posterior limits of another partly free pleonite 9
4a Posterior margin of pleotelson notched 5
4b Posterior margin of pleotelson tapering to a rounded tip .... *Synidotea nodulosa*
5a Anterior margin of cephalon either with a pair of distinct spines or divided by notches into sharp-tipped projections 6
5b Anterior margin of cephalon neither with a pair of distinct spines nor divided by notches into sharp-tipped projections 8
6a Pleon decidedly longer than wide; dorsal surface of each pereonite with 3 tubercles, one of which is on the midline (the cephalon also has 3 tubercles, plus a pair of spines close to the midline at the anterior margin) .... *Synidotea pettiboneae*
6b Pleon about as wide as long; dorsal surface of each pereonite not with 3 tubercles, but with either a slight transverse ridge or with several tubercles, none of which is on the midline 7
7a Dorsal surface of each pereonite with a slight but distinct transverse ridge (on most pereonites, this is near the posterior border), but without tubercles; anterolateral corners of the cephalon not extending beyond the two more nearly median projections of the cephalon .... *Synidotea bicuspida*
7b Dorsal surface of each pereonite with several tubercles, all of which are lateral to the midline; anterolateral corners of the cephalon prominent, extending considerably beyond the 2 more nearly median projections of the cephalon .... *Synidotea ritteri*
8a Pereonite 4 definitely wider than the preceding and succeeding pereonites .... *Synidotea nebulosa*
8b Pereonites 2-7 all about the same width .... *Synidotea angulata*
9a Palp of maxilliped consisting of 4 articles 10
9b Palp of maxilliped consisting of 5 articles 13
10a Posteriormost portion of pleotelson either angular or coming to a distinct projection, definitely not concave 11
10b Posteriormost portion of pleotelson concave, but with a slight median projection .... *Idotea (Idotea) rufescens*
11a Only the coxal plate of pereonite 7 reaching the posterior limit of the pereonite .... *Idotea (Idotea) fewkesi* (fig. 17.17)
11b Coxal plates of pereonites 6 and 7 reaching the posterior limits of these pereonites 12
12a Posterior portion of pleotelson angular, but the tip not extended as a distinct projection; lateral margins of pleotelson not concave .... *Idotea (Idotea) urotoma* (fig. 17.18)
12b Posterior portion of pleotelson tipped with a distinct projection; lateral margins of pleotelson often concave (the margins between the tip and the posterolateral corners may also be concave) .... *Idotea (Idotea) ochotensis*
13a Posterior margin of pleotelson concave .... *Idotea (Pentidotea) resecata* (fig. 17.21)

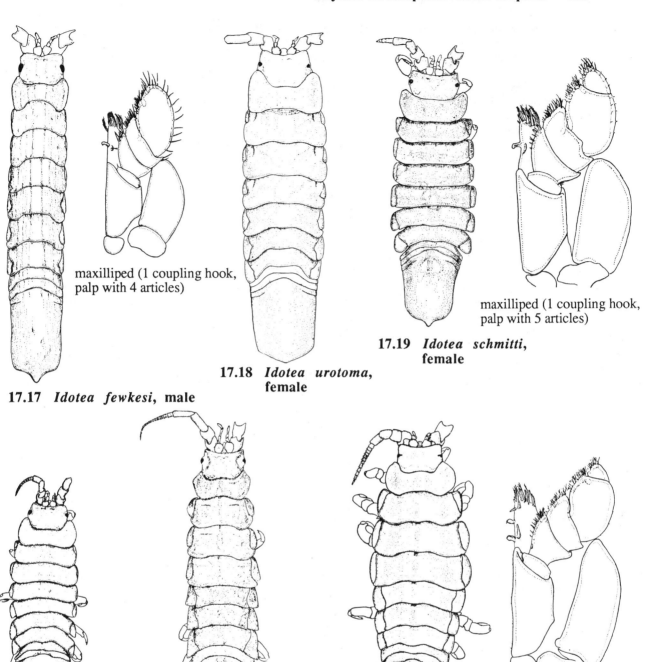

17.17 *Idotea fewkesi*, male

17.18 *Idotea urotoma*, female

17.19 *Idotea schmitti*, female

17.20 *Idotea wosnesenskii*, male

17.21 *Idotea resecata*, male

17.22 *Idotea stenops*, male

13b	Posterior margin of pleotelson convex, often with an apical projection	14
14a	Posterolateral margin of coxal plate of pereonite 7 acute	15
14b	Posterolateral margin of coxal plate of pereonite 7 bluntly rounded	*Idotea (Pentidotea) schmitti* (fig. 17.19)
15a	Pereonite 1 only slightly wider (not more one-tenth) than the cephalon; body slender, all pereonites about the same width	16
15b	Pereonite 1 decidedly wider (at least one-tenth) than the cephalon; pereonites near the middle of the body wider than the more anterior and posterior pereonites	17
16a	Frontal process of cephalon (visible in dorsal view) broadly triangular, not extending forward beyond the tip of the frontal lamina (this plate lies between the bases of the antennae 2 and is visible in ventral view)	male *Idotea (Pentidotea) montereyensis*
16b	Frontal process slender, sharp-tipped, extending forward considerably farther than the tip of the frontal lamina	*Idotea (Pentidotea) kirchanskii*
17a	Width of pleotelson about equal to that of the cephalon (lateral margins of pleotelson straight or slightly concave; blunt apical projection of pleotelson conspicuous)	*Idotea (Pentidotea) aculeata*
17b	Pleotelson wider than the cephalon	18
18a	Pleonite 1 decidedly narrower laterally than along the midline; eyes kidney-shaped (common intertidal species)	*Idotea (Pentidotea) wosnesenskii* (fig. 17.20)
18b	Pleonite 1 not significantly narrower laterally than along the midline; eyes not kidney-shaped	19
19a	Length less than 2 cm; apex of frontal process without a median notch; eyes almost circular; apical projection of pleotelson inconspicuous	female *Idotea (Pentidotea) montereyensis*
19b	Length up to about 5.5 cm; apex of frontal process with a median notch; eyes distinctly elongated; apical projection of pleotelson prominent	*Idotea (Pentidotea) stenops* (fig. 17.22)

One species, marked with an asterisk, is not in the key.

Family Arcturidae

Idarcturus hedgpethi Menzies, 1951

Family Idoteidae

Idotea (Idotea) fewkesi Richardson, 1905 (fig. 17.17)
Idotea (Idotea) ochotensis Brandt, 1851
Idotea (Idotea) rufescens Fee, 1926
Idotea (Idotea) urotoma Stimpson, 1864 (fig. 17.18)
Idotea (Pentidotea) aculeata Stafford, 1913
Idotea (Pentidotea) kirchanskii Miller & Lee, 1970
Idotea (Pentidotea) montereyensis Maloney, 1933
**Idotea (Pentidotea) obscura* Rafi, 1972
Idotea (Pentidotea) resecata Stimpson, 1857 (fig. 17.21)
Idotea (Pentidotea) schmitti Menzies, 1950 (fig. 17.19)
Idotea (Pentidotea) stenops Benedict, 1898 (fig. 17.22)
Idotea (Pentidotea) wosnesenskii (Brandt, 1851) (fig. 17.20)
Saduria entomon (Linnaeus, 1767)
Synidotea angulata Benedict, 1897
Synidotea bicuspida (Owen, 1839)
Synidotea nebulosa Benedict, 1897
Synidotea nodulosa (Krøyer, 1848)
Synidotea pettiboneae Hatch, 1947
Synidotea ritteri Richardson, 1904

Suborder Epicaridea

Roland Bourdon

Butler, T. H. 1964. Redescription of the parasitic isopod *Holophryxus alaskensis* Richardson and a note on its synonymy. J. Fish. Res. Bd. Canada, 21:971-6.
Muscatine, L. 1956. A new entoniscid (Crustacea: Isopoda) from the Pacific coast. J. Wash. Acad. Sci., 46:122-6.
Richardson, H. 1905a. Isopods from the Alaska salmon investigation. Bull. U. S. Bur. Fish., 24:211-21.
------. 1905b. A monograph on the isopods of North America. Bull. 54, U. S. Nat. Mus. lii + 727 pp.
------. 1908. On some isopods of the family Dajidae from the northwest Pacific, with descriptions of a new genus and two new species. Proc. U. S. Nat. Mus., 33:689-96.
------. 1910. Isopods collected in the northwest Pacific by the U. S. Bureau of Fisheries Steamer "Albatross" in 1906. Proc. U. S. Nat. Mus., 37:75-129.

1a Protandric hermaphrodites in which the last larval stage (cryptoniscus), after functioning as a free-living male, becomes transformed into a parasitic female; this loses all or most appendages and becomes hypertrophied in such a way that the body has 2 to several lobes (section Cryptoniscina)   2
1b Sexes separate and both parasitic, the male being much smaller and more isopodlike than the female (there is usually a male closely associated with each female) (section Bopyrina)   3
2a Parasitic in barnacles (*Balanus glandula*, *Chthamalus dalli*); adult female conserving only the cephalon and some of the more anterior segments of the cryptoniscus, the rest of the body consisting of several lobes; basal article of antenna 1 and coxal plates of male toothed
Hemioniscus balani
2b Hyperparasitic on the rhizocephalans *Peltogaster paguri* and *Peltogasterella gracilis*, which are parasites of hermit crabs; adult female, after a period of completely internal parasitism, becomes constricted into 2 nearly spherical portions, the anterior one being buried in the host, the posterior one being external; basal article of antenna 1 and coxal plates of male smooth
Liriopsis pygmaea
3a Endoparasitic (not visible externally) in the visceral cavity of the crabs *Hemigrapsus oregonensis* and *H. nudus* (body of female bent into a V, the cephalon divided into 2 hemispheres, the pereopods reduced to stubs, the pleon with 5 pairs of wrinkled lateral lobes; male with only 6 pairs of pereopods)   *Portunion conformis*
3b Ectoparasitic, either located in the branchial chamber of the crustacean host, attached to the dorsal surface, or attached to the underside of the abdomen   4
4a Attached to the dorsal surface of the host, which is a decapod or mysid (female symmetrical, with only 5 pairs of pereopods, these grouped around the oral region, and with most of the posterolateral portion covered by a pair of enlarged oostegites   5
4b In the branchial chamber or under the abdomen of decapod crustaceans   7
5a Parasitic on the shrimp *Pasiphaea pacifica*; female with all pereonites and pleonites fused or so nearly fused that they are indistinct (male with pleonites fused)   *Holophryxus alaskensis*
5b Parasitic on mysids (possibly other planktonic crustaceans); female with at least some pereonites and pleonites distinct dorsally   6
6a Female with all pereonites and pleonites distinct dorsally (male with all pleonites distinct) (parasitic on the mysid *Eucopia australis*, collected in very deep water, and perhaps associated with other mysids)   *Arthophryxus beringanus*
6b Female with pereonites 1-3 and pleonites distinct dorsally (type specimen found in a collection of mysids from very deep water)   *Prophryxus alascensis*
7a Female symmetrical or only slightly asymmetrical, with all pereonites distinct, and with 5 pairs of nearly equal oostegites; female with 7 pairs of pereopods and generally with 6 distinct pleonites   8
7b Abdominal parasites of shrimps of the genera *Lebbeus*, *Spirontocaris*, *Eualus*, and *Heptacarpus*; female very asymmetrical, with pereonites not completely separated dorsally; on

the swollen side, the oostegites are large and form a closed brood pouch, and generally only pereopod 1 persists; on the nearly straight side, oostegites 2-4 are reduced to small plates, and there are 7 pereopods; pleon with 5 pleonites, 4 pairs of leaflike lateral plates, and small uniramous pleopods; male with fused pleonites   *Hemiarthrus abdominalis*

8a   Parasitic on the underside of the abdomen of the thalassinid *Upogebia pugettensis* (pleon of female without lateral plates, but with well developed biramous pleopods and uniramous uropods; pleon of male with lateral plates)   *Phyllodurus abdominalis*

8b   Parasitic in the branchial chamber of thalassinids, hermit crabs, galatheids, and shrimps of the families Crangonidae, Hippolytidae, and Pandalidae   9

9a   On the thalassinid *Callianassa californiensis*; pleon of female with 6 pairs of lateral plates, which are long and thin and whose posterior margins are fringed by slender outgrowths (brood pouch of female closed; pleon of male with elongated lateral plates, but these are not fringed)   *Ione cornuta*

9b   On hermit crabs, galatheids, or shrimps; pleon either with 5 pairs of lateral plates, with rudimentary plates, or without any plates   10

10a Pleon with 5 pairs of lamellar and smooth lateral plates; brood pouch closed; pleonites of male fused or not fused   11

10b Pleon either without lateral plates or with rudimentary plates; brood pouch open; pleonites of male fused   13

11a On the galatheid *Munida quadrispina* (see also choice 12b); female with 2 pairs of lateral lamellae on the posteroventral border of the cephalon; coxal plates of the pereon well developed; pleopods visible when the animal is viewed from the dorsal side; pleotelson elongated and swollen near its tip, and bearing biramous uropods; pleonites of male fused
                                                                                                           Munidion parvum

11b Female with 1 pair of lateral lamellae on the posteroventral border of the cephalon; pereon without coxal plates; pleopods not visible when the animal is viewed from the dorsal side; pleotelson short, rounded or truncate at its tip, and bearing uniramous uropods; pleonites of male not fused   12

12a On hermit crabs; endopodites of pleopods of female much longer than the exopodites, their surfaces more or less tuberculated   *Pseudione giardi* (fig. 17.24)

12b On the galatheid *Munida quadrispina* (see also choice 11a); endopodites of pleopods of female only slightly longer than the exopodites, their surfaces smooth
                                                                                         Pseudione galacanthae (fig. 17.25)

13a On shrimps of the family Crangonidae (*Crangon, Argis*); female with well developed biramous pleopods and uniramous uropods   *Argeia pugettensis* (fig. 17.23)

13b On shrimps of the families Hippolytidae (*Spirontocaris, Eualus, Heptacarpus*) and Pandalidae; female either without pleopods or with pleopods reduced to tubercles, and without uropods   *Bopyroides hippolytes*

Family Bopyridae

Argeia pugettensis Dana, 1853 (fig. 17.23)
Bopyroides hippolytes (Krøyer, 1838)
Hemiarthrus abdominalis (Krøyer, 1840)
Ione cornuta Bate, 1864
Munidion parvum Richardson, 1904
Phyllodurus abdominalis Stimpson, 1857
Pseudione galacanthae Hansen, 1897 (fig. 17.25)
Pseudione giardi Calman, 1898 (fig. 17.24)

Family Dajidae

Arthophryxus beringanus Richardson, 1908
Holophryxus alaskensis Richardson, 1905
Prophryxus alascensis Richardson, 1909

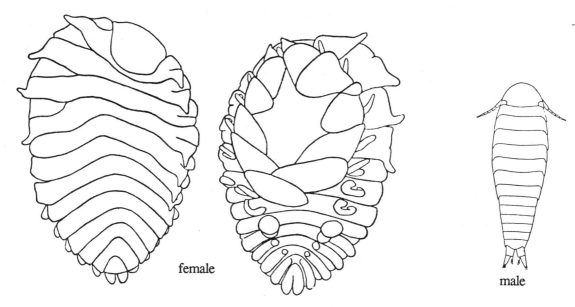

17.23 *Argeia pugettensis*

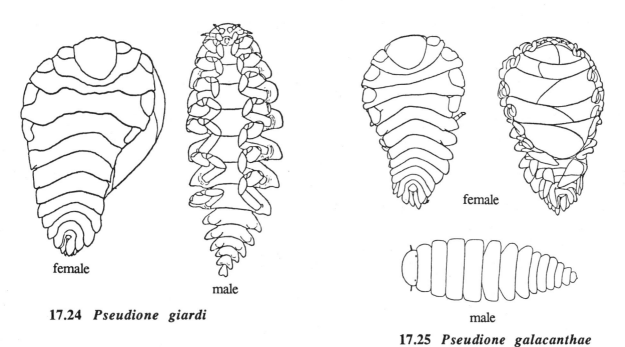

17.24 *Pseudione giardi*

17.25 *Pseudione galacanthae*

Family Entoniscidae

Portunion conformis Muscatine, 1956

Family Hemioniscidae

Hemioniscus balani (Buchholz, 1866)

Family Liriopsidae

Liriopsis pygmaea (Rathke, 1843)

Suborder Oniscoidea

Miller, M. A. 1936. California isopods of the genus *Porcellio* with descriptions of a new species and a new subspecies. Univ. Calif. Publ. Zool., 41:165-72.
Van Name, W. G. 1936. American land and fresh-water isopod Crustacea. Bull. Amer. Mus. Nat. Hist., 71:1-535.

1a Flagellum of antenna 2 with more than 10 articles (but there may be as few as 6 articles in immature specimens); up to about 2.5 cm long; found in rock crevices (and sometimes on pilings) just above the high tide line ... 2
1b Flagellum of antenna 2 with not more than 4 articles; less than 2 cm long; not typically found in rock crevices (generally on uppermost portions of sandy beaches or truly terrestrial) ... 3
2a Basal article of uropod about as broad as long ... *Ligia* (*Ligia*) *pallasii*
2b Basal article of uropod several times as long as broad (rarely found as far north as Oregon) ... *Ligia* (*Megaligia*) *occidentalis*
3a Flagellum of antenna 2 with 3 or 4 articles; tracheal system absent; under debris on sandy beaches, near uppermost limits of intertidal zone ... 5
3b Flagellum of antenna 2 with 2 articles; tracheal system present in exopodites of pleopods 1 and 2; truly terrestrial, though occasionally found close to the uppermost limits of the intertidal zone ... 4
4a Capable of rolling up into a ball; antenna 2 not extending posteriorly beyond pereonite 2; pleotelson blunt ... *Armadillidium vulgare*
4b Not capable of rolling up into a ball; antenna 2 extending posteriorly beyond pereonite 2; pleotelson pointed ... *Porcellio scaber*
5a Flagellum of antenna 2 with 3 articles ... 6
5b Flagellum of antenna 2 with 4 articles ... 7
6a Pleonites becoming gradually narrower than the pereonites; cephalon with large and tuberculate median and anterolateral lobes; antenna 2 not extending beyond pereonite 1 ... *Alloniscus perconvexus*
6b Pleonites becoming abruptly narrower than the pereonites; cephalon rounded in front and with only small lateral lobes, if any; antenna 2 long, extending beyond pereonite 2 ... *Littorophiloscia richardsonae*
7a Anterior margin of cephalon deeply 3-lobed, the median lobe forming a rostrum; basal article of uropod expanded, forming a large lobe lateral to the endopodite and nearly reaching the distal tip of the latter ... *Armadilloniscus holmesi*
7b Anterior margin of cephalon 2-lobed, without a rostrumlike median lobe; basal article of uropod not expanded, the endopodite inserted upon it in a posterolateral position ... *Detonella papillicornis*

Family Armadillidiidae

Armadillidium vulgare (Latreille, 1804)

Family Ligiidae

Ligia (*Ligia*) *pallasii* Brandt, 1833
Ligia (*Megaligia*) *occidentalis* Dana, 1853

Family Oniscidae

Alloniscus perconvexus Dana, 1854
Littorophiloscia richardsonae Holmes & Gay, 1909

Family Porcellionidae

Porcellio scaber Latreille, 1804

Family Syphacidae

Armadilloniscus holmesi Arcangeli, 1933
Detonella papillicornis Richardson, 1904

18

PHYLUM ARTHROPODA: SUBPHYLUM CRUSTACEA: CLASS MALACOSTRACA: SUBCLASS PERACARIDA: ORDER AMPHIPODA

In the keys for amphipods, the term "head" (i. e., the head plus one thoracic segment) is equivalent to "cephalon" as used in peracaridan orders covered in Chapter 17.

Key to Suborders

1a   Pereon consisting of 7 pereonites, all of which have well developed pereopods; abdomen not vestigial; body neither especially slender nor resembling that of a praying mantis   2
1b   Pereon consisting of 6 visible pereonites, some of which have only vestigial percopods; abdomen vestigial; body (except for species found on whales) usually slender and often resembling that of a praying mantis   Suborder Caprellidea
2a   Eyes generally large, occupying most of the surface of the head; coxae of pereopods small and often fused with the body; usually planktonic, associated with jellyfishes, or occupying the tunic of dead salps   Suborder Hyperiidea
2b   Eyes usually present and conspicuous, but not so large as to occupy most of the surface of the head; coxae of pereopods well developed, and usually expanded into platelike structures; mostly benthic, but a few species are planktonic   Suborder Gammaridea

Suborder Gammaridea

Craig P. Staude

Although gammaridean amphipods are of great ecological importance, most of the numerous species in our region are difficult to identify. The key provided here has been written for the nonspecialist and is based primarily on features that are fairly obvious. Reference to mouthparts and the need for dissection are minimized. As a rule, two or three contrasting characters are used in each couplet, so that it may be possible to identify a specimen even if a critical appendage is missing.

    Certain details that are not easily visible under a dissecting microscope should be observed with a compound microscope. For this, it may be advantageous to mount the whole specimen on a depression slide, or, if a standard slide is used, to support the coverglass. It is often advisable to remove the urosome and mount it separately so that the telson can be rotated by carefully moving the coverglass. If further dissection is required, an individual pereopod may be detached by pulling on the coxa of the appendage. Fine forceps and probes tipped with insect pins are useful tools. Barnard (1969b) and Bousfield (1973) explain procedures that are employed in more thorough studies of amphipod morphology.

    Terminology is illustrated by figures or defined in the glossary. Figure 18.1 shows the main features of amphipod structure. Note that the first two pereopods are generally modified as grasping appendages called gnathopods (figs. 18.2-18.8), and that all pereopods, with a few exceptions, consist of 7 articles (figs. 18.9-18.11). Various conditions of the telson are illustrated in Figures 18.12-18.15. Throughout the key it is important to distinguish flexible setae from rigid spines.

    Some characters change with age. Adult amphipods often have more spines or setae on a particular part of the body than juveniles do. It is always best to use mature specimens when using the key.

    Sexual dimorphism is pronounced in some families. Many couplets therefore refer to male or female characters. Females can be recognized by the leaflike oostegites that cover the sternal brood chamber within which embryos develop. Even if a mature female is not brooding, she will have oostegites next to the gills at the bases of coxae 2-5; the oostegites may, however, be partly concealed by the coxae. Males sometimes have larger eyes, larger gnathopods, and longer antennae than females do.

After arriving at a name in the key, one should turn to the species list in order to determine that the habitat given agrees with that at the collection site. Whenever possible, it is a good idea to confirm your identification by consulting a complete description of the species in one of the appropriate monographs listed below.

The taxonomy of some groups of gammaridean amphipods is still in a state of flux. In the last few years, several new genera and many new species have been described from the northeastern Pacific region by E. L. Bousfield and his associates at the National Museum of Canada. It is expected that many more will be described in the near future, and that some determinations that have been made in the past will be changed.

Alderman, A. L. 1936. Some new and little known amphipods of California. Univ. Calif. Publ. Zool., 41:53-74.

Barnard, J. L. 1962a. Benthic marine Amphipoda of southern California: families Aoridae, Photidae, Ischyroceridae, Corophiidae, Podoceridae. Pacific Naturalist, 3:1-72.

------. 1962b. Benthic marine Amphipoda of southern California: families Tironidae to Gammaridae. Pacific Naturalist, 3:73-115.

------. 1962c. Benthic marine Amphipoda of southern California: families Amphilochidae, Leucothoidae, Stenothoidae, Argissidae, Hyalidae. Pacific Naturalist, 3:116-63.

------. 1964. Marine Amphipoda of Bahia de San Quintin, Baja California. Pacific Naturalist, 4:55-139.

------. 1965. Marine Amphipoda of the family Ampithoidae from southern California. Proc. U. S. Nat. Mus., 118(3522):1-46.

------. 1966. Part V. Systematics: Amphipoda. Submarine canyons of southern California. Allan Hancock Pac. Exped., 27:1-166.

------. 1969a. Gammaridean Amphipoda of the rocky intertidal of California: Monterey Bay to La Jolla. Bull. 258, U. S. Nat. Mus. 230 pp.

------. 1969b. The families and genera of marine gammaridean Amphipoda. Bull. 271, U. S. Nat. Mus. 535 pp.

------. 1971. Gammaridean Amphipoda from a deep-sea transect off Oregon. Smithsonian Contrib. Zool., no. 61. 86 pp.

------. 1979. Littoral gammaridean Amphipoda from the Gulf of California and the Galapagos Islands. Smithsonian Contrib. Zool., no. 271. 149 pp.

Bousfield, E. L. 1973. Shallow-water Gammaridean Amphipoda of New England. Ithaca: Cornell University Press. 312 pp.

------. 1979. The amphipod superfamily Gammaroidea in the northeastern Pacific region: systematics and distributional ecology. Bull. Biol. Soc. Wash., 3: 297-357.

------. 1981. Evolution in north Pacific coastal marine amphipod crustaceans. In Scudder, G. G. E., & J. L. Reveal (eds.), Evolution Today, Proc. 2nd. Internat. Congr. System. Evol. Biol., pp. 69-89.

------. 1983. An updated phyletic classification and paleohistory of the Amphipoda. In Schram, F. R. (ed.), Crustacean Phylogeny. Rotterdam: Balkema, pp. 257-77.

Conlan, K. E. 1983. The amphipod superfamily Corophioidea in the northeastern Pacific region. 3. Family Isaeidae: systematics and distributional ecology. Nat. Mus. Canada, Publ. Nat. Sci., no. 4. 73 pp.

Conlan, K. E., & E. L. Bousfield. 1982. Studies on amphipod crustaceans of the northeastern Pacific region. 2. Family Ampithoidae. Nat. Mus. Canada, Publ. Biol. Oceanogr., 10:41-75.

Hurley, D. E. 1963. Amphipoda of the family Lysianassidae from the west coast of North and Central America. Occas. Pap. Allan Hancock Found., no. 25. 160 pp.

Jarrett, N. E., & E. L. Bousfield. 1982. Studies on the amphipod crustaceans of the northeastern Pacific region. 4. Family Lysianassidae, genus *Hippomedon*. Nat. Mus. Canada, Publ. Biol. Oceanogr., 10:103-28.

Laubitz, D. 1977. A revision of the genera *Dulichia* Krøyer and *Paradulichia* Boeck (Amphipoda, Podoceridae). Can. J. Zool., 55:942-82.

Phylum Arthropoda: Order Amphipoda

Mills, E. L. 1962. Amphipod crustaceans of the Pacific coast of Canada. II. Family Oedicerotidae. Nat. Mus. Canada, Nat. Hist. Pap., 15:1-21.

Otte, G. 1975. A laboratory key for the identification of *Corophium* species (Amphipoda, Corophiidae) of British Columbia. Environ. Canada, Fish. Mar. Serv. Tech. Rep. no. 519. 19 pp.

Staude, C. P., J. W. Armstrong, R. M. Thom, & K. K. Chew. 1977. An illustrated key to the intertidal gammaridean Amphipoda of central Puget Sound. Univ. Wash., College of Fisheries, Contrib. no. 466. 27 pp.

Choosing One of the Four Main Keys

Gammaridea have been assigned to four main keys on the basis of the presence or absence of the third uropod, and its structure if it is present. The four arrangements are described and illustrated below.

Uropod 3 with a single ramus which never greatly exceeds the length of the peduncle (if there is a second ramus, it is scarcely noticeable)
           Proceed to Key I

Uropod 3 biramous; both rami shorter than the peduncle   Proceed to Key II

Uropod 3 biramous; at least one ramus equal to or longer than the peduncle   Proceed to Key III

Uropod 3 absent or lost   Proceed to Key IV

Key I

Uropod 3 essentially uniramous (if there is a second ramus, it is scarcely noticeable); the obvious ramus never more than twice the length of the peduncle.

1a Article 4 of antenna 2 extraordinarily thickened and bearing large ventral spines or teeth; urosome dorsoventrally flattened; ramus of uropod 3 flattened and paddlelike (fig. 18.16) (*Corophium*)   2

Phylum Arthropoda: Order Amphipoda 349

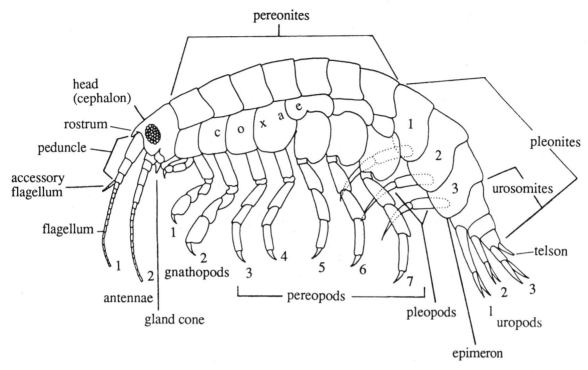

18.1 Body parts of a gammaridean amphipod

1b   Article 4 of antenna 2 not especially thickened and without spines or teeth; urosome laterally compressed; ramus of uropod 3 somewhat conical or bladelike (figs. 18.17-18.25) 8
2a   Urosomal segments distinctly separate (fig. 18.26) 3
2b   Urosomal segments fused (fig. 18.27) 5
3a   Both sexes with a large curved tooth at the lower distal end of article 4 of antenna 2, but lacking a small secondary tooth within the concave margin of that tooth
       .... *Corophium spinicorne* Stimpson, 1857
3b   Male with a large curved tooth at the lower distal end of article 4 of antenna 2, and also bearing a small tooth within the concave margin of that tooth (fig. 18.26); female bearing articulated spines (instead of a large distal tooth) on the ventral margin of article 4 of antenna 2 4
4a   Article 1 of antenna 1 rather flat, the proximal portion of its ventral surface smooth; mature male with article 3 of antenna 2 often greatly exceeding the length of article 1; article 4 of antenna 2 of female bearing only 2 ventral spines (1 proximal and 1 distal)
       .... *Corophium salmonis* Stimpson, 1857 (fig. 18.26)
4b   Article 1 of antenna 1 with a proximal toothlike process in males and armed with a row of about 4 spines in females (often hidden by antenna 2); antenna 1 and 2 generally more setose than in *C. salmonis*; article 3 of antenna 2 of male about as long as article 1; article 4 of antenna 2 of female bearing 3 pairs of spines along its ventral margin
       .... *Corophium brevis* Shoemaker, 1949
5a   Article 1 of antenna 1 of mature individuals armed ventrally with a row of 6 or more spines; head of male with acute lateral lobes; female bearing a dense row of spines along the ventral margin of article 4 of antenna 2 .... *Corophium crassicorne* Bruzelius, 1859
5b   Article 1 of antenna 1 rarely with more than 4 ventral spines; lateral lobes of head of male not acute; article 4 of antenna 2 of female with spines widely set, in pairs or singly, but not in a continuous row 6

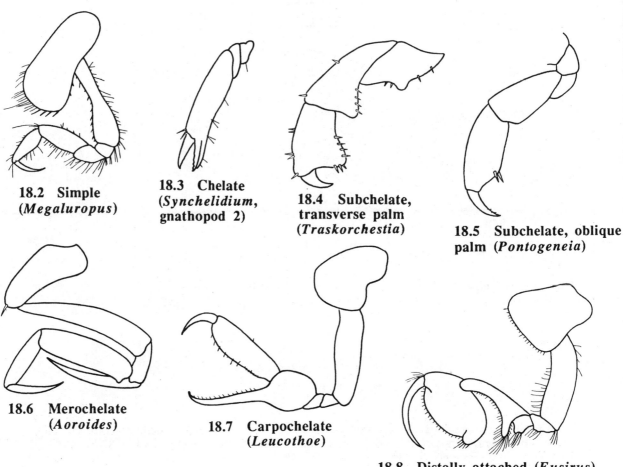

18.2 Simple (*Megaluropus*)
18.3 Chelate (*Synchelidium*, gnathopod 2)
18.4 Subchelate, transverse palm (*Traskorchestia*)
18.5 Subchelate, oblique palm (*Pontogeneia*)
18.6 Merochelate (*Aoroides*)
18.7 Carpochelate (*Leucothoe*)
18.8 Distally attached (*Eusirus*)

Types of gnathopods (gnathopod 1, unless otherwise noted)

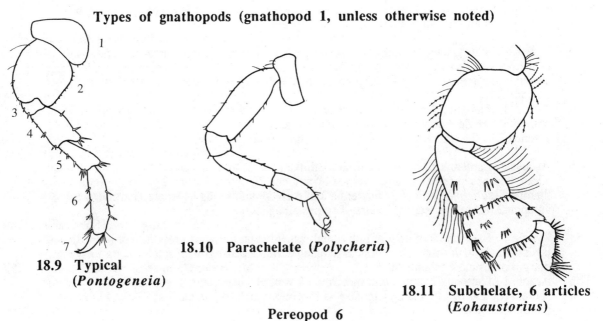

18.9 Typical (*Pontogeneia*)
18.10 Parachelate (*Polycheria*)
18.11 Subchelate, 6 articles (*Eohaustorius*)

Pereopod 6

6a Mature male with distal tooth of article 4 of antenna 2 directed medially; article 4 of antenna 2 of female bearing 3 single spines, these widely spaced along the ventral margin
Corophium baconi Shoemaker, 1934

6b Mature male with tooth of article 4 of antenna 2 directed forward; subterminal ventral spines of article 4 of antenna 2 of female arranged in pairs 7

7a Peduncle of uropod 1 with 3 (rarely 2) distal spines along inner margin; male rostrum much shorter than the lateral lobes; article 4 of antenna 2 of female with 7 ventral spines (3 proximal pairs and a single terminal spine) *Corophium acherusicum* Costa, 1857 (fig. 18.27)

7b Peduncle of uropod 1 with 1 distal spine (rarely 2 spines) on inner margin; male rostrum about as long as the lateral lobes; article 4 of antenna 2 of female with 5 ventral spines (2 proximal pairs and a single terminal spine) *Corophium insidiosum* Crawford, 1939

8a Coxa 1 much reduced and coxa 4 greatly enlarged (often more than 3 times the width of coxa 2); ramus of uropod 3 with 2 distinct articles (fig. 18.17) Stenothoidae

8b Coxae 1-4 nearly equal in size (although becoming slightly wider posteriorly); obvious ramus of uropod 3 with a minute second article (fig. 18.18) or uniarticulate (figs. 18.19-18.25) 9

9a Eyes borne on prominent anteriorly-directed lateral lobes of head; proximal posterior margin of coxa 4 without a deep concavity; peduncle of uropod 3 slender and distinctly longer than the telson; telson thick and fleshy, not cleft 10

9b Eyes not borne on lateral lobes; proximal posterior margin of coxa 4 with a deep concavity in its posteroproximal portion; peduncle of uropod 3 stout, never greatly exceeding the length of the telson; telson sometimes thickened, but usually thin and cleft 18

10a Coxae 3 and 4 longer than wide; uropod 3 with minute inner ramus (usually hidden in lateral view), large outer ramus bearing a minute second article with setae at its apex (fig. 18.18); telson without denticulate pads; male gnathopod 2 subchelate (*Photis*) 11

10b Coxae 3 and 4 wider than long; uropod 3 truly uniramous, the ramus without setae but with small apical denticles (fig. 18.19); telson with 2 denticulate pads; male gnathopod 2 carpochelate (as in *Leucothoe*, fig. 18.7) (*Ericthonius*) 17

11a Palm of gnathopod 2 of male with a bifid process directed toward the medial side of the dactyl *Photis bifurcata* Barnard, 1962 (fig. 18.29)

11b Palm of gnathopod 2 of male with a concavity defined posteriorly by a thumb-like tooth (this tooth neither bifid nor directed medially) 12

12a Dactyl of gnathopod 2 of male with a callus, tooth, or protuberance on its inner margin 13

12b Dactyl of gnathopod 2 of male with inner margin smooth and undeveloped 15

13a Inner margin of dactyl of gnathopod 2 of male with small spines embedded in irregular calluses *Photis pachydactyla* Conlan, 1983 (fig. 18.30)

13b Inner margin of dactyl of gnathopod 2 of male with a single smooth tooth at its midpoint 14

14a Tooth of dactyl of gnathopod 2 of male rounded; palm of gnathopod 2 of male with a wide concavity which is more than twice as wide as the dactylar tooth; palms of gnathopods 1 and 2 of female without a distinct concavity *Photis brevipes* Shoemaker, 1942 (fig. 18.28)

14b Tooth of dactyl of gnathopod 2 of male square; palm of gnathopod 2 of male with a narrow concavity which is only slightly wider than the dactylar tooth; palms of gnathopods 1 and 2 of female distinctly concave *Photis parvidons* Conlan, 1983 (fig. 18.31)

15a Thumb of gnathopod 2 of fully mature male reaching beyond the hinge point of the dactyl
Photis conchicola-oligochaeta group (fig. 18.32)

15b Thumb of gnathopod 2 of male not reaching the hinge point of the dactyl (fig. 18.33) 16

16a Article 5 of gnathopod 1 of male about as long as article 6, its posterior margin not lobate; palm of gnathopod 1 of male slightly convex *Photis lacia* Barnard, 1962 (fig.18.33)

16b Article 5 of gnathopod 1 of male distinctly shorter than article 6, its posterior margin lobate (especially in male); palm of gnathopod 1 of male concave
Photis macinerneyi Conlan, 1983 (fig.18.33)

17a Article 5 of gnathopod 2 of male with bifid distal tooth *Ericthonius* cf. *brasiliensis* Dana, 1853

17b Article 5 of gnathopod 2 of male with a simple distal tooth *Ericthonius hunteri* Bate, 1862

18a Ramus of uropod 3 vestigial, scarcely one-fourth as long as the peduncle (fig. 18.20) 19

18b Ramus of uropod 3 distinct, at least one-half as long as the peduncle (figs. 18.21-18.25) 20

19a Rostrum small but distinct; posterior margins of pereopods 5-7 and uropods 1-2 with many long, crowded setae, but without spines *Proboscinotus loquax* (Barnard, 1967)

18.12 Entire (*Calliopius*)

18.13 Notched, apical spines (*Oligochinus*)

18.14 Cleft, lateral spines (*Accedomoera*)

18.15 Cleft, lyre-shaped (*Pontogeneia* cf. *ivanovi*)

Telson

19b Without a rostrum, but dorsal part of anterior portion of head with a shallow concavity; pereopods 5-7 and uropods 1-2 bearing only sparse, short setae among a few spines
Najna spp.
20a Antenna 1 longer than the peduncle of antenna 2; telson bearing no more than 2 stiff spines 21
20b Antenna 1 shorter than the peduncle of antenna 2; telson with more than 2 spines (Talitridae, "beach hoppers") 27
21a Antenna 1 slightly longer than antenna 2; uropod 3 bearing a minute inner ramus (hidden in lateral view) (fig. 18.21); telson bearing 2 stiff spines *Parallorchestes* spp.
21b Antenna 1 shorter than, or about as long as antenna 2; uropod 3 truly uniramous (figs. 18.22 and 18.23); telson naked, or bearing setae or small, flexible spines 22
22a Antenna 1 of female only slightly shorter than antenna 2 (in males, it is distinctly shorter than antenna 2); apex of uropod 3 bearing only setae or flexible spinules (fig. 18.22); telson not cleft to its base (*Allorchestes*) 23
22b Antenna 1 of both sexes distinctly shorter than antenna 2 (fig. 18.34); apex of uropod 3 bearing stiff spines (fig. 18.23); telson cleft to its base 24
23a Gnathopod 1 of male with dactyl weakly inflated and bifid at the tip; palm without teeth; pereopod spines long and sharp; pleonites not dorsally carinate
Allorchestes angusta Dana, 1856 group
23b Gnathopod 1 of male with dactyl much inflated and simply pointed at the tip; palm toothed; pereopod spines short and blunt; pleonites dorsally carinate
Allorchestes bellabella Barnard, 1974
24a Flagellum of antenna 2 long (with 20 or more articles in adult specimens); uropod 1 with an enlarged peduncular spine at the base of the rami (this spine lateral to the rami); telson naked
Hyale frequens Stout, 1913 (fig. 18.34) group
24b Flagellum of antenna 2 short (with less than 20 articles in adult specimens); uropod 1 with or without an enlarged peduncular spine at the base of the rami (when present, this spine is between the rami); telson bearing slender spinules 25
25a Antenna 2 of male without plumose setae; dactylar setae of pereopods stout; distal peduncular spine of uropod 1 not distinctly larger than the next proximal spine of the peduncle
Hyale pugettensis (Dana, 1852)
25b Antenna 2 of male with plumose setae; dactylar setae of pereopods slender; distal peduncular spine of uropod 1 distinctly larger than the next proximal spine of the peduncle 26
26a Plumose setae present on the proximal portion of the posterior margin of article 5 of antenna 2 of male (some plumose setae present on this article in both male and female); inner margin of dactyls nearly smooth; distal peduncular spine of uropod 1 very large (at least one-third the length of the rami) *Hyale plumulosa* (Stimpson, 1857)
26b Plumose setae absent from the proximal portion of posterior margin of article 5 of antenna 2 of male (some plumose setae present on the medial and distal margins of this article of the male, but not in the female); inner margin of dactyls with a row of small but distinct bumps; distal peduncular spine of uropod 1 not attaining one-third the length of the rami
Hyale anceps (Barnard, 1969)

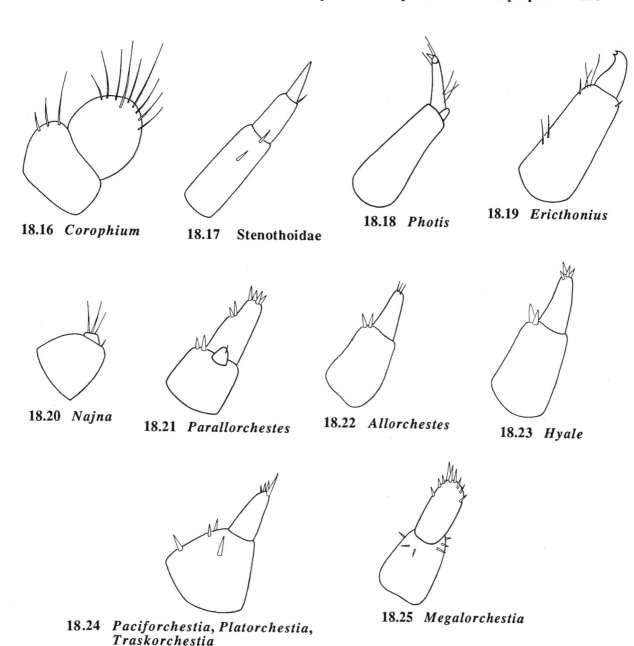

18.16 *Corophium*
18.17 Stenothoidae
18.18 *Photis*
18.19 *Ericthonius*
18.20 *Najna*
18.21 *Parallorchestes*
18.22 *Allorchestes*
18.23 *Hyale*
18.24 *Paciforchestia, Platorchestia, Traskorchestia*
18.25 *Megalorchestia*

Uniramous uropod 3

27a Gnathopod 1 subchelate (especially in males), its dactyl not longer than the width of article 6; pereopod 6 not exceeding the length of pereopod 7; ramus of uropod 3 shorter than the peduncle and tapering to a point (fig. 18.24) ... 28

27b Gnathopod 1 simple, its dactyl longer than the width of article 6; pereopod 6 slightly longer than pereopod 7; ramus of uropod 3 at least as long as the peduncle, not tapering distally (fig. 18.25) (*Megalorchestia*) ... 31

28a Peduncle of uropod 1 with enlarged interramal spine; pleopod 3 about two-thirds the length of pleopods 1 and 2 ... *Paciforchestia klawei* (Bousfield, 1961)

28b Peduncle of uropod 1 lacking an enlarged interramal spine; pleopods almost equal in length ... 29

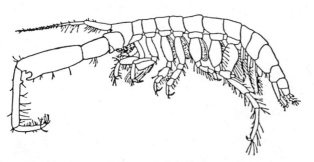
18.26 *Corophium salmonis*, male

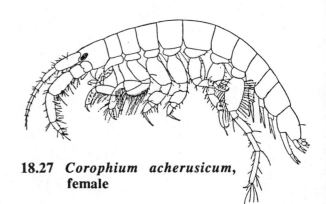
18.27 *Corophium acherusicum*, female

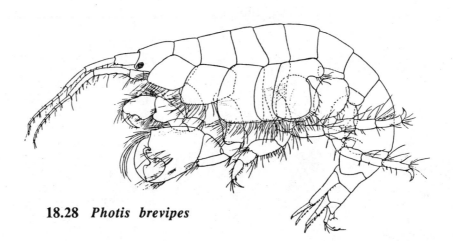
18.28 *Photis brevipes*

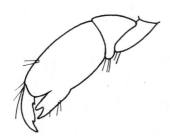
18.29 *Photis bifurcata*

18.30 *Photis pachydactyla*

18.31 *Photis parvidons*

18.32 *Photis conchicola*, *Photis oligochaeta*

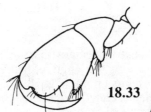
18.33 *Photis lacia*, *Photis macinerneyi*

Gnathopod 2 of male *Photis*

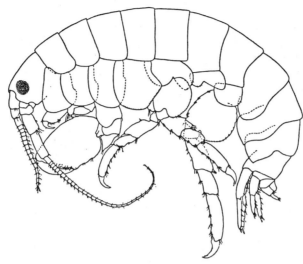

18.34 *Hyale frequens*

29a Flagellum of antenna 2 with about 6 articles; lateral margin of outer ramus of uropod 1 without spines   *Platorchestia chathamensis* Bousfield, 1982
29b Flagellum of antenna 2 with more than 10 articles; lateral margin of outer ramus of uropod 1 with spines   30
30a Pleopods slender, their rami with 4 to 7 articles; posterior margin of article 4 of gnathopod 1 of male without a tubercle   *Traskorchestia georgiana* (Bousfield, 1958)
30b Pleopods stout, their rami with 7 to 10 articles; article 4 of gnathopod 1 of male bearing a small posterior tubercle (fig. 18.4)   *Traskorchestia traskiana* (Stimpson, 1857)
31a Flagellum of antenna 2 shorter than peduncle; bases of peduncles of pleopods only slightly expanded; inner margin of outer ramus of uropod 2 without spines
                                                                                                                                                                                           Megalorchestia pugettensis (Dana, 1853)
31b Flagellum of antenna 2 not shorter than the peduncle (in males it is much longer than the peduncle); bases of peduncles of pleopods nearly twice as wide as their distal ends; inner margin of outer ramus of uropod 2 with spines   32
32a Article 5 of gnathopod 1 of female with a posterodistal tubercle; ventral margins of epimera 2-3 each bearing several small spines; rami of pleopods less than half the length of the peduncles (the antennae are pinkish orange in life)   *Megalorchestia californiana* Brandt, 1851
32b Article 5 of gnathopod 1 of female without a tubercle; ventral margins of epimera without spines; rami of pleopods at least one-half the length of the peduncles
                                                                                                                                                                                           Megalorchestia columbiana (Bousfield, 1958)

Key II

Uropod 3 biramous; both rami shorter than the peduncle.

1a   Outer ramus of uropod 3 with 1 or 2 curved apical spines or teeth (figs. 18.35-18.37)   2
1b   Neither ramus of uropod 3 with curved apical spines or teeth (apical spines, if present, are small and straight) (figs. 18.38-18.40)   15
2a   Outer ramus of uropod 3 with 2 curved apical spines; inner ramus of uropod 3 flattened, with setae as well as spines (fig. 18.37); outer lobe of lower lip with a deep cleft (Ampithoidae)   3
2b   Outer ramus of uropod 3 with a single hooked apical spine and some smaller proximal teeth; inner ramus of uropod 3 conical, with one or more spines but no setae; outer lobe of lower lip broadly rounded, not cleft   14

3a	Accessory flagellum present and consisting of about 3-5 articles *Cymadusa uncinata* (Stout, 1912) (fig. 18.41)	
3b	Accessory flagellum absent or consisting of a single minute article	4
4a	Palm of gnathopod 1 nearly transverse (slightly oblique to nearly chelate); article 2 of pereopods 3 and 4 ovate; uropod 1 with a distal peduncular tooth extending between the rami (*Peramphithoe*)	5
4b	Palm of gnathopod 1 distinctly oblique; article 2 of pereopods 3 and 4 rectangular; uropod 1 without a distal peduncular tooth (*Ampithoe*)	9
5a	Adults 2-3 cm in length; proximal flagellar articles of antenna 2 typical, not fused; gnathopod 2 of male only slightly enlarged, if at all	6
5b	Adults rarely over 1 cm; proximal flagellar articles of antenna 2 usually fused into a single article which is distinctly longer than the subsequent articles; gnathopod 2 of male distinctly enlarged	7
6a	Article 5 of gnathopod 1 distinctly longer than article 6; gnathopod 2 of male not obviously different from gnathopod 1; pereopod 7 longer than pereopod 6 *Peramphithoe humeralis* (Stimpson, 1864) (fig. 18.48)	
6b	Articles 5 and 6 of gnathopod 1 nearly equal in length; article 6 of gnathopod 2 of male about twice as broad as in gnathopod 1; pereopods 6 and 7 nearly equal in length *Peramphithoe mea* (Gurjanova, 1938) (fig. 18.49)	
7a	Antenna 2 nearly as long as antenna 1; article 6 of gnathopod 2 of male rectangular *Peramphithoe plea* (Barnard, 1965) (fig. 18.51)	
7b	Antenna 2 scarcely three-fourths as long as antenna 1; article 6 of gnathopod 2 of male ovate	8
8a	Proximal 5-6 articles of flagellum of antenna 2 fused to form a single unit; palm of gnathopod 2 of male without a large distal tooth *Peramphithoe lindbergi* (Gurjanova, 1938) (fig.18.50)	
8b	Proximal flagellar articles of antenna 2 fused in pairs; palm of gnathopod 2 of male with a large distal tooth *Peramphithoe tea* (Barnard, 1965) (fig. 18.52)	
9a	Posteroventral corner of epimera 2 and 3 with a blunt but distinct tooth	10
9b	Posteroventral corner of epimera 2 and 3 rounded or forming a right angle, but without a distinct tooth	11
10a	Antenna 1 longer than antenna 2; article 6 of gnathopod 2 of mature male nearly rectangular, becoming slightly chelate in older specimens *Ampithoe lacertosa* Bate, 1858 (fig. 18.45)	
10b	Antenna 1 shorter than antenna 2; article 6 of gnathopod 2 of mature male ovate, its palm with a deep excavation between the thumb and dactylar hinge *Ampithoe simulans* Alderman, 1936 (fig. 18.42)	
11a	Article 6 of gnathopod 2 of male rectangular, its palm nearly transverse and somewhat convex	12
11b	Article 6 of gnathopod 2 of male ovate, its palm strongly oblique and concave or incised	13
12a	Ventral margin of article 5 of antenna 2 with many dense long setae; palm of gnathopod 2 of mature male rather oblique and slightly convex *Ampithoe plumulosa* Shoemaker, 1938 (fig.18.46)	
12b	Ventral margin of article 5 of antenna 2 with a few short setae; palm of gnathopod 2 of mature male transverse and with a nearly square median tooth *Ampithoe valida* Smith, 1873 (fig. 18.47)	
13a	Antennae and uropods with close-set, long setae (setae of antennal peduncles longer than twice the width of each article); gnathopod 2 of male with a narrow excavation between the thumb and dactylar hinge *Ampithoe sectimanus* Conlan & Bousfield, 1982 (fig. 18.43)	
13b	Antennae and uropods with sparse, short setae (nearly all setae of antennal peduncles shorter than the width of each article); gnathopod 2 of male with a wide, shallow excavation between the thumb and dactylar hinge *Ampithoe dalli* Shoemaker, 1938 (fig. 18.44)	
14a	Outer ramus of uropod 3 with 2 teeth proximal to the curved apical spine; gnathopod 2 of female and juveniles with a concave palm; article 6 of gnathopod 2 of male bearing a thumblike process *Jassa* spp. (fig. 18.35)	
14b	Outer ramus of uropod 3 with 1 or 2 rows of denticles proximal to the curved apical spine; palm of gnathopod 2 of female and juveniles not concave; article 6 of gnathopod 2 of male narrow and curved, without a thumb *Ischyrocerus anguipes* (Krøyer, 1838) (fig. 18.36) group	

Phylum Arthropoda: Order Amphipoda 357

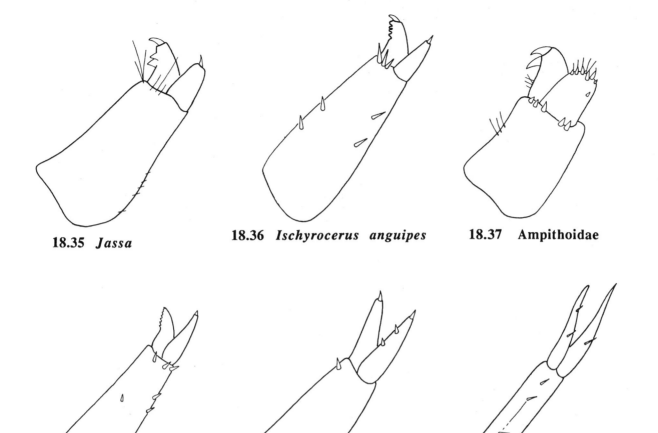

18.35 *Jassa* 18.36 *Ischyrocerus anguipes* 18.37 Ampithoidae

18.38 *Microjassa litotes* 18.39 *Ischyrocerus serratus* 18.40 Amphilochidae

Biramous uropod 3 with rami shorter than the peduncle

15a Telson short, thick and fleshy 16
15b Telson long and thin 18
16a Pereonites 6-7 and pleonites 1-3 each with 1-2 dorsal teeth; outer ramus of uropod 3 without denticles or serrations *Ischyrocerus serratus* Gurjanova, 1938 (fig. 18.39)
16b Pereonites and pleonites without dorsal teeth; outer ramus of uropod 3 with denticles or fine serrations 17
17a Eyes usually light when preserved in alcohol; coxa 1 about half as long as coxa 2; outer ramus of uropod 3 with a single row of fine serrations (fig. 18.38); apex of telson with 1 pair of thin spinules *Microjassa* spp.
   Probably includes *M. litotes* Barnard, 1954 (fig. 18.38).
17b Eyes usually dark when preserved in alcohol; coxa 1 nearly equal in length to coxa 2; outer ramus of uropod 3 with 2 parallel rows of denticles; apex of telson with 1 to 3 pairs of large spines *Ischyrocerus* sp.

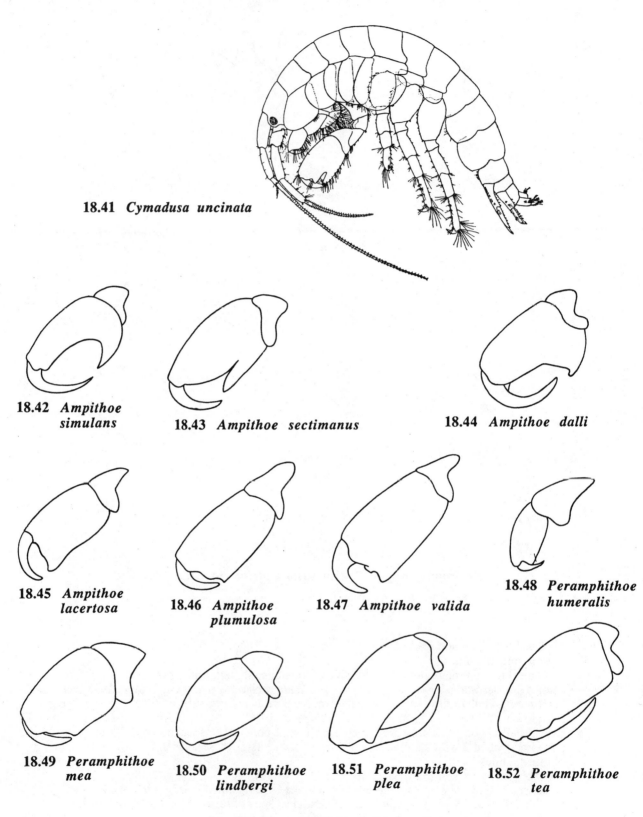

18.41 *Cymadusa uncinata*

18.42 *Ampithoe simulans*

18.43 *Ampithoe sectimanus*

18.44 *Ampithoe dalli*

18.45 *Ampithoe lacertosa*

18.46 *Ampithoe plumulosa*

18.47 *Ampithoe valida*

18.48 *Peramphithoe humeralis*

18.49 *Peramphithoe mea*

18.50 *Peramphithoe lindbergi*

18.51 *Peramphithoe plea*

18.52 *Peramphithoe tea*

Ampithoidae, gnathopod 2 of male

Phylum Arthropoda: Order Amphipoda 359

Proceed to Key III A if

18.53 or 18.54

If not, continue to one of the following choices:

Key III B if

18.55 and 18.56

Key III C if

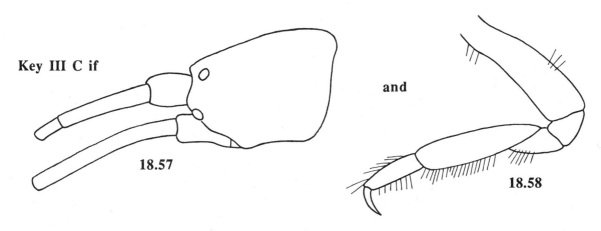

18.57 and 18.58

Key III D if

18.59

Key III E if

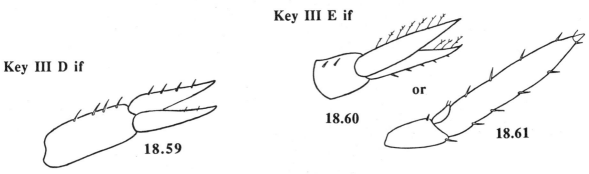

18.60 or 18.61

Picture key illustrating choices A to E of Key III

18a Gnathopods weakly subchelate, article 5 not projecting distally along the posterior margin of article 6; some pleonites dorsally toothed or serrated; telson with a terminal notch or cleft
Melphidippidae
18b Gnathopods strongly subchelate, article 5 projecting distally along the posterior margin of article 6; pleonites dorsally smooth; telson without a notch or cleft   19
19a Rostrum small (shorter than half the length of article 1 of antenna 1); coxa 1 "normal," not overlapped by coxa 2; gnathopod 1 carpochelate, the lobe of article 5 projecting distally as far as the hinge of the dactyl, article 6 without a distinct palm   *Leucothoe* sp. (fig.18.7)
19b Rostrum large (longer than half the length of article 1 of antenna 1); coxa 1 reduced, largely overlapped by coxa 2; gnathopod 1 subchelate, the lobe of article 5 not projecting as far as the hinge of the dactyl, article 6 without a distinct palm (Amphilochidae)   20
20a Lobe of article 5 of gnathopod 2 not projecting more than halfway along the posterior margin of article 6   *Amphilochus litoralis* Stout, 1912
20b Lobe of article 5 of gnathopod 2 projecting along the entire posterior margin to the corner of the palm of article 6   21
21a Article 4 of gnathopod 2 with only a single small spine at the distal apex; mandibular molar large and serrated   *Gitanopsis vilordes* Barnard, 1962
21b Article 4 of gnathopod 2 with 2 spines along the posterior margin in addition to an apical spine or some setae; mandibular molar small and smooth   22
22a Apex of article 4 of gnathopod 2 with a spine and some setae; anterior margin of article 6 projecting distally beyond the hinge of the dactyl   *Amphilochus picadurus* Barnard, 1962
22b Apex of article 4 of gnathopod 2 usually with a single spine; anterior margin of article 6 not projecting distally   *Amphilochus neapolitanus* Della Valle, 1893

Key III

Uropod 3 biramous, at least one of the rami as long as or longer than the peduncle (in certain species that have an exceptionally long outer ramus the inner ramus is so small by comparison that it may be overlooked in a lateral view).

1a Rostrum enlarged into a hood-shaped or saberlike process that is at least half as long as article 1 of antenna 1 (fig. 18.53 or 18.54)   Subkey III A
1b Rostrum either without a projection or with only a small point that is less than half as long as article 1 of antenna 1   2
2a Peduncle of antenna 1 short and stout (combined length of articles 2 and 3 usually less than half the length of article 1); article 3 of gnathopod 2 distinctly longer than article 4 (figs. 18.55 and 18.56)   Subkey III B
2b Peduncle of antenna 1 not especially stout; article 3 of gnathopod 2 equal to or shorter than article 4   3
3a Eye composed of 2 or 8 widely separated, clear lenses (in addition to an internal pigmentary mass); gnathopod 2 simple and very slender, article 5 about twice as long as article 6 (figs. 18.57 and 18.58)   Subkey III C
3b Lenses of eye clustered in a multifaceted mass or absent altogether; gnathopod 2 usually subchelate (rarely simple), article 5 less than twice as long as article 6   4
4a Neither ramus of uropod 3 distinctly longer than the peduncle (fig. 18.59)   Subkey III D
4b One or both rami of uropod 3 at least 1.5 times as long as the peduncle (fig. 18.60 or 18.61)   Subkey III E

Subkey III A

Rostrum enlarged into a hood-shaped or saberlike projection that is at least half as long as article 1 of antenna 1.

1a One or more pleonites dorsally toothed or with a raised keel (fig. 18.62 and 18.63)   2
1b Pleonites dorsally smooth and undeveloped (occasionally with a small rounded hump on

	pleonites 3 or 4, as in some mature male phoxocephalids) (fig. 18.64)	13
2a	Rostrum dorsoventrally compressed (its sagittal depth much less than its lateral breadth); urosome with a dorsal hooked tooth that is directed anteriorly *Rhepoxynius vigitegus* (Barnard, 1971)	
2b	Rostrum not dorsoventrally compressed (its sagittal depth not much less than its lateral breadth); dorsal projections of the pleonites never directed anteriorly	3
3a	Rostrum blunt, thick and immense, longer than the first 3 articles of antenna 1 *Pleustes depressa* Alderman, 1936 (fig. 18.65)	
3b	Rostrum acutely pointed or very narrow, rarely exceeding the length of the first article of antenna 1	4
4a	Head very bulbous, its dorsal margin very convex over the eye; main eyes fused as a single dorsal mass; coxa 4 much smaller than coxa 3	5
4b	Head not very bulbous, its dorsal margin straight or slightly convex over the eye; eyes never fused as a single dorsal mass; coxa 4 not much smaller than coxa 3	6
5a	With a pair of small eye lenses ventral to, and widely separated from, the main eye mass; gnathopods simple *Tiron biocellata* Barnard, 1962 (fig. 18.62)	
5b	Without eye lenses separate from the main eye mass; gnathopods subchelate *Syrrhoe longifrons* Shoemaker, 1964	
6a	Gnathopod 1 strongly subchelate, article 5 wide (its width at least half its length), article 5 usually much shorter than article 6	7
6b	Gnathopod 1 weakly subchelate or minutely chelate, article 5 narrow (its width about one-fourth its length), article 5 not much shorter than article 6	11
7a	Gnathopods of moderate size, posterior margin of article 5 not projecting as a narrow lobe; pleonites 1 and 2 with a single dorsal ridge; urosomites 2 and 3 fused (*Atylus*)	8
7b	Gnathopods large, article 5 projecting as a narrow posterior lobe alongside article 6 (fig. 18.66); pleonites 1 and 2 dorsally tridentate; urosomites not fused (*Rhachotropis*)	10
8a	Eye small; gnathopods rather slender; posteroventral corner of article 2 of pereopod 7 not projecting alongside article 3; urosomite 1 with a dorsal ridge, this ridge not projecting as a large tooth *Atylus levidensus* Barnard, 1956	
8b	Eye medium to large; gnathopods stout; posteroventral corner of article 2 of pereopod 7 projecting distally alongside article 3; urosomite 1 with a dorsal ridge that is elevated into a distinct tooth	9
9a	Spines of gnathopods stout and peglike; dorsal surface of each urosomite with a single blunt keel-like tooth, but without an anterior notch; rami of uropod 3 of male stout and spinose *Atylus collingi* (Gurjanova, 1938)	
9b	Spines of gnathopods moderately slender, not peglike; dorsal surface of each urosomite with a small notch preceding an acute tooth (giving the appearance of 2 teeth on each segment); rami of uropod 3 of male long and setose *Atylus tridens* (Alderman, 1936) (fig. 18.63)	
10a	Urosomite 1 with a large acute dorsal tooth; telson cleft for about one-twelfth of its length *Rhachotropis clemens* Barnard, 1967 (fig. 18.66)	
10b	Urosomite 1 without a dorsal tooth; telson cleft for about half its length *Rhachotropis oculata* (Hansen, 1887)	
11a	Mouthparts normal (rectangularly grouped); gnathopod 1 weakly subchelate; some pleonites dorsally ridged but without distinct teeth *Bruzelia tuberculata* G. O. Sars, 1895	
11b	Mouthparts grouped together to form a cone; gnathopod 1 minutely chelate; some pleonites with dorsal teeth (Acanthonotozomatidae)	12
12a	Article 6 of gnathopod 2 narrow, not widening distally; with pairs of dorsal teeth at the posterior margins of pleonites 1-3 *Coboldus* sp.	
12b	Article 6 of gnathopod 2 broad, widening distally to twice its proximal width; with a single rounded dorsal tooth on pleonite 3 *Odius kelleri* Bruggen, 1907	
13a	Rostrum laterally compressed into a thin bladelike projection that emerges between the first articles of antennae 1 (fig. 18.54); pereopods 5, 6, and 7 similar in size and shape *Pontogeneia* cf. *rostrata* Gurjanova, 1938	
13b	Rostrum laterally expanded so that its proximal portion overlies the first articles of antennae 1; pereopods 5, 6, and 7 distinctly unequal in size or shape	14
14a	Accessory flagellum absent; eyes (when present) forming a single dorsal mass; pereopod 7 much longer than pereopod 6; telson not cleft (Oedicerotidae)	15

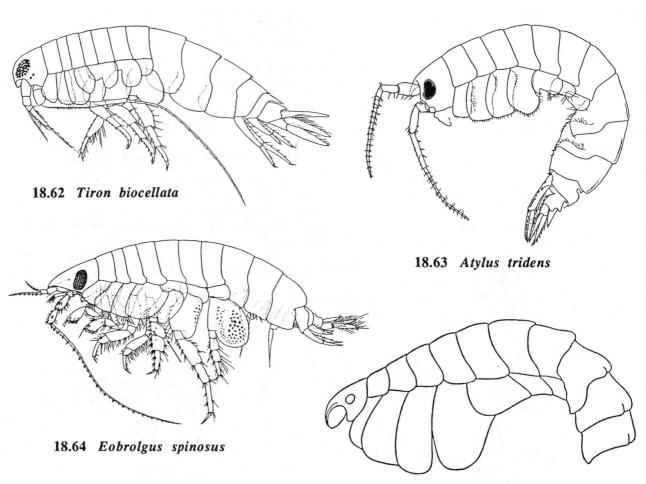

18.62 *Tiron biocellata*

18.63 *Atylus tridens*

18.64 *Eobrolgus spinosus*

18.65 *Pleustes depressa*

14b Accessory flagellum present, multiarticulate; eyes (when present) separate and laterally placed; pereopod 6 much longer than pereopod 7; telson cleft (Phoxocephalidae) 18
15a Gnathopod 2 chelate (fig. 18.3) 16
15b Gnathopod 2 subchelate (figs. 18.67 and 18.68) 17
16a Palm of gnathopod 1 oblique; article 2 of pereopod 7 without long setae on its posterior margin .... *Synchelidium shoemakeri* Mills, 1962
16b Palm of gnathopod 1 transverse; article 2 of pereopod 7 with long setae on its posterior margin .... *Synchelidium rectipalmum* Mills, 1962
17a Gnathopods 1 and 2 similar, posterodistal lobe of article 5 not projecting alongside article 6 .... *Westwoodilla caecula* (Bate, 1856) (fig. 18.67)
17b Gnathopods dissimilar, posterodistal lobe of article 5 projecting alongside article 6 (in gnathopod 2, this extends for nearly the full length of the posterior margin of article 6) .... *Monoculodes* spp.
    Monoculodes zernovi Gurjanova, 1936 (fig. 18.68) and *M. spinipes* Mills, 1962 have been reported.
18a Article 3 of pereopod 5 about as wide as the widest point of article 2; posteroventral corner of epimeron 3 projecting as a sharp tooth beyond its posterior margin 19
18b Article 3 of pereopod 5 half as wide as the widest point of article 2; posteroventral corner of

	epimeron 3 rounded or acute, but rarely projecting as a long tooth (such a tooth present only in *Eyakia robustus*)	20
19a	Eyes absent in both sexes; article 2 of antenna 2 not extended as a long bladelike process; article 2 of pereopod 7 without long setae *Harpiniopsis fulgens* Barnard, 1960	
19b	Eyes present in male, small or absent in female; article 2 of antenna 2 projecting anteriorly as a long bladelike process; article 2 of pereopod 7 with long setae *Heterophoxus oculatus* (Holmes, 1908)	
20a	Rostrum very long, the head thus as long as the first 4 pereonites combined; eyes absent; ramal spines of uropods 1 and 2 crowded near the apices *Mandibulophoxus gilesi* Barnard, 1957	
20b	Rostrum of moderate length, the head rarely longer than the first 3 pereonites; eyes present (minute or absent in female of *R. daboius*); spines on distal portions of rami of uropods 1 and 2 sparse or absent	21
21a	Article 5 of gnathopods less than one-fourth the length of article 6 and cryptic (in gnathopod 2, not reaching the posterior margin, as if squeezed out by article 4); article 6 of gnathopod 1 rectangular	22
21b	Article 5 of gnathopods never less than half the length of article 6, well exposed along the posterior margin in both gnathopods; article 6 of gnathopod 1 ovate or triangular	24
22a	Article 6 of gnathopod 1 only slightly longer than wide, its palm oblique; inner rami of uropods 1 and 2 with about 3 setae; mandibular molar large and ridged *Phoxocephalus homilis* Barnard, 1960	
22b	Article 6 of gnathopod 1 twice as long as wide, palm transverse or slightly chelate; inner rami of uropods 1 and 2 naked or with only 1 seta; mandibular molar small and unridged	23
23a	Posterior margin of article 6 of gnathopod 1 prolonged so as to appear slightly chelate; anterior and posterior margins of article 6 of gnathopod 2 nearly parallel, and the palm of this article transverse *Metaphoxus fultoni* (Scott, 1890) (fig. 18.69)	
23b	Palm of article 6 of gnathopod 1 transverse; article 6 of gnathopod 2 distally expanded, its palm oblique *Metaphoxus frequens* Barnard, 1960	
24a	Rostrum narrow, abruptly tapering in front of the eyes (lateral margins concave when viewed from above)	25
24b	Rostrum broad, gradually tapering in front of the eyes (lateral margins slightly convex when viewed from above)	32
25a	Coxae 1-3 with a small cusp at the posteroventral corner; telson with mid-dorsal spines *Grandifoxus grandis* (Stimpson, 1856)	
25b	Coxae 1-3 without cusps; mid-dorsal part of telson naked or with setae (*Rhepoxynius*)	26
26a	Urosomites 1 and 2 partially coalesced and bearing a dorsal, anteriorly-directed tooth *Rhepoxynius vigitegus* (Barnard, 1971)	
26b	Urosomites clearly articulated and without dorsal tooth	27
27a	Article 2 of pereopod 7 with 2 large posterior teeth *Rhepoxynius bicuspidatus* (Barnard, 1960)	
27b	Article 2 of pereopod 7 with more than 2 posterior teeth	28
28a	Epistome with a distinct anterior cusp (more easily seen by placing animal on its back and sliding a black insect pin between the second antennae and the mouthparts)	29
28b	Epistome without an anterior cusp (rarely with a small anterior bump)	31
29a	Article 2 of pereopod 7 with 5 or more posterior teeth; peduncle of uropod 2 with about 4-7 dorsal spines; tips of telson with short spines and setules (lacking long setae) *Rhepoxynius abronius* (Barnard, 1960)	
29b	Article 2 of pereopod 7 with 3-4 posterior teeth; peduncle of uropod 2 with about 2-3 dorsal spines; tips of telson with long setae	30
30a	Eyes of female minute or absent; epimeron 2 with a cluster of 3 facial setae above its posteroventral corner; teeth on article 2 of pereopod 7 small *Rhepoxynius daboius* (Barnard, 1960)	
30b	Eyes of female small but distinct; epimeron 2 without setae above its posteroventral corner (although setae are present along the ventral margin); teeth on basis of pereopod 7 medium to large *Rhepoxynius variatus* (Barnard, 1960)	
31a	Article 2 of pereopod 7 with 3-4 moderately large teeth, these teeth becoming larger distally; inner ramus of uropod 3 of adult female not more than half the length of the outer ramus *Rhepoxynius tridentatus* (Barnard, 1954)	

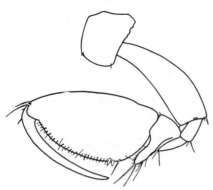

18.66 *Rhachotropis clemens*

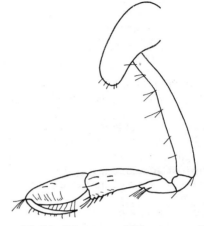

18.67 *Westwoodilla caecula*

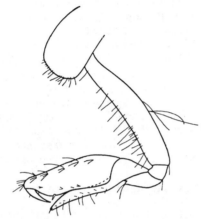

18.68 *Monoculodes zernovi*

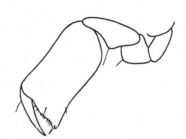

18.69 *Metaphoxus fultoni*

Gnathopod 2

31b Article 2 of pereopod 7 with 4-5 small teeth, these teeth varying randomly in size; inner ramus of uropod 3 of adult female more than half the length of the outer ramus
 .... *Rhepoxynius heterocuspidatus* (Barnard, 1960)
32a Article 1 of antenna 2 projecting forward as a small cusp (lift or remove coxa 1 to expose it); posterior margin of epimeron 2 with a vertical row of several long setae (*Foxiphalus*) 33
32b Article 1 of antenna 2 not projecting as a cusp; posterior margin of epimeron 2 naked or with short setules 35
33a Ventral margin of epimera 2 and 3 rather straight and projecting at the posterior corner as a small tooth; posterior and ventral margins of epimera 2 and 3 with dense rows of setae; peduncle of uropods 1 and 2 lined with setae .... *Foxiphalus major* (Barnard, 1960)
33b Ventral margin of epimera 2 and 3 somewhat convex, the corner not projecting; marginal setae not dense (they are especially sparse along the ventral margin of epimeron 3); peduncle of uropod 1 and 2 with spines 34
34a Epistome not projecting, or with only a small anterior bump; telson with some stout spines proximal to the apex .... *Foxiphalus obtusidens* (Alderman, 1936)
34b Epistome projecting as a long cusp or finger; telson usually lacking spines proximal to the apex .... *Foxiphalus cognatus* (Barnard, 1960) and *F. similis* (Barnard, 1960)
35a Article 2 of antenna 1 with setae widely distributed along its ventral margin; article 2 of

pereopod 5 tapering distally, the distal end only slightly wider than article 3; epimeron 3 with a setose lateral ridge that is extended obliquely to form a posteroventral tooth
Eyakia robustus (Holmes, 1908)

35b Article 2 of antenna 1 with ventral setae confined to its distal end; article 2 of pereopod 5 not significantly tapered, its distal end twice as wide as article 3; epimeron 3 without lateral setae and lacking a posteroventral tooth 36

36a Article 4 of pereopod 5 wider than long; epimeron 3 with 3-5 posterior setae
Eobrolgus spinosus (Holmes, 1905) (fig. 18.64)

36b Article 4 of pereopod 5 longer than wide; epimeron 3 without any setae
Paraphoxus oculatus G. O. Sars, 1879

Subkey III B

Peduncle of antenna 1 short and stout (combined length of article 2 and 3 usually less than half the length of article 1); article 3 of gnathopod 2 distinctly longer than article 4.

1a Gnathopod 1 distinctly chelate (fig. 18.3), the thumb (posterior corner of the palm) of article 6 extending well beyond the hinge of the dactyl 2

1b Gnathopod 1 simple (fig. 18.2) or subchelate (figs. 18.4 and 18.5), sometimes slightly chelate, but the thumb never extending much beyond the hinge of the dactyl 4

2a Dactyl and palm of gnathopod 1 outlining a wide circular gap; the dactyl and thumb touching only at the tips; telson deeply cleft *Opisa tridentata* Hurley, 1963

2b Inner margins of dactyl and palm of gnathopod 1 nearly parallel, touching throughout their length when closed; telson not cleft 3

3a Accessory flagellum small but erect and usually visible in a lateral view; eye usually present; tip of thumb of gnathopod 1 curving posteriorly in the same direction as the dactyl
Prachynella lodo Barnard, 1964

3b Accessory flagellum cryptic, not visible in a lateral view; eye absent; tip of thumb of gnathopod 1 curving slightly anteriorly toward the dactyl   *Pachynus* cf.*barnardi* Hurley, 1963

4a Ventral margin of epimeron 3 extending well beyond the posterior margin to form a large tooth at the posteroventral corner (fig.18.70) 5

4b Ventral margin of epimeron 3 forming nearly a right angle with the posterior margin, the posteroventral corner rounded or angular, sometimes with a small point but without a large tooth (fig. 18.71) (if tooth size is intermediate, check *Acidostoma*, choice 20a) 12

5a Eyes present or absent; inner ramus of uropod 2 bladelike, with spines (if any) restricted to the proximal end of the dorsal margin 6

5b Eyes present; inner ramus of uropod 2 bearing a distinct spine at or beyond the midpoint of its dorsal margin 9

6a Article 4 of antenna 1 about as long as article 1; pereopod 7 with a gill, but this is reduced; tips of telson tapering to a point and usually with only 1-2 apical spines   *Hippomedon* spp.

6b Article 4 of antenna 1 distinctly shorter than article 1; pereopod 7 without a gill; tips of telson truncated and usually with 3-5 apical spines or setae 7

7a Article 4 of antenna 1 shorter than either article 2 or 3; pereopod 7 longer than 6; with a deep incision above the corner tooth of epimeron 3
Psammonyx longimerus Jarrett & Bousfield, 1982

7b Article 4 of antenna 1 about as long as the combined length of articles 2 and 3; pereopod 6 longer than 7; without an incision above the corner tooth of epimeron 3 (*Wecomedon*) 8

8a Antenna 1 of female about four-fifths as long as antenna 2; anterior margin of article 2 of gnathopod 1 sparsely setose; corner tooth of epimeron 3 short and stout
Wecomedon wecomus (Barnard, 1971)

8b Antenna 1 of female about equal in length to antenna 2; anterior margin of article 2 of gnathopod 1 densely setose; corner tooth of epimeron 3 long and slender
Wecomedon similis Jarrett & Bousfield, 1982

9a Coxa 1 reduced and partly hidden by coxa 2; palm of gnathopod 1 very oblique and poorly set off from the posterior margin so that it appears simple (compare figs. 18.2 and 18.5); article 5 of gnathopod 1 slightly longer than article 6   *Schisturella cocula* Barnard, 1966

366 Phylum Arthropoda: Order Amphipoda

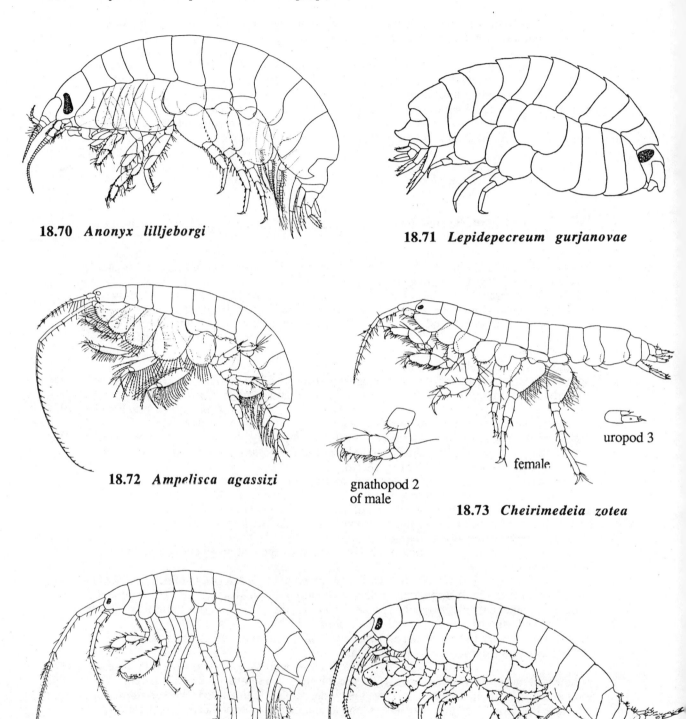

18.70 *Anonyx lilljeborgi*

18.71 *Lepidepecreum gurjanovae*

18.72 *Ampelisca agassizi*

gnathopod 2 of male

female

uropod 3

18.73 *Cheirimedeia zotea*

18.74 *Melita dentata*

calceolus, enlarged

18.75 *Eogammarus confervicolus*

9b Coxa 1 not reduced and not hidden by coxa 2; palm of gnathopod 1 transverse (fig. 18.4); article 5 of gnathopod 1 not longer than article 6 10
10a Eye orange; coxa 1 not widened distally; spine at the base of the dactyls of pereopods 3 and 4 not distally hooked (observe under compound microscope)
Orchomene decipiens (Hurley, 1963)
10b Eye reddish brown; coxa 1 widened distally; spine at the base of the dactyls of pereopods 3 and 4 distally hooked (*Anonyx*) 11
11a Article 6 of pereopods 3 and 4 with paired rows of long setae and short spines along the posterior margin; spine on inner ramus of uropod 2 small, and the ramus not constricted distal to the spine *Anonyx* cf. *laticoxae* Gurjanova, 1962
11b Article 6 of pereopods 3 and 4 with a row of paired, long setae (but no spines) along the posterior margin; spine on inner ramus of uropod 2 large and the ramus somewhat constricted distal to the spine *Anonyx* cf. *lilljeborgi* Boeck, 1870 (fig. 18.70)
12a Telson cleft for more than half of its length (figs. 18.14 and 18.15) 13
12b Telson not cleft (fig. 18.12), or cleft for less than half of its length (fig. 18.13) 19
13a Pereonite 1 dorsally convex, forming a distinct hump just behind the head (especially in the male); article 2 of pereopod 5 with a long posterior tooth that projects beyond the distal end of article 5 *Cyphocaris challengeri* Stebbing, 1888
13b Pereonite 1 not forming a dorsal hump; article 2 of pereopod 5 without a prominent tooth that projects beyond article 5 14
14a Coxa 1 tapering distally, much shorter than coxa 2, and concealed by coxa 2; palm of gnathopod 1 indistinctly set off from the posterior margin so that it appears simple; pereopods parachelate (not quite so pronounced as in *Polycheria*, fig. 18.10)
Aristias pacificus Schellenberg, 1936
14b Coxa 1 not tapering distally, only slightly shorter than coxa 2, and not concealed by coxa 2; palm of gnathopod 1 clearly set off by spines, distinctly subchelate; pereopods normal, not parachelate 15
15a Urosomite 1 with an erect dorsal keel which is usually very angular; pereopods 3 and 4 with a pair of small spines at the base of the dactyl (these spines are distally hooked; observe with a compound microscope) 16
15b Urosomite 1 without a dorsal keel, but often with a rounded protuberance; pereopods 3 and 4 with 1 or 2 simple (not hooked) spines at the base of the dactyl 18
16a Article 1 of antenna 1 not distally expanded over article 2
Orchomene pacifica (Gurjanova, 1938)
16b Article 1 of antenna 1 distally expanded into a pronounced tooth that extends dorsally over article 2 (fig. 18.71) (*Lepidepecreum*) 17
17a Eyes present; only urosomite 1 with a dorsal keel
Lepidepecreum gurjanovae Hurley, 1963 (fig.18.71)
17b Eyes absent; all pereonites and pleonites 1-3, as well as urosomite 1, with a dorsal tooth or keel *Lepidepecreum garthi* Hurley, 1963
18a Eye distinctly broadened ventrally; lateral lobe of head evenly rounded and parallel to the anterior margin of the eye; urosomite 1 with a large dorsal protuberance
Orchomene obtusa (G. O. Sars, 1890)
18b Eye only slightly broadened ventrally; lateral lobe of head somewhat angular, but with rounded corners; urosomite 1 with a slight protuberance, or none at all
Orchomene cf. *pinguis* (G. O. Sars, 1895)
19a Gnathopod 1 simple (fig. 18.2) 20
19b Gnathopod 1 subchelate or slightly chelate (fig. 18.3-18.5) 21
20a Peduncle of uropod 2 with an upright dorsal keel; inner ramus of uropod 2 without a long spine near its apex; uropod 3 very short, hardly exceeding the length of the peduncle of uropod 2 *Acidostoma* sp.
20b Peduncle of uropod 2 without a dorsal keel; inner ramus of uropod 2 with a long spine near its apex; uropod 3 extending well beyond the peduncle of uropod 2
Lysianassa holmesi (Barnard, 1955)
21a Eyes very elongated, nearly meeting dorsally; lateral lobe of head broadly rounded throughout; coxa 1 widened distally; article 6 of gnathopod 1 about twice as long as article 5
Koroga megalops Holmes, 1908

21b Eyes ovate or reniform, not dorsally elongated; lateral lobe of head expanded acutely in front of eye, although apically rounded; coxa 1 not widened distally; article 6 of gnathopod 1 about as long as article 5 22
22a Gnathopod 1 chelate but only slightly so; telson usually triangular (especially in the female) but variable; apical portion of the telson extending beyond the distal spines, apex minutely notched .... *Allogaussia recondita* Stasek, 1958
22b Gnathopod 1 subchelate, the palm transverse; telson nearly square, its small distal spines truly apical and separated by a shallow concavity .... *Orchomene* sp.

Subkey III C

Eye composed of 2 or 8 clear lenses that are widely separated (in addition to an internal pigmented mass); gnathopod 2 simple and very slender, article 5 about twice as long as article 6. (Families Ampeliscidae and Argissidae).

1a Eye composed of 8 small lenses, these arranged in pairs around the periphery of the eye; accessory flagellum present and consisting of 2 articles; coxae 1 to 3 successively shorter, but coxa 4 longer than coxa 1 .... *Argissa hamatipes* (Norman, 1869)
1b Eye composed of 2 lenses, 1 dorsolateral and 1 anteroventral (fig. 18.57); accessory flagellum absent; coxae 1 to 4 nearly equal or slightly increasing in length (Ampeliscidae) 2
2a Article 2 of pereopod 7 about twice as wide as article 3, not distally expanded; article 6 of pereopod 7 less than half as long as article 5 .... *Haploops tubicola* Lilljeborg, 1856
2b Article 2 of pereopod 7 at least 3 times as wide as article 3, distally expanded; article 6 of pereopod 7 not noticeably shorter than article 5 3
3a Posterior lobe of article 2 of pereopod 7 with setae extending along the ventral margin up to its junction with article 3; article 6 of pereopod 7 with several groups of spines along the posterior margin; preserved specimens usually with dark pigment, especially around the eyes (*Byblis*) 4
3b Posterior lobe of article 2 of pereopod 7 lacking setae near its junction with article 3; article 6 of pereopod 7 with posterior spines restricted to the distal corner (proximal posterior margin without spines); preserved specimens usually without dark pigment (*Ampelisca*) 5
4a Antenna 1 extending beyond the end of the peduncle of antenna 2; antenna 2 about as long as the body; coxae 2 and 3 distally tapered and only slightly serrated along their ventral margins; outer ramus of uropod 1 longer than inner ramus .... *Byblis veleronis* Barnard, 1954
4b Antenna 1 just reaching the end of the peduncle of antenna 2; antenna 2 about two-thirds as long as the body; coxae 2 and 3 not noticeably tapered, their ventral margins distinctly serrated; rami of uropod 1 nearly equal in length .... *Byblis millsi* Dickinson, 1983
5a Outer ramus of uropod 2 with a long spine at its apex; article 5 of pereopod 7 shorter than article 6 (but it is nearly equal to article 6 in *A. brevisimulata*); epimeron 3 with a distinct tooth at the posteroventral corner 6
5b Outer ramus of uropod 2 without a long apical spine; article 5 of pereopod 7 nearly equal in length to article 6; epimeron 3 usually lacking a tooth at the posteroventral corner (there is a small tooth in *A. hancocki*) 10
6a Article 5 of pereopod 7 nearly equal in length to article 6; posterior margin of epimeron 3 with a marked convexity that often extends posteriorly beyond the posteroventral tooth; telson without apical notches (best observed when mounted separately) .... *Ampelisca brevisimulata* Barnard, 1954
6b Article 5 of pereopod 7 shorter than article 6; posterior margin of epimeron 3 with a slight or moderate convexity that does not extend beyond the posteroventral tooth; telson with one or more apical notches 7
7a Dorsal and ventral margins of distal half of head nearly parallel; dorsal keel of urosomite 1 thin and usually rounded in lateral view .... *Ampelisca cristata* Holmes, 1908
7b Dorsal and ventral margins of the head not parallel; dorsal keel of urosomite 1 thick and angular in lateral view 8
8a Posterior lobe of article 4 of pereopod 7 large, nearly covering the posterior margin of article 5; anterior margin of article 5 of pereopod 7 notched .... *Ampelisca pugetica* Stimpson, 1864

Phylum Arthropoda: Order Amphipoda

8b Posterior lobe of article 4 of pereopod 7 small, reaching about half way along article 5; article 5 without an anterior notch 9

9a Ventral margin of head slightly concave just behind the ventral eye lens; anterodorsal margin of head slightly bulbous and projecting over the attachment of antenna 1; antenna 1 slightly longer than the peduncle of antenna 2 *Ampelisca careyi* Dickinson, 1982

9b Ventral margin of head not concave (corneal eyes occasionally absent); anterodorsal margin of head not bulbous and not projecting over the attachment of antenna 1; antenna 1 shorter than the peduncle of antenna 2 *Ampelisca unsocalae* Barnard, 1960

10a Urosomite 1 with a prominent dorsal keel; peduncle of uropod 1 stout, nearly twice as wide as the peduncle of uropod 2; both tips of telson with a row of 3-5 setae *Ampelisca agassizi* (Judd, 1896) (fig. 18.72)

10b Urosomite 1 with a slight dorsal keel or low protuberance; peduncle of uropod 1 usually only slightly wider than that of uropod 2 (although moderately wide in *A. hancocki*); both tips of telson bearing a single small spine (occasionally accompanied by a seta) 11

11a Posterior margin of article 5 of pereopod 7 much shorter than the anterior margin, which is deeply notched; posteroventral corner of epimeron 3 rounded *Ampelisca fageri* Dickinson, 1982

11b Posterior and anterior margins of article 5 of pereopod 7 nearly equal, and the anterior margin without a deep notch; posteroventral corner of epimeron 3 forming a right angle or with a tooth 12

12a Posterior lobe of article 4 of pereopod 7 reaching slightly more than half way along article 5; posteroventral corner of epimeron 3 with a prominent tooth projecting below a concave posterior margin *Ampelisca hancocki* Barnard, 1954

12b Posterior lobe of article 4 of pereopod 7 reaching less than half way along article 5; posteroventral corner of epimeron 3 forming a right angle or with an inconspicuous tooth below a nearly straight posterior margin *Ampelisca lobata* Holmes, 1908

Subkey III D

Neither ramus of uropod 3 distinctly longer than the peduncle (rostrum not developed; article 3 of gnathopod 2 short; lenses of the eye clustered or eye absent altogether).

1a Telson thin and laminar, cleft or not cleft 2
1b Telson thick and fleshy, usually not cleft but sometimes with a slight notch 8
2a Urosomite 1 with a dorsal keel; urosomites 2 and 3 fused *Dexamonica reduncans* Barnard, 1957
2b Urosomite 1 without a dorsal keel; urosomites 2 and 3 distinctly separate 3
3a Telson cleft for at least half its length, tips of telson with numerous stout spines; gnathopod 2 of male greatly enlarged (about 4 times as large as gnathopod 1) (not recorded with certainty north of California) *Elasmopus* spp.
3b Telson not cleft or cleft for about one-fourth its length, apex of telson with no more than 2 small spines (fig. 18.12 or 18.13); gnathopod 2 of male not much larger than gnathopod 1 4
4a Eye absent or indistinct; article 5 of gnathopod 2 longer than article 6; pereopod 7 much longer than pereopod 6 (Oedicerotidae, with short rostrum) 5
4b Eye distinct; article 6 of gnathopod 2 longer than article 5; pereopods 6 and 7 nearly equal in length 7
5a Palm of gnathopods poorly distinguished from the posterior margin of article 6; tip of telson slightly concave *Bathymedon pumilus* Barnard, 1962
5b Palm of gnathopods defined by a large spine at the posterior corner; tip of telson not concave 6
6a Posterior lobe of article 5 of gnathopod 1 not extending along article 6 (fig. 18.67); article 2 of pereopod 7 rectangular, its posterior margin with only sparse setae *Bathymedon flebilis* Barnard, 1967
6b Posterior lobe of article 5 of gnathopod 1 extending distally along article 6 (fig. 18.66 or 18.68); article 2 of pereopod 7 somewhat triangular, its posterior margin decidedly setose Oedicerotidae, unidentified

7a	Coxae 1-4 with spines along their posterior margins; telson notched about one-fourth its length and with a spine at each tip (fig. 18.13) *Oligochinus lighti* Barnard, 1969	
7b	Coxae 1-4 without spines along their posterior margins; telson not cleft, its margin with 6-8 small setules *Paracalliopiella pratti* (Barnard, 1954)	
8a	Rami of uropod 3 naked except for a small spine embedded in the tip of each ramus *?Megamphopus* sp.	
8b	Rami of uropod 3 bearing several spines or setae	9
9a	Eye set back from lateral lobe of head; gnathopod 1 of male merochelate (fig. 18.6) and much larger than gnathopod 2	10
9b	Eye centered on lateral lobe of head; gnathopod 1 of male subchelate and smaller than gnathopod 2	15
10a	Accessory flagellum of antenna 1 composed of 2 articles, small but distinctly visible in lateral view; enlarged coxa 1 of male circular in outline; distal peduncular spine of uropods 1 and 2 more than half the length of the rami *Columbaora cyclocoxa* Conlan & Bousfield, 1982	
10b	Accessory flagellum vestigial, not apparent in lateral view; enlarged coxa 1 of male somewhat triangular in outline (fig. 18.6); distal peduncular spine of uropods 1 and 2 about one-third the length of the rami (*Aoroides*)	11
11a	Outer ramus of uropod 3 usually with apical setae only (rarely with 1-2 small spines); posterior margin of article 2 of enlarged gnathopod 1 of male without setae (although setae present on lateral margin); article 2 of gnathopod 1 of female with anterodistal setae which are longer than the width of this article	12
11b	Outer ramus of uropod 3 with 1-3 small proximal spines in addition to the apical setae; posterior margin of article 2 of enlarged gnathopod 1 of male setose; article 2 of gnathopod 1 of female with minute anterodistal setae	14
12a	Body diffusely pigmented, except for light patches on head and pereonites 6 and 7; article 5 of gnathopod 1 of male with only one distal group of short setae along anterior margin, this article wider than article 2 *Aoroides columbiae* Walker, 1898	
12b	Head and all pereonites pigmented in discrete spots; article 5 of gnathopod 1 of male setose along its anterior margin, this article about as wide as article 2	13
13a	Body pigment usually concentrated in posteroventral corners of pereonites 1-5; article 5 of gnathopod 1 of male with more than 7 groups of setae along its anterior margin *Aoroides inermis* Conlan & Bousfield, 1982	
13b	Body pigment not concentrated in corners of pereonites; article 5 of gnathopod 1 of male with 5-7 groups of setae along its anterior margin *Aoroides intermedius* Conlan & Bousfield, 1982	
14a	Body diffusely pigmented, except for light patches on head and pereonites 6 and 7; article 2 of pereopod 7 about twice as wide as articles 3-6, with sparse, short setae along its posterior margin; articles 2 and 3 of gnathopod 1 of male with crowded setae that are longer than the width of each article *Aoroides exilis* Conlan & Bousfield, 1982	
14b	Head and pereonites pigmented with discrete spots; article 2 of pereopod 7 about 3 times as wide as articles 3-6, with crowded, long setae along its posterior margin; articles 2 and 3 of gnathopod 1 of male with sparse setae that are shorter than the width of each article *Aoroides spinosus* Conlan & Bousfield, 1982	
15a	Posterior margins of urosomites 1 and 2 with a tooth and long seta on each side of the dorsal midline; epimeron 3 with a convex posterior margin above a small posteroventral tooth; telson notched and with 2 stout spines	16
15b	Urosomites 1 and 2 without dorsal teeth or long setae; epimeron 3 rounded or quadrate, but without a small distinct tooth at the posteroventral corner; telson not notched and with only slender spines or setae	18
16a	Lateral lobe of head acute, but not tapering to an extremely sharp-pointed tooth; palm of gnathopod 2 of both sexes with a broad, uniform concavity; coxa 7 not enlarged in male *Gammaropsis ellisi* Conlan, 1983	
16b	Lateral lobe of head tapering to an extremely sharp-pointed tooth; palm of gnathopod 2 of both sexes not broadly concave (palm of male *G. thompsoni* with 2 small notches); coxa 7 enlarged in male	17
17a	Coxa 1 about as long as coxa 2; article 2 of gnathopod 1 of male with a posterodistal lobe;	

palm of gnathopod 2 of male oblique and broadly convex
Gammaropsis shoemakeri Conlan, 1983

17b Coxa 1 distinctly shorter than coxa 2 (especially in the male); article 2 of gnathopod 1 of male without a posterodistal lobe; palm of gnathopod 2 of male nearly transverse and with 2 small notches *Gammaropsis thompsoni* (Walker, 1898)

18a Accessory flagellum of antenna 1 absent or consisting of 1 cryptic article; eye medium to large; epistome with a sharp projection; article 2 of pereopod 5 of male with a posterodistal notch (very slight in *Podoceropsis chionoecetophila*) 19

18b Accessory flagellum of antenna 1 with 2 or more distinct articles; eye medium to small; epistome not developed; article 2 of pereopod 5 of male without a posterodistal notch 21

19a Eye medium; coxae 1-5 increasing in length, their ventral margins distinctly setose; posterior margin of article 2 of pereopods 5-7 with long setae; article 2 of pereopod 5 of male with a very slight posterodistal notch *Podoceropsis chionoecetophila* Conlan, 1983

19b Eye large; coxae 1-5 nearly equal in length, their ventral margins with only sparse short setae; posterior margin of article 2 of pereopods 5-7 with only sparse short setae; article 2 of pereopod 5 of male with a deep posterodistal notch 20

20a Antenna 1 shorter than antenna 2; article 6 of gnathopod 2 of mature male rectangular, the dactyl more than half as long as article 6 *Kermystheus ociosa* Barnard, 1962

20b Antennae 1 and 2 about equal in length; article 6 of gnathopod 2 of male ovate, the dactyl about half as long as article 6 *Podoceropsis barnardi* Kudryashov & Tsvetkova, 1975

21a Lateral lobe of head acute; coxae 1-4 long, fully covering article 2 of pereopods 1-4; posterior margin of article 2 of pereopods 5-7 without long setae; ventral margin of epimeron 2 without long setae; outer ramus of uropod 3 with only short spines
Gammaropsis spinosa (Shoemaker, 1942)

21b Lateral lobe of head rounded or quadrate; coxae 1-4 short, extending less than half way along article 2 of pereopods 1-4; posterior margin of article 2 of pereopods 5-7 with long setae; ventral margin of epimeron 2 with long setae; outer ramus of uropod 3 with 1 or more apical setae 22

22a Article 5 of gnathopod 2 of male shorter than article 6; article 4 of pereopod 3 with no more than 4 small clusters of setae along its anterior margin; inner ramus of uropod 3 usually about half as long as outer ramus (fig. 18.73) (except in *C. similicarpa*) (*Cheirimedeia*) 23

22b Article 5 of gnathopod 2 of male longer than article 6; article 4 of pereopod 3 with setae along its entire anterior margin; inner ramus of uropod 3 nearly as long as outer ramus (*Protomedeia*) 25

23a Inner ramus of uropod 3 about as long as outer ramus; anterior margin of article 5 of gnathopod 2 of female with a prominent convexity, and this article wider than article 6
Cheirimedeia similicarpa Conlan, 1983

23b Inner ramus of uropod 3 about half as long as outer ramus; anterior margin of article 5 of gnathopod 2 of female without a prominent convexity, and this article not distinctly wider than article 6 24

24a Article 6 of gnathopods 1 and 2 gradually tapered to a short transverse palm (exclusive of any posterior spines), the palm greatly overlapped by the dactyl; posteroventral corner of epimeron 3 rounded *Cheirimedeia zotea* (Barnard, 1962) (fig. 18.73)

24b Article 6 of gnathopods 1 and 2 not tapered proximal to the palm, palm oblique and nearly as long as the posterior margin, the dactyl not greatly overlapping the palm; posteroventral corner of epimeron 3 quadrate *Cheirimedeia macrocarpa* subsp. *americana* Conlan, 1983

25a Antenna 1 about 1.5 times the length of antenna 2; articles 5 and 6 of gnathopod 2 of male very wide, about 4 times the width of articles 5 and 6 of gnathopod 1; coxa 5 of male about equal in length to coxa 4 *Protomedeia grandimana* Bruggen, 1905

25b Antenna 1 only slightly longer than antenna 2; articles 5 and 6 of gnathopod 2 of male not more than twice as wide as articles 5 and 6 of gnathopod 1; coxa 5 of male distinctly longer than coxa 4 26

26a Article 5 of gnathopods 1 and 2 of male about 1.5 times the length of article 6; palm of gnathopod 2 of male with a large tooth at the corner which reaches to the hinge point of the dactyl *Protomedeia penates* Barnard, 1966 and *P. prudens* Barnard, 1966

26b Article 5 of gnathopods 1 and 2 of male about equal in length to article 6; palm of gnathopod 2 of male with an articulated spine at the corner which fails to reach the hinge point of the dactyl *Protomedeia articulata* Barnard, 1962

Subkey III E

Uropod 3 biramous, with one or both rami at least 1.5 times as long as the peduncle (rostrum not especially developed; article 3 of gnathopod 2 short; eyes typical or absent).

1a	One or both gnathopods simple (fig. 18.2)	2
1b	Neither gnathopod simple (the palm is distinct, although it may be shorter than the dactyl)	9
2a	Eyes large and distinct; article 5 of gnathopod 2 with a posterodistal lobe projecting somewhat alongside article 6 (similar to fig. 18.66); rami of uropod 3 broad and paddlelike *Megaluropus* sp.	
2b	Eyes absent or indistinct, generally not recognizable in preserved specimens; article 5 of gnathopod 2 without a posterodistal lobe projecting alongside article 6; rami of uropod 3 slender or foliaceous, but not paddlelike	3
3a	Articles 4 and 5 of antenna 2 slender and only slightly setose; pereopods 6 and 7 slender; pereopod 6 not subchelate (Pardaliscidae)	4
3b	Articles 4 and 5 of antenna 2 broad and with many long setae; pereopods 6 and 7 broad; pereopod 6 appearing subchelate (fig. 18.11) (*Eohaustorius*)	5
4a	Mouthparts clustered in a conical bundle; gnathopods with article 6 longer than article 5, article 6 proximally expanded *Halicella halona* Barnard, 1971	
4b	Mouthparts not conically arranged; gnathopods with article 5 longer than article 6, article 6 not proximally expanded *Pardalisca tenuipes* G. O. Sars, 1893	
5a	Article 2 of pereopod 7 with a tooth at its proximal posterior corner; article 5 of adult pereopod 7 with 3 groups of spines along the anterior margin	6
5b	Article 2 of pereopod 7 lacking a posteroproximal tooth; article 5 of adult pereopod 7 with 2 groups of spines along the anterior margin	7
6a	Article 6 of adult pereopod 5 with 3 groups of spines along the posterior margin; posteroproximal tooth on article 2 of pereopod 7 stout and recurved *Eohaustorius washingtonianus* (Thorsteinson, 1941)	
6b	Article 6 of adult pereopod 5 with 2 groups of spines along posterior margin; posteroproximal tooth on article 2 of pereopod 7 triangular and only slightly developed *Eohaustorius brevicuspis* Bosworth, 1973	
7a	Article 7 of gnathopod 1 as long as article 4 *Eohaustorius sencillus* Barnard, 1962	
7b	Article 7 of gnathopod 1 half as long as article 4	8
8a	Article 5 of pereopod 4 with groups of spines all along the posterior margin; article 6 (which is the terminal one, there being no dactyl) of pereopod 5 longer than article 5; article 2 of pereopod 7 with a rounded projection on its posteroproximal corner *Eohaustorius sawyeri* Bosworth, 1973	
8b	Article 5 of pereopod 4 without spines along its posteroproximal margin; article 6 of pereopod 5 approximately equal in length to article 5; article 2 of pereopod 7 with proximal posterior corner smoothly rounded *Eohaustorius estuarius* Bosworth, 1973	
9a	Gnathopods with article 6 distally attached to a long anterodistal lobe of article 5 at a point close to the dactylar hinge (fig. 18.8)	10
9b	Gnathopods not as above	11
10a	Peduncle of antenna 1 shorter than the head; accessory flagellum of article 1 reduced to a small scale fused to article 3 of antenna 1; telson not notched, its apex rounded *Pleusirus secorrus* Barnard, 1969	
10b	Peduncle of antenna 1 longer than the head; accessory flagellum of article 1 small but distinctly articulated; telson with an apical notch *Eusirus* spp.	
11a	Accessory flagellum distinct, multiarticulate and usually visible in a lateral view	12
11b	Accessory flagellum lacking or indistinct (it is sometimes represented by a small scale or lobe)	27
12a	Rami of uropod 3 nearly equal in length	13
12b	Inner ramus of uropod 3 very short, less than one-half the length of the outer ramus	18
13a	Urosome without dorsal teeth or spines (*Maera*)	14
13b	Urosome with dorsal teeth or spines	16
14a	Palm of gnathopod 2 of male with 1 or 2 deep notches among the teeth; rami of uropod	

	3 less than twice the length of the peduncle; each lobe of telson with two or more stout subapical spines that clearly extend beyond the apex	*Maera simile* Stout, 1913
14b	Palm of gnathopod 2 of male with a row of spines and teeth; rami of uropod 3 more than twice the length of the peduncle; each lobe of the telson with 1 or more slender spines or setae that scarcely extend beyond the apex, if at all	15
15a	Eye dark; article 2 of pereopods 5-7 twice as long as wide; epimeron 3 with ventral margin convex to such a degree that the corner tooth occurs near the midpoint of the posterior margin	*Maera danae* Stimpson 1853
15b	Eye inconspicuous; article 2 of pereopods 5-7 three times longer than wide; tooth of epimeron 3 occurring at the typical posteroventral corner	*Maera loveni* (Bruzelius, 1859)
16a	Urosomite 2 with several dorsal spines	*Lagunogammarus setosus* (Dementieva, 1931)
16b	Urosomite 2 with a single dorsal tooth	17
17a	Posterior margin of epimeron 3 smooth; urosomite 1 with several dorsal spines	*Anisogammarus pugettensis* (Dana, 1853)
17b	Posterior margin of epimeron serrate; urosomite 1 with a single dorsal tooth	*Ceradocus spinicaudus* (Holmes, 1908) (fig. 18.80)
18a	Urosome with dorsal teeth (occasionally with spines) (*Melita*)	19
18b	Urosome with dorsal spines, but without teeth	23
19a	Urosomite 1 with fewer than 3 dorsal teeth	20
19b	Urosomite 1 with 3 or more dorsal teeth	22
20a	Urosomite 1 without a dorsal tooth; gnathopod 1 of male with dactyl reduced and closing on the broad medial surface of article 6	*Melita oregonensis* Barnard, 1954
20b	Urosomite 1 with a single dorsal tooth; dactyl of gnathopod 1 of male either normal (fig. 18.74) or reduced	21
21a	Dactyl of gnathopod 1 of male reduced and closing on the broad medial surface of article 6; palm of gnathopod 2 of male with a densely setose ridge lateral to the dactyl	*Melita sulca* (Stout, 1913)
21b	Dactyl of gnathopod 1 of male typical, closing on the palm; palm of gnathopod 2 of male without a ridge and only sparsely setose	*Melita desdichada* Barnard, 1962
22a	All pleonites with dorsal teeth; urosomite 1 with 5 or more dorsal teeth	*Melita dentata* (Krøyer, 1842) (fig. 18.74)
22b	Pleonites 1-3 dorsally without dorsal teeth; urosomite 1 with 3 dorsal teeth (there are variants with minute secondary teeth)	*Melita californica* Alderman, 1936
23a	Pleonites 1-3 with spines or setae along their dorsal posterior margins	24
23b	Pleonites 1-3 without dorsal spines or setae	25
24a	Pleonites 1-3 usually with dorsal setae only; inner ramus of uropod 1 with 2 marginal spines	*Ramellogammarus ramellus* (Weckel, 1907)
24b	Pleonites 1-3 each with 1-2 dorsal spines (plus setae); inner ramus of uropod 1 with 3-4 marginal spines	*Ramellogammarus vancouverensis* Bousfield, 1979
25a	Uropods 1 and 2 short, not reaching beyond the peduncle of uropod 3; outer ramus of uropod 3 appearing to be uniarticulate (article 2 minute); epimera (especially epimeron 2) with posteroventral corners rounded or forming a right angle	*Locustogammarus levingsi* Bousfield, 1979
25b	Uropods 1 and 2 long, reaching beyond the peduncle of uropod 3; outer ramus of uropod 3 with a small but distinct article 2; posteroventral corner of each epimeron acute and bearing a spine (*Eogammarus*)	26
26a	Articles 4 and 5 of antenna 2 with 3-4 subterminal groups of setae along the posterior margin; flagellum of female lacking calceoli; apices of telson each with 2 spines (plus setae)	*Eogammarus oclairi* Bousfield, 1979
26b	Articles 4 and 5 of antenna 2 with 2 (rarely 3) subterminal groups of setae along the posterior margin; flagellum of female with minute calceoli; apices of telson each with a single spine (plus setae)	*Eogammarus confervicolus* (Stimpson, 1856) (fig. 18.75)
27a	Telson not cleft (fig. 18.12)	28
27b	Telson cleft (figs. 18.14 and 18.15)	35
28a	Articles 5 and 6 of gnathopod 2 very linear, 6-8 times longer than wide	*Oradarea longimana* (Boeck, 1871)
28b	Articles 5 and 6 of gnathopod 2 ovate or rectangular, rarely more than 3 times longer than wide	29

374 Phylum Arthropoda: Order Amphipoda

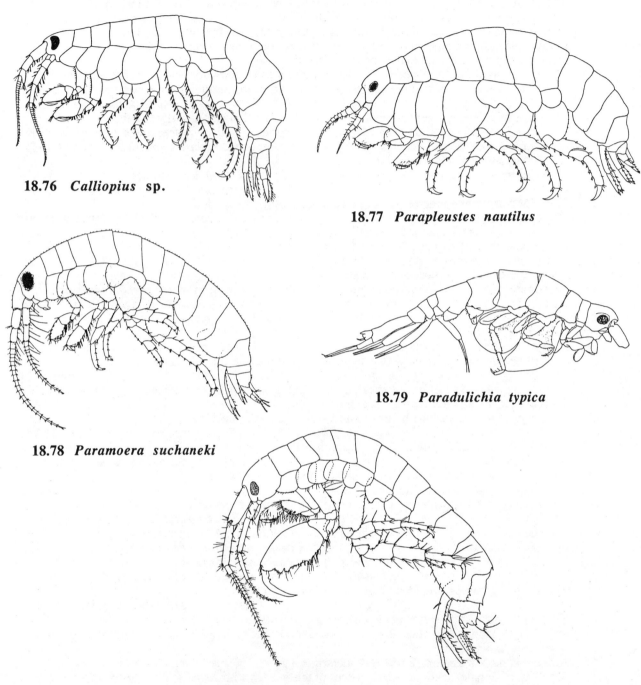

18.76 *Calliopius* sp.

18.77 *Parapleustes nautilus*

18.78 *Paramoera suchaneki*

18.79 *Paradulichia typica*

18.80 *Ceradocus spinicaudus*

29a	Lower distal corner of article 3 of antenna 1 forming a tooth that extends along the proximal flagellar articles	*Calliopius* spp. (fig. 18.76)
29b	Article 3 of antenna 1 distally not forming a prominent tooth (Pleustidae, in part)	30
30a	Corner of epimeron 3 forming a stout tooth that extends well beyond the posterior margin; mandibular molar large and with ridges (*Pleusymtes*)	31
30b	Corner of epimeron 3 forming a right angle, either without a tooth, or with a small tooth that does not extend much beyond the posterior margin; mandibular molar small and without ridges (*Parapleustes*; caution: undescribed species may occur)	32
31a	Posteroventral corners of coxae 1-3 with a minute notch and seta; posterior margin of epimeron 3 convex	*Pleusymtes subglaber* (Barnard & Given, 1960)
31b	Posteroventral corners of coxae 1-3 with a deep notch and curved tooth; posterior margin of epimeron 3 straight	*Pleusymtes* sp.
32a	Antennae 1 and 2 not longer than the head and first 3 pereonites combined; eye small; uropods short, the outer ramus of uropod 3 not reaching much beyond the tip of the telson	*Parapleustes nautilus* Barnard, 1969 (fig. 18.77)
32b	Antennae 1 and 2 longer than head and first 3 pereonites combined; eye moderate to large; uropods long, outer ramus of uropod 3 reaching well beyond the tip of the telson	33
33a	Article 6 of gnathopods nearly rectangular, their palms only slightly oblique; palm of gnathopod 2 equal to about one-third the length of the posterior margin of article 6	*Parapleustes oculatus* (Holmes, 1908)
33b	Article 6 of gnathopods ovate, their palms very oblique; palm of gnathopod 2 equal to more than half of the length of the posterior margin of article 6	34
34a	Gnathopods of adults with palms straight and not clearly distinct from the posterior margin of article 6; palms densely setose and lacking a cusp at their midpoint	*Parapleustes den* Barnard, 1969
34b	Gnathopods of adults with palms convex and clearly distinct from the posterior margin of article 6; palm sparsely setose and with a small cusp at its midpoint	*Parapleustes pugettensis* (Dana, 1853) group
35a	Pereopods 3-7 parachelate (fig. 18.10); urosomites 2 and 3 fused	*Polycheria osborni* Calman, 1898
35b	Pereopods 3-7 not parachelate (fig. 18.9); urosomites 2 and 3 distinctly separate (superfamily Eusiroidea)	36
36a	Gland cone of article 2 of antenna 2 closely appressed to article 3, not prominent in lateral view; posterior margin of article 6 of gnathopod 1 with 2-3 irregular groups of setae; apices of telson without spines or setae	37
36b	Gland cone of article 2 of antenna 2 prominent in lateral view (fig. 18.1); posterior margin of article 6 of gnathopod 1 with 3 or more uniform, comblike groups of setae; apices of telson with spines or setae (*Paramoera*)	41
37a	Accessory flagellum composed of a single small article, hidden against the medial surface of article 4 of antenna 1 (if article 6 of gnathopods is greatly enlarged, as in fig. 18.66, see *Rhachotropis* in Subkey III A, couplet 10); coxae 1-2 with large spines along their posterior margin; telson with a single small spine on each lateral margin (fig. 18.14)	*Accedomoera vagor* Barnard, 1969
37b	Accessory flagellum absent; posterior margins of coxae 1-2 without large spines, but sometimes with setae or small spines; telson without spines, but sometimes with small setules (*Pontogeneia*)	38
38a	Rostrum reaching about one-fourth the length of article 1 of antenna 1; eye black (in alcohol); apices of telson rounded	*Pontogeneia intermedia* Gurjanova, 1938
38b	Rostrum reaching more than one-third the length of article 1 of antenna 1; eye red, brown, or yellow (in alcohol); apices of telson pointed	39
39a	Rostrum reaching half the length of article 1 of antenna 1; gnathopods with article 5 nearly equal in length to article 6; epimeron 3 with a distinct tooth at its corner; inner margins of the cleft of the telson straight, diverging only slightly at the tips	*Pontogeneia* cf. *rostrata* Gurjanova, 1938
39b	Rostrum reaching less than half the length of article 1 of antenna 1; gnathopods with article 5 shorter than article 6; corner of epimeron 3 obtusely angular, the tooth indistinct; inner margins of the cleft of the telson convex or sinuous, clearly diverging at the tips	40

40a	Posterior margin of epimeron 3 slightly convex, broadly rounded; telson lyre-shaped, the inner margins of the cleft somewhat sinuously convex (fig. 18.15) *Pontogeneia* cf. *ivanovi* Gurjanova, 1951
40b	Posterior margin of epimeron 3 decidedly convex, obtusely angular; inner margins of the cleft of the telson convex *Pontogeneia inermis* (Krøyer, 1838)
41a	Uropod 3 with several plumose setae along the inner margins of the rami (fig. 18.60) 42
41b	Uropod 3 without plumose setae 45
42a	Anteroventral corner of the head developed as a broad plate that extends anteriorly almost as far as the lateral lobe of the head; article 2 of pereopods 5-7 conspicuously serrated along the posterior margin; epimeron 3 with a distinct recurved tooth at the posteroventral corner *Paramoera serrata* Staude (in press)
42b	Anteroventral corner of the head obviously notched, not extending anteriorly as far as the lateral lobe of the head; article 2 of pereopods 5-7 minutely serrated along the posterior margin; epimeron 3 without a recurved tooth at the posteroventral corner 43
43a	Gland cone with 1-3 stout apical spines (plus a setae in females); posterior margin of article 6 of gnathopod 2 with 3-5 groups of setae; palm of gnathopod 2 of male very oblique (palm longer than posterior margin of article 6); posterior margin of article 6 of pereopods 3-4 with a row of groups of spines (2-3 spines in each group) *Paramoera columbiana* Bousfield, 1958
43b	Gland cone with apical setae only (no spines); posterior margin of article 6 of gnathopod 2 with 6-10 groups of setae; palm of gnathopod 2 of male moderately oblique (palm shorter than posterior margin of article 6); posterior margin of article 6 of pereopods 3-4 with a row of spine/seta pairs (1 spine and 1 seta in each pair) 44
44a	Gland cone with 2 apical setae; outer ramus of uropod 2 shorter than inner ramus; uropod 3 extending well beyond the end of uropod 1; telson with 2 setae at the apex of each lobe; length to 7 mm *Paramoera mohri* Barnard, 1952
44b	Gland cone with 3-4 apical setae (especially in adults); outer ramus of uropod 2 longer than, or equal to, the inner ramus; uropod 3 not extending much beyond the end of uropod 1; telson with 2-3 apical spines and 1 subapical spine on each lobe; length to 13 mm *Paramoera suchaneki* Staude (in press) (fig. 18.78)
45a	Eye small and black; gland cone with 2 small apical spinules; article 6 of gnathopod 1 broad (length about 1.5 times the width); posterior margin of epimeron 3 serrate, with 6-7 small notches; telson cleft for about half its length, each apex truncate and with 3-4 setae 46
45b	Eye not as described in choice 45a; gland cone with a single apical seta; article 6 of gnathopod 1 long (length at least twice the width); posterior margin of epimeron 3 with a single notch above a small corner tooth; telson cleft for more than half its length, each apex pointed and with 1-2 setae 47
46a	Eye about one-third the depth of the lateral lobe of the head; article 6 of the gnathopods widest proximal to the midpoint; gill absent from pereopod 7; peduncle of uropod 1 with more than 10 spines in the dorsolateral row, and with an isolated distoventral spine *Paramoera bucki* Staude (in press)
46b	Eye about half the depth of the lateral lobe; article 6 of the gnathopods widest at or distal to the midpoint; gill present on pereopod 7; peduncle of uropod 1 with less than 10 spines in the dorsolateral row; lacking an isolated distoventral spine *Paramoera carlottensis* Bousfield, 1958
47a	Eye small and white; antennae and pereopods 3-4 nearly naked, only sparsely setose (there may be a distal spine on article 1 of antenna 1); apices of telson each with 2 small setae *Paramoera leucophthalma* Staude (in press)
47b	Eye large and black; antennae and pereopods 3-4 with several setae or spines; apices of telson each with a single long seta and a minute setule *Paramoera bousfieldi* Staude (in press)

Key IV

Uropod 3 absent or lost.

1a	Coxae small, usually not overlapping one another; urosomite 1 much longer than deep, at least 4 times the length of urosomite 2; uropod 3 absent or vestigial and scarcely noticeable (Podoceridae)	2
1b	Coxae moderate to large, overlapping one another; urosomite 1 not much longer than deep, less than 3 times the length of urosomite 2; uropod 3 present in life, but routinely lost in collection and preservation (thus a "socket" in the urosomite may be apparent; this part of the key deals only with species that are especially likely to lose the third uropod)	6
2a	Pereonites 6-7 and pleonites 1-2 with dorsal keels; uropod 3 composed of a minute bladelike single article	*Podocerus cristatus* (Thomson, 1879) group
2b	None of the pereonites or pleonites with a dorsal keel; uropod 3 absent	3
3a	Uropod 2 vestigial, uniramous; gnathopod 2 of male approximately equal to gnathopod	*Paradulichia typica* Boeck, 1870 (fig. 18.79)
3b	Uropod 2 well developed, biramous; gnathopod 2 of male much larger than gnathopod 1	4
4a	With 3 pairs of coxal gills; article 2 of pereopods 3-4 expanded, with a decidedly convex anterior margin; pereopods 5-7 not much longer than pereopods 3-4	*Dyopedos* spp.
4b	With 4 pairs of coxal gills; article 2 of pereopods 3-4 narrow, with a nearly straight anterior margin; pereopods 5-7 much longer than pereopods 3-4	5
5a	Article 5 of pereopod 7 approximately equal in length to article 6; article 2 of gnathopod 2 of female equal in length to article 6	*Dulichia rhabdoplastis* McCloskey, 1970
5b	Article 5 of pereopod 7 one and a half times as long as article 6; article 2 of gnathopod 2 of female longer than article 6	*Dulichia* spp.
6a	Rostrum conspicuous (rarely reduced to a small point); urosomite 1 without dorsal teeth; telson not cleft	7
6b	Rostrum inconspicuous; urosomite 1 with 1 or more dorsal teeth; telson slightly notched or deeply cleft	9
7a	Eyes when present fused as a single dorsal mass; coxa 1 neither reduced nor hidden by the following coxae; pereopod 7 about twice as long as pereopod 6 (Oedicerotidae)	8
7b	Eyes not dorsally fused; coxa 1 reduced (much shorter than coxa 3) and hidden by the coxae that follow; pereopod 7 about as long as pereopod 6 (Amphilochidae)	Key II, couplet 20
8a	Rostrum bulbous; eyes present	Subkey III A, couplet 15
8b	Rostrum reduced to a small point; eye absent or indistinct	Subkey III D, couplet 5
9a	Eyes on lateral bulges of head; coxae short and overlapping only slightly; telson slightly notched (similar to fig. 18.13)	Melphidippidae
9b	Eyes not on lateral bulges; coxae large and markedly overlapping; telson deeply cleft	10
10a	Gnathopod 1 simple (fig. 18.2); coxa 3 shorter than coxa 2; anterior margins of coxae setose	*Megaluropus* sp.
10b	Gnathopod 1 subchelate; coxa 3 nearly as long as coxa 2; anterior margins of coxae without setae	11
11a	Epimeron 3 serrate, with numerous irregular teeth above and below its posteroventral corner	*Ceradocus spinicaudus* (Holmes, 1908) (fig. 18.80)
11b	Epimeron 3 typically with a single large tooth at its posteroventral corner (*Melita*)	Subkey III E, couplet 19

The checklist is arranged according to the phylogenetic scheme of Bousfield (1983). A few of the species in the checklist are not in the keys. They are indicated by an asterisk.

Superfamily Eusiroidea

Family Pontogeneiidae

Accedomoera vagor Barnard, 1969. Intertidal and shallow subtidal; on algae and mixed sediments.
Paramoera bousfieldi Staude (in press). Intertidal; sometimes near freshwater seepage; in mixed sediment

(especially cobbles).
Paramoera bucki Staude (in press). Intertidal in freshwater seepage and in the tidal region of streams; in gravel.
Paramoera carlottensis Bousfield, 1958. Intertidal; especially in low-salinity tidepools and seepage; mixed sediment.
Paramoera columbiana Bousfield, 1958. Low intertidal; especially in situations of low salinity; in gravel and other sediments.
Paramoera leucophthalma Staude (in press). Subtidal; in gravel and fine sediment.
Paramoera mohri Barnard, 1952. Intertidal (rarely subtidal); in gravel.
Paramoera serrata Staude (in press). Low intertidal; in coarse sand and mixed sediment.
Paramoera suchaneki Staude (in press) (fig. 18.78). Intertidal; in gravel and cobbles or in mussel beds.
Pontogeneia inermis (Krøyer, 1838). Habitat uncertain; known on this coast only from a single dubious specimen.
Pontogeneia intermedia Gurjanova, 1938. Intertidal and shallow subtidal; on algae and various sediments.
Pontogeneia cf. *ivanovi* Gurjanova, 1951 (fig. 18.15). Low intertidal and shallow subtidal; mixed sediments (especially sand); not in complete agreement with Gurjanova's description.
Pontogeneia cf. *rostrata* Gurjanova, 1938. Low intertidal and subtidal; on algae and various sediments; not in complete agreement with Gurjanova's description.

Family Calliopiidae

Calliopius spp. (fig. 18.76). Low intertidal to deep subtidal; on algae or mixed sediment and around docks; somewhat pelagic; there are probably 2 or 3 species in local waters, perhaps all undescribed.
Paracalliopiella pratti (Barnard, 1954). Low intertidal and subtidal; on algae, mixed sediment, and especially seagrasses; variants with atypical antennae.
Oligochinus lighti Barnard, 1969. Low intertidal; in mixed sediments and among algae.
Oradarea longimana (Boeck, 1871). Subtidal (sometimes deep); mixed mud, sand, and shell (possibly commensal).

Family Eusiridae

Eusirus spp. Deep subtidal; on fine sediment and probably also pelagic; one species close to *E. longipes* Boeck, 1861, is found locally, others expected.
Rhachotropis clemens Barnard, 1967 (fig. 18.66). Deep subtidal; on fine sediment and probably also pelagic.
Rhachotropis oculata (Hansen, 1887). Deep subtidal; on fine sediment and probably also pelagic.

Superfamily Oedocerotoidea

Family Oedicerotidae

Bathymedon flebilis Barnard, 1967. Subtidal; fine sediment.
Bathymedon pumilus Barnard, 1962. Subtidal; fine sediment.
Monoculodes spp. Low intertidal to deep subtidal; fine sediment; *M. zernovi* Gurjanova, 1938 (fig. 18.68) and *M. spinipes* Mills 1962 have been reported (see Mills, 1962), but a number of variants make this genus confusing.
Synchelidium rectipalmum Mills, 1962. Low intertidal and subtidal; sandy sediment.
Synchelidium shoemakeri Mills, 1962. Low intertidal to deep subtidal; fine sediment.
Westwoodilla caecula (Bate, 1856) (fig. 18.67). Low intertidal to deep subtidal; fine sediment.
Unidentified sp. Deep subtidal; fine sediment, probably undescribed.

Superfamily Leucothoidea

Family Pleustidae

It is likely that this family is represented by several undescribed species in addition to those listed.

Parapleustes den Barnard, 1969. Intertidal; exposed rocky beaches.
Parapleustes nautilus Barnard, 1969 (fig. 18.77). Intertidal, exposed rocky beaches.
Parapleustes oculatus (Holmes, 1908). Subtidal, habitat poorly known.
Parapleustes pugettensis (Dana, 1853) group. Low intertidal to subtidal; on various substrata, and sometimes commensal; a group of species, incompletely described.
Pleusirus secorrus Barnard, 1969. Low intertidal and subtidal; cobbles.
Pleustes depressa Alderman, 1936 (fig. 18.65). Subtidal; on algae attached to rock surfaces, and on eelgrass.
Pleusymtes subglaber (Barnard & Given, 1960). Subtidal; sand; genus synonymous with *Pleusymptes* Barnard and *Sympleustes* Barnard.
Pleusymtes sp. Shallow subtidal; sandy(?) sediment; probably undescribed.

Family Amphilochidae

Amphilochus litoralis Stout, 1912. Low intertidal; probably commensal.
Amphilochus neapolitanus Della Valle, 1893. Low intertidal; probably commensal.
Amphilochus picadurus Barnard, 1962. Low intertidal; probably commensal.
Gitanopsis vilordes Barnard, 1962. Low intertidal; probably commensal.

Family Leucothoidae

Leucothoe sp. (fig. 18.7). Low intertidal and subtidal; probably commensal; distinct from *L. alata* and *L. spinicarpa*, and probably undescribed.

Family Stenothoidae

Metopa cistella, *Metopella* ?*carinata*, *Proboloides* sp., and *Stenula* spp. have been reported locally. Low intertidal to deep subtidal; often commensal with anemones, hydroids, and sea pens; a poorly known group whose species are difficult to identify due to their small size and the need to examine mouthpart structure.

Superfamily Talitroidea

Family Hyalidae

Allorchestes angusta Dana, 1856 group (at least 4 species). Intertidal and shallow subtidal; ranging into water of reduced salinity; on various substrata and among drift algae or wood chips.
Allorchestes bellabella Barnard, 1974. Intertidal (and also planktonic).
Hyale anceps (Barnard, 1969). Low intertidal; rocky beaches with algae.
Hyale frequens Stout, 1913 (fig. 18.34) group (about 10 species). Mid-intertidal to shallow subtidal; on various substrata with algae.
Hyale pugettensis (Dana, 1852). High intertidal tidepools; possibly synonymous with *H. californica* Barnard, 1969.
Hyale plumulosa (Stimpson, 1857). Low intertidal; mixed sediment (especially cobbles) with algae.
Parallorchestes spp. Intertidal to shallow subtidal; usually on rocky beaches with algae; Bousfield (1981) indicates that there are 12 species of *Parallorchestes*, including *P. ochotensis* (Brandt, 1851) and some undescribed species, along the North Pacific Rim.

Family Dogielinotidae

Proboscinotus loquax (Barnard, 1967). Intertidal; burrowing in sand on beaches of the outer coast.

Family Najnidae

Najna spp. Low intertidal and shallow subtidal; on *Alaria* and other algae, burrowing into stipes; Bousfield (1981) indicates that there are 10 species of *Najna*, including *N. kitamati* Barnard, 1979 (*N. ?consiliorum* of Barnard, 1962c), along the North Pacific Rim.

Family Talitridae

Megalorchestia californiana Brandt, 1851. High intertidal; sandy beaches of the open coast.
Megalorchestia columbiana (Bousfield, 1958). High intertidal; sandy beaches (occasionally in brackish situations).
Megalorchestia pugettensis (Dana, 1853). High intertidal; coarse to fine sand; open coast to protected estuaries.
Paciforchestia klawei (Bousfield, 1961). High intertidal; coarse sand and gravel (habitat incompletely known).
Platorchestia chathamensis Bousfield, 1982. High intertidal; among driftwood logs; known from a single specimen collected near Victoria, British Columbia.
Traskorchestia georgiana (Bousfield, 1958). High intertidal; coarse sand and gravel beaches.
Traskorchestia traskiana (Stimpson, 1857). High intertidal; widely distributed, but largely associated with gravel and rocky beaches.

Superfamily Phoxocephaloidea

Family Phoxocephalidae

Eobrolgus spinosus (Holmes, 1905) (fig. 18.64). Intertidal and shallow subtidal; fine sediment (especially sandy mud).
Eyakia robustus (Holmes, 1908). Subtidal; fine sediment.
Foxiphalus cognatus (Barnard, 1960) and *F. similis* (Barnard, 1960). Low intertidal to deep subtidal; fine sediment; taxonomy needs to be studied.
Foxiphalus major (Barnard, 1960). Subtidal; fine sediment.
Foxiphalus obtusidens (Alderman, 1936). Low intertidal and subtidal; fine sediment.
Grandifoxus grandis (Stimpson, 1856). Intertidal and shallow subtidal; sand; synonymous with *Paraphoxus milleri* Thorsteinson, 1941.
Harpiniopsis fulgens Barnard, 1960. Deep subtidal; fine sediment.
Heterophoxus oculatus (Holmes, 1908). Subtidal (sometimes deep); mud.
Mandibulophoxus gilesi Barnard, 1957. Deep subtidal; fine sediment.
Metaphoxus frequens Barnard, 1960. Deep subtidal; fine sediment.
Metaphoxus fultoni (Scott, 1890) (fig. 18.69). Deep subtidal; fine sediment.
Paraphoxus oculatus G. O. Sars, 1879. Subtidal (sometimes deep); fine sediment.
Phoxocephalus homilis Barnard, 1960. Deep subtidal; fine sediment.
Rhepoxynius abronius (Barnard, 1960). Shallow subtidal; fine sediment (especially sandy mud).
Rhepoxynius bicuspidatus (Barnard, 1960). Deep subtidal; fine sediment.
Rhepoxynius daboius (Barnard, 1960). Subtidal; fine sediment.
Rhepoxynius heterocuspidatus (Barnard, 1960). Subtidal; fine sediment.
Rhepoxynius tridentatus (Barnard, 1954). Low intertidal and subtidal; fine sediment.
Rhepoxynius variatus (Barnard, 1960). Subtidal; fine sediment.
Rhepoxynius vigitegus (Barnard, 1971). Subtidal; fine sediment.

Family Urothoidae

Urothoe spp. Deep fjords; there are perhaps 2 or 3 species.

Superfamily Lysianassoidea

Family Lysianassidae

Acidostoma sp. Subtidal (sometimes deep); on soft sediment; possibly commensal.
Allogaussia recondita Stasek, 1958. Intertidal; commensal in the gut of sea anemones; not reported north of Oregon.
Anonyx cf. *laticoxae* Gurjanova, 1962. Shallow to deep subtidal; soft sediment; some local populations mature at an unusually small size; possibly an undescribed species or pair of species.
Anonyx cf. *lilljeborgi* Boeck, 1870 (fig. 18.70). Shallow to deep subtidal; soft sediment; uropod 2 not very constricted; probably a new species.
Aristias pacificus Schellenberg, 1936. Subtidal; commensal with brachiopods and ascidians; possibly synonymous with *A. veleronis* Hurley, 1963; determination uncertain.
Cyphocaris challengeri Stebbing, 1888. Deep subtidal; pelagic.
Hippomedon spp. Subtidal (sometimes deep); soft sediment; undescribed species are expected (see Jarrett & Bousfield, 1982).
Koroga megalops Holmes, 1908. Subtidal (sometimes deep); soft sediment.
Lepidepecreum garthi Hurley, 1963. Deep subtidal; soft sediment.
Lepidepecreum gurjanovae Hurley, 1963 (fig. 18.71). Shallow to deep subtidal; on various substrata (kelp holdfasts to soft sediment); undescribed species are expected.
Lysianassa holmesi (Barnard, 1955). Subtidal (sometimes deep); soft sediment; previously in the genus *Aruga*.
Opisa tridentata Hurley, 1963. Deep subtidal; soft sediment.
Orchomene sp. Intertidal; possibly commensal with anemones; similar to *Allogaussia recondita* Stasek, 1958.
Orchomene decipiens (Hurley, 1963). Deep subtidal; soft sediment.
Orchomene obtusa (G. O. Sars, 1890). Subtidal (sometimes deep); epibenthic and on soft sediment; abundant in waters of British Columbia, but not yet reported in Washington.
Orchomene pacifica (Gurjanova, 1938). Subtidal (sometimes deep); on various substrata.
Orchomene cf. *pinguis* (G. O. Sars, 1895). Low intertidal and subtidal; mixed sediment.
Pachynus cf. *barnardi* Hurley, 1963. Subtidal (sometimes deep); soft sediment; absence of eyes and structure of accessory flagellum do not agree with Hurley's description.
Prachynella lodo Barnard, 1964. Subtidal; mixed sediment.
Psammonyx longimerus Jarrett & Bousfield, 1982. Subtidal (sometimes deep); sandy sediment.
Schisturella cocula Barnard, 1966. Deep subtidal; soft sediment.
Wecomedon similis Jarrett & Bousfield, 1982. Intertidal and subtidal; soft sandy sediment.
Wecomedon wecomus (Barnard, 1971). Low intertidal to deep subtidal; soft sandy sediment.

Superfamily Synopioidea

Family Synopiidae

Bruzelia tuberculata G. O. Sars, 1895. Deep subtidal; soft sediment.
Syrrhoe longifrons Shoemaker, 1964. Deep subtidal; soft sediment.
Tiron biocellata Barnard, 1962 (fig. 18.62). Low intertidal to deep subtidal; various sediments.

Family Argissidae

Argissa hamatipes (Norman, 1869). Subtidal (sometimes deep); soft sediment; possibly a group of undescribed species.

Superfamily Stegocephaloidea

Family Acanthonotozomatidae

Coboldus sp. Subtidal; rocky substrata, especially with algae (e. g., corallines and kelp holdfasts), possibly commensal; possibly synonymous with ?*Panoploea hedgpethi* Barnard, 1969.
Odius kelleri Bruggen, 1907. Subtidal; rocky substrata, especially with algae.

Family Lafystiidae

*Some new genera and species recently reported.

Superfamily Pardaliscoidea

Family Pardaliscidae

This is a poorly sampled group; additional species should be expected.

Halicella halona Barnard, 1971. Subtidal (sometimes deep); soft sediment.
**Pardalisca cuspidata* Krøyer, 1842. Deep subtidal; soft sediment.
Pardalisca tenuipes G. O. Sars, 1893. Deep subtidal; soft sediment.

Family Stilipedidae

**Stilipes* sp. Deep fjords.

Superfamily Dexaminoidea

Family Atylidae

Atylus collingi (Gurjanova, 1938). Shallow subtidal; sand and gravel (especially with eelgrass); euryhaline.
Atylus levidensus Barnard, 1956. Low intertidal and subtidal; various sediments (especially sand).
Atylus tridens (Alderman, 1936) (fig. 18.63). Low intertidal and subtidal; associated with sand, eelgrass, and rocky bottoms; occasionally pelagic.

Family Dexaminidae

Dexamonica reduncans Barnard, 1957. Subtidal; soft sediments.
Polycheria osborni Calman, 1898. Low intertidal; burrowing into compound ascidians that colonize firm substrata.

Superfamily Ampeliscoidea

Family Ampeliscidae

Ampelisca agassizi (Judd, 1896) (fig. 18.72). Low intertidal to subtidal; tube-building in soft sediment.
Ampelisca brevisimulata Barnard, 1954. Subtidal (sometimes deep); tube-building in soft sediment.
Ampelisca careyi Dickinson, 1982. Subtidal (sometimes deep); tube-building in soft sediment; Dickinson recently distinguished *A. careyi* from *A. macrocephala* Lilljeborg, 1842; some local specimens display characters that seem to be intermediate between those typical of the two species.

Ampelisca cristata Holmes, 1908. Subtidal (sometimes deep); tube-building in coarse sand.
Ampelisca fageri Dickinson, 1982. Intertidal and subtidal; tube-building in mixed sand and boulders.
Ampelisca hancocki Barnard, 1954. Subtidal (sometimes deep); tube-building in soft sediment.
Ampelisca lobata Holmes, 1908. Subtidal; tube-building in mixed sand and rock, often associated with plants.
Ampelisca pugetica Stimpson, 1864. Subtidal; tube-building in sand.
Ampelisca unsocalae Barnard, 1960. Subtidal (sometimes deep); tube-building in very fine sediment.
Byblis millsi Dickinson, 1983. Subtidal (sometimes deep); tube-building in soft sediment; ramal spines of uropod 1 very small in local specimens; other species expected in depths of 200 m or more.
Byblis veleronis Barnard, 1954. Subtidal (sometimes deep); tube-building in soft sediment; other species expected in depths of 200 m or more.
Haploops tubicola Lilljeborg, 1856. Deep subtidal; tube-building in soft sediment.

Superfamily Pontoporeioidea

Family Haustoriidae

Eohaustorius brevicuspis Bosworth, 1973. Shallow subtidal; sand.
Eohaustorius estuarius Bosworth, 1973. Estuarine; sand.
Eohaustorius sawyeri Bosworth, 1973. Shallow subtidal; sand.
Eohaustorius sencillus Barnard, 1962. Shallow subtidal; sand.
Eohaustorius washingtonianus (Thorsteinson, 1941). Intertidal and shallow subtidal; sand.

Superfamily Gammaroidea

Family Anisogammaridae

Anisogammarus pugettensis (Dana, 1853). Intertidal and subtidal; various substrata, but especially associated with eelgrass, algae, and deposits of wood chips.
Eogammarus confervicolus (Stimpson, 1856) (fig. 18.75). Estuarine, intertidal, and subtidal; various substrata, but especially associated with sedges, eelgrass, algae, and deposits of wood chips.
Eogammarus oclairi Bousfield, 1979. Estuarine, intertidal, and shallow subtidal; various substrata; characters may intergrade with those of *E. confervicolus,* making identification difficult.
Locustogammarus levingsi Bousfield, 1979. Estuarine and intertidal; cobble and shingle beaches.
Ramellogammarus ramellus (Weckel, 1907). Stream mouths and high intertidal; coarse sand, stones, and wood debris.
Ramellogammarus vancouverensis Bousfield, 1979. Stream mouths and high intertidal; coarse sand, stones, and wood debris.

Family Gammaridae

Lagunogammarus setosus (Dementieva, 1931). Estuarine, intertidal, and subtidal; fine sediments.

Superfamily Melphidippoidea

Family Melphidippidae

Melphidippella and *Melphissana* have been reported locally; species of this family are rare in Washington waters, and poorly known; deep subtidal; in soft sediment.

Megaluropus Family Group

Megaluropus sp. Intertidal and subtidal; associated with algae, but also planktonic; probably undescribed.

Superfamily Hadzioidea

Family Melitidae

Ceradocus spinicaudus (Holmes, 1908) (fig. 18.80). Intertidal and subtidal; cobbles.
Elasmopus spp. Intertidal; associated with algal cover on rocks; not confirmed north of California.
Maera danae Stimpson, 1853. Shallow to deep subtidal; fine sediment to gravel.
Maera loveni (Bruzelius, 1859). Subtidal; mud.
Maera simile Stout, 1913. Shallow subtidal; associated with algal cover on rocks.
Melita californica Alderman, 1936. Intertidal to deep subtidal; cobbles to fine sediment; some subtidal individuals may belong to an undescribed species.
Melita dentata (Krøyer, 1842) (fig. 18.74). Low intertidal to deep subtidal; on various substrata.
Melita desdichada Barnard, 1962. Low intertidal and subtidal; soft sediment.
Melita oregonensis Barnard, 1954. Intertidal; associated with algal cover on rocks.
Melita sulca (Stout, 1913). Low intertidal to deep subtidal; often associated with algal cover on rocks.

Superfamily Corophioidea

Family Ampithoidae

Ampithoe dalli Shoemaker, 1938 (fig. 18.44). Intertidal and shallow subtidal; algae and eelgrass.
Ampithoe lacertosa Bate, 1858 (fig. 18.45). Intertidal and shallow subtidal; algae and eelgrass.
Ampithoe plumulosa Shoemaker, 1938 (fig. 18.46). Intertidal and shallow subtidal; algae and surfgrass; rare north of California.
Ampithoe sectimanus Conlan & Bousfield, 1982 (fig. 18.43). Low intertidal; exposed rocky beaches with algae.
Ampithoe simulans Alderman, 1936 (fig. 18.42). Low intertidal; various substrata, associated with algae and eelgrass.
Ampithoe valida Smith, 1873 (fig. 18.47). Low intertidal and shallow subtidal; usually on soft sediment with algae or eelgrass, somewhat estuarine.
Cymadusa uncinata (Stout, 1912) (fig. 18.41). Low intertidal and shallow subtidal; builds plant-debris nests at the base of boulders on exposed beaches, also associated with kelp and surfgrass.
Peramphithoe humeralis (Stimpson, 1864) (fig. 18.48). Low intertidal and subtidal; curls blades of kelp and eelgrass to form a tube.
Peramphithoe lindbergi (Gurjanova, 1938) (fig. 18.50). Low intertidal and shallow subtidal; eelgrass and algal holdfasts.
Peramphithoe mea (Gurjanova, 1938) (fig. 18.49). Subtidal; eelgrass.
Peramphithoe plea (Barnard, 1965) (fig. 18.51). Shallow subtidal; kelp holdfasts.
Peramphithoe tea (Barnard, 1965) (fig. 18.52). Intertidal and subtidal; algae.

Family Isaeidae

There are still many taxonomic problems in this family, despite the useful paper of Conlan (1983).

Cheirimedeia macrocarpa subsp. *americana* Conlan, 1983. Low intertidal; brackish and marine sandflats.
Cheirimedeia similicarpa Conlan, 1983. Subtidal; shelly sediments.
Cheirimedeia zotea (Barnard, 1962) (fig. 18.73). Low intertidal to deep subtidal; mixed sediments.
Gammaropsis ellisi Conlan, 1983. Low intertidal and subtidal; on algae and sponges.
Gammaropsis shoemakeri Conlan, 1983. Low intertidal and subtidal; on algae and hydroids.

Gammaropsis spinosa (Shoemaker, 1942). Low intertidal and subtidal; on algae, sponges, and polychaete tubes; reported north and south of Washington.
Gammaropsis thompsoni (Walker, 1898). Low intertidal and subtidal; on various substrata, but especially among encrusting animals and in algal holdfasts.
Kermystheus ociosa Barnard, 1962. Subtidal; sand and gravel; apparently synonymous with *Podoceropsis augustimana* Conlan, 1983.
?*Megamphopus* sp. Low intertidal and subtidal; sand, possibly associated with eelgrass; an undescribed species of uncertain generic affiliation; referred to as "near *Podoceropsis inaequistylis*" by Staude et al., 1977.
Photis bifurcata Barnard, 1962 (fig. 18.29). Intertidal to deep subtidal; usually on soft sediment.
Photis brevipes Shoemaker, 1942 (fig. 18.28). Low intertidal to deep subtidal; in various sediments, but especially sand.
Photis conchicola Alderman, 1936 (fig. 18.32). Intertidal and subtidal; rocky beaches with algae and surfgrass, often pagurid-like, living in empty gastropod shells; differing from *P. oligochaeta* only by its more setose coxae.
Photis lacia Barnard, 1962 (fig. 18.33). Subtidal; soft sediments.
Photis macinerneyi Conlan, 1983 (fig. 18.33). Low intertidal and subtidal; sand.
Photis oligochaeta Conlan, 1983 (fig. 18.32). Low intertidal and subtidal; sand and gravel; differing from *P. conchicola* only by its less setose coxae, a character which is size-related, according to Conlan (1983).
Photis pachydactyla Conlan, 1983 (fig. 18.30). Low intertidal and subtidal; hard substratum, and occasionally in empty barnacle shells.
Photis parvidons Conlan, 1983 (fig. 18.31). Low intertidal and subtidal; sandy sediment.
Podoceropsis barnardi Kudryashov & Tsvetkova, 1975. Subtidal; mixed sediment, especially sand; not reported south of Vancouver Island.
Podoceropsis chionoecetophila Conlan, 1983. Deep subtidal; commensal in egg masses of the crab *Chionoecetophila tanneri*; reported only from Alaska and Oregon.
Protomedeia articulata Barnard, 1962. Low intertidal to deep subtidal; soft sediments.
Protomedeia grandimana Bruggen, 1905. Low intertidal to deep subtidal; soft sediments; not reported south of Vancouver Island.
Protomedeia penates Barnard, 1966 and *P. prudens* Barnard, 1960. Subtidal (sometimes deep); soft sediments; Conlan (1983) suggests that the 2 species are synonymous.

Family Ischyroceridae

There are many undescribed species. Those on the Pacific coast are being studied by Conlan.

Ischyrocerus anguipes (Krøyer, 1838) (fig. 18.36) group. Low intertidal and subtidal; tube-building on various substrata.
Ischyrocerus serratus Gurjanova, 1938 (fig. 18.39). In beds of *Mytilus californianus* on exposed rocky shores; assignment to genus doubtful.
Ischyrocerus sp. Low intertidal and subtidal; tube-building on various substrata.
Jassa spp. (fig. 18.35). Low intertidal and subtidal; tube-building on various substrata; Conlan will soon describe new species from this coast.
Microjassa spp. Low intertidal and subtidal; in algal holdfasts; Conlan has recognized at least 3 species, one of which is probably *M. litotes* Barnard, 1954 (fig. 18.38).

Family Aoridae

Aoroides columbiae Walker, 1898. Low intertidal to deep subtidal; mixed sediment with algae.
Aoroides exilis Conlan & Bousfield, 1982. Low intertidal and subtidal; on various sediments, but especially with algae and eelgrass.
Aoroides inermis Conlan & Bousfield, 1982. Low intertidal and subtidal; sand.
Aoroides intermedius Conlan & Bousfield, 1982. Low intertidal and subtidal; sand and gravel, especially with algae and eelgrass.
Aoroides spinosus Conlan & Bousfield, 1982. Low intertidal and subtidal; on various substrata, but

especially with algae and debris.

Columbaora cyclocoxa Conlan & Bousfield, 1982. Low intertidal and subtidal; exposed rocky beaches with algae.

Family Cheluridae

*Chelura terebrans Philippi, 1839. Associated with wood-boring isopods of the genus *Limnoria*; presence north of California not confirmed.

Family Corophiidae

Corophium acherusicum Costa, 1857 (fig. 18.27). Intertidal and subtidal; tube-building on sediment, algae, and eelgrass.
Corophium baconi Shoemaker, 1934. Intertidal and subtidal; tube-building in soft sediment; reported from Bering Sea and California.
Corophium brevis Shoemaker, 1949. Intertidal and subtidal; tube-building on various substrata.
Corophium crassicorne Bruzelius, 1859. Subtidal; tube-building in soft sediment.
Corophium insidiosum Crawford, 1939. Intertidal and subtidal; tube-building in soft sediment.
Corophium salmonis Stimpson, 1857 (fig. 18.26). Intertidal and subtidal; tube-building in soft sediment, especially in estuarine situations.
Corophium spinicorne Stimpson, 1857. Intertidal and subtidal; tube-building in soft sediment, primarily in freshwater.
Ericthonius brasiliensis Dana, 1853. Subtidal; forming mats of muddy tubes on various substrata.
Ericthonius hunteri Bate, 1862. Subtidal; forming mats of muddy tubes on various substrata.
*Grandidierella japonica Stephensen, 1938. Intertidal and subtidal; soft-sediments; probably introduced with the oyster *Crassostrea gigas*.

Family Podoceridae

Dulichia spp. Shallow to deep subtidal; on various substrata, but especially epibiotic; undescribed species are to be expected.
Dulichia rhabdoplastis McCloskey, 1970. Subtidal; commensal on the spines of the sea urchin *Strongylocentrotus franciscanus* and also occurring on soft sediment.
Dyopedos spp. Shallow to deep subtidal; on various substrata, but especially epibiotic; undescribed species are to be expected.
Paradulichia typica Boeck, 1870 (fig. 18.79). Shallow to deep subtidal; on various substrata, but especially epibiotic.
Podocerus cristatus (Thomson, 1879) group. Subtidal; on various substrata, but especially epibiotic; until this group is revised, Barnard (1979) hesitates to identify species unless they have been collected at the type localities.

Suborder Caprellidea

Dougherty, E. C., & J. E. Steinberg. 1953. Notes on the skeleton shrimps (Crustacea: Caprellidae) of California. Proc. Biol. Soc. Wash., 66:39-50.
Laubitz, D. R. 1970. Studies on the Caprellidae (Crustacea, Amphipoda) of the American North Pacific. Nat. Mus. Canada Publ. Biol. Oceanogr., no. 1. 89 pp.
------. 1972. The Caprellidae (Crustacea, Amphipoda) of Atlantic and Arctic Canada. Nat. Mus. Canada Publ. Biol. Oceanogr., no. 4. 82 pp.
Leung, Y.-M. 1967. An illustrated key to the species of whale-lice (Amphipoda, Cyamidae), ectoparasites of Cetacea, with a guide to the literature. Crustaceana, 12:279-91.
McCain, J. C. 1968. The Caprellidae (Crustacea: Amphipoda) of the western North Atlantic. Bull. 278, U. S. Nat. Mus. vi +147 pp.

------. 1969. A new species of caprellid (Crustacea: Amphipoda) from Oregon. Proc. Biol. Soc. Wash., 82:507-10.
------. 1970. Familial taxa within the Caprellidea (Crustacea: Amphipoda). Proc. Biol. Soc. Wash., 82:837-42.
McCain, J. C., & J. E. Steinberg, 1970. Amphipoda I, Caprellidea I, Fam. Caprellidae. Crustaceorum Catalogus, part 2. 78 pp.
Marelli, D. C. 1981. New records for Caprellidae in California, and notes on a morphological variant of *Caprella verrucosa* Boeck, 1871. Proc. Biol. Soc. Wash., 94:654-62.
Martin, D. M. 1977. A survey of the family Caprellidae (Crustacea, Amphipoda) from selected sites along the northern California coast. Bull. So. Calif. Acad. Sci., 76:146-67.

The key does not include the species of Cyamidae, which live on whales, dolphins, and porpoises. These are listed, however, and all of them are dealt with in the work of Leung (1967).

1a	With gills on pereonites 2-4; mandible without a molar process	2
1b	With gills on pereonites 3 and 4 only; mandible with a molar process	3
2a	Pereopods 3 and 4 with 1 article, pereopod 5 with 6 articles; abdomen with 5 apparent segments	*Cercops compactus*
2b	Pereopods 3 and 5 with 3 articles, pereopod 4 with 1 article; abdomen small and without obvious segmentation	*Perotripus brevis* (fig. 18.82)
3a	With vestigial pereopods on pereonites 3 and 4; mandible with a palp	4
3b	Without pereopods on pereonites 3 and 4; mandible without a palp	7
4a	Pereopods 3 and 4 consisting of 2 articles; antenna 2 without long setae	5
4b	Pereopods 3 and 4 consisting of only 1 article; antenna 2 with long setae on the larger articles	6
5a	Head with a median dorsal spine (in the male) or a knob (in the female); pereopod 5 consisting of 6 articles; gnathopod 2 of male attached to the anterior half of pereonite 2	*Deutella californica* (fig. 18.83)
5b	Head without a median dorsal spine or knob; pereopod 5 consisting of 3 articles; gnathopod 2 of male attached to the posterior half of pereonite 2	*Mayerella banksia*
6a	Next-to-last article of antenna 2 about 4 times as long as wide, with setae not appreciably longer than its width; short spines on pereonites directed more nearly anteriorly than laterally	*Tritella laevis* (fig. 18.84)
6b	Next-to-last article of antenna 2 about 6 times as long as wide, with some setae about twice as long as its width; short spines on pereonites directed more nearly laterally than anteriorly	*Tritella pilimana*
7a	Width of propodus of gnathopod 2 about four-fifths the length of this article; length of adults about 4 mm; with a pair of grasping spines on the proximal half of the propodus of pereopods 5-7 (found on the surface of the sea star *Henricia leviuscula*, as well as on hydroids and other organisms)	*Caprella greenleyi*
7b	Width of propodus of gnathopod 2 less than two-thirds the length of this article; length of adults generally greater than 5 mm; with a single grasping spine or tooth on the propodus of pereopods 5-7	8
8a	With a ventral spine between the bases of gnathopods 2	9
8b	Without a ventral spine between the bases of gnathopods 2	12
9a	With a prominent, anteriorly directed spine on the dorsal side of the head; propodus of gnathopod 2 of male about 4 times as long as wide	*Caprella californica*
9b	Without a spine on the dorsal side of the head; propodus of gnathopod 2 of male only about twice as long as wide	10
10a	Pereonite 5 with a small but distinct lateral projection on both sides of its anterior half (there are also projections on the posterior half); with a conspicuous, anterolaterally directed spine lateral to the base of gnathopod 2	*Caprella equilibra* (fig. 18.87)
10b	Pereonite 5 without a projection on both sides of its anterior half; if a spine is present lateral to the base of gnathopod 2, it is small and not directed anterolaterally	11
11a	Dactyl of gnathopod 2 decidedly setose; without a spine lateral to the base of gnathopod 2	*Caprella pilidigitata*

11b Dactyl of gnathopod 2 not setose; with a small spine lateral to the base of gnathopod 2
Caprella mendax (fig. 18.86)

12a Head without any dorsal spines, knobs, or tubercles 13
12b Head with 1 or a pair of dorsal spines, knobs, or tubercles 18
13a Propodus of pereopods 5-7 slender, its inner edge nearly straight and with a tooth near the middle; dactyl of gnathopod 2 of male decidedly setose; basis of gnathopod 2 of male much longer than the propodus
Caprella gracilior
13b Propodus of pereopods 5-7 rather stout, much of its inner edge at least slightly concave (except perhaps in *C. striata*) and with a tooth decidedly proximal to its middle; dactyl of gnathopod 2 of male not setose; basis of gnathopod 2 of male shorter than the propodus 14
14a Propodus of gnathopod 2 with 2 small spines at the base of the large grasping spine; pereonite 1 of male about 3 times as long as the head
Caprella irregularis
14b Propodus of gnathopod 2 with only 1 small spine at the base of the large grasping spine; pereonite 1 of male not more than twice as long as the head 15
15a Pereonites (at least the more posterior ones) with dorsal spines or tubercles; antenna 1 more than half the length of the body in the female, more than two-thirds the length of the body in the male 16
15b Pereonites without dorsal spines or tubercles; antenna 1 less than half the length of the body 17
16a Flagellum of antenna 1 longer than the peduncle; antenna 2 of male about as long as the peduncle of antenna 1 (subtidal)
Caprella striata
16b Flagellum of antenna 1 shorter than the peduncle; antenna 2 of male shorter than the combined length of the first 2 articles of antenna 1
Caprella alaskana
17a Outline of gills nearly circular; gnathopod 2 of male setose; gnathopod 2 of female attached near the middle of pereonite 2 (not likely to be found in our region)
Caprella drepanochir
17b Outline of gills elongated; gnathopod 2 of male not setose; gnathopod 2 of female attached near the anterior end of pereonite 2 (common)
Caprella laeviuscula (fig. 18.88)
18a Head with a single dorsal spine or knob 19
18b Head with a pair of dorsal spines or tubercles 23
19a Dorsal spine on head an anteriorly directed, triangular projection 20
19b Dorsal projection on head a knob that is directed upward 22
20a Tubercles on dorsal surface of pereonites small or absent (if present at all, lacking on pereonite 1); gnathopod 2 of male attached near the anterior end of pereonite 2
Caprella angusta
20b Tubercles on dorsal surface of pereonites distinct, present on all pereonites; gnathopod 2 of male attached near the middle of pereonite 2 21
21a Tubercles on pereonites large; propodus of gnathopod 2 of male shorter than pereonite 2
Caprella verrucosa
21b Tubercles on pereonites small; propodus of gnathopod 2 of male as long as pereonite 2
Caprella incisa (fig. 18.85)
22a Head and pereon covered with large and small tubercles; antenna 2 with some long setae (these are more than twice as long as the width of the articles from which they originate); gnathopod 2 and much of the body of the male setose
Caprella pustulata
22b Head and pereon with only a few tubercles; antenna 2 without especially long setae (the setae are not more than twice as long as the width of the articles from which they originate); gnathopod 2 and body of male not setose
Caprella borealis
23a Head with a pair of tubercles 24
23b Head with a pair of rather sharp spines 25
24a Posterior pereonites with large dorsal spines; length of pereonite 1 of male about the same as that of the head
Caprella ferrea
24b Posterior pereonites with small dorsal tubercles; pereonite 1 of male about 1.5 times as long as the head
Caprella rudiuscula
25a Flagellum of antenna 1 longer than the peduncle; pereonites 1 and 2 of male not more than twice as long as wide
Metacaprella anomala (fig. 18.89)
25b Flagellum of antenna 1 shorter than the peduncle; pereonites 1 and 2 of male more than twice as long as wide
Metacaprella kennerlyi

Phylum Arthropoda: Order Amphipoda

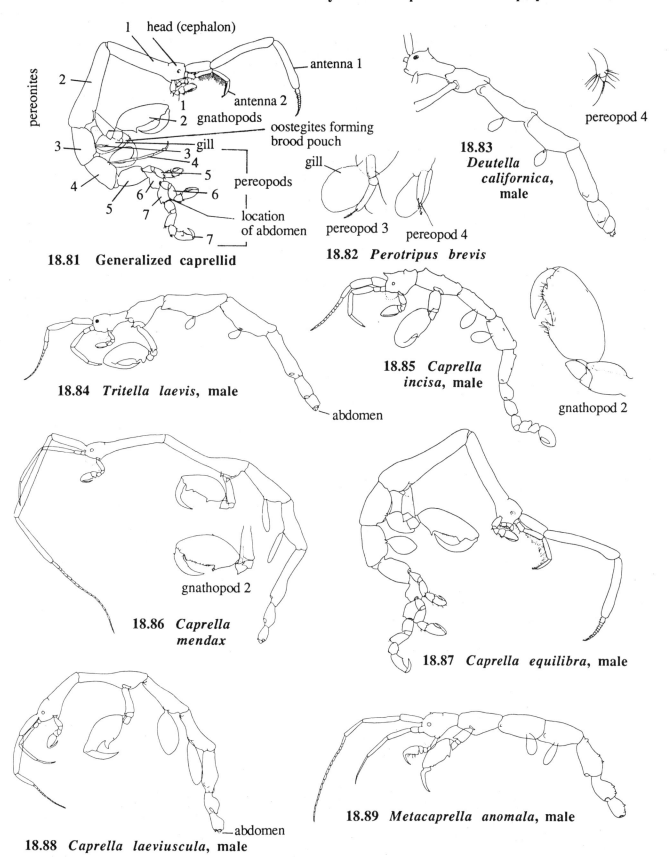

18.81 Generalized caprellid
18.82 *Perotripus brevis*
18.83 *Deutella californica*, male
18.84 *Tritella laevis*, male
18.85 *Caprella incisa*, male
18.86 *Caprella mendax*
18.87 *Caprella equilibra*, male
18.88 *Caprella laeviuscula*, male
18.89 *Metacaprella anomala*, male

Infraorder Caprellida

Two species, marked with an asterisk, are not in the key; they are found only in deep water (see Laubitz, 1970).

Family Phtisicidae

Perotripus brevis (La Follette, 1915) (fig. 18.82)

Family Aeginellidae

Deutella californica Mayer, 1890 (fig. 18.83)
Mayerella banksia Laubitz, 1970
Tritella laevis Mayer, 1903 (fig. 18.84)
Tritella pilimana Mayer, 1890

Family Caprogammaridae

Cercops compactus Laubitz, 1970

Family Caprellidae

Caprella alaskana Mayer, 1903
Caprella angusta Mayer, 1903
Caprella borealis Mayer, 1903
Caprella californica Stimpson, 1857
Caprella drepanochir Mayer, 1890
Caprella equilibra Say, 1818 (fig. 18.87)
Caprella ferrea Mayer, 1903
Caprella gracilior Mayer, 1903
Caprella greenleyi McCain, 1969
Caprella incisa Mayer, 1903 (Fig. 18.85)
Caprella irregularis Mayer, 1890
Caprella laeviuscula Mayer, 1903
Caprella mendax Mayer, 1903 (fig. 18.86)
Caprella pilidigitata Laubitz, 1970
Caprella pustulata Laubitz, 1970
Caprella rudiuscula Laubitz, 1970
Caprella striata Mayer, 1903
**Caprella ungulina* Mayer, 1903
Caprella verrucosa Boeck, 1872
Metacaprella anomala (Mayer, 1903) (fig. 18.89)
Metacaprella kennerlyi (Stimpson, 1864)
**Pseudoliropus vanus* Laubitz, 1970

Infraorder Cyamida

Family Cyamidae

Cyamus boopis (Lütken, 1870). On *Megaptera novaeangliae* (humpback whale) and *Physeter catodon* (sperm whale).
Cyamus catodontis Margolis, 1954. On *Physeter catodon*.
Cyamus ceti (Linnaeus, 1754). On *Eschrichtius glaucus* (gray whale).

Cyamus kessleri Brandt, 1872. On *Eschrichtius glaucus*.
Cyamus sp. On *Berardius bairdi* (Baird beaked whale).
Isocyamus delphinii (Guérin-Ménéville, 1837). On porpoises and dolphins.
Neocyamus physeteris (Pouchet, 1888). On *Physeter catodon* and on *Phocoenoides dalli* (Dall porpoise)
Platycyamus sp. On *Berardius bairdi*.

Suborder Hyperiidea

About 30 species of hyperiidean amphipods, some of which are restricted to deep water, have been reported from our region. Most of them are independent planktonic organisms. A few, however, are associated with jellyfishes, and the remarkable *Phronima sedentaria* (Forskål, 1775), often observed in quiet water around floating docks, lives in empty tunics of salps.

Several genera of the largest family (Hyperiidae) are dealt with in detail by Bowman (1973). This work, and others listed below, will lead one to additional pertinent literature concerning the amphipods of this suborder.

Bowman, T. E. 1973. Pelagic amphipods of the genus *Hyperia* and closely related genera (Hyperiidea: Hyperiidae). Smithsonian Contr. Zool., no. 136. 76 pp.

Bowman, T. E., & H. E. Gruner. 1973. The families and genera of Hyperiidea (Crustacea: Amphipoda). Smithsonian Contr. Zool., no. 146. iv + 64 pp.

Brusca, G. J. 1981. Annotated keys to the Hyperiidea (Crustacea: Amphipoda) of North American coastal waters. Tech. Rep. Allan Hancock Found., no. 5. iii + 76 pp.

Shih, C.-T. 1969. The systematics and biology of the family Phronimidae (Crustacea: Amphipoda). Dana Reports, 74:1-100.

PHYLUM ARTHROPODA: SUBPHYLUM CRUSTACEA: CLASS MALACOSTRACA: SUBCLASS EUCARIDA

Order Euphausiacea

Except at night, when certain species migrate closer to the surface, euphausiaceans are not likely to be collected except by use of large plankton nets towed at depths of 100 m or more. The works of Banner (1950) and Boden, Johnson, & Brinton (1955) deal with morphological characters of species found in our region, and should be used for making identifications. Brinton (1962) has summarized the geographic distribution of Pacific euphausiaceans.

Most crustaceans of this group are restricted to oceanic waters. The 2 species most likely to be found in Puget Sound, the San Juan Archipelago, Strait of Georgia, and contiguous inland waters are *Euphausia pacifica* Hansen, 1911 (without a rostrum) and *Thysanoessa raschii* M. Sars, 1863 (with a slender rostrum).

Banner, A. H. 1950. A taxonomic study of the Mysidacea and Euphausiacea (Crustacea) of the northeastern Pacific. Part III. Euphausiacea. Trans. Royal Canad. Inst., 28(1949):1-63.

Boden, B. P., M. W. Johnson, & E. Brinton. 1955. The Euphausiacea (Crustacea) of the North Pacific. Bull. Scripps Inst. Oceanogr., Univ. Calif., 6:287-400.

Brinton, E. 1962. The distribution of Pacific euphausiids. Bull. Scripps Inst. Oceanogr., Univ. Calif., 8:51-270.

Order Decapoda

Rathbun, M. J. 1904. Decapod crustaceans of the northwest coast of North America. Harriman Alaska Exped., 10:1-210.

Schmitt, W. L. 1921. The marine decapod Crustacea of California. Univ. Calif. Publ. Zool., 23:1-470.

"Natantia"--Shrimps

A few of the shrimps of our region--those in the families Penaeidae and Sergestidae--are placed in the decapod suborder Dendrobranchiata. The majority, however, are assigned to the suborder Pleocyemata, which includes thalassinids, crabs, hermit crabs, lobsters, and related crustaceans. It is convenient, nevertheless, to deal with shrimps of both suborders in one section of this book.

Butler, T. H. 1980. Shrimps of the Pacific coast of Canada. Can. Bull. Fish. Aquat. Sci., no. 202. 280 pp.

1a   Pereopod 3 chelate; epimera of abdominal segment 2 overlapped by those of segment 1; abdomen without a sharp bend; primarily oceanic, and not likely to be found except in deep water .... Suborder Dendrobranchiata

1b Pereopod 3 not chelate; epimera of abdominal segment 1 overlapped by those of segment 2; abdomen with a sharp bend in some families, but not in others; includes many littoral and shallow-water species, as well as some that are mostly restricted to deep water
　　　　　　　　　　　　　　　　　　　　　　　　　　　　　　　　Suborder Pleocyemata

Suborder Dendrobranchiata

1a Pereopods 1-3 obviously chelate　　　　　　　　　　　　　　　　Family Penaeidae
1b Pereopod 1 not chelate, and pereopods 2 and 3 with small, inconspicuous chelae
　　　　　　　　　　　　　　　　　　　　　　　　　　　　　　　　Family Sergestidae

Family Penaeidae

1a Curved transverse grooves on the carapace interrupted by the median ridge; telson with 2 pairs of stout, movable spines　　　　　*Bentheogennema borealis* (Rathbun, 1902)
1b Curved transverse grooves on the carapace not interrupted by the median ridge; telson with only 1 pair of stout, movable spines　　*Bentheogennema burkenroadi* Krygier & Wasmer, 1975

The following species have been reported for the region, but are likely to be taken only in deep water and only in the southern part.

Gennadas incertas Balss, 1925
Gennadas propinquus Rathbun, 1906
Gennadas tinayrei Bouvier, 1906
Hemipenaeus spinidorsalis Bate, 1881

Family Sergestidae

1a With a prominent spine on the dorsal part of the carapace just behind each eye, and with a similar spine on both sides of the carapace at a slightly more posterior level
　　　　　　　　　　　　　　　　　　　　　　　　　　　Eusergestes similis (Hansen, 1903)
1b Without the spines described in couplet 1a　　　　　　*Sergia tenuiremis* (Krøyer, 1855)

Petalidium subspinosum Burkenroad, 1937. Has been reported.

Suborder Pleocyemata

Infraorder Caridea

1a Pereopods 1-5 with exopodites　　　　　　　　　　　　　　　　　　　　　　　　2
1b None of the pereopods with exopodites　　　　　　　　　　　　　　　　　　　　3
2a Pereopods 1 and 2 longer and obviously more stout than the others　　Family Pasiphaeidae
2b Pereopods 1 and 2 shorter than the others, and not obviously more stout　Family Oplophoridae
3a Pereopod 1 subchelate; carpus of pereopod 2 not subdivided into several units
　　　　　　　　　　　　　　　　　　　　　　　　　　　　　　　　Family Crangonidae
3b Pereopod 1 not subchelate; carpus of pereopod 2 subdivided into several units　　　　4
4a Carpus of pereopod 2 subdivided into more than 7 units; rostrum prominent, and with movable dorsal spines　　　　　　　　　　　　　　　　　　　　Family Pandalidae
4b Carpus of pereopod 2 subdivided into 3-7 units; rostrum, if present, without movable dorsal spines　　　　　　　　　　　　　　　　　　　　　　　　　　　　　　　5

394   Phylum Arthropoda: Order Decapoda

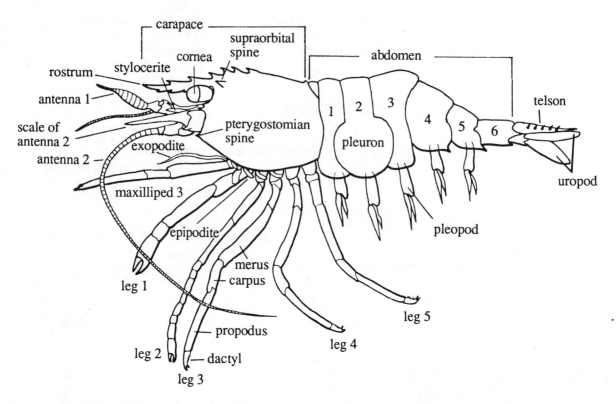

19.1 Diagram of a hippolytid shrimp, showing terminology used in identification of decapod crustaceans

5a	Rostrum absent; eyes at least partly covered dorsally by the carapace	Family Alpheidae
5b	Rostrum present; eyes not covered dorsally by the carapace	Family Hippolytidae

Family Oplophoridae

Butler, T. H. 1980. Shrimps of the Pacific coast of Canada. Can. Bull. Fish. Aquat. Sci., no. 202. 280 pp.

Man, J. G. de. 1920. The Decapoda of the Siboga Expedition. Part IV. Families Pasiphaeidae, Stylodactylidae, Hoplophoridae, Nematocarcinidae, Thalassocaridae, Pandalidae, Psalidopodidae, Gnathophyllidae, Processidae, Glyphocrangonidae and Crangonidae. Siboga-Expeditie, 39a3 (livr. 87). 318 pp.

Stevens, B. A., & F. A. Chace, Jr. 1965. The mesopelagic caridean shrimp *Notostomus japonicus* Bate in the northeastern Pacific. Crustaceana, 8:277-84.

1a	Abdominal segment 6 with a median dorsal ridge (segments 1-5 or 2-5 may also have a ridge)	2
1b	Abdominal segment 6 without a dorsal median ridge (in *Systellaspis*, segment 3, and sometimes segment 4, also have a median ridge)	3
2a	All abdominal segments with a median dorsal ridge; carapace with a prominent lateral ridge that runs its full length on both sides	*Notostomus japonicus* Bate, 1888
2b	Abdominal segment 1 without a median dorsal ridge; carapace without a prominent lateral ridge that runs its full length on both sides	*Acanthinephyra curtirostris* Wood-Mason, 1891
3a	Eyes almost without pigment; none of the abdominal segments with a median dorsal ridge	4

Phylum Arthropoda: Order Decapoda 395

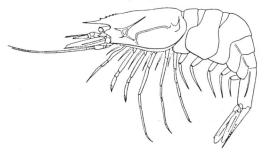

19.2 Oplophoridae: *Hymenodora frontalis*, female

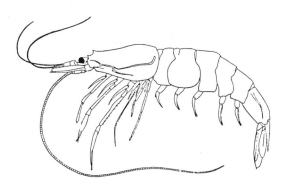

19.3 Pasiphaeidae: *Pasiphaea pacifica*, female

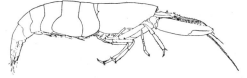

19.4 Alpheidae: *Betaeus harrimani*

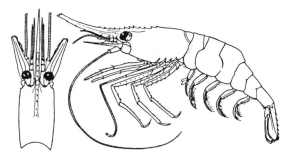

19.5 Pandalidae: *Pandalus danae*, male

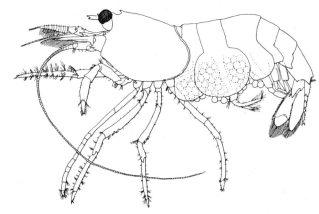

19.6 Hippolytidae: *Heptacarpus pugettensis*, female

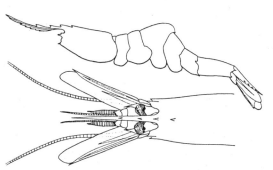

19.7 *Heptacarpus flexus*, female

scale of antenna 2

19.8 *Heptacarpus paludicola*

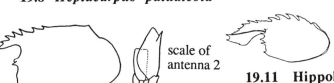

scale of antenna 2

19.11 Hippolytidae: *Spirontocaris lamellicornis*

19.10 *Heptacarpus kincaidi*, female

19.9 *Heptacarpus brevirostris*

3b   Eyes dark, with considerable pigment; abdominal segment 3, and sometimes 4, with a median dorsal ridge 7
4a   Telson with a bluntly rounded, setose (but not spinose) distal portion that projects beyond the last pair of lateral spines, which are much longer than the other lateral spines
Hymenodora acanthitelsonis Wasmer, 1972
4b   Telson without a rounded distal portion that projects beyond the last pair of lateral spines (the tip of the telson, moreover, has several spines) 5
5a   Rostrum extending only slightly beyond the cornea of the eye 6
5b   Rostrum extending up to or beyond the distal end of the peduncle of antenna 1
Hymenodora frontalis Rathbun, 1902 (fig. 19.2)
6a   Maxilliped 2 with a podobranch on the epipodite; carapace with a V-shaped groove (the apex of the V directed ventrally) on both sides near the middle of its length
Hymenodora gracilis Smith, 1886
6b   Maxilliped 2 without a podobranch on the epipodite; carapace without a V-shaped groove on both sides   *Hymenodora glacialis* (Buchholz, 1874)
7a   Rostrum about half as long as the rest of the carapace; abdominal segment 5 without a median dorsal spine at its posterior edge   *Systellaspis braueri* (Balss, 1914)
7b   Rostrum more than half as long as the rest of the carapace; abdominal segment 5 with a median dorsal spine at its posterior edge   *Systellaspis cristata* (Faxon, 1893)

Family Pasiphaeidae

Butler, T. H. 1980. Shrimps of the Pacific coast of Canada. Can. Bull. Fish. Aquatic Sci., no. 202. 280 pp.

Man, J. G. de. 1920. The Decapoda of the Siboga Expedition. Part IV. Families Pasiphaeidae, Stylodactylidae, Hoplophoridae, Nematocarcinidae, Thalassocaridae, Pandalidae, Psalidopodidae, Gnathophyllidae, Processidae, Glyphocrangonidae and Crangonidae. Siboga-Expeditie, 39a3 (livr. 87). 318 pp.

1a   With a prominent lateral spine on both sides of the carapace at or close to the anterior end; without a distinct rostrum at the anterior end of the carapace, but with a large, anteriorly directed median spine close to the anterior end 2
1b   Without a prominent lateral spine on both sides of the carapace at or close to the anterior end; with a distinct rostrum projecting from the anterior end of the carapace
Parapasiphae sulcatifrons Smith, 1884
2a   Lateral spines of the carapace slightly behind the anterior edge of the latter; mostly transparent when alive, but generally spotted or tinged with red; length up to about 8 cm
Pasiphaea pacifica Rathbun, 1902 (fig. 19.3)
2b   Lateral spines of the carapace at the anterior edge of the latter; mostly crimson when alive; length up to about 21 cm   *Pasiphaea tarda* Krøyer, 1845

Family Alpheidae

Hart, J. F. L. 1964. Shrimps of the genus *Betaeus* on the Pacific coast of North America with description of three new species. Proc. U. S. Nat. Mus., 115:433-66.

1a   Frontal edge of carapace with an obvious indentation; propodus of cheliped (exclusive of the fixed claw) about as wide as long   *Betaeus setosus* Hart, 1964
1b   Frontal edge of carapace slightly convex or nearly straight, definitely without an indentation; propodus of cheliped more than twice as long as wide
Betaeus harrimani Rathbun, 1904 (fig. 19.4)

Family Pandalidae

Butler, T. H. 1980. Shrimps of the Pacific coast of Canada. Can. Bull. Fish. Aquat. Sci., no. 202. 280 pp.

1a	Antenna 1 about twice as long as the carapace (including the rostrum)	2
1b	Antenna 1 about as long as the carapace, or only slightly longer	3
2a	Distal half of rostrum with spines on its dorsal edge	*Pandalopsis dispar* Rathbun, 1902
2b	Distal half of rostrum without spines on its dorsal edge (restricted to deep water)	*Pandalopsis ampla* Bate, 1888
3a	Abdominal segment 3 somewhat compressed, and with a median dorsal ridge that forms a spine decidedly anterior to the posterior edge of the segment	4
3b	Abdominal segment 3 not compressed, without a median dorsal ridge and without a spine decidedly anterior to the posterior edge	6
4a	Posterior edges of abdominal segments 3 and 4 without a median spine or sharp projection	*Pandalus borealis* Krøyer, 1860
4b	Posterior edges of abdominal segments 3 and 4 without a median spine or sharp projection (segment 3 may, however, have an obtuse or rounded projection)	5
5a	Distal half of the rostrum with dorsal spines	*Pandalus jordani* Rathbun, 1902
5b	Distal half of the rostrum without dorsal spines	*Pandalus goniurus* Stimpson, 1860
6a	Only the anterior half of the carapace (exclusive of the rostrum) with median dorsal spines	7
6b	Median dorsal spines extending posteriorly behind the middle of the carapace (exclusive of the rostrum)	8
7a	Abdominal segment 6 about 3 times as long as wide	*Pandalus tridens* Rathbun, 1902
7b	Abdominal segment 6 less than twice as long as wide	*Pandalus platyceros* Brandt, 1851
8a	With 17-21 dorsal median spines on the carapace (exclusive of the rostrum)	*Pandalus hypsinotus* Brandt, 1851
8b	With fewer than 15 (usually 8-12) dorsal median spines on the carapace (exclusive of the rostrum)	9
9a	Rostrum at least 1.5 times as long as the rest of the carapace	*Pandalus gurneyi* Stimpson, 1871
9b	Rostrum less than 1.5 times as long as the rest of the carapace	10
10a	Telson with 3-5 pairs of small lateral spines; distal half of the blade of the exopodite (scale) of antenna 2 narrower than the adjacent thickened axis; outer margin of blade distinctly concave	*Pandalus stenolepis* Rathbun, 1902
10b	Telson with 6 pairs of small lateral spines; distal half of the exopodite (scale) of antenna 2 not narrower than the adjacent thickened axis; outer margin of the scale nearly straight	*Pandalus danae* Stimpson, 1857 (fig. 19.5)

Family Hippolytidae

Gregory C. Jensen

Butler, T. H. 1971. *Eualus berkeleyorum* n. sp., and records of other caridean shrimps (Order Decapoda) from British Columbia. J. Fish. Res. Bd. Canada, 28:1615-20.

------. 1980. Shrimps of the Pacific coast of Canada. Can. Bull. Fish. Aquat. Sci., no. 202. 280 pp.

Chace, F. A., Jr. 1951. The grass shrimps of the genus *Hippolyte* from the west coast of North America. J. Wash. Acad. Sci., 41:35-9.

Hayashi, K. I. 1977. Studies on the hippolytid shrimps from Japan--VI. The genus *Spirontocaris* Bate. J. Shimonoseki Univ. Fish., 25:155-86.

------. 1979. Studies on the hippolytid shrimps from Japan--VII. The genus *Heptacarpus* Holmes. J. Shimonoseki Univ. Fish., 28:11-32.

Holthuis, L. B. 1947. The Decapoda of the Siboga Expedition. Part IX. The Hippolytidae and Rhynchocinetidae collected by the Siboga and Snellius Expeditions with remarks on other species. Siboga-Expeditie, 39a8 (livr. 140). 100 pp.

------. 1955. The recent genera of the caridean and stenopodidean shrimps (Class Crustacea, Order Decapoda, Supersection Natantia) with keys for their determination. Zool. Verhandel. Leiden, no. 26. 157 pp.

Jensen, G. C. 1983. *Heptacarpus pugettensis*, a new hippolytid shrimp from Puget Sound, Washington. J. Crust. Biol., 3:314-20.

Rathbun, M. J. 1904. Decapod crustaceans of the northwest coast of North America. Harriman Alaska Exped., 10:1-210.

Schmitt, W. L. 1921. The marine decapod Crustacea of California. Univ. Calif. Publ. Zool., 23:1-470.

Wicksten, M. K., & T. H. Butler. 1983. Description of *Eualus lineatus* new species, with a redescription of *Heptacarpus herdmani* (Walker) (Caridea: Hippolytidae). Proc. Biol. Soc. Wash., 96:1-6.

1a	Carpus of leg 2 consisting of 7 articles	2
1b	Carpus of leg 2 consisting of 3 articles	*Hippolyte clarki* Chace, 1951
2a	Supraorbital spines present	30
2b	Supraorbital spines absent	3
3a	Maxilliped 3 with an exopodite	21
3b	Maxilliped 3 without an exopodite	4
4a	Dactyls of legs 3-5 with bifid tips	6
4b	Dactyls of legs 3-5 with simple tips	5
5a	Pleuron of abdominal segment 4 with a ventral spine; leg 3 with an epipodite	*Heptacarpus stimpsoni* Holthuis, 1947
5b	Pleuron of abdominal segment 4 without a ventral spine; leg 3 without an epipodite	*Heptacarpus herdmani* (Walker, 1898)
6a	Rostrum short, not reaching beyond article 1 of antenna 1	7
6b	Rostrum reaching to or beyond article 2 of antenna 1	9
7a	Rostrum reaching to or beyond the cornea of the eye	8
7b	Rostrum strongly downcurved, not reaching the cornea	*Heptacarpus taylori* (Stimpson, 1857)
8a	Leg 3 with an epipodite; distal portion of article 1 of antenna 1 with 3-5 small dorsal spines	*Heptacarpus brevirostris* (Dana, 1852) (fig. 19.9)
8b	Leg 3 without an epipodite; distal portion of article 1 of antenna 1 with a single dorsal spine	*Heptacarpus pugettensis* Jensen, 1983 (fig. 19.6)
9a	Pterygostomian spine absent	10
9b	Pterygostomian spine present	11
10a	Rostrum with 5-7 ventral teeth and reaching beyond the peduncle of antenna 1 by at least half its length; all legs without epipodites	*Heptacarpus stylus* (Stimpson, 1864)
10b	Rostrum with 1-3 ventral teeth and, at most, barely reaching beyond the peduncle of antenna 1; leg 1 with an epipodite	*Heptacarpus littoralis* Butler, 1980
11a	Pleuron of abdominal segment 4 with a ventral spine	12
11b	Pleuron of abdominal segment 4 without a ventral spine	15
12a	Leg 2 with an epipodite	13
12b	Leg 2 without an epipodite	14
13a	Length of the rostrum about equal to that of the rest of the carapace and reaching the end of the scale of antenna 2	*Heptacarpus paludicola* (Holmes, 1900) (fig. 19.8)
13b	Length of the rostrum shorter than that of the rest of the carapace and not reaching the end of the scale of antenna 2	*Heptacarpus pictus* (Stimpson, 1871)
14a	Maxilliped 3 reaching beyond the scale of antenna 2 by about half the length of its distal article	*Heptacarpus moseri* (Rathbun, 1902)
14b	Maxilliped 3, at most, barely reaching beyond the scale of antenna 2	*Heptacarpus sitchensis* (Brandt, 1851)
15a	Maxilliped 3 with an epipodite	16
15b	Maxilliped 3 without an epipodite	*Heptacarpus tenuissimus* Holmes, 1900
16a	Leg 1 with an epipodite	17
16b	Leg 1 without an epipodite	18
17a	Leg 3 with an epipodite; rostrum broad and horizontal or descending with respect to the rest of the carapace	*Heptacarpus carinatus* Holmes, 1900

17b Leg 3 without an epipodite; rostrum slender and ascending with respect to the rest of the carapace *Heptacarpus flexus* (Rathbun, 1902) (fig. 19.7)
18a With 4-7 dorsal teeth on the carapace and rostrum, the anteriormost tooth decidedly farther forward than the cornea of the eye 19
18b Usually with 3 dorsal teeth on the carapace and rostrum, the base of the anteriormost tooth behind, or even with, the cornea of the eye *Heptacarpus tridens* (Rathbun, 1902)
19a Abdominal segment 6 longer than the telson *Heptacarpus decorus* (Rathbun, 1902)
19b Abdominal segment 6 shorter than the telson 20
20a Tip of rostrum nearly bifid due to the presence of a small, ventral, subterminal tooth; article 1 of antenna 1 lacking a dorsal spine; maxilliped 3 reaching to or beyond the tip of the scale of antenna 2 *Heptacarpus kincaidi* (Rathbun, 1902) (fig. 19.10)
20b Rostrum not appearing bifid; article 1 of antenna 1 with a distal, dorsal spine; maxilliped 3 reaching to about the middle of the scale of antenna 2 *Heptacarpus camtschaticus* (Stimpson, 1860)
21a Abdominal segments 3-6 with sharp dorsal spines *Eualus barbatus* (Rathbun, 1899)
21b Abdominal segments lacking dorsal spines 22
22a Leg 1 without an epipodite 23
22b Leg 1 with an epipodite 24
23a Rostrum oval, shorter than the carapace *Eualus macrophthalmus* (Rathbun, 1902)
23b Rostrum bladelike, longer than the carapace *Eualus biunguis* (Rathbun, 1902)
24a Rostrum shorter than the carapace 25
24b Rostrum as long as, or longer than, the carapace 28
25a Dactyls of legs 3-5 with simple tips 26
25b Dactyls of legs 3-5 with bifid tips 27
26a Pleuron of abdominal segment 4 with a ventral spine *Eualus avinus* (Rathbun, 1899)
26b Pleuron of abdominal segment 4 without a ventral spine *Eualus berkeleyorum* Butler, 1971
27a Distal portion of article 1 of antenna 1 with 3 dorsal spines; rostrum not reaching article 2 of antenna 1 *Eualus lineatus* Wicksten & Butler, 1983
27b Distal portion of article 1 of antenna 1 with a single dorsal spine; rostrum not reaching article 2 of antenna 1 *Eualus pusiolus* (Krøyer, 1841)
28a Distal half of rostrum with dorsal spines 29
28b Distal half of rostrum without dorsal spines *Eualus fabricii* (Krøyer, 1841)
29a Pleuron of abdominal segment 4 with a ventral spine; scale of antenna 2 shorter than the carapace (rostrum excluded) *Eualus suckleyi* (Stimpson, 1864)
29b Pleuron of abdominal segment 4 without a ventral spine; scale of antenna 2 as long as or longer than the carapace (rostrum excluded) *Eualus townsendi* (Rathbun, 1902)
30a Maxilliped 3 with an exopodite; with 2 or more supraorbital spines 31
30b Maxilliped 3 without an exopodite; with only 1 supraorbital spine 39
31a Dorsal carapace spines with spiny anterior margins *Spirontocaris prionota* (Stimpson, 1864)
31b Dorsal carapace spines without spiny anterior margins 32
32a Leg 3 without an epipodite 33
32b Leg 3 with an epipodite 35
33a Pleuron of abdominal segment 4 with a ventral spine *Spirontocaris synderi* Rathbun, 1902
33b Peuron of abdominal segment 4 without a spine 34
34a Leg 2 with an epipodite *Spirontocaris holmesi* Holthuis, 1947
34b Leg 2 without an epipodite *Spirontocaris sica* Rathbun, 1902
35a Dactyls of legs 3-5 with bifid tips 36
35b Dactyls of legs 3-5 with simple tips *Spirontocaris lamellicornis* (Dana, 1852) (fig. 19.11)
36a Posteriormost carapace spine near the middle of the carapace 37
36b Posteriormost carapace spine on the posterior third of the carapace 38
37a Carapace with 2 supraorbital spines; rostrum with a midrib extending to the anteriormost point *Spirontocaris ochotensis* (Brandt, 1851)
37b Carapace usually with 3 supraorbital spines; dorsal and ventral blades of the rostrum exceeding the tip of the midrib *Spirontocaris truncata* Rathbun, 1902
38a Dactyls of legs 3-5 stout, with 5-7 spines *Spirontocaris arcuata* Rathbun, 1902
38b Dactyls of legs 3-5 slender, with 8-11 spines *Spirontocaris spina* (Sowerby, 1805)
39a Rostrum longer than the eye 40
39b Rostrum shorter than the eye *Lebbeus* sp.

40a	Pleura of abdominal segments 2 and 3 terminating in sharp points	
		Lebbeus groenlandicus (Fabricius, 1775)
40b	Pleura of abdominal segments 2 and 3 rounded	41
41a	Leg 3 with an epipodite	42
41b	Leg 3 without an epipodite (live specimens with purple, yellow, and red bands)	
		Lebbeus grandimanus (Brazhnikov, 1907)
42a	Rostrum with 2 or 3 ventral teeth; stylocerite not reaching beyond article 1 of antenna 1	
		Lebbeus washingtonianus (Rathbun, 1902)
42b	Rostrum with 1 ventral tooth; stylocerite reaching article 2 of antenna 1	
		Lebbeus schrencki (Brazhnikov, 1907)

Family Crangonidae

Gregory C. Jensen

Butler, T. H. 1980. Shrimps of the Pacific coast of Canada. Can. Bull. Fish. Aquat. Sci., no. 202. 280 pp.

Kuris, A. M., & J. T. Carlton. 1977. Description of a new species, *Crangon handi*, and new genus, *Lissocrangon*, of crangonid shrimps (Crustacea: Caridea) from the California coast, with notes on adaptation in body shape and coloration. Biol. Bull., 153:540-59.

Rathbun, M. J. 1904. Decapod crustaceans of the northwest coast of North America. Harriman Alaska Exped., 10:1-210.

Schmitt, W. L. 1921. The marine decapod Crustacea of California. Univ. Calif. Publ. Zool., 23:1-470.

1a	Rostrum more than half as long as the rest of the carapace; with only 4 pairs of legs	
		Paracrangon echinata Dana, 1852
1b	Rostrum, if present, less than half as long as the rest of the carapace; with 5 pairs of legs	2
2a	Dactyl of leg 5 broad and flattened; eyestalks parallel and shielded by a hood that consists of the antennal spines, postorbital spines, and rostrum	3
2b	Dactyl of leg 5 not broad and flattened; eyestalks not parallel and not shielded by a hood consisting of spines	7
3a	Carapace with 2 median spines in addition to the spine that forms the rostrum	4
3b	Carapace with 3 or 4 median spines in addition to the spine that forms the rostrum	5
4a	Longitudinal dorsal ridges (carinae) on abdominal segment 6 extended beyond the posterior edge of the segment as spines	*Argis dentata* (Rathbun, 1902)
4b	Longitudinal dorsal ridges on abdominal segment 6 not extended beyond the posterior edge of the segment as spines	*Argis lar* (Owen, 1839)
5a	Longitudinal dorsal ridges (carinae) on abdominal segment 6 extended beyond the posterior edge of the segment as spines	*Argis alaskensis* (Kingsley, 1882)
5b	Longitudinal dorsal ridges on abdominal segment 6 not extended beyond the posterior edge of the segment as spines	6
6a	Abdominal segments 1-4 with a longitudinal dorsal ridge; pleuron of abdominal segment 4 with a small posteroventral spine	*Argis crassa* (Rathbun, 1899)
6b	Abdominal segments 1-4 without a longitudinal dorsal ridge; pleuron of abdominal segment 4 without a posteroventral spine	*Argis levior* (Rathbun, 1902)
7a	Carapace with a median dorsal spine or spines	8
7b	Carapace without median spines	*Lissocrangon stylirostris* (Holmes, 1900)
8a	Submedian spines present on carapace	9
8b	Submedian spines absent on carapace	13
9a	Carapace with 2 submedian spines on both sides	*Mesocrangon munitella* (Walker, 1898)
9b	Carapace with 1 submedian spine on both sides	10
10a	Pleura of abdominal segments 1-3 with ventrally directed spines	
		Metacrangon spinosissima (Rathbun, 1902)

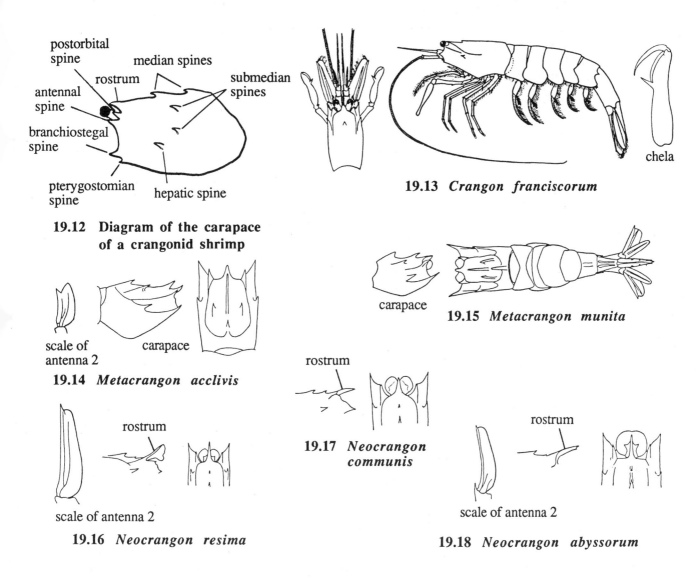

19.12 Diagram of the carapace of a crangonid shrimp

19.13 *Crangon franciscorum*

19.14 *Metacrangon acclivis*

19.15 *Metacrangon munita*

19.16 *Neocrangon resima*

19.17 *Neocrangon communis*

19.18 *Neocrangon abyssorum*

10b Pleura of abdominal segments 1-3 without ventrally directed spines 11
11a Anterior median carapace spine extending beyond the bases of the orbits of the eyes, and larger than the posterior median carapace spine
.... *Metacrangon acclivis* (Rathbun, 1902) (fig. 19.14)
11b Anterior median carapace spine not extending beyond the bases of the orbits of the eyes, and not larger than the posterior median carapace spine 12
12a Abdominal segments 3-5 with a prominent longitudinal dorsal ridge (carina); spine of scale of antenna 2 reaching to or beyond the tip of the lamella
.... *Metacrangon variabilis* (Rathbun, 1902)
12b Abdominal segments 3-5 without a prominent longitudinal dorsal ridge; tip of the lamella of the scale of antenna 2 reaching beyond the spine
.... *Metacrangon munita* (Dana, 1852) (fig. 19.15)
13a Carapace obviously sculptured 14
13b Carapace smooth 15
14a Carapace with 1 or 2 median spines; outer margin of scale of antenna 2 concave
.... *Rhynocrangon alata* (Rathbun, 1902)

14b Carapace with 3 or 4 median spines; outer margin of scale of antenna 2 convex
Sclerocrangon boreas (Phipps, 1774)
15a Carapace with 2 median spines 16
15b Carapace with 1 median spine 18
16a Rostrum ascending at an angle of about 45° and extending beyond the eyes; rostrum, in adult, with a large ventral expansion   *Neocrangon resima* (Rathbun, 1902) (fig. 19.16)
16b Rostrum ascending at an angle of 30° or less and not extending beyond the eyes; rostrum without a ventral expansion 17
17a Tip of rostrum rounded; eyes not contiguous
Neocrangon communis (Rathbun, 1899) (fig. 19.17)
17b Tip of rostrum acute; eyes large and contiguous
Neocrangon abyssorum (Rathbun, 1902) (fig. 19.18)
18a Dorsal surface of abdominal segment 6 with 2 prominent longitudinal ridges (carinae)
Crangon dalli Rathbun, 1902
18b Dorsal surface of abdominal segment 6 without longitudinal ridges 19
19a Abdominal segment 5 with a spine on the upper posterolateral margin of both sides 20
19b Abdominal segment 5 without spines on the upper posterolateral margins 21
20a Dactyl of leg 1 reaching the base of the fixed finger when the chela is closed; propodus of leg 1 of male 4.5 times as long as wide
Crangon franciscorum subsp. *franciscorum* Stimpson, 1856 (fig. 19.13)
20b Dactyl of leg 1 crossing the base of the fixed finger when the chela is closed; propodus of leg 1 of male 5.5 times as long as wide
Crangon franciscorum subsp. *angustimana* Rathbun, 1902
21a Ventral surface of abdominal segment 6 with a median groove (sulcus) 22
21b Ventral surface of abdominal segment 6 without a median groove   *Crangon alba* Holmes, 1900
22a Flagella of antenna 1 equal in length; scale of antenna 2 not more than twice as long as wide (not known to occur north of California)   *Crangon handi* Kuris & Carlton, 1977
22b Inner flagellum of antenna 1 longer than the outer flagellum; scale of antenna 2 more than twice as long as wide 23
23a Spine of scale of antenna 2 extending beyond the tip of the lamella, which is narrow; scale of antenna 2 as long as, or longer than, the telson   *Crangon alaskensis* Lockington, 1877
23b Spine of scale of antenna 2 reached or exceeded by the tip of the lamella, which is a rounded lobe; scale of antenna 2 shorter than the telson   *Crangon nigricauda* Stimpson, 1856

"Reptantia"

Thalassinids, crabs, hermit crabs, and other non-shrimp decapods are placed in the suborder Pleocyemata. This group also includes the caridean shrimps dealt with in the preceding section. Thus the heading for Pleocyemata is repeated.

Suborder Pleocyemata

Infraorder Thalassinidea

Hart, J. F. L. 1982. Crabs and Their Relatives of British Columbia. Handbook 40, British Columbia Provincial Museum, Victoria. 266 pp.
Stevens, B. A. 1928. Callianassidae from the west coast of North America. Publ. Puget Sound Biol. Station, 6:315-69.

1a Rostrum conspicuous, at least twice as long as wide, with 5 or 6 teeth on each side (burrowing in muddy substrata at depths generally greater than 100 m)   Family Axiidae
1b Rostrum either a small tooth or a broad process that is divided into 3 teeth 2
2a Rostrum a small tooth, not hairy; chelipeds chelate and decidedly unequal (in mature or nearly mature specimens)   Family Callianassidae

19.19 Callianassidae: *Callianassa californiensis*

19.20 Galatheidae: *Munida quadrispina*

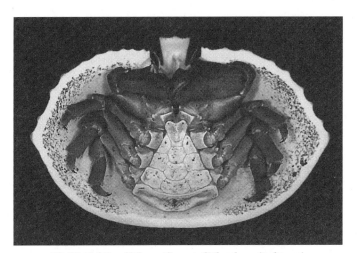

19.21 Lithodidae: *Cryptolithodes sitchensis*

19.22 Paguridae: *Discorsopagurus schmitti*

2b Rostrum a broad process divided into 3 teeth and hairy dorsally; chelipeds subchelate (the dactyl bends back to close against the propodus itself, rather than against a clawlike process of the latter) and nearly equal . . . . Family Upogebiidae

Family Axiidae

Four widely distributed species have been reported in our region. All burrow in mud at considerable depths. None is likely to be found in water shallower than 100 m. A key and descriptions are found in Hart (1982).

Axiopsis spinulicauda (Rathbun, 1902)
Calastacus stilirostris Faxon, 1893
Calocaris investigatoris (Anderson, 1896) (fig. 19.23)
Calocaris quinqueseriatus (Rathbun, 1902)

Family Upogebiidae

Upogebia pugettensis (Dana, 1852). Common intertidal species that burrows in muddy sand and muddy gravel.

Family Callianassidae

1a Rostrum bluntly rounded; chela of larger cheliped, when closed, with a conspicuous gap between the dactyl and the clawlike process of the propodus; carpus of smaller cheliped decidedly wider than the merus; common intertidally in muddy sand
Callianassa californiensis Dana, 1854 (fig. 19.19)
1b Rostrum sharp; chela of larger cheliped, when closed, without a conspicuous gap between the dactyl and the clawlike process of the propodus; carpus of smaller cheliped not much wider than the merus; locally common intertidally and subtidally (to about 50 m) in muddy sand
Callianassa gigas Dana, 1852 (fig. 19.24)

Callianopsis goniophthalma (Rathbun, 1901). Found in mud at depths greater than 400 m (see Hart, 1982).

Infraorder Anomura

Haig, J. 1960. The Porcellanidae (Crustacea Anomura) of the eastern Pacific. Allan Hancock Pacific Exped., 24. 440 pp.
Hart, J. F. L. 1937. Larval and adult stages of British Columbia Anomura. Canad. J. Res., D, 15:179-220.
------. 1971. New distribution records of reptant decapod Crustacea, including descriptions of three new species of *Pagurus*, from the waters adjacent to British Columbia. J. Fish. Res. Bd. Canada, 28:1527-44.
------. 1982. Crabs and Their Relatives of British Columbia. Handbook 40, British Columbia Provincial Museum, Victoria. 266 pp.
McLaughlin, P. A. 1974. The hermit crabs (Crustacea Decapoda, Paguridea) of northwestern North America. Zool. Verhandel. Leiden, no. 130. 396 pp.
Makarov, V. V. 1938. [Crustacea. Anomura]. *In* Fauna SSSR, n. s., no. 16, 10. x + 324 pp. (in Russian). (English translation, 1962. Jerusalem: Israel Program for Scientific Translations.)
Schmitt, W. L. 1921. The marine decapod Crustacea of California. Univ. Calif. Publ. Zool., 23:1-470.
Stevens, B. A. 1925. Hermit crabs of Friday Harbor, Washington. Publ. Puget Sound Biol. Station, 3:273-309.

1a Body as a whole almost ovoid; leg 1 not chelate or subchelate; telson twice as long as wide; found only on exposed sandy beaches, burrowing just under the surface
Superfamily Hippoidea: Family Hippidae

Phylum Arthropoda: Order Decapoda

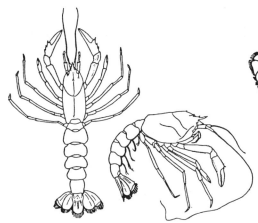

19.23 Axiidae: *Calocaris investigatoris*

19.24 Callianassidae: *Callianassa gigas*

19.26 *Pagurus hirsutiusculus*

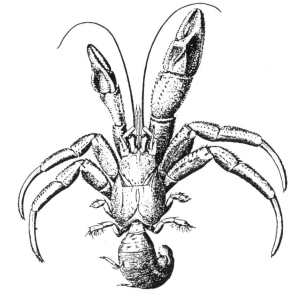

19.25 Paguridae: *Pagurus tanneri*

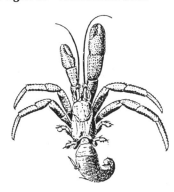

19.27 *Pagurus granosimanus*

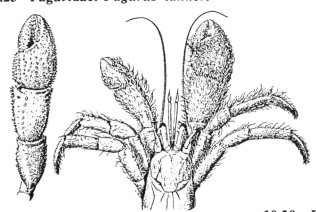

19.28 *Pagurus capillatus*

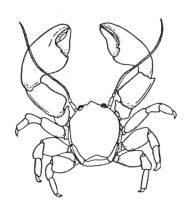

19.29 Porcellanidae: *Petrolisthes cinctipes*

1b	Body as a whole not ovoid; leg 1 chelate or subchelate; telson, if obvious, not twice as long as wide; generally in rocky or stony situations, on muddy bottoms, or living in permanent burrows in mud or muddy sand	2
2a	Uropods present and constituting part of a well developed tail fan (even in the case of crablike species in which the abdomen is permanently folded against the underside of the thorax)	3
2b	Uropods, if present, not constituting part of a well developed tail fan	5
3a	Body somewhat lobsterlike, the abdomen bent under itself but not folded tightly against the underside of the thorax	4
3b	Body crablike, the abdomen permanently folded against the underside of the thorax Superfamily Galatheoidea: Family Porcellanidae	
4a	Telson bent under the abdomen; not likely to be found in situations shallower than 900 m Superfamily Galatheoidea: Family Chirostylidae	
4b	Telson not bent under the abdomen; found in shallow water, as well as at considerable depths Superfamily Galatheoidea: Family Galatheidae	
5a	Uropods present, hooklike, and with a rasplike area on each ramus; abdomen soft, unsegmented, not folded against the underside of the thorax, generally twisted (but sometimes straight); living in snail shells, worm tubes, or cavities in sponge masses	6
5b	Uropods absent; abdomen either firm or soft, segmented or unsegmented, folded against the underside of the thorax, not twisted; not living in snail shells, worm tubes, or cavities in sponge masses Superfamily Paguroidea: Family Lithodidae	
6a	Bases of third maxillipeds as close together as those of the first 2 pairs; both sexes with paired pleopods (male with 2 pairs, female with 1 pair); with 13 pairs of gills; chelipeds similar and nearly equal (includes 2 species of hermit crabs found subtidally) Superfamily Paguroidea: Family Diogenidae	
6b	Bases of third maxillipeds much more widely separated than those of the first 2 pairs; either without paired pleopods, or with paired pleopods only in the male; with 11 pairs of gills; chelipeds generally dissimilar and unequal	7
7a	Male without paired pleopods; female with paired gonopores (all intertidal hermit crabs and most subtidal hermit crabs belong in this group) Superfamily Paguroidea: Family Paguridae	
7b	Male with paired first and second pleopods; female with a gonopore on the left side only (includes a rare hermit crab found in deep water) Superfamily Paguroidea: Family Parapaguridae	

Superfamily Paguroidea

Family Lithodidae

1a	Abdomen mostly soft (but the telson may be calcified, and there may be a few other calcified plates), and not tightly applied to the underside of the thorax	2
1b	Abdomen largely or completely covered by calcified plates and tightly pressed against the underside of the thorax	6
2a	Abdomen rather thin and flat; chelipeds nearly equal (the right cheliped only slightly larger than the left); carapace and legs with protuberances that resemble scales *Placetron wosnessenskii* Schalfeew, 1892	
2b	Abdomen thick; chelipeds decidedly unequal; dactyl of chelipeds generally shorter than the palm of the chela; carapace and legs without protuberances that resemble scales (they may, however, have spines, setae, and tubercles)	3
3a	Rostrum, from its base to its tip, studded with stout spines *Acantholithodes hispidus* (Stimpson, 1860)	
3b	Rostrum without spines (but it may be setose and its tip is sharp)	4
4a	Cephalothorax and legs much flattened dorsoventrally, and very setose; upper surface of chelipeds with spines; anterior margins of most articles of legs 2-4 with large, flattened spines (the spines are so broad that they give the margins of the articles from which they originate a serrated appearance)	5
4b	Cephalothorax and legs not much flattened dorsoventrally and not especially setose; upper surface of chelipeds with prominent, granular tubercles, but without obvious spines; anterior	

	margins of articles of legs 2-4 without large spines (there are, however, some setae, as well as tubercles and small spines) *Oedignathus inermis* (Stimpson, 1860)
5a	Length of rostrum about twice its width at the base; upper surface of palm of right chela usually with 4 rows of spines running lengthwise *Hapalogaster mertensii* Brandt, 1850
5b	Length of rostrum less than twice its width at the base; upper surface of right chela with 3 rows of spines running lengthwise *Hapalogaster grebnitzkii* Schalfeew, 1892
6a	Carapace with lateral expansions that completely conceal the legs when the animal is viewed from above 7
6b	Carapace without lateral expansions that completely conceal the legs when the animal is viewed from above 8
7a	Propodus of chelipeds smooth; rostrum broader distally than proximally; abdominal plates without raised margins *Cryptolithodes sitchensis* Brandt, 1853 (fig. 19.21)
7b	Propodus of chelipeds with coarse tubercles; rostrum generally broader proximally than distally; abdominal plates with raised margins *Cryptolithodes typicus* Brandt, 1849
8a	Outline of carapace, in dorsal view, similar to that of an equilateral triangle (there are blunt or sharp-tipped spines along the lateral margins, however); posterior half of dorsal surface of carapace with a pronounced concavity or semicircular depression 9
8b	Outline of carapace, in dorsal view, rounded posterolaterally, not at all similar to that of an equilateral triangle; posterior half of dorsal surface of carapace with neither a pronounced concavity nor a semicircular depression 10
9a	Rostrum divided into 2 blunt lobes; most spines on legs blunt, papillate; legs with few if any conspicuous setae; posterior half of carapace with an extensive concavity, this consisting essentially of 2 depressions, 1 on each side of the midline, the concavity as a whole bordered by large tubercles; lateral margins of carapace with stout, blunt spines *Phyllolithodes papillosus* Brandt, 1849
9b	Rostrum not divided into 2 blunt lobes; spines on legs sharp-tipped; legs with many long, curved setae (most of these are longer than the spines); posterior half of carapace with a semicircular depression, this not bordered by prominent tubercles; lateral margins of carapace with sharp-tipped spines *Rhinolithodes wosnessenskii* Brandt, 1849
10a	Carapace bumpy, due to the presence of large, granular tubercles; legs 2-4 stout, not appreciably longer than the width of the carapace 11
10b	Carapace spiny, but without large tubercles; legs 2-4 decidedly longer than the width of the carapace 12
11a	Tubercles on chelipeds and other legs spinelike; posterior edge of carpus of chelipeds excavated into a deep sinus (when this is in apposition to a shallow sinus on the anterior edge on the carpus of leg 2, a circular hole is formed) *Lopholithodes foraminatus* (Stimpson, 1862)
11b	Tubercles on chelipeds and other legs mostly rounded and blunt; posterior edge of carpus of chelipeds not excavated into a sinus *Lopholithodes mandtii* Brandt, 1849
12a	Most of the integument of the abdomen calcified; median plates of abdominal segments 3-5 replaced by a membranous area that has calcareous nodules 13
12b	Integument of the abdomen not extensively calcified; median plate of abdominal segments 3-5 distinct 15
13a	Abdominal segment 2 covered by 5 distinct plates; rostrum ending in a single, sharp-tipped projection (not likely to be found south of northern British Columbia, and mostly at depths greater than 50 m) *Paralithodes camtschatica* (Tilesius, 1815)
13b	Plates of abdominal segment 2 completely or partly fused; rostrum ending in 2 diverging spines (thus essentially bifid) or in a spine that is flanked by 2 other spines (thus almost trifid) 14
14a	Length of rostrum about 4 times its width at the base; rostrum terminating in a pair of diverging spines, without dorsal spines, but with a pair of large lateral spines near its middle (not likely to be found at depths less than 250 m) *Lithodes couesi* Benedict, 1894
14b	Length of rostrum about twice its width at the base; rostrum with some dorsal spines as well as with a pair of lateral spines flanking the terminal projection (not likely to be found at depths less than 75 m) *Lithodes aequispina* Benedict, 1894
15a	Carapace (including rostrum) slightly longer than wide; right cheliped not decidedly larger than the left cheliped; legs 2-4 appreciably flattened (not likely to be found at depths less than 1200 m) *Paralomis verrilli* (Benedict, 1894)

Phylum Arthropoda: Order Decapoda

15b Carapace about as broad as long; right cheliped usually decidedly longer than the left cheliped; legs 2-4 not appreciably flattened (not likely to be found at depths less than 800 m)
Paralomis multispina (Benedict, 1894)

Family Paguridae

The majority of the pagurids found in this region are subtidal. The following species, however, are intertidal, and unless otherwise noted they occur in Puget Sound and contiguous inland waters as well as on the open coast: *Pagurus beringanus, P. hirsutiusculus* (fig. 19.26), *P. granosimanus* (fig. 19.27), *P. hemphilli* (primarily open coast), and *P. samuelis* (only open coast).

Other pagurids that are primarily subtidal, but that are occasionally encountered intertidally, are *Discorsopagurus schmitti* (fig. 19.22), *Elassochirus gilli, Pagurus armatus, P. caurinus, P. quaylei* (open coast), and *P. kennerlyi*.

1a Carapace smooth, and only a portion of it (shield) calcified; basal portions of eyescales not covered by the carapace, visible from above; leg 4 subchelate; male with pleopods, but these are unpaired 2
1b Carapace with dorsal spines, and almost all of it calcified; basal portions of eyescales covered by the carapace, not visible from above; male without pleopods
Labidochirus splendescens (Owen, 1839)
2a Abdomen not coiled; uropods symmetrical; in tubes of sabellariid or serpulid polychaetes, or in shells of scaphopod molluscs (rarely shells of gastropod molluscs) 3
2b Abdomen coiled; uropods asymmetrical; in shells of gastropods or in sponge masses 4
3a Palm of right chela wider at the level where the dactyl and fixed claw originate than at its proximal end; right and left halves of telson partly separated by a groove; in shells of scaphopod molluscs; subtidal off the open coast *Orthopagurus minimus* (Holmes, 1900)
3b Palm of right chela little if any wider at the level where the dactyl and fixed claw originate than at its proximal end; right and left halves of telson not separated by a groove; almost always in tubes of sabellariid and serpulid polychaetes (rarely in shells of gastropods); subtidal, but occurring in inland waters as well as off the open coast
Discorsopagurus schmitti (Stevens, 1925) (fig. 19.22)
4a Carpus of right cheliped at least slightly longer than wide; neither the carpus nor the propodus of the right cheliped much flattened or with 1 or both margins thinned further; eyescales without a deep median furrow 5
4b Carpus of right cheliped as wide or wider than long; either the carpus or propodus, or both, much flattened and with 1 or both margins thinned further; eyescales with a deep median furrow 22
5a Dactyl of legs 2 and 3 obviously twisted in relation to the propodus, and with 2 reddish brown stripes 6
5b Dactyl of legs 2 and 3 not twisted in relation to the propodus, and without stripes 8
6a Dactyl and propodus of chelipeds spiny dorsally and marginally, but with relatively few spines on the ventral side *Pagurus armatus* (Dana, 1851)
6b Dactyl and propodus of chelipeds with about as many tubercles or small spines on the ventral surface as on the dorsal surface 7
7a Dorsal surface of dactyl of legs 2 and 3 with a single distinct longitudinal groove; merus and carpus of chelipeds without an opalescent sheen *Pagurus aleuticus* (Benedict, 1892)
7b Dorsal surface of dactyl of legs 2 and 3 with longitudinal rows of small spines or tubercles, these rows separated by 2 shallow, indistinct grooves; merus and carpus of chelipeds with an opalescent sheen *Pagurus ochotensis* Brandt, 1851
8a Dorsal surface of palm of left chela with a prominent ridge or crest near the midline, and with a distinct concavity lateral to this; subtidal, and found only in deep water off the open coast 9
8b Dorsal surface of palm of left chela without a prominent ridge or crest near the midline (but there may be a slight eminence), and without a concavity; includes intertidal and subtidal species, some of which occur in inland waters 11
9a Right chela about 2.5 times as long as wide; the nearly triangular elevation on the dorsal surface of the palm of the right chela continuing onto the fixed claw as a rather sharp, spiny crest *Pagurus tanneri* (Benedict, 1892) (fig. 19.25)

9b	Right chela not more than twice as long as wide; the nearly triangular elevation on the dorsal surface of the palm of the right chela continuing onto the fixed claw as a low, spiny or tuberculate ridge	10
10a	Apex of the nearly triangular elevation on the dorsal surface of the palm of the right chela forming a prominent spiny or tuberculate projection	*Pagurus cornutus* (Benedict, 1892)
10b	Apex of the nearly triangular elevation on the dorsal surface of the palm of the right chela not forming a prominent projection	*Pagurus confragosus* (Benedict, 1892)
11a	Ventral surface of merus of right cheliped with 1 or 2 prominent tubercles (these are about twice as large as the other tubercles); includes intertidal and subtidal species	12
11b	Ventral surface of merus of right cheliped without prominent tubercles; strictly subtidal species	17
12a	Ventral surface of merus of right cheliped with 2 prominent tubercles	13
12b	Ventral surface of merus of right cheliped with only 1 prominent tubercle (and in *Pagurus hirsutiusculus*, a common intertidal species, this is often obscure)	15
13a	Ventrolateral surface of dactyl and propodus of left leg 3 without spines or tubercles, or the propodus with only 1 or 2 spines on its distal portion; legs, including chelipeds, olive-green, with small blue dots	*Pagurus granosimanus* (Stimpson, 1858) (fig. 19.27)
13b	Ventrolateral surface of dactyl and propodus of left leg 3 with 1 or more irregular rows of small spines or tubercles; color of legs not olive-green with small blue dots	14
14a	Shield of carapace about as wide as long; chelae with small spines rather than tubercles or granules; dorsal surface of carpus of left cheliped with a single row of stout spines; rostrum generally rounded; legs 2 and 3 with a red band at the articulation of the dactyl and propodus; not restricted to the open coast, but found only subtidally and in the lowest part of the intertidal region	*Pagurus beringanus* (Benedict, 1892)
14b	Shield of carapace decidedly longer than wide; chelae with tubercles or granules; dorsal surface of carpus of left cheliped with 2 rows of rather short spines; rostrum acute; dactyl of legs 2 and 3 bright blue; restricted to the open coast, where it occurs intertidally	*Pagurus samuelis* (Stimpson, 1857)
15a	Chelae with irregular rows of rather stout spines, these partly obscured by tufts of long setae; antennae 2 without white spots (reddish brown in color); with a white band at the articulation of the dactyl and propodus of legs 2 and 3	*Pagurus caurinus* Hart, 1971
15b	Chelae with closely spaced tubercles or granules, these often inconspicuous (scattered tufts of setae may also be present); with or without white spots on antennae 2 ; with or without a light band at the articulation of the dactyl and propodus of legs 2 and 3	16
16a	Ventral side of carpus of right cheliped greatly swollen; shield of carapace decidedly longer than wide; antennae 2 orange-red, without white spots; legs, on the whole, dark red, with minute yellow spots, but without a light band at the articulation of the dactyl and propodus; on the open coast, where it is restricted to the lowest part of the intertidal region and the shallow subtidal region	*Pagurus hemphilli* (Benedict, 1892)
16b	Ventral side of carpus of right cheliped not greatly swollen; shield of carapace about as wide as long, or wider than long; antennae 2 largely green, with white spots; legs greenish brown, legs 2 and 3 with white or bluish bands at the articulation of the dactyl and propodus; on the open coast and in inland waters, mostly in the mid-intertidal region	*Pagurus hirsutiusculus* (Dana, 1851) (fig. 19.26)
17a	Dactyl of left chela bowed to the extent that when it is closed against the fixed claw, there is a prominent gap between them; eyescale usually tipped with 2 to 5 spines	*Pagurus quaylei* Hart, 1971
17b	Dactyl of left chela not bowed to the extent that there is a prominent gap between it and the fixed claw; eyescales terminating in a single spine or merely acute	18
18a	Carpus of left cheliped about 3 times as long as wide, and usually appreciably longer than the merus	19
18b	Carpus of left cheliped about 2 times as long as wide, and about the same length as the merus	20
19a	Carpus of right cheliped about twice as long as wide; left chela more nearly triangular than elongate-ellipsoidal, and with a double row of divergent spines on the eminence near the midline of the dorsal surface of the palm; rostrum not pronounced, with a blunt tip; distal part of the merus of chelipeds without a white band	*Pagurus stevensae* Hart, 1971

19b Carpus of right cheliped about 1.5 times as long as wide; left chela more nearly elongate-ellipsoidal than triangular, and with a single row of large spines on the eminence near the midline of the dorsal surface of the palm; rostrum pronounced, triangular and sharp-tipped; distal part of the merus of chelipeds with a distinct white band   *Pagurus dalli* (Benedict, 1892)

20a Dorsal side of the base of the cornea with a tuft of setae; eyestalks (including the cornea) about 5 times as long as wide; distal portion of the merus of chelipeds with a white band; antenna 2 with light and dark bands   *Pagurus kennerlyi* (Stimpson, 1864)

20b Dorsal side of the base of the cornea without a tuft of setae; eyestalks (including the cornea) generally not more than 4 times as long as wide; distal portion of the merus of chelipeds without a white band; antenna 2 not banded   21

21a Ventral margin of dactyl of legs 2 and 3 with a row of small corneous tubercles; chelae so hairy that the spines or tubercles are nearly hidden; found in inland waters as well as on the open coast (antenna 2 grayish tan)   *Pagurus capillatus* (Benedict, 1892) (fig. 19.28)

21b Ventral margin of dactyl of legs 2 and 3 with a row of sharp corneous spines; chelae not so hairy that the spines or tubercles are nearly hidden; restricted to the open coast
     Pagurus setosus (Benedict, 1892)

22a Propodus of right cheliped, in dorsal view, slightly wider than the carpus; dorsal surface of propodus of right cheliped roughened by tubercles and spines; margins of merus of legs (especially chelipeds) bright grayish blue (elsewhere on the legs, the color ranges from cream to red)   *Elassochirus tenuimanus* (Dana, 1851)

22b Propodus of right cheliped, in dorsal view, narrower than the carpus; dorsal surface of propodus of right cheliped nearly smooth; prevailing coloration of legs orange to red, without grayish blue on the margins of the merus   23

23a Dorsal surface of carpus of right cheliped with 1 or more irregular rows of small spines; merus and carpus of chelipeds with a lavender-purple area
     Elassochirus cavimanus (Miers, 1879)

23b Dorsal surface of carpus of right cheliped without small spines; chelipeds rather uniformly orange to red   *Elassochirus gilli* (Benedict, 1892)

Family Parapaguridae

Parapagurus pilosimanus subsp. *benedicti* de Saint Laurent, 1972. Taken only in very deep water off the coast.

Family Diogenidae

1a Eyestalks rather stout, their length equal to about two-thirds to three-fourths the width of the anterior portion of the carapace (common subtidal species)
     Paguristes turgidus (Stimpson, 1857)

1b Eyestalks slender, their length equal to, or slightly greater than, the width of the anterior portion of the carapace   *Paguristes ulreyi* Schmitt, 1921

Superfamily Hippoidea

Family Hippidae

Emerita analoga (Stimpson, 1857). Occasionally found on sandy beaches of the outer coast of Oregon, Washington, and Vancouver Island; gregarious and moving up and down the beach with the tide; the populations are derived form planktonic larvae originating farther south, and do not persist.

Superfamily Galatheoidea

Family Galatheidae

1a Eyes with dark pigment, faceted; exopodite of maxilliped terminating in a flagellum; sometimes fairly common at depths greater than about 20 m
Munida quadrispina Benedict, 1902 (fig. 19.20)
1b Eyes without dark pigment, not faceted; exopodite of first maxilliped not terminating in a flagellum; not likely to be taken except in deep water on the continental shelf
Munidopsis quadrata (Faxon, 1893)

Family Porcellanidae

1a Chelipeds equal or nearly equal, flattened, with a finely granulated surface, not covered with plumose hairs 2
1b Chelipeds decidedly unequal, thick, roughened by tubercles, and covered with plumose hairs 3
2a Carpus of chelipeds about twice as long as wide, with parallel anterior and posterior margins
Petrolisthes eriomerus Stimpson, 1871
2b Carpus of chelipeds about 1.5 times as long as wide, with anterior and posterior margins not parallel *Petrolisthes cinctipes* (Randall, 1839) (fig. 19.29)
3a Telson with 7 plates; carpus of chelipeds with 3 or 4 uneven, serrated teeth on the anterior margin *Pachycheles pubescens* Holmes, 1900
3b Telson with 5 plates; carpus of chelipeds with a broad, nearly triangular lobe on the anterior margin *Pachycheles rudis* Stimpson, 1860

Infraorder Brachyura

Chace, F. A., Jr. 1951. The oceanic crabs of the genera *Planes* and *Pachygrapsus*. Proc. U. S. Nat. Mus., 101:65-103.
Hart, J. F. L. 1982. Crabs and Their Relatives of British Columbia. Handbook 40, British Columbia Provincial Museum, Victoria. 266 pp.
Menzies, R. J. 1948. A revision of the brachyuran genus *Lophopanopeus*. Occas. Pap. Allan Hancock Found., no. 4. 45 pp.
Rathbun, M. J. 1904. Decapod crustaceans of the northwest coast of North America. Harriman Alaska Exped., 10:1-210.
------. 1918. The grapsoid crabs of America. Bull. 97, U. S. Nat. Mus. xxii +461 pp.
------. 1925. The spider crabs of America. Bull. 129, U. S. Nat. Mus. xx + 613 pp.
------. 1930. The cancroid crabs of America of the families Euryalidae, Portunidae, Atelecyclidae, Cancridae, and Xanthidae. Bull. 152, U. S. Nat. Mus. xvi + 609 pp.
------. 1937. The oxystomatous and allied crabs of America. Bull. 166, U. S. Nat. Mus. vi + 278 pp.
Schmitt, W. L. 1921. The marine decapod Crustacea of California. Univ. Calif. Publ. Zool., 23:1-470.
Wells, W. W. 1928. Pinnotheridae of Puget Sound. Publ. Puget Sound Biol. Station, 6:283-314.
------. 1940. Ecological studies on the pinnotherid crabs of Puget Sound. Univ. Wash. Publ. Oceanogr., 2:19-50.

1a Aggregate of mouthparts tapering anteriorly, thus forming a triangle (carapace broader than long, and with a large, laterally directed tooth originating at the widest part)
Family Calappidae
1b Aggregate of mouthparts not tapering anteriorly 2
2a With a prominent rostrum that is divided into 2 processes (these may be partly or wholly fused) Family Majidae
2b Without a rostrum 3

19.30 Majidae: *Chionoecetes bairdi*

19.31 Majidae: *Scyra acutifrons*

19.32 Grapsidae: *Hemigrapsus oregonensis*

3a   Anterior margin of carapace, between the eyes, usually toothed (not toothed, however, in the family Xanthidae); lateral margins of carapace distinctly toothed   4
3b   Anterior margin of carapace, between the eyes, not toothed; lateral margins of carapace either toothed or not toothed   6
4a   Anterior margin of carapace, between the eyes, without teeth, but generally with a slight notch; anterolateral margins of carapace with 3 or more blunt teeth; abdomen of male with 4 segments plus telson   Family Xanthidae
4b   Anterior margin of carapace, between the eyes, with several distinct teeth; anterolateral

	margins of carapace with at least 6 teeth; abdomen of male with 6 segments plus telson	5
5a	Carapace more or less five-angled; marginal teeth with several spines; setae stiff, club-shaped	
		Family Atelecyclidae
5b	Carapace broadly oval; each marginal tooth with not more than 1 spine; setae, when present, not club-shaped	Family Cancridae
6a	Carapace almost rectangular, with distinct teeth on the lateral margins; free-living	
		Family Grapsidae
6b	Carapace oval or nearly circular, without teeth on the lateral margins; closely associated with other invertebrates (living in worm tubes, ascidians, mantle cavity of bivalve molluscs, etc.)	
		Family Pinnotheridae

Section Oxystomata

Family Calappidae

Mursia gaudichaudi (Milne Edwards, 1837). Found subtidally (30-400 m) as far north as southern Oregon, and there is one record for British Columbia.

Family Majidae

1a	Rostrum about half the length of the rest of the carapace, consisting largely of 2 slender, spinelike processes (these may be fused medially for much of their length)	2
1b	Rostrum not more than one-third or two-fifths the length of the rest of the carapace, consisting of 2 broad, flattened processes	3
2a	Spinelike processes of rostrum nearly parallel, at least in the proximal half; dorsal surface of carapace with a large, sharp tooth behind each eye, but otherwise rather smooth	
		Oregonia gracilis Dana, 1851 (fig. 19.33)
2b	Spinelike processes of rostrum diverging proximally as well as distally; dorsal surface of carapace without a prominent tooth behind the orbit of each eye, but otherwise rather spiny	
		Chorilia longipes Dana, 1851 (fig. 19.34)
3a	Carapace, including the rostrum, about as wide as, or wider than, long	4
3b	Carapace, including the rostrum, decidedly longer than wide	6
4a	Carapace, viewed from above, nearly hexagonal; lateral margins of carapace extended into thin, winglike sheets, each with 2 sharp-tipped lobes; dorsal surface of carapace nearly smooth; intertidal as well as subtidal *Mimulus foliatus* Stimpson, 1860 (fig. 19.35)	
4b	Carapace, viewed from above, broadly oval to nearly circular; lateral margins of carapace not extended into winglike sheets with lobes; dorsal surface of carapace with tubercles or spines, or both; strictly subtidal	5
5a	Rostrum sharply inclined upward; dorsal surface of carapace spiny; generally at depths greater than 1000 m *Chionoecetes angulatus* Rathbun, 1893	
5b	Rostrum not sharply inclined upward; dorsal surface of carapace roughened by tubercles, but not especially spiny; at depths as shallow as 20 m	
		Chionoecetes bairdi Rathbun, 1893 (fig. 19.30)
6a	Carapace with a prominent, sharp projection, directed outward or slightly forward on both sides of the body near or behind the middle	7
6b	Carapace without a prominent, sharp projection on both sides of the body near or behind the middle (any such projections are limited to the anterior half of the body)	9
7a	Dorsal surface of carapace almost smooth; distance between the eyes less than about one-third the greatest width of the carapace *Pugettia producta* (Randall, 1839) (fig. 19.36)	
7b	Dorsal surface of carapace with tubercles; distance between the eyes about one-half the greatest width of the carapace	8
8a	Dorsal surface of merus of cheliped with some tubercles, but not with a distinct ridge; posterolateral expansions of carapace much broader basally than the next expansions anterior to them *Pugettia richii* Dana, 1851	

414   Phylum Arthropoda: Order Decapoda

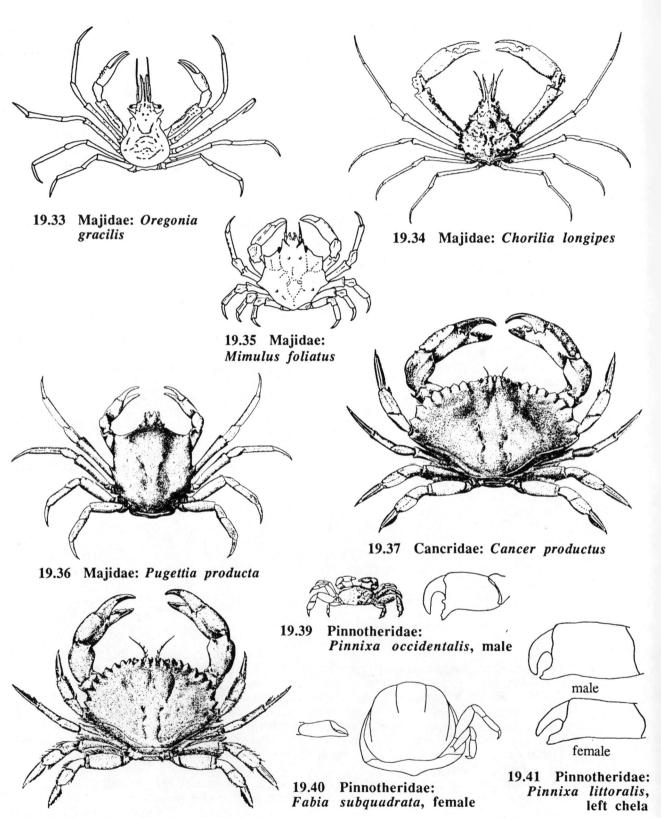

19.33 Majidae: *Oregonia gracilis*
19.34 Majidae: *Chorilia longipes*
19.35 Majidae: *Mimulus foliatus*
19.36 Majidae: *Pugettia producta*
19.37 Cancridae: *Cancer productus*
19.38 Cancridae: *Cancer antennarius*
19.39 Pinnotheridae: *Pinnixa occidentalis*, male
19.40 Pinnotheridae: *Fabia subquadrata*, female
19.41 Pinnotheridae: *Pinnixa littoralis*, left chela

8b   Dorsal surface of merus of cheliped with an irregularly toothed ridge; posterolateral expansions of carapace only slightly broader than the next expansions anterior to them
Pugettia gracilis Dana, 1851

9a   Outline of body, as viewed from above, lyre-shaped, owing to a toothed expansion of the carapace on both sides of the anterior half; rostral processes widest at the base   10

9b   Outline of body, as viewed from above, nearly triangular, becoming narrowed gradually toward the eyes; rostral processes widest near the middle
Scyra acutifrons Dana, 1851 (fig. 19.31)

10a   Length of carapace, including rostrum, about 1.5 times the width; at depths of 500 m or more
Oregonia bifurca Rathbun, 1902

10b   Length of carapace, including rostrum, about one and one-fifth times the width; in shallow water as well as at considerable depths   *Hyas lyratus* Dana, 1851

Section Cancridea

Family Atelecyclidae

Telmessus cheiragonus (Tilesius, 1815). Generally greenish or brownish yellow, hairy and with a carapace width of up to 10 cm; found in beds of eelgrass and on masses of algae; common from Puget sound northward, and although the range extends to California, not often encountered south of Washington.

Family Cancridae

1a   Tip of dactyl and fixed claw of the chelipeds dark, sometimes nearly black   2

1b   Tip of dactyl and fixed claw of the chelipeds not dark   5

2a   Dorsal surface of carapace smooth or slightly rough to the touch (like fine sandpaper), but not setose or obviously tuberculated; carapace at least 1.5 times as wide as long; width of carapace commonly exceeding 6 cm   3

2b   Much of the dorsal surface of the carapace with low tubercles, granular elevations, or setae; carapace less than 1.5 times as wide as long; width of carapace not exceeding 6 cm   4

3a   Dorsal surface of the carapace slightly rough to the touch (like fine sandpaper); propodus of chelipeds without obvious tubercles; ventral surface, in living or freshly collected specimens, with red spots; width of carapace up to 10 cm (not likely to be found north of Oregon)
Cancer antennarius Stimpson, 1856 (fig. 19.38)

3b   Dorsal surface of carapace not rough to the touch, propodus of chelipeds usually with a few tubercles; ventral surface without red spots; width of carapace up to 18 cm (common throughout our region)   *Cancer productus* Randall, 1839 (fig. 19.37)

4a   Dorsal surface of carapace obviously setose and with granular elevations; dorsal surface of carpus, propodus, and dactyl of chelipeds with spiny ridges, but without prominent tubercles; width of carapace up to 6 cm; subtidal   *Cancer branneri* Rathbun, 1898

4b   Dorsal surface of carapace with tubercles and sometimes also setose; dorsal surface of carpus, propodus, and dactyl of chelipeds without spiny ridges, but with prominent tubercles; width of carapace up to about 5 cm; common intertidal and subtidal species, generally nestling in holes, large barnacle shells, kelp holdfasts, etc.
Cancer oregonensis (Dana, 1852)

5a   Carapace widest at the 10th in the series of teeth that begins lateral to the eye; carpus, propodus, and dactyl of chelipeds with spiny ridges; posterolateral margins of carapace without a distinct tooth just behind the tooth that marks the widest point of the carapace; width of the carapace up to 23 cm   *Cancer magister* Dana, 1852

5b   Carapace widest at the 9th in the series of teeth that begins lateral to the eye; carpus, propodus, and dactyl of chelipeds without spiny ridges; posterolateral margins of carapace with a distinct tooth just behind the tooth that marks the widest point of the carapace; width of the carapace up to 12 cm   *Cancer gracilis* Dana, 1852

Section Brachyrhyncha
Family Xanthidae

1a Carpus of chelipeds smooth; carpus of legs 2-5 with 1 slight tubercle or crest; intertidal, mostly under rocks resting on gravel   *Lophopanopeus bellus* subsp. *bellus* (Stimpson, 1860)
1b Carpus of chelipeds with prominent tubercles; carpus of legs 2-5 with 2 prominent tubercles or crests; subtidal in our area, in sand, mud, or gravel
    Lophopanopeus bellus subsp. *diegensis* Rathbun, 1900

Family Grapsidae

1a Merus of legs 2-5 much flattened, used for swimming; male without a tuft of hairs on the propodus of the cheliped; oceanic crabs, only occasionally washed onto shore with floating debris or other animals   2
1b Merus of legs 2-5 not markedly flattened; male with a tuft of hairs on the propodus of the cheliped; common intertidal crabs   3
2a Carapace decidedly longer than wide; lateral margins convex   *Planes cyaneus* Dana, 1852
2b Carapace only slightly longer than wide; lateral margins almost straight, but converging slightly posteriorly   *Planes marinus* Rathbun, 1914
3a Dorsal surface of carapace with flat transverse ridges, and with red or purple transverse lines on a blackish green background; carapace with 2 teeth on the anterolateral margin (not likely to be found north of Oregon)   *Pachygrapsus crassipes* Randall, 1839
3b Dorsal surface without flat transverse ridges, and without transverse lines; carapace with 3 teeth on the anterolateral margin   4
4a Legs fringed with abundant hairs; prevailing coloration generally grayish green; chelipeds without conspicuous purple spots   *Hemigrapsus oregonensis* (Dana, 1851) (fig. 19.32)
4b Legs not obviously hairy; prevailing coloration generally reddish; chelipeds with conspicuous purple spots   *Hemigrapsus nudus* (Dana, 1851)

Family Pinnotheridae

1a Width of carapace less than 1.5 times the length   2
1b Width of carapace at least 1.5 times the length   6
2a Carapace calcified and rather firm   3
2b Carapace soft   4
3a Hairs on legs 3 and 4 featherlike and much longer than those on the other legs; dactyls of walking legs decidedly curved; generally living in the mantle cavity of bivalve molluscs (*Modiolus modiolus*, *Mytilus californianus*, *M. edulis*, *Tresus capax*, *Mya arenaria*, *Astarte compacta*, *Cyclocardia ventricosa*, *Crenella decussata*, *Kellia suborbicularis*), but sometimes free-living   mature males and females of *Fabia subquadrata* (Dana, 1851)
    For ovigerous females of this species see choice 4a.
3b Hairs on legs 3 and 4 not appreciably longer than those on the other legs; dactyls of walking legs only slightly curved; living in burrows of *Callianassa* and *Upogebia*
    Scleroplax granulata Rathbun, 1897
4a Dorsal surface of the carapace with a longitudinal groove extending posteriorly from the orbit of each eye; dactyls of walking legs decidedly curved (in the mantle cavity of various bivalve molluscs; see choice 3a)
    subadults and ovigerous females of *Fabia subquadrata* (Dana, 1851) (fig. 19.40)
4b Dorsal surface of the carapace without a longitudinal groove extending posteriorly from the orbit of each eye; dactyls of walking legs nearly straight   5
5a Dactyl of leg 5 only slightly longer than the dactyl of legs 3 and 4; dorsal surface of carapace with a pair of small tubercles just behind the middle (carapace of male with fine setae) (in the ascidians *Corella willmeriana*, *Ascidia paratropa*, and *A. ceratodes*)
    Pinnotheres taylori Rathbun, 1918

5b Dactyl of leg 5 about one-fifth or one-fourth longer than the dactyls of legs 3 and 4; dorsal surface of carapace without a pair of tubercles just behind the middle (in the ascidians *Ascidia paratropa, Halocynthia aurantium,* and *H. igaboja*; also in the mantle cavity of the bivalve *Hinnites giganteus*) .... *Pinnotheres pugettensis* Holmes, 1900
6a Corneous tips of dactyls of legs 2-5 decidedly curved 7
6b Corneous tips of dactyls of legs 2-5 nearly straight 8
7a Outer margin of eye orbits (in anterior view) rounded; male with a gap between the dactyl and fixed finger of the chela, but female without this gap (mature individuals, including gravid females, in the bivalve *Tresus capax*; immature individuals in many bivalves, and sometimes in hosts other than bivalves) .... *Pinnixa faba* (Dana, 1851)
7b Outer margin of eye orbits (in anterior view) somewhat acute; both sexes with a gape between the dactyl and fixed finger of the chela (mature individuals in the bivalve *Tresus capax*; immature individuals in many bivalves, and sometimes in hosts other than bivalves) .... *Pinnixa littoralis* Holmes, 1894 (fig. 19.41)
8a Propodus of legs 3-5 longer than the dactyl (it is also about 3 times as wide as the dactyl) (in tubes of terebellid polychaetes, including *Eupolymnia heterobranchia*) .... *Pinnixa tubicola* Holmes, 1894
8b Propodus of legs 3-5 not longer than the dactyl (and not usually more than twice as wide as the dactyl, except in *Pinnixa schmitti*) 9
9a Inner margin of dactyl of chela with a distinct tooth; carapace about twice as wide as long, rounded laterally (in tubes of polychaetes of the genus *Abarenicola*) .... *Pinnixa eburna* Wells, 1928
9b Inner margin of dactyl of chela without a distinct tooth; carapace more than twice as wide as long, pointed laterally 10
10a Dorsal surface of carapace bumpy (in burrows of the subtidal echiuran *Echiurus echiurus* subsp. *alaskanus*, and apparently also free-living) .... *Pinnixa occidentalis* Rathbun, 1893 (fig. 19.39)
10b Dorsal surface of carapace smooth (in burrows of *Upogebia* and *Callianassa*, but also associated with the holothurian *Leptosynapta clarkii*, the ophiuroid *Amphiodia urtica*, and perhaps other invertebrates) .... *Pinnixa schmitti* Rathbun, 1918

20
PHYLA
PHORONIDA, BRACHIOPODA, ENTOPROCTA

PHYLUM PHORONIDA

Emig, C.-C. 1971a. Remarques sur la systématique des Phoronidea X. Notes sur l'écologie, la morphologie et la taxonomie de *Phoronis ijimai* et de *P. vancouverensis*. Mar. Biol., 8:154-9.
------. 1971b. Taxonomie et systématique des Phoronidiens. Bull. Mus. Natl. Hist. Nat., Paris, Zool., 8:473-568.
Marsden, J. C. R. 1959. Phoronidea from the Pacific coast of North America. Can. J. Zool., 37:87-111.
Pixell, H. L. M. 1912. Two new species of the Phoronidea from Vancouver Island. Q. J. Micr. Sci., 58:257-84.
Torrey, H. B. 1901. On *Phoronis pacifica*, sp. nov. Biol. Bull., 2:283-8.
Zimmer, R. L. 1967. The morphology and function of accessory reproductive glands in the lophophores of *Phoronis vancouverensis* and *Phoronopsis harmeri*. J. Morphol., 121:159-78.

1a With a distinct groove separating the lophophore from the trunk; length commonly exceeding 6 cm; tubes typically buried in sand or in a mixture of sand and gravel *Phoronopsis harmeri*
1b Without a distinct groove separating the lophophore from the trunk; length not exceeding 5 cm; attached to rock or wood, burrowing in calcareous substrata, or embedded in the mixture of mud and mucus lining the burrows of *Upogebia pugettensis* 2
2a Length usually 2-5 cm; typically in clumps of several to many specimens attached to rock (especially sandstone and limestone) or wood, the tubes covered with silt (intertidal, and also on floats and logs whose location is permanent) *Phoronis vancouverensis*
2b Length not more than 5 mm; burrowing in calcareous shells or in limestone, or embedded in the mixture of mud and mucus lining the burrows of *Upogebia* 3
3a Burrowing in shells of barnacles (especially *Balanus nubilus*) and bivalve molluscs (especially *Pododesmus cepio* and *Hinnites giganteus*), but also found in limestone (mostly subtidal) *Phoronis ovalis*
3b Embedded in the mixture of mud and mucus lining the burrows of *Upogebia* *Phoronis pallida*

Family Phoronidae

See the note concerning *Phoronopsis pacifica*, which is not in the key.

Phoronis ovalis Wright, 1856
Phoronis pallida (Schneider, 1862)
Phoronis vancouverensis Pixell, 1912
Phoronopsis harmeri Pixell, 1912
Phoronopsis pacifica (Torrey, 1901). This was the first species to be described from the Pacific coast of North America. Torrey's 8 specimens came from widely separated localities: Humboldt Bay, in northern California, and Puget Sound. The Puget Sound specimens almost certainly belonged to what we now call *Phoronopsis harmeri*; the specimens from Humboldt Bay may have belonged to *P. harmeri*, which occurs at least as far south as Coos Bay, Oregon, to *P. viridis* Hilton, 1930, which is common at least as far north as Sonoma County, California, or to an undescribed species.

Type specimens have not survived and Torrey's description is not sufficiently detailed to enable us to be certain that the material he studied belonged to just one species. If it can be established that the only *Phoronopsis* in Humboldt Bay is *P. harmeri*, then it would probably be logical to give *P. pacifica* priority. If, on the other hand, *P. viridis* or an undescribed species prevails in Humboldt Bay, it would be best to keep *P. pacifica* out of consideration in future taxonomic studies.

PHYLUM BRACHIOPODA

The key includes species that occur intertidally or subtidally at depths of less than 200 m. For several other species that occur in very deep water, consult the paper of Bernard (1972). The only common intertidal species is *Terebratalia transversa*. It is extremely variable, not only in the shape of its shell but also in the extent to which the valves are sculptured. In some specimens, the valves have prominent ribs; in others, they are nearly smooth.

Bernard, F. R. 1972. The living Brachiopoda of British Columbia. Syesis, 5:73-82.

1a	Pedicle (attachment stalk) absent; shell flattened (the thickness is less that half the width); lower valve firmly cemented to rock (thus this species could be confused with certain bivalve molluscs, such as *Pododesmus cepio* or small oysters)	*Crania californica* (fig. 20.1)
1b	Pedicle present; shell not especially flattened (the thickness is at least half the width); lower valve free of the substratum	2
2a	Shell dark gray or nearly black; posterior beak of the larger valve rather sharply pointed, its apex not obliterated by the opening through which the pedicle passes	*Hemithiris psittacea*
2b	Shell yellowish, brownish, or reddish; posterior beak of the larger valve not pointed, and its apex may be partly or wholly obliterated by the opening through which the pedicle passes	3
3a	Shell usually wider than long; margins of the valves at the anterior (broader) end with pronounced complementary undulations	*Terebratalia transversa*
3b	Shell usually longer than wide; margins of valves at the anterior (broader) end without pronounced complementary undulations (they may not, however, be perfectly straight, and they may have small teeth)	4
4a	Anterior margins of valves with regularly spaced small teeth; shell not translucent; valves with delicate, radiating striations, but usually without concentric growth lines; apex of the larger valve completely obliterated by the opening through which the pedicle passes	*Terebratulina unguicula*
4b	Anterior margins of valves without regularly spaced teeth; shell somewhat translucent; valves with crowded concentric growth lines, but without radiating striations; apex of the larger valve not completely obliterated by the opening through which the pedicle passes	*Laqueus californianus* (fig. 20.2)

Class Inarticulata

Order Neotremata

Family Craniidae

Crania californica Berry, 1921 (fig. 20.1). Subtidal off the open coast, and mostly on large rocks.

420   Phylum Entoprocta

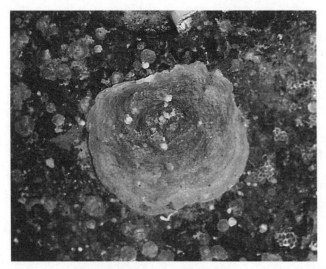

20.1 *Crania californica*

20.2 *Laqueus californianus*

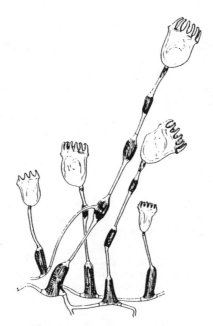

20.3 *Barentsia benedeni*

Class Articulata
Order Rhynchonellida
Family Hemithyrididae

Hemithiris psittacea (Dillwyn, 1817)

Order Terebratulida

Suborder Terebratulidina

Family Terebratulidae

Terebratulina unguicula (Carpenter, 1864) (fig. 20.2)

Suborder Terebratellidina

Family Laqueidae

Terebratalia transversa (Sowerby, 1846)
Laqueus californianus (Koch, 1848)

PHYLUM ENTOPROCTA

Emscherman, P. 1972. *Loxokalypus socialis* gen. et sp. nov. (Kamptozoa, Loxokalypodidae fam. nov.), ein neuer Kamptozoentyp aus dem nordlichen Pazifischen Ozean. Ein Vorschlag zur Neufassung der Kamptozoen Systematik. Mar. Biol., 12:237-54.
Mariscal, R. N. 1965. The adult and larval morphology and life history of the entoproct *Barentsia gracilis* (M. Sars, 1835). J. Morphol., 116:311-38. (Concerns *B. benedeni*.)
Nielsen, C. 1964. Studies on Danish Entoprocta. Ophelia, 1:1-76.
------. 1966. Some Loxosomatidae (Entoprocta) from the Atlantic coast of the United States. Ophelia, 3:249-75.
Osburn, R. C. 1953. Bryozoa of the Pacific coast of America. Part 3. Cyclostomata, Ctenostomata, Entoprocta, and addenda. Allan Hancock Pacific Exped., 14:613-841.
Robertson, A. 1900. Studies in Pacific coast Entoprocta. Proc. Calif. Acad. Sci., ser. 3, Zool., 2:323-48.
Ryland, J. S. 1971. Two species of *Loxosomella* (Entoprocta) from west Norway. Sarsia, 1:31-8.
Wieser, W. 1959. Free-living nematodes and other small invertebrates of Puget Sound beaches. Univ. Wash. Publ. Biol., 19. x + 179 pp.

1a	Zooids separate from one another, not connected by stolons or united basally into a cluster; generally attached to other organisms	2
1b	Zooids forming colonies, either connected by stolons or united basally into a cluster; free-living or attached to other organisms	4
2a	Base of stalk differentiated into a muscular disk by which the animal is attached (attachment is not permanent, however; the animal is capable of moving about to some extent)	*Loxosoma davenporti*
2b	Base of stalk not differentiated into a muscular disk	3
3a	With a horseshoe-shaped flange partly encircling the calyx, below the base of the ring of 8 tentacles (the flange is most prominent on the anal side, where the longest tentacles are located); attached to bryozoans	*Loxosomella nordgaardi*
3b	Without a horseshoe-shaped flange below the base of the ring of tentacles; attached to hosts other than bryozoans	*Loxosomella* spp.
4a	Producing buds at the base of the stalk, thereby forming a cluster of zooids that remain attached to one another; on the body surface of the polychaete *Glycera nana* (Glyceridae)	*Loxokalypus socialis*
4b	Zooids connected by stolons; free-living or attached to other organisms, but not to polychaetes	5

5a	Stalks thick, without muscular enlargements where they join the stolon	6
5b	Stalks thin, with muscular enlargements where they join the stolon (and sometimes with 1 or more muscular enlargements between the base and the calyx)	7
6a	Orientation of tentacles in a plane that is decidedly oblique to the long axis of the stalk; both stalk and calyx strongly muscularized	*Myosoma spinosa*
6b	Tentacles oriented at right angles to the long axis of the stalk; neither stalk nor calyx strongly muscularized	*Pedicellina cernua*
7a	Stalks with 1 or more muscular enlargements (in addition to those at their bases) that enable them to bend sharply	8
7b	Stalks without muscular enlargements (other than those at their bases)	9
8a	Stalks with pores, and branching by producing new individuals at the points where the muscular enlargements are located	*Barentsia ramosa*
8b	Stalks without pores, and rarely branching	*Barentsia benedeni* (fig. 20.3)
9a	Stalks with numerous pores; height of individual zooids up to about 4 mm	*Barentsia misakiensis*
9b	Stalks with few or no pores; height of individual zooids not often exceeding 1 mm	*Barentsia gracilis*

One species, marked with an asterisk, is not in the key.

Order Solitaria

Family Loxosomatidae

Loxosoma davenporti Nickerson, 1898
Loxosomella nordgaardi Ryland, 1964. On bryozoans, including *Tegella*.
Loxosomella spp. Small species, some or all of which may be on various hosts, including the sipunculan *Phascolosoma agassizi*, the polychaetes *Diopatra ornata* (Onuphidae) and *Gattyana cirrosa* (Polynoidae).

Order Coloniales

Family Loxokalypodidae

Loxokalypus socialis Emscherman, 1972. On the body surface of the polychaete *Glycera nana* (Glyceridae).

Family Pedicellinidae

Myosoma spinosa Robertson, 1900
Pedicellina cernua (Pallas, 1771)

Family Barentsiidae

Barentsia benedeni (Foettinger, 1887) (fig. 20.3)
Barentsia gracilis (M. Sars, 1835)
Barentsia misakiensis (Oka, 1895)
Barentsia ramosa (Robertson, 1900)
*Barentsia robusta Osburn, 1955

21

PHYLUM BRYOZOA (ECTOPROCTA)

Anne Bergey and David Denning

Hayward, P. J., & J. S. Ryland. 1980. British Ascophoran Bryozoans. Synopses of the British Fauna (new series), no. 14. London & New York: Academic Press. 312 pp.

Kliuge (Kluge), G. A. 1962. [Bryozoans of the northern seas of the USSR.] Opredeliteli po Faune SSSR, 76. 584 pp. (in Russian). (English translation, 1975. New Delhi: Amerind Publishing Co.)

McCain, K. W., & J. R. P. Ross. 1974. Annotated faunal list of cheilostome Ectoprocta of Washington state. Northwest Sci., 48:9-16.

Nielsen, C. 1981. On morphology and reproduction of '*Hippodiplosia*' *insculpta* and *Fenestrulina malusii* (Bryozoa, Cheilostomata). Ophelia, 20:91-125.

O'Donoghue, C. H. 1925. Notes on certain Bryozoa in the collection of the University of Washington. Publ. Puget Sound Biol. Sta., 5:15-23.

O'Donoghue, C. H., & E. O'Donoghue. 1923. A preliminary list of the Bryozoa (Polyzoa) from the Vancouver Island region. Contr. Canad. Biol. Fish., n. s., 1:143-201.

------. 1925. List of Bryozoa from the vicinity of Puget Sound. Publ. Puget Sound Biol. Sta., 5:91-108.

------. 1926. A second list of the Bryozoa (Polyzoa) from the Vancouver Island region. Contr. Canad. Biol. Fish., n. s., 3:49-131.

Osburn, R. C. 1950. Bryozoa of the Pacific coast of America. Part 1, Cheilostomata--Anasca. Allan Hancock Found. Pac. Exped., 14:1-269.

------. 1952. Bryozoa of the Pacific coast of America. Part 2, Cheilostomata--Ascophora. Allan Hancock Found. Pac. Exped., 14:271-611.

------. 1953. Bryozoa of the Pacific coast of America. Part 3, Cyclostomata, Ctenostomata, Entoprocta and addenda. Allan Hancock Found. Pac. Exped., 14:613-841.

Osburn, R. C., & J. D. Soule. 1953. Suborder Ctenostomata. *In* Osburn, R. C., Bryozoa of the Pacific coast of America. Part 3, Cyclostomata, Ctenostomata, Entoprocta and addenda. Allan Hancock Found. Pac. Exped., 14:726-58.

Pinter, P. A. 1973. The *Hippothoa hyalina* (L.) complex with a new species from the Pacific coast of California. *In* Larwood, G. P., Living and Fossil Bryozoa, pp. 437-46. London & New York: Academic Press, Inc.

Powell, N. A. 1970. *Schizoporella unicornis*--an alien bryozoan introduced into the Strait of Georgia. J. Fish. Res. Bd. Canada, 27:1847-53.

Robertson, A. 1905. Non-incrusting chilostomatous Bryozoa of the west coast of North America. Univ. Calif. Publ. Zool., 2:235-344.

------. 1908. The incrusting chilostomatous Bryozoa of the west coast of North America. Univ. Calif. Publ. Zool., 4:253-344.

------. 1910. The cyclostomatous Bryozoa of the west coast of North America. Univ. Calif. Publ. Zool., 6:225-84.

Ross, J. R. P. 1970. Keys to the recent cyclostome Ectoprocta of marine waters of northwest Washington state. Northwest Sci., 44:154-69.

Ross, J. R. P., & K. W. McCain. 1976. *Schizoporella unicornis* (Ectoprocta) in coastal waters of the northwestern United States and Canada. Northwest Sci., 50:160-71.

Ryland, J. S., & P. J. Hayward. 1977. British Anascan Bryozoans. Synopses of the British Fauna (new series), no. 10. London & New York: Academic Press. 188 pp.

Schopf, T. J. M., K. O. Collier, & B. O. Bach. 1980. Relation of the morphology of stick-like bryozoans at Friday Harbor, Washington, to bottom currents, suspended matter, and depth. Paleobiology, 6:466-76.

Yoshioka, P. M. 1982. Predator-induced polymorphism in the bryozoan *Membranipora membranacea* (L.). J. Exp. Mar. Biol. Ecol., 61:233-42.

1a Zooecia calcified to at least some extent; apertures of zooecia either simple openings or closed by a lidlike operculum, but not closed by a constriction
                Class Stenolaemata: Order Cyclostomata and Class Gymnolaemata: Order Cheilostomata
1b Zooecia not calcified, either united into a fleshy, leathery, or gelatinous mass or connected by stolons; apertures of zooecia closed by a constriction
                Class Gymnolaemata: Order Ctenostomata

Class Stenolaemata: Order Cyclostomata and Class Gymnolaemata: Order Cheilostomata

Cyclostomes, which are in the minority, are indicated by the abbreviation Cycl. All other species in the key are cheilostomes.

1a	Colony erect or recumbent (with leaflike lobes loosely attached to the substratum), but not truly encrusting	2
1b	Colony encrusting (but the distal portions of the zooecia may be raised up)	47
2a	Colony consisting mostly of stiff, two-layered folds	3
2b	Colony not consisting of two-layered folds	6
3a	Most of the frontal of the zooecia with numerous large pores	4
3b	Frontal of zooecia with pores only at the margins (a portion of the frontal usually elevated as an umbo, to which ridges extend from the margins; a small avicularium generally present on the slope of the umbo; see also choice 130b)	*Rhamphostomella costata* (fig. 21.66)
4a	Avicularia usually present, these interzooecial and large (in some colonies, however, avicularia are absent) (frontal of zooecia in larger colonies with pores arranged in several regular rows; see also choice 65a)	*Lyrula hippocrepis* (fig. 21.56)
4b	Avicularia absent	5
5a	Ovicells absent; aperture of zooecia with a thin, raised collar and a central sharp tooth inside the proximal margin (see also choice 110a)	*Cheilopora praelonga*
5b	Ovicells present, these large, globular, without pores, but with radiating ribs; proximal rim of aperture of zooecia raised into a peaked mound (between the aperture and the mound is an oval depression that could be mistaken for a pore or avicularium; see also choice 115a)	*Hippodiplosia insculpta* (fig. 21.59)
6a	Colony recumbent, flexible, attached loosely to the substratum by hairlike projections (the colony may form broad lobes or fingerlike branches) (this choice leads to a few cheilostomes in which the frontal is not calcified)	7
6b	Colony erect (this choice leads to cyclostomes, some cheilostomes with a calcified frontal, and some cheilostomes with a noncalcified frontal)	11
7a	With interzooecial avicularia	8
7b	Without interzooecial avicularia	9
8a	With numerous slender spines on the margins of the zooecia, the spines curving inward over the uncalcified frontal; mandible of avicularia very long, broad at its base, tapering to a thin, curved point	*Hincksina pallida* (fig. 21.26)
8b	Spines lacking, or small and located at the distal corners of the zooecia; avicularia rare, the base of each one nearly square and the mandible small	*Chartella membranaceo-truncata*
9a	Zooecia closely connected; colony divided into broad lobes resembling those of certain lichens; zooecia usually (but not always) with several slender spines curving for a short distance over the frontal	*Dendrobeania lichenoides* (fig. 21.50)
9b	Zooecia loosely connected; lobes of colony variable, but usually split into narrower lobes peripherally; spines on zooecia large and stout	10
10a	Spines at edges of zooecia long, sometimes curving over the entire frontal	*Dendrobeania longispinosa* (fig. 21.51)

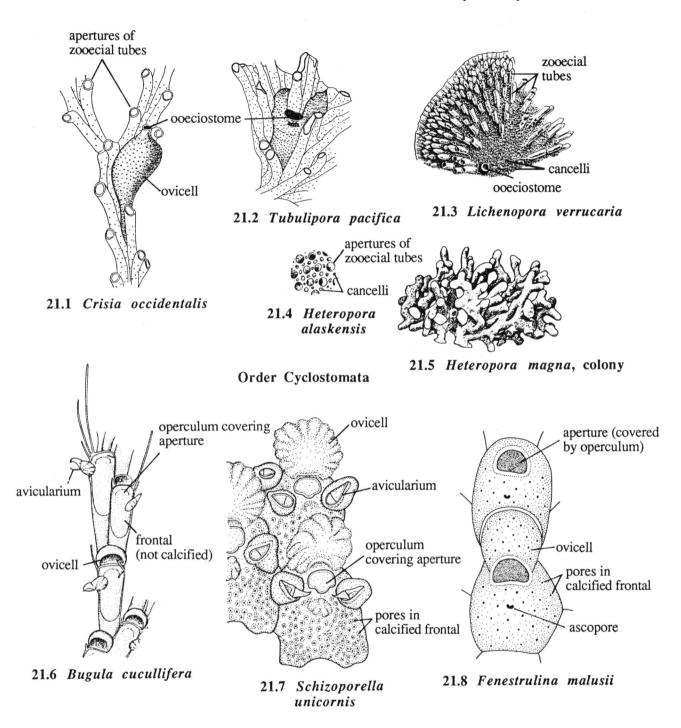

Terminology used in identification of cyclostome and cheilostome bryozoans

10b Spines at edges of zooecia not so long as to curve over the entire frontal
Dendrobeania laxa (fig. 21.49)
11a Colony a flattened cup attached to the substratum by a narrow stalk and small disk, and with numerous bundles of zooecial tubes (12-20 zooecia per bundle radiating outward from the edge of the cup) *Discocytis canadensis* (Cycl.)
11b Colony branching or lacy in appearance, not cuplike 12
12a Colony a stiff, lacy network (live specimens usually pale orange or salmon-orange)
Phidolopora labiata (fig. 21.69)
12b Colony not in the form of a lacy network 13
13a Cheilostomes in which at least half of the frontal of the zooecia is not calcified 14
13b Either cheilostomes in which most of the frontal of the zooecia is calcified, or cyclostomes in which the zooecia (generally tubular) are calcified 30
14a Vibracula present on the upper suface of the colony 15
14b Vibracula absent 17
15a Vibracular chamber large, with an elongate seta; scutum may or may not be present; colony not regularly jointed 16
15b Vibracular chamber small or absent, with a short seta (usually shorter than a zooecium); scutum always present, small, ranging from a curved spine to a paddle-shaped structure; colony regularly jointed and bushy *Scrupocellaria californica* (fig. 21.37)
16a Scutum present, large; giant frontal avicularia often present (these are nearly as large as the zooecia) *Caberea boryi*
16b Scutum absent; giant avicularia absent *Caberea ellisi* (fig. 21.36)
17a Scutum present, curving over the frontal 18
17b Scutum absent 21
18a Colony jointed, usually with 3 zooecia in each internode 19
18b Colony jointed, usually with 5 or more zooecia in each internode 20
19a Scutum attached considerably below the middle of the noncalcified portion of the frontal; frontal avicularia absent; ovicells with pores *Tricellaria occidentalis* (fig. 21.39)
19b Scutum attached at the middle, or above the middle, of the noncalcified portion of the frontal; frontal avicularia present; ovicells without pores *Tricellaria ternata*
20a Scutum attached considerably below the middle of the noncalcified portion of the frontal; ovicells with a few pores *Tricellaria praescuta* (fig. 21.40)
20b Scutum attached at the middle, or above the middle, of the noncalcified portion of the frontal; ovicells without pores *Tricellaria erecta* (fig. 21.38)
21a Colony branching from the upper portion of a long, jointed stalk 22
21b Colony not branching from the upper portion of a long, jointed stalk 24
22a Zooecia with 4 or 5 long, jointed spines on the outer distal corner *Caulibugula ciliata*
22b Zooecia with fewer than 4 spines 23
23a Distal end of zooecia rounded, with 2 or 3 long, jointed outer spines and 1 inner spine; nearly half of the frontal not calcified *Caulibugula occidentalis* (fig. 21.47)
23b Distal end of zooecia markedly angled, with 1 prominent outer spine (there may, however, be some other small spines); at least three-fourths of the frontal not calcified
Caulibugula californica (fig. 21.46)
24a Zooecia small (about 0.4 mm long), attached to one another in pairs, back-to-back (colony inconspicuous) *Synnotum aegyptiacum* (fig. 21.41)
24b Zooecia not attached to one another in pairs (if the colony is biserial, the zooecia are more nearly end-to-end than back-to-back) 25
25a Colony multiserial, sometimes partly recumbent; both lateral and frontal avicularia present 26
25b Colony biserial or multiserial, always erect; only lateral avicularia present 27
26a Avicularia long, the beak and mandible curved in such a way that they meet only at their tips; spines on the frontal absent or poorly developed *Dendrobeania curvirostrata* (fig. 21.48)
26b Avicularia short, the beak and mandible capable of touching for most of their length; with definite spines curving over the frontal *Dendrobeania murrayana* (fig. 21.52)
27a Colony forming spiral whorls (zooecia arranged biserially; avicularia usually of 2 sizes, attached near the middle of the zooecia; ovicells globular) *Bugula californica* (fig. 21.42)
27b Colony not forming spiral whorls 28
28a Zooecia arranged multiserially; ovicells absent; avicularia attached near the middle of the zooecial margin *Bugula pugeti* (fig. 21.44)

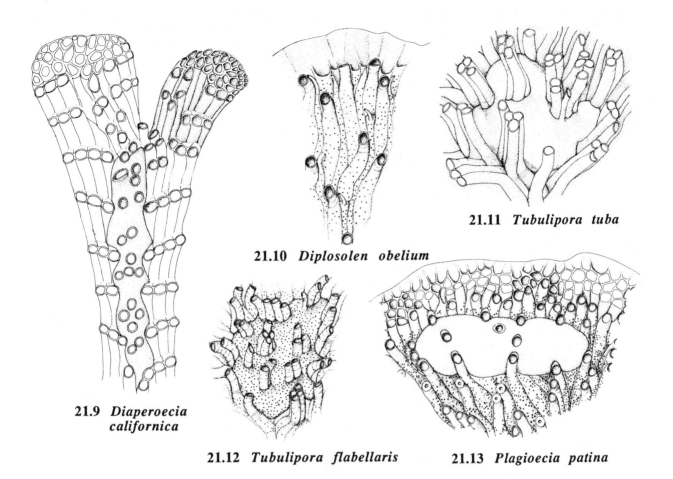

21.9 *Diaperoecia californica*
21.10 *Diplosolen obelium*
21.11 *Tubulipora tuba*
21.12 *Tubulipora flabellaris*
21.13 *Plagioecia patina*

28b Zooecia arranged biserially; ovicells present; avicularia attached near the base of the zooecial margin 29
29a Avicularia large, longer than the width of the zooecia from which they originate; ovicell hoodlike and with a broad aperture *Bugula cucullifera* (figs. 21.6 and 21.45)
29b Avicularia not especially large; ovicell reduced, caplike *Bugula pacifica* (fig. 21.43)
30a Colony erect, solid, neither flexible nor jointed 31
30b Colony consisting of calcified units connected by sclerotized, flexible joints 38
31a Branches of colony generally cylindrical; apertures of the zooecia distributed more or less evenly around the branches 32
31b Branches somewhat flattened; apertures of the zooecia circular, concentrated on one side of each branch, usually in groups of 4 or 5 (colonies variable; the branches may become attached to the substratum or become fused to form reticulated masses)
    Diaperoecia californica (fig. 21.9) (Cycl.)
32a Zooecia indistinct, embedded within a cylindrical matrix 33
32b Zooecia more or less distinct, flask-shaped, attached to one another for most of their length, but each with a long, tubular, raised peristome (there is a pair of avicularia on the rim of the peristome of at least some zooecia) *Lagenipora punctulata* (fig. 21.70)
33a Apertures of zooecia distinct and with a proximal, U-shaped sinus; apertures separated from one another by at least the width of one aperture; operculum present; avicularia present 34
33b Apertures of zooecia circular, without a sinus; apertures often tightly packed, raised slightly above, or level with, the surface of the branch; operculum absent; avicularia absent 36

34a	Avicularia about as large as the aperture, single, on the midline distal to the aperture (colony tall and stout)	*Myriozoum coarctatum* (fig. 21.71)
34b	Avicularia much smaller than the aperture, sometimes single, either on the midline or to one side of the midline, and distal to the aperture	35
35a	Avicularia on or very near the midline, usually single, but occasionally paired; colony tall, stout	*Myriozoum subgracile*
35b	Avicularia almost always paired, separated by a distance about equal to the width of the aperture, symmetrically placed with respect to the midline; colony tall, slender	*Myriozoum tenue*
36a	Branches up to 5 mm in diameter	37
36b	Branches about 2 mm in diameter (color light, often yellow or with yellow-tipped branches; branches rarely anastomosing)	*Heteropora alaskensis* (fig. 21.4) (Cycl.)
37a	Color often gray-purple; branches rarely anastomosing; zooecia protruding only slightly, if at all, above the surface of the colony	*Heteropora magna* (fig. 21.5) (Cycl.)
37b	Color usually yellow to gray, branches occasionally with pinkish tips; branches frequently anastomosing; peristomes of zooecia sometimes distinctly protruding above the surface of the branches	*Heteropora pacifica* (Cycl.)
38a	Zooecia somewhat boxlike, embedded in the matrix of the colony, the frontals of the zooecia forming the surface of the colony; aperture approximately semicircular, with an operculum	39
38b	Units of the colony formed by tubular zooecia lying against one another and becoming fused; aperture circular, without an operculum	41
39a	Internodes of colony about 1 cm long, elliptical in transverse section; zooecia not hexagonal; avicularia with a triangular mandible	*Microporina borealis* (fig. 21.35)
39b	Internodes of colony up to 2.5 cm long, circular in transverse section; zooecia distinctly hexagonal; avicularia with a semicircular mandible	40
40a	Avicularium broader than the proximal portion of the zooecium with which it is associated, the mandible (which may be up to 0.2 mm wide) brown	*Cellaria mandibulata*
40b	Avicularium about as broad as the zooecium with which it is associated, the mandible colorless	*Cellaria diffusa*
41a	Zooecia arranged in biserial rows	42
41b	Zooecia arranged in uniserial rows	45
42a	Ooeciostome curved or bent forward	43
42b	Ooeciostome straight	44
43a	Ooeciostome long, slender, bent forward, its aperture circular	*Crisia pugeti* (fig. 21.16) (Cycl.)
43b	Ooeciostome short, its aperture elliptical	*Crisia serrulata* (Cycl.)
44a	Branches of the colony curved inward or with spikelike projections at their tips; internodes short	*Crisia occidentalis* (fig. 21.1) (Cycl.)
44b	Branches straight; internodes long	*Crisia maxima* (Cycl.)
45a	Without spines (ovicell attached for the entire length of an internode; ooeciostome terminal)	*Filicrisia franciscana* (fig. 21.18) (Cycl.)
45b	With long spines	46
46a	With only 1 zooecium in each internode	*Crisidia cornuta* (fig. 21.17) (Cycl.)
46b	With at least 2 zooecia in each internode (fertile internodes have 3-5 zooecia)	*Bicrisia edwardsiana* (fig. 21.15) (Cycl.)
47a	Colony consisting of creeping, adherent, stolonlike zooecia, the distal ends of which rise in erect tubes that have an operculum at the aperture	*Aetea* spp.
47b	Colony not stolonlike	48
48a	Colony consisting of simple calcified tubes fused together (secondary calcification may obscure the tubes, but portions of them are almost always visible at the surface of the colony); apertures usually circular, without an operculum; with neither avicularia nor spines (in *Disporella fimbriata*, the tubes end in sharp points, but these are not spines)	49
48b	Colony not consisting of calcified tubes (the zooecia are more complex than simple tubes, and have an aperture on the frontal); aperture variable, but with an operculum; either spines or avicularia, or both, may be present	62
49a	Colony uniserial, branching dichotomously at nearly right angles	*Stomatopora granulata* (Cycl.)
49b	Colony multiserial or massive	50

Phylum Bryozoa

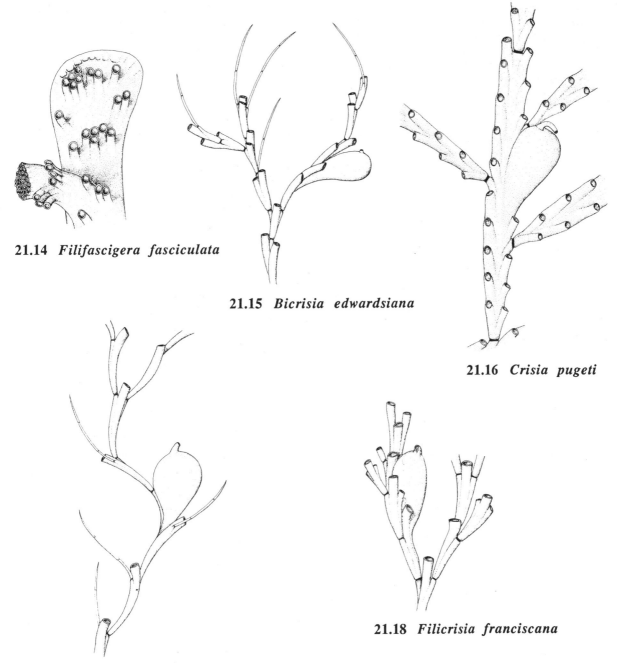

21.14 *Filifascigera fasciculata*

21.15 *Bicrisia edwardsiana*

21.16 *Crisia pugeti*

21.18 *Filicrisia franciscana*

21.17 *Crisidia cornuta*

50a	Colony disk-shaped	51
50b	Colony not disk-shaped (fanlike colonies may, however, grow outward to the extent that they become almost circular)	54
51a	Colony large, up to 10 cm in diameter, complex, composed of subcolonies with 8-12 radially arranged bundles of tubules (color usually deep purple) *Disporella separata* (Cycl.)	
51b	Colony small, simple, not composed of subcolonies	52
52a	Ends of zooecial tubes flaring and ending in 2-5 sharp points; pores on the surface of the colony (not the raised apertures of the zooecial tubes) nearly circular, not so crowded that	

	they nearly touch one another *Disporella fimbriata* (Cycl.)
52b	Ends of zooecial tubes neither flaring nor ending in 2 or more sharp points; pores (or calcified porelike structures) on the surface of the colony irregular in form and size, and sometimes so crowded as to nearly touch one another ... 53
53a	Elevated tubular portions of the zooecia arranged in radiating, uniserial rows, the tubes adhering to one another up to their tips *Lichenopora novae-zelandiae* (Cycl.)
53b	Elevated tubular portions of the zooecia not arranged in regular, radiating rows over the entire colony (the tubes in the central portion may be so arranged, however), and not adhering to one another up to their tips *Lichenopora verrucaria* (fig. 21.3) (Cycl.)
54a	Colony consisting of straplike branches that adhere tightly to the substratum for most of their length ... 55
54b	Colony growing outward in such a way that it produces one or more fan-shaped lobes ... 58
55a	Zooecial tubes erect and in bundles; colony in the form of narrow, creeping fingers, the basal width of these relatively constant *Filifascigera fasciculata* (fig. 21.14) (Cycl.)
55b	Zooecial tubes separate, not in bundles; colony with straplike branches that vary in width (they are usually 2-8 zooecia wide) ... 56
56a	Branches of the colony short, often anastomosing to form a network, the entire colony always adherent; ovicells near the ends of the branches, their ooeciostomes much shorter than the peristomes of the zooecia *Proboscina incrassata* (Cycl.)
56b	Branches of the colony generally not anastomosing, the tips of the branches sometimes erect; ovicells varying with respect to their position on the branches, but their ooeciostomes are nearly as long as the peristomes of the zooecia ... 57
57a	Ovicells always on short, erect branch tips *Diaperoecia intermedia* (Cycl.)
57b	Ovicells on branches that adhere to the substratum *Diaperoecia johnstoni* (Cycl.)
58a	Colony with zooecia of 2 distinct sizes, the smaller tubes being about one-fourth the diameter of the larger tubes, and usually adnate on the larger tubes; ovicells inflated; ooeciostome isolated, not attached to a peristome, more or less central on the ovicell *Diplosolen obelium* (fig. 21.10) (Cycl.)
58b	Zooecial tubes generally of 1 size; ovicells variable, not necessarily inflated; ooeciostome usually at the edge of the ovicell and may be attached to the side of a peristome ... 59
59a	Peristomes never arranged in bundles or rows; ooeciostome short, located on the distal edge of the ovicell, free from the peristome, its aperture circular, 60-80 µm in diameter *Plagioecia patina* (fig. 21.13) (Cycl.)
59b	Peristomes often arranged in bundles or rows; ooeciostome at the side of a zooecial tube, its aperture not circular ... 60
60a	Colony less than 5 mm in diameter; ooeciostome short, angled sharply away from the base of a peristome (usually encrusting on algae) *Tubulipora pacifica* (fig. 21.2) (Cycl.)
60b	Colony up to 8 mm in diameter; ooeciostome long, partly adherent and generally parallel to a peristome ... 61
61a	Peristomes at the distal margins of the colony always in bundles of 2-30, the bundles usually in single or double rows, rarely arranged radially; ooeciostome large, flaring at its tip, the aperture elliptical *Tubulipora tuba* (fig. 21.11) (Cycl.)
61b	Peristomes usually fused together in such a way that they form short rows, rather than bundles, the rows often arranged radially; ooeciostome narrow, compressed distally, its aperture therefore slitlike *Tubulipora flabellaris* (fig. 21.12) (Cycl.)
62a	Pores on the frontal of the zooecia arranged in a series of parallel rows (the pores are formed by incomplete fusion of spines that extend across the zooecia; in some species there are pores on the ovicells but not on the zooecia, so be sure you are looking at zooecia) ... 63
62b	Zooecia with or without pores on the frontal (if pores are present, they are not in distinct rows) ... 69
63a	With avicularia (these may be irregularly distributed and are sometimes absent from the entire colony, or at least from portions of it) ... 64
63b	Without avicularia ... 66
64a	Avicularia (minute and setose) located on the frontal, one on each side of the aperture; zooecium with 6-8 spines uniting to form an umbo in the center of the frontal *Puellina setosa* (fig. 21.54)
64b	Avicularia interzooecial; zooecia without a central umbo ... 65
65a	Avicularia nearly as long as the zooecia, the mandible long and spatulate; aperture lyre-

shaped; ovicells absent (older zooecia often secondarily calcified)
Lyrula hippocrepis (fig. 21.56)
65b Avicularia much smaller than the zooecia, the mandible pointed; aperture semicircular, surrounded by about 5 spines; ovicells present   *Colletosia radiata* (fig. 21.53)
66a Frontal spines fused for their entire length, forming complete and regular series of pores   67
66b Frontal spines not fusing completely on the midline, so that uneven gaps rather than regular pores are formed   *Cribrilina corbicula*
67a With an umbo on the proximal border of the aperture; 2 stout spines present on each side of the aperture and 2 similar ones situated at the distal corners; ovicells small
Cribrilina annulata (fig. 21.55)
67b Without a pronounced umbo on the proximal border of the aperture; if apertural spines are present, there are only 2; ovicells prominent   68
68a Pores arranged in transverse series across the frontal; a short, often bifurcate spine sometimes present on both sides of the frontal   *Reginella furcata* (fig. 21.57)
68b Pores arranged in series radiating from the center of the frontal; frontal glossy, without spines next to the aperture (in older specimens, the proximal border of the aperture is often thickened to form a low umbo)   *Reginella nitida* (fig. 21.58)
69a Less than half of the frontal calcified, leaving a large membranous area either exposed or protected only by spines   70
69b More than three-fourths of the frontal calcified   89
70a Zooecia without distinct spines (in certain species, however, there are some minute spines on the frontal, at least at its distal corners; do not confuse a pointed tooth on the midline with a spine)   71
70b Zooecia with relatively stout spines along the margins (the spines may be absent, however, in some portions of the colony)   77
71a Avicularia absent   72
71b Avicularia present, but these are interzooecial   75
72a Zooecia rectangular   73
72b Zooecia oval   74
73a Zooecia elongate rectangles with thin walls (the margins often slightly serrated); occasionally with hollow spines or knobs at the distal corners (sclerotized spinules may be present on the frontal; colonies usually encrusting on large algal fronds)
Membranipora membranacea (fig. 21.20)
73b Zooecia usually elongate rectangles (but sometimes short and broad), their walls heavily calcified and granular; proximal corners often closed off to form 2 triangular regions that may be membranous or knoblike and that may fuse to form a single knob
Conopeum reticulum (fig. 21.19)
74a Calcification variable, extending from the proximal margin half way up the frontal; with a single, stout, pointed tooth arising from the distal margin of the calcified frontal; operculum white, semicircular, calcified; ovicells absent   *Electra crustulenta* subsp. *arctica* (fig. 21.21)
74b Frontal entirely membranous, with wide, crenulated walls; a minute spine sometimes present on both sides of the aperture; operculum sclerotized; ovicells present   *Alderina brevispina*
75a Interzooecial avicularia with elongate mandibles   76
75b Interzooecial avicularia with small, triangular mandibles, the avicularium chambers nearly square   *Ellisina levata*
76a Avicularium chambers diamond-shaped; mandible mostly filiform, with a triangular base; ovicells porous, granular   *Copidozoum tenuirostre* (fig. 21.30)
76b Avicularium chambers oval, nearly as long as the zooecia; mandible large (about 1 mm long), with a slender central rib and with winglike expansions on both sides; ovicells without pores (operculum of zooecia large, with a conspicuous yellow border)   *Hincksina alba* (fig. 21.25)
77a Lateral spines branching, resembling miniature antlers, distal spines straight or occasionally bifid   78
77b Spines not branching   79
78a Avicularia large, interzooecial, the mandible long, narrow, pointed, and curved at the tip; noncalcified portion of the frontal occupying more than half of the frontal surface
Copidozoum protectum (fig. 21.29)

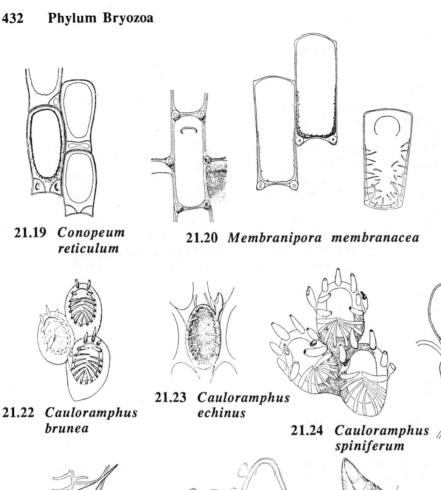

21.19 *Conopeum reticulum*

21.20 *Membranipora membranacea*

21.21 *Electra crustulenta* subsp. *arctica*

21.22 *Cauloramphus brunea*

21.23 *Cauloramphus echinus*

21.24 *Cauloramphus spiniferum*

21.25 *Hincksina alba*

21.26 *Hincksina pallida*

21.27 *Callopora armata*

21.28 *Callopora horrida*

21.29 *Copidozoum protectum*

21.30 *Copidozoum tenuirostre*

21.31 *Doryporella alcicornis*

21.32 *Tegella armifera*

21.33 *Tegella robertsonae*

78b Avicularia small, on the proximal portion of the frontal, the mandible short, triangular; noncalcified portion of the frontal small, occupying less than half of the frontal surface
Doryporella alcicornis (fig. 21.31)
79a Avicularia stalked, tall, slender, with a narrow base, distributed more or less sparsely among the lateral spines (avicularia are sometimes lacking altogether, however); ovicells either absent or so deeply embedded that they are not evident 80
79b Avicularia not stalked; ovicells prominent 82
80a Zooecia separated by deep or wide grooves 81
80b Zooecia crowded, not separated by grooves   *Cauloramphus spiniferum* (fig. 21.24)
81a With 2 pairs of stout and erect distal spines, the other spines pointed and curved markedly over the membranous frontal (spines usually brown)   *Cauloramphus brunea* (fig. 21.22)
81b With a pair of distal spines directed forward, the other spines curving slightly over the frontal (all spines slender)   *Cauloramphus echinus* (fig. 21.23)
82a Zooecia with long spines only on the portion distal to the aperture; usually with an avicularium distal to the aperture; ovicells with 1 or 2 small triangular avicularia; membranous portion of the frontal occupying less than two-thirds of the surface 83
82b Zooecia with spines lateral to the aperture; small avicularia interzooecial or close to the lateral or proximal margins of the aperture; avicularia on the ovicells variable, but sometimes large, oblong, pointed; membranous portion of the frontal occupying more than two-thirds of the surface 84
83a Zooecia about 0.7-0.85 mm long; spines straight, extending forward from the distal rim; colony encrusting, but loosely attached   *Chapperiella patula*
83b Zooecia about 0.6 mm long; most nearly proximal pair of spines curving down over the frontal; colony tightly encrusting on a variety of substrata, including small worm tubes
Chapperiella condylata
84a Frontal entirely membranous 85
84b Frontal with a narrow calcified shelf extending inward from the rim 87
85a Only proximal avicularia present, these single or paired 86
85b Both proximal and lateral avicularia present (inner margin of the zooecial rim faintly toothed)
Callopora armata (fig. 21.27)
86a Zooecia separated by wide grooves, the deepest portions of these with large pores
Callopora circumclathrata
86b Grooves between zooecia narrow and without pores   *Callopora horrida* (fig. 21.28)
87a With small avicularia lateral to the aperture   *Tegella armifera* (fig. 21.32)
87b Without lateral avicularia 88
88a Zooecia without erect, tubular spines, but usually with 1-3 curved spines bending over the frontal; ovicells prominent   *Tegella aquilirostris*
88b Zooecia with an erect, tubular spine on both sides, just proximal to the aperture (there are also usually some curved spines bending over the frontal, considerably proximal to the aperture); ovicells low, not prominent   *Tegella robertsonae* (fig. 21.33)
89a Zooecia with pores distributed over most of the frontal (be sure you are looking at the zooecia, not the ovicells, which may have pores when the zooecia do not) 90
89b Zooecia either without pores or with pores in a line around the edge of the frontal 119
90a With an ascopore (a distinct pore on the midline proximal to the aperture; it is the opening to the ascus, a compensation sac beneath the calcified frontal) 91
90b Without an ascopore 97
91a Without avicularia; frontal pores present between the ascopore and the aperture 92
91b With avicularia, although these are absent on some zooecia; ascopore close to the aperture, there being no frontal pores between them 93
92a Frontal inflated only slightly and more or less evenly   *Fenestrulina malusii* (fig. 21.8)
92b Frontal with a conspicuous umbo just proximal to the ascopore
Fenestrulina malusii subsp. *umbonata*
93a Avicularia usually single (occasionally paired), located proximal to the ascopore 94
93b Avicularia usually paired (occasionally single), lateral to the ascopore 96
94a Avicularia large, the mandible slender, up to 1 mm long   *Microporella vibraculifera*
94b Avicularia small, the mandible much shorter than 1 mm 95
95a With 3 umbones (1 central, the others lateral to the aperture); ascopore near the aperture
Microporella umbonata

434 Phylum Bryozoa

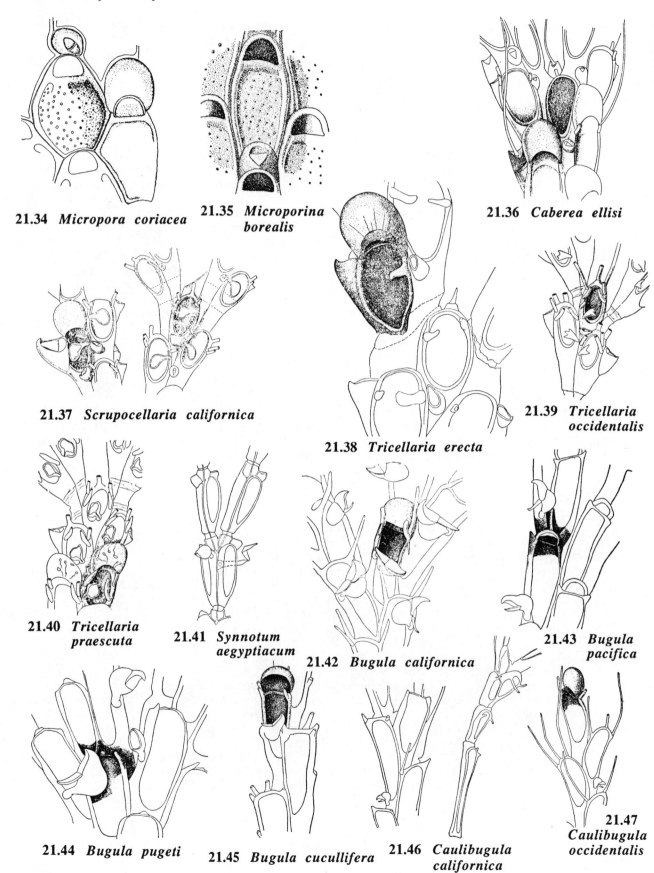

21.34 *Micropora coriacea*
21.35 *Microporina borealis*
21.36 *Caberea ellisi*
21.37 *Scrupocellaria californica*
21.38 *Tricellaria erecta*
21.39 *Tricellaria occidentalis*
21.40 *Tricellaria praescuta*
21.41 *Synnotum aegyptiacum*
21.42 *Bugula californica*
21.43 *Bugula pacifica*
21.44 *Bugula pugeti*
21.45 *Bugula cucullifera*
21.46 *Caulibugula californica*
21.47 *Caulibugula occidentalis*

Phylum Bryozoa 435

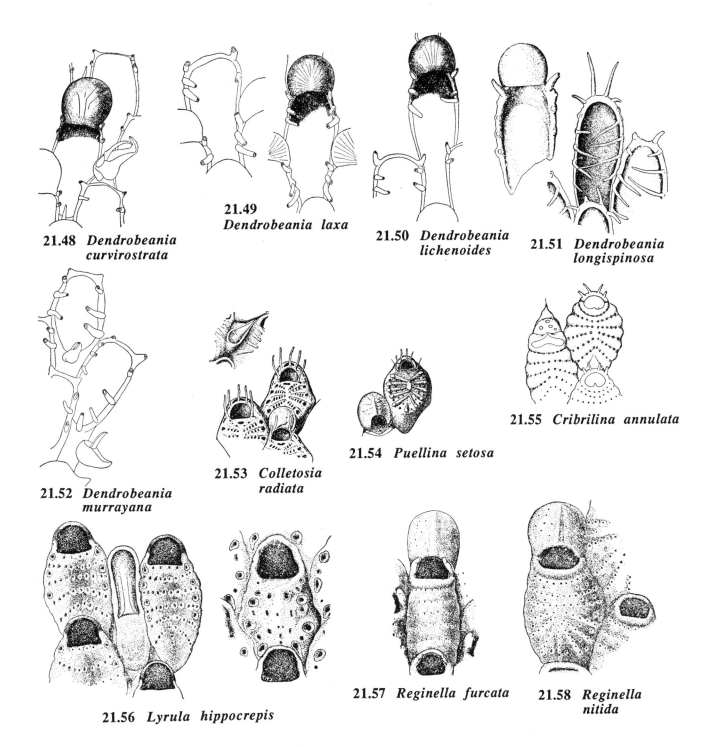

21.48 *Dendrobeania curvirostrata*
21.49 *Dendrobeania laxa*
21.50 *Dendrobeania lichenoides*
21.51 *Dendrobeania longispinosa*
21.52 *Dendrobeania murrayana*
21.53 *Colletosia radiata*
21.54 *Puellina setosa*
21.55 *Cribrilina annulata*
21.56 *Lyrula hippocrepis*
21.57 *Reginella furcata*
21.58 *Reginella nitida*

436   Phylum Bryozoa

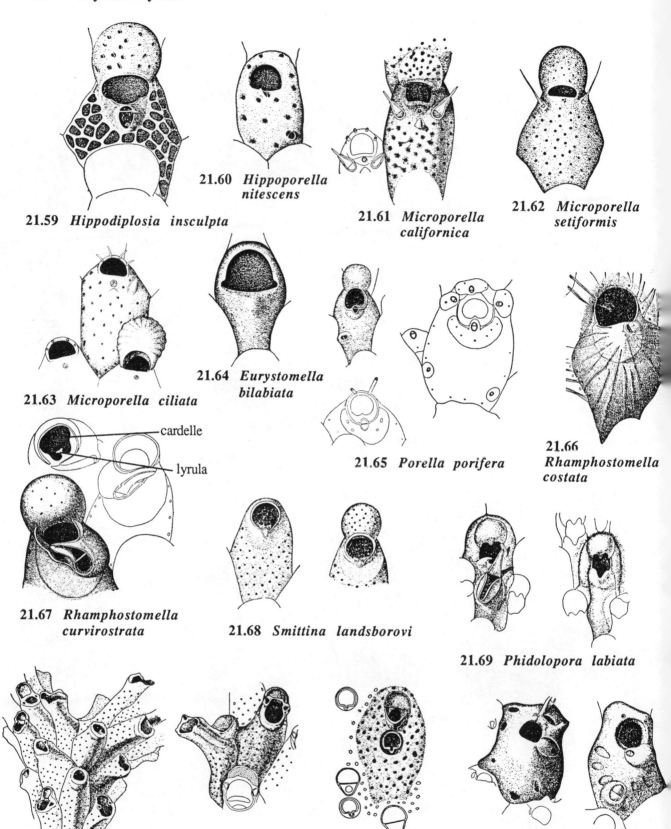

21.59 *Hippodiplosia insculpta*
21.60 *Hippoporella nitescens*
21.61 *Microporella californica*
21.62 *Microporella setiformis*
21.63 *Microporella ciliata*
21.64 *Eurystomella bilabiata*
cardelle
lyrula
21.65 *Porella porifera*
21.66 *Rhamphostomella costata*
21.67 *Rhamphostomella curvirostrata*
21.68 *Smittina landsborovi*
21.69 *Phidolopora labiata*
21.70 *Lagenipora punctulata*
21.71 *Myriozoum coarctatum*
21.72 *Lepraliella bispina*

95b With 1 umbo; ascopore rather widely separated from the aperture
Microporella ciliata (fig. 21.63)
96a Bases of avicularia small, narrow; mandible elongated and slender, but less than half as long as the zooecium *Microporella setiformis* (fig. 21.62)
96b Bases of avicularia broad; mandible triangular *Microporella californica* (fig. 21.61)
97a With avicularia 98
97b Without avicularia 110
98a Avicularia interzooecial *Micropora coriacea* (fig. 21.34)
98b Avicularia located on the frontal 99
99a Avicularia paired, one on each side of the aperture, and sometimes on the raised lip of the aperture 100
99b Avicularia not arranged as described in choice 99a 103
100a Zooecia flask-shaped; frontal with pores over its entire surface; peristome thin, flaring, sometimes with short spines around its border; ovicell high on the distal side of the peristome, its semicircular front with pores but without radiating ribs *Lagenipora socialis*
100b Zooecia globular, especially in older colonies; frontal without pores in the central area; peristome thick, without spines; semicircular front of ovicel with radially arranged ribs separating slitlike pores (the pores may disappear with further calcification) 101
101a Colony encrusting on rock or shell, occasionally rising up to form stout protuberances; color dark orange to pale peach; numerous zooecia in the colony 0.7 mm long, and not many less than 0.55 mm long *Costazia ventricosa*
101b Colony small, usually almost globular, on algae or worm tubes; color white; zooecia not often more than 0.65 mm long 102
102a Some zooecia with a median avicularium proximal to the aperture in addition to the ones on both sides of the aperture *Costazia robertsoniae*
102b With only 2 avicularia, these on both sides of the aperture *Costazia costazi*
103a Avicularia long and pointed, their length about equal to the diameter of the aperture
Schizoporella unicornis (fig. 21.7)
103b Avicularia small, their length much less than the diameter of the aperture 104
104a Aviculaium on the midline, proximal to the aperture (in some species, however, a second avicularium may be found elsewhere on the frontal) 105
104b Avicularium lateral to the midline *Codonellina cribriformis*
105a Ovicells absent (a small avicularium is sometimes present proximal to the aperture)
Cryptosula pallasiana
105b Ovicells present 106
106a Ovicell without pores (median apertural avicularium often directed laterally)
Hippoporella nitescens (fig. 21.60)
106b Ovicell with pores 107
107a Zooecia less than 0.5 mm long; avicularium mounted on a small umbo some distance proximal to the aperture *Schizomavella auriculata*
107b Zooecia more than 0.65 mm long; avicularium located at the proximal margin of the aperture 108
108a Aperture wider than high; with a large avicularium on a smooth, triangular area proximal to the aperture *Hippodiplosia reticulato-punctata*
108b Aperture either nearly circular or higher than wide; avicularium small 109
109a Mature zooecia with a raised rim around the distal half of the aperture, but not around the proximal half (in younger zooecia, however, the rim around the aperture is nearly complete); an avicularium rarely present just proximal to the aperture, where there is no rim; frontal surface coarsely granular or ribbed; color of live colonies frequently dull orange or brown *S mittina cordata*
109b Aperture completely encircled by a raised rim; an avicularium present proximal to the aperture, becoming fused with the rim; frontal smooth to slightly granular; color light
Smittina landsborovi (fig. 21.68)
110a Aperture with a pointed proximal tooth (ovicells absent) *Cheilopora praelonga*
110b Aperture without a pointed proximal tooth 111
111a With a U-shaped or V-shaped sinus on the proximal rim of the aperture 112
111b Without a U-shaped or V-shaped sinus on the rim of the aperture 115
112a Ovicells absent; zooecia large, up to 0.8 mm long; frontal becoming roughened by

	secondary calcification	*Stomachetosella cruenta*
112b	Ovicells present; zooecia small to large; frontal often granular, but not highly calcified or rough	113
113a	Ovicells without pores (the ovicells are, moreover, deeply embedded) *Stomachetosella limbata*	
113b	Ovicells with 1 or more pores	114
114a	Ovicells with only 1 pore (the ovicells are deeply embedded and have only a small exposed surface); zooecia more than 0.5 mm long, their lateral margins angular	*Stomachetosella sinuosa*
114b	Ovicells with many pores; zooecia less than 0.5 mm long, their lateral margins straight, which gives them a rectangular appearance	*Schizoporella linearis* subsp. *inarmata*
115a	Ovicells without pores, but with conspicuous radial ribs; proximal rim of the aperture raised up in an umbo, a deep transverse depression separating the peak of the umbo from the aperture (colony usually yellow to yellow-brown, growing in double sheets)	*Hippodiplosia insculpta* (fig. 21.59)
115b	Ovicells with pores; frontal with neither an umbo nor a depression	116
116a	Ovicells large, globular, often inflated in such a way as to form 3 bulbous lobes, but lacking a distinct frontal area; dwarf zooecia present at the distal ends of normal zooecia	*Trypostega claviculata*
116b	Ovicells with a distinct, porous frontal area; without dwarf zooecia	117
117a	Frontal of ovicells forming an elongate triangle	*Dakaria pristina*
117b	Frontal of ovicells circular or semicircular	118
118a	Frontal of ovicells circular, distinctly separate from the zooecial aperture	*Dakaria dawsoni*
118b	Frontal of ovicells somewhat semicircular, contiguous with the zooecial aperture	*Dakaria ordinata*
119a	Avicularia absent	120
119b	Avicularia present	123
120a	Zooecia hyaline, glossy, and usually corrugated transversely; aperture with a U-shaped proximal sinus	121
120b	Zooecia not hyaline; aperture without a U-shaped proximal sinus	122
121a	Colony multiserial, usually forming a monolayer over the substratum but sometimes forming nodular encrustations; pores sometimes present between the zooecia; ovicells large, with pores, borne on dwarf zooecia	*Hippothoa hyalina*
121b	Colony uniserial, with lateral branches; ovicells smooth, globular, without pores	*Hippothoa divaricata*
122a	With a large umbo proximal to the aperture; aperture bordered by 6-8 long, erect spines; lyrula well developed; frontal with pores near the margins	*Mucronella ventricosa*
122b	Without an umbo proximal to the aperture; aperture not bordered by spines; lyrula absent; frontal smooth, without pores (colony distinctly rose-pink; operculum rose-red, its outline shaped like that of a narrow-brimmed derby hat seen in side view)	*Eurystomella bilabiata* (fig. 21.64)
123a	Avicularia interzooecial, large, with a brown, spoon-shaped mandible (a small avicularium also generally present at the proximal edge of the aperture)	*Holoporella brunnea*
123b	Without interzooecial avicularia (the avicularia are on the frontal)	124
124a	With an avicularium on the midline, adjacent to the proximal edge of the aperture	125
124b	Avicularia, if adjacent to the proximal edge of the aperture, not located on the midline	128
125a	Avicularium chamber small, narrow, shaped like a truncated cone, curving forward over the deep, U-shaped sinus; ovicells with pores	*Lacerna fistulata*
125b	Avicularium chamber not shaped like a truncated cone; ovicells without pores	126
126a	A small frontal avicularium usually present in addition to the one that is adjacent to the proximal edge of the aperture; chamber of the avicularium that is adjacent to the aperture large, inflated, extending transversely for most of the width of the aperture, and with 2-6 pores	*Porella porifera* (fig. 21.65)
126b	Only 1 avicularium present (this adjacent to the proximal edge of the aperture); avicularium chamber usually without pores	127
127a	Avicularium chamber prominent, with an umbo, but not extending laterally around the aperture; frontal margin of the zooecia with a few pores, these becoming occluded with age	*Porella concinna*

127b Avicularium chamber considerably inflated, extending laterally around the aperture; pores along the frontal margin of the zooecia large, conspicuous   *Porella columbiana*
128a With an avicularium at the proximal edge of the aperture   129
128b Without an avicularium at the proximal edge of the aperture (1 or 2 avicularia are present, but they are not adjacent to the edge of the aperture)   132
129a Rostrum and mandible of the avicularium not overhanging the aperture   130
129b Rostrum and mandible of the avicularium overhanging the aperture   131
130a Avicularium small, fitting into the rimlike peristome, the mandible elongate-triangular; cardelles present in the aperture   *Rhamphostomella cellata*
130b Avicularium of moderate size, located on a large umbo, the mandible spatulate; cardelles absent (lyrula present; frontal with prominent ribs running from the margins to the umbo)   *Rhamphostomella costata* (fig. 21. 66)
131a Rim of the aperture in the form of a high collar; rostrum of the avicularium that is adjacent to the proximal edge of the aperture large, laterally curved; cardelles present; without any avicularia other than the one adjacent to the aperture; frontal without pronounced radiating ribs   *Rhamphostomella curvirostrata* (fig. 21.67)
131b Rim of the aperture not raised up as a collar; rostrum of the avicularium that is adjacent to the proximal edge of the aperture sharply hooked at the tip; cardelles absent; in older specimens, frontal avicularia other than the one adjacent to the aperture are present; frontal with pronounced radiating ribs   *Rhynchozoon tumulosum*
132a Rostrum of the avicularium or avicularia directed proximally, though often at an angle to the long axis of the zooecium   133
132b Rostrum of the avicularia directed distally   134
133a Frontal with 1 row of pores along the margin; aperture oval; avicularia conspicuous, with a long, narrow rostrum   *Gemelliporella inflata*
133b Frontal reticulated, with 1-3 rows of pores along the margin; proximal border of the aperture with a broad, V-shaped sinus; avicularia elongate, lateral to the aperture (they may be paired) (operculum of zooecia light brown, sclerotized)   *Hippomonavella longirostrata*
134a With ovicells; with 2 or 3 different types of avicularia   135
134b Without ovicells; avicularia of 1 type, small, 1 on each side of the aperture (frontal smooth, with large pores around the margin)   *Umbonula arctica*
135a Aperture with a thin raised collar, a sinus scarcely noticeable or absent; young zooecia with 2-4 spines at the distal margin; ovicells with pores   136
135b Aperture without a raised collar, but with a deep, V-shaped sinus; zooecia without spines; ovicells without pores (with a pair of small avicularia lateral to, or slightly proximal to, the sinus of the aperture)   *Stephanosella vitrea*
136a Frontal with 1 to several distinct nodules or tuberosities; young zooecia with only 2 spines on the distal margin   *Parasmittina collifera*
136b Frontal without tuberosities; young zooecia with 2-4 spines on the distal margin   *Parasmittina trispinosa*

Class Gymnolaemata: Order Ctenostomata

1a Zooecia connected by creeping stolons (in *Immergentia* and *Penetrantia*, the colonies are embedded in calcareous shells of molluscs and barnacles)   2
1b Zooecia united into a fleshy, leathery, or gelatinous mass that is erect or sheetlike   6
2a Colony embedded in a calcareous shell, the apertures of the zooids about flush with the surface   *Immergentia* sp. and *Penetrantia* sp.
2b Colony not embedded in a calcareous shell   3
3a Zooecia attached to stolons by slender stalks   *Triticella pedicellata*
3b Zooecia attached directly to stolons, without obvious stalks   4
4a Each zooecium expanded at the base, joined to several stolons; gizzard absent (surface covered with a fine layer of silt)   *Nolella stipata*
4b Each zooecium attached to a single stolon; gizzard present or absent   5
5a Zooecia 0.3-0.5 mm long, the proximal one-third of each one usually adhering to the stolon   *Buskia nitens*

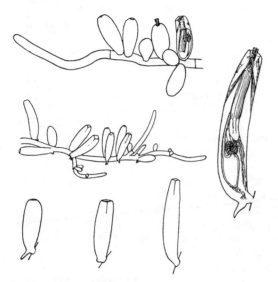

21.73 *Bowerbankia gracilis*

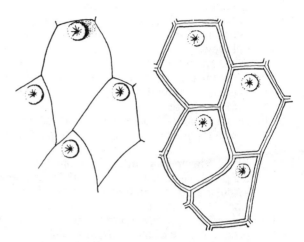

21.74 *Alcyonidium polyoum*

5b Zooecia 1-2.2 mm long, distinct from the stolon for their entire length (gizzard present)
Bowerbankia gracilis (fig. 21.73)
6a Colony an erect, club-shaped structure with an annulated stalk *Clavopora occidentalis*
6b Colony not club-shaped and without an annulated stalk 7
7a Colony generally erect, rarely encrusting, consisting of flattened lobes bearing many forked spines (color pale tan to dark brown) *Flustrellidra corniculata*
7b Colony erect or encrusting, without spines 8
8a Colony erect, lobed, attached by a short stalk to a basal disk *Alcyonidium pedunculatum*
8b Colony encrusting, forming a gelatinous coating over the substratum
Alcyonidium polyoum (fig. 21.74)

Class Stenolaemata

Order Cyclostomata

Suborder Tubuliporina

Family Oncousoeciidae

Proboscina incrassata (Smitt, 1866)
Stomatopora granulata (Milne-Edwards, 1836)

Family Diastoporidae

Diaperoecia californica (d'Orbigny, 1852) (fig.21.9)
Diaperoecia intermedia (O'Donoghue, 1923)
Diaperoecia johnstoni (Heller, 1867)
Diplosolen obelium (Johnston, 1838) (fig.21.10)
Plagioecia patina (Lamarck, 1816) (fig. 21.13)

Family Tubuliporidae

Tubulipora flabellaris (Fabricius, 1780) (fig.21.12)
Tubulipora pacifica Robertson, 1910 (fig. 21.2)
Tubulipora tuba (Gabb & Horn, 1862) (fig. 21.11)

Family Frondiporidae

Filifascigera fasciculata (Hincks, 1880) (fig. 21.14)

Suborder Articulata

Family Crisiidae

Bicrisia edwardsiana (d'Orbigny, 1839) (fig. 21.15)
Crisia maxima Robertson, 1910
Crisia occidentalis Trask, 1857 (fig. 21.1)
Crisia pugeti Robertson, 1910 (fig. 21.16)
Crisia serrulata (Gabb & Horn, 1862)
Crisidia cornuta (Linnaeus, 1758) (fig. 21.17)
Filicrisia franciscana (Robertson, 1910) (fig. 21.18)

Suborder Cancellata

Family Cytididae

Discocytis canadensis O'Donoghue, 1926

Suborder Cerioporina

Family Heteroporidae

Heteropora alaskensis (Borg, 1933) (fig. 21.4)
Heteropora magna O'Donoghue, 1923 (fig. 21.5)
Heteropora pacifica Borg, 1933

Suborder Rectangulata

Family Lichenoporidae

Disporella fimbriata (Busk, 1875)
Disporella separata Osburn, 1953
Lichenopora novae-zelandiae (Busk, 1875)
Lichenopora verrucaria (Fabricius, 1780) (fig. 21.3)

Class Gymnolaemata
Order Ctenostomata
Suborder Euctenostomata

Family Flustrellidridae

Flustrellidra corniculata (Smitt, 1871)

Family Alcyonidiidae

Alcyonidium pedunculatum Robertson, 1902
Alcyonidium polyoum (Hassall, 1841) (fig. 21.74)

Family Clavoporidae

Clavopora occidentalis (Fewkes, 1889)

Family Triticellidae

Triticella pedicellata (Alder, 1857)

Family Arachnidiidae

Nolella stipata Gosse, 1855

Family Immergentiidae

Immergentia sp.

Family Vesiculariidae

Bowerbankia gracilis Leidy, 1855 (fig. 21.73)

Family Buskiidae

Buskia nitens Alder, 1857

Family Penetrantiidae

Penetrantia sp.

Order Cheilostomata
Suborder Anasca
Family Aeteidae

Aetea spp.

Family Membraniporidae

Conopeum reticulum (Linnaeus, 1767) (fig. 21.19)
Membranipora membranacea (Linnaeus, 1767) (fig. 21.20)

Family Electridae

Electra crustulenta subsp. *arctica* Borg, 1931 (fig. 21.21)

Family Flustridae

Chartella membranaceo-truncata (Smitt, 1867)

Family Hincksinidae

Cauloramphus brunea Canu & Bassler, 1930 (fig. 21.22)
Cauloramphus echinus (Hincks, 1882) (fig. 21.23)
Cauloramphus spiniferum (Johnston, 1832) (fig. 21.24)
Ellisina levata (Hincks, 1882)
Hincksina alba (O'Donoghue, 1923) (fig. 21.25)
Hincksina pallida (Hincks, 1884) (fig. 21.26)

Family Alderinidae

Alderina brevispina (O'Donoghue, 1926)
Callopora armata O'Donoghue, 1926 (fig. 21.27)
Callopora circumclathrata (Hincks, 1881)
Callopora horrida (Hincks, 1880) (fig. 21.28)
Copidozoum protectum (Hincks, 1884) (fig. 21.29)
Copidozoum tenuirostre (Hincks, 1880) (fig. 21.30)
Doryporella alcicornis (O'Donoghue, 1923) (fig. 21.31)
Tegella aquilirostris (O'Donoghue, 1923)
Tegella armifera (Hincks, 1880) (fig. 21.32)
Tegella robertsonae O'Donoghue, 1926 (fig. 21.33)

Family Chapperiellidae

Chapperiella condylata Canu & Bassler, 1930
Chapperiella patula (Hincks, 1881)

Family Microporidae

Micropora coriacea (Esper, 1791) (fig. 21.34)
Microporina borealis (Busk, 1855) (fig. 21.35)

Family Cellariidae

Cellaria diffusa Robertson, 1905
Cellaria mandibulata Hincks, 1882

Family Scrupocellariidae

Caberea boryi (Audouin, 1826)
Caberea ellisi (Fleming, 1814) (fig. 21.36)
Scrupocellaria californica Trask, 1857 (fig. 21.37)
Tricellaria erecta (Robertson, 1900) (fig. 21.38)
Tricellaria occidentalis (Trask, 1857) (fig. 21.39)
Tricellaria praescuta Osburn, 1950 (fig. 21.40)
Tricellaria ternata (Solander, 1786)

Family Epistomiidae

Synnotum aegyptiacum (Audouin, 1826) (fig. 21.41)

Family Bicellariellidae

Bugula californica Robertson, 1905 (fig. 21.42)
Bugula cucullifera Osburn, 1912 (figs. 21.6 and 21.45)
Bugula pacifica Robertson, 1905 (fig. 21.43)
Bugula pugeti Robertson, 1905 (fig. 21.44)
Caulibugula californica (Robertson, 1905) (fig. 21.46)
Caulibugula ciliata (Robertson, 1905)
Caulibugula occidentalis (Robertson, 1905) (fig. 21.47)
Dendrobeania curvirostrata (Robertson, 1905) (fig. 21.48)
Dendrobeania laxa (Robertson, 1905) (fig. 21.49)
Dendrobeania lichenoides (Robertson, 1905) (fig. 21.50)
Dendrobeania longispinosa (Robertson, 1905) (fig. 21.51)
Dendrobeania murrayana (Johnston, 1847) (fig. 21.52)

Family Cribrilinidae

Colletosia radiata (Moll, 1803) (fig. 21.53)
Cribrilina annulata (Fabricius, 1780) (fig. 21.55)
Cribrilina corbicula (O'Donoghue, 1923)
Lyrula hippocrepis (Hincks, 1882) (fig. 21.56)
Puellina setosa (Waters, 1899) (fig. 21.54)
Reginella furcata (Hincks, 1884) (fig. 21.57)
Reginella nitida Osburn, 1950 (fig. 21.58)

PHYLUM ECHINODERMATA

Class Crinoidea

The only crinoid commonly collected in our region is *Florometra serratissima*. It occurs at depths as shallow as 30 m, but is rather local in distribution. Other species listed below are restricted to deep water.

Clark, A. H. 1907. Descriptions of new species of recent unstalked crinoids from the North Pacific Ocean. Proc. U. S. Nat. Mus., 33:69-84.
Clark, A. H., & A. McG. Clark. 1967. A monograph of the existing crinoids. Vol. 1. The comatulids. Part 5. Suborders Oligophreata (concluded) and Macrophreata. Bull. 82, U. S. Nat. Mus. xiv + 860 pp.
Mladenov, P. V., & F. S. Chia. 1983. Development, settling behavior, metamorphosis and pentacrinoid feeding and growth of the feather star *Florometra serratissima*. Mar. Biol., 73:309-23.

Order Hyocrinida
Family Hyocrinidae

Ptilocrinus pinnatus A. H. Clark, 1907

Order Bourgeticrinida
Family Bathycrinidae

Bathycrinus pacificus A. H. Clark, 1907

Order Comatulida
Family Pentametrocrinidae

Pentametrocrinus sp.

Family Antedonidae

Antedon sp.
Florometra serratissima (A. H. Clark, 1907)
Psathyrometra fragilis A. H. Clark, 1908
Retiometra alascana A. H. Clark, 1936

Class Asteroidea

Alton, M. S. 1961. Bathymetric distribution of sea stars (Asteroidea) off the northern Oregon coast. J. Fish. Res. Bd. Canada, 23:1673-1714.

Chia, F. S. 1966. Systematics of the six-rayed sea star *Leptasterias* in the vicinity of San Juan Island, Washington. Systematic Zool., 15:300-6.

D'iakonov, A. M. 1950. [Sea stars of the seas of the USSR.] Opredeliteli po Faune SSSR, no. 34. 203 pp. (in Russian). (English translation, 1968, by Israel Program for Scientific Translations, Jerusalem.)

Fisher, W. K. 1911-1930. Asteroidea of the north Pacific and adjacent waters. Bull. 76, U. S. Nat. Mus. Part 1 (1911), vi + 419 pp; Part 2 (1928), iii + 245 pp.; Part 3 (1930), iii + 356 pp.

Lambert, P. 1978a. New geographic and bathymetric records for some northeast Pacific asteroids (Echinodermata: Asteroidea). Syesis, 11:61-4.

------. 1978b. British Columbia Marine Faunistic Survey Report: Asteroids from the Northeast Pacific. Canada Fisheries and Marine Service Technical Report No. 773. 23 pp.

------. 1981. The Sea Stars of British Columbia. Handbook 39, British Columbia Provincial Museum, Victoria, 153 pp.

Mauzey, K. P., C. Birkeland, & P. K. Dayton. 1968. Feeding behavior of asteroids and escape responses of their prey in the Puget Sound region. Ecology, 49:603-19.

Verrill, A. E. 1914. Monograph of the shallow-water starfishes of the North Pacific coast from the Arctic Ocean to California. Harriman Alaska Expedition, 14. Part 1 (text), xii + 408 pp; Part 2 (plates), 222 pp.

The key includes all species likely to be found intertidally or subtidally to a depth of about 200 m. For deep-water sea stars that do not seem to fit any of the choices, consult the handbook of Lambert (1981), which is the most useful work for our region.

1a	With 5 or 6 rays (except in specimens that have lost a ray, or that have regenerated extra rays after being injured)	2
1b	With 8 to 20 or more rays	24
2a	With 6 rays; diameter rarely greater than 11 cm	*Leptasterias hexactis*
2b	With 5 rays; diameter frequently greater than 11 cm	3
3a	Rays bordered by conspicuous marginal plates that are much larger than any other ossicles visible at the surface (in *Hippasteria spinosa*, each plate bears a single stout spine, and in *Luidia foliolata* each plate bears several spines)	4
3b	Rays not bordered by conspicuous marginal plates (small marginal plates may be present, but these are not likely to be noted without magnification)	11
4a	Marginal plates not visible when the animal is viewed from above (they must be observed from the side), and each plate with several spines (tube feet pointed, without suckers)	*Luidia foliolata* (fig. 22.1)
4b	Marginal plates visible when the animal is viewed from above, and the plates either without spines or with only 1 or 2 stout spines	5
5a	Each marginal plate with 1 or 2 spines	6
5b	Marginal plates without spines	7
6a	Color of aboral surface orange to vermilion; aboral surface with some conspicuous bivalved, unstalked pedicellariae; most marginal plates with 2 spines	*Hippasteria spinosa* (fig. 22.3)
6b	Color of aboral surface gray or yellowish; aboral surface without pedicellariae; each marginal plate with a single spine	*Ctenodiscus crispatus*
7a	Tube feet pointed; diameter up to about 9 cm	*Leptychaster pacificus*
7b	Tube feet tipped with suckers; diameter generally greater than 9 cm	8
8a	Rays so short that the body appears almost pentagonal (diameter of the disk at least three-fourths of the total diameter)	9
8b	Rays not so short that the body appears almost pentagonal	10
9a	Flat-topped plates of aboral surface with 4 to 12 marginal units surrounding 1 to 3 central units	*Ceramaster arcticus*

Phylum Echinodermata 449

22.1 *Luidia foliolata*

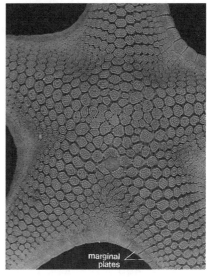

22.2 *Mediaster aequalis*, **aboral surface**

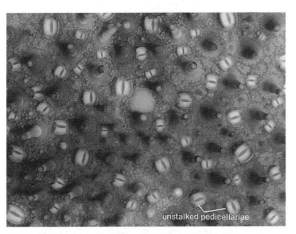

22.3 *Hippasteria spinosa*, **aboral surface**

22.4 *Pteraster tesselatus*

22.5 *Crossaster papposus*

22.6 *Asterina miniata*

9b   Flat-topped plates of aboral surface with 12 to 15 marginal units surrounding 4 to 12 central units   *Ceramaster patagonicus*

10a   Aboral surface with circular paxillae that consist of numerous granulelike spines, all of which are similar   *Gephyreaster swifti*

10b   Aboral surface with circular to oval or hexagonal, flat-topped plates, each consisting of about 25 marginal granules and a central mass of slightly different and less markedly flattened granules (with an odor like that of exploded gunpowder)   *Mediaster aequalis* (fig. 22.2)

11a   Aboral surface smooth and slippery, with few if any exposed spines   12

11b   Aboral surface rough or gritty due to exposed spines or plates   13

12a   Thickness equal to about one-third the diameter; aboral surface with a conspicuous and slightly elevated central opening (this leads into a spongy cavity that lies between the body wall proper and a superficial membrane); madreporite not visible; color of aboral surface usually pale orange or brownish yellow; without an odor like that of exploded gunpowder, but generally secreting a large amount of nearly gelatinous mucus, especially when first collected; subtidal   *Pteraster tesselatus* (fig. 22.4)

12b   Thickness less than one-third the diameter; aboral surface without a conspicuous and elevated central opening (there is no spongy cavity covered by a membrane distinct from the rest of the body wall); madreporite evident, though sunken; aboral surface usually reddish brown, with some patches of gray or purple; with an odor like that of exploded gunpowder, but not secreting a large amount of nearly gelatinous mucus; intertidal and subtidal   *Dermasterias imbricata*

13a   Diameter of disk about one-third to one-half the total diameter (found on the west coast of Vancouver Island and in California, but rarely if ever encountered in Washington or Oregon)   *Asterina miniata* (fig. 22.6)

13b   Diameter of disk much less than one-third the total diameter   14

14a   Pedicellariae absent; aboral surface often with the texture of fine sandpaper when rubbed with a finger, but not noticeably spiny when viewed without magnification   15

14b   Pedicellariae present; aboral surface noticeably spiny   20

15   Our representatives of the genus *Henricia*, couplets 15 to 19, need study. Although all intertidal and nearly all shallow subtidal specimens will key to *H. leviuscula*, this species, as presently understood, is extremely variable. Several subspecies were proposed by Fisher (1911, 1930) and by Verrill (1914), but they intergrade and no attempt will be made here to separate them.

15a   Plates forming the ridges on the aboral surface with a single row of conical spines (1-4 spines per plate), these largely covered with tissue so that they may not at first be distinct (rare subtidal species; only a few specimens known)   *Henricia asthenactis*

15b   Plates forming the ridges on the aboral surface bearing minute, scattered spines or clusters of prominent spines, or modified as spiny paxillae, all of these elements exposed and readily apparent with magnification   16

16a   Diameter of meshlike areas on the aboral surface appreciably greater than the width of the ridges that separate them (the ridges bear minute, scattered spines or clusters of spines whose length is commonly 1 to 1.5 mm)   17

16b   Diameter of meshlike areas on the aboral surface no greater than, and generally less than, the diameter of the spiny paxillae that separate them   18

17a   Ridges separating the meshlike areas of the aboral surface with scattered, minute spines (aboral surface yellow to brick red; rare subtidal species)   *Henricia aspera*

17b   Ridges separating the meshlike areas of the aboral surface with radiating clusters of sharp spines (these commonly 1 to 1.5 mm long, and 2-9 in a cluster) (aboral surface mostly white; rare subtidal species)   *Henricia longispina*

18a   Marginal plates prominent and arranged in 3 distinct rows; rays not thickened so much at the base that they are separated from one another by a crease that extends into the disk (aboral surface usually orange, brick-red, or brown, but sometimes cream or some other color; color not always uniform, and there may be mottling, gray patches, etc.)   19

18b   Marginal plates not prominent and not forming distinct rows; rays usually noticeably thickened at the base and separated from one another by a crease that extends into the disk (color off-white to orange: subtidal)   *Henricia sanguinolenta*

19a   Genital pores (located between the rays) slightly aboral with respect to the margin of the disk; diameter frequently exceeding 8 cm; color uniformly orange or orange-red, or with a disk of a different color than the rays (the disk is sometimes mottled, but the rays are not mottled); females not known to brood young (common intertidal and subtidal species) *Henricia leviuscula*

19b Genital pores slightly oral with respect to the margin of the disk; diameter rarely greater than 5 cm; disk and rays irregularly mottled; females brooding young under oral surface in winter or early spring (intertidal and subtidal) *Henricia* sp.
Apparently an undescribed species; believed to be conspecific with the "brooding variety" of *H. leviuscula* reported by Fisher (1911).

20a Spines on upper parts of aboral surface generally not higher than 2 mm, typically in irregular clusters or arranged in networklike patterns, though some may be single and some may be in short rows; most pedicellariae in raised, hemispherical cushions that have a central spine 21

20b Spines on upper parts of aboral surface mostly 3-5 mm long, typically well separated and more inclined to form poorly defined rows than clusters or networks; most pedicellariae concentrated in raised, hemispherical cushions that have a central spine 23

21a Diameter of disk not more than one-sixth the total diameter; rays typically broadest a short distance away from the disk; jaws of the straight pedicellariae not forked at their tips into a pair of blades; color extremely variable, may be gray, greenish, brown, orange, or red, but not likely to be purple or pink *Evasterias troschelii*

21b Diameter of disk usually greater than one-fifth the total diameter; rays typically broadest where they join the disk; jaws of the straight pedicellariae forked at their tips into a pair of unequal blades; color generally orange, orange-ochre, brown, purple, or some shade of pink 22

22a Spines on upper parts of aboral surface usually arranged in a networklike pattern; color orange, orange-ochre, brown, or purple; diameter not often exceeding 25 cm; intertidal and subtidal *Pisaster ochraceus*

22b Spines on upper parts of aboral surface usually single or clustered, but sometimes arranged in a networklike pattern; color some shade of pink; diameter commonly exceeding 25 cm, and sometimes exceeding 40 cm; almost strictly subtidal *Pisaster brevispinus*

23a Prevailing coloration of aboral surface some shade of red, often concentrated in blotches or bands that contrast with a light background; teeth on jaws of crossed pedicellariae essentially a series of small serrations, none of them fanglike; intertidal and subtidal
Orthasterias koehleri

23b Prevailing coloration of aboral surface dark brown, olive, or gray, without any obvious reddish tones; teeth on jaws of crossed pedicellariae long, the pair at the tip of each jaw fanglike; subtidal *Stylasterias forreri*

24a With 8-16 rays (except in very small individuals); without pedicellariae; body not flabby; diameter not exceeding 50 cm 25

24b With 20-24 rays (except in very small individuals); with pedicellariae; body flabby when separated from the substratum; diameter up to 80 cm *Pycnopodia helianthoides*

25a Paxillae of aboral surface rather widely separated and giving rise to long, slender spines of uneven length (thus the aboral surface appears prickly) *Crossaster papposus* (fig. 22.5)

25b Paxillae of aboral surface crowded, giving rise only to short, blunt projections (thus the aboral surface is rather smooth) 26

26a Diameter of disk usually about one-fourth the total diameter; aboral surface usually some shade of orange or pink, with a grayish blue streak radiating from the center of the disk to the tip of each ray; intertidal and subtidal *Solaster stimpsoni*

26b Diameter of disk usually about one-third the total diameter; aboral surface usually orange, brown, or gray, without conspicuous darker streaks radiating from the center of the disk 27

27a Rays (of which there are 8-10) typically narrowing rapidly just after leaving the base, thus becoming slender rather quickly; aboral surface usually orange; subtidal and extremely rare
Solaster paxillatus

27b Rays typically tapering rather evenly from the base; aboral surface usually orange, brown, or gray; common subtidal species 28

28a With 7-13 rays (usually 8-11); paxillae of aboral surface almost contiguous; aboral surface usually pale orange, sometimes orange-red *Solaster endeca*

28b With 8-16 rays (usually 11 or 12); paxillae of aboral surface separated from one another for a

distance about equal to their diameter; aboral surface usually gray or brown, but sometimes orange, and occasionally showing a mottled pattern in which there are 2 shades of brown
Solaster dawsoni

Species marked with an asterisk are not in the key. With few exceptions, they are restricted to deep water. Most of them are dealt with, however, in the handbook of Lambert (1981) and in the monographs of Fisher (1911, 1928, 1930) and D'iakonov (1950).

Order Platyasterida

Family Luidiidae

Luidia foliolata Grube, 1866 (fig. 22.1)

Order Paxillosida

Suborder Diplozonina

Family Astropectinidae

*Dipsacaster anoplus Fisher, 1910
*Dipsacaster borealis Fisher, 1910
*Dytaster gilberti Fisher, 1905
*Leptychaster anomalus Fisher, 1906
*Leptychaster arcticus (M. Sars, 1851)
*Leptychaster inermis (Ludwig, 1905)
Leptychaster pacificus Fisher, 1906
*Psilaster pectinatus (Fisher, 1905)
*Thrissacanthias penicillatus (Fisher, 1905)

Suborder Cribellina

Family Goniopectinidae

Ctenodiscus crispatus (Retzius, 1805)

Family Porcellanasteridae

*Eremicaster gracilis (Sladen, 1883)
*Eremicaster pacificus (Ludwig, 1905)

Suborder Notomyotina

Family Benthopectinidae

*Benthopecten acanthonotus Fisher, 1911
*Benthopecten claviger Fisher, 1910
*Benthopecten mutabilis Fisher, 1910
*Luidiaster dawsoni (Verrill, 1880)
*Nearchaster aciculosus (Fisher, 1910)
*Pectinaster agassizi subsp. *evoplus* (Fisher, 1910)

Order Valvatida

Suborder Granulosina

Family Goniasteridae

Ceramaster arcticus (Verrill, 1909)
**Ceramaster clarki* Fisher, 1910
**Ceramaster japonicus* (Sladen, 1889)
Ceramaster patagonicus (Sladen, 1889)
**Cryptopeltaster lepidonotus* Fisher, 1904
**Hippasteria californica* Fisher, 1905
Hippasteria spinosa Verrill, 1909 (fig. 22.3)
Mediaster aequalis Stimpson, 1857 (fig. 22.2)
**Pseudarchaster dissonus* Fisher, 1910
**Pseudarchaster parelii* (Düben & Koren, 1846)

Family Asterinidae

Asterina miniata (Brandt, 1835) (fig. 22.6). Long known as *Patiria miniata*.

Family Poraniidae

**Poraniopsis inflata* (Fisher, 1906)

Family Asteropseidae

Dermasterias imbricata (Grube, 1857)

Family Radiasteridae

Gephyreaster swifti (Fisher, 1905)

Order Spinulosida

Suborder Eugnathina

Family Solasteridae

**Crossaster borealis* Fisher, 1906
Crossaster papposus (Linnaeus, 1767) (fig. 22.5)
**Lophaster furcifer* subsp. *vexator* (Fisher, 1911)
**Lophaster furcilliger* Fisher, 1905
**Heterozonias alternatus* (Fisher, 1906)
Solaster dawsoni Verrill, 1880
Solaster endeca (Linnaeus, 1771)
Solaster paxillatus Sladen 1889
Solaster stimpsoni Verrill, 1880

Family Korethrasteridae

Peribolaster biserialis Fisher, 1905

Family Pythonasteridae

Asthenactis fisheri Alton, 1966

Family Pterasteridae

*?*Cryptaster* sp.
**Diplopteraster multipes* (M. Sars, 1865)
**Hymenaster perissonotus* Fisher, 1910
**Hymenaster quadrispinosus* Fisher, 1904
**Pteraster jordani* Fisher, 1904
**Pteraster militaris* (O. F. Müller, 1776)
Pteraster tesselatus Ives, 1888 (fig. 22.4)

Suborder Leptognathina

Family Echinasteridae

Henricia aspera Fisher, 1906
Henricia asthenactis Fisher, 1910
Henricia leviuscula (Stimpson, 1857)
Henricia longispina Fisher, 1910
**Henricia polyacantha* Fisher, 1911
Henricia sanguinolenta (O. F. Müller, 1776)
Henricia sp. Broods young under oral surface.

Order Forcipulatida

Suborder Asteriadina

Family Asteriidae

**Ampheraster chiroplus* Fisher, 1928
**Ampheraster marianus* (Ludwig, 1905)
**Anteliaster coscinactis* Fisher, 1928
**Anteliaster megatretus* Fisher, 1928
Evasterias troschelii (Stimpson, 1862)
Leptasterias hexactis (Stimpson, 1862)
**Lethasterias nanimensis* (Verrill, 1914)
Orthasterias koehleri (de Loriol, 1897)
**Pedicellaster magister* Fisher, 1923
Pisaster brevispinus (Stimpson, 1857)
Pisaster ochraceus (Brandt, 1835)
Pycnopodia helianthoides (Brandt, 1835)
**Rathbunaster californicus* Fisher, 1906
Stylasterias forreri (de Loriol, 1887)
**Tarsaster alaskanus* Fisher, 1928

Order Zorocallida

Family Zoroasteridae

Myxoderma platyacanthum (H. L. Clark, 1913)
Myxoderma sacculatum Fisher, 1904
Zoroaster evermanni Fisher, 1904
Zoroaster ophiurus Fisher, 1904

Order Euclasterida

Family Brisingidae

Astrocles actinodetus Fisher, 1917
Astrolirus panamensis (Ludwig, 1905)
Brisingella exilis (Fisher, 1928)
Craterobrisinga synaptoma Fisher, 1917
Freyella microplax (Fisher, 1917)
Freyellaster fecundus (Fisher, 1905)

Class Ophiuroidea

Fewer than half of the species of ophiuroids that have been reported from the region are likely to be collected at depths shallower than 200 m. Those restricted to deep water are omitted from the key. They are listed, however, and some of them can be identified with the help of the more nearly complete key of Kyte (1969) and the works of H. L. Clark (1911) and D'iakonov (1967).

Clark, H. L. 1911. North Pacific ophiurans in the collection of the United States National Museum. Bull. 75, U. S. Nat. Mus. xvi + 302 pp.
D'iakonov, A. M. 1967. [Ophiuroids of seas of the USSR.] Opredeliteli po Faune SSSR, 55. 136 pp. (in Russian).
Kyte, M. A. 1969. A synopsis and key to the recent Ophiuroidea of Washington State and southern British Columbia. J. Fish. Res. Bd. Canada, 26:1727-41.
------. 1982. *Ophiacantha abyssa*, new species, and *Ophiophthalmus displasia* (Clark), a suggested new combination in the ophiuroid family Ophiacanthidae (Echinodermata: Ophiuroidea) from off Oregon, USA. Proc. Biol. Soc. Wash., 95:505-8.

1a Arms repeatedly dichotomously branched; plates of arms and disk concealed by a thick skin; diameter of disk sometimes exceeding 5 cm; strictly subtidal
Gorgonocephalus eucnemis (fig. 22.8)
1b Arms not branched; plates of arms and disk usually distinct, although those on the disk may be small or partly obscured by spines; diameter of disk not exceeding 2 cm; intertidal or subtidal 2
2a Plates on upper surface of arms separated from one another by small supplementary plates; coloration usually with some bright red or purplish red, this typically concentrated in streaks or blotches
Ophiopholis aculeata
2b Plates on upper surface of arms not separated from one another by small supplementary plates; coloration usually not including bright red or purplish red streaks or blotches (in *Ophiura leptoctenia*, which is strictly subtidal, the aboral surface of the disk may be rose, violet, or red, with a few streaks; some other species are pinkish, but these are not likely to have prominent markings) 3
3a Aboral surface of disk with conspicuous marginal notches into which the arms are inserted, each notch bordered on both sides by an arm comb 4

456   Phylum Ecinodermata

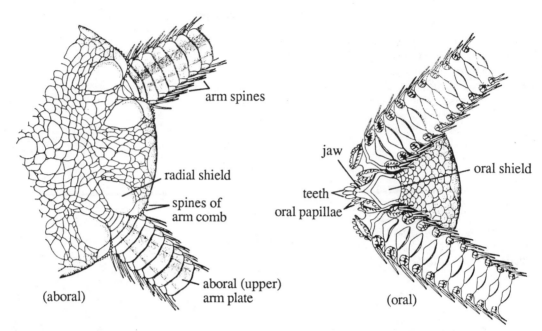

22.7 *Ophiura sarsi.* Terminology used in identification of ophiuroids.

22.8 *Gorgonocephalus eucnemis*, oral surface

3b	Aboral surface of disk without marginal notches into which the arms are inserted, and without arm combs	6
4a	Oral shields (the largest plates on the oral surface, situated between the bases of the arms) widest on the side nearer the margin of the disk; spines of arm combs at least 5 times as long as wide, with sharp tips; prevailing color of aboral surface of disk rose, violet, or red	*Ophiura leptoctenia*
4b	Oral shields widest near their middle portions; spines of arm combs usually not more than 3 times as long as wide, with blunt or truncate tips; prevailing color of aboral surface generally some shade of gray, bluish gray, or greenish gray, rarely reddish	5
5a	Spines of arm combs contiguous and with truncate tips	*Ophiura lütkeni*
5b	Spines of arm combs well separated and with bluntly rounded tips	*Ophiura sarsi* (fig. 22.7)
6a	With 4 oral papillae on both sides of each jaw	7
6b	With 3 oral papillae on both sides of each jaw	8
7a	Plates on upper surface of arms nearly circular, hardly wider than long, and barely in contact with one another even near the bases of the arms; interbrachial areas of oral surface of disk (the areas between the arm insertions) without scales	*Amphioplus strongyloplax*
7b	Plates on upper surface of arms nearly triangular (but with truncated corners), decidedly wider than long and in contact with one another at least near the bases of the arms; interbrachial areas of oral surface of disk with small scales	*Amphioplus macraspis*
8a	Oral papillae of the pair farthest from the tip of the jaw decidedly longer and wider than those of the other 2 pairs	9
8b	All 3 pairs of oral papillae about the same size	10
9a	Length of arms not over 4.5 times the diameter of the disk; longest arm spines usually shorter than the width of the arm units from which they originate; ovoviviparous (the young develop in the bursal pockets)	*Amphipholis squamata*
9b	Length of arms 7 or 8 times the diameter of the disk; longest arm spines longer than the width of the arm units from which they originate; oviparous (discharging eggs instead of brooding young in the bursal pockets)	*Amphipholis pugetana*
10a	Arm spines blunt, markedly flattened	*Amphiodia occidentalis*
10b	Arm spines pointed, either not flattened or only slightly flattened (oval in cross-section)	11
11a	Arm spines not at all flattened (circular in cross-section)	*Amphiodia urtica*
11b	Arm spines slightly flattened (oval in cross section)	*Amphiodia periercta*

Species marked with an asterisk are not in the key. Most of them are found only at depths greater than 200 m.

Order Phrynophiurida

Suborder Euryalina

Family Gorgonocephalidae

Gorgonocephalus eucnemis Müller & Troschel, 1842 (fig. 22.8)

Family Asteronychidae

*Asteronyx loveni Müller & Troschel, 1842

Family Asteroschematidae

*Asteroschema sublaeve Lütken & Mortensen, 1899

Suborder Ophiomyxina

Family Ophiomyxidae

*Ophioscolex corynetes (H. L. Clark, 1911)

Order Ophiurida

Suborder Laemophiurina

Family Ophiacanthidae

*Ophiacantha abyssa Kyte, 1982
*Ophiacantha bathybia H. L. Clark, 1911
*Ophiacantha rhacophora H. L. Clark, 1911
*Ophiacantha trachybactra (H. L. Clark, 1911)
*Ophiolimna bairdi (Lyman, 1883)
*Ophiophthalmus cataleimmoidus (H. L. Clark, 1911)
*Ophiophthalmus displasia (H. L. Clark, 1911)
*Ophiophthalmus eurypoma H. L. Clark, 1911
*Ophiophthalmus normani (Lyman, 1879)

Suborder Gnathophiurina

Family Ophiactidae

Ophiopholis aculeata (Linnaeus, 1767)
*Ophiopholis bakeri McClendon, 1909
*Ophiopholis longispina H. L. Clark, 1911

Family Amphiuridae

*Amphilepas patens Lyman, 1879
Amphiodia occidentalis (Lyman, 1860)
Amphiodia periercta H. L. Clark, 1911
Amphiodia urtica (Lyman, 1860)
Amphioplus macraspis (H. L. Clark, 1911)
Amphioplus strongyloplax (H. L. Clark, 1911)
Amphipholis pugetana (Lyman, 1860)
Amphipholis squamata (delle Chiaje, 1829)
*Amphiura carchara H. L. Clark, 1911
*Amphiura diomedeae Lütken & Mortensen, 1899
*Amphiura polyacantha Lütken & Mortensen, 1899
*Dougaloplus gastracanthus (Lütken & Mortensen, 1899)

Family Ophiocomidae

*Ophiopteris papillosa (Lyman, 1875)

Suborder Chilophiurina

Family Ophioleucidae

Ophioleuce oxycraspedon Baranova, 1954

Family Ophiuridae

Amphiophiura bullata subsp. *pacifica* Litvinova, 1971
Amphiophiura ponderosa (Lyman, 1878)
Amphiophiura superba Lütken & Mortensen, 1899
Ophiomusium glabrum Lütken & Mortensen, 1899
Ophiomusium jolliensis McClendon, 1909
Ophiomusium lymani Wyville Thompson, 1873
Ophiura bathybia H. L. Clark, 1911
Ophiura cryptolepas H. L. Clark, 1911
Ophiura flagellata (Lyman, 1878)
Ophiura irrorata (Lyman, 1878)
Ophiura leptoctenia H. L. Clark, 1911
Ophiura lütkeni (Lyman, 1860)
Ophiura sarsi Lütken, 1855 (fig. 22.7)

Class Echinoidea

Jensen, M. 1974. The Strongylocentrotidae (Echinoidea), a morphologic and systematic study. Sarsia, 57:113-48.

McCauley, J. E., & A. G. Carey, Jr. 1967. Echinoidea of Oregon. J. Fish. Res. Bd. Canada, 24:1385-1401.

Mironov, A. N. 1973. [New deep-sea species of sea urchins of the genus *Echinocrepis* and the distribution of the family Pourtalesiidae (Echinoidea, Meridosternina).] Trudy Inst. Okeanol., 91: 240-7. (in Russian).

------. 1975. [Deep-sea urchins (Echinodermata, Echinoidea) collected during the 14th cruise of the r/v 'Akademik Kurchatov.'] Trudy Inst. Okeanol., 103:281-8 (in Russian, with English summary).

Mortensen, T. 1928-51. A Monograph of the Echinoidea. 5 vols. København: C. A. Reitzel.

Swan, E. F. 1953. The Strongylocentrotidae (Echinoidea) of the northeast Pacific. Evolution, 7:269-73.

1a A sand dollar--body disk-shaped, its diameter several times greater than the thickness *Dendraster excentricus*
1b Body not disk-shaped, its diameter not more than twice the thickness 2
2a A heart urchin--the body nearly heart-shaped when viewed from the oral or aboral side; ambulacra in distinct furrows, one of which is much deeper than the others; symmetry essentially bilateral, the mouth at one end of the oral surface (close to the deepest ambulacral furrow) and the anus at the margin of the test on the opposite side *Brisaster latifrons*
2b Sea urchins--body tomato-shaped, circular in outline when viewed from the oral or aboral side; ambulacra not in distinct furrows; symmetry almost perfectly radial, the mouth at the center of the oral surface and the anus at the center of the aboral surface 3
3a Prevailing general color greenish or whitish, definitely not red, purple, or orange-pink 4
3b Prevailing color red, purple, or orange-pink 5
4a Prevailing color of spines whitish, although some, especially near the center of the aboral surface, may be greenish or reddish; tube feet usually either lighter than the spines or about the same color, rarely much darker; diameter of test not often greater than 6 cm (but sometimes nearly 8 cm); cleaned test mostly white, but with light green to reddish tints in the

region of the periproct; primary spines with 17-26 lengthwise wedges; usually with 6 or 7 pore pairs in each arc; oral spines white, with reddish brown or brown tips, the color persisting for several days after specimens have been preserved in formalin; strictly subtidal and usually at depths greater than 30 m ..... *Strongylocentrotus pallidus*

4b Prevailing color of spines pale green; tube feet darker than the spines, and usually purple; diameter of test frequently more than 6 cm (maximum about 8 cm); cleaned test decidedly greenish; primary spines with 26-36 lengthwise wedges; with 5 or 6 pore pairs in each arc; oral spines slightly purplish, sometimes with white tips, becoming distinctly purple after specimens have been preserved in formalin; intertidal and subtidal
..... *Strongylocentrotus droebachiensis*

5a Prevailing color of living animal orange-pink; cleaned test pale orange-pink, extremely fragile; subtidal and mostly at depths greater than 100 m ..... *Allocentrotus fragilis*

5b Prevailing color of living animal red or purple; cleaned test usually gray or pale purple, not extremely fragile; intertidal and subtidal 6

6a Prevailing color usually bright red, reddish purple, or maroon, although the larger spines of lighter individuals may be rose and the smaller spines may be almost white; diameter of test up to about 15 cm; spines up to 7 cm long; intertidal and subtidal
..... *Strongylocentrotus franciscanus*

6b Prevailing coloration purple; diameter of test not exceeding 9 cm; spines rarely more than 2.5 cm long; largely intertidal on rocky shores that have considerable wave action, but subtidal to some extent ..... *Strongylocentrotus purpuratus*

Species marked with an asterisk are not in the key. Nearly all of them are restricted to very deep water, and some are primarily northern. A few have not yet been recorded from Oregon, Washington, or British Columbia, but their known distribution indicates that they should be expected.

Subclass Euechinoidea

Order Echinothurioida

Family Echinothuriidae

*ractical*Sperosoma biseriatum* Döderlein, 1901
**Sperosoma giganteum* A. Agassiz & H. L. Clark, 1907

Order Echinoida

Family Strongylocentrotidae

Allocentrotus fragilis (Jackson, 1912)
Strongylocentrotus droebachiensis (O. F. Müller, 1776)
Strongylocentrotus franciscanus (A. Agassiz, 1863)
Strongylocentrotus pallidus (G. O. Sars, 1871)
Strongylocentrotus purpuratus (Stimpson, 1857)

Order Clypeasteroida

Suborder Scutellina

Family Dendrasteridae

Dendraster excentricus (Eschscholtz, 1831)

Order Holasteroida

Family Urechinidae

*Urechinus loveni (A. Agassiz, 1898)

Family Pourtalesiidae

*Ceratophysa valvaecristata Mironov, 1975
*Cystocrepis setigera (A. Agassiz, 1898)
*Echinocrepis rostrata Mironov, 1973
*Pourtalesia tanneri A. Agassiz, 1898
*Pourtalesia thomsoni Mironov, 1975
*Rodocystis rosea (A. Agassiz, 1879)

Order Spatangoida

Suborder Hemiasterina

Family Schizasteridae

Brisaster latifrons (A. Agassiz, 1898)

Family Aeropsidae

*Aeropsis fulva (A. Agassiz, 1898)

Class Holothuroidea

Clark, H. L. 1901. The holothurians of the Pacific coast of North America. Zool. Anz., 24:162-71.
------. 1907. The apodous holothurians--A monograph of the Synaptidae and Molpadidae. Smithson. Contrib. Knowl., 35:1-231.
------. 1922. The holothurians of the genus Stichopus. Bull. Mus. Comp. Zool. Harvard, 65:39-74.
------. 1924. Some holothurians from British Columbia. Can. Field Nat., 38:54-7.
Deichmann, E. 1937. The Templeton Crocker Expedition. IX. Holothurians from the Gulf of California, the west coast of Lower California, and Clarion Island. Zoologica, 22:161-76.
------. 1938. New holothurians from the western coast of North America, and some remarks on the genus Caudina. Proc. New England Zool. Club, 16:103-15.
Edwards, C. L. 1907. The holothurians of the North Pacific coast of North America collected by the Albatross in 1903. Proc. U. S. Nat. Mus., 33:49-68.
Heding, S. G. 1928. Synaptidae. Papers from Dr. Th. Mortensen's Pacific Expedition 1914-16, no. 46. Vidensk. Medd. Dansk Naturhist. Foren. København, 85:105-324.
Lambert, P. 1984. British Columbia Marine Faunistic Survey Report: Holothurians from the Northeast Pacific. Can.Tech. Rep. Fish. Aquatic Sci., no. 1234. 30 pp.
------. 1986. Northeast Pacific holothurians of the genus Parastichopus with a description of a new species, Parastichopus leukothele (Echinodermata). Can. J. Zool., 64:2266-72.

1a Upper surface of body with numerous large, fleshy, pointed projections; buccal tentacles moplike; length commonly greater than 25 cm ... 2
1b Upper surface of body without fleshy, pointed projections; buccal tentacles bushy, pinnately branched, or unbranched, but not moplike; length rarely exceeding 25 cm ... 3

22.9 *Parastichopus californicus*

22.10 *Cucumaria piperata*

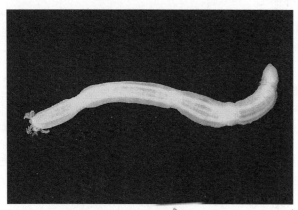

22.11 *Leptosynapta clarki*

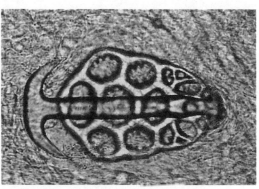

22.12 *Leptosynapta clarki*, ossicles of body wall (the anchor-shaped ossicle protrudes externally)

2a Color usually reddish brown or orange-brown, sometimes mottled; length up to about 50 cm; primarily subtidal, but also low intertidal *Parastichopus californicus* (fig. 22.9)

2b Color reddish orange with small white papillae and rust-colored patches; length up to about 38 cm; strictly subtidal *Parastichopus leukothele*

3a Without tube feet other than the buccal tentacles; buccal tentacles either unbranched or pinnately branched 4

3b With tube feet on at least part of the general body surface; buccal tentacles bushy 7

4a Buccal tentacles unbranched, fingerlike; body wall thick, without anchor-shaped ossicles that project externally; aboral end of the body tapering to a "tail" 5

4b Buccal tentacles pinnately branched; body wall thin, with anchor-shaped ossicles that project externally (these ossicles are microscopic, but they engage one's skin when the animal is handled); body slender, but not tapering to a "tail" 6

5a Body elongated, with a long tail; color generally pinkish, purplish, or silvery gray; intertidal or subtidal, in rather clean sand *Paracaudina chilensis*

5b Body stout, with a short tail; color mostly reddish brown or reddish purple; subtidal, in mud *Molpadia intermedia*

6a Usually with 12 (sometimes 11 or 13) buccal tentacles; body surface with raised papillae; color almost uniformly pale pink; average diameter of oocytes about 400 μm; primarily

	intertidal in muddy sand and gravel *Leptosynapta clarki* (figs. 22.11 and 22.12)
6b	Usually with 10 (sometimes 9 or 11) buccal tentacles; body surface without raised papillae; color white with dark red spots (more spots on the upper surface and near the anterior end than elsewhere); shallow subtidal *Leptosynapta transgressor*
	L. roxtona Heding, 1928, inadequately described, is doubtfully distinct.
7a	Lower surface of body a flattened "sole" with 3 double rows of tube feet; upper surface covered by large calcareous plates — 8
7b	Lower surface of body not flattened to form a "sole;" upper surface not covered by calcareous plates — 10
8a	Calcareous plates on upper surface with small nodules readily visible at 10x magnification (caution: the nodules, to which the sandpaperlike texture of the plates is due, break off readily if scraped); greatest diameter of plates not often exceeding 5 mm — 9
8b	Calcareous plates on upper surface without small nodules; greatest diameter of plates in larger specimens (at least 5 cm long) sometimes exceeding 1 cm, and usually at least 5 mm in smaller specimens (general color reddish brown; tentacles red; common shallow water species, and sometimes intertidal) *Psolus chitonoides*
9a	Length up to 3 cm; general color pinkish or lavender; greatest diameter of calcareous plates about 3 mm (common subtidal species) *Psolidium bullatum*
9b	Length up to 6 cm; general color not pinkish or lavender; greatest diameter of calcareous plates in larger specimens (3-6 cm long) about 5 mm *Psolus squamatus*
10a	Length commonly 20-25 cm, but sometimes attaining 35 cm; general body color usually reddish brown, sometimes with a purplish tinge, but the buccal tentacles bright orange or orange-brown, and the introvert about the same color; common intertidal and subtidal species, usually in crevices and between rocks *Cucumaria miniata*
	See note under *C. fallax*, choice 16a.
10b	Length rarely exceeding 20 cm (and in most species, other than *Havelockia benti* and *Cucumaria fallax*, not often exceeding 10 cm); general body color not reddish brown, and the buccal tentacles and introvert not bright orange or orange-brown (they may, however, be pale orange or pale brown) — 11
11a	General body color (the buccal tentacles and introvert may be different) white, cream, yellowish, pale orange, pinkish, brown, or tan, but neither dark nor conspicuously speckled with dark pigment (*Havelockia benti*, whose color ranges from pale orange to brown, often has small dots of brown pigment, but these contribute to a nearly uniform coloration rather than to conspicuous speckles) — 12
11b	Body either dark (often blackish) or conspicuously speckled with dark pigment — 19
12a	Body tapered at both ends (the posterior end may look like a tail) and curved at least slightly (sometimes the body is nearly U-shaped) — 13
12b	Body not obviously tapered at either end and not curved — 15
13a	Skin with the texture of fine sandpaper; perforated ossicles of body wall triangular to oval (length up to 10 cm, but commonly 5 or 6 cm) *Pentamera pseudocalcigera*
13b	Skin rather smooth to the touch, not at all like sandpaper; perforated ossicles of body wall diamond-shaped, circular, or star-shaped — 14
14a	Body plump, decidedly curved, and with the posterior portion tapering quickly to a nipplelike tail; perforated ossicles of body wall circular to star-shaped and with an obvious central spine *Pentamera populifera*
14b	Body not especially plump, and only slightly curved, the posterior portion not tapering so quickly that there is a tail; most ossicles of the body wall diamond-shaped, with 4 large perforations in the central portion (there are also a few smaller, circular plates that have minute spinelets on one surface) *Pentamera lissoplaca*
15a	Tube feet not restricted to the ambulacra; general body color light brown, but sometimes pale orange; length up to 20 cm *Havelockia benti*
15b	Tube feet restricted to the ambulacra (although the rows of tube feet may be irregular); general body color white, yellowish, pale orange, pale brown, or tan; length not often exceeding 10 cm (except in *Cucumaria fallax*, whose length may reach 15 cm) — 16
16a	General body color pale brown, tan, or yellowish; length up to 15 cm; body slimy when the animal is alive (introvert white or whitish; buccal tentacles faintly pinkish or yellowish; subtidal, mostly under rocks in cobble) *Cucumaria fallax*

This species could be confused with small specimens of *C. miniata*, choice 10a, although the latter is not so slimy as *C. fallax*, and its tube feet are proportionately thicker. The eggs of *C. fallax* are tan, whereas those of *C. miniata* are green.

16b General body color white to pale orange (not brown or tan); length rarely exceeding 10 cm; body not slimy 17

17a Length not often more than 3 cm; perforated ossicles of body wall with numerous knobs and spines on one surface (rare subtidal species) *Pentamera trachyplaca*

17b Length commonly 5-10 cm; perforated ossicles of body wall without knobs and spines (common intertidal and subtidal species) 18

18a Tube feet slender (when the animal is relaxed), their length usually less than the width of the interambulacral areas; color nearly pure white; body wall thin; animal often covering itself with pieces of algae, bits of shell, and other foreign material *Eupentacta pseudoquinquesemita*

18b Tube feet rather stout (when animal is relaxed), their length usually greater than the width of the interambulacral areas; color white to pale yellow or pale orange; body wall tough; animal not often covering itself with foreign matter *Eupentacta quinquesemita*

19a Prevailing color whitish or yellowish, with purple, brown, or black speckles *Cucumaria piperata* (fig. 22.10)

19b Color almost uniformly grayish, grayish brown, or blackish (albinistic specimens occur, however) 20

20a With many tube feet between the ambulacra; ossicles of body wall button-shaped, with 4 holes; length up to about 5 cm; subtidal *Cucumaria lubrica*

20b With few if any tube feet between the ambulacra; ossicles of body wall neither button-shaped nor with 4 holes; length up to about 2.5 cm; intertidal, and usually in beds of *Mytilus californianus* *Cucumaria pseudocurata*

Species marked with an asterisk are not in the key. All of them are subtidal, and most are restricted to deep water. Some have been collected in the region only once or twice.

Order Dendrochirotida

Family Psolidae

Psolidium bullatum Ohshima, 1915
Psolus chitonoides H. L. Clark, 1901
Psolus squamatus (Koren, 1844)

Family Sclerodactylidae

Eupentacta pseudoquinquesemita Deichmann, 1938
Eupentacta quinquesemita (Selenka, 1867)

Family Phyllophoridae

Pentamera lissoplaca (H. L. Clark, 1924)
Pentamera populifera (Stimpson, 1864)
Pentamera pseudocalcigera Deichmann, 1938
Pentamera trachyplaca (H. L. Clark, 1924)
*Pentamera spp. At least 2 species, probably undescribed.
Havelockia benti (Deichmann, 1937). Generally called *Thyone benti*.

Family Cucumariidae

Abyssocucumis albatrossi Cherbonnier, 1947
Cucumaria fallax Ludwig, 1894
**Cucumaria frondosa* subsp. *japonica* Mortensen, 1932
Cucumaria lubrica H. L. Clark, 1901
Cucumaria miniata (Brandt, 1835)
Cucumaria piperata (Stimpson, 1864) (fig. 22.10)
Cucumaria pseudocurata Deichmann, 1938
*Cucumaria vegae Théel, 1886
**Duasmodactyla commune* (Forbes, 1852)

Order Dactylochirotida

Family Ypsilothuridae

**Sphaerothuria bitentaculata* Ludwig, 1893

Order Aspidochirotida

Family Stichopodidae

Parastichopus californicus (Stimpson, 1857) (fig. 22.9)
Parastichopus leukothele Lambert, 1986

Family Synallactidae

**Mesothuria murrayi* (Théel, 1886)
*Paelopatides sp.
**Pseudostichopus mollis* Théel, 1886
**Pseudostichopus nudus* Ohshima, 1915
**Pseudostichopus villosus* Théel, 1886
**Synallactes gilberti* Ohshima, 1915
**Zygothuria lactea* (Théel, 1886)

Order Elasipodida

Family Deimatidae

**Oneirophanta mutabilis* Théel, 1879

Family Laetmogonidae

**Laetmogone wyvillethompsoni* Théel, 1882
**Laetmophasma fecundum* Ludwig, 1894
**Pannychia moseleyi* Théel, 1882

Family Elpidiidae

*Amperima rosea (Perrier, 1896)
*Amperima sp. Similar to *A. naresi* (Théel, 1882)

Capheira mollis Ohshima, 1915
Peniagone gracilis (Ludwig, 1894)
*Peniagone sp.
Scotoplanes clarki Hansen, 1975
Scotoplanes globosa Théel, 1879

Family Psychropotidae

Benthodytes sanguinolenta Théel, 1882
Benthodytes incerta Théel, 1893
Psychropotes longicaudata Théel, 1882
Psychropotes raripes Ludwig, 1893

Order Apodida

Family Synaptidae

Leptosynapta clarki Heding, 1928 (figs. 22.11 and 22.12)
Leptosynapta roxtona Heding, 1928
Leptosynapta transgressor Heding, 1928
Protankyra duodactyla H. L. Clark, 1907
Protankyra pacifica Ludwig, 1894

Family Chiridotidae

Chiridota albatrossii Edwards, 1907
Chiridota nanaimensis Heding, 1928

Family Myriotrochidae

Myriotrochus bathibius H. L. Clark, 1920
Myriotrochus giganteus H. L. Clark, 1920
*Myriotrochus sp.

Order Molpadiida

Family Molpadiidae

Ceraplectana trachyderma H. L. Clark, 1908
Molpadia borealis M. Sars, 1859
Molpadia granulosa Ludwig, 1893
Molpadia intermedia (Ludwig, 1894)
Molpadia musculus Risso, 1826

Family Caudinidae

Paracaudina chilensis (J. Müller, 1850)

23

PHYLA UROCHORDATA, HEMICHORDATA, CHAETOGNATHA

PHYLUM UROCHORDATA

Class Ascidiacea

Charles C. Lambert, Gretchen Lambert, and Eugene N. Kozloff

Abbott, D. P., & W. Trason. 1968. Two new colonial ascidians from the west coast of North America. Bull. So. Calif. Acad. Sci., 67:143-53.
Berrill, N. J. 1950. The Tunicata. With an Account of the British Species. London: Ray Society. iii + 354 pp.
Huntsman, A. G. 1912a. Ascidians from the coasts of Canada. Trans. Canad. Inst., 1911:111-48.
------. 1912b. Holosomatous ascidians from the coast of western Canada. Contrib. Canad. Biol., 1906-10:103-85.
Lambert, G., C. C. Lambert, & D. P. Abbott. 1981. *Corella* species in the American Pacific NW: distinction of *C. inflata* Huntsman, 1912 from *C. willmeriana* Herdman, 1898 (Ascidiacea, Phlebobranchia). Can. J. Zool., 59:1493-1504.
Newberry, A. T. 1984. *Dendrodoa* (*Styelopsis*) *abbotti*, sp. nov. (Styelidae, Ascidiacea) from the Pacific coast of the United States, and its impact on some gonadal criteria of its genus and subgenus. Proc. Calif. Acad. Sci., 43:239-48.
Ritter, W. E. 1900. Some ascidians from Puget Sound, collections of 1896. Ann. New York Acad. Sci., 12:589-616.
------. 1913. The simple ascidians from the northeastern Pacific in the collection of the United States National Museum. Proc. U. S. Nat. Mus., 45:427-505.
Ritter, W. E., & R. A. Forsyth. 1917. Ascidians from the littoral zone of southern California. Univ. Calif. Publ. Zool., 16:439-512.
Tokioka, T. 1963. The outline of the Japanese ascidian fauna as compared with that of the Pacific coast of North America. Publ. Seto Mar. Biol. Lab., 11:131-56.
------. 1967. Pacific Tunicata of the United States National Museum. Bull. 251, U. S. Nat. Mus. 247 pp.
Van Name, W. G. 1945. The North and South American ascidians. Bull. Amer. Mus. Nat. Hist., 84:1-476.

1a	Solitary ascidians (not reproducing by budding) or social ascidians (reproducing by budding and connected, at least initially, by stolons or sheets of tissue and tunic material), but not embedded in a common tunic	2
1b	Compound ascidians, the several to many zooids of the colony embedded in a common tunic, which is usually gelatinous in texture	33
2a	Solitary ascidians (if the animals are in clusters, it is because they have settled on one another, not because they have reproduced by budding); individuals usually more than 1 cm in diameter	3
2b	Social ascidians (members of an aggregation connected, at least initially, by stolons or sheets of tissue and tunic material); members usually less than 1 cm in diameter	28
3a	Tunic transparent or translucent (in *Molgula pacifica*, however, considerable foreign material is incorporated into the translucent tunic)	4

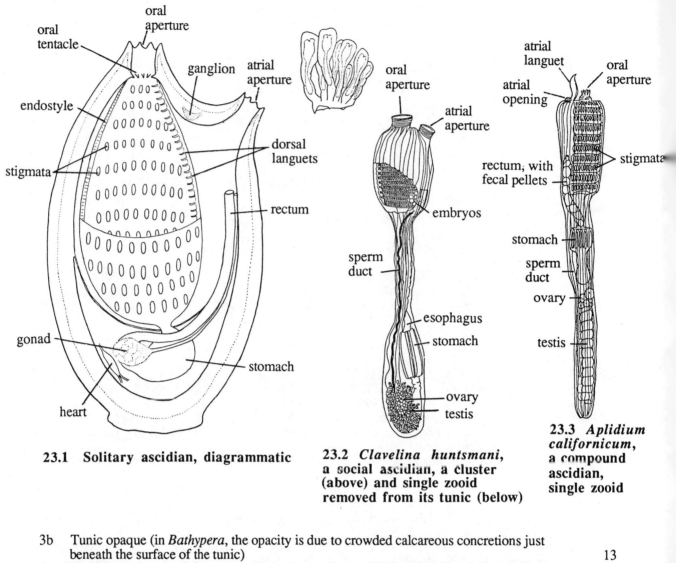

23.1 Solitary ascidian, diagrammatic

23.2 *Clavelina huntsmani*, a social ascidian, a cluster (above) and single zooid removed from its tunic (below)

23.3 *Aplidium californicum*, a compound ascidian, single zooid

3b	Tunic opaque (in *Bathypera*, the opacity is due to crowded calcareous concretions just beneath the surface of the tunic)	13
4a	Both the oral and atrial apertures situated on a flattened disk consisting of several distinct plates	5
4b	Oral and atrial apertures not situated on a flattened disk that consists of several distinct plates	6
5a	Plates of disk usually showing concentric growth lines; muscle strands connecting the 2 central plates of the disk not visible through the tunic; no intermediary plates between the central and marginal plates; diameter of disk often exceeding 1.5 cm; intertidal, subtidal, and on floats .... *Chelyosoma productum* (fig. 23.8)	
5b	Plates of disk without concentric growth lines; muscle strands connecting the two central plates of the disk visible through the tunic in preserved specimens; 1-3 intermediary plates between the central and marginal plates; diameter of disk rarely exceeding 1.5 cm; subtidal .... *Chelyosoma columbianum* (fig. 23.9)	
6a	Not permanently attached to a firm substratum; tunic covered with short, hairlike projections .... *Molgula pugetiensis*	
6b	Attached to a firm substratum; tunic smooth or with papillae, but without hairlike projections	7
7a	Body low, somewhat flattened	8
7b	Body taller than wide	9
8a	Atrial aperture usually near the end of the anterior third of the body; with about 25-35 oral tentacles around the opening to the pharynx; common subtidally and on floats, and	

	occasionally found intertidally	*Ascidia callosa*
8b	Atrial aperture usually near the middle of the body; with about 150-200 oral tentacles around the opening to the pharynx; strictly subtidal in our region (in California, found also intertidally, and often common on floats and pilings)	*Ascidia ceratodes*
9a	Tunic with considerable foreign material, even pieces of algae, incorporated into it; atrial siphon about twice as long as the oral siphon; both siphons orange-red or pinkish red; height up to about 2 cm (open coast, intertidal and subtidal)	*Molgula pacifica*
9b	Tunic generally free of foreign material; oral and atrial siphons about the same length; siphons not some shade of red; height of mature specimens usually more than 2 cm	10
10a	Tunic with scattered large papillae	*Ascidia paratropa*
10b	Tunic smooth or irregularly wrinkled	11
11a	Body much taller than wide; tunic transparent, but tinted yellow-green; longitudinal muscle bands distinctly visible beneath the tunic; oral and atrial apertures borne on short siphons	*Ciona intestinalis*
11b	Body not much taller than wide; tunic transparent and colorless; longitudinal muscle bands not obvious beneath the tunic; oral and atrial apertures not borne on distinct siphons	12
12a	Rectum less than half the height of the body; a portion of the atrium expanded into a pocket in which embryos are brooded	*Corella inflata*
12b	Rectum more than three-fourths the height of the body; atrium not expanded into a pocket for brooding embryos (this species does not brood)	*Corella willmeriana*
13a	Surface of tunic underlain by a stratum of conspicuous, closely spaced calcareous concretions (body usually not as tall as wide; diameter up to about 4 cm, but generally much smaller; color usually grayish white, sometimes pinkish; subtidal)	*Bathypera ?ovoida*
	B. ovoida was described from southern California. The *Bathypera* of our region may be a separate species.	
13b	Surface of tunic not underlain by a stratum of conspicuous calcareous concretions	14
14a	Tunic covered with spinelike projections (these are usually branched)	15
14b	Tunic smooth or wrinkled, but without spinelike projections	17
15a	Body attached to the substratum by a distinct stalk; spinelike projections generally branched, but the branches not arranged in circles (low intertidal and subtidal; common on floats in some parts of our region)	*Boltenia villosa* (fig 23.6)
15b	Body not attached to the substratum by a distinct stalk; branches of spinelike projections arranged in one or more circles	16
16a	Spinelike projections almost completely obscuring the rest of the tunic, each projection encircled by several rings of recurved, thornlike secondary spinelets; height up to about 10 cm	*Halocynthia igaboja*
16b	Spinelike projections not obscuring the rest of the tunic, each projection with a single irregular circle of several (usually 4-8) secondary spinelets; height not often exceeding 4 cm	*Boltenia echinata* (fig. 23.11)
17a	Both the oral and atrial apertures situated on a flattened disk that consists of several distinct plates	18
17b	Oral and atrial apertures not situated on a flattened disk that consists of several distinct plates	19
18a	Plates of disk usually showing concentric growth lines; muscle strands connecting the 2 central plates not visible through the tunic; no intermediary plates between the central and marginal plates; diameter of disk often exceeding 1.5 cm; low intertidal and subtidal, common on floats	*Chelyosoma productum* (fig. 23.8)
18b	Plates of disk without concentric growth lines; muscle strands connecting the 2 central plates visible through the tunic in preserved specimens; 1-3 intermediary plates between the central and marginal plates; diameter of disk rarely exceeding 1.5 cm; subtidal	*Chelyosoma columbianum* (fig. 23.9)
19a	Body elongated, attached to the substratum by a rather small area of the tunic, and sometimes with a distinct stalk	20
19b	Body not elongated, attached to the substratum by a rather large area of the tunic	22
20a	With a slender, furrowed stalk that is about half the total height of the body; restricted to the open coast	*Styela montereyensis* (fig. 23.4)
20b	Without a slender, furrowed stalk; not restricted to the open coast	21

470 Phylum Urochordata

23.4 *Styela montereyensis*

23.5 *Pyura haustor*

23.6 *Boltenia villosa*

23.7 *Metandrocarpa taylori*

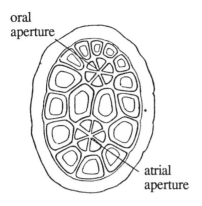
23.8 *Chelyosoma productum*

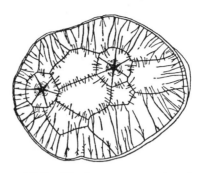
23.9 *Chelyosoma columbianum*

23.10 *Pyura mirabilis*

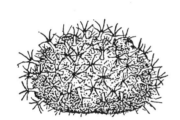
23.11 *Boltenia echinata*

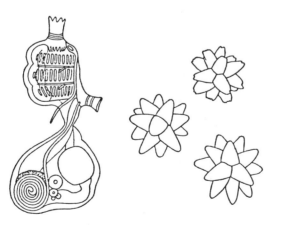
23.12 *Tridemnum opacum*, a zooid and calcareous concretions

23.13 *Didemnum albidum*, calcareous concretions

21a Body barrel-shaped; tunic peach-colored, smooth or with inconspicuous furrows and wrinkles; height up to about 6 cm .... *Halocynthia aurantium*
21b Body elongate, cylindrical, shaped much like a short cucumber; tunic brown or brownish red, with conspicuous lengthwise wrinkles; height not often more than 4 cm .... *Styela gibbsii*
22a Body broadened in such a way that the oral and atrial apertures are widely separated and on opposite sides of the body .... *Pyura mirabilis* (fig. 23.10)
22b Body not broadened in such a way that the oral and atrial apertures are widely separated and on opposite sides of the body 23
23a Entire animal bright pinkish red or orange-red; tunic smooth and shiny .... *Cnemidocarpa finmarkiensis*
23b Animal not entirely red (if some bright red coloration is present, it is restricted to the areas surrounding the oral and atrial apertures); tunic usually wrinkled to at least some extent, not so smooth as to be shiny 24
24a Body taller than wide; siphons red (tunic tough, leathery, coarsely wrinkled, mostly orange-brown; height up to about 5 cm; common intertidal and subtidal species) .... *Pyura haustor* (fig. 23.5)
24b Body low, approximately hemispherical or appreciably flatter than hemispherical; siphons not red 25
25a Greatest diameter about 1.2 cm; animal, when relaxed, about one-fourth as high as its greatest diameter, but contracting to a nearly flat disk when disturbed; siphonal apertures closing down to simple slits when disturbed (color mostly translucent gray, tinted with ochre or brownish pink) .... *Dendrodoa abbotti*
25b Greatest diameter often 2 cm or more; animal, when relaxed, approximately hemispherical, not capable of contracting to a nearly flat disk when disturbed; siphonal apertures not closing down to simple slits when disturbed 26
26a Attached to the substratum by the ventral side of the body; tunic finely wrinkled and studded with small, translucent tubercles; tunic dull brown .... *Styela coriacea*
26b Attached to the substratum by the left side of the body, or partly by the left side and partly by the ventral side; tunic rather smooth, but often encrusted with foreign material or small organisms; tunic nearly colorless to light brown or greenish 27
27a Atrial aperture usually near the end of the anterior third of the body; with about 25-35 oral tentacles around the opening to the pharynx; common subtidally and on floats, and occasionally intertidal .... *Ascidia callosa*
27b Atrial aperture usually near the middle of the body; with about 150-200 oral tentacles around the opening to the pharynx; strictly subtidal in our region (in California, found also intertidally, and often common on floats and pilings) .... *Ascidia ceratodes*
28a Individuals not much, if any, taller than wide 29
28b Individuals much taller than wide 31
29a Individuals translucent, greenish, attached to the substratum by a small portion of the tunic .... *Perophora annectens*
29b Individuals opaque and usually orange-red or brick-red (sometimes yellowish), attached to the substratum by a considerable portion of the tunic 30
30a Contiguous individuals joined by slender stolons or by thin sheets of tunic material, the connections often indistinct (on rocks, intertidal and subtidal) .... *Metandrocarpa taylori* (fig. 23.7)
30b Contiguous individuals joined to the extent that they nearly resemble zooids of a compound ascidian (usually on kelp, but also on rocks, intertidal and subtidal) .... *Metandrocarpa dura*
31a Individuals widely separated, less than 1 cm tall when mature; pharynx yellow-orange .... *Pycnoclavella stanleyi*
31b Individuals clustered, usually more than 1 cm tall when mature; pharynx if pigmented, not yellow-orange (caution: the larvae are yellow) 32
32a Dorsal lamina and endostyle bright fluorescent pink; margins of oral and atrial apertures not lobed .... *Clavelina huntsmani* (fig. 23.2)
32b Dorsal lamina and endostyle not fluorescent pink; margin of oral aperture irregularly lobed, the atrial aperture with 2 lips, each divided into 3 small lobes .... *Clavelina* sp.
33a Colonies forming thin encrusting sheets, thick encrusting slabs (often irregular and lobed), or low mounds attached for most of their diameter to the substratum (the colony may be very lumpy, but it does not consist of club-shaped, mushroomlike, leaflike, or globular masses

	that have distinct stalks)	34
33b	Colonies forming club-shaped, mushroomlike, leaflike, or globular masses, usually with distinct stalks (even when the masses are moundlike, they have narrow stalks)	43
34a	Colony forming encrustations not more than 3 mm thick and not lumpy (in *Diplosoma macdonaldi*, however, the common atrial apertures are on prominent elevations)	35
34b	Colony forming cakes, low mounds, or slabs (these often irregular, lobed, and lumpy) 5 mm to 1 cm or more in thickness	37
35a	Colony transparent, colorless, gray, olive, or tan, up to about 3 mm thick; tunic soft and gelatinous, lacking calcareous concretions (but it may have scattered flecks of white pigment) *Diplosoma macdonaldi*	
35b	Colony opaque, whitish, only about 1 or 2 mm thick; tunic firm, containing many-rayed, globular calcareous concretions (the degree of firmness and whiteness of the tunic varies according to the density of the concretions)	36
36a	Pharynx with 4 rows of stigmata; calcareous concretions with stubby rays that are not as much as one-fourth the total diameter of the concretions and that usually have bluntly rounded tips *Didemnum albidum* (fig. 23.13) In *D. carnulentum*, which may also occur in our region (it is primarily southern), the rays of the concretions, though less than one-fourth the total diameter of the concretions, are generally pointed rather than bluntly rounded.	
36b	Pharynx with 3 rows of stigmata; calcareous concretions with rays that are about one-fourth to one-third the total diameter of the concretions and that have pointed tips *Trididemnum opacum* (fig. 23.12) In *T. strangulatum*, which probably occurs in our region, the concretions resemble those of *Didemnum albidum* in having short, bluntly rounded rays.	
37a	Tunic with large, densely packed bladder cells and with disk-shaped calcareous concretions (both visible at a magnification of 15x) (color whitish tan or pink) *Cystodytes lobatus*	
37b	Tunic without bladder cells or disk-shaped calcareous concretions	38
38a	Most or all zooids arranged in systems (the individual zooids of a system have their oral apertures at the surface, but their atrial apertures join a cavity that has a single opening)	39
38b	Zooids not arranged in systems (the oral and atrial apertures of each zooid are at the surface)	41
39a	Tunic tough, leathery, encrusted with sand and with sand embedded in it; zooids with 2 body regions; pharynx of zooids with 3 rows of stigmata *Archidistoma psammion*	
39b	Tunic gelatinous or fleshy, sometimes encrusted with sand but not with sand embedded in it; zooids with 3 body regions; pharynx of zooids with at least 8 rows of stigmata	40
40a	Colonies forming smooth or irregular sheets 1-3 cm thick; zooids tan, yellowish, or orange-brown; pharynx with 7-15 (usually 8-12) rows of stigmata *Aplidium californicum* (fig. 23.3)	
40b	Colonies forming flat-topped slabs up to 5 cm thick, often massive and sometimes encrusted with sand; zooids generally red or orange-brown; pharynx with 12-16 (usually 13-15) rows of stigmata *Aplidium solidum*	
41a	Colonies somewhat globular, up to 10 cm in diameter and 2.5 cm thick (tunic white, opaque; portions of the zooids near the surface bright red, so the colony as a whole appears pinkish white and speckled with red dots) *Archidistoma molle*	
41b	Colonies forming thick sheets or slabs from which several to many lobes (these with broad bases) or knobs project	42
42a	Colony usually pale yellow, owing to yellow zooids in the semi-transparent test, without purple pigment granules; lobes or knobs sometimes accumulating considerable sand, especially in their basal portions *Archidistoma ritteri*	
42b	Colony usually purple or lavender (but sometimes gray), owing to small purple pigment granules near the surface of the test and in the zooids (in the zooids, the pigment is often concentrated at the edges of the 3 rows of stigmata); upper parts of the colony free of sand, although there may be some sand in the basal portion *Archidistoma* sp.	
43a	Pharynx with 4 rows of stigmata; colonies not sand-encrusted	44
43b	Pharynx with 5 or more rows of stigmata; colonies often sand-encrusted	45
44a	Colony consisting of club-shaped or mushroomlike masses, sometimes broad mounds (but these have narrow stalks); color ranging from pale orange to dark purplish red; found in protected situations, especially on floats and pilings, as well as on the open coast *Distaplia occidentalis*	

44b Colony consisting of a cluster of leaflike or paddlelike lobes; color ranging from cream to light orange-brown; restricted to the open coast *Distaplia smithi*
45a Zooids arranged in systems (the individual zooids of a system have their oral apertures at the surface, but their atrial apertures join a cavity that has a single opening) 46
45b Zooids not arranged in systems (the oral and atrial apertures of each zooid are at the surface) 50
46a Pharynx with 5 rows of stigmata; tunic encrusted and impregnated with sand *Aplidium arenatum*
46b Pharynx with at least 10 rows of stigmata; tunic, if encrusted with sand, not also impregnated with sand 47
47a Pharynx usually with 17-21 rows of stigmata (young zooids, however, may have fewer) (lobes of colony usually globular and often somewhat flattened above, encrusted with sand, not brightly colored) *Aplidium propinquum*
47b Pharynx with 10-16 rows of stigmata 48
48a Pharynx with 14-16 rows of stigmata; colonies encrusted with sand (zooids up to 4 cm long, parallel to one another for much of their length but diverging as they approach the surface; tunic orange, zooids red in life) *Synoicum parfustis*
This species is not known, with certainty, to occur north of California; see list of species for comments on other species of *Synoicum* that may be expected in our region.
48b Pharynx with 10-12 rows of stigmata; lobes of colony encrusted with sand only at the base, if at all 49
49a Atrial languet usually with a pair of lateral lobes arising near its base (when these lobes are well developed, the languet is tridentate); restricted to the open coast *Aplidium glabrum*
49b Atrial languet without a pair of lateral lobes; found in protected situations as well as on the open coast *Aplidium* sp.
50a Colony heavily encrusted with sand (colony consisting of several to many tightly packed, capitate lobes; zooids orange-brown, with 7-10 rows of stigmata; stomach wall with ridges) *Ritterella aequalisiphonis*
50b Colony lightly, if at all, encrusted with sand 51
51a Lobes of colony round-topped, up to 3 cm high, usually reddish; pharynx of zooids with 10-13 rows of stigmata; stomach wall with tubercles rather than ridges *Ritterella rubra*
51b Lobes of colony often flat-topped, up to 4 cm high, usually orange; pharynx of zooids usually with 8-10 rows of stigmata; stomach wall with ridges *Ritterella pulchra*

Species indicated by an asterisk are not in the key. Most of them are rare in our region.

Order Enterogona

Suborder Aplousobranchia

Family Clavelinidae

Archidistoma molle (Ritter, 1900)
Archidistoma psammion (Ritter & Forsyth, 1917)
Archidistoma ritteri (Van Name, 1945)
Archidistoma sp.
Clavelina huntsmani Van Name, 1931 (fig. 23.2)
Clavelina sp. Similar to *C. concrescens* Hartmeyer, 1924.
Cystodytes lobatus (Ritter, 1900)
Distaplia occidentalis Bancroft, 1899
Distaplia smithi Abbott & Trason, 1968
Pycnoclavella stanleyi Berrill & Abbott, 1949

Family Polyclinidae

Aplidium arenatum (Van Name, 1945)
Aplidium californicum (Ritter & Forsyth, 1917) (fig. 23.3)
Aplidium glabrum (Verrill, 1871)
Aplidium propinquum (Van Name, 1945)
Aplidium solidum (Ritter & Forsyth, 1917)
Aplidium sp.
**Euherdmania claviformis* (Ritter, 1903). Not recorded with certainty north of California.
Ritterella aequalisiphonis (Ritter & Forsyth, 1917)
Ritterella pulchra (Ritter, 1901)
Ritterella rubra Abbott & Trason, 1968
Synoicum parfustis Ritter & Forsyth, 1917
*Synoicum spp. Several northern species may be expected. One of them is *S. jordani* (Ritter, 1899), described from a specimen collected in the Pribilof Island group. Van Name (1945) stated that material from the Bering Sea, which seemed to fit the description of *S. jordani*, could not be distinguished from *S. pulmonaria* (Ellis & Solander, 1786). (Zooids of this well known species, which is found in the North Atlantic, North Sea, and contiguous regions, have about 19 or 20 rows of stigmata). The matter is complicated by the fact that 2 other poorly described species have been reported from the northern Pacific: *S. cymosum* Redikorzev, 1927 and *S. kincaidi* (Ritter, 1899). Furthermore, "*Synoicum* sp. A" and "*Synoicum* sp. B," collected at Ucluelet, on the west coast of Vancouver Island, were mentioned by Huntsman (1912a), but never described. It should be noted that not all species of *Synoicum* will fit under choice 33b; some are moundlike and have no tendency to form stalked lobes.

Family Didemnidae

Didemnum albidum (Verrill, 1871) (fig. 23.13)
Didemnum carnulentum Ritter & Forsyth, 1917
Diplosoma macdonaldi Herdman, 1886
Trididemnum opacum (Ritter, 1907) (fig. 23.12)
Trididemnum strangulatum (Ritter, 1901)

Order Phlebobranchia

Family Cionidae

Ciona intestinalis (Linnaeus, 1767)

Family Perophoridae

Perophora annectens Ritter, 1893

Family Corellidae

Chelyosoma columbianum Huntsman, 1912 (fig. 23.9)
Chelyosoma productum Stimpson, 1864 (fig. 23.8)
Corella inflata Huntsman, 1912
Corella willmeriana Herdman, 1898

Family Ascidiidae

Ascidia callosa Stimpson, 1852
Ascidia ceratodes (Huntsman, 1912)
Ascidia paratropa (Huntsman, 1912)
*Ascidia prunum O. F. Müller, 1776. Northern, but has been found in British Columbia; *Ascidiopsis nanaimoensis* Huntsman, 1912 is a synonym.

Order Stolidobranchia

Family Styelidae

*Botryllus sp.
Cnemidocarpa finmarkiensis (Kiaer, 1893)
Dendrodoa abbotti Newberry, 1984
Metandrocarpa dura (Ritter, 1896)
Metandrocarpa taylori Huntsman, 1912 (fig. 23.7)
**Pelonaia corrugata* Goodsir & Forbes, 1841. Introduced to British Columbia from Japan.
*Styela clavata (Pallas, 1774)
Styela coriacea (Alder & Hancock, 1848)
Styela gibbsii (Stimpson, 1864)
Styela montereyensis (Dall, 1872) (fig. 23.4)
*Styela truncata Ritter, 1901. *Katatropa uclueletensis* Huntsman, 1912 and *K. vancouverensis* Huntsman, 1912 are synonyms.

Family Pyuridae

Bathypera ?ovoida (Ritter, 1907)
Boltenia echinata (Linnaeus, 1767) (fig. 23.11)
Boltenia villosa (Stimpson, 1864) (fig. 23.6)
Halocynthia aurantium (Pallas, 1787)
Halocynthia igaboja Oka, 1906
Pyura haustor (Stimpson, 1864) (fig. 23.5)
Pyura mirabilis (von Drasche, 1884) (fig. 23.10)
*Pyura spp.

Family Molgulidae

*Molgula cooperi (Huntsman, 1912)
*Molgula oregonia Ritter, 1913
Molgula pacifica (Huntsman, 1912)
Molgula pugetiensis Herdman, 1898

Class Thaliacea

Order Salpida

The salps are pelagic animals that are almost completely restricted to the open ocean. Some species, however, are occasionally washed up on the shore, especially on the outer coast, or carried by currents into Puget Sound and waters adjacent to it.

The paper of Yount (1954), concerned with the rich fauna of salps found in the central part of the Pacific Ocean, provides a key to nearly all valid species, a few of which occur in our region. The other works listed in the bibliography will be helpful for confirming identifications.

Apstein, C. 1901. Salpidae, Salpen. Nordisches Plankton, Zool., 2, Abt. III:5-10.
Fraser, J. H. 1947. Thaliacea--I. Family: Salpidae. Fiches d'Identification du Zooplanction (Conseil Permanent International pour l'Exploration de la Mer), Sheet 9. 4pp.
Metcalf, M. M. 1919. The Salpidae collected by the United States Fisheries Steamer "Albatross," in Philippine waters, during the years 1908 and 1909. Bull. 100, U. S. Nat. Mus., vol. 2, part 2. 193 pp.
Yount, J. L. 1954. The Taxonomy of the Salpidae (Tunicata) of the central Pacific Ocean. Pacific Sci., 9:276-330.

Order Doliolida

Doliolids are rarely taken except in plankton tows made well away from the coast. Any species likely to be encountered in our area can probably be identified with the aid of the references listed.

Borgert, A. Die nordischen Dolioliden. Nordisches Plankton, Zool., 2, Abt. III:1-4.
Fraser, J. H. 1947. Thaliacea--II. Family: Doliolidae. Fiches d'Identification du Zooplancton (Conseil Permanent International pour l'Exploration de la Mer), Sheet 10. 4pp.

Class Larvacea

Berrill, N. J. 1950. The Tunicata. With an Account of the British Species. London: Ray Society. iii + 354 pp.
Essenberg, C. E. 1926. Copelata from the San Diego region. Univ. Calif. Publ. Zool., 28:399-521.
Fenaux, R. 1966. Synonymie et distribution géographique des appendiculaires. Bull. Inst. Oceanogr. Monaco, 66:1-23.
Tokioka, T. 1960. Studies on the distribution of appendicularians and some thaliaceans of the north Pacific, with some morphological notes. Publ. Seto Mar. Biol. Lab., 8:351-443.

1a Trunk elongate, more than one-third (usually about one-half) the length of the tail
Fritillaria borealis
1b Trunk ovoid, less than one-third (usually about one-fourth or one-fifth) the length of the tail   2
2a Profile of dorsal surface of trunk somewhat irregular and with a distinct hump near the mouth; with 2 spindle-shaped subchordal cells on the right side of the notochord near the middle of the tail; present in the plankton throughout the year, especially common from spring to autumn   *Oikopleura dioica*
2b Profile of dorsal surface of trunk rather smooth and without a distinct hump near the mouth; with about 15-25 large cuboidal subchordal cells along the notochord in the terminal third of the tail; present in the plankton from fall to spring, but most common in winter
Oikopleura labradoriensis

Family Fritillariidae

Fritillaria borealis Lohmann, 1896

Family Oikopleuridae

Oikopleura dioica Fol, 1872
Oikopleura labradoriensis Lohmann, 1892

PHYLUM HEMICHORDATA

Class Enteropneusta

Several enteropneusts have been found in our region, but only one has been named. This is *Glossobalanus berkeleyi* Willey, 1931, described from an incomplete specimen collected near Nanaimo. A species of *Saccoglossus*, characterized by an orange proboscis, is abundant intertidally in Willapa Bay, Washington, and a similar species occurs in North Bay, at Cape Arago, Oregon. Chuckanut Bay, near Bellingham, Washington, is another locality from which an intertidal enteropneust has been reported. There are also subtidal species, one of them being rather abundant in coarse sand near Bamfield, Vancouver Island.

A systematic study of Pacific coast enteropneusts, in which some of the species mentioned above are to be named and described, is in preparation.

Willey, A. 1931. *Glossobalanus berkeleyi*, a new enteropneust from the west coast. Trans. Royal Soc. Canada, Sect. V, ser. 3, 24:19-28.

PHYLUM CHAETOGNATHA

Only two species of chaetognaths, *Sagitta elegans* and *Eukrohnia hamata*, are common in coastal waters of our region. The key includes two others that are likely to be found in small numbers, and the list includes a few deep-water chaetognaths that may occasionally be encountered in plankton collections made near the surface.

Fraser, J. H. 1957. Chaetognaths. Fiches d'Identification du Zooplancton (Conseil Permanent International pour l'Exploration de la Mer), no. 1 (revised).
Lea, H. E. 1955. The chaetognaths of western Canadian coastal waters. J. Fish. Res. Bd. Canada, 12:593-617.
Tokioka, T. 1965. The taxonomical outline of the chaetognaths. Publ. Seto Mar. Biol. Lab., 12:335-57.

1a With 2 completely separate lateral fins on both sides of the body   2
1b With a long single fin or 2 incompletely separate fins on both sides of the body   3
2a Anterior lateral fins slightly longer than the posterior lateral fins, and reaching nearly to the level of the posterior edge of the ventral ganglion   *Sagitta decipiens*
2b Anterior lateral fins slightly shorter than the posterior lateral fins, and terminating at a distance equal to more than half their length behind the posterior edge of the ventral ganglion   *Sagitta elegans*
3a With 2 incompletely separate lateral fins on both sides of the body; anus anterior to the tail septum; without oil droplets in the intestine   *Sagitta scrippsae*
3b With a single long lateral fin on both sides of the body; anus at the tail septum; with conspicuous oil droplets in the intestine   *Eukrohnia hamata*

Species marked with an asterisk are not in the key. They are generally restricted to deep water.

Class Sagittoidea

Order Phragmophora

Family Eukrohniidae

Eukrohnia hamata Möbius, 1875

Order Aphragmophora

Family Sagittidae

**Sagitta bierii* Conant, 1896
Sagitta decipiens Fowler, 1905
Sagitta elegans Verrill, 1873
**Sagitta friderici* Ritter-Zahony, 1911
**Sagitta maxima* Conant, 1896
**Sagitta minima* Grassi, 1881
Sagitta scrippsae Alvariño, 1962. This name applies to Pacific coast specimens that have previously been referred to *S. lyra* Krohn, 1853.

GLOSSARY

Aboral. Opposite the end or side on which the mouth is located
Accessory flagellum. In certain Crustacea, a branch of antenna 1
Acicle. A spinelike projection
Acicula (also spelled aciculum). In polychaete annelids, a chitinous needlelike structure that provides internal support for the lobe of a parapodium (it usually does not protrude)
Acicular seta. In polychaete annelids, a stout seta that resembles an acicula, but that is like other setae in that it can be protruded
Acontium. In some sea anemones, a free filament that originates from the edge of each septum that subdivides the cavity of the gut
Adductor muscle scar. On the shell of a bivalve mollusc, the scar of a large muscle that pulls the valves together (there is usually an anterior and a posterior adductor muscle)
Ambulacrum. In echinoderms, a region along which the tube feet (podia) are arranged (in asteroids, each ambulacrum is a prominent groove bordered by tube feet)
Ampulla (in oligochaete annelids). *See* Sperm trap
Antenna. In polychaete annelids, a slender sensory appendage arising from the prostomium; in insects, millipedes, and centipedes, the first head appendage (*see also* Antenna 1, Antenna 2)
Antenna 1. In Crustacea, the first head appendage; also called antennule
Antenna 2. In Crustacea, the second head appendage; also called antenna, when the antenna 1 is referred to as the antennule
Article. In arthropods, a unit ("segment") of a jointed appendage
Atrial languet. In zooids of compound ascidians, a tonguelike process (sometimes divided) arising from the lip of the atrial aperture
Avicular uncinus. A short, stout seta that has a hooklike form, there being one large tooth and several small ones distal to it
Avicularium. In bryozoans, a specialized zooid that resembles the beak of a bird (the movable jaw is homologous to the operculum of a feeding zooid)

Basis. In Crustacea, the second article of an appendage (if the appendage is biramous, the basis is the article that bears the endopodite and exopodite); in Nemertea, a mass of granules to which the functional stylet is attached
Bifid. Divided into 2 lobes or teeth
Biramous. Divided into 2 branches (in Crustacea, the two branches [endopodite, exopodite] of a biramous appendage originate on article 2 [the basis])
Body whorl. In the coiled shell of a gastropod mollusc, the youngest (and usually the largest) whorl
Branchia. A structure that functions as a gill

Caecum. A blind pouch; a diverticulum
Calceolus. In amphipod crustaceans, a sensory structure (a disk-shaped seta) on antenna 1 or 2
Calyx. In Entoprocta, the cuplike portion of the body in which the gut lies and from which the tentacles arise
Cancellate sculpture. In shells of gastropod molluscs, a pattern of sculpture formed by the crossing of axial ribs by spiral ridges; in bivalve molluscs, a pattern formed by the crossing of radial ribs by concentric ridges or lines
Capillary seta. A slender simple seta
Capitate. With a knob or swelling at the tip
Capitulum. In stalked barnacles, the armored portion within which the appendages and most of the viscera are located
Carapace. In some malacostracan Crustacea, a continuous covering (often hardened to at least some extent) over the head and thorax; in barnacles, the aggregation of plates that protects the rest of the body

Cardelles. In cheilostome Bryozoa, toothlike projections at the edge of the aperture, serving as points of attachment for the operculum

Cardinal teeth. In the shell of a bivalve mollusc, the teeth that radiate from the part of the hinge immediately adjacent to the umbo

Carina. A ridge; in barnacles, one of the plates of the carapace (see figs. 16.1 and 16.2)

Carinate. With a prominent ridge

Carpochelate. In amphipod crustaceans, a condition of a gnathopod in which the carpus (article 5) is elongated to form a thumb against which the dactyl (article 7) closes

Carpopropodus. In Mysidacea, a multiarticulate portion of the endopodite of a thoracopod, representing the carpus plus propodus

Carpus. In the thoracic appendages (legs, pereopods) of malacostracan crustaceans, article 5

Cephalic. Pertaining to the head region

Cephalon. In arthropods, the head, derived from the first 6 embryonic somites; the term is often applied (as in the case of isopods) to what is essentially a short cephalothorax, derived from the head plus 1 thoracic somite

Cerata. In opisthobranch gastropod molluscs, fleshy processes of the dorsum, containing diverticula of the digestive gland

Chela. In Crustacea, a pincer in which the dactyl serves as the movable claw and the propodus serves as the stationary claw

Chelate. Provided with a chela (*see also* Subchelate)

Cheliped. A leg (pereopod) provided with a chela

Chondrophore. In the valve of the shell of a bivalve mollusc, a plate or shelf to which an internal hinge ligament is attached

Chromatophore. A structure (sometimes a single cell) in which pigment is concentrated

Cirrus. In general, a soft appendage, usually fingerlike or tentaclelike; in barnacles, one of the biramous thoracic appendages used for collecting food

Clavus. In nudibranch gastropods, the distal portion of a rhinophore, often club-shaped and somewhat resembling a gill

Clitellum. In oligochaete annelids, a glandular thickening of the epidermis of the genital segments

Coelomocytes. Cells that circulate freely in the coelom

Colloblast. In Ctenophora, a "glue cell," used in capture of prey (colloblasts are typically concentrated on the tentacles, but are sometimes found on other parts of the body)

Comb row. In Ctenophora, each of the 8 meridional rows of ctenes, or "combs," which consist of long locomotor cilia

Copulatory bursa. In oligochaete annelids, an involution of the ventral body wall around the male pore(s), often unpaired and on the midline

Corbula. In certain hydroids, a basketlike structure within which the gonophores are concentrated

Cormidium. In siphonophores, a lateral branch of the stem that trails the swimming bell (it consists initially of a gastrozooid, and this eventually produces one or more gonophores)

Coupling hook. On a maxilliped of an isopod crustacean, a small, hooklike projection on the medial margin

Coxa. In Crustacea, the first article of an appendage (if an epipodite is present, it is borne on the coxa; in amphipods, the coxa is expanded into a broad plate)

Coxal plate. In amphipod Crustacea, a coxa that is expanded into a broad plate

Crenate or crenulate. Having a scalloped margin

Cryptocyst. In cheilostome Bryozoa, a calcified shelf beneath the peripheral portion of the frontal membrane

Cuticular penis sheath (of oligochaete annelids). *See* Penis

Dactyl. In Crustacea, the terminal article, usually clawlike, of a thoracic appendage

Dextral. Referring to the right, or something turned or coiled toward the right

Distal (as opposed to proximal). In describing appendages or other outgrowths, referring to the portion farthest from the point of origin

Dorsal lamina. In ascidians, a narrow membrane running along the mid-dorsal line of the pharynx

Dorsum. The upper surface of an animal; the back

Egg sac (in oligochaete annelids). A diverticulum of the septum of the ovarian segment (extending posteriorly from the clitellum), in which eggs are stored

Elytron. In certain families of polychaete annelids, a shieldlike plate on the dorsal surface
Endopodite. In Crustacea, the inner (medial) branch of a biramous appendage
Endostyle. In ascidians, a grooved band of ciliated and glandular tissue running along the mid-ventral line of the pharynx
Epibenthic. Living at or close to the surface of a bottom deposit
Epimeron (or epimeral plate). In Crustacea, a flattened lateral extension of a segment, directed laterally or ventrolaterally
Epipodite. In certain Crustacea, a lateral extension of the coxa of a leg (pereopod), usually functioning as a gill
Eudoxid. In siphonophores, a cormidium that has become detached from the stem on which it was produced and that may survive for a time as a free-living individual
Exopodite. In Crustacea, the outer (lateral) branch of a biramous appendage
Eye scale. In decapod Crustacea, a flattened projection at the base of the eyestalk

Filiform. Slender and threadlike
Flagellum. In certain Crustacea, the slender distal portion of an antenna; the term is more generally used to indicate a vibratile protoplasmic extension of a cell, involved in locomotion, attachment, or creating water currents
Frontal. In cheilostome Bryozoa, the surface of a zooecium on which the aperture is located

Gastrozooid. In colonial Hydrozoa and some other cnidarians, a member specialized for feeding; a feeding polyp
Girdle. In chitons, the portion of the mantle that borders the 8 shell plates
Gland cone. In amphipod crustaceans, a conical ventral process of article 2 of antenna 2 (it contains the duct of the antennal gland)
Gnathopod. In amphipods and certain other Crustacea, one or both of the first 2 legs (pereopods), usually differing from the remaining legs in being chelate or subchelate
Gonangium. In Hydrozoa, a polyp specialized for production of medusae
Gonophore. In Hydrozoa, a structure within which gametes are produced (the term is commonly applied to a medusa that remains attached; *see also* Medusoid); in siphonophores, a medusiform sexual member of a colony
Gonopore. A genital pore
Gonotheca. In Hydrozoa, a hardened outer covering (perisarc) enclosing a gonangium
Gymnocyst. In cheilostome Bryozoa, a calcified portion of the frontal wall, proximal and lateral to the membranous portion

Hair seta. In oligochaete annelids, a seta that is longer and more flexible than other setae (a nodulus, moreover, is never present; hair setae are found only in the dorsal setal bundles of some Tubificidae and Naididae)
Heterogomph. In polychaete annelids, referring to compound setae in which the 2 lobes at the end of the proximal portion are decidedly unequal
Homogomph. In polychaete annelids, referring to compound setae in which the 2 lobes at the end of the proximal portion are equal
Hydranth. In colonial hydroids, a feeding polyp; a gastrozooid
Hydrotheca. In hydroids, a hardened covering (perisarc) around a hydranth

Introvert. In sipunculans, a portion of the body that can be retracted by being pulled into itself
Ischium. In Crustacea, article 3 of a leg (pereopod)

Keel. A prominent ridge, such as that forming the keel of a boat

Labrum. An unpaired upper lip (not an appendage), which in arthropods is just anterior to the mandibles
Lamella. A thin, platelike structure
Lanceolate. Tapering gradually to a point
Lateral teeth. In the shell of a bivalve mollusc, any teeth that may be present on the hinge plate anterior or posterior to the cardinal teeth

Lithocyst. In hydrozoan medusae, a structure concerned with balance, consisting of a crystalline mass within a chamber containing sensory cells
Lophophore. In bryozoans, brachiopods, and phoronids, a fold or ridge on the body wall that bears ciliated tentacles used in collecting food and in respiration
Lyrula. In cheilostome Bryozoa, a median tooth or shelf on the proximal border of the aperture

Madreporite. A perforated calcareous ossicle ("sieve plate") on the aboral surface of echinoid and asteroid echinoderms, permitting water to enter the water-vascular system (a simplified madreporite is found on the oral surface of some ophiuroids)
Mandible. In Crustacea, one of the third pair of appendages of the head (the first pair of appendages associated with the mouth)
Manubrium. In medusae, a stalk on which the mouth is located
Maxilla. In Crustacea, a member of the fourth and fifth pairs of appendages of the head (maxillae 1, maxillae 2), which follow the mandibles
Maxilliped. In Crustacea, an anterior thoracic appendage that functions as a mouthpart (the number of pairs of maxillipeds varies according to the group of Crustacea)
Medusoid. In the broad sense, a medusa, but the term is more commonly applied to a hydrozoan medusa that produces gametes while remaining attached to the polyp that produced it, and that is usually much modified
Meridional canal. In Ctenophora, each of the 8 digestive canals that lie beneath the comb rows
Merochelate. In amphipod curstaceans, a condition of a gnathopod in which the merus (article 4) is extended as a thumb against which the dactyl (article 7) closes
Merus. In Crustacea, article 4 of a thoracic appendage
Mesoglea. In Cnidaria, the jellylike material between the gut and epidermis (especially prominent in medusae)
Modified genital seta (in oligochaete annelids). *See* Penial seta, Spermathecal seta
Molar. In the mandibles of some Crustacea and other arthropods, a stout process specialized for grinding or crushing food
Moniliform. Appearing to consist of a series of beads

Nectophore. In siphonophores, a swimming bell
Nematocyst. In cnidarians, a secreted, intracellular capsule that contains an inverted thread (upon appropriate stimulation, the thread is everted and used in prey capture or defense)
Nematophore. In certain hydroids, a polyp specialized for production of nematocysts
Nephridiopore. In annelids, the external opening of an excretory organ (protonephridium or metanephridium), or of a compound structure that consists partly of an excretory organ
Neuroseta. In polychaete annelids, one of the setae arising from the lower lobe (neuropodium) of a parapodium
Nodulus. In oligochaete annelids, a swelling on a seta, at the point where it emerges from the setal sac
Notoseta. In polychaete annelids, one of the setae arising from the upper lobe (notopodium) of a parapodium
Nuchal. Referring to the posterior part of the head or to the neck

Ocellus. A simple eye; an eyespot
Oeciostome. In cyclostome Bryozoa, the aperture of a communal ovicell
Ommatidium. In arthropods, a unit of a compound eye
Operculum. A flap or trap door, such as is used for closing the aperture of a snail shell, the aperture of a bryozoan zooecium, or the hydrotheca of a hydroid
Osculum. In sponges, an opening by which water leaves the colony
Ossicle. In echinoderms, a calcareous skeletal structure, often platelike or spinelike
Ovicell. In Bryozoa, a structure in which eggs are brooded

Pallial line. In bivalve molluscs, a line that marks the attachment of the edge of the mantle to the shell
Pallial sinus. In bivalve molluscs, an indentation of the pallial line, indicating the area into which the siphons can be withdrawn
Palm. In Crustacea, the broadened portion of the propodus (article 6) of a chelate or subchelate appendage

Palp. In certain polychaete annelids, a sensory outgrowth (sometimes consisting of 2 units) of the ventral or frontal side of the prostomium; in some other polychaetes, a long sensory outgrowth of the first true segment; in Crustacea and other arthropods, the slender distal portion of a mandible or maxilla, which functions as a sensory structure

Palpon. In Siphonophora, a nonfeeding member of the colony, lacking a mouth and having a single, unbranched tentacle (also called dactylozooid)

Papilla. A small, fleshy projection of the body wall or of some other structure

Parachelate. In amphipod crustaceans, a condition of a gnathopod or pereopod in which the propodus (article 6) is broadened into a tooth against which the dactyl (article 7) closes (the dactyl usually extends well beyond the tooth)

Paragnath. In certain polychaete annelids, such as members of the family Nereidae, one of the small teeth on the outside of the everted proboscis

Parapodium. In polychaete annelids, a fleshy flap on each side of at least certain segments (the setae arise from the parapodia)

Paxilla. In some asteroid echinoderms, a columnar calcareous ossicle on the body surface (the top of a paxilla, usually nearly flat, is typically covered by small ossicles)

Pectinate. Comblike

Pectinate seta (in oligochaete annelids). A bifid dorsal seta that has a few spinelike teeth between the 2 primary teeth

Pedicellaria. In asteroid and echinoid echinoderms, a pincerlike structure (usually stalked) on the body surface (in asteroids, the pedicellariae commonly have 2 jaws, but sometimes are formed by a cluster of ossicles; in echinoids of our region, the pedicellariae have 3 jaws)

Peduncle. A stalk; in the case of some Crustacea, applied to the proximal portion of an antenna 1 or antenna 2, as distinct from the flagellum of the appendage

Penial seta. In oligochaete annelids, a ventral seta, usually different from other setae, associated with the male genital pore (there may be more than one penial seta)

Penis. A male copulatory organ; in oligochaete annelids, a permanent fold of the body wall, often enclosed by a cuticular sheath, around the male genital complex

Peptonephridia. In oligochaete annelids (especially Enchytraeidae), elongated spongy structures that extend from the posterodorsal part of the pharynx; sometimes called salivary glands, post-pharyngeal bulbs, or esophageal peptonephridia

Pereon. In Crustacea, the complex of thoracic segments from which the legs (pereopods, gnathopods) originate (it does not include segments that are fused to the head, and whose appendages [maxillipeds] function as mouthparts)

Pereonite. A segment of the pereon

Pereopod. An appendage (leg) arising from a segment of the pereon

Perfoliate. Referring to a structure in which the stalk is encircled by one or more continuous flanges

Periostracum. In molluscs, the organic material, often fibrous, on the outside of the shell

Periproct. In sea urchins, the area immediately surrounding the anus (it is usually somewhat membranous)

Peristome. In cyclostome Bryozoa, the portion of a tubular zooecium that projects beyond the surface of the colony; in cheilostome Bryozoa, an elevated rim around the aperture of a zooecium

Peristomium. In polychaete annelids, the first true segment, directly behind the prostomium, on which the mouth is located (additional segments may be fused with it, and the complex as a whole may bear tentacular cirri; the cirri are derived from parapodia, and must not be confused with antennae, which are restricted to the prostomium)

Pharyngeal glands. Glands closely associated with the pharynx (in some oligochaete annelids [Enchytraeidae], the cell bodies of the glands extend posteriorly from the dorsal side of the eversible pharynx and may be seen around the gut in segments 4, 5, and 6)

Pleon. In certain Crustacea, especially Peracarida, the abdomen

Pleonite. A segment of the pleon

Pleopod. In Crustacea, a paired biramous appendage arising from the ventral side of an abdominal segment (pleonite), used in swimming, copulation, tending eggs, and other functions

Pleotelson. In certain Crustacea, such as isopods, a complex formed by fusion of the telson with one or more segments of the pleon

Proboscis. An extensile or eversible structure that is used in feeding, but that is not part of the gut (an eversible pharynx, such as is characteristic of many polychaete annelids, is here not considered to be a proboscis); in some oligochaete annelids, the proboscis is the prostomium; in echiurans, it is a substantial preoral lobe; in nemerteans, it is a complex structure everted by hydrostatic pressure

Propodus. In Crustacea, article 6 (next to last article) of a leg (pereopod, gnathopod)

Prostomium. In polychaete annelids, the head lobe, dorsal and anterior to the mouth (if antennae or anteroventral palps are present, these originate from the prostomium)

Protandric. Referring to an animal that is first male, then female

Pseudohydrotheca. In certain hydroids, an extension of the perisarc that encloses at least a part of a hydranth but that differs from a true hydrotheca in being of inconsistent and somewhat irregular form

Pseudopenis. In oligochaete annelids, a type of penis that consists of only one layer (atrial), rather than two layers (atrial and epidermal)

Pygidium. In annelids, the posteriormost unit of the body (not a true segment), on which the anus is located

Rachis. The main stem of a branched structure, especially one that is featherlike

Radiole. In sabellid, serpulid, and spirorbid polychaetes, one of several to many featherlike structures at the anterior end of the body (they are outgrowths of the peristomium)

Ramus. A branch

Rhinophore. In opisthobranch gastropod molluscs, a member of a pair of outgrowths, sometimes with a sheath and often elaborate, arising from the dorsal part of the head

Rhopalium. In scyphozoan medusae, a structure concerned with equilibrium, consisting of a fleshy outgrowth of the margin of the bell, weighted by a mass of crystals, that contacts a sensory lobe

Rostrum. In Crustacea generally, an anteriorly directed prolongation of the head or carapace; in barnacles, one of the plates of the carapace (see figs. 16.1 and 16.2)

Scale (of antenna 2). In decapod Crustacea, the exopodite of antenna 2

Scutum. In barnacles, one of a pair of plates that form part of the carapace (see figs. 16.1 and 16.2)

Seminal vesicle. A portion of the male reproductive system in which sperm are stored; in oligochaete annelids, a sperm sac formed by distension of the septa on the anterior and/or posterior sides of the testis segment

Setigerous. Bearing setae

Setose. Studded with setae; bristly

Setule. In Crustacea and other arthropods, a small seta

Shield (of carapace). In hermit crabs, the anterior, calcified portion of the carapace

Sinistral. Referring to something that is left-handed, or turned or coiled to the left

Sinus. In cheilostome Bryozoa, an incision, usually rounded, in the proximal portion of the edge of the aperture

Siphonoglyph. In Anthozoa, a ciliated groove running the length of the pharynx (there may be one or two of these, and they are usually evident at the level of the mouth)

Sperm collar. In oligochaete annelids, a portion of the male duct to which mature sperm become attached before copulatory transfer

Sperm funnel. In some oligochaete annelids (Enchytraeidae), the thickened, glandular part of the vas deferens, which supports the sperm collar (it is located anterior to the septum between segments 11 and 12); in other oligochaetes, synonymous with sperm collar (see above) and located on the septum between segments 10 and 11 (Tubificidae) or on the septum between segments 4 and 5 or 5 and 6 (Naididae)

Sperm morula. In oligochaete annelids, a syncytial mass of spermatocytes, floating in the coelom

Sperm trap. In some oligochaete annelids (Tubificidae), an abrupt ($90°$) bend in the spermatheca (the bend distinctly separates the ampulla from the rest of the spermatheca)

Spermatheca. An organ that receives sperm from the partner during copulation (it may store the sperm for some time, but releases them when the eggs are ready to be fertilized; synonymous with seminal receptacle, which is generally preferred)

Spermathecal seta. In oligochaete annelids, a ventral seta, associated with the spermatheca, that is usually different from other setae

Spermatophore. A package of sperm, transferred as a unit during copulation

Statocyst. A type of sensory structure concerned with equilibrium (usually a vesicle containing a secreted hard body)

Stigmata. In ascidians, the perforations of the pharynx

Stolon. A "runner," creeping over or through the substratum, from which new individuals or members of a colony are budded

Stylet. A spinelike or needlelike structure, such as is present on the proboscis of some nemerteans or on the penis of some polyclad flatworms

Subchelate. In Crustacea, the condition of a gnathopod or pereopod in which the dactyl folds back against the palm of the propodus (article 6) to form a pincer (differing from chelate in that the propodus does not have a distal prolongation against which the dactyl closes)

Telson. In Crustacea, a terminal flap or plate attached to the sixth abdominal segment (in certain groups, such as the Isopoda, it may be indistinct from one or more of the abdominal segments [*see* Pleotelson])

Tentacle sac. In ctenophores, the sac into which a tentacle may be withdrawn

Tentacular bulb. In hydrozoan medusae, a swelling at the base of a tentacle

Tentillae. In ctenophores and certain cnidarians, a branch tentacle

Terete. Approximately circular in cross-section

Tergum. In kinorhynchs and arthropods, the dorsal portion of an interrupted ring of cuticle that encircles a segment; in barnacles, one of two plates that form part of the carapace (see figs. 16.1 and 16.2)

Theca. An outer covering, generally hardened to some extent

Thoracopod. In Crustacea, an appendage of the thorax

Torus. A parapodial lobe (neuropodium or notopodium) that is reduced to a small ridge and that bears a series of uncini (stout, hooked setae)

Tracheal system. In insects and certain Crustacea, a system of tubes for breathing air

Trepan. In some syllid polychaetes, a ring of hardened teeth at the anterior end of the pharynx

Triturative. Specialized for chewing or grinding

Trochoid. Resembling the shell of a gastropod mollusc of the family Trochidae

Tunic. The outer covering, often rather thick, of a solitary or social ascidian; also the matrix in which the several to many zooids of a compound ascidian are embedded

Umbilicus. In gastropod molluscs, a pit at the base of the shell, leading into the pillar around which the whorls are spiralled

Umbo (plural umbones). In the shell of a bivalve mollusc, the oldest portion of each valve, near the hinge, often elevated or somewhat beaklike

Uniarticulate. Descriptive of an appendage or ramus that consists of a single article

Uniramous. Unbranched, referring to a crustacean appendage in which only the exopodite or endopodite of the biramous type of appendage persists

Uropod. In malacostracan Crustacea, a member of 1 or more pairs of appendages arising from the posterior part of the abdomen (pleon) (there is only 1 pair in most malacostracans, but amphipods have 3 pairs)

Urosome. In amphipod Crustacea, the complex formed by the last 3 segments of the abdomen (pleon); these segments bear the 3 pairs of uropods

Urosomite. In amphipod Crustacea, one of the last 3 segments of the abdomen (pleon)

Vas deferens. A sperm duct; in oligochaete annelids, specifically a duct that carries sperm from the sperm funnel to the atrium or some other part of the male reproductive system, such as the penial bulb (Enchytraeidae)

Velum. In hydrozoan medusae, a circular membrane extending inward from the margin of the bell

Vibraculum. In a few cheilostome Bryozoa, a specialized individual consisting largely of a slender, vibratile projection

Zooecium. In bryozoans, the boxlike or tubelike "house" secreted by a zooid

ADDITIONS AND CORRECTIONS

2. Phylum Porifera

Page
14   Between 3a and 3b, add *?Psammopemma* sp. This has the appearance of a layer of agglutinated sand, with pores.
    Following 4b add *Spongionella* sp. This has reticulate spongin fibers, but is unlike *Dysidea fragilis* in that the fibers do not contain sand.
    5a   Color rose or red   *Aplysilla* similar to *glacialis*
          (An ivory or light tan sponge that resembles it is *Pleraplysilla* sp.)
    5b   For *Aplysilla polyraphis* read *Chelonaplysilla polyraphis*.
25   For the second 107a read 107b
       115a After *Mycale macginitiei* add *Mycale bamfieldense* Reiswig & Kaiser, 1989. This has micranthoxeas (very small, spiny oxeas) in addition to the spicules characteristic of *M. macginitiei*.
28   In Family Tethyidae
        For *Tethya* close to *aurantia* (Pallas, 1766) read *Tethya californiana* (de Laubenfels, 1932).
29   In Family Mycalidae add *Mycale bamfieldense* Reiswig & Kaiser, 1989. Intertidal.
31   In Family Dysideidae
        Dysidea fragilis. Subtidal.
        Add *Spongionella* sp. Intertidal.
    In Family Aplysillidae replace the species listed with
        Aplysilla similar to *glacialis* of de Laubenfels, 1930. Intertidal and subtidal.
        Chelonaplysilla polyraphis (de Laubenfels, 1930). Intertidal.
        Pleraplysilla sp. Intertidal.
    Add, at the bottom of the page, Undetermined Family, with *Psammopemma* sp. Intertidal. (It has the appearance of an agglutinated layer of sand, with pores.)

3. Phylum Cnidaria

32   In Hydromedusae add the following references:
        Brinckmann-Voss, A. 1980. A new species of the genus *Sarsia* (Hydrozoa, Corynidae) from Vancouver Island and Puget Sound. Life Sci. Occas. Pap. Royal Ontario Mus., 34:1-4.
        Brinckmann-Voss, A. 1989. *Sarsia cliffordi* n. sp. (Cnidaria, Hydrozoa, Anthomedusae) from British Columbia, with distribution records and evaluation of related species. Can. J. Zool., 67:685-91.
        Miller, R. L. 1982. Identification of sibling species within the "*Sarsia tubulosa* complex" at Friday Harbor, Washington (Hydrozoa: Anthomedusae). J. Exp. Mar. Biol. Ecol., 62:153-72.
34   Fig. 3.2. For *Obelia dichotoma* read *Obelia longissima*.
35   17a *Cladonema californicum* usually has 9 tentacles, and these are branched once or twice; one branch ends in a sucker, the others have swollen clusters of nematocysts. *Cladonema radiatum* has 8-10 tentacles, and these are branched several times; the lower 1-4 branches end in suckers, the upper 4-6 branches have swollen clusters of nematocysts.
       31a For *Bougainvillia ramosa* read *Bougainvillia muscus*.
36   43b With 8-16 large tentacles etc.
37   Fig. 3.8. For *Bougainvillia ramosa* read *Bougainvillia muscus*.
38   46a For *Obelia dichotoma* read *Obelia longissima*.
       46b For *Obelia geniculata* read *Obelia dichotoma*.

Additions and Corrections

39	49a For *Phialidium gregarium* read *Clytia gregaria*.
	49b For *Phialidium lomae* read *Clytia lomae*.
41	In Family Corynidae add the following species:
	Sarsia apicula (Murbach & Shearer, 1902)
	Sarsia eximia (Allman, 1859)
	Sarsia cliffordi Brinckmann-Voss, 1989
	Sarsia japonica (Nagao, 1962)
	Sarsia sp., not yet described (see Miller, 1982)
42	In Family Cladonematidae add *Cladonema radiatum* Dujardin, 1843. Apparently introduced from Europe; now abundant in beds of eelgrass in Padilla Bay, Skagit County, Washington.
	In Family Bougainvilliidae for *Bougainvillia ramosa* (van Beneden, 1844) read *Bougainvillia muscus* (Allman, 1863)
43	In Family Mitrocomidae delete *Tiaropsidium* and *Tiaropsis*.
	After Family Mitrocomidae add Family Tiaropsidae, with genera *Tiaropsidium* and *Tiaropsis*.
	In Family Campanulariidae
	Add *Obelia bidentata* Clark, 1875.
	For *Phialidium gregarium* read *Clytia gregaria*.
	For *Phialidium lomae* read *Clytia lomae*.
44	Family Proboscidactylidae. Some authors now place this family in the Suborder Athecata (Anthomedusae).
	Delete Order Trachylina. The suborders Trachymedusae and Narcomedusae are now considered to be within the Order Hydroida.
	In Family Rhopalonematidae, add *Benthocodon pedunculata* (Bigelow, 1913).
	Under Hydroid Polyps add the following references:
	Brinckmann-Voss, A., D. M. Lickey, & C. E. Mills. 1993. *Rhysia fletcheri* (Cnidaria, Hydrozoa, Rhysiidae), a new species of colonial hydroid from Vancouver Island (British Columbia, Canada) and the San Juan Archipelago (Washington, U.S.A.). Can. J. Zool., 71:401-6.
	Calder, D. R. 1988. Shallow-water hydroids of Bermuda: the Athecatae. Life Sci. Contrib. Royal Ontario Mus., 148:1-107.
	------. 1991. Shallow-water hydroids of Bermuda: the Thecatae, exclusive of Plumularioidea. Royal Ontario Mus., Life Sci. Contrib., 154:1-140.
	Petersen, K. W. 1990. Evolution and taxonomy in capitate hydroids and medusae (Cnidaria: Hydrozoa). Zool. J. Linnean Soc., 100:101-231.
45	6b For *Rhysia* sp. read *Rhysia fletcheri*.
47	Fig. 3.39. For *Bougainvillia ramosa* read *Bougainvillea muscus*.
48	14a For *Tubularia crocea* read *Ectopleura crocea*.
	17a For *Tubularia marina* read *Ectopleura marina*.
	25b polyp releasing medusae with 5-10 branched tentacles
	Cladonema californicum or *C. radiatum*
50	43b This leads to *Plumularia* ssp. (choice 52a), but in this genus the hydrotheca is not large enough to accommodate the hydranth when it contracts.
	46b operculum consisting of 1-4 flaps
51	Fig. 3.43. For *Obelia geniculata* read *Obelia dichotoma*.
52	Fig. 3.52. For *Obelia dichotoma* read *Obelia longissima*.
56	Revised key to species of *Obelia* (based on corrections of Calder, 1991)
	64a Rim of hydrotheca even, not wavy or toothed; usually on brown algae, but sometimes on animals or inert substrata (hydranths on short branches of the erect stems)
	Obelia dichotoma (fig. 3.43)
	64b Rim of hydrotheca wavy or toothed; usually on animals or inert substrata, less often on algae 65
	65a Rim of hydrotheca with 8-11 prominent teeth; stems (these arising from creeping stolons) usually not branched except for production of individual hydranths, but variable in this respect; perisarc not darkening as the colony ages *Obelia bidentata*
	65b Rim of hydrotheca wavy, but not toothed; stems long (sometimes exceeding 50 cm) and extensively branched; perisarc of older portions of the colony becoming dark brown or blackish *Obelia longissima* (fig. 3.52)

Additions and Corrections

57 In Family Tubulariidae
For *Tubularia crocea* read *Ectopleura crocea.*
For *Tubularia marina* Torrey, 1912 read *Ectopleura marina* (Torrey, 1912).

58 After Family Boreohydridae, add Family Candelabridae and *Candelabrum* sp., reported from Sonoma County, California, and from Cape Arago, Coos County, Oregon.
In Family Cladonematidae add *Cladonema radiatum* Dujardin, 1843. Medusae of this species are common among eelgrass at Padilla Bay, Skagit County, Washington.
In Family Rhysiidae, for *Rhysia* sp. read *Rhysia fletcheri* Brinckmann-Voss, Lickey, & Mills, 1993.
In Family Bougainvilliidae
For *Bougainvillia* sp. (fig. 3.39) read *Bougainvillia muscus* (Allman, 1863) (fig. 3.39).
Add *Bougainvillia* spp.

59 In Family Mitrocomidae
Delete *Tiaropsidium* and *Tiaropsis*
After Family Mitrocomidae add Family Tiaropsidae, with genera *Tiaropsidium* and *Tiaropsis.*

60 In Family Campanulariidae
For *Obelia dichotoma* (Linnaeus, 1758) (fig. 3.52) read *Obelia longissima* (Pallas, 1766) (fig. 3.52).
For *Obelia geniculata* read *Obelia dichotoma.*

61 Some authorities place the Family Proboscidactylidae under Suborder Athecata.

63 In Family Agalmidae, add **Nanomia bijuga* (delle Chiaje, 1841). This name has been applied to specimens collected off California, whereas specimens collected in the Northwest have generally been identified as *N. cara*. Whether both species occur along the Pacific coast needs to be determined.

67 To references for Order Stauromedusae, add
Larson, R. J., & D. G. Fautin. 1989. Stauromedusae of the genus *Manania* (=*Thaumatoscyphus*) (Cnidaria, Scyphozoa) in the northeast Pacific, including description of new species *Manania gwilliami* and *Manania handi*. Can. J. Zool., 67:1543-9.

New key to Order Stauromedusae

1a Calyx with poorly developed lobes; at least the outer tentacles with cushionlike swellings at their bases; coronal muscle (the muscle that encircles the bell near its edge) not interrupted (Family Depastridae) 2
1b Calyx with 8 well developed marginal lobes; none of the tentacles with cushionlike swellings at their bases; coronal muscle interrupted at each tentacle cluster, thus consisting of 8 sections (Family Haliclystidae) 5
2a With 12 tentacle clusters; with only 1 canal extending lengthwise throughout the stalk
Manania hexaradiata (Broch, 1907)
2b With 8 tentacle clusters; with 4 canals extending lengthwise throughout the stalk 3
3a Stalk gradually flaring into the calyx; color mostly yellow-green, but with 8 nearly white bands on the calyx; found in quiet waters, and rare
Manania handi Larson & Fautin, 1989
3b Stalk sharply distinct from the calyx; color usually cream to tan or some shade of red; on the open coast, in areas of considerable wave action 4
4a Color of calyx and stalk usually reddish, but varying from tan to magenta (with white subumbrellar nematocyst clusters) *Manania gwilliami* Larson & Fautin, 1989
4b Color of calyx and stalk usually cream to light tan (gut evident in the calyx as a dark brown, pinnately branched structure; rare)
Manania distincta (Kishinouye, 1910)
5a Marginal anchors (these located between tentacle clusters) expanded into broad trumpet-shaped cups and with conspicuous stalks; gonads extending into the lobes for only about half their length *Haliclystus salpinx* Clark, 1863 (fig. 3.82)
5b Marginal anchors egg-shaped, not expanded into broad cups, and with inconspicuous stalks; gonads nearly reaching the ends of the lobes
Haliclystus stejnegeri Kishinouye, 1899

Additions and Corrections

68   To references for Subclass Alcyonaria add the following:
Verseveldt, J., & L. P. van Ofwegen. 1992. New and redescribed species of *Alcyonium* Linnaeus, 1758 (Anthozoa: Alcyonacea). Zool. Mededel. (Leiden), 66:1-15.

69   In Family Alcyonidae
For the first ?*Alcyonium* sp. read *Alcyonium rudyi* Verseveldt & van Ofwegen, 1992. This species, forming small, whitish to pale salmon colonies up to about 3.5 mm thick, is common in some intertidal areas on the open coast. It is found on rock and shell rubble.

For Stonlonifera read Stolonifera.

73   To references for Order Actiniaria add
Fautin, D. G., Bucklin, A., & C. Hand. 1989. Systematics of sea anemones belonging to the genus *Metridium* (Coelenterata: Actiniaria), with a description of *M. giganteum*, new species. Wasmann J. Biol., 47:77-85.

75   9a For *Metridium senile* read *Metridium senile* subsp. *fimbriatum* Verrill, 1865.
9b For *Metridium* sp. read *Metridium giganteum*.

78   In Family Metridiidae
For *Metridium senile* (Linnaeus, 1767) read *Metridum senile* subsp. *fimbriatum* (Verrill, 1865).
For *Metridium* sp. read *Metridium giganteum* Fautin, Bucklin, & Hand, 1989.

4. Phyla Ctenophora, Orthonectida, Dicyemida

79   To references for Ctenophora add the following:
Arai, M. N. 1988. *Beroe abyssicola* Mortensen, 1927: a redescription. Contrib. Nat. Sci., 9:1-7.
Mackie, G. O., C. E. Mills, & C. L. Singla. 1988. Structure and function of the prehensile tentilla of *Euplokamis* (Ctenophora, Cydippida). Zoomorphology, 107:319-37.
Matsumoto, G. I. 1988. A new species of lobate ctenophore, *Leucothea pulchra* sp. nov., from the California Bight. J. Plankton Res., 10:301-11.
Mills, C. E. 1987. Revised classification of the genus *Euplokamis* Chun, 1880 (Ctenophora: Cydippida: Euplokamidae n. fam.), with a description of the new species *Euplokamis dunlapae*. Can. J. Zool., 65:2661-8.

Revised key to species of *Beroe*

2a   Meridional digestive canals with branching diverticula in a layer just below the body surface   3

2b   Meridional digestive canals without diverticula just below the body surface, although there are a few diverticula directed inward toward the pharynx (with spots of reddish or golden brown pigment)   *Beroe gracilis*

3a   Body more or less cylindrical or only slightly flattened   4

3b   Body decidedly flattened in the same plane as the pharynx (less than one-third as thick as wide)   5

4a   Meridional canals and pharyngeal canals branched; all comb rows from one-half to two-thirds of the body length; gonads in diverticula of the meridional canals; pharynx colorless to pink, purple, deep claret red, or nearly black   *Beroe abyssicola*

4b   Meridional canals branched, but pharyngeal canals not branched; all comb rows about three-fourths of the body length; gonads in the walls of the meridional canals; pharynx colorless or slightly yellow in small animals to deep pink in large specimens   *Beroe cucumis*

5a   Aboral end rounded; usually with a prominent orange or red pigment spot on both sides of the pharynx near the middle of the body; meridional canals branching, but the branches not anastomosing; comb rows less than two-thirds the length of the body, with the central two rows on both sides noticeably shorter the others (rare, probably almost strictly oceanic)   *Beroe mitrata*

5b   Aboral end pointed; body without a prominent orange or red pigment spot on both sides of the pharynx, but often with a general pink pigmentation that is darker under the comb rows; meridional canals branching extensively, the branches anastomosing; comb rows nearly as long as the body and all of approximately equal length   *Beroe forskalii*

80 In Family Pleurobrachiidae add *Hormiphora* sp. This species, apparently more common in California than in our region, is colorless, very transparent, and a little larger and more elongate than *Pleurobrachia bachei*. It has long, straight tentacle sheaths, and the tentacles emerge close to the aboral end.
In Family Bolinopsidae, for *Leucothea* sp. read *Leucothea pulchra* Matsumoto, 1988.
In Family Beroidae, the following species are now known to occur in the region:
  Beroe abyssicola Mortensen, 1927
  Beroe cucumis Fabricius, 1780
  Beroe forskalii Milne-Edwards, 1841
  Beroe gracilis Künne, 1939
  Beroe mitrata (Moser, 1907)

82 To references for Orthonectida add
  Kozloff, E. N. 1992. The genera of the phylum Orthonectida. Cah. Biol. Mar., 33:377-406.
  ------. 1993. Three new species of *Stoecharthrum* (phylum Orthonectida). Cah. Biol. Mar., 34:523-34.
  ------. 1994. The structure and origin of the plasmodium of *Rhopalura ophiocomae* (phylum Orthonectida). Acta Zool., 75:191-9.
In Family Rhopaluridae add, after *Rhopalura ophiocomae*: Develops in hypertrophied muscle cells in the wall of the gut and bursae.
In Family Intoshiidae add the following:
  Stoecharthrum fosterae Kozloff, 1993. Parasitic in *Mytilus trossulus* (Mollusca, Bivalvia, Mytilidae).
  Stoecharthrum burresoni Kozloff, 1993. Parasitic in *Ascidia callosa* (Urochordata, Ascidiacea).
To references for Family Dicyemidae add the following:
  Bogolepova-Dobrokhotova, I. I. 1963. [The current classification of the dicyemids.] Parazitol. Sbornik, 21:259-71 (in Russian).
1a Calotte much broader than the rest of the body, and also distinctly flattened
        Dicyemodeca
1b Calotte neither conspicuously broader than the rest of the body nor flattened   2
2a Calotte with 4 propolar cells and 4 metapolar cells; vermiform embryos without an abortive second axial cell   *Dicyema*

83 Under *Octopus dofleini*, for *Conocyema deca* McConnaughey, 1957 read *Dicyemodeca deca* (McConnaughey, 1957).

5. Phyla Platyhelminthes, Gnathostomulida

89 To references for Order Neorhabdocoela add
  Hyra, G. S. 1993. *Genostoma kozloffi* sp. nov. and *G. inopinatum* sp. nov. (Turbellaria: Neorhabdocoela: Genostomatidae) from leptostracan crustaceans of the genus *Nebalia*. Cah. Biol. Mar., 34:111-26.
  Schell, S. C. 1986. *Graffilla pugetensis* n. sp. (Order Neorhabdocoela: Graffillidae), a parasite in the pericardial cavity of the bent-nose clam, *Macoma nasuta* (Conrad, 1837). J. Parasit., 72:748-54.
  Westervelt, C. A., Jr., & E. N. Kozloff. 1992. Two new species of *Syndesmis* (Turbellaria: Neorhabdocoela; Umagillidae) from the sea urchins *Strongylocentrotus droebachiensis* and *Allocentrotus fragilis*. Cah. Biol. Mar., 33:115-24.

90 To Suborder Dalyellioida add Family Genostomatidae, with *Genostoma kozloffi* Hyra, 1993. On *Nebalia pugetensis* (Crustacea, Leptostraca).
In Family Graffillidae add *Graffilla pugetensis* Schell, 1986. In the pericardial cavity of *Macoma nasuta*.
In Family Umagillidae replace *Syndesmis* spp. with
  Syndesmis inconspicua Westervelt & Kozloff, 1992. In the intestine of *Allocentrotus fragilis*.
  Syndesmis neglecta Westervelt & Kozloff, 1992. In the intestine of *Strongylocentrotus droebachiensis*.

492 Additions and Corrections

90   In Family Umagillidae, for *Collastoma pacifica* read *Collastoma pacificum* (emendation of spelling of species name).

6. Phylum Nemertea

94   To references add the following:
   Kozloff, E. N. 1991. *Malacobdella siliquae* sp. nov and *Malacobdella macomae* sp. nov., commensal nemerteans from bivalve molluscs on the Pacific coast of North America. Can. J. Zool., 69:1612-8.

96   10a 14
     10b 11
     13a *Lineus* sp. is probably *L. rubescens* Coe, 1904.

97   23a The species commensal with *Macoma secta* and *M. nasuta* is *Malacobdella macomae* Kozloff, 1991; the species commensal with *Siliqua patula* is *Malacobdella siliquae* Kozloff, 1991.

98   35a 36
     35b .... *Zygonemertes virescens*

100 See addendum for page 97 concerning *Malacobdella macomae* and *M. siliquae*.

8. Phylum Annelida: Class Polychaeta

115 32b 33
117 Fig. 8.29. Delete *Nothria conchylega*. (The drawing is of *Diopatra cuprea*, an Atlantic species.)
120 43a .... ; usually without eyes (but eyes present in some species); etc.
121 55b 56
   In Family Phyllodocidae add the following references:
   Kravitz, M. J., & H. R. Jones. 1979. Systematics and ecology of benthic Phyllodocidae (Annelida: Polychaeta) off the Columbia River, U.S.A. Bull. So. Calif. Acad. Sci., 78:1-19.
   Pleijel, F. 1991. Phylogeny and classification of the Phyllodocidae (Polychaeta). Zool. Scripta, 20:225-61.
   ------. 1988. *Phyllodoce* (Polychaeta, Phyllodocidae) from northern Europe. Zool. Scripta, 17:141-53
   ------ 1993. Polychaeta Phyllodocidae. Mar. Invertebr. Scandinavia, 8:1-158.

123 11b For *Phyllodoce (Genetyllis) castanea* (Marenzeller, 1879) read *Nereiphylla castanea* (Marenzeller, 1879).
124 20a For *Eulalia (Bergstroemia) nigrimaculata* Moore, 1909 read *Bergstroemia nigrimaculata* (Moore, 1909).
   Add the following species of Phyllodocidae:
   Phyllodoce hartmanae Blake & Weldon, 1977
   Phyllodoce multipapillata Kravitz & Jones, 1979
   Phyllodoce longipes (Kinberg, 1866)
   Eteone spilotus Kravitz & Jones, 1979
   Eteone columbianus Kravitz & Jones, 1979
   Eteone fauchaldi Kravitz & Jones, 1979
   Eteone barbata (Malmgren, 1865)

126 In Family Glyceridae add the following reference:
   Hilbig, B. 1992. New polychaetous annelids of the families Nereididae, Hesionidae, and Nephtyidae from the Santa Maria Basin, California, with a redescription of *Glycera nana* Johnson, 1901. Proc. Biol. Soc. Washingon, 105:709-22.
   3a For *Glycera capitata* Ørsted, 1843 read *Glycera nana* Johnson, 1901.
   In Family Hesionidae add the following reference:
   Fournier, J. A. 1991 New species of *Microphthalmus* (Polychaeta: Hesionidae) from the Pacific Northwest. Bull. Mar. Sci., 48208-13.
   Westheide, W & G. Purschke. 1992. *Microphthalumus simplicichaetosus* (Annelida: Polychaeta), a new hesionid from the northwestern American Pacific coast with exclusively simple setae. Proc. Biol. Soc. Washington, 105:132-5.

Additions and Corrections 493

126   3b   For *Podarkeopsis brevipalpa* (Hartmann-Schröder, 1959) read *Podarkeopsis glabrus* (Hartman, 1961)
          Add the following species:
            Heteropodarke sp. (possibly *H. heteromorpha* Hartmann-Schröder, 1962)
            Microphthalmus coustalini Fournier, 1991
            Microphthalmus hystrix Fournier, 1991
            Microphthalmus simplicichaetosus Westheide & Purschke, 1992

129   In Family Pilargidae add the following reference:
          Fitzhugh, K, & P. S. Wolf. 1990. Gross morphology of the brain of pilargid polychaetes: taxonomic and systematic implications. Amer. Mus. Novitates, no. 2992.
       In Family Syllidae add the following references and species:
          Garwood, P. R., 1991. Reproduction and classification of the family Syllidae (Polychaeta). Ophelia Suppl. 5:81-88.
          Riser, N. W., 1991. An evaluation of taxonomic characters in the genus *Sphaerosyllis* (Polychaeta: Syllidae). Ophelia Suppl. 5:209-18.
          Autolytus alexandri Malmgren, 1867

131   11a   For *Sphaerosyllis hystrix* Claparède, 1863 read *Sphaerosyllis* sp.
       11b   For *Sphaerosyllis pirifera* Claparède, 1868 read *Sphaerosyllis californiensis* Hartman, 1966.
       13b   For *Exogone* sp. read *Exogone* (*Exogone*) ?*naidina* Ørsted, 1845.

133   In Nereididae add the following references:
          Fitzhugh, K., 1987. Phylogenetic relationships within the Nereidae (Polychaeta): implications at the subfamily level. Bull. Biol. Soc. Washington, 7:174-83.
          Glasby, C. J., 1991. Phylogenetic relationships in the Nereididae (Annelida: Polychaeta), chiefly in the subfamily Gymnonereidinae, and the monophyly of the Namanereidinae. Bull. Mar. Sci., 48:559-73.

137   In Family Polynoidae add the following references and species:
          Pettibone, M. H. 1993. Scaled polychaetes (Polynoidae) associated with ophiuroids and other invertebrates and review of species referred to *Malmgrenia* McIntosh and replaced by *Malmgreniella* Hartman, with descriptions of new taxa. Smithsonian Contr. Zool., no. 538.
          ------. 1986. Review of the Iphioninae (Polychaeta: Polynoidae) and revision of *Iphione cimex* Quatrefages, *Gattyana deludens* Fauvel, and *Harmothoe iphionelloides* Johnson (Harmothoinae). Smithsonian Contrib. Zool., no. 428.
          Arcteobia anticostiensis (McIntosh, 1874)
          Arcteobia spinielytris Annenkova, 1937
          Malmgreniella bansei Pettibone, 1993
          Malmgreniella liei Pettibone, 1993
          Malmgreniella berkeleyorum Pettibone, 1993
          Malmgreniella nigralba (E. Berkeley, 1923)
          Malmgreniella scriptoria (Moore, 1910)

139   16b   For *Gattyana iphionelloides* read *Gaudichaudius iphionelloides*.

140   For Family Polyodontidae read Family Pholoidae (Polyodontidae, Peisidicidae) and add the the following reference:
          Pettibone, M. H. 1992. Contribution to the polychaete family Pholoidae Kinberg. Smithsonian Contrib. Zool., 532. iii + 24 pp.
       For *Peisidice aspera* Johnson, 1897 read *Pholoides asperus* (Johnson, 1897); delete "is the only species known to occur in the region."
       From Family Sigalionidae, remove *Pholoe minuta* (Fabricius, 1780) (fig. 8.89) and transfer it to Family Pholoidae (replacing Polyodontidae, p. 140).
       In Family Sigalionidae add the following reference and species:
          Mackie, A. S. Y., & S. J. Chambers. 1990. Revision of the type species of *Sigalion*, *Thalenessa* and *Eusigalion* (Polychaeta: Sigalionidae). Zool. Scripta, 19:39-56.
          Sigalion mathildae Audouin & Milne-Edwards (in Cuvier, 1830)

141   In Family Euphrosinidae
          Kudenov, J, D. 1987. Review of the primary species characters for the genus *Euphrosine* (Polychaeta: Euphrosinidae). Bull. Biol. Soc. Washington, 7:184-93.

494 Additions and Corrections

141 In Family Onuphidae add the following reference:
   Paxton, H. 1986. Generic revision and relationships of the family Onuphidae (Annelida: Polychaeta). Rec. Austral. Mus., 38:1-74.
   4a Delete (fig. 8.29). (This figure shows a species of *Diopatra*.)
142 Fig. 8.95. For *Schistomeringos rudolphi* read *Dorvillea* (*Schistomeringos*) *longicornis*.
143 In Family Eunicidae
   2b For *Eunice aphroditois* read *Eunice ?aphroditois*.
   In Family Lumbrineridae
   5a For *Lumbrineris lagunae* Fauchald, 1970 read *Eranno lagunae* (Fauchald, 1970).
   In Family Arabellidae (Some authorities now use Oenonidae in place of Arabellidae.)
144 2a For *Arabella iricolor* read *Arabella ?iricolor*.
   In Family Dorvilleidae add the following reference:
      Orensanz, J. M. 1990. The eurnicemorph polychaete annelids from Antarctic and Subantarctic seas. Antarctic Res. Ser., 52:1-183.
   1a *Ophryotrocha vivipara* Banse,1963 (fig. 8.96) is now placed in a separate family, Iphitimidae.
   4a For *Schistomeringos moniliceras* (Moore, 1909) read *Dorvillea* (*Dorvillea*) *moniliceras* (Moore, 1909).
   4b For *Schistomeringos pseudorubrovittata* (E. Berkeley, 1927) read *Dorvillea* (*Dorvillea*) *pseudorubrovittata* E. Berkeley, 1927.
   7a For *Schistomeringos caeca* read *Parougia caeca*.
   6a For *Schistomeringos annulata* (Moore, 1909) read *Dorvillea* (*Schistomeringos*) *annulata* Moore, 1906.
   7a For *Schistomeringos caeca* read *Parougia caeca*.
   8a For *Schistomeringos japonica* read *Dorvillea* (*Schistomeringos*) *japonica*.
   8b For *Schistomeringos rudolphi* (delle Chiaje, 1828) read *Dorvillea* (*Schistomeringos*) *longicornis* (Ehlers, 1901).
147 In Family Spionidae add the following references:
   Blake, J. A., & N. J. Maciolek, 1987. A redescription of *Polydora cornuta* Bosc (Polychaeta: Spionidae) and designation of a neotype. Bull. Biol. Soc. Washington, 7:11-15.
   Fournier, J. & Levings, C. D. 1982. Polychaetes recorded near two pulp mills on the coast of northern British Columbia: a preliminary taxonomic and ecological account. Syllogeus, 40:1-91.
   Maciolek, N. J . 1987. New species and records of *Scololepis* (Polychaeta: Spionidae) from the east coast of North America, with a review of the subgenera. Bull. Biol. Soc. Washington, 7:16-40.
   ------. 1990 A redescription of some species belonging to the genera *Spio* and *Microspio* (Polychaeta: Annelida) and descriptions of three new species from the northwestern Atlantic Ocean. J. Nat. Hist. 24:1109-41.
   3a For neuropodia read notopodia; for *Polydora* (*Boccardia*) *hamata* Webster, 1879 read *Boccardiella hamata* (Webster, 1879).
   14a For *Polydora* (*Polydora*) *ligni* Webster, 1879 read *Polydora* (*Polydora*) *cornuta* Bosc, 1802.
   Add the following species:
      Polydora (*Polydora*) *caulleryi* Mesnil, 1897
      Polydora (*Polydora*) *nuchalis* Woodwick, 1953
      Prionospio lighti Maciolek, 1985
      Pygospio californica Hartman, 1936
150 In Family Paraonidae
   10a For *Tauberia gracilis* (Tauber, 1879) read *Levinsenia gracilis* (Tauber, 1879).
151 In Family Cirratulidae add the following reference:
   Blake, J. A., 1991. Revision of some genera and species of Cirratulidae (Polychaeta) from the western North Atlantic. Ophelia Suppl. 5:17-30.
   6b For *Tharyx secundus* Banse & Hobson, 1968 read *Aphelochaeta secunda* (Banse & Hobson, 1968)
   6b For *Tharyx serratisetis* Banse and Hobson, 1968 read *Aphelochaeta serratiseta* (Banse & Hobson, 1968).

151	7a	For *Tharyx parvus* E. Berkeley, 1929 read *Aphelochaeta parva* (E. Berkeley, 1929).
	7b	For *Tharyx multifilis* Moore, 1909 read *Aphelochaeta multifilis* (Moore, 1909).

151 Add the following species:
   ?*Aphelochaeta monilaris* (Hartman, 1960)
   Monticellina tesselata (Hartman, 1960)

152 In Family Cossuridae:
   1a   For *Cossura soyeri* Laubier, 1961 read *Cossura pygodactylata* Jones, 1956.
   Add *Cossura modica* Fauchald and Hancock, 1981.

154 In Family Capitellidae add the following references:
   Eckelbarger, K. J., & J. P. Grassle. 1987 Interspecific variation in genital spine, sperm, and larval morphology in six sibling species of *Capitella*. Bull. Biol. Soc. Washington, 7:62-76.
   Grassle, J. P., E. E. Gelfman, & S. W. Mills. 1987. Karyotypes of *Capitella* sibling species, and of several species in the related genera *Capitellides* and *Capitomastus* (Polychaeta). Bull. Biol. Soc. Washington, 7:77-88.
   Warren, L. M., 1991. Problems in capitellid taxonomy. The genera *Capitella*, *Capitomastus*, and *Capitellides* (Polychaeta). Ophelia Suppl. 5:275-82.
   1a Worms referred to *Capitella capitata* form a complex of sibling species. Names have not yet been proposed for these.
   In Family Arenicolidae
   1b For *Branchiomaldane vincentii* Langerhans, 1881 read *Branchiomaldane simplex* Berkeley & Berkeley, 1932.
   Arenicolid polychates producing gelatinous, stalked egg masses have been found in Boundary Bay, British Columbia, and in Willapa Bay, Washington. These worms are probably either *Arenicola cristata* Stimpson, 1856 or *A. brasiliensis* Nonato, 1958.

155 In Family Maldanidae add the following references:
   Wilson, W. H., Jr. 1983 Life-history evidence for sibling species in *Axiothella rubrocincta* (Polychaeta: Maldanidae). Mar. Biol, 76:297-300.
   Light, W. J. H., 1991. Systematic revision of the genera of the polychaete subfamily Maldaninae Arwidsson. Ophelia Suppl. 5:133-46.
   5a For *Asychis similis* read *Chirimia similis*.
   6a For *Asychis disparidenta* Moore, 1904 read *Metasychis disparidenta* (Moore, 1904).
   7b For *Asychis biceps* read *Chirimia biceps*. Two subspecies are recognized: *biceps* (M. Sars, 1861) and *lacera* (Moore, 1906)

157 21b *Axiothella "rubrocincta"* is perhaps a complex of sibling species.

159 In Family Ampharetidae add the following reference:
   Holthe, T. 1986. Polychaeta Terebellomorpha. Mar. Invertebr. Scandinavia, 7: 3-192.
   17a For *Amphisamytha bioculata* (Moore, 1906) read *Mooresamytha bioculata* (Moore, 1906).

160 In Family Terebellidae add the following references:
   Holthe, T. 1986. Polychaeta Terebellomorpha. Mar. Invertebr. Scandinavia, 7:3-192.
   Safronova, M. A. 1984. [On polychaete worms of the genus *Pista* (Polychaeta, Terebellidae) from the Pacific.] Zoologicheskii Zhurnal, 63:983-92 (in Russian, with English summary).
   ------ (Saphronova). 1991. Redescription of some species of *Scionella* Moore, 1903, with a review of the genus and comments on some species of *Pista* Malmgren, 1866 (Polychaeta: Terebellidae). Ophelia Suppl. 5:239-48.
   Smith, R. I., 1992. Three nephromixial patterns in polychaete species currently assigned to the genus *Pista* (Annelida, Terebellidae). J. Morph., 213:365-93.
   16a For *Scionella estevanica* Berkeley & Berkeley, 1942 read *Pista estevanica* (Berkeley & Berkeley, 1942).
   Revised key to species of *Pista*
   20a Gills (2 pairs) club-shaped, the swollen terminal portion of each one consisting of many short, crowded filaments (length up to about 9 cm)   *Pista cristata* (O. F. Müller, 1776)
   20b Gills (2 or 3 pairs) either elongated or bushy, definitely not club-shaped   21
   21a Gills (3 pairs) elongated and trailing, with many clusters of short filaments
         Pista moorei Berkeley & Berkeley, 1942
   21b Gills (2 or 3 pairs) bushy, not elongated   22

Additions and Corrections

160   22a   With 2 pairs of gills (those of the first pair much shorter than those of the second pair); tube consisting of mud and sand (length up about 4 cm)   *Pista brevibranchiata* Moore, 1923

         22b   With 3 pairs of gills; tube somewhat parchmentlike, with embedded sand grains   23

         23a   Opening of tube expanded into a broad, triangular or nearly circular lobe, this with a fringe of outgrowths; length up to about 35 cm; in mud and sand flats
                                                       Pista pacifica Berkeley & Berkeley, 1942

         23b   Opening of tube with many filamentous outgrowths that form a loosely woven plug; length up to about 20 cm; in rocky habitats (often among roots of *Phyllospadix*)
                                                       Pista elongata Moore, 1909

       Add the following species:
           Pista bansei Safronova, 1988
           Pista wui Safronova, 1988

163   In Family Sabellidae add the following references:
           Fitzhugh, K. 1989. A systematic revision of the Sabellidae-Caobangiidae-Sabellongidae complex (Annelida: Polychaeta). Bull. Amer. Mus. Nat. Hist., 192:1-104.
           Fitzhugh, K. 1991. Further revisions of the Sabellidae subfamilies and cladistic relationships among the Fabriciinae (Annelida: Polychaeta). Zool. J. Linnean Soc., 102:305-32.
           Smith, R. I., 1991. Relationships within the order Sabellida. Ophelia Suppl. 5:248-60.
       14a   For *Sabella (Demonax) media* read *Demonax medius*.
       Add *Demonax rugosus* (Moore, 1904).

164   In Family Serpulidae
        7a   For *Apomatus geniculatus* read *Apomatopsis geniculata*.

165   In Family Spirorbidae add the following references:
           Knight-Jones, P., E. W. Knight-Jones, & G. Buzhinskaya. 1991. Distribution and interrelationships of northern spirorbid genera. Bull. Mar. Sci., 48:189-97.
           Hess, H. C., 1993. The evolution of parental care in brooding spirorbid polychaetes: the effect of scaling constraints. Amer. Nat., 141:577-96.

166   10a   For *Pileolaria (Jugaria) quadrangularis* read *Jugaria quadrangularis*.
       11a   For *Pileolaria (Jugaria) similis* read *Jugaria similis*.
       11b   For *Sinistrella verruca* read *Bushiella verruca*.
       12a   For *Sinistrella abnormis* read *Bushiella abnormis*.
       12b   For *Sinistrella media* (Pixell, 1912?) read *Protoleodora asperata* (Bush, 1904).
       Add the following species:
           Paradexiospira (Paradexiospira) violacea (Levinsen, 1883)
           Pileolaria (Nidularia) dalestraughanae Vine, 1992
           Pileolaria (Pileolaria) berkeleyana (Rioja, 1942)

9. Phylum Annelida: Classes Oligochaeta, Hirudinoidea

178   In Family Piscicolidae, in the list of leeches and their hosts
         line 2   For Soleidae, read Pleuronectidae.
         line 3   For anomuran read brachyuran.
         line 15   For Soleidae read Pleuronectidae.

179   Following Unidentified species, for *Allocentrotus pallidus* read *Allocentrotus fragilis*.

12. Phylum Mollusca: Class Gastropoda

195   29b   For Rissoidae read Rissoidae and Barleeidae.
197   42a   Shell sculpturing limited to spiral ridges
208   19b   For *Lottia painei* Lindberg, 1987 read *Lottia* sp. (not yet described).
210   For Family Rissoidae read Family Rissoidae and Family Barleeidae. (*Alvania* and *Onoba* are assigned to the former; *Barleeia* to the latter.)
       In Family Barleeiidae add the following reference:
           Ponder, W. F. 1983. Review of the genera of the Barleeidae (Mollusca: Gastropoda: Rissoacea). Rec. Austral. Mus., 35: 231-81.

Additions and Corrections 497

210   In Family Rissoidae add the following reference:
         Ponder, W. F. 1985. A review of the genera of the Rissoidae (Mollusca: Gastropoda: Rissoacea). Rec. Austral. Mus., Suppl. 4:1-221.
     4a   For *Alvania carpenteri* Weinkauff, 1885 read *Onoba* (*Onoba*) *carpenteri* (Weinkauff, 1885).
     4b   5
     5b   Ponder (1985) indicates that *Alvania compacta* (Carpenter, 1864) may be a synonym of *A. acutelirata* (Carpenter, 1864). He also considers *A. rosana* Bartsch, 1911 to be a separate species.

217   In Family Velutinidae
     1a   Aperture oval, considerably higher than wide; periostracum generally with small ridges that diverge as they approach the edge of the aperture (associated with the solitary ascidians *Styela gibsii*, *Pyura haustor*, and *Chelyosoma productum*)
               Velutina plicatilis O. F. Müller, 1776
     1b   Aperture about as wide as high; periostracum without spiral ridges
               Velutina prolongata Carpenter, 1864

223   In Family Buccinidae add *Volutharpa ampullacea* (Middendorff, 1848) (fig. 12.73).
     In the list of species, change *strigillosum* to *strigillatum*.

225   2a   Delete this choice. (*Volutharpa ampullacea* is in the Family Buccinidae, p. 223.)
     In the list of names following choice 8b, for *rophius* read *trophius*.

226   12a   For (Dall, 1919) read (Dall, 1891).
     In the list of species following choice 12b, for *N. humboldtiana* A. G. Smith, 1968 read *N. humboldtiana* A. G. Smith, 1971.

229   10a   11
     10b   12
     15a   For *Kurtziella plumbea* (Hinds, 1843) read *Kurtziella crebricostata* (Carpenter, 1864).

232   47b   Delete this choice.

237   In Order Cephalaspidea add the following reference:
         Gibson, G. D., & F.-S. Chia. 1989. Description of a new species of *Haminoea*, *Haminoea callidegenita* (Mollusca: Opisthobranchia), with a comparison with two other *Haminoea* species found in the northeast Pacific. Can. J. Zool., 67:914-2.
     Revised key for species of *Haminaea* (formerly *Haminoea*)
     3a   Widest part of aperture (in the lower third of the shell) more than half the height of the shell
               4
     3b   Widest part of aperture not more than half the height of the shell (animal dark greenish brown, with brown and yellow spots; usually on *Phyllospadix* in rocky intertidal habitats)
               Haminaea virescens
     4a   Cephalic shield deeply divided; animal reddish brown, with brown, white, and orange dots (in lagoons and bays, often on *Ulva* and *Chaetomorpha*)   *Haminaea callidegenita*
     4b   Cephalic shield not divided; animal brown, with cream, dark brown, and black dots (in bays, on *Zostera*)   *Haminaea vesicula*
     In Family Atyidae (Haminaeidae) add *Haminaea callidegenita* Gibson & Chia, 1989.
     Add Family Diaphanidae, with **Diaphana californica* Dall, 1919. Now known to occur as far north as the Olympic Peninsula, Washington.
     Add Family Runcinidae, with **Runcina macfarlandi* Gosliner, 1991. This has been found on *Cladophora* in high intertidal pools on the coast of central Oregon.

240   For *Cylichna alba* (Carpenter, 1864) read *Cylichna alba* (Brown, 1827).
     For *Cylichna attonsa* (Brown, 1927) read *Cylichna attonsa* (Carpenter, 1864).
     In Family Pleurobranchidae add **Pleurobranchaea californica* MacFarland, 1966. This has been found as far north as Port Orford, Oregon; subtidal.

241   In Order Gymnosomata add Family Cliopsidae, with *Cliopsis krohni* Troschel, 1854.

243   To references for Order Nudibranchia add the following:
         Behrens, D. W. 1991. Pacific Coast Nudibranchs. Monterey, California: Sea Challengers. vi + 107 pp.
         Goddard, J. H. R. 1984. The opisthobranchs of Cape Arago, Oregon, with notes on their biology and a summary of benthic opisthobranchs known from Oregon. Veliger, 27:143-63.
         ------. 1990. Additional opisthobranch mollusks from Oregon, with a review of deep-water records and observations on the fauna of the south coast. Veliger, 33:230-7.

498 Additions and Corrections

243   Gosliner, T. M. 1991. Four new species and a new genus of opisthobranch gastropods from the Pacific coast of North America. Veliger, 34:272-90.

           Millen, S. V. 1983. Range extensions of opisthobranchs in the northeastern Pacific. Veliger, 25:383-6.

           ------. 1986. Northern, primitive tergipedid nudibranchs, with a description of a new species from the Canadian Pacific. Can. J. Zool., 64:1356-62.

           ------. 1987. The nudibranch genus *Adalaria*, with a description of a new species from the northeastern Pacific. Can. J. Zool., 65:2696-2702.

           ------, & J. W. Nybakken. 1991. A new species of *Corambe* (Nudibranchia: Doridoidea) from the northeastern Pacific. J. Molluscan Studies, 57, Suppl.:209-15.

           Rivest, B. 1984. Copulation by hypodermic injection in the nudibranchs *Palio zosterae* and *P. dubia* (Gastropoda, Opisthbranchia). Biol. Bull. 167:543-54.

245   For Eolidacea read Aeolidacea.

250   In Family Corambidae add *Corambe thompsoni* Millen & Nybakken, 1991.

        In Family Onchidorididae add the following species:

           Acanthodoris lutea MacFarland, 1925. This has been found as far north as Cape Arago, Oregon, where it feeds on a species of *Alcyonidium* (Bryozoa).

           Acanthodoris rhodoceras Cockerell in Cockerell & Elliot, 1905. Now known to occur in Oregon; feeds on a species of *Alcyonidium*.

           Adalaria jaunae Millen, 1987 (replacing *Adalaria* sp.). Feeds on bryozoans.

        In Family Polyceridae add the following species:

           Polycera atra MacFarland, 1905. Found as far north as Coos Bay, Oregon; feeds on bryozoans.

           Triopha maculata MacFarland, 1905. Now known to occur as far north as Bamfield, British Columbia.

        In Family Actinocyclidae, after *Hallaxa chani*, add the following note: This species feeds on a sponge of the genus *Halisarca*, not on *Didemnum carnulentum*.

251   4a   Delete this choice. (*Tritonia exsulans* is a synonym of *Tritonia diomedea*.)

253   In Family Tritoniidae

        Delete *Tritonia exsulans*.

        Tritonia festiva is known to feed on *Alcyonium rudyi*.

254   3b   *Janolus barbarensis* probably does not occur north of California.

257   To Family Tergipedidae add the following:

           Catriona rickettsi Behrens, 1984. Has been found with *Tubularia* (Hydrozoa) in southern Oregon.

           Cuthona flavovulta (MacFarland, 1966). Found as far north as the Olympic Penisula, Washington.

           Cuthona fulgens (MacFarland, 1966). Found as far north as the Olympic Peninsula.

           Cuthona lagunae (O'Donoghue, 1926). Known to occur as far north as Curry County, Oregon.

           Cuthona punicea Millen, 1986

           Tenellia adspersa (Nordmann, 1845). Has been found in Coos Bay, Oregon; feeds on *Tubularia*.

258   In Order Basommatophora add Family Trimusculidae, with *Trimusculus reticulatus* (Sowerby, 1835). This species has been found as far north as the Olympic Peninsula, Washington.

13. Phylum Mollusca: Class Bivalvia

264   30a   .... *Serripes groenlandicus* (Fig. 13.31)

270   To references for Family Mytilidae add the following:

        Seed, R. 1992. Systematics, evolution, and distribution of mussels belonging to the genus *Mytilus*. Amer. Malacol. Bull., 9:123-37.

    3b   For *Mytilus edulis* Linnaeus, 1758 read *Mytilus trossulus* Gould, 1850. (True *M. edulis* is believed to be restricted to Atlantic shores. Distinctions between these species are based primarily on biochemical characteristics, rather than on visible features.

274   In Family Thyasiridae

    1b   Outline of valves nearly circular; etc.

Additions and Corrections

278   6b   This choice should perhaps be deleted, because *Clinocardium fucanum* seems not to be distinct from *C. blandum* (choice 4a).

279   4b   For *nucleoides* read *nuculoides*.

Some species of *Macoma* that have been left out of the key are known to occur in shallow subtidal habitats, and certain of them are occasionally found intertidally. A key to all species that may be expected is therefore provided here. Common intertidal macomas are adequately covered by couplets 5-8 of the original key.

5a   Portion of the shell anterior to the umbones about 1.5 times as long as the posterior portion (most of the periostracum polished and transparent, not conspicuously eroding; length less than 3 cm; relatively rare, and not likely to be encountered at depths of less than 10 meters)   6

5b   Portion of the shell anterior to the umbones not more than 1.3 times as long as the posterior portion   7

6a   Posterior slope of the shell with a concavity where the ligament is attached; length up to 2 cm   *Macoma (Psammacoma) yoldiformis* Carpenter, 1864

6b   Posterior slope of the shell without a concavity where the ligament is attached; length up to 3 cm   *Macoma (Macoma) moesta* (Deshayes, 1855)

7a   Ligament wide, only about one-fourth as long as the posterior slope of the shell (shell up to about 8 cm long, about 1.3 times as long as high; persisting portions of periostracum polished; usually in rather clean sand, often common intertidally   *Macoma (Rexithaerus) secta* (Conrad, 1837)

7b   Ligament not especially wide, and decidedly more than one-fourth as long as the posterior slope of the shell   8

8a   Periostracum somewhat polished, generally yellowish, except where darkened (length up to nearly 5 cm, and 1.4 times the height (mostly subtidal)   *Macoma (Rexithaerus) expansa* Carpenter, 1864

8b   Periostracum not polished, rather fibrous, and usually conspicuously eroded   9

9a   Posterior slope of shell, when viewed from the side, almost perfectly straight, and with an obvious lengthwise flattening or concavity when viewed from above (the ligament occupies about half of the flattening or concavity) (length up to about 3.5 cm; periostracum gray-green, much of it persisting; probably strictly subtidal)   *Macoma (Macoma) elimata* Dunnill & Coan, 1968

9b   Posterior slope of shell not almost perfectly straight, and without a lengthwise flattening or concavity where the ligament is attached   10

10a   Shell not so much as 1.25 times as long as high (posterior portions of both valves bent slightly to the right; length up to about 3 cm; probably strictly subtidal)   *Macoma (Macoma) obliqua* (Sowerby, 1817)

10b   Shell from 1.25 to 1.5 times as long as high   11

11a   Posterior portions of both valves obviously bent to the right   12

11b   Posterior portions of valves essentially straight   15

12a   Ventral margins of valves evenly rounded (length up to 7 cm; probably strictly subtidal)   *Macoma (Macoma) lipara* Dall, 1916

12b   Ventral margins of valves not evenly rounded (a considerable portion of the posterior half may be nearly straight or even slightly concave)   13

13a   Pallial sinus of left valve reaching the scar of the anterior adductor muscle (but the pallial sinus of the right valve does not reach the corresponding scar) (length up to 6.5 cm; common intertidal and subtidal species)   *Macoma (Heteromacoma) nasuta* (Conrad, 1837)

13b   Pallial sinus of left valve not reaching the scar of the anterior adductor muscle   14

14a   Length up to 8 cm; ligament conspicuous, nearly half the length of the posterior slope   *Macoma (Macoma) brota* Dall, 1916

14b   Length up to about 2.5 cm; ligament narrow, inconspicuous, not more that one-third the length of the posterior slope of the shell   *Macoma (Psammacoma) carlottensis* Whiteaves, 1880

15a   Anteriormost portion of pallial sinus not turning abruptly backward before joining the pallial line   16

500 Additions and Corrections

279   15b   Pallial sinus, after reaching its anteriormost point, turning abruptly backward and joining the pallial line at a narrow angle (length up to nearly 6 cm; probably strictly subtidal)
                                          Macoma (*Macoma*) *calcarea* (Gmelin, 1791)

        16a   Length rarely exceeding 2 cm; color white, yellowish, pink, or blue (common intertidal and shallow subtidal species)   *Macoma* (*Macoma*) *balthica* (Linnaeus, 1758)

        16b   Length up to about 4.5 cm; color (except for the periostracum) white (common intertidal and shallow subtidal species)
                                          Macoma (*Heteromacoma*) *inquinata* (Deshayes, 1855)

        In the list of species for *eliminata* read *elimata*.

288   In Family Cuspidariidae for *Cuspidaria* (*Cuspidaria*) *glacialis* read *Cuspidaria* (*Cuspidaria*) *subglacialis*.

14. Phylum Mollusca: Classes Scaphopoda, Cephalopoda

289   To references for Class Scaphopoda add the following:
        Shimek, R. L. 1989. Shell morphogenesis and systematics: a revision of the slender, shallow-water *Cadulus* of the northeastern Pacific (Scaphopoda: Gadilida). Veliger, 32:233-46.
        ------, & G. Moreno. 1996. A new species of eastern Pacific *Fissidentalium* (Mollusca: Scaphopoda). Veliger, 39:71-82.

290   In Family Dentaliidae
        For *Dentalium pretiosum* Sowerby, 1860 read *Antalis pretiosum* (Sowerby, 1860).
        For *Dentalium rectius* Carpenter, 1864 read *Rhabdus rectius* (Carpenter, 1864), now in Family Rhabdidae (below).
        Add the following species:
           Dentalium vallicolens Raymond, 1904
           Dentalium neohexagonum (Pilsbry & Sharp, 1897)
           Fissidentalium erosum Shimek & Moreno, 1996.
           Fissidentalium actiniophorum Shimek (in press)
           Fissidentalium megathyris (Dall, 1890)
           Siphonodentalium quadrifissatum (Pilsbry & Sharp, 1898)
        (Not all of the above have been found in the region, but at least some of them are to be expected.)
        After Family Dentaliidae, add Family Rhabdidae and *Rhabdus rectius* (Carpenter, 1864).
      In Family Gadilidae
        For *Cadulus aberrans* Whiteaves, 1887 read *Gadila aberrans* (Whiteaves, 1887).

291   To references for Class Cephalopoda, add:
        Nesis, K. R., 1987. Cephalopods of the World: Squids, Cuttlefishes, Octopuses, and Allies. Neptune City, New Jersey: THF Publications, Inc. 351pp.

15. Phylum Tardigrada and Phylum Arthropoda: Subphyla Chelicerata, Uniramia

301   Under Order Acarida, add
        Traskorchestianoetus spiceri Fain & Colloff, 1990 and *T. brevipes* Fain & Colloff, 1990 are parasitic on *Traskorchestia traskiana* (Amphipoda) in British Columbia, and perhaps also in Washington and Oregon. They are assigned to the family Histiostomatidae. The authors do not mention *Gammaridacarus orchestoideae*, mentioned above, which may be closely related, even if it has been placed in a different family.
      To references for Order Acarida, add:
        Fain, A., & M. J. Colloff. 1990. A new genus and two new species of mites (Acari, Histiostomatidae) phoretic on *Traskorchestia traskiana* (Stimpson, 1857) (Crustacea: Amphipoda) from Canada. J. Nat. Hist., 24:667-72.

16. Phylum Arthropoda: Subphylum Crustacea: Classes Branchiopoda, Copepoda, Ostracoda, Cirripedia

303   To references for Family Podonidae add the following:
         Della Croce, N. 1974. Cladocera. Fiches d'Identification du Zooplancton. Conseil Permanent pour l'Exploration de la Mer, no. 143 (supersedes Rhamner, 1939).
         Smirnov, N. N., & B. V. Timms. 1893. A revision of the Australian Cladocera (Crustacea). Rec. Austral. Mus., Suppl. 1:1-130.

315   2a   For *Mitella polymerus* read *Pollicipes polymerus*.

316   In Family Scalpellidae
         For *Mitella polymerus* Sowerby, 1833 read *Pollicipes polymerus* (Sowerby, 1833).

317   5a   Tubes within wall plates without transverse septa, at least in the upper part of the wall (expose the tubes by filing the plates, then blow out the powder); etc.
       5b   Tubes within wall plates with transverse septa; etc.

17. Phylum Arthropoda: Subphylum Crustacea: Class Malacostraca: Subclasses Phyllocarida, Peracarida (in part)

321   To references for Order Mysidacea add the following:
         Kathman, R. D., W. C. Austin, J. C. Saltman, & J. D. Fulton. 1986. Identification Manual to the Mysidacea of the northeast Pacific. Can. Spec. Publ. Fish. Aquat. Sci., no. 93:1-411.

325   To references for Order Cumacea add the following:
         Gladfelter, W. B. 1975. Quantitative distribution of shallow-water Cumacea from the vicinity of Dillon Beach, California, with descriptions of five new species. Crustaceana, 29:241-51.
         Watling, L., & L. McCann. 1996. Taxonomic Atlas of the Benthic Fauna of the Santa Maria Basin and Western Santa Barbara Channel. Cumacea. Santa Barbara: Santa Barbara Museum of Natural History.

328   In Family Lampropidae add or change the following:
         Hemilamprops californicus (Zimmer, 1907)
         Lamprops obfuscatus (Gladfelter, 1975)
         Lamprops serratus Hart, 1930
         Lamprops tomalesi Gladfelter, 1975
         Lamprops triserratus (Gladfelter, 1975)
         Mesolamprops dillonensis Gladfelter, 1975
         For *Lamprops carinata* read *Lamprops carinatus*.
         For *Lamprops serrata* read *Lamprops serratus*.
         Lamprops fuscatus G. O. Sars, 1865 (*L. fuscata*) should perhaps be deleted.
    In Family Diastylidae, add or change the following:
         Anchicolurus occidentalis (Calman, 1912)
         Diastylis abbotti Gladfelter, 1975
         Diastylis crenellata Watling & McCann, 1996
         Diastylis quadriduplicata Watling & McCann, 1996
         Diastylis santamariaensis Watling & McCann, 1996
         Diastylis sentosa Watling & McCann, 1996
         Diastylis serratocostata Watling & McCann, 1996
         Leptostylis abditis Watling & McCann, 1996
         Leptostylis calva Watling & McCann, 1996
         For *Diastylis dawsoni* read *Diastylopsis dawsoni*.
         Delete *Diastylis rathkei* (Krøyer, 1841) and *Leptostylis villosa* G. O. Sars, 1869.
    In Family Leuconidae add or change the following:
         Eudorella redacticruris Watling & McCann, 1996
         Delete *Eudorella tridentata*. (It is a synonym of *Eudorella pacifica*.)
         Leucon (Diaphonoleucon) declivis Watling & McCann, 1996
         Leucon (Leucon) falcicostata Watling & McCann, 1996
         Nippoleucon hinumensis (Gamo, 1967) In estuaries, introduced; originally described from Japan.

502 Additions and Corrections

328   *Leucon fulvus* G. O. Sars, 1864 and *Leucon nasica* Krøyer, 1841 should perhaps be deleted.
In Family Nannastacidae add
    Campylaspis biplicata Watling & McCann, 1996
    Campylaspis blakei Watling & McCann, 1996
    Campylaspis canaliculata Zimmer, 1936
    Campylaspis maculinodulosa Watling & McCann, 1996
    Cumella (Cumella) californica Watling & McCann, 1996
    Cumella (Cumella) morion Watling & McCann, 1996
    Procampylaspis caenosa Watling & McCann, 1996
In Family Bodotriidae add or change the following:
    Cyclaspis nubila Zimmer, 1936
    Delete family Colurostylidae and *Colurostyis occidentalis* (the latter changed to *Anchicolurus occidentalis* and transferred to Diastylidae).

337   In Family Eurycopidae, for *Acanthomunnopsis milleri* Wilson, 1982 read *Baeonectes improvisus* Wilson, 1982.
In Family Munnopsidae, for *Baeonectes improvisus* Wilson, 1982 read *Acanthomunnopsis milleri* Wilson, 1982.

338   To references for Suborder Valvifera add the following:
    Rafi, F., & D. R. Laubitz. 1990. The Idoteidae (Crustacea: Isopoda: Valvifera) of the shallow waters of the northeastern Pacific Ocean. Can. J. Zool., 658:2649-87.
    7a   For *Synidotea bicuspida* read *Synidotea consolidata*.
    12b *Idotea (Idotea) ochotensis* is believed not to occur in our region.

340   17a For *Idotea (Pentidotea) aculeata* read *Idotea (Pentidotea) recta*.
In Family Idoteidae add or change the following:
    For *Idotea (Pentidotea) aculeata* Stafford, 1913 read *Idotea (Pentidotea) recta* Rafi & Laubitz 1990.
    Add *Idotea (Idotea) aleutica* Gurjanova, 1933.
    Delete *Idotea (Pentidotea) ochotensis* Brandt, 1851.
    For *Synidotea bicuspida* (Owen, 1839) read *Synidotea consolidata* Stimpson, 1856.
    Add *Synidotea cornuta* Rafi & Laubitz, 1990, *Synidotea minuta* Rafi & Laubitz, 1990, and *Edotia sublitoralis* (Menzies & Barnhart, 1959).

18. Phylum Arthropoda: Subphylum Crustacea: Class Malacostraca: Subclass Peracarida: Order Amphipoda

347   To references for Suborder Gammaridea add the following:
Barnard, J. L., & C. M. Barnard. 1981. The amphipod genera *Eobrolgus* and *Eyakia* (Crustacea: Phoxocephalidae) in the Pacific Ocean. Proc. Biol. Soc. Wash., 94: 295-313.
------, & ------. 1982. Revision of *Foxiphalus* and *Eobrolgus* (Crustacea: Amphipoda: Phoxocephalidae) from American oceans. Smithsonian Contrib. Zool., 372: 1-35.
------, & G. S. Karaman. 1991. The families and genera of marine gammaridean Amphipoda (except marine gammaroids). Rec. Austral, Mus., Suppl. 13, Part I (pp.1-417) and Part II (pp.419-866).
Blake, J. A., L. Watling, & P. H. Scott (eds.). 1995. Taxonomic Atlas of the Santa Maria Basin and Western Santa Barbara Channel, Vol. 12, The Crustacea, Part 3. Amphipoda. Santa Barbara: Santa Barbara Museum of Natural History. xiv + 251 pp.
Bousfield, E. L., & E. A. Hendrycks, 1994. The amphipod superfamily Leucothoidea on the Pacific coast of North America. Family Pleustidae: Subfamily Pleustinae. Systematics and biogeography. Amphipacifica, 1(2.:3-69.
------, & ------. 1995a. The amphipod superfamily Eusiroidea in the North American Pacific region. I. Family Eusiridae: systematics and distributional ecology. Amphipacifica, 1(4):3-59.
------, & ------. 1995b. The amphipod family Pleustidae on the Pacific coast of North America: part III. Subfamilies Parapleustinae, Dactylopleustinae, and Pleusirinae. Systematics and distributional ecology. Amphipacifica, 2(1): 65-133.

347 ------, & P. M. Hoover. 1995. The amphipod superfamily Pontoporeioidea on the Pacific coast of North America. I. Family Haustoriidae. Genus *Eohaustorius* J.L. Barnard: systematics and distributional ecology. Amphipacifica, 2(1):35-63.

------, & Z. Kabata. 1988. Amphipoda, *In* Margolis, L., & Z. Kabata (eds.), Guide to the Parasites of Pacific Fishes of Canada. Part II. Crustacea. Can. Spec. Publ. Fish. Aquat. Sci., 101:149-63.

------, & J. A. Kendall. 1994. The amphipod superfamily Dexaminoidea on the North American Pacific Coast; Families Atylidae and Dexaminidae: Systematics and distributional ecology. Amphipacifica, 1(3):3-66.

------, & C.-t. Shih. 1994. The phyletic classification of amphipod crustaceans; problems in resolution. Amphipacifica, 1(3): 76-134.

------, & C. P. Staude. 1994. The impact of J. L. Barnard on North American Pacific amphipod research: a tribute. Amphipacifica, 1(1): 3-16.

Conlan, K. E.. 1990. Revision of the crustacean amphipod genus *Jassa* Leach (Corophioidea: Ischyroceridae). Canadian J. Zool., 68: 2031-75.

------. 1995. Thumbing doesn't always make the genus: revision of *Microjassa* Stebbing (Corophioidea: Ischyroceridae). Bull. Mar. Sci., 57:333-77.

Coyle, K. O. 1982. The amphipod genus *Grandifoxus* Barnard (Gammaridea: Phoxocephalidae) in Alaska. J. Crust. Biol., 2: 430-50.

Jarrett, N. E., & E. L. Bousfield. 1994a. The amphipod superfamily Phoxocephaloidea on the Pacific coast of North America. Family Phoxocephalidae. Part I. Metharpiniidae, new subfamily. Amphipacifica, 1(1):58-140.

------, & ------. 1994b. The amphipod superfamily Phoxocephaloidea on the Pacific coast of North America. Family Phoxocephalidae. Part II. Subfamilies Pontharpiniinae, Parharpiniinae, Brolginae, Phoxocephalinae, and Harpiniinae. Systematics and distributional ecology. Amphipacifica, 1(2): 71-150.

Meyers, A. A., & D. McGrath. 1984. A revision of the north-east Atlantic species of *Ericthonius* (Crustacea: Amphipoda). J. Mar. Biol. Assoc. United Kingdom, 64: 379-400.

Moore, P. G. 1992. A study on amphipods from the superfamily Stegocephaloidea Dana 1852 from the northeastern Pacific region: systematics and distributional ecology. J. Nat. Hist., 26:905-36.

Staude, C. P. 1995. The amphipod genus *Paramoera* Miers (Gammaridea: Eusiroidea: Pontogeneiidae) in the eastern North Pacific. Amphipacifica, 1(4):61-102.

350 On fig. 18.9 (typical amphipod leg) number the articles from 1 (top of figure) to 7 (bottom of figure)

351 17b For *Erichthonius hunteri* Bate 1862 read *Ericthonius rubricornis* (Stimpson, 1853).

361 10a For *Rhachotropis clemens* Barnard. 1967 read *Rhacotropis barnardi* Bousfield & Hendrycks, 1995.

12a For *Coboldus* sp. read *Iphimedia rickettsi* (Shoemaker, 1931).

12b For *Odius kelleri* Bruggen, 1907 read *Cryptodius kelleri* (Brüggen, 1907).

362 Fig. 18.65 For *Pleustes depressa* read *Thorlaksonius depressus*

363 19a For *Heterophoxus oculatus* (Holmes, 1908) read *Heterophoxus* spp.

20a For *Mandibulophosus gilesi* Barnard, 1957 read *Mandibulophoxus* spp.

22a For *Phoxocephalus homilis* (Barnard, 1960) read *Cephalophoxoides homilis* (Barnard, 1960).

25a For *Grandifoxus grandis* (Stimpson, 1856) read *Grandifoxus* spp.

27a For *Rhepoxynius bicuspidatus* (Barnard, 1960) read *Rhepoxynius barnardi* Jarrett & Bousfield, 1994.

30b For *Rhepoxynius variatus* (Barnard, 1960) read *Rhepoxynius boreovariatus* Jarrett & Bousfield, 1994.

364 Fig. 18.69 For *Metaphoxus fultoni* read *Parametaphoxus quaylei*.

33a For *Foxiphalus major* (Barnard, 1960) read *Majoxiphalus* spp.

34a For *Foxiphalus obtusidens* (Alderman, 1936) read *Foxiphalus* spp.

365 35a For *Eyakia robustus* (Holmes, 1908) read *Eyakia robusta* group.

504   Additions and Corrections

365	36a	For *Eobrolgus spinosus* (Holmes, 1905) read *Eobrolgus chumashi* Barnard & Barnard, 1981.
	36b	For *Paraphoxus oculatus* G. O. Sars, 1879 read *Paraphoxus* spp.
367	16a	For *Orchomene pacifica* read *Orchomene pacificus*.
	18a	For *Orchomene obtusa* read *Orchomene obtusus*.
	20b	For *Lysianassa holmesi* (Barnard, 1955) read *Aruga holmesi* Barnard, 1955.
368	22a	For *Allogaussia recondita* Stasek, 1958 read *Orchomene recondita* (Stasek, 1958).
369	2a	For *Dexamonica reduncans* Barnard, 1957 read *Guernea reduncans* (Barnard, 1957).
370	8a	For *?Megamphopus* sp. read *Gammaropsis* sp.
	16a	*Gammaropsis (Gammaropsis) ellisi*
371	17a	*Gammaropsis (Gammaropsis) shoemakeri*
	18a	*Gammaropsis chionoecetophila*
	19a	*Gammaropsis (Podoceropsis) chionoecetophila*
	20a	*Gammaropsis (Podoceropsis) ociosa*
	20b	*Gammaropsis (Podoceropsis) barnardi*
	21a	*Gammaropsis (Gammaropsis) spinosa*
	25a	For Bruggen, 1905 read Brüggen, 1906.
372	4a	For *Halicella halona* Barnard, 1971 read *Rhynohalicella halona* (Barnard, 1971).
	4b	For 1893 read 1895.
	10b	For *Eusirus* spp. read *Eusirus columbianus* Bousfield & Hendrycks, 1995.
374	Fig. 18.77	For *Parapleustes nautilus* read *Micropleustes nautilus*.
375	32a	For *Parapleustes nautilus* (Barnarad, 19969) read *Micropleustes* spp.
	33a	For *Parapleustes oculatus* (Holmes, 1908) read *Chromopleustes* spp.
	34a	For *Parapleustes den* Barnard, 1969 read *Gnathopleustes* spp.
	34b	For *Parapleustes pugettensis* (Dana, 1853) group read *Gnathopleustes* and *Trachypleustes* spp.
	35a	For *Polycheria osborni* Calman, 1898 read *Polycheria* spp.
376	42a	*Paramoera serrata* Staude, 1995
	44b	*Paramoera suchaneki* Staude, 1995
	46a	*Paramoera bucki* Staude, 1995
	47a	*Paramoera leucophthalma* Staude, 1995
	47b	*Paramoera bousfieldi* Staude, 1995

377   In Family Pontogeneidae add
          Paramoera bousfieldi Staude, 1995
378   *Paramoera bucki* Staude, 1995
          Paramoera leucophthalma Staude, 1995
          Paramoera serrata Staude, 1995
          Paramoera suchaneki Staude, 1995
      In Family Eusiridae
          For *Eusirus* ssp. read *Eusirus columbianus* Bousfield & Hendrycks, 1995.
          For *Rhachotropis clemens* Barnard, 1967 read *Rhacotropis barnardi* Bousfield & Hendrycks, 1995.
          Add *Rhacotropis conlanae* Bousfield and Hendrycks, 1995 and *R. miniata* Bousfield & Hendrycks, 1995. Both deep subtidal, on fine sediment, and probably also pelagic.
379   In Family Pleustidae
          For *Parapleustes den* Barnard, 1969 read *Gnathopleustes den* (Barnard, 1969).
          For *Parapleustes nautilus* Barnard, 1969 read *Micropleustes nautilus* (Barnard, 1969).
          For *Parapleustes oculatus* read *Chromopleustes oculatus*.
          For *Parapleustes pugettensis* group read *Gnathopleustes pugettensis* group.
          For *Pleustes depressa* Alderman, 1936 read *Thorlaksonius depressus* (Alderman, 1936).
      Add the following species:
          Chromopleustes lineatus Bousfield & Hendrycks. Low intertidal to subtidal, on rocks with algae and surfgrass.
          Dactylopleustes echinoides Bousfield & Hendrycks, 1995. Low intertidal to subtidal, possibly commensal with sea urchins.
          Gnathopleustes polychaetus Bousfield & Hendrycks, 1995. Low intertidal to subtidal, on rocks with algae.
          Gnathopleustes serratus Bousfield & Hendrycks, 1995. Low intertidal on rocky shores.

379 *Gnathopleustes simplex* Bousfield & Hendrycks, 1995. Low intertidal and subtidal, on rocks with sponges and algae.
Gnathopleustes trichodus Bousfield & Hendrycks, 1995. Subtidal, habitat unknown.
Micropleustes nautiloides Bousfield & Hendrycks, 1995. Intertidal and shallow subtidal, on algal mats and seagrass.
Pleustes constantinus Bousfield & Hendrycks, 1995. Shallow subtidal, habitat unknown.
Pleustes victoriae Bousfield & Hendrycks, 1994. Low intertidal to subtidal; habitat unknown.
Thorlaksonius borealis Bousfield & Hendrycks, 1994. Low intertidal and subtidal, on rocks with algae.
Thorlaksonius brevirostris Bousfield & Hendrycks, 1994. Low intertidal and subtidal, on algae and seagrass.
Thorlaksonius carinatus Bousfield & Hendrycks, 1994. Shallow subtidal, on rocks with algae.
Thorlaksonius grandirostris Bousfield & Hendrycks, 1994. Low intertidal, on rocks with seagrass; probably mimics a snail.
Thorlaksonius subcarinatus Bousfield & Hendrycks, 1994. Low intertidal and subtidal, on rocks with algae.
Thorlaksonius truncatus Bousfield & Hendrycks, 1994. Shallow subtidal, on sand with drift algae.
Trachypleustes trevori Bousfield & Hendrycks, 1995. Low intertidal, on rocks with algae.
Trachypleustes vancouverensis Bousfield & Hendrycks, 1995. Low intertidal rocks with algae.

380 In Family Phoxocephalidae
For *Eobrolgus spinosus* (Holmes, 1905) read *Eobrolgus chumashi* Barnard and Barnard, 1981.
For *Foxiphalus major* read *Majoxiphalus major*.
For *Heterophoxus oculatus* (Holmes, 1908) read *Paraphoxus gracilis* Jarrett & Bousfield, 1994.
For *Metaphoxus frequens* Barnard, 1960 read *Parametaphoxus quaylei* Jarrett & Bousfield, 1994
For *Phoxocephalus homilis* Barnard, 1960 read *Cephalophoxoides homilis* (Barnard, 1960)
For *Rhepoxynius variatus* ((Barnard, 1960) read *Rhepoxynius boreovariatus* Jarrett & Bousfield, 1994
Add the following species:
Foxiphalus aleuti Barnard & Barnard, 1982. In sand, subtidal to deep.
Foxiphalus falciformis Jarrett & Bousfield, 1994. In sand, low intertidal.
Foxiphalus fucaximeus Jarrett & Bousfield, 1994. In sand, low intertidal.
Foxiphalus xiximeus Barnard & Barnard, 1982. In sand, low intertidal.
Grandifoxus aciculatus Coyle, 1982. Subtidal to deep, in fine sediment.
Grandifoxus dixonensis Jarrett & Bousfield, 1994. Deep, in fine sediment.
Grandifoxus lindbergi (Gurjanova, 1953) In sand, shallow subtidal.
Grandifoxus longirostris (Gurjanova, 1953) In sand, subtidal.
Heterophoxus affinis (Holmes, 1908). In fine sediment, subtidal to deep.
Heterophoxus conlanae Jarrett & Bousfield, 1994. In fine sediment, subtidal.
Heterophoxus ellisi Jarrett & Bousfield, 1994. In fine sediment, subtidal.
Majoxiphalus maximus Jarrett & Bousfield, 1994. In fine sediment, subtidal.
Mandibulophoxus alaskensis Jarrett & Bousfield, 1994. In fine sediment, low intertidal to subtidal.
Mandibulophoxus mayi Jarrett & Bousfield, 1994. In fine sediment, low intertidal to subtidal.
Paraphoxus communis Jarrett & Bousfield, 1994. In mixed sediment, low intertidal to shallow subtidal.
Paraphoxus pacificus Jarrctt & Bousfield, 1994. In mixed sediment, low intertidal to subtidal.

Additions and Corrections

380
    Paraphoxus similis Jarrett & Bousfield, 1994. Subtidal, in mixed sediment.
    Rhepoxynius barnardi Jarrett & Bousfield, 1994. In fine sediment, subtidal.

381 In Superfamily Lysianassoidea add Family Cyphocaridae, with *Cyphocaris challengeri* Stebbing, 1888. Deep subtidal; pelagic.
    In Family Lysianassidae
        For *Acidostoma* sp. read *Acidostoma hancocki* Hurley, 1963. Subtidal (sometimes deep), on soft sediment; possibly commensal.
        Delete *Cyphocaris challengeri* Stebbing, 1888. (It has been tranferred to Family Cyphocaridae, above.)
        For *Lysianassa holmesi* (Barnard, 1955) read *Aruga holmesi* Barnard, 1955.

382 Under Superfamily Stegocephaloidea add *Family Stegocephalidae with *Stegocephalexia penelope* Moore, 1992. Subtidal.
    In Family Acanthonotozomatidae
        For *Coboldus* sp. read *Iphimedia rickettsi* (Shoemaker, 1931). Subtidal on rocky substrata, especially with coralline algae and holdfasts of kelps; possibly commensal; similar to *Coboldus hedgpethi* from California.
        For *Odius kelleri* Brüggen, 1907 read *Cryptodius kelleri* (Brüggen, 1907). Subtidal, on rocky substrata, especially with algae; on Amchitka Island, Alaska, this species co-occurs with a similar species, *Imbrexodius oclairi* Moore, 1992.
    In Family Lafystiidae add
        *Members of this family are fish parasites. (See Bousfield and Kabata, 1988, cited above).
    In Family Pardaliscidae
        For *Halicella halona* Barnard, 1971 read *Rhynohalicella halona* (Barnard, 1971).
        Add *Pardaliscella symmetrica* Barnard, 1959. Deep subtidal, soft sediment.
    In Family Atylidae add
        Atylus borealis Bousfield and Kendall, 1994. Shallow subtidal sand and eelgrass.
        Atylus georgianus Bousfield and Kendall, 1994. Shallow subtidal sand and eelgrass.
    In Family Dexaminidae
        For *Dexamonica reduncans* Barnard, 1957 read *Guernea reduncans* (Barnard, 1957).
        Add *Polycheria carinata* Bousfield & Kendall, 1994. Low intertidal to subtidal, commensal on ascidians and sponges.
        Add *Polycheria mixillae* Bousfield & Kendall, 1994. Low intertidal to subtidal, commensal on sponges.

383 In Superfamily Pontoporeioidea add *Family Pontoporeiidae, with *Pontoporeia femorata* Krøyer, 1842. Shallow subtidal, soft sediment.
    In Family Melphidippidae, for *Melphissana* read *Melphisana*.

384 In Family Isaeidae add
    Gammaropsis (*Gammaropsis*) *ellisi* Conlan, 1983.
    Gammaropsis (*Gammaropsis*) *shoemakeri* Conlan, 1983
    Gammaropsis (*Gammaropsis*) *spinosa* (Shoemaker, 1942)
    Gammaropsis (*Gammaropsis*) *thompsoni* (Walker, 1898)

385 For *Kermystheus ociosa* Barnard, 1962 read *Gammaropsis* (*Podoceropsis*) *ociosa* (Barnard, 1962).
    For ?*Megamphopus* sp. read *Gammaropsis* (*Podoceropsis*) sp.
    For *Podoceropsis barnardi* read *Gammaropsis* (*Podoceropsis*) *barnardi* .
    For *Podoceropsis chionoecetophila* read *Gammaropsis* (*Podoceropsis*) *chionoecetophila*.
    Add *Photis macrotica* Barnard, 1962. In soft sediments, subtidal to deep.
    In Family Ischyroceridae add
        * *Jassa borowskyae* Conlan, 1990. Exposed rocky shores, on algae and surfgrass.
        Jassa morinoi Conlan, 1990. Low intertidal, on rocks and algae.
        Jassa oclairi Conlan, 1990. Low intertidal and subtidal, on algae and sponges.
        Jassa shawi Conlan, 1990. Low intertidal and subtidal, on hard substrata and sponges.
        Jassa slatteryi Conlan, 1990. Low intertidal and subtidal, on algae and hydroids.
        Jassa staudei Conlan, 1990. Low intertidal and subtidal, on rocks and algae.
        Microjassa barnardi Conlan, 1995. Subtidal, soft sediment.
        Microjassa bousfieldi Conlan, 1995. Subtidal to deep, soft sediment.

386 In Family Corophiidae
For *Erichtonius hunteri* Bate, 1862 read *Ericthonius rubricornis* (Stimpson, 1853).
To references for Suborder Caprellidea add the following:
Watling, L. 1995. The Suborder Caprellidea. *In* Blake, J. A., L. Watling, & P. H. Scott, Taxonomic Atlas of the Benthic Fauna of the Santa Maria Basin and Western Santa Barbara Channel, vol. 12. The Crustacea Part 3. The Amphipoda, pp.223-240.

19. Phylum Arthropoda: Subphylum Crustacea: Class Malacostraca: Subclass Eucarida

392 To references for Order Decapoda add the following:
Jensen, G. 1995. Pacific Coast Crabs and Shrimps. Monterey, California: Sea Challengers. vii + 87 pp.
To "Natantia"--Shrimps add Family Palaemonidae, with *Palaemon macrodactylus* Rathbun, 1902. This estuarine species, introduced from southeast Asia, became established in San Francisco Bay about 1950, and is now also known to be present in Coos Bay, Oregon and Willapa Bay, Washington.

393 In Family Sergestidae, for *Petalidium subspinosum* Burkenroad, 1937 read *Petalidium suspiriosum* Burkenroad, 1937.

397 In Family Pandalidae add the following reference:
Squires, H. J. 1992. Recognition of *Pandalus eous* Makarov, 1935, as a Pacific species not a variety of the Atlantic *Pandalus borealis* Krøyer, 1838 (Decapoda, Caridea). Crustaceana, 63:257-62.

Revision of key beginning with couplet 3
3a Posterior margin of abdominal segment 4 with a sharp median spine
Pandalus eous Makarov, 1935
3b Posterior margin of abdominal segment 4 without a median spine   4
4a Distal half of the scale of antenna 2 narrower than the spine
Pandalus stenolepis Rathbun, 1902
4b Distal half of the blade of the scale of antenna 2 not narrower than the spine   5
5a Telson with 8-13 pairs of dorsolateral spines; color uniformly pink, with no bands or other markings on legs or body   *Pandalus jordani* Rathbun, 1902
5b Telson with 4-6 pairs of dorsolateral spines; legs and/or body with bands or other markings   6
6a Distal half of rostrum with dorsal spines   7
6b Distal half of rostrum without dorsal spines   8
7a Only the anterior half of the carapace (exclusive of the rostrum) with median dorsal spines
Pandalus platyceros Brandt, 1851
7b Median dorsal spines extending posteriorly beyond the middle of the carapace (exclusive of the rostrum)   *Pandalus hypsinotus* Brandt, 1851
8a Abdominal segment 3 with a median lobe or spine; rostrum bifid
Pandalus goniurus Stimpson, 1860
8b Abdominal segment 3 without a median lobe or spine; rostrum trifid   9
9a Median dorsal spines extending slightly posterior to the middle of the carapace (exclusive of the rostrum); dactyl of leg 5 about one-third as long as the propodus; abdomen with irregular brown or red stripes   *Pandalus danae* Stimpson, 1857
9b Only the anterior half of the carapace (exclusive of the rostrum) with median dorsal spines; dactyl of leg 5 about one-fifth the length of the propodus; abdomen uniformly pink
Pandalus tridens Rathbun, 1902

In Family Hippolytidae add the following references:
Jensen, G. C. 1987. A new species of the genus *Lebbeus* (Caridea: Hippolytidae) from the northeastern Pacific. Bull. So. Calif. Acad. Sci., 86:89-94.
Wicksten, M. K. 1990. Key to the hippolytid shrimp of the eastern Pacific Ocean. Fish. Bull. U. S., 88:587-98.

398 7b *Heptacarpus taylori* is a southern species that probably does not occur in our region.
13b *Heptacarpus pictus* may be a synonym of *Heptacarpus sitchensis* (choice 14b).
399 27a and 27b Delete: rostrum not reaching article 2 of antenna 1.
31a and 31b For spiny read serrated.

Additions and Corrections

399   38a   Pleura of abdominal segments 1-3 with broadly rounded ventral margins
Spirontocaris arcuata Rathbun, 1902
      38b   Pleura of abdominal segments 1-3 with pointed ventral margins
smaller specimens of *Spirontocaris lamellicornis* (Dana, 1852)
Note: *Spirontocaris spina* (originally in choice 38b) probably does not occur in our region.
39b For *Lebbeus* sp. read *Lebbeus catalepsis* Jensen, 1987.

400   In Family Hippolytidae add *Lebbeus catalepsis* Jensen, 1987. This northern species has been found in the Strait of Juan de Fuca.

402   In Family Crangonidae add *Crangon stylirostris* Holmes, 1900.
      2a   To the characters of Family Callianassidae, add: with 3 pairs of pleopods

403   2b   To the characters of Family Upogebiidae, add: with 4 pairs of pleopods

404   In Family Callianassidae
      1a   For *Callianassa californiensis* Dana, 1854 read *Neotrypaea californiensis* (Dana, 1854).
      1b   For *Callianassa gigas* Dana, 1852 read *Neotrypaea gigas* (Dana, 1852).

407   13a   For *camtschatica* read *camtschaticus*.

415   Revised key to Family Cancridae
      1a   Dactyl and fixed claw of the chelipeds dark, often black   2
      1b   Dactyl and fixed claw of the chelipeds not dark   6
      2a   Walking legs setose along their entire length; 5 teeth on edge of carapace between the eyes variable in size, shape, or spacing   3
      2b   Propodus and carpus of walking legs nearly devoid of setae; 5 teeth on edge of carapace between the eyes of similar size, shape, and spacing
Cancer productus Randall, 1839 (fig. 19.37)
      3a   Anterolateral and posterolateral margins of the carapace meeting in a smooth curve that has a rounded outline when viewed from above (common intertidal and subtidal species, often nestling in holes, large barnacle shells, kelp holdfasts, etc.)
Cancer oregonensis (Dana, 1852)
      3b   Outline of carapace with a distinct angle where the anterolateral and posterolateral margins meet (mostly found in areas on or near the open coast)   4
      4a   Carapace widest at the 8th tooth behind the eye; carapace up to 18 cm wide (ventral surface of living and fresh specimens with conspicuous red spots)
Cancer antennarius Stimpson, 1856 (fig. 19.38)
      4b   Carapace widest at the 9th tooth behind the eye; carapace less than 6 cm wide   5
      5a   Dorsal margin of dactyl of chelipeds with a series of large, sharp, reddish spines; 10th tooth behind the eye small but distinct (carapace up to 6 cm wide; primarily subtidal)
Cancer branneri Rathbun, 1898
      5b   Dorsal margin of dactyl of chelipeds sometimes with sharp tubercles, but without large spines; 10th tooth behind the eye represented at most by a slight irregularity in the posterolateral margin (rare north of Oregon, but has been found in northern Washington)
Cancer jordani Rathbun, 1900
      6a   Carapace widest at the 10th tooth behind the eye; propodus and dactyl of chelipeds with serrated dorsal margins; carapace up to 23 cm wide   *Cancer magister* Dana, 1852
      6b   Carapace widest at the 9th tooth behind the eye; propodus and dactyl of chelipeds with some spines on the dorsal margin, but not serrate; carapace less than 12 cm wide
Cancer gracilis Dana, 1852

416   In Family Xanthidae, add *Rhithropanopeus harrisii* (Gould, 1841). This is an estuarine Atlantic species that became established in San Francisco Bay about 1940. It has since been found as far north as Coos Bay, Oregon.. The general body color is dull green, and the dactyls and fixed claws are white-tipped.

In Family Grapsidae
Pachygrapsus crassipes has been found as far north as the Olympic Peninsula, Washington. It is rarely seen north of Oregon, however.

20. Phyla Phoronida, Brachiopoda, Entoprocta

421 Correct list of references for Phylum Entoprocta as follows:
Emschermann .... Kamptozoensystematik
O'Donoghue, C. H., & E. O'Donoghue. 1923. A preliminary list of Bryozoa (Polyzoa) from the Vancouver Island Region. Contrib. Can. Biol. Fish., n. s., 1:143-201.
O'Donoghue, C. H., & E. O'Donoghue. 1926. A second list of the Bryozoa (Polyzoa) from the Vancouver Island Region. Contrib. Can. Biol. Fish., n. s., 3:49-131.
Ryland, J. S. 1961.
Delete Wieser, W. 1959.

4b ..... ; free-living or attached to other organisms, but not on the body surface of polychaetes (some colonial species are often found on polychaete tubes, however)

422 6a ..... ; both stalk and calyx strongly muscularized, and with conspicuous cuticular spines
6b .... ; neither stalk nor calyx strongly muscularizd, and not bearing any spines *Pedicellina* sp.
7b Stalks without muscular enlargements (other than those at their bases) *Barentsia robusta*
8a Stalks short (usually less than 2 mm) and delicate, lacking conspicuous cuticular pores 9
8b Stalks tall and robust, bearing conspicuous cuticular pores 10
9a Stalks with several muscular enlargements, thus with a beaded appearance *Barentsia benedeni* (fig. 20.3)
9b Stalks with only 1 or 2 muscular enlargements *Barentsia parva*
10a Stalks generally with 4-6 muscular enlargements, from many of which erect branches arise, making the colonies very bushy *Barentsia ramosa*
10b Stalks generally with 2-4 muscular enlargements, without branches, the colonies therefore not bushy *Barentsia gracilis* var. *nodosa*

Revised checklist of species
Family Loxosomatidae
Loxosoma davenporti Nickerson, 1898
Loxosomella nordgaardi Ryland, 1951. On bryozoans, including *Tegella*.
Loxosomella spp. Small species found on invertebrate hosts, including the sipunculans *Phascolosoma agassizi*, *Golfingia pugettensis*, and *Themiste pyroides*, and the polychaetes *Diopatra ornata*, *Gattyana cirrosa*, *Lepidonotus squamatus*, *Eunoe oerstedi*, and *Lumbrineris* sp.
Family Loxokalypodidae
Loxokalypus socialis Emschermann, 1972. On the body surface of *Glycera nana*.
Family Pedicellinidae
Myxosoma spinosa Robertson, 1900
Pedicellina sp.
Family Barentsiidae
Barentsia benedeni (Foettinger, 1887) (fig. 20.3). Introduced, and found only in bays.
Barentsia gracilis var. *nodosa* (Lomas, 1886) sensu O'Donoghue & O'Donoghue, 1926
Barentsia parva (O'Donoghue & O'Donoghue, 1923)
Barentsia ramosa (Robertson, 1900)
Barentsia robusta Osburn, 1953

21. Phylum Bryozoa (Ectoprocta)

423 To references for Phylum Bryozoa, add
O'Donoghue, C. H., & E. O'Donoghue. 1926. A second list of the Bryozoa (Polyzoa) from the Vancouver Island Region. Contrib. Canad. Biol. Fish., 3:49-131.
Reed, C. G. 1988. The reproductive biology of the gymnolaemate bryozoan *Bowerbankia gracilis* (Ctenostomata: Vesiculariidae). Ophelia, 29:1-23.
Thorpe, J. P., & J. E. Winston. 1984. On the identity of *Alcyonidium gelatinosum* (Linnaeus, 1761) (Bryozoa: Ctenostomata) J. Nat. Hist., 18:853-60.
------. 1986. On the identity of *Alcyonidium diaphanum* Lamouroux, 1813 (Bryozoa: Ctenostomata). J. Nat. Hist. 20:845-8.

510 Additions and Corrections

424   5b   After *Hippodiplosia insculpta* delete (fig. 21.59).

426   The term scutum, as used in couplets 15 to 20, applies to a modified spine that overhangs the frontal area. It may be oval, elliptical, paddle-shaped, fingerlike, or forked.

436   For *Hippodiplosia insculpta* (fig. 21.59) read *Hippodiplosia reticulato-punctata*.

437   108a   After *Hippodiplosia reticulato-punctata* add (fig. 21.59).

438   115a   After *Hippodiplosia insculpta* delete (fig. 21.59).

       121a   For *Hippothoa hyalina* read *Celleporella hyalina*.

440   Fig. 21.74   For *Alcyonidium polyoum* read *Alcyonidium gelatinosum*.

       8b   For *Alcyonidium polyoum* read *Alcyonidium gelatinosum*.

442   In Family Alcyonidiidae
       For *Alcyonidium polyoum* (Hassall, 1841) read *Alcyonidium gelatinosum* (Linnaeus, 1761)
       Add **Alcyonidium albidum*.
   In Family Vesiculariidae, add:
       **Bowerbankia gracilis* Leidy, 1855 var. *aggregata* O'Donoghue & O'Donoghue, 1926

443   In Family Membraniporidae
       Add *Conopeum tenuissimum* (Canu). Established in Coos Bay, Oregon, to which it was introduced with Atlantic oysters.

445   In Family Hippothoidae, for *Hippothoa hyalina* (Linnaeus, 1758) read *Celleporella hyalina* (Linnaeus, 1767).
   In Family Schizoporellidae
       After *Hippodiplosia insculpta* (Hincks, 1882) delete (fig. 21.59).
       After *Hippodiplosia reticulato-punctata* (Hincks, 1877) add (fig. 21.59).

22. Phylum Echinodermata

448   To references for Class Asteroidea add the following:
       Kwast, K. E., D. W. Foltz, & W. B. Stickle. 1990. Population genetics and systematics of the *Leptasterias hexactis* (Echinodermata: Asteroidea) species complex. Mar. Biol., 105:477-89.

       2a   Specimens keying to *Leptasterias hexactis* are now believed to belong to three recognizable entities. They are distinguished to some extent by biochemical properties, but the following visible features will usually enable one to separate them. *Leptasterias hexactis* (Stimpson, 1862): small pedicellariae (as distinguished from larger pedicellariae) especially abundant around the aboral spines, generally embedded in tissue either at the base or about midway on the spines, and forming a characteristically wreathlike arrangement; color of aboral surface usually dark olive-green or indigo. *Leptasterias epichlora* (Brandt, 1835): small pedicellariae few and randomly arranged around the aboral spines; color of aboral surface commonly indigo, blue-gray, or dark green, usually mottled. *Leptasterias aequalis* (Stimpson, 1862) (believed to be a hybrid of *L. hexactis* and *L. epichlora*): small pedicellariae usually numerous and randomly arranged around the aboral spines; color of aboral surface extremely variable, ranging from olive-green, indigo, or gray to coral-red or orange (in our region, the last two colors not often noted in the other two species).

451   19a   Specimens keying to *Henricia leviuscula* belong to this species and to what may be an undescribed species or a hybrid. *Henricia leviuscula* reaches a diameter of about 10 cm and its aboral surface is uniformly orange. Specimens of about the same size, but orange-red or pinkish in color, are perhaps distinct. Another color morph reaches a diameter of about 15 cm; it is mostly orange or reddish and has gray or lavender patches where the rays join the disk.

454   In Family Echinasteridae, note that specimens keying to *Henricia leviuscula* belong to this species and to at least one other undescribed species. See note concerning choice 19a, page 451 (above).
   In Family Asteriidae add
       Leptasterias epichlora (Brandt, 1835)
       Leptasterias aequalis (Stimpson, 1862) (considered to be a hybrid of *L. hexactis* and *L. epichlora*)

460   4a   To the description of *Strongylocentrotus pallidus* add: ridges on spines with nearly flat surfaces, without obvious sculpturing;

460	4b	To the description of *Strongylocentrotus droebachiensis* add: ridges on spines with rounded surfaces, and with periodic sculpturings (these somewhat fan-shaped)
461		To references for Class Holothuroidea add the following:

      Kirkendale, L., & P. Lambert. 1995, *Cucumaria pallida*, a new species of sea cucumber from the northeast Pacific Ocean (Echinodermata, Holothuroidea). Can. J. Zool., 73:542-51.

      Lambert, P. 1990. A new combination and synonymy for two subspecies of *Cucumaria fisheri* Wells (Echinodermata: Holothuroidea). Proc. Biol. Soc. Washington, 103:913-21.

      Panning, A. 1949. Versuch einer Neuordnung der Familie Cucumariidae (Holothuroidea: Dendrochirota). Zool. Jahrb., Abt. System., 78:404-70.

463   16a   For *Cucumaria fallax* read *Cucumaria pallida*.

465   In Family Cucumariidae

      For *Cucumaria fallax* Ludwig, 1894 read *Cucumaria pallida* Kirkendale & Lambert, 1995.

      Add **Pseudocnus astigmatus* (Wells, 1924). Similar to *Cucumaria piperata* in general appearance, but is not speckled and has different ossicles (see Lambert, 1990).

23. Phyla Urochordata, Hemichordata, Chaetognatha

467   To references for Class Ascidiacea, add:

      Lambert, G. 1989. A new species of the compound ascidian *Eudistoma* (Ascidiacea, Polycitoridae) from the northeastern Pacific. Can. J. Zool., 67:2700-3.

      Young, C. M., & E. Vazquez. 1995. Morphology, larval development, and distribution of *Bathypera feminalba* n. sp. (Ascidiacea, Pyuridae), a deep-water ascidian from the fjords and sounds of British Columbia. Invert. Biol., 114:89-106.

469   13a   Surface of tunic underlain by a stratum of conspicuous, closely spaced calcareous spicules (these with a long central spine and several lateral spines) (body usually not as tall as wide; diameter up to about 4 cm, but generally much smaller; color usually grayish white, sometimes pinkish; subtidal)   *Bathypera feminalba*

473   35a   For *Diplosoma macdonaldi* read *Diplosoma listerianum*.

      42b   For *Archidistoma* sp. read *Eudistoma purpuropunctatum*.

474   In Family Clavelinidae replace *Archidistoma* sp. with *Eudistoma purpuropunctatum* Lambert, 1989.

475   In Family Polyclinidae delete the note concerning **Euherdmania claviformis*. This species has recently been found at Friday Harbor, Washington.

      In Family Didemnidae replace *Diplosoma macdonaldi* Herdman, 1886 with *Diplosoma listerianum* (Milne-Edwards, 1841)

    For Order Phlebobranchia read Suborder Phlebobranchia, and directly above this add Order Pleurogona.

476   In Family Ascidiidae, add the following notes:

      Ascidia callosa Stimpson, 1852. Formerly abundant in the San Juan Archipelago and the Vancouver Island region, but now extremely rare.

      Ascidia ceratodes (Huntsman, 1912). Known only from Saanich Inlet, Vancouver Island.

    After the list of species in Family Ascidiidae, add Order Pleurogona.

    For Order Stolidobranchia read Suborder Stolidobranchia.

    In Family Styelidae, add

      *Botryllus spp. and **Botrylloides* spp. Recent introductions, becoming abundant.

      *Styela clava Herdman, 1881. Has been found at French Creek Marina, Vancouver Island.

    In Family Pyuridae, for *Bathypera ?ovoida* (Ritter, 1907) read *Bathypera feminalba* Young & Vazquez, 1995.

    In Family Molgulidae add **Molgula manhattensis* (DeKay, 1843).

478   To references under Class Enteropneusta, add:

      King, G. M., Giray, C., & I. Kornfield. 1994. A new hemichordate, *Saccoglossus bromphenolosus* (Hemichordata: Enteropneusta: Harrimaniidae) from North America. Proc. Biol. Soc. Washington, 107:383-90.

512 Additions and Corrections

478   To the text under Class Enteropneusta, add:

       In external morphology and certain biochemical features, specimens of the *Saccoglossus* from Willapa Bay appear to be indistinguishable from the Atlantic *S. bromphenolosus*. It must be stated, however, that until the internal morphology of both enteropneusts has been studied carefully, it is not advisable to jump to the conclusion that they are really identical.

     In Phylum Chaetognatha, add the following references:

       Bieri, R., 1991. Six new genera in the chaetognath family Sagittidae. Gulf Res. Rep., 8:221-5.

       ------. 1991. Systematics of the Chaetognatha. *In* Bone, Q., H. Kapp, & A. C. Pierrot-Bults, The Biology of Chaetognaths, pp. 122-36. Oxford, New York, and Tokyo: Oxford University Press.

2a   Add . . . . ; intestine relatively narrow, not filling the trunk cavity

2b   Add . . . . ; intestine hypertrophied, almost completely filling the trunk cavity (the boundaries between the enlarged intestinal cells form lines that appear to partition the trunk cavity transversely)

3a   anus widely separated from the tail septum

3b   anus not noticeably separated from the tail septum

479   For Order Phragmophora read Order Monophragmophora

     In Family Sagittidae

     For *Sagitta bieri* Conant, 1896 read *Serratosagitta bierii* (Alvariño, 1961).

     For *Sagitta decipiens* Fowler, 1905 read *Decipisagitta decipiens* (Fowler, 1905).

     For *Sagitta friderici* Ritter-Zahony, 1911 read *Tenuisagitta friderici* (Ritter-Zahony, 1911).

     For *Sagitta maxima* Conant, 1896 read *Pseudosagitta maxima* (Conant, 1896).

     For *Sagitta minima* Grassi, 1881 read *Decipisagitta minima* (Grassi, 1881)

     For *Sagitta scrippsae* Alvariño, 1962 read *Pseudosagitta scrippsae* (Alvariño, 1962). This is perhaps a synonym of *Pseudosagitta lyra* (Krohn, 1853).

     Note: not all specialists accept the recent changes in generic names.

INDEX

This index includes the entries in Additions and Corrections as well as those in the first printing. Bold-face numerals refer to pages on which figures are located.

Abarenicola claparedi subsp.
   oceanica, 155
  claparedi subsp. *vagabunda*, 155
  pacifica, 155
Abietinaria anguina, 53
  filicula, **53**
  pulchra, **53**
  variabilis, **53**
  spp., **46**, 50, 60
Abra profundorum, 280
Abraliopsis felis, 294
Abyssocucumis albatrossi, 465
Acanella sp., 69
Acanthamunnopsis milleri, 337
Acanthascus platei, 13
Acanthina spirata, 199, **220**, 221
Acanthinephyra curtirostris, 394
Acanthochitonidae, 192
Acanthochitonina, 192
Acanthodoris armata, 250
  atrogriseata, 250
  brunnea, **244**, 248, 250
  hudsoni, 248, 250
  lutea, 498
  rhodoceras, 498
  nanaimoensis, 248, 250
  pilosa, 248, 250
Acanthogorgiidae, 69
Acantholithodes hispidus, 406
Acanthomunnopsis milleri, 502
Acanthomysis columbiae, 323, 324
Acanthonotozomatidae, 382, 506
Acarida, 300-1, 500
Acarnus erithacus, 20, 21, 29
Accedomoera, **352**
  vagor, 375, 377
Acerotisa alba, 87
Acetabulosoma kozloffi, 314
Acharax johnsoni, 265
Achelia alaskensis, 299
  chelata, 299
  discoidea, 299
  echinata, 299
  gracilipes, 299
  latifrons, 299
  nudiuscula, **298**, 299
Acidostoma sp., 367, 381, 506
  hancocki, 506
Acila castrensis, **263**, 266

Acmaea mitra, 205
Acmaeidae, 194, 205
Acochlidiacea, **234**, 236
Acoelida, 84
Acotylea, 88
Acrocirridae, 114, 152
Acrocirrus columbianus, 152
  heterochaetus, 152
  occipitalis, 152
Acrothoracica, 319
Acteonidae, 195, 238
Actiniaria, 72-8, 490
Actiniidae, 77
Actinocyclidae, 250, 498
Actinostolidae, 77-8
Adalaria jaunae, 498
  sp., 248, 250, 498
Admete californica, 228
  gracilior, 198, **215**, 228
Adocia gellindra, 22, 31
  spp., 22, 31
Adula californiensis, 270
  diegensis, 270
  falcata, 270
Aega symmetrica, 332, 334
Aegidae, 334
Aegina citrea, **40**, 41, 44
Aeginellidae, 390
Aeginidae, 44
Aegires albopunctatus, 246, 250
Aeolidacea, **244**, 255-8, 498
Aeolidea papillosa, **249**, 255, 258
Aeolididae, 258
Aequorea victoria, 35, 43, 56, 61
Aequoreidae, 43, 61
Aeropsidae, 461
Aeropsis fulva, 461
Aetea spp., 428, 443
Aeteidae, 443
Aforia circinata, **215**, 229
Agalmidae, 63, 489
Aglaja ocelligera, **234**, 238, 240
Aglajidae, 240
Aglantha digitale, 35, **40**, 44
Aglaophamus rubella subsp.
  anops, 135
Aglaophenia latirostris, **55**
  struthionides, **51**, **55**
  spp., 56, 60
Aglaopheniidae, 60

Agriodesma saxicola, **286**, 287
Akentrogonida, 319
Akentrogonidae, 319
Aktedrilus oregonensis, 175
Alciopa reynaudii, 124, **125**
Alciopidae, 110, **116**, 124-6
Alcyonacea, 68, 69-70
Alcyonaria, 68-71, 490
Alcyonidiidae, 442, 510
Alcyonidium albidum, 510
  gelatinosum, 510
  pedunculatum, 440, 442
  polyoum, **440**, 442, 510
Alcyoniidae, 69, 490
Alcyoniina, 68-70
Alcyonium rudyi, 490
  sp., 69
Alderia modesta, 242
Alderina brevispina, 431, 443
Alderinidae, 443
Aldisa albomarginata, 247, 250
  cooperi, 247, 251
  sanguinea, 247, 251
  tara, 247, 251
Aldisidae, 250-1
Alepas pacifica, 316
Algamorda subrotundata, 209, **211**
Alia carinata, **224**, 226
  gausapata, 226
  permodesta, **224**, 226
  tuberosa, **224**, 226
Alienacanthomysis macropsis, 323,324
Allocentrotus fragilis, 460, 497
Allogaussia recondita, 368, 381, 504
Alloniscus perconvexus, 344, 345
Allopora petrograpta, 61
  porphyra, 61
  venusta, 61
  verrilli, 61
Allorchestes, **353**
  angusta, 352, 379
  bellabella, 352, 379
Alpheidae, 394, **395**, 396
Alvania acutelirata, 497
  carpenteri, 210, 497
  compacta, 210, 497
  rosana, 497
  sanjuanensis, 210

Amaeana occidentalis, 160
Amage anops, 159
Amauropsis purpurea, 216
Amblyops abbreviata, 322, 324
Amblyosyllis sp., 132
Americanaplana fernaldi, 92
Ammothea hilgendorfi, 297, 299
Ammotheidae, 299
Ammothella tuberculata, 297, 299
  sp., 297, 299
Ampelisca agassizi, **366**, 369, 382
  brevisimulata, 368, 382
  careyi, 369, 382
  cristata, 368, 383
  fageri, 369, 383
  hancocki, 369, 383
  lobata, 369, 383
  pugetica, 368, 383
  unsocalae, 369, 383
Ampeliscidae, 382-3
Ampeliscoidea, 382-3
Amperima rosea, 465
  sp., 465
Ampharete acutifrons, 159
  finmarchica, 159
Ampharete goesi subsp.
  brazhnikovi, 159
  goesi subsp. *goesi*, 159
  labrops, 158
Ampharetidae, 115, **119**, 158-9, 495
Ampheraster chiroplus, 454
  marianus, 454
Amphichaeta sp., 171
Amphicteis glabra, 159
  mucronata, 159
  scaphobranchiata, 159
Amphictenidae, 158
Amphilepas patens, 458
Amphilochidae, **357**, 379
Amphilochus litoralis, 360, 379
  neapolitanus, 360, 379
  picadurus, 360, 379
Amphinema platyhedos, 33, 42
Amphinomida, 141
Amphinomidae, 114, **117**, 141
Amphiodia occidentalis, 457, 458
  periercta, 457, 458
  urtica, 457, 458
Amphiophiura bullata subsp.
  pacifica, 459
  ponderosa, 459
  superba, 459
Amphioplus macraspis, 457, 458
  strongyloplax, 457, 458
Amphipholis pugetana, 457, 458
  squamata, 457, 458
Amphipoda, 346-91, 502-7
Anchicolurus occidentalis, 501
  occidentalis, 502
Amphiporidae, 100
Amphiporus angulatus, 100

  angulatus, 98
  bimaculatus, 95, 98, 100
  cruentatus, 98, 100
  formidabilis, 98, 100
  imparispinosus, 98, 100
  rubellus, 98, 100
  tigrinus, 98, 100
Amphisamytha bioculata, 159, 495
Amphissa columbiana, 226
  reticulata, **224**, 226
  versicolor, **224**, 226
Amphitrite cirrata, 161
Amphiura carchara, 458
  diomedeae, 458
  polyacantha, 458
Amphiuridae, 458
Amphoriscidae, 10
Ampithoe dalli, 356, **358**, 384
  lacertosa, 356, **358**, 384
  plumulosa, 356, **358**, 384
  sectimanus, 356, **358**, 384
  simulans, 356, **358**, 384
  valida, **356**, **358**, 384
Ampithoidae, 357, 358, 384
Anaata brepha, 23, 30
  spongigartina, 23, 30
Anasca, 443-4
Anaspidea, 233, 240
Anatoma baxteri, 199
  crispata, 199
Anchinoidae, 30
Ancula pacifica, 246, 250
Angulosaccus tenuis, 318
Anisodoris lentiginosa, 251
  nobilis, 247, 251
Anisogammaridae, 383
Anisogammarus pugettensis, 373, 383
Annelida, 109-79, 492-6
Anobothrus gracilis, 158
Anomalodesmata, 287
Anomiidae, 259, 274
Anomura, 404-11
Anonyx laticoxae, 367, 381
  lilljeborgi, **366**, 367, 381
Anopla, 99
Anoplodactylus erectus, 300
  viridintestinalis, 299, 300
Anoplodium hymanae, 90
Antalis pretiosum, 500
Antedon sp., 447
Antedonidae, 447
Anteliaster coscinactis, 454
  megatretus, 454
Anthessius nortoni, 306
Anthoarcuata graceae, 23, 30
Anthomedusae, 41-2, 57-9
Anthopleura artemisia, 75, 77
  elegantissima, 75, 77
  xanthogrammica, 75, 77
Anthoptilidae, 70
Anthoptilum grandiflorum, 70

Anthothela pacifica, 70
Anthothelidae, 70
Anthozoa, 68-78
Anthuridae, 331
Anthuridea, 331
Antinoella macrolepida, 139
Antipatharia, 71
Antipathes sp., 71
Antipathidae, 71
Antipathina, 71
Antiplanes thalea, 229
  perversa, 229, **231**
  voyi, 229
Aoridae, 385-6
Aoroides, **350**
  columbiae, 370, 385
  exilis, 370, 385
  inermis, 370, 385
  intermedius, 370, 385
  spinosus, 370, 385
Aphelochaeta monilaris, 495
  multifilis, 495
  parva, 495
  secunda, 495
  serratiseta, 495
Aphrocallistes vastus, 10, **12**, 13
Aphrocallistidae, 13
Aphrodita japonica, 136
  longipalpa, 136
  negligens, 136
  parva, 136
  refulgida, 137
Aphroditacea, 136-40
Aphroditidae, 112, **116**, 136-7
Apistobranchidae, 115, **117**, 145
Apistobranchus ornatus, 145
  tullbergi, **117**, 145
Aplacophora, 184-5
Aplidium arenatum, 474, 475
  californicum, **468**, 473, 475
  glabrum, 474, 475
  propinquum, 474, 475
  solidum, 473, 475
  sp., 474, 475
Aplousobranchia, 474-5
Aplysiidae, 240
Aplysilla glacialis, 14, 31, 487
  polyraphis, 14, 31, 487
Aplysillidae, 31, 487
Aplysiopsis smithi, 242
Apodida, 466
Apomatopsis geniculata, 496
Apomatus geniculatus, 164, 496
  timmsii, 164
Arabella iricolor, **142**, 144, 494
  semimaculata, 144
Arabellidae, 121, **142**, 143-4, 494
Arachnida, 300
Arachnidiidae, 442
Archaeobalanidae, 318
Archaeogastropoda, 199-204
Archaeomysis grebnitzkii, 321, 324

Archaeopulmonata, 258
Archiannelids, 166-9
Archidistoma molle, 473, 474
   psammion, 473, 474
   ritteri, 473, 474
   sp., 473, 474, 511
Archidorididae, 251
Archidoris montereyensis, 247, 251
   odhneri, 248, 251
Architeuthidae, 294
Architeuthis japonica, 294
Archoophora, 84-8
Arcidae, 269
Arcoida, 269
Arcteobia anticostiensis, 493
   spinielytris, 493
Arctomelon stearnsii, 227
Arctonoe fragilis, **138**, 140
   pulchra, 140
   vittata, 140
Arcturidae, 340
Arenicola brasiliensis, 495
   cristata, 495
   marina, 154
Arenicolidae, **118**, 121, 154-5, 495
Argeia pugettensis, 342, **343**
Argilloecia conoidea, 314
   cylindrica, 314
Arginula bella, 200
Argis alaskensis, 400
   crassa, 400
   dentata, 400
   lar, 400
   levior, 400
Argissa hamatipes, 368, 381
Argissidae, 381
Argulidae, 310
Arguloida, 310
Argulus borealis, 310
   pugettensis, 310
Arhynchite pugettensis, 180
Aricidea assimilis, 150
   lopezi, 150
   minuta, 150
   neosuecica, 150
   quadrilobata, **117**, 150
   ramosa, 150
   wassi, 150
Aristias pacificus, 367, 381
Armadillidiidae, 344
Armadillidium vulgare, 344
Armadilloniscus holmesi, 344, 345
Armandia brevis, 153
Armina californica, 245, **249**, 254
Arminacea, 245, 254-5
Arminidae, 254
Arndtanchora sp., 23, 30
Artacama conifera, 159
Arthophryxus beringanus, 341, 342

Arthropoda, 296-417
Articulata, 420-1, 441
Aruga holmesi, 504
Asabellides lineata, 159
   sibirica, 159
Asbestopluma occidentalis, 16, 29
Ascidia callosa, 469, 472, 476, 511
   ceratodes, 469, 472, 476, 511
   paratropa, 469, 476
   prunum, 476
Ascidiacea, 467-76, 511-2
Ascidicolidae, 305-6
Ascidiidae, 476, 511
Asclerocheilus beringianus, 153
Ascophora, 445-6
Ascothoracica, 319
Asellota, 331, 336-7
Aspidochirotida, 465-6
Assiminea californica, 195, 210
Assimineidae, 195, 210
Astarte compacta, 276
   undata, 276
   esquimalti, 276, **277**
Astartidae, 264, 276
Asteriadina, 454
Asteriidae, 454, 511
Asterina miniata, **449**, 450, 453
Asterinidae, 453
Asteroidea, 448-55, 510-1
Asteronychidae, 457
Asteronyx loveni, 457
Asteropella sp., 312
Asteropseidae, 453
Asteroschema sublaeve, 457
Asteroschematidae, 457
Asthenactis fisheri, 454
Astraea gibberosa, 201, **202**
Astrocles actinodetus, 455
Astrolirus panamensis, 455
Astropectinidae, 452
Asychis biceps, 155, 495
   disparidenta, 155, 495
   lacera, 155
   rubrocincta, 495
   similis, 155, 495
Atelecyclidae, 413, 415
Athecanephria, 182
Athecata, 41-2, 57-9
Atlantia gaudichaudi, 208
Atlantiidae, 208
Atyidae, 238, 382
Atylidae, 506
Atylus borealis, 506
   collingi, 361, 382
   levidensus, 361, 382
   tridens, 361, **362**, 382
Aulactinia incubans, **74**, 75, 77
Aurelia aurita, 65, 67
   limbata, 65, 67
Austrorhynchus californicus, 91
   pacificus, 91
Autolytus alexandri, 493

   cornutus, 129
   fasciatus, **116**
   magnus, 129
   prismaticus, 129
   trilineatus, 129
   verrilli, 129
Axiidae, 402, 403-4, **405**
Axinella sp., 19, 28
Axinellida, 28
Axinellidae, 28
Axinopsida serricata, 264, **271**, 274
Axinulus redondoensis, 274
Axiopsis spinulicauda, 404
Axiothella rubrocincta, **156**, 157, 495
Axocielita originalis, 24, 30

Bacescuella labeosa, 175
Baeonectes improvisus, 337, 502
Balanidae, 318
Balanomorpha, 314, 316-8
Balanophyllia elegans, 72
Balanus balanus, 317, 318
   crenatus, 317, 318
   glandula, **315**, 317, 318
   improvisus, 317, 318
   nubilus, 317, 318
   rostratus, 317, 318
Balcis columbiana, 218
   montereyensis, 218
Balgetia pacifica, 90
Balticina californica, 71
   septentrionalis, 71
Bankia setacea, 285
Barantolla americana, 154
Barentsia benedeni, **420**, 422, 509
   gracilis, 422
      var. *nodosa,* 509
   misakiensis, 422
   parva, 509
   ramosa, 422, 509
   robusta, 422, 509
Barentsiidae, 422
Barleeia acuta, 210
   haliotiphila, 210
   subtenuis, 210
Barleeidae, 497
Barnea subtruncata, 284
Baseodiscidae, 99
Baseodiscus princeps, 96, 99
Basibranchia, 183
Basommatophora, 258, 498
Bathyarca nucleator, 269
Bathybembix bairdi, 203
Bathycrinidae, 447
Bathycrinus pacificus, 447
Bathycuma, 326, 327
Bathydrilus litoreus, **176**, 177
   torosus, 175
Bathyleberis sp., 312
Bathymedon flebilis, 369, 378

pumilus, 369
Bathypera feminalba, 511
  ovoida, 469, 476, 511
Batillaria zonalis, 197, **211**, 212
Batillipedidae, 296
Batillipes mirus, 296
Bdellidae, 301
Bdellonemertea, 100
Bentheogennema borealis, 393
  burkenroadi, 393
Benthocodon pedunculata, 488
Benthoctopus abruptus, 295
  profundum, 295
Benthodytes incerta, 466
  sanguinolenta, 466
Benthopecten acanthonotus, 452
  claviger, 452
  mutabilis, 452
Benthopectinidae, 452
Bergstroemia nigrimaculata, 492
Beringius crebricostatus, 225
  eyerdami, **222**, 225
  kennicotti, **222**
Beroe abyssicola, 79, 81, 490, 491
  cucumis, 79, 81, 490, 491
  forskali, 490
  gracilis, 490, 491
  mitrata, 490, 491
  sp., 79, 81
Beroida, 81
Beroidae, 81, 491
Berryteuthis anonychus, 294
  magister, 293, 294
Berthella californica, 233, 240
Betaeus harrimani, **395**, 396
  setosus, 396
Betapista dekkerae, 160
Bicellariellidae, 444
Bicidium aequoreae, 77
Bicrisia edwardsiana, 428, **429**, 441
Biemna rhadia, 25, 29
Biemnidae, 29
Bimeria spp., 50, 58
Bittium attenuatum, 212, **214**
  eschrichtii, 212, 214
  munitum, 212, **214**, 215
Bivalvia, 259-88, 499-500
Boccardiella hamata, 494
Bodotriidae, 328, 502
Bolinopsidae, 81, 491
Bolinopsis infundibulum, 80, 81
Bolitaenidae, 295
Boltenia echinata, 469, **471**, 476
  villosa, 469, **470**, 476
Bomolochidae, 307-8
Bonelliidae, 180
Bonelloinea, 180
Bonneviella spp., 57, 60
Bonneviellidae, 60
Bopyridae, 342
Bopyroides hippolytes, 342
Boreohydridae, 49, 58, 489

Botrylloides spp., 511
Botryllus sp., 476, 511
Bougainvillia muscus, 487, 488, 489
  principis, 36, 42
  ramosa, 35, **37**, 42, **47**, 487, 488
  superciliaris, 36, 42
  spp., **46**, 49, 58, 489
Bougainvilliidae, 42, 58, 488, 489
Bourgeticrinida, 447
Bowerbankia gracilis, **440**, 442
  var. *aggregata*, 510
Brachionidae, 105
Brachionus plicatilis, 105
Brachiopoda, 419-21
Brachyrhyncha, 416-7
Brachyura, 411-7
Brada sachalina, 152
  villosa, 152
Branchellion lobata, 178
Branchiomaldane simplex, 495
  vincentii, 154
Branchiopoda, 303, 501
Branchiura, 310
Brania brevipharyngea, 129, **130**
Breslauilla relicta, 90
Briarosaccus callosus, 319
Brinkmanniella palmata, 91
Brisaster latifrons, 459, 461
Brisingella exilis, 455
Brisingidae, 455
Bruzelia tuberculata, 361, 381
Bryozoa, 423-46
Buccinidae, 198, 223, 497
Buccinum aleuticum, 223
  castaneum, 223
  diplodetum, 223
  glaciale, 223
  planeticum, 223
  plectrum, 222, 223
  scalariforme, 223
  strigillatum, 497
  strigillosum, 223, 497
Bugula californica, 426, **434**, 444
  cucullifera, **425**, 427, **434**, 444
  pacifica, 427, **434**, 444
  pugeti, 426, **434**, 444
Bursovaginoidea, 93
Bushiella abnormis, 496
  verruca, 496
Buskia nitens, 439, 442
Buskiidae, 442
Byblis millsi, 368, 383
  veleronis, 368, 383
Bythotiara depressa, 39, 42
  huntsmani, 34, 42, 48, 59

Caberea boryi, 426, 444
  ellisi, 426, **434**, 444
Cadlina flavomaculata, 248, 250
  luteomarginata, 248, 250
  modesta, 248, 250

Cadulus aberrans, 289, 290, 500
  californicus, 289, 290
  tolmiei, 289, **290**
Caecidae, 193, 212
Caesaromysis hispida, 321, 324
Calanoida, 303-4
Calappidae, 411, 413
Calastacus stilirostris, 404
Calcarea, 6, 7-10
Calcaronea, 9-10
Calcigorgia spiculifera, 69
Calcinea, 9
Caligidae, 309
Calinaticina oldroydi, 216
Callianassa californiensis, **403**, 404, 508
  gigas, 404, **405**, 508
Callianassidae, 402, **403**, 404, **405**, 508
Callianopsis goniophthalma, 404
Calliobdella knightjonesi, 178
Calliopiidae, 378
Calliopius, **352**
  spp., **374**, 375, 378
Callioplanidae, 88
Calliostoma annulatum, 203
  bernardi, 204
  canaliculatum, **202**, 204
  ligatum, 204
  platinum, 203
  variegatum, 204
Calliotropis carlotta, 204
  ceratophora, 204
Callistochiton crassicostatus, 187, 192
Callistochitonidae, 192
Callogorgia kinoshitae, 70
Callopora armata, **432**, 433, 443
  circumclathrata, 433, 443
  horrida, **432**, 433, 443
Calocaris investigatoris, 404, **405**
  quinqueseriatus, 404
Calycella spp., 56, 60
Calycophorae, 65
Calycopsidae, 42, 59
Calycopsis nematophora, 35, 42
Calyptogena gigas, 280
  kilmeri, 280
  pacifica, 280
Calyptraea fastigiata, 213
Calyptraeidae, 193, 213, 216
Campanularia ritteri, **52**
  spp., 56, 57, 60
Campanulariidae, 43, 56, 488, 489
Campanulinidae, 56, 60
Campylaspis, 326, 327
  biplicata, 502
  blakei, 502
  canaliculata, 502
  hartae, 328
  maculinodulosa, 502
  rubromaculata, 328
  rufa, 328

Campyloderes sp., 106, 108
Cancellaria crawfordiana, 198, 228
Cancellariidae, 198, 228
Cancellata, 441
Cancer antennarius, **414**, 415, 509
  branneri, 415, 508
  gracilis, 415, 508
  jordani, 508
  magister, 415, 508
  oregonensis, 415, 509
  productus, **414**, 415, 508
Cancridae, 413, **414**, 415, 508-9
Cancridea, 415
Capheira mollis, 466
Candelabridae, 489
Candelabrum sp., 489m
Capitella capitata, **111**, **146**, 154, 495
Capitellida, 154
Capitellidae, **111**, 121, **146**, 154, 495
Caprella alaskana, 388, 390
  angusta, 388, 390
  borealis, 388, 390
  californica, 387, 390
  drepanochir, 388, 390
  equilibra, 387, **389**, 390
  ferrea, 388, 390
  gracilior, 387, 390
  greenleyi, 388, 390
  incisa, 388, **389**, 390
  irregularis, 388, 390
  laeviuscula, 388, **389**, 390
  mendax, 388, **389**, 390
  pilidigitata, 387, 390
  pustulata, 388, 390
  rudiuscula, 388, 390
  striata, 388, 390
  ungulina, 390
  verrucosa, 388, 390
Caprellida, 390
Caprellidae, 390
Caprellidea, 346, 386-91, 507
Caprogammaridae, 390
Carcinonemertes epialti, 97, 100
  errans, 97, 100
Carcinonemertidae, 100
Cardiidae, 262, 264, 276-8
Cardiomya californica, **286**, 288
  curta, 288
  oldroydi, 288, **286**
  pectinata, 288
  planetica, **286**, 288
Carditidae, 276
Carenzia inermis, 204
Cargoa vancouverensis, 250
Caridea, 393-402
Carinaria cristata, 208
  japonica, 208
  lamarcki, 208
  latidens, 208
Carinariidae, 208
Carinoma mutabilis, 96, 99

Carinomidae, 99
Caryophyllia alaskensis, 72
Caryophylliidae, 72
Caryophylliina, 72
Catablema multicirrata, 39, 42
  nodulosa, 41, 42
Catriona columbiana, 257
  rickettsi, 498
Caudinidae, 466
Caudofoveata, 184-5
Caulibugula californica, 426, **434**, 444
  ciliata, 426, 444
  occidentalis, 426, **434**, 444
Caulleriella alata, 151
  hamata, 151
Cauloramphus brunea, **432**, 433, 443
  echinus, **432**, 433, 443
  spiniferum, **432**, 433, 443
Cavolinia gibbosa, 241
  tridentata, 241
Cavoliniidae, 241
Cecina manchurica, 195, 210, **211**
Cellaria diffusa, 428, 444
  mandibulata, 428, 444
Cellariidae, 444
Celleporella hyalina, 510
Celleporidae, 446
Centroderidae, 108
Cephalaspidea, 194, 233, **234**, 237-40, 497-8
Cephalophoxoides homilis, 503, 505
Cephalopoda, 291-5, 500
Cephalothricidae, 99
Cephalothrix pacifica, 95, 99
Ceractinomorpha, 28-31
Ceradocus spinicaudus, 373, **374**, 377, 384
Ceramaster arcticus, 448, 453
  clarki, 453
  japonicus, 453
  patagonicus, 450, 453
Ceraplectana trachyderma, 466
Ceratocephale loveni, 133
Ceratopera axi, 91
  pilifera, 91
Ceratophysa valvaecristata, 461
Ceratostoma foliatum, 219
  inornatum, 219, **220**, 221
Cercops compactus, 387, 390
Cerebratulus albifrons, 96, 99
  californiensis, 97, 99
  herculeus, 97, 99
  longiceps, 97, 99
  marginatus, 97, 99
  montgomeryi, 96, 99
  occidentalis, 97, 99
Ceriantharia, 68, 71
Cerianthidae, 71
Ceriantipatharia, 68, 71
Cerioporina, 441

Cerithiidae, 197, 212
Cerithiopsidae, 197, 213
Cerithiopsis columna, 213, **214**
  signa, 213
  stejnegeri, 213
Cestoplana sp., 85, 88
Cestoplanidae, 88
Chaetodermatida, 184-5
Chaetognatha, 478-9, 512
Chaetonotida, 104
Chaetonotidae, 104
Chaetonotus testiculophorus, 104
Chaetopleura gemma, 187, 192
Chaetopleuridae, 192
Chaetopterida, 149
Chaetopteridae, 115, **117**, 149
Chaetopterus variopedatus, 149
Chaetozone acuta, 151
  setosa, 151
  spinosa, 151
Chama arcana, 276
Chamidae, 259, 276
Chapperiella condylata, 433, 443
  patula, 433, 443
Chapperiellidae, 443
Chartella membranaceo-truncata, 424, 443
Cheilonereis cyclurus, **112**, 133, **134**
Cheilopora praelonga, 424, 437, 446
Cheiloporinidae, 446
Cheilostomata, 424-39, **425**, 443-6
Cheirimedeia macrocarpa subsp. *americana*, 371, 384
  similicarpa, 371, 384
  zotea, **366**, 371, 384
Chelicerata, 296-300, 501
Chelonethida, 300
Chelonaplysilla polyraphis, 487
Chelophyes appendiculata, 63, 65
Chelura terebrans, 386
Cheluridae, 386
Chelysoma columbianum, 468, 469, **471**, 475
  productum, 468, 469, **471**, 475
Chevroderma whitlachi, 185
Childia groenlandica, 84
Childiidae, 84
Chilophiurina, 459
Chionoecetes angulatus, 413
  bairdi, **412**, 413
Chiridota albatrossii, 466
  nanaimensis, 466
Chiridotidae, 466
Chirimia biceps, 495
  similis, 495
Chirostylidae, 406
Chiroteuthidae, 295
Chiroteuthis calyx, 295
Chitonina, 191-2
Chlamys behringiana, 273

hastata, **263**, 273
jordani, 273
rubida, 273
Chloeia entypa, 141
pinnata, 141
Chondracanthidae, 308
Chone aurantiaca, 164
duneri, **162**, 164
ecaudata, 163
infundibuliformis, 164
magna, 164
mollis, 164
Chonelasma calyx, 10, **12**, 13
tenerum, 13
Chorilia longipes, 413, **414**
Choristes carpenteri, 210
coani, 210
Choristida, 27
Choristidae, 210
Chromodoridae, 250
Chromopleustes lineatus, 504
oculatus, 504
spp., 504
Chrysaora fuscescens, 67
melanaster, **66**, 67
Chrysogorgiidae, 69
Chrysopetalidae, **112**, 140
Chrysopetalum occidentale, **112**, 140
Chthamalidae, 317
Chthamalus, **315**
dalli, 316, 317
Cidarina cidaris, 203, **215**
Ciliocincta sabellariae, 82
Ciocalypta penicillus, 19, 28
Ciona intestinalis, 469, 475
Cionidae, 475
Circeis armoricana, 165
spirillum, 165
Cirolana harfordi, 332, **333**, 334
Cirolanidae, 334
Cirrata, 295
Cirratulida, 150-1
Cirratulidae, **111**, 115, 120, **146**, 150-1, 495
Cirratulus cirratus, 151
spectabilis, 151
Cirriformia spirabranchia, 151
Cirripedia, 314-9, 501
Cirrophorus branchiatus, 150
lyra, 150
Cirroteuthidae, 295
Cirroteuthis muelleri, 295
Cladocarpus spp., 56, 60
vancouverensis, **55**
Cladocera, 303
Cladonema californicum, 35, 42, 48, 58, 487, 488
Cladonema radiatum, 487, 488, 489
Cladonematidae, 42, 58, 488, 489
Cladorhizidae, 29
Clathriidae, 30

Clathrina blanca, 9
coriacea, 9
spp., 9
Clathrinida, 9
Clathrinidae, 9
Clathromangelia interfossa, 229
Clausidiidae, 306
Clausidium vancouverense, 306
Clavelina huntsmani, **468**, 472, 474
sp., 472, 474
Clavelinidae, 474, 511
Clavidae, 58
Clavopora occidentalis, 440, 442
Clavoporidae, 442
Clavularia moresbii, 69
spp., 69
Clavulariidae, 69
Cleistocarpidae, 67
Clinocardium blandum, **277**, 278, 499
californiense, 278
ciliatum, 278
fucanum, 278, 499
nuttallii, 278
Clio polita, 241
pyramidata, 241
recurva, 241
Cliona argus, 14, 28
celata subsp. *californiana*, 14, 28
lobata, 14, 28
sp., 20
warreni, 14, 28
Clione limacina, **234**, **239**, 241
Clionidae, 28, 241
Cliopsidae, 497
Cliopsis krohni, 497
Clistosaccidae, 318
Clistosaccus paguri, 318
Clymenella torquata, 157
Clymenura columbiana, 155
Clypeasteroida, 460
Clytia gregaria, 488
johnstoni, **52**
lomae, 488
spp., 56, 60
Cnemidocarpa finmarkiensis, 472, 476
Cnidaria, 32-78, 487-90
Coboldus sp., 361, 382, 503, 506
Cocculina cowani, 204
Cocculinidae, 204
Codonellina cribriformis, 437, 446
Coelogynopora cochleare, 92
falcaria, 92
frondifera, 92
nodosa, 92
scalpri, 92
Coelogynoporidae, 92
Coleoidea, 293
Coleoptera, 302
Collastoma kozloffi, 90
pacifica, 90, 492

pacificum, 492
Collembola, 302
Colletosia radiata, 431, **435**, 444
Colobomatus kyphosus, 307
Coloniales, 422
Colossendeidae, 299-300
Colossendeis angusta, 299
colossea, 299
tenera, 300
Colotrachelus careyi, 204
Columbaora cyclocoxa, 370, 386
Columbellidae, 196, 198, 199, 226
Columbiaemysis ignota, 324
Colurostylidae, 328, 502
Colurostylis, 326
occidentalis, 328, 502
Colus adonis, **222**, 225
halli, **222**, 225
spitzbergensis, 225
trophius, 497
Comatulida, 447
Compsomyax subdiaphana, 282
Conaea, 306
Conchocele bisecta, **271**, 274
excavata, 274
Conchoderma auritum, 316
virgatum, 316
Conchoecia alata subsp. *minor*, 312
elegans, 312
spinirostris, 312
Conocyema, 82
deca, 83, 491
Conopeum reticulum, 431, **432**, 443
tenuissimum, 510
Convolutidae, 84
Cooperella subdiaphana, 265, 283
Cooperellidae, 265, 283
Copepoda, 303-10
Copidozoum protectum, 431, **432**, 443
tenuirostre, 431, **432**, 443
Corallimorpharia, 72, 78
Corallimorphidae, 78
Corallimorphus sp., 78
Corambe pacifica, 246, 250
thompsoni, 498
Corambidae, 250, 498
Cordagalma cordiformis, 62, 63
Cordylophora caspia, 48, 58
Corella inflata, 469, 475
willmeriana, 469, 475
Corellidae, 475
Corolla spectabilis, 241
Coronopharynx pusillus, 90
Coronula diadema, 317
reginae, 317
Coronulidae, 317, 386
Corophiidae, 386, 507
Corophioidea, 384-6
Corophium, **353**

acherusicum, 351, **354**, 386
baconi, 351, 386
brevis, 349, 386
crassicorne, 349, 386
insidiosum, 351, 386
salmonis, 349, **354**, 386
spinicorne, 349, 386
Corycaeidae, 306
Corycaeus, 306
Corymorpha sp., 47, 57
Corymorphidae, 41-2, 57
Corynactis californica, 72, 78
Coryne sp., 48, 58
Corynidae, 41, 58, 488
*Cossura longocirrat*a, **146**, 152
modica, 495
pygodactyla, 495
soyeri, 152, 495
Cossurida, 152
Cossuridae, 121, **146**, 152, 495
Costazia costazi, 437, 446
robertsoniae, 437, 446
ventricosa, 437, 446
Cotylea, 88
Cranchia scabra, 295
Cranchiidae, 295
Crangon alaskensis, 402
alba, 402
dalli, 402
franciscorum, **401**
franciscorum subsp.
angustimana, 402
handi, 402
nigricauda, 402
stylirostris, 508
Crangonidae, 393, 400-2
Crania californica, 419, **420**
Craniella spinosa, 18, 27
villosa, 18, 27
Craniidae, 419
Crassicardia crassidens, 276
Crassonemertes robusta, 101
Crassostrea gigas, 272
virginica, 272
Cratenemertidae, 100
Craterobrisinga synaptoma, 455
Crenella decussata, 272
Crepidula adunca, 213, **214**
dorsata, 213
fornicata, 216
nummaria, 216
perforans, **214**, 216
Creseis virgula, 241
Cribellina, 452
Cribrilina annulata, 431, **435**, 444
corbicula, 431, 444
Cribrilinidae, 444
Cribrinopsis fernaldi, 75, 77
williamsi, 77
Crimora coneja, 250
Crinoidea, 447
Crisia maxima, 428, 441
occidentalis, **425**, 428, 441

pugeti, 428, **429**, 441
serrulata, 428, 441
Crisidia cornuta, 428, **429**, 441
Crisiidae, 441
Crossaster borealis, 453
papposus, **449**, 451, 453
Crossota sp., 35, 44
Crucigera irregularis, 164
zygophora, 164
Crustacea, 303-417, 501-9
Cryptaster sp., 454
Cryptobranchia concentrica, 205
Cryptochiton stelleri, 185, 192
Cryptodius kelleri, 504
Cryptodonta, 265-6
Cryptolaria spp., 57, 60
Cryptolepas rachianecti, 318
Cryptolithodes sitchensis, **403**, 407
typicus, 407
Cryptomya californica, 283
Cryptonemertes actinophila, 100
Cryptopeltaster lepidonotus, 453
Cryptosula pallasiana, 437, 446
Crystallophrisson sp., 184
Crystallophrissonidae, 184
Ctenodiscus crispatus, 448, 452
Ctenodrilida
Ctenodrilidae, 120, **146**, 151
Ctenodrilus serratus, 120, **146**, 151
Ctenophora, 79-81, 490-1
Ctenostomata, 424, 439-40, 442
Cucumaria fallax, 463, 465, 511
frondosa subsp. *japonica*, 465
lubrica, 464, 465
miniata, 463, 465
pallida, 511
piperata, **462**, 464, 465
pseudocurata, 464, 465
vegae, 465
Cucumariidae, 465, 511
Cultellidae, 262, 278-9
Cumacea, 325-8, 501-2
Cumanotidae, 257
Cumanotus beaumonti, 255, 257
Cumella, 326, 327, 502
californica, 502
morion, 502
vulgaris, 328
Cumingia californica, 280
Cunina sp., 41, 44
Cuninidae, 44
Cuspidaria apodema, 288
cowani, 288
filatovae, 288
glacialis, 288
subglacialis, 500
variola, 288
Cuspidariidae, 260, 288
Cuthona abronia, 256, 257
albocrusta, 256, 257
cocoachroma, 257, 258

concinna, 255, 258
flavovulta, 498
fulgens, 498
lagunae, 498
punicea, 498
divae, 255, 258
nana, 255, 258
pustulata, 258
Cyamida, 390-1
Cyamidae, 390-1
Cyamus boopis, 390
catodontis, 390
ceti, 390
kessleri, 391
sp., 391
Cyanea capillata, **66**, 67
Cyaneidae, 67
Cyathoceras quaylei, 72
Cyathura carinata, 331
Cyclaspis, 326, 327
nubila, 502
Cyclocardia crebricostata, 276, **277**
ventricosa, 276
Cyclopecten argenteus, 273
bistriatus, 273
carlottensis, 273
knudseni, 273
squamiformis, 274
Cyclorhagida, 106-8
Cyclostomata, 424-30, 440-1
Cyclostremella concordia, 235, 236
Cyclostremellidae, **196**, 235, 236
Cydippida, 81
Cylichna alba, 238, 240, 497
attonsa, 238, **239**, 240, 497
Cylichnella cerealis, 238, 240
culcitella, 238, **239**, 240
eximia, 240
harpa, 238, 240
Cylichnidae, 240
Cylindrolberididae, 312
Cymadusa uncinata, 356, **358**, 384
Cymatiidae, 194, 217
Cymbuliidae, 241
Cymodoce japonica, 333, 334
Cymothoidae, 334
Cyphocaris challengeri, 367, 381, 506
Cyprididinidae, 312
Cystocrepis setigera, 461
Cystodytes lobatus, 473, 474
Cytharella victoriana, 229
Cythere alveolivalva, 313
lutea, 313
unifalcata, 313
Cythereis arachis, 313
aurita, 313
dunelmensis, 313
filoplumosa, 313
glauca, 313

longiductus, 313
montereyensis, 313
pacifica, 313
polita, 313
serridentata, 313
Cytheridae, 313
Cytherideidae, 313
Cytherois pusilla, 314
Cytherura clathrata, 313
Cytheruridae, 313
Cytididae, 441

Dacrydium pacificum, 272
rostriferum, 272
Dactylochirotida, 465
Dactylopleustes echinoides, 504
Dajidae, 342
Dakaria dawsoni, 438, 445
ordinata, 438, 445
pristina, 438, 445
Dalyellioida, 90, 491
Decamastus gracilis, 154
Decapoda, 392-417, 507-9
Decipisagitta decipiens, 512
minima, 512
Deimatidae, 465
Delectopecten randolphi, 273
vancouverensis, 273
Demonax medius, 496
rugosus, 496
Demospongiae, 7, 13-31
Dendraster excentricus, 459, 460
Dendrasteridae, 460
Dendrobeania curvirostrata, 426, **435**, 444
laxa, 426, **435**, 444
lichenoides, 424, **435**, 444
longispinosa, 424, **435**, 444
murrayana, 426, **435**, 444
Dendrobranchiata, 392, 393
Dendroceratida, 31
Dendrochirotida, 464-5
Dendrochiton flectens, **186**, 190, 192
semiliratus, **186**, 190, 192
Dendrodoa abbotti, 472, 476
Dendronotacea, **244**, 245, 251-4
Dendronotidae, 253
Dendronotus albopunctatus, 252, 253
albus, 253
dalli, 253
diversicolor, 253
frondosus, 253
iris, 252, 253
rufus, 252, 253
subramosus, **244**, 252, 253
Dendrophylliidae, 72
Dendrophylliina, 72
Dendropoma lituella, 212
Dentaliida, 290
Dentaliidae, 290, 500

Dentalium agassizi, 289, 290
neoxagonum, 500
pretiosum, 289, **290**, 500
rectius, 289, **290**, 500
Depastridae, 489
Dermasterias imbricata, 450, 453
Desmophyes annectens, 63, 65
Desmophyllum cristagalli, 72
Desmote inops, 90
Desmoxyidae, 28
Detonella papillicornis, 344, 345
Deutella californica, **389**, 390
Dexaminidae, 382, 506
Dexaminoidea, 382
Dexamonica reduncans, 369, 382, 504, 506
Diacria trispinosa, 241
Diaperoecia californica, **427**, 440
intermedia, 430, 440
johnstoni, 430, 440
Diaphana brunnea, 239
Diaphanidae, 239
Diaphorodoris lirulatocauda, 248, 250
Diarthrodes cystoecus, 305
Diasterope sp., 312
Diastoporidae, 440
Diastylidae, 328, 501-2
Diastylis, 326
abbotti, 501
alaskensis, 328
aspera, 328
bidentata, 328
crenellata, 501
dalli, 328
dawsoni, 328, 501
nucella, 328
paraspinulosa, 328
pellucida, 328
quadriduplicata, 501
rathkei, 328, 501
santamariaensis, 501
sentosa, 501
serratocostata, 502
umatillensis, 328
Diastylopsis, 326
dawsoni, 501
Diatomovora amoena, 84
Diaulula sandiegensis, 247, 251
Dictyociona asodes, 23, 30
Dicyema, 82, 491
apollyoni, 83
Dicyemennea, 82
abreida, 83
brevicephala, 83
brevicephaloides, 83
filiformis, 83
Dicyemida, 82-3
Dicyemidae, 82-3
Dicyemodeca deca, 491
Didemnidae, 475, 511
Didemnum albidum, **471**, 473, 475

carnulentum, 475
Dimophyes arctica, 63, 65
Dinonemertidae, 101
Dinophilida, 168
Dinophilidae, 167, 168
Dinophilus kincaidi, 168
Diodora aspera, 200
Diogenidae, 406, 410
Diopatra ornata, 141, **142**
Diphasia spp., 50, 60
Diphyidae, 65
Dipleurosoma typicum, 35, 43
Dipleurosomatidae, 43
Diplodonta impolita, 275
orbella, **271**, 275
Diplopteraster multipes, 454
Diplosolen obelium **427**, 430, 440
Diplosoma listerianum, 511
macdonaldi, 473, 475, 511
Diplostraca, 303
Diplozonina, 452
Dipsacaster anoplus, 452
borealis, 452
Diptera, 302
Dirona albolineata, **249**, 254
aurantia, 254
picta, 254
Dironidae, 254
Disacanthomysis dybowskii, 323, 324
Discocytis canadensis, 426, 441
Discodorididae, 251
Discodoris heathi, 247, 251
Discordiprostatus longisetosus, 175, **177**
Discorsopagurus schmitti, **403**, 408
Discurria insessa, 206
Disporella fimbriata, 430, 441
separata, 429, 441
Distaplia occidentalis, 473, 474
smithi, 474
Diurodrilus ankeli, 168
Dodecaceria concharum, **146**, 151
fewkesi, 151
Dogielinotidae, 380
Doliolida, 477
Doridacea, **244**, 245-51
Doridella steinbergae, 246, 250
Doridicola ptilosarci, 307
Dorididae, 251
Doris odonoghuei, 251
Dorvillea japonica, 494
longicornis, 494
longicornis, 494
moniliceras, 494
pseudorubrovittata, 494
Dorvilleidae, 117, 120, **142**, 144, 494
Dorvillea annulata, 494
Doryporella alcicornis, **432**, 433, 443

Doto amyra, 252, 254
  columbiana, 252, 254
  kya, 252, 254
Dotoidae, 254
Dougaloplus gastracanthus, 458
Drilonereis falcata subsp. *minor*, 144
  longa, **142**, 144
  nuda, 144
Dryodora glandiformis, 81
Duasmodactyla commune, 465
Dulichia rhabdoplastis, 377, 386
  spp., 377, 386
Duplacorhynchus major, 91
Dynamena operculata, **54**
  spp., 50, 60
Dynamenella benedicti, 334, 336
  glabra, 334, 336
  sheareri, 334, 336
Dyopedos spp., 377, 386
Dysidea fragilis, 14, **21**, 31, 487
Dysideidae, 31, 487
Dysponetus pygmaeus, 140
Dytaster gilberti, 452

Echinasteridae, 454, 510
Echiniscoides sigismundi, 296
Echiniscoididae, 296
Echinocrepis rostrata, 461
Echinoderes kozloffi, 106, **107**
  pennaki, 106
  sp., 106
Echinoderidae, 106-7
Echinodermata, 447-66, 510-1
Echinoida, 460
Echinoidea, 459-61
Echinothuriidae, 460
Echinothurioida, 460
Echiura, 180-1
Echiuridae, 180-1
Echiuroinea, 180-1
Echiurus echiurus subsp. *alaskanus*, 180
Ectopleura crocea, 488, 489
  marina, 488, 489
Ectoprocta, 423-46
Ectyomyxilla parasitica, 26, 29
Edotia sublitoralis, 502
Edwardsia sipunculoides, 73, 77
Edwardsiidae, 77
Eirene mollis, 39, 43
Eirenidae, 43, 61
Elasipodida, 465
Elasmopus spp., 369, 384
Elassochirus cavimanus, 410
  gilli, 408, 410
  tenuimanus, 410
Electra crustulenta subsp. *arctica*, 431, **432**, 443
Electridae, 443
Ellisina levata, 431, 443
Elpidiidae, 465-6
Elysia hedgpethi, **234**, 242, 243

Elysiidae, 243
Emerita analoga, 410
Empleconia vaginata, 269
Emplectonema gracile, 98, 100
  purpuratum, 97, 100
Emplectonematidae, 100
Enchytraeidae, 171, 172-4
Enchytraeus kincaidi, 173
  multiannulatus, 173
Enopla, 100-1
Enoploteuthidae, 294
Enterogona, 474-2
Enteropneusta, 478, 511
Enteroxenos parastichopoli, 219
Entoconchidae, 219
Entodesma pictum, 287
Entoniscidae, 343
Entoprocta, 421-2
Eobrolgus chumashi, 504
  spinosus, **362**, 365, 380, 504, 505
Eogammarus confervicolus, **366**, 373, 383
  oclairi, 373, 383
Eohaustorius, **350**
  brevicuspis, 372, 383
  estuarius, 372, 383
  sawyeri, 372, 383
  sencillus, 372, 383
  washingtonianus, 372, 383
Eolidacea, 245, 444498
Eperetmus typus, 36, **38**, 41,43
Epiactis fernaldi, 76, 77
  lisbethae, 76, 77
  prolifera, 76, 77
  ritteri, 76, 77
Epicaridea, 330, 341-4
Epidiopatra hupferiana subsp. *monroi*, 141
Epistomiidae, 444
Epitoniidae, 195, 217-8
Epizoanthidae, 78
Epizoanthus scotinus, 72, 78
Eranno lagunae ., 494
Eremicaster gracilis, 452
  pacificus, 452
Ergasilidae, 307-8
Ericthonius, **353**
  brasiliensis, 351, 386
  hunteri, 351, 386, 503, 507
  rubricornis, 503, 507
Errinopora pourtalesii, 61
Esperiopsidae, 29
Eteone barbata, 492
  columbianus, 492
  californica, 123
  fauchaldi, 492
  longa, **122**, 123
  pacifica, 122
  spetsbergensis, 122
  spilotus, 492
  tuberculata, 122
Eualus avinus, 399

  barbatus, 399
  berkeleyorum, 399
  biunguis, 399
  fabricii, 399
  lineatus, 399
  macrophthalmus, 399
  pusiolus, 399
  suckleyi , 399
  townsendi, 399
Eubranchidae, 257
Eubranchus misakiensis, 257
  olivaceus, 257
  rustyus, 257
  sanjuanensis, 257
Eucarida, 392-417, 507-9
Euchone analis, 163
  hancocki, 163
  incolor, 163
Eucladocera, 303
Euclasterida, 455
Euclymene reticulata, 157
  sp., 157
  zonalis, 157
Euctenostomata, 442
Eudendriidae, 59
Eudendrium californicum, **46**
  spp., 49, 59
Eudistoma purpuropunctatum, 511
Eudistylia catharinae, 162
  polymorpha, 162
  vancouveri, **162**
Eudorella, 326, 328
  pacifica, 328, 501
  redacticruris, 501
  tridentata, 328, 501
Eudorellopsis, 326, 328
  integra, 328
Euechinoidea, 460-1
Eugnathina, 453-4
Euherdmania claviformis, 475, 511
Eukrohnia hamata, 478, 479
Eukrohniidae, 479
Eulalia bilineata, 124
  levicornuta, 124
  longicornuta, 124
  macroceros, 124
  nigrimaculata, 124, 492
  parvoseta, 124
  quadrioculata, 124
  sanguinea, **111**, 124
  tubiformis, **122**, 124
  viridis, 124
Eulecithophora, 89-93
Eulima micans, **214**, 218
  randolphi, 218
  rutila, 218
  thersites, **214**, 218
Eulimidae, 195, 218
Eumastia sitiens, 19, 28
Eunice aphroditois, 143, 494
  valens, 143
Eunicida, 141-4
Eunicidae, **117**, 120, 143, 494

Eunoe depressa, **138**, 139
  nodosa, 139
  oerstedi, 139
  senta, 139
  uniseriata, 139
Eupentacta pseudoquinquesemita, 464
  quinquesemita, 464
Euphausia pacifica, 392
Euphausiacea, 392
Euphilomedes carcharodonta, 312
  longiseta, 312
  producta, 312
Euphrosine arctia, 141
  bicirrata, 141
  heterobranchia, 141
  hortensis, 141
Euphrosinidae, 114, **117**, 141, 494
Euphysa flammea, 33, 34, **40**, 41
  japonica, 34, 41
  ruthae, 47, 57
  tentaculata, 33, **40**, 41
  spp., **37**, 47, 57
Euphysidae, 41, 57
Euplokamidae, 81
Euplokamis dunlapae, **80**, 81
Eupolymnia heterobranchia, 161
Euretidae, 13
Euryalina, 457
Eurycopidae, 337, 502
Eurylepta aurantiaca, **86**, 87, 88
  leoparda, 88
Euryleptidae, 88
Eurystomella bilabiata, **436**, 438, 445
Eurystomellidae, 445
Eusergestes similis, 393
Eusiridae, 378, 504
Eusiroidea, 377-8
Eusirus columbianus, 504
  spp., 350, 372, 378, 504
Eusyllis
  blomstrandi, 132
  japonica, 133
  magnifica, 133
Euthecosomata, 241
Eutonina indicans, **38**, 39, 43, 56, 61
Euzonus mucronatus, 153, **146**
  williamsi, 153
Evadne nordmanni, 303
Evasterias troschelii, 451, 454
Exacanthomysis alaskensis, 323, 324
  davisi, 323, 324
Excirolana kincaidi, 332, 334
  linguifrons, 332, 334
  vancouverensis, 332, 334
Exilioidea rectirostris, 197, **222**, 225
Exogone lourei, **112**, 131
  molesta, **130**, 131

  verugera, 131
Exosphaeroma amplicauda, 334, 336
  crenulatum, 334, 336
  inornata, 334, 336
  naidina, 493
  rhomburum, 334, 336
  sp., 493
Eyakia robusta group, 503
  robustus, 365, 380, 503

Fabia subquadrata, **414**, 416
Fabricia oregonica, 163
  sabella, **162**, 163
Fabriciola berkeleyi, 163
Facelinidae, 258
Fallacohospes inchoatus, 90
Farranula, 306
Fartulum occidentale, 212
Fecampiidae, 90
Fenestrulina malusii, **425**, 433, 445
  malusii subsp. *umbonata*, 433, 445
Filellum serpens, **53**
  spp., 57, 60
Filicrisia franciscana, 428, **429**, 441
Filifascigera fasciculata, **429**, 430, 441
Fiona pinnata, 255, 258
Fionidae, 258
Fissidentalium actiniophoru, 500
  erosu, 500
  megathyris, 500
Fissurellidae, 193, **196**, 200
Fissurellidea bimaculata, 200
Fissurisepta pacifica, 200
Flabellifera, 331, 332-6
Flabelligera affinis, 152
Flabelligerida, 152
Flabelligeridae, 114, **118**, 152
Flabellina fusca, 256, 257
  iodinea, 255, 257
  pricei, 256, 257
  trilineata, 256, 257
  verrucosa, 256, 257
Flabellinidae, 257
Florometra serratissima, 447
Flustrellidra corniculata, 440, 442
Flustrellidridae, 442
Flustridae, 443
Foersteria, 59
  purpurea, 39, 43
Forcepia japonica, 26, 29
Forcipulatida, 454
Foxiphalus aleuti, 505
  cognatus, 364, 380
  falciformis, 505
  fucaximeus, 505
  major, 364, 380, 503, 505
  obtusidens, 364, 380, 503
  similis, 364, 380

  xiximeus, 505

Freemania litoricola, **86**, 87, 88
Freyella microplax, 455
Freyellaster fecundus, 455
Fritillaria borealis, 477
Fritillariidae, 477
Frondiporidae, 441
Funiculina parkeri, 70
Funiculinidae, 70
Fusinidae, 198, 227
Fusinus barbarensis, 227
  harfordi, 227, **231**,
  monksae, 227
Fusitriton oregonensis, 194, 217

Gadila aberrans, 500
Gadilida, 290
Gadilidae, 290, 500
Galathealinum brachiosum, 182
Galatheidae, **403**, 406, 411
Galatheoidea, 406, 411
Galeommatidae, 275
Galiteuthis phyllura, 295
Gammaridacarus orchestoideae, 301, 500
Gammaridae, 383
Gammaridea, 346-86, 502-7
Gammaroidea, 383
Gammaropsis barnardi, 504, 506
  chionoecetophila, 504, 506
  ellisi, 370, 384, 504, 507
  ociosa, 504, 507
  shoemakeri, 371, 384, 504, 506
  spinosa, 371, 385, 504, 506
  thompsoni, 371, 385, 506
  sp., 504, 506
Gari californica, 265, 280
Garveia annulata, **47**, 50, 58
  groenlandica, 50, 58
Gastrodelphyidae, 307
Gastrodelphys, 307
Gastropoda, 193-258, 497-8
Gastropteridae, 240
Gastropteron pacificum, 238, 240
Gastrotricha, 103-4
Gattyana ciliata, 139
  cirrosa, **138**, 139
  iphionelloides, **138**, 139, 493
  treadwelli, 139
Gaudichaudius iphionelloides, 493
Gemelliporella inflata, 439, 445
Gennadas incertas, 393
  propinquus, 393
  tinayrei, 393
Genostoma kozloffi, 491
Genostomatidae, 491
Geodia mesotriaena, 18, 19, **21**, 27
Geodiidae, 27
Geodinella robusta, 19, 25, 27
Geomackiea zephyrolata, 36, 42
Gephyreaster swifti, 450, 453

Gersemia rubiformis, 69
Gigantocypris agassizii, 312
Gitanopsis vilordes, 360, 379
Glans carpenteri, 262, 276
Glossobalanus berkeleyi, 478
Glycera americana, 126, **127**
   capitata, **116**, 126, 492
   convoluta, 127
   gigantea, 126
   nana, 492
   robusta, 127
   tenuis, 126
   tesselata, 127
Glyceracea, 126-8
Glyceridae, 110, **116**, 126-**127**, 492
Glycinde armigera, 128
   picta, 128
   polygnatha, 128
Glycymerididae, 260, 269
Glycymeris, **261**
   corteziana, 269
   subobsoleta, 269
Gnathophiurina, 458
Gnathopleustes den, 504
   polychaetus, 504
   pugettensis group, 504
   serratus, 504
   simplex, 505
   trichodus, 505
   spp., 504
Gnathostomula karlingi, 93
Gnathostomulida, 93
Gnathostomulidae, 93
Gnorimosphaeroma insulare, 334, **335**, 336
   noblei, 334, 336
   oregonense, 334, **335**, 336
Golfingia margaritacea, 182
   pugettensis, 181, 182
   vulgaris, 181, 182
Golfingiidae, 182
Gonatidae, 294
Gonatopsis borealis, 294
Gonatus berryi, 294
   californiensis, 294
   madokai, 294
   onyx, 294
   oregonensis, 294
   pyros, 294
   ursabrunae, 294
Goniada annulata, 128
   brunnea, **116**, 128
   maculata, **127**, 128
Goniadidae, 110, **116**, **127**-8
Goniasteridae, 453
Goniodorididae, 250
Gonionemus vertens, 36, 43, 49, 61
Goniopectinidae, 452
Gonothyraea clarki, **52**
   spp., 56, 60
Gorgonocephalidae, 457

Gorgonocephalus eucnemis, 455, **456**, 457
Graffilla pugetensis, 491
Graffillidae, 90, 491
Grammaria spp., 57, 60
Grandifoxus aciculatus, 505
   dixonensis, 505
   grandis, 363, 380, 503
   lindbergi, 505
   longirostris, 505
   spp., 503
Grania incerta, 172
   paucispina, 172
Grantia comoxensis, 10
   compressa, 8, 10
Grantiidae, 10
Granulina margaritula, 194, **215**, 228
Granulosina, 453
Grapsidae, **412**, 413, 416, 509
Gymnocrater baxteri, 204
Gymnolaemata, 424-439, 442-6
Gymnomorpha, 234, 258
Gymnosomata, 233, **234**, 241, 497
Gymnosomina, 241
Gyratrix hermaphroditus, 91
   proaviformis, 91

Habepegris washingtonia, 272
Hadromerida, 27-8
Hadzioidea, 384
Haeckelia rubra, 81
Haeckeliidae, 81
Halacaridae, 300
Halcampa crypta, 73, 77
   decementaculata, 73, 77
Halcampidae, 77
Halcampoides purpurea, 73, 77
Halcampoididae, 77
Haleciidae, 59
Halecium, **46**, 59
   kofoidi, **52**
   labrosum, **52**
   muricatum, **52**
   spp., 50
Halicardia perplicata, 287
Halicella halona, 372, 382, 504, 506
Halichondria bowerbanki, 24, 28
   panicea, 22, 24, 28
   spp., 24, 28
Halichondriida, 28-9
Halichondriidae, 28
Haliclona ecbasis, 22, 24, 31
   permollis, 22, 24, 31
Haliclonidae, 31
Haliclystidae, **66**, 67, 489
Haliclystus salpinx, **66**, 67, 489
   stejnegeri, 67, 489
Halimedusa typus, 36, 42, 49, 59
Halimedusidae, 42, 59
Haliophasma geminata, 331

Haliotidae, 194, 199
Haliotis cracherodii, 199
   kamtschatkana, 199
   rufescens, 199
   walallensis, 199
Haliplanella lineata, 75, 78
Haliplanellidae, 78
Halisarca sacra, 14, 31
Halisarcidae, 31
Halistylus pupoideus, 201
Halitholus pauper, 41, 42
   sp., **37**, 41, 42
Hallaxa chani, 247, 250, 498
Halobisium occidentale, 300
Haloclavidae, 77
Halocynthia aurantium, 472, 476
   igaboja, 469, 476
Halocyprida, 312
Halocyprididae, 312
Halocypridina, 312
Halosydna brevisetosa, 139
Hamacanthidae, 29
Hamigera lundbecki, 19, 30
Haminaea callidegenita, 497
   vesicula, 497
   virescens, 497
Haminoea vesicula, 238, **239**, 497
   virescens, 238, 239, 497
Hanleya oldroydi, 191
Hanleyidae, 191
Hansenulus trebax, 310
Hapalogaster grebnitzkii, 407
   mertensii, 407
Haploops tubicola, 368, 383
Haplosclerida, 31
Haplosyllis spongicola subsp. *spongicola*, 131
Harbansus mayeri, 312
Harmothoe extenuata, 137
   fragilis, 137
   imbricata, **112**, 137, **138**
   lunulata, 137, **138**
   multisetosa, 137
Harpacticoida, 304-5
Harpiniopsis fulgens, 363, 380
Hataia parva, 48, 58
Haustoriidae, 383
Havelockia benti, 463, 464
Hebella spp., 57, 60
Hedgpethia californica, 300
Hedylopsidae, 236
Hedylopsis sp., **234**, 236
Helicoptilum rigidum, 70
Hemectyon hyle, 18, 28
Hemiarthrus abdominalis, 342
Hemiasterina, 461
Hemichordata, 478, 511
Hemicyclops subadhaerens, 306
   thysanotus, 306
Hemicythere bicarina, 313
Hemicytheridae, 313
Hemigrapsus nudus, 416
   oregonensis, **412**, 416

Hemilamprops, 326
   californicus, 501
   gracilis, 328
Hemioniscidae, 343
Hemioniscus balani, 341, 343
Hemipenaeus spinidorsalis, 393
Hemipodus borealis, 126, **127**
Hemiptera, 302
Hemithiris psittacea, 419, 420
Hemithyrididae, 420
Henricia aspera, 450, 454
   asthenactis, 450, 454
   leviuscula, 451, 454, 510
   longispina, 450, 454
   polyacantha, 454
   sanguinolenta, 450, 454
   sp., 451, 454
Heptabrachia ctenophora, 182
Heptacarpus brevirostris, **395**, 398
   camtschaticus, 399
   carinatus, 398
   decorus, 399
   flexus, **395**, 399
   herdmani, 398
   kincaidi, **395**, 399
   littoralis, 398
   moseri, 398
   paludicola, **395**, 398
   pictus, 398, 507
   pugettensis, **395**, 398
   sitchensis, 398, 507
   stimpsoni, 398
   stylus, 398
   taylori, 398, 507
   tenuissimus, 398
   tridens, 399
Hermadion truncata, 137
Hermaea oliviae, 242
   vancouverensis, 242
Hermannella, 307
Hermissenda crassicornis, **244**, 256, 258
Herpyllobiidae, 309
Hesionidae, **116**, 120, 128
Hesionura coineaui subsp. *difficilis*, **122**, 123
Hesperonoe adventor, 139
   complanata, 137, **138**
Heterodonta, 274-85
Heteromastus filiformis, 154
   filobranchus, **146**, 154
Heteromysis odontops, 321, 324
Heteronemertea, 99
Heterophoxus affinis, 505
   conlanae, 505
   ellisi, 505
   oculatus, 363, 380, 503, 505
   spp., 503
Heteropoda, 208
Heteropodarke heteromorpha, 493
   sp., 493
Heteropora alaskensis, **425**, 428, 441

   magna, **425**, 428, 441
   pacifica, 428, 441
Heteroporidae, 441
Heterotiara anonyma, 39, 42
Heterozonias alternatus, 453
Hexactinellida, 6, 10-3
Hexactinosa, 13
Hexadella sp., 14, 31
Hexasterophora, 13
Hiatella arctica, 284
Hiatellidae, 262, 265, 284
Higginsia higgini, 22, 28
Hincksina alba, 431, **432**, 443
   pallida, 424, **432**, 443
Hincksinidae, 443
Hinnites giganteus, 260, 273
Hippasteria californica, 453
   spinosa, 448, **449**, 453
Hippidae, 404, 410
Hippodiplosia insculpta, 424, **436**, 438, 445, 510
   reticulato-punctata, 437, 445, 510
Hippoidea, 404, 410
Hippolyte clarki, 398
Hippolytidae, 394, **395**, 397-400, 508
Hippomedon spp., 365, 381
Hippomonavella longirostrata, 439, 445
Hipponicidae, 193, 213
Hipponix cranioides, 213
Hippoporella nitescens, **436**, 437, 445
Hippoporinidae, 445
Hippothoa divaricata, 438, 445
   hyalina, 438, 445, 510
Hippothoidae, 445, 510
Hirudinea, 178-9
Hirudinoidea, 178-9, 496
Histioteuthidae, 294
Histioteuthis dofleini, 294
Hobsonia florida, 159
Holasteroida, 461
Holaxonia, 69-70
Holmesiella anomala, 322, 324
Holmesimysis costata, 323, 324
   nuda, 323, 324
   nudensis, 323, 324
   sculpta, 323, 324
   sculptoides, 323, 324
Hololepida magna, 139
Holophryxus alaskensis, 341, 342
Holoporella brunnea, 438, 446
Holothuroidea, 461-6, 511
Homalopoma lacunatum, 201
   luridum, 201, 202
   subobsoletum, 201
Homalorhagida, 108
Homoscleromorpha, 26
Homoscleromorphida, 26
Hopkinsia rosacea, 245, 250

Hoplonemertea, 100
Hormathiidae, 78
Hormiphora cucumis, 80, 81
   sp., 491
Humilaria kennerlyi, **281**, 282
Huxleyia munita, 266
Hyale, **353**
   anceps, 352, 379
   frequens, 352, 355, 379
   plumulosa, 352, 379
   pugettensis, 352, 379
Hyalidae, 379
Hyalopecten neoceanus, 274
Hyalospongia, 10
Hyas lyratus, 415
Hybocodon prolifer, 33, 41, **47**, 48, 57
Hyboscolex pacificus, 153
Hydractinia aggregata, 49, 58
   laevispina, 49, 58
   milleri, 49, 58
   sp., 49, 58
Hydractiniidae, 49, 58
Hydrallmania spp., 50, 60
Hydrodendron spp., 50, 59
Hydroid polyps, 44-61, 488
Hydroida, 41-4, 57-61, 488
Hydromedusae, 32-44, 487-8
Hydrozoa, 32-65
Hymedesanisochela rayae, 21, 25, 30
Hymedesmia spp., 23, 26, 30
Hymedesmiidae, 30
Hymenamphiastra cyanocrypta, 22, 23, 30
Hymenanchora sp., 26, 30
Hymenaster perissonotus, 454
   quadrispinosus, 454
Hymendectyon lyoni, **21**, 26, 29
Hymeniacidon perleve, 22, 29
   sinapium, 24, 29
   ungodon, 22, 24, 29
   sp., 29
Hymeniacidonidae, 29
Hymenodora acanthitelsonis, 396
   frontalis, **395**, 396
   glacialis, 396
Hyocrinida, 447
Hyocrinidae, 447
Hyperiidea, 346, 391

Ianiropsis analoga, 336, 337
   kincaidi, 336, 37
   magnocula, **335**, 336, 337
   tridens, **335**, 336, 337
Idanthyrsus armatus, **119**, 158
   ornamentatus, 158
Idarcturus hedgpethi, 338, 340
Idotea aculeata, 340, 502
   aleutica, 502
   fewkesi, 338, **339**, 340
   kirchanskii, 340
   montereyensis, 340

obscura, 340
ochotensis, 338, 340, 502
urotoma, 338, **339**, 340
recta, 502
resecata, 338, **339**, 340
rufescens, 338, 340
schmitti, **339**, 340
stenops, **339**, 340
wosnesenskii, **339**, 340
Idoteidae, 340, 502
Ilyanassa obsoleta, 198, 227
Imbrexodius oclairi, 506
Immergentia sp., 439, 442
Immergentiidae, 442
Inarticulata, 419
Incirrata, 295
Insecta, 301-2
Intoshiidae, 82, 492
Inusitatomysis insolita, 321, 324
Invenusta paracnida, 92
Ione cornuta, 342
Iophon chelifer var. *californiana*, 18, 29
  piceus var. *pacifica*, 25, 29
Iothia lindbergi, 205
Iphimedia rickettsi, 503, 506
Iphitimidae, 494
Isaeidae, 384-5, 506
Ischnochiton abyssicola, 191
  interstinctus, 187, 191
  trifidus, 187, 191
Ischnochitonidae, 191-2
Ischyroceridae, 385, 508
Ischyrocerus anguipes, 356, **357**, 385
  serratus, **357**, 385
  sp., 357
Iselica obtusa, 235, 236
  ovoidea, 235, 236, **237**
Isididae, 69
Isocirrus longiceps, 157
Isocyamus delphinii, 391
Isopoda, 330-45, 502
Itaipusa bispina, 91
  curvicirra, 91
Itaspiella armata, 92

Jaeropsididae, 337
Jaeropsis setosa, 336, 337
Janiralata occidentalis, 336, 337
  solasteri, 336, 337
Janiridae, 337
Janolus barbarensis, 254, 255, 498
  fuscus, 254, 255
Janthina prolongata, 218
Janthinidae, 218
Janua pagenstecheri, **165**
Japatella diaphana, 295
Jasmineira pacifica, **162**, 163
Jassa borowskyae, 506
  morinoi, 506
  oclairi, 506
  shawi, 506

  slatteryi, 506
  staudei, 506
  spp., 356, **357**, 385
Johanssonia sp., 178
Jones *amaknakensis*, 23, 26, 30
Jonesia rostrata, 313
Jugaria quadrangularis., 496
  similis, 496

Kaburakia excelsa, 85, 88
Kalyptorhynchia, 91
Katadesmia gibbsii, 268
Katharina tunicata, 186, 192
Kefersteinia cirrata, 128
Kellia suborbicularis, 275
Kelliella galatheae, 280
Kelliellidae, 280
Kelliidae, 264, 275
Kentrogonida, 318-9
Keratella cochlearis subsp. *tecta*, 105
Kermystheus ociosa, 371, 385, 507
Kinorhyncha, 105-8
Kinorhynchus cataphractus, 106, 108
  ilyocryptus, 106, **107**, 108
Koinocystididae, 91
Kophobelemnidae, 70
Kophobelemnon affine, 70
  biflorum, 70
  hispidum, 70
Korethrasteridae, 454
Koroga megalops, 367, 381
Kronborgia pugetensis, 90
Kurtzia arteaga, 229
  variegata, 232
Kurtziella crebricostata, 497, *plumbea*, 229, 497

Labidochirus splendescens, 408
Lacerna fistulata, 438, 445
Lacuna marmorata, 209
  porrecta, 209
  variegata, 209, **211**
  vincta, 209, **211**
Lacunidae, 195, 208-9
Laelaptidae, 301
Laemophiurina, 458
Laeodiceidae, **46**
Laetmogone wyvillethompsoni, 465
Laetmogonidae, 465
Laetmonice pellucida, 136
Laetmophasma fecundum, 465
Lafoea dumosa, **46, 53**
  spp., 57, 60
Lafoeidae, 60
Lafystiidae, 382
Lagenipora punctulata, 427, **436**, 446
  socialis, 437, 446

Lagunogammarus setosus, 373, 383
Laila cockerelli, 246, 250
Lamellaria diegoensis, 217
Lamellariidae, 217
Lamellibrachia barhami, 183
Lamellibrachiida, 183
Lamellibrachiidae, 183
Lamellisabella coronata, 183
  zachsi, 183
Lamellisabellidae, 183
Lampropidae, 328, 501
Lamprops, 326
  beringi, 328
  carinata, 328, 501
  fuscata, 328, 501
  fuscatus, 501
  krasheninnikovi, 328
  obfuscatus, 501
  serrata, 328, 501
  serratus, 501
  tomalesi, 501
Lanassa venusta subsp. *venusta*, 160
Lanice sp., 161
Laodicea sp., 36, 43
Laodiceidae, 43, 59
Laonice cirrata, 148
  pugettensis, 148
Laonome kröyeri, 162
Laphania boecki, 160
Laqueidae, 421
Laqueus californianus, 419, **420**, 421
Larvacea, 477
Lasaea subviridis, 275, **277**
Lasaeidae, 265, 275
Laternula limicola, 283
Laternulidae, 283
Latrunculia sp., 18, 28
Latrunculiidae, 28
Laxosuberites sp., 24, 27
Leachia dislocata, 295
Leaena abranchiata, 160
Lebbeus catalepsis, 508
  grandimanus, 400
  groenlandicus, 400
  schrencki, 400
  washingtonianus, 400
  sp., 399, 508
Leitoscoloplos panamensis, 145
  pugettensis, 145
Lensia baryi, 63, 65
  conoidea, 63, 65
Lepadidae, 316
Lepadomorpha, 314, 315
Lepas, **315**
  anatifera, 316
  fascicularis, 316
  hilli, 316
  pacifica, 316
Lepeta caeca, 205
Lepetidae, 193, 204, 205-6

Lepidasthenia berkeleyae, 139
   longicirrata, 139
Lepidepecreum garthi, 367, 381
   gurjanovae, **366**, 367, 381
Lepidochitona dentiens, 187, 192
   fernaldi, 187, 192
   hartwegii, 192
Lepidochitonidae, 192
Lepidonotus squamatus, **111**, 137
Lepidopleurina, 191
Lepidozona cooperi, **188**, 189, 191
   mertensii, **188**, 189, 191
   retiporosa, 187, **188**, 191
   scabricostata, 190, 192
   willetti, **188**, 190, 192
Lepraliella bispina, **436**, 446
Leptasterias aequalis, 510
   epichlora, 510
   hexactis, 448, 454, 510
Leptochelia savignyi, 329, **330**
Leptochiton nexus, 187, 191
   rugatus, 187, 191
Leptochitonidae, 191
Leptocuma, 326, 327
Leptognathia gracilis, 329
Leptognathina, 454
Leptomedusae, 43-4, 59
Leptoplana vesiculata, 87, 88
Leptoplanidae, 88
Leptostraca, 320
Leptostylis, 326
   abditis, 502
   calva, 502
   villosa, 328, 502
Leptosynapta clarki, **462**, 463, 466
   roxtona, 466
   transgressor, 463, 466
Leptychaster anomalus, 452
   arcticus, 452
   inermis, 452
   pacificus, 448, 452
Lernaeopodidae, 309
Lethasterias nanimensis, 454
Leucandra heathi, 8, 10
   levis, 8, 10
   pyriformis, 8, 10
   taylori, 8, 10
Leucilla nuttingi, 8, 10
Leuckartiara foersteri, 39, 42
   spp., 39, 42, 49, 59
Leucon, 326, 327
   declivis, 501
   falcicostata, 501
   fulvus, 328, 502
   nasica, 328, 501
Leuconidae, 328, 501
Leucopsila stylifera, 10
Leucosolenia eleanor, **7**, 9
   nautilia, 9
   spp., 9
Leucosoleniida, 9

Leucosoleniidae, 9
Leucothea pulchra, 491
   sp., 80, 81, 491
Leucothoe spp., 360, **350**, 379
Leucothoidae, 379
Leucothoidea, 379
Leuroleberis sharpei, 312
Levinsenia gracilis, 495
   rectangulata, 178
Lichenopora novae-zelandiae, 430, 441
   verrucaria, **425**, 430, 441
Lichenoporidae, 441
Lichomolgidae, 306-7
Lictorella, 57
Ligia occidentalis, 344
   pallasii, 344
Ligiidae, 344
Limacina helicina, 233, 240, 241
Limacinidae, 241
Limalepeta caecoides, 206
Limatula attenuata, 272
   saturna, 272
   subauriculata, **271**, 272
   vancouverensis, 272
Limidae, 260, 272
Limnodriloides monothecus, 175
   victoriensis, 175
Limnomedusae, 43-4, 61
Limnoria algarum, 332, 334, **335**
   lignorum, 332, 334, **335**
   tripunctata, 332, 334, **335**
Limnoriidae, 334
Limoida, 272
Limopsidae, 269
Limopsis akutanica, 269
   dalli, 269
Lineidae, 99
Lineus bilineatus, 96, 99
   pictifrons, 96, 99
   ruber, 99
   rubescens, 99, 492
   sp., 96, 99, 492
   vegetus, 96, 99
Liocyma fluctuosa, 282
Liponema brevicornis, 77
Liponematidae, 77
Liriopsidae, 344
Liriopsis pygmaea, 341, 344
Lironeca californica, 333, 334
   vulgaris, 333, 334
Lirularia lirulata, **202**, 203
   parcipicta, 203, **215**
   succincta, **202**, 203, **215**
Lissocrangon stylirostris, 400
Lissodendoryx firma, **21**, 26, 30
   sp., 26, 30
Listriolobus hexamyotus, 181
Lithodes aequispina, 407
   couesi, 407
Lithodidae, **403**, 406-8
Lithophaga plumula, 270
Littorina keenae, 209, **211**

   plena, 209
   scutulata, 209
   sitkana, 209
Littorinidae, 195, 209
Littorophiloscia richardsonae, 344, 345
Lobata, 81
Locustogammarus levingsi, 373, 383
Loimia medusa, 161
Loliginidae, 294
Loligo opalescens, **292**, 294
Lophaster furcifer subsp. *vexator*, 453
   furcilliger, 453
Lophelia californica, 72
Lopholithodes foraminatus, 407
   mandtii, 407
Lophopanopeus bellus subsp. *bellus*, 416
   bellus subsp. *diegensis*, 416
Lottia alveus, 206
   asmi, 207, **211**
   digitalis, **202**, 207
   gigantea, 206
   instabilis, 206
   limatula, 207, **211**
   ochracea, 207
   painei, 208
   pelta, 207, 208
   strigatella, 207, **211**
   triangularis, 206, 207
Lottiidae, 194, **196**, 205, 206-8
Loxoconcha dentiarticula, 314
   fragilis, 314
   tenuiungula, 314
Loxoconchidae, 314
Loxokalypodidae, 422
Loxokalypus socialis, 421, 422, 509
Loxosoma davenporti, 421, 422, 509
Loxosomatidae, 422
Loxosomella nordgaardi, 421, 422, 509
   spp., 421, 422, 509
Loxothylacus panopaei, 319
Lubbockia, 306
Lucina tenuisculpta, 262, **271**, 274
Lucinidae, 262, 264, 274
Lucinoma annulata, 264, **271**, 274
Luidia foliolata, 448, **449**, 452
Luidiaster dawsoni, 452
Luidiidae, 452
Lumbricillus annulatus, 174
   curtus, 174
   lineatus, 174
   mirabilis, 174
   pagenstecheri, 174
   qualicumensis, 174
   tsimpseanis, 174

tuba, 174
Lumbrineridae, **112**, **117**, 121, **142**, 143, 494
Lumbrineris bicirrata, **142**, 143
  californiensis, 143
  cruzensis, 143
  inflata, 143
  japonica, 143
  lagunae, 143, 494
  latreilli, **112**, **142**, 143
  limicola, 143
  luti, 143
  pallida, 143
  similabris, 143
  zonata, 143
Lycastopsis sp., 133
Lyonsia californica, 287, **286**
Lyonsiella parva, 287
Lyonsiidae, 262, 287
Lyrula hippocrepis, 424, 431, **435**, 444
Lysianassa holmesi, 367, 381, 504
Lysianassidae, 381, 506
Lysianassoidea, 381, 506
Lysilla loveni, 160
Lysippe labiata, 159
Lyssacinosa, 13

Macclintockia scabra, **202**, 205, 207
Macoma balthica, 279, 500
  brota, 279, 499
  calcarea, 279, 500
  carlottensis, 280, 499
  elimata, 279, 499
  expansa, 280, 499
  inquinata, 279, 500
  moesta, 279, 499
  lipara, 499
  nasuta, 279, 499
  obliqua, 279, 499
  secta, 279, 499
  yoldiformis, 279, 499
Macrochaeta pege, 152
Macroclymene sp., 157
Macrodasyida, 104
Macrodasyidae, 104
Macrodasys cunctatus, 104
Mactridae, 265, 278
Madreporaria, 72
Maera danae, 373, 384
  loveni, 373, 384
  simile, 373, 384
Magelona berkeleyi, 149
  hobsonae, 149
  longicornis, 149
  sacculata, 149
Magelonida, 149
Magelonidae, 115, **118**, 149
Majidae, 411, **412**, 413, **414**, 415
Majoxiphalus maximus, 505
  spp., 504

Malacobdella macomae, 492
  siliquae, 492
  spp., 97, 100
Malacobdellidae, 100
Malacoceros fuliginosus, 148
  glutaeus, 148
Malacostraca, 320-417
Maldane glebifex, 155
Maldane sarsi subsp. *sarsi*, 155
Maldanella harai, 157
  robusta, 157
Maldanidae, 112, **118**, 155, **156**, 157, 495
Malletia faba, 268
  pacifica, 268
  talama, 269
  truncata, 269
Malletiidae, 266, 268-9
Malmgreniella bansei, 493
  berkeleyorum, 493
  liei, 493
  nigralba, 493
  scriptoria, 493
Malmiana diminuta, 178
  virida, 178
Manania distincta, 489
  gwilliami, 489
  hexaradiata, 489
Manayunkia aestuarina, 163
Mandibulophoxus alaskensis, 505
  gilesi, 363, 380, 503
  mayi, 505
  spp., 503
Margarites beringensis, 203
  marginatus, **202**, 203
  pupillus, **202**, **215**, 203
  rhodia, 203, **215**
  simblus, 204
Marginellidae, 194, 228
Maricola, 93
Marionina appendiculata, 173
  charlottensis, 173
  glandulifera, 172, **173**
  klaskisharum, 172
  neroutsensis, 172, 174
  sjaelandica, 174
  southerni, 172
  subterranea, 172, **173**
  vancouverensis, 173
Marphysa stylobranchiata, 143
Marseniidae, 194, 217
Marsenina rhombica, 217
  stearnsii, 217
Marseniopsis sharonae, 217
Marsipobdella sacculata, 178
Mayerella banksia, 387, 390
Mediaster aequalis, **449**, 450, 453
Mediomastus ambiseta, 154
  californiensis, 154
  capensis, 154
Megacrenella columbiana, 272
Megalomma splendida, **119**, 161
Megalorchestia, 353

  californiana, 355, 380
  columbiana, 355, 380
  pugettensis, 355, 380
Megaluropus spp., **350**, 372, 377, 384
Megamphopus sp., 370, 385, 504, 506
Megasurcula carpenteri, 229
Meiomenia swedmarki, 185
Meiomeniidae, 185
Melampidae, 194, 258
Melanellidae, 218
Melanochlamys diomedea, 238, 240
Melibe leonina, 252, 253
Melicertidae, 43
Melicertum octocostatum, 35, **40**, 43
Melinna cristata, 158
  elisabethae, 158
Melita californica, 373, 384
  dentata, **366**, 373, 384
  desdichada, 373, 384
  oregonensis, 373, 384
Melita sulca, 373, 384
Melitidae, 384
Melphidippella, 383
Melphidippidae, 360, 377, 383, 506
Melphidippoidea, 383
Melphisana, 506
Melphissana, 383, 506
Membranipora membranacea, **249**, 431, **432**, 443, 510
Membraniporidae, 443
Merriamum oxeota, 19, 26, 30
Mertensia sp., 80, 81
Mertensiidae, 81
Mesochaetopterus taylori, **117**, 149
Mesocrangon munitella, 400
Mesogastropoda, 208-19
Mesolamprops, 326
  dillonensis, 501
  californiensis, 328
Mesonerilla sp., 167
Mesothuria murrayi, 465
Metacaprella anomala, 388, **389**, 390
  kennerlyi, 388, 390
Metacopina, 314
Metacrangon acclivis, **401**
  munita, **401**
  spinosissima, 400
  variabilis, 401
Metandrocarpa dura, 472, 476
  taylori, **470**, 472, 476
Metaphoxus frequens, 363, 380, 505
  fultoni, 363, **364**, 380, 503
Metasychis disparidenta, 495
Meterythrops robusta, 322, 324
Metopa cistella, 379

Metopella carinata, 379
Metridiidae, 78
Metridium giganteum, 490
   senile, 75, 78, 490
      subsp. *fimbriatum*, 490
      sp., **74**, 75, 78, 490
Micranellum crebricinctum, **211**, 212
Microcerberidae, 331
Microcerberidea, 331
Microcerberus abbotti, 331
   abbotti subsp. *juani*, 331
Microciona microjoanna, 23, 30
   primitiva, 30
   prolifera, 19, 23, 30
Microglyphis estuarinus, 238
Microjassa barnardi, 507
   bousfieldi, 507
   litotes, **357**
   spp., 357, 385
Micronereis nanaimoensis, 133
Microphthalmus coustalini, 493
   hystrix, 493
   simplicichaetosus, 493
Micropleustes nautiloides, 505
   nautilus, 504, 505
   spp., 504
Micropodarke dubia, 128
Micropora coriacea, **434**, 437, 444
Microporella californica, **436**, 437, 445
   ciliata, **436**, 437, 445
   setiformis, **436**, 437, 445
   umbonata, 433, 445
   vibraculifera, 433, 445
Microporellidae, 445
Microporidae, 444
Microporina borealis, 428, **434**, 444
Micrura alaskensis, 96, 99
   verrilli, 96, 99
   wilsoni, 96, 99
Mimulus foliatus, 413, **414**
Miontodiscus prolongatus, 276
Mitella polymerus, 315, 316, 501
Mitrocoma, 59
   cellularia, **38**, 39, 43
Mitrocomella, 59
   polydiademata, **38**, 39, 43
Mitrocomidae, 43, 56, 59, 488, 489
Mitromorpha gracilior, 229
Modiolus modiolus, 270
   rectus, 270
Mohnia frielei, 225
Molgula cooperi, 476
   manhattensis, 511
   oregonia, 476
   pacifica, 469, 476
   pugetiensis, 468, 476
Molgulidae, 476, 511
Mollusca, 184-295, 497-500
Molpadiida, 466

Molpadia borealis, 466
   granulosa, 466
   intermedia, 462, 466
   musculus, 466
Molpadiidae, 466
Monobrachium parasiticum, 45, 61
Monoculodes spinipes, 372, 378
   zernovi, **364**, 372, 378
   spp., 362, 378
Monogononta, 105
Monophragmophora, 512
Monopylephorus cuticulatus, 174
   parvus, 174
   rubroniveus, 174
Monostilifera, 100-1
Monstrilla canadensis, 310
   helgolandica, 310
   longiremis, 310
   spinosa, 310
   wandelii, 310
Monstrillidae, 310
Monstrilloida, 310
Montacutidae, 264, 275-6
Monticellina tesselata, 495
Mooreonuphis stigmatis, 141
Mooresamytha bioculata, 495
Mopalia ciliata, **189**, 191, 192
   cirrata, 190, 192
   cithara, 192
   egretta, 192
   hindsii, **189**, 191, 192
   imporcata, 191, 192
   laevior, 190, 192
   lignosa, **189**, 190, 192
   muscosa, 190, 192
   phorminx, 192
   porifera, 191, 192
   sinuata, 190, 192
   spectabilis, 191, 192
   swanii, 191, 192
Mopaliidae, 192
Moroteuthis robusta, 293, 294
Mucronella ventricosa, 438, 446
Muggiaea atlantica, 63, **64**, 65
Munida quadrispina, **403**, 411
Munidion parvum, 342
Munidopsis quadrata, 411
Munna chromatocephala, **335**, 337
   fernaldi, 337
   ubiquita, 335, 337
Munnidae, 337
Munnogonium tillerae, 337
Munnopsidae, 337, 502
Muricidae, **196**, 198, 219-21
Mursia gaudichaudi, 413
Musculista senhousia, 270
Musculus discors, **271**, 272
   niger, 272
   taylori, 272
Musellifer sublitoralis, 104
Mya arenaria, **263**, 284

   truncata, 284
Mycale adhaerens, 14, 16, **21**, 26, 29
   bamfieldensis, 487
   bellabellensis, 16, 29
   hispida, 26, 29
   macginitiei, 25, 29, 487
   richardsoni, 25, 29
   toporoki, 25, 29
Mycalecarmia lobata, 29
Mycalidae, 29, 487
Myicolidae, 306
Myidae, 264, 283-4
Myina, 283-4
Myodocopa, 312
Myodocopida, 312
Myoida, 283-4
Myonera tillamookensis, 288
Myopsida, 294
Myosoma spinosa, 422
Myriochele oculata, 158
Myriotrochidae, 466
Myriotrochus bathibius, 466
   giganteus, 466
   sp., 466
Myriozoidae, 446
Myriozoum coarctatum, 428, **436**, 446
   subgracile, 428, 446
   tenue, 428, 446
Mysella tumida, 264, 275, **277**
Mysidacea, 320-4, 501
Mysidae, 324
Mysidella americana, 321, 324
Mysina, 324
Mysis litoralis, 321, 324
Mystides borealis, 123
Mytilicola orientalis, 308
Mytilicolidae, 308
Mytilidae, 261, 270-2, 498
Mytilimeria nuttalli, 287
Mytiloida, 270-2
Mytilus californianus, 270
   edulis, 270, 498
   trossulus, 498
Myxicola infundibulum, 163
Myxilla behringensis, 23, 30
   incrustans, 16, **21**, 26, 30
   lacunosa, 19, 20, 30
Myxillidae, 29-30
Myxoderma platyacanthum, 455
   sacculatum, 455
Myzostomida, 166
Myzostomidae, 166
Myzostomum pseudogigas, 166
Myxosoma spinosa, 422, 509

Nacellidae, 194, 205, 206, 207
Nacellina, 204-8
Naididae, 171-2
Naineris dendritica, **111**, 145
   quadricuspida, 145
   uncinata, 145

Nais communis, 172
  elinguis, 172
  variabilis, 172
Najna spp., 352, **353**, 380
Najnidae, 380
Nannastacidae, 328, 502
Nanomia cara, 62, 63, **64,** 489
  bijuga, 489
Narcomedusae, 44, 488
Nassariidae, 197, 198, 227
Nassarius fossatus, 227
  fraterculus, **224**, 227
  mendicus, **224**, 227
  perpinguis, 227
  rhinites, 227
Natantia, 392-402, 507-8
Natica clausa, **214**, 216
Naticidae, 195, 216
Nearchaster aciculosus, 452
Nebalia pugettensis, 320
Nebaliidae, 320
Nectonemertes mirabilis, 101
Nectonemertidae, 101
Nellobia eusoma, 180
Nematoda, 102-3
Nematoplana nigrocapitula, 92
Nematoplanidae, 92
Nematostella vectensis, 73, 77
Nemertea, 94-101, 492
Nemertopsis gracilis, 98, 100
Nemocardium centifilosum, 262, **277**, 278
Neoadmete circumcincta, 228
  modesta, 198, 228
Neoamphitrite edwardsi, 161
  groenlandica, 161
  robusta, 161
Neocrangon abyssorum, **401**, 402
  communis, **401**, 402
  resima, **401**, 402
Neocyamus physeteris, 391
Neoesperiopsis digitata, 18, 29
  infundibula, 16, 29
  rigida, 18, 29
  vancouverensis, 18, 29
Neogastropoda, 219-32
Neoleprea californica, 161
  japonica, 161
  spiralis, 161
Neoloricata, 191-2
Neomenia dalyelli, 185
Neomeniamorpha, 185
Neomeniidae, 185
Neomolgus littoralis, 301
Neomysis kadiakensis, 323, 324
  mercedis, 322, 324
  rayii, 322, 324
Neorhabdocoelida, 89-91, 491
Neotremata, 419
Neotrypaea californiensis, 508
  gigas, 508
Neoturris breviconis, **37**, 39, 42
  spp., 49, 59

Nephtheidae, 69
Nephtyidacea, 135-6
Nephtyidae, **116**, 120, **134**, 135-6
Nephtys assignis, 136
  brachycephala, 136
  caeca, 136
  caecoides, **134**, 136
  californiensis, 136
  ciliata, 136
  cornuta subsp. *cornuta*, **134**, 135
  cornuta subsp. *franciscana*, 135
  ferruginea, 136
  longosetosa, **134**, 135
  punctata, 135
  rickettsi, 136
Neptunea amianta, 226
  humboldtiana, 226, 497
  ithia, 226
  lyrata, 226
  phoenicia, **222**, 226
  pribiloffensis, 226
  smirnia, **224**, 226
  stilesi, 226
  tabulata, 198, **224**, 226
Neptuneidae, 197-9, 225
Nereidacea, 128-35
Nereidae, **111**, **112**, 120, 133-5, **134**, 493
Nereiphylla castanea, 492
Nereis brandti, 135
  eakini, 135
  grubei, 133
  limnicola, 135
  neoneanthes, 133, **134**
  paucidentata, 135
  pelagica, 135
  procera, 135
  vexillosa, 133, **134**
  virens, 135
  wailesi, 133
  zonata, 135
Nerilla antennata, 168
  digitata, **167**, 168
  inopinata, 168
Nerillida, 167-8
Nerillidae, **167**
Nesion arcticum, 93
Nesionidae, 93
Netastoma rostrata, 284
Neverita lamonae, 216
Nexilidae, 93
Nexilis epichitonius, 93
Nicolea zostericola, 161
Nicomache lumbricalis, **156**, 157
  personata, 157
Nicon moniloceras, 133
Nicothoidae, 309-10
Ninoe sp., 143
Nippoleucon hinumensis, 501
Nipponnemertes pacificus, 98, 100

  punctatulus, 100
Nitidiscala caamanoi, **215**, 218
  catalinae, 218
  catalinensis, 218
  hindsii, 218
  indianorum, **214**, 218
  sawinae, 218
  tincta, **214**, 218
Niveotectura funiculata, 206
Nolella stipata, 439, 442
Nootkadrilus compressus, 177
  gracilisetosus, 177, 178
  hamatus, 178
  verutus, 177
Notaspidea, 233, 240
Notholca striata, 105
Nothria conchylega, 141, 492
Notocaryoplanella glandulosa, 92
Notocirrus californiensis, 144
Notodelphyidae, 305-6
Notomastus giganteus, 154
  lineatus, 154
  tenuis, **146**, 154
  variegatus, 154
Notomyotina, 452
Notophyllum imbricatum, 123
  tectum, 123
Notoplana atomata, 87, 88
  celeris, 87, 88
  inquieta, 87, 88
  inquilina, 87, 88
  longastyletta, 87, 88
  natans, 87, 88
  rupicola, 88
  sanguinea, 87, 88
  sanjuania, **86**, 87, 88
Notoproctus pacificus, 157
Notostomobdella cyclostoma, 178
Notostomus japonicus, 394
Novobranchus pacificus, 159
Nucella canaliculata, **220**, 223
  emarginata, **222**, 223
  lamellosa, 198, **220**, 223
  lima, **220**, 223
Nucellidae, 198, 199, 221-3
Nucinellidae, 266
Nucula cardara, 266
  carlottensis, 266
  chrysocoma, 266
  darela, 266
  linki, 266
  tenuis, 266
Nuculana amiata, 268
  cellulita, 266
  conceptionis, **267**, 268
  extenuata, 268
  fossa, **267**, 268
  gomphoidea, 268
  hamata, 266, **267**
  leonina, 268
  liogona, 268
  lomaensis, 268
  minuta, 266

penderi, 268
pernula, 268
spargana, 268
 taphria, **267**, 268
 tenuisculpta, 268
Nuculanidae, 266-8
Nuculidae, 266
Nuculoida, 260, 266
Nudibranchia, 234-5, 243-58, 497-8
Nuttallina californica, 187, 192
Nymphon, **298**
 grossipes, 299, 300
 longitarse, 300
 pixellae, 299, 300
 rubrum, 300
Nymphonidae, 300
Nymphopsis spinosissima, 297, **298**, 299
Nynantheae, 77-8

Obelia, **46**
 bidentata, 56, 60, 488
 dichotoma, 34, 38, 43, **52**, 56, 60, 487, 488
 geniculata, 38, 43, **51**, 56, 60, 487, 488
 longissima, 487, 488
Oceanobdella pallida, 178
Ocenebra interfossa, **220**, 221
 lurida, **220**, 221
 orpheus, **215**, 221
 painei, **215**, 221
 sclera, 221
 triangulata, 221
Octocorallia, 68
Octopoda, 295
Octopodidae, 295
Octopoteuthidae, 294
Octopoteuthis deletron, 294
Octopus dofleini, **292**, 293, 295
 leioderma, 291, **292**, 295
 rubescens, 293, 295
Odius kelleri, 361, 382, 504, 506
Odontogena borealis, 275
Odontosyllis fulgurans subsp.
 japonica, 132
 parva, 132
 phosphorea, 132
Odostomia, 235
 angularis, 236
 barkleyensis, 236, **237**
 canfieldi, 236
 cassandra, 236
 chinooki, 236
 columbiana, **214**, 236
 cypria, 236
 engbergi, 236
 grippiana, 236
 kennerlyi, 236
 nuciformis, 236
 oregonensis, 236
 quadrae, 236, **237**

 satura, 236
 spreadboroughi, 236
 tacomaensis, 236
 tenuisculpta, 236
 vancouverensis, 236,
 willetii, 236
 youngi, 236, **237**
Oedicerotidae, 369, 378
Oedignathus inermis, 407
Oedocerotoidea, 378
Oegopsida, 294
Oenonidae, 494
Oenopota alaskensis, 229, 232
 alitakensis, 230
 babylonia, 230
 crebricostata, 232
 elegans, 230, 232
 excurvata, 230, 231
 fidicula, 230, **231**
 harpa, 232
 harpularia, 230
 akrausei, 232
 kyskana, 232
 levidensis, 229
 maurelli, 230
 pleurotomaria, 230, 232
 popovia, 230
 pyramidalis, 232
 reticulata, 232
 rosea, 232
 sculpturata, 230
 solida, 232
 tabulata, 230
 turricula, 230
 viridula, 230
Oerstedia spp., 98, 101
Oikopleura dioica, 477
 labradoriensis, 477
Oikopleuridae, 477
Oithona, 305
Oithonidae, 305
Okenia vancouverensis, 246
Olea hansineensis, 242, 243
Oleidae, 243
Oligochaeta, 170-8
Oligochinus, **352**
 lighti, 370, 378
Olindiasidae, 43, 61
Olivella baetica, 228, 231
 biplicata, 228, **231**
 pycna, 228
Olividae, 197, 228
Ommastrephes bartrami, 294
Ommastrephidae, 294
Oncaea, 306
Oncaeidae, 306
Onchidella borealis, 234, 258
 carpenteri, 258
Onchidiacea, 234, 258
Onchidorididae, 250, 498
Onchidoris bilamellata, 248, 250
 muricata, **249**, 250
Oncousoeciidae, 440

Oneirophanta mutabilis, 465
Oniscidae, 345
Oniscoidea, 331, 344-5
Onoba carpenteri, 497
Onuphidae, 494
Onuphis elegans, 141
 geophiliformis, 141
 iridescens, 141
Onychoteuthidae, 294
Onychoteuthis borealijaponica, 294
Opalia borealis, 217
 montereyensis, 218
Ophelia limacina, 153
Opheliida, 153
Opheliidae, **118**, 121, **146**, 153
Ophelina acuminata, **146**, 153
 breviata, 153
Ophiacantha abyssa, 458
 bathybia, 458
 rhacophora, 458
 trachybactra, 458
Ophiacanthidae, 458
Ophiactidae, 458
Ophidonais serpentina, 171
Ophiocomidae, 458
Ophiodermella cancellata, 229
 inermis, 229, **231**
Ophiodromus pugettensis, **116**, 128
Ophioleuce oxycraspedon, 459
Ophioleucidae, 459
Ophiolimna bairdi, 458
Ophiomusium glabrum, 459
 jolliensis, 459
 lymani, 459
Ophiomyxidae, 458
Ophiomyxina, 458
Ophiopholis aculeata, 455, 458
 bakeri, 458
 longispina, 458
Ophiophthalmus cataleimmoidus, 458
 displasia, 458
 eurypoma, 458
 normani, 458
Ophiopteris papillosa, 458
Ophioscolex corynetes, 458
Ophiura bathybia, 459
 cryptolepas, 459
 flagellata, 459
 irrorata, 459
 leptoctenia, 457, 459
 lütkeni, 457, 459
 sarsi, **456**, 457, 459
Ophiurida, 458
Ophiuridae, 459
Ophiuroidea, 455
Ophlitaspongia pennata, 24, 30
Ophryotrocha vivipara, **142**, 144, 494
Opisa tridentata, 365, 381
Opisthobranchia, 194, 195, 232-58

Opisthoteuthidae, 295
Opisthoteuthis californiana, 295
Oplophoridae, 393, 394, **395**-6
Oradarea longimana, 373, 378
Orbiniella nuda, 145
Orbiniida, 144
Orbiniidae, **111**, **117**, 121, 144
Orchomene decipiens, 367, 381
  obtusa, 367, 381, 504
  obtusus, 504
  pacifica, 367, 381, 504
  pacificus, 504
  pinguis, 367, 381
  recondita, 504
  sp., 368, 381
Oregonia bifurca, 415
  gracilis, 413, **414**
Oregoniplana opisthopora, 93
Orina sp., 25, 31
Oriopsis gracilis, 163
  minuta, **162**, 163
Orthasterias koehleri, 451, 454
Orthonectida, 81-2, 491
Orthopagurus minimus, 408
Orthoplana kohni, 92
Orthopyxis spp., 43, **51**, 56, 60
Ostracoda, 310-14
Ostrea lurida, 272
Ostreidae, 259, 272
Ostreina, 272
Ostreobdella papillata, 178
Ostreoida, 272
Otocelididae, 84
Otocelis luteola, 84
Otoplanidae, 92
Otopsis longipes, 129
Ototyphlonemertes americana, 97, 100
Ototyphlonemertidae, 100
Owenia fusiformis, **118**, 158
Oweniida, 158
Oweniidae, 110, **118**, 158
Oxystomata, 413-5
Oxyurostylis, 326, 327

Pachastrellidae, 27
Pachycerianthus fimbriatus, 71, **74**
Pachychalina spp., 22, 31
Pachycheles pubescens, 411
  rudis, 411
Pachygrapsus crassipes, 416, 508
Pachynus barnardi, 365, 381
Pacifacanthomysis nephrophthalma, 323, 324
Pacifides psammophilus, 93
Paciforchestia, **353**
  klawei, 353, 380
Paelopatides sp., 465
Paguridae, **403**, **405**, 406, 408-10
Paguristes turgidus, 410
  ulreyi, 410
Paguroidea, 406-10

Pagurus aleuticus, 408
  armatus, 408
  beringanus, 408, 409
  capillatus, **405**, 410
  caurinus, 408, 409
  confragosus, 409
  cornutus, 409
  dalli, 410
  granosimanus, **405**, 408, 409
  hemphilli, 408, 409
  hirsutiusculus, **405**, 408, 409
  kennerlyi, 408, 410
  ochotensis, 408
  quaylei, 408, 409
  samuelis, 408, 409
  setosus, 410
  stevensae, 409
  tanneri, **405**, 408
Palaemon macrodactylus, 507
Palaemonidae, 507
Palaeonemertea, 99
Palaeotaxodonta, 266-9
Paleanotus bellis, 140
Palio zosterae, 246, 250
Pancolus californiensis, 329
Pandalidae, 393, **395**, 397
Pandalopsis ampla, 397
  dispar, 397
Pandalus borealis, 397
  danae, **395**, 397, 507
  eous, 507
  goniurus, 397, 507
  gurneyi, 397
  hypsinotus, 397, 507
  jordani, 397, 507
  platyceros, 397
  stenolepis, 397, 507
  tridens, 397, 507
Pandaridae, 309
Pandea rubra, 39, 42
Pandeidae, 42, 49, 59
Pandora bilirata, 287
  filosa, **286**, 287
  glacialis, 287
  punctata, 287
  wardiana, 287
Pandoridae, 260, 287
Pannychia moseleyi, 465
Panomya, **261**
  beringiana, 265
  chrysis, 284
Panope abrupta, 265, 284
Pantachogon sp., 35, 44
Pantinonemertes californiensis, 97, 101
Paracalliopiella pratti, 370, 378
Paracaudina chilensis, 462, 466
Paracrangon echinata, 400
Paractinostola faeculenta, 77
Paracyathus stearnsi, 72
Paradexiospira vitrea, 165
  violacea, 496
Paradoxostoma cuneata, 314

  fraseri, 314
  striungulum, 314
Paradoxostomatidae, 314
Paradulichia typica, **374**, 377, 386
Paragorgia pacifica, 70
Paragorgiidae, 70
Paralaeospira malardi, 165
Paralithodes camtschatica, 407, 508
  camtschaticus, 508
Parallorchestes spp., 352, **353**, 379
Paralomis multispina, 408
  verrilli, 407
Parametaphoxus quaylei, 503, 505
Paramoera bousfieldi, 376, 377, 504
  bucki, 376, 378, 504
  carlottensis, 376, 378
  columbiana, 376, 378
  leucophthalma, 376, 378, 504
  mohri, 376, 378
  serrata, 376, 378
  suchaneki, 376, **374**, 378, 504
Paranais frici, 172
  litoralis, 172
Paranemertes peregrina, 97, 100
  sanjuanensis, 97, 100
Paraonella platybranchia, 150
  spinifera, 150
Paraonidae, 114, **117**, 121, 150, 495
Parapaguridae, 406, 410
Parapagurus pilosimanus subsp. *benedicti*, 410
Parapasiphae sulcatifrons, 396
Paraphoxus communis, 506
  oculatus, 365, 380, 504
  pacificus, 506
  similis, 506
Parapleustes den, 375, 379, 504, 505
  nautilus, **374**, 375, 379, 504
  oculatus, 375, 379, 504, 505
  pacificus, 506
  pugettensis, 375, 379
  pugettensis group, 504, 505
  similis, 506
Paraprionospio pinnata, **146**, 148
Parasitiformes, 300, 301
Parasmittina collifera, 439, 446
  trispinosa, 439, 446
Parastennella sp., 70
Parasterope sp., 312
Parastichopus californicus, **462**, 465
  leukothele, 462, 465
Paratanaidae, 329
Paratomella unichaeta, 84
Paratomellidae, 84
Paraturbanella intermedia, 104
Pardalisca cuspidata, 382
  tenuipes, 372, 382

Pardaliscella symmetrica, 506
Pardaliscidae, 382, 506
Pardaliscoidea, 382
Parergodrilidae, 120, 151
Paresperella psila, 25, 29
Parotoplana pacifica, 92
Parougia caeca, 494
Parvamussium alaskensis, **271**, 273
Pasiphaea pacifica, **395**, 396
Pasiphaea tarda, 396
Pasiphaeidae, 393, **395**, 396
Patellogastropoda, 194, 204-8
Patinopecten caurinus, 273
Paxillosida, 452
Peachia quinquecapitata, 73, 77
Pectinaria californiensis, 158
    granulata, 158
    moorei, 158
Pectinariidae, 115, **119**, 158
Pectinaster agassizi subsp.
    evoplus, 452
Pectinidae, 260, 273-4
Pectinina, 273-4
Pedicellaster magister, 454
Pedicellina cernua, 422
    sp., 509
Pedicellinidae, 422, 509
Peisidice aspera, 114, 140, 493
Peisidicidae, 140, 493
Pelagiidae, 67
Pelagonemertes brinkmanni, 101
    joubini, 101
Pelonaia corrugata, 476
Peltogaster boschmae, 319
    paguri, 319
Peltogasterella gracilis, 319
Peltogastridae, 319
Penaeidae, 393-4
Penares cortius, 19, 20, 27
Penetrantia sp., 439, 442
Penetrantiidae, 442
Peniagone gracilis, 466
    sp., 466
Penitella conradi, 285
    gabbii, 285
    penita, **281**, 285
    turnerae, 285
Pennatula phosphorea, 71
Pennatulacea, 68-71
Pennatulidae, 71
Pennellidae, 309
Pentamera lissoplaca, 463, 464
    populifera, 463, 464
    pseudocalcigera, 463, 464
    trachyplaca, 464
    spp., 464
Pentametrocrinidae, 447
Pentametrocrinus sp., 447
Peracarida, 320-91, 501-7
Paramoera bucki, 504
Peramphithoe humeralis, 356, **358**, 384

    lindbergi, 356, **358**, 384
    mea, 356, **358**, 384
    plea, 356, **358**, 384
    tea, 356, **358**, 384
Peribolaster biserialis, 454
Perigonimus repens, **47**
    spp., **46**, 59
Perinereis monterea, 135
Perophora annectens, 472, 475
Perophoridae, 475
Perotripus brevis, 387, **389**, 390
Petalidium subspinosum, 393, 507
    suspiriosum, 507
Petaloconchus compactus, 212
Petaloproctus tenuis subsp.
    borealis, 157
    tenuis subsp. *tenuis*, 157
Petricola carditoides, 265, **281**, 283
    pholadiformis, 262, 283
Petricolidae, 262, 265, 283
Petrolisthes cinctipes, **405**, 411
    eriomerus, 411
Petrosiida, 31
Petrosiidae, 31
Pettiboneia pugettensis, 144
Phacellophora camtschatica, **66**, 67
Phakettia beringensis, 16, 28
Phallodrilus tempestatis, 177
Phascolosoma agassizii, 181, 182
Phascolosomatidae, 182
Pherusa inflata, 152
    negligens, 152
    plumosa, **118**, 152
Phialidium gregarium, **38**, 39, 43, 488
    lomae, 39, 43, 488
Phidolopora labiata, 426, **436**, 446
Philichthyidae, 307
Philine bakeri, 238, 240
Philinidae, 240
Philobrya setosa, **267**, 269
Philobryidae, 261, 269
Philomedes dentata, 312
Philomedidae, 312
Philosyrtis sanjuanensis, 92
Phlebobranchia, 475-6, 511
Pholadidae, 260, **261**, 284-5
Pholadina, 284-5
Pholadomyoida, 287
Pholoe minuta, 140, **142**
Pholoidae, 493
Pholoides asperus, 493
Phoronida, 418-9
Phoronidae, 418-9
Phoronis ovalis, 418
    pallida, 418
    vancouverensis, 418
Phoronopsis harmeri, 418
    pacifica, 418
Photis, 353

    bifurcata, 351, **354**, 385
    brevipes, 351, **354**, 385
    conchicola, **354**, 385
    conchicola-oligochaeta, 351
    lacia, 351, **354**, 385
    macinerneyi, 351, **354**, 385
    macrotica, 506
    oligochaeta, **354**, 385
    pachydactyla, 351, **354**, 385
    parvidons, 351, **354**, 385
Phoxichilidiidae, 300
Phoxichilidium femoratum, **298**, 299, 300
Phoxocephalidae, 380, 505
Phoxocephaloidea, 380-1
Phoxocephalus homilis, 363, 380, 503, 505
Phragmophora, 479, 512
Phronima sedentaria, 391
Phrynophiurida, 457-8
Phtisicidae, 390
Phylactellidae, 446
Phyllaplysia taylori, 233, 240
Phyllocarida, 320
Phyllochaetopterus claparedii, **117**, 149
    prolifica, 149
Phyllodoce castanea, 123
    citrina, 123
    hartmanae, 492
    groenlandica, 124
    longipes, 492
    maculata, **122**, 123
    madeirensis, 123
    medipapillata, 123
    multipapillata, 492
    mucosa, 123
    multiseriata, 123
    polynoides, **122**, 123
    williamsi, 123
Phyllodocida, 121-40
Phyllodocidacea, 121-6
Phyllodocidae, **111**, 120-4, **122**, 492
Phyllodurus abdominalis, 342
Phyllolithodes papillosus, 407
Phyllophoridae, 464
Phylloplana viridis, 85, 88
Physonectae, 63-5
Physophora hydrostatica, 62, 65
Physophoridae, 65
Phytia myosotis, 194, 258
Pilargiidae, **116**, 119, 129
Pilargis berkeleyae, **116**, 129
Pileolaria berkeleyana, 496
    dalestraughanae, 496
    potswaldi, 165
    quadrangularis, **165**, 166, 496
    similis, **165**, 166, 496
Pinnixa eburna, 417
    faba, 417
    littoralis, **414**, 417
    occidentalis, **414**, 417

schmitti, 417
tubicola, 417
Pinnotheres pugettensis, 417
taylori, 416
Pinnotheridae, 413, **414**, 416-7
Pionosyllis gigantea, 132
uraga, 132
Piromis eruca, 152
Pisaster brevispinus, 451, 454
ochraceus, 451, 454
Piscicola sp., 179
Piscicolidae, 178-9, 496
Pista bansei, 496
brevibranchiata, 161, 496
cristata, **156**, 160, 495
elongata, 161, 496
estevanica, 495
moorei, 160, 4955
pacifica, 160, 496
wui, 496
Placetron wosnessenskii, 406
Placida dendritica, **234**, 242, 243
Placiphorella rufa, 192
velata, **189**, 190, 192
Plagioecia patina, **427**, 430, 440
Plakina brachylopha, 25, 26
trilopha, 25, 27
sp., 25, 27
Plakinidae, 26
Planes cyaneus, 416
marinus, 416
Planktonemertidae, 101
Platorchestia, **353**
chathamensis, 355, 380
Platyasterida, 452
Platycyamus sp., 391
Platyhelminthes, 84-93, 491
Platynereis bicanaliculata, 135
Platyodon cancellatus, **281**, 284
scaber, 288
Pleocyemata, 393-402
Pleopis polyphemoides, 303
Pleraplysilla sp., 487
Pleurobrachia bachei, **80**, 81, 491
Pleurobrachiidae, 81, 491
Pleurobranchaea californica, 497
Pleurobranchidae, 240
Pleurogona, 511
Pleurogoniidae, 337
Pleurogonium rubicundum, 337
Pleurotomariina, 199
Pleusirus secorrus, 372, 379, 382
Pleustes constantinus, 505
depressa, 361, **362**, 379, 503, 504
victoriae, 505
Pleustidae, 379, 505
Pleusymtes sp., 375, 379
subglaber, 375, 379
Plexauridae, 69
Plicifusus griseus, 198, 225
Plocamia karykina, 24, 30
Plocamiidae, 30

Plocamilla illgi, 22, 30
lambei, 23, 30
Ploima, 105
Plotocnide borealis, 34, 41
Plotohelmis tenuis, **125**, 126
Plotonemertes adhaerens, 101
Plumularia lagenifera, 55
setacea, **46**, **51**, **55**
spp., 50, 60, 488
Plumulariidae, 60
Pluribursaeplana illgi, 92
Pneumoderma atlanticum, 241
Pneumodermatidae, 241
Pneumodermopsis macrochira, 241
Podarkeopsis brevipalpa, 128, 493
glabrus, 493
Podoceridae, 386
Podoceropsis barnardi, 371, 385, 506
chionoecetophila, 371, 385, 506
cristatus, 377, 386
Podocopa, 313-4
Podocopida, 313-4
Podocopina, 313-4
Pododesmus cepio, 259, 274
Podon leuckarti, 303
Podonidae, 303, 501
Podotuberculum hoffmanni, 19, 30
Poecillastra rickettsi, 27
Poecilosclerida, 29-30
Poecilostomatoida, 306-8
Pogonophora, 182-3
Policordia alaskana, 287
Polinices draconis, 216
lewisii, **214**, 216
pallidus, 216
Pollicipes polymerus, 501
Polybrachia canadensis, 182
Polybrachiidae, 182
Polycera atra, 498
tricolor, 246, 250
Polyceridae, 250, 498
Polychaeta, 109-69, 492
Polycheria, **350**
carinata, 506
mixillae, 506
osborni, 375, 382, 504
spp., 504
Polycirrus californicus, 160
Polycladida, 85-8
Polyclinidae, 475, 511
Polycystididae, 91
Polycystis hamata, 91
Polydora alloporis, 147
armata, 147
brachycephala, 147
cardalia, 147
caulleryi, 494
columbiana, 147
commensalis, 147
cornuta, 494
giardi, 147
hamata, 147, 494

kempi subsp. *japonica*, 147
ligni, 147, 494
limicola, 147
nuchalis, 494
polybranchia, 147
proboscidea, **146**, 147
pugettensis, 147
pygidialis, 147
quadrilobata, 147
socialis, **146**, 148
spongicola, 147
websteri, 147
Polyeunoa tuta, 139
Polygordiida, 168
Polygordiidae, 167, 168
Polygordius, 168
Polymastia pachymastia, 18, 20, 27
pacifica, 27
Polymastiidae, 27
Polynoe canadensis, 137, **138**
gracilis, 137
Polynoidae, **111**, **112**, 114, 137-40, **138**, 493
Polyodontidae, 114, 140, 493
Polyorchidae, 42
Polyorchis penicillatus, 36, **37**, 42
Polyplacophora, 185-92
Polyrhabdoplana posttestis, 92
Polystilifera, 101
Pontocyprididae, 314
Pontocypris clemensi, 314
Pontogeneia ivanovi, **350**, **352**, 376, 378
rostrata, 361, 375, 378
inermis, 376, 378
intermedia, 375, 378
Pontogeneiidae, 377-8, 504
Pontoporeioidea, 383, 506
Poraniidae, 453
Poraniopsis inflata, 453
Porcellanasteridae, 452
Porcellanidae, **405**, 406, 411
Porcellio scaber, 344, 345
Porcellionidae, 345
Porella columbiana, 439, 446
concinna, 438, 446
porifera, **436**, 438, 446
Porifera, 6-31, 487
Poromya beringiana, 288
canadensis, 288
leonina, 288
malespinae, 288
tenuiconcha, 288
Poromyidae, 288
Portlandia dalli, 268
Portunion conformis, 341, 343
Potamididae, 197, 212-3
Potamilla intermedia, 163
myriops, 163
neglecta, 162
occelata, **162**, 163

Pourtalesia tanneri, 461
  thomsoni, 461
Pourtalesiidae, 461
Prachynella lodo, 365, 381
Praxillella affinis subsp. *affinis*, 156
  affinis subsp. *pacifica*, 156
  gracilis, **156**
  praetermissa, **156**, 157
Praxillura maculata, 157
Praya dubia, 63, 65
  reticulata, 63, 65
Prayidae, 65
Prianos problematicus, 24, 29
Priapulida, 108
Priapulidae, 108
Priapulomorpha, 108
Priapulus caudatus, 108
Primnoa willeyi, 70
Primnoidae, 70
Prionospio lighti, 494
  multibranchiata, 148
  steenstrupi, 148
  sp., 148
Proales spp., 105
Proalidae, 105
Proboloides sp., 379
Proboscidactyla flavicirrata, 35, **38**, 44, 45, 61
Proboscidactylidae, 44, 61, 488
Proboscina incrassata, 430, 440
Proboscinotus loquax, 351, 380
Procampylaspis caenosa, 502
Procerodes pacifica, 93
Procerodidae, 93
Prochaetodermatidae, 185
Proclea graffii, 160
Prolecithophora, 89
Promesostoma hymanae, 91
  infundibulum, 91
Promesostomidae, 91
Proneomysis wailesi, 323, 324
Propeamussidae, 260, 273, 274
Propeamussium malpelonium, 274
Prophryxus alascensis, 341, 342
Proseriata, 91-2
Prosobranchia, 199-32
Prosorhochmidae, 101
Prosorhochmus spp., 98, 101
Prosuberites sp., 25, 27
Protankyra duodactyla, 466
  pacifica, 466
Protoleodora asperata, 496
Protoariciella oligobranchia, 145
Protodorvillea gracilis, 144
Protodrilida, 168
Protodrilidae, 167, 168
Protodriloides chaetifer, 168
Protodrilus flabelliger, 168
Protohydra leuckarti, 45, **47**, 59
Protohydridae, 59
Protolaeospira capensis, 166
  eximia, 166

Protomedeia articulata, 371, 385
  grandimana, 371, 385
  penates, 371, 385
  prudens, 371, 385
Protopelagonemertidae, 101
Protoptilidae, 70
Protothaca staminea, 282
  tenerrima, **263**, 280
Protula pacifica, 164
Provorticidae, 90
Psammobiidae, 265, 280
Psammonyx longimerus, 365, 381
Psammopemma sp., 487
Psathyrometra fragilis, 447
Psephidia lordi, 282
  ovalis, 282
Pseudanthessiidae, 307
Pseudanthessius latus, 307
Pseudarchaster dissonus, 453
  parelii, 453
Pseudaxinella rosacea, 22, 28
Pseudevadne tergestina, 303
Pseudione galacanthae, 342, **343**
  giardi, 342, **343**
Pseudoceros canadensis, 87, 88
Pseudocerotidae, 88
Pseudochama exogyra, 276
Pseudochitinopoma occidentalis, 164
Pseudocnus astigmatus, 511
Pseudococculinidae, 204
Pseudoliropus vanus, 390
Pseudolubbockia, 306
Pseudomma berkeleyi, 322, 324
Pseudopythina compressa, 264, 276
  rugifera, 264, 275
Pseudosagitta lyra, 512
  maxima, 512
  scrippsae, 512
Pseudoscorpionida, 300
Pseudostichopus mollis, 465
  nudus, 465
  villosus, 465
Pseudostylochus burchami, 87, 88
  ostreophagus, 87, 88
Pseudosuberites spp., 25, 27
Pseudotanaidae, 329
Pseudotanais oculatus, 329
Pseudothecosomata, 241
Psilaster pectinatus, 452
Psolidae, 464
Psolidium bullatum, 463, 464
Psolus chitonoides, 463, 464
  squamatus, 463, 464
Psychropotes longicaudata, 466
  raripes, 466
Psychropotidae, 466
Pteraster jordani, 454
  militaris, 454
  tesselatus, **449**, 450, 454
Pterasteridae, 454
Pteriomorphia, 269-74

Pterotrachea coronata, 208
  hippocampus, 208
  minuta, 208
  scutata, 208
Pterotracheidae, 208
Ptilocrinus pinnatus, 447
Ptilosarcus gurneyi, 70, 71
Ptychogastria polaris, 35, 44
Ptychogastriidae, 44
Ptychogena, 59
  lactea, 36, 43
Puellina setosa, 430, **435**, 444
Pugettia gracilis, 415
  producta, 413, **414**
  richii, 413
Pulmonata, 193, 194, 258
Pulsellidae, 290
Pulsellum salishorum, 289, 290
Puncturella cooperi, 200
  cucullata, 200
  decorata, 200
  expansa, 200
  galeata, 200
  multistriata, 200
  rothi, 200
Pycnoclavella stanleyi, 472, 474
Pycnogonida, 296
Pycnogonidae, 300
Pycnogonum rickettsi, 297, 300
  stearnsi, 297, **298**, 300
Pycnophyes sanjuanensis, 106, 108
Pycnophyidae, 108
Pycnopodia helianthoides, 451, 454
Pygospio californica, 494
  elegans, 148
Pyramidellacea, 233, 235-6
Pyramidellidae, 195, 235, 236
Pyrenidae, 226
Pythonasteridae, 454
Pyura haustor, **470**, 472, 476
  mirabilis, **471**, 472, 476
  spp., 476
Pyuridae, 476, 511

Questa caudicirra, 121, 150
Questidae, 121, 150

Rachiglossa, 219-8
Radiasteridae, 453
Radiceps sp., 69
Ramellogammarus ramellus, 373, 383
  vancouverensis, 373, 383
Raphidophallus actuosus, 84
Raspailiidae, 28
Rathbunaster californicus, 454
Rathkea octopunctata, 35, **40**, 42, 49, 58
Rathkeidae, 42, 58
Rectangulata, 441
Reginella furcata, 431, **435**, 444

nitida, 431, **435**, 444
Reniera mollis, 24, 31
   sp., 22, 24, 31
Reptantia, 402-17
Reteporidae, 446
Retiometra alascana, 447
Retusidae, 239
Rhabdidae, 500
Rhabdocalyptus dawsoni, 12, 13
Rhabdus rectius, 500
Rhachotropis barnardi, 503, 504
   conlanae, 505
   clemens, 361, **364**, 378, 503, 504
   miniata, 505
   oculata, 361, 378
Rhamphidonta retifera, 275
Rhamphostomella cellata, 439, 446
   costata, 424, **436**, 439, 446
   curvirostrata, **436**, 439, 446
Rhepoxynius abronius, 363, 380
   barnardi, 503, 506
   bicuspidatus, 363, 380, 503
   boreovariatus, 503, 505
   daboius, 363, 380
   heterocuspidatus, 364, 380
   tridentatus, 363, 380
   variatus, 363, 380, 503, 505
   vigitegus, 361, 363, 380
Rhinolithodes wosnessenskii, 407
Rhithropanopeus harrisii, 508
Rhizocaulus verticillatus, **53**, 56, 60
Rhizocephala, 318-9
Rhizodrilus pacificus, 175
Rhizogeton sp., 48, 58
Rhodine bitorquata, 155
Rhopalonematidae, 44, 488
Rhopalura ophiocomae, 82, 491
Rhopaluridae, 82, 491
Rhynchobdellida, 178-9
Rhynchonellida, 420
Rhynchonerella angelini, **125**, 126
   gracilis, **125**, 126
Rhynchozoon tumulosum, 439, 446
Rhynocrangon alata, 401
Rhynohalicella halona, 504, 506
Rhysia fletcheri, 488
   sp., 45, 58, 488
Rhysiidae, 58
Rictaxis punctocaelatus, 195, 238, **239**
Ridgeia phaeophiale, 183
   piscesae, 183
Ridgeiidae, 183
Rissoidae, 195, 210, 497
Rissoina newcombiana, 195, 210
Rissoinidae, 195, 210
Ritterella aequalisiphonis, 474, 475

   pulchra, 474, 475
   rubra, 474, 475
Rocinela angustata, 333, 334
   belliceps, 333, 334
   propodialis, 333, 334
   tridens, 332, 334
Rodocystis rosea, 461
Rossellidae, 13
Rossia pacifica, **292**, 293
Rostanga pulchra, 247, 251
Rostangidae, 251
Rotifera, 104-5
Roya spongotheres, 240
Rutiderma rostratum, 312
Rutidermatidae, 312

Sabella crassicornis, 163
   media, 163, 496
   pacifica, **162**, 163
Sabellacheres, 307
Sabellaria cementarium, 158
Sabellariidae, 115, **119**, 158
Sabellastarte sp., 162
Sabellida, 161-4
Sabellidae, 114, **119**, 161, **162**-4, 496
Sabelliphilidae, 307
Sabia conica, 213
Sabinella ptilocrinicola, 218
Saccocirridae, **167**, 169
Saccocirrus eroticus, 169
   sonomacus, **167**, 169
Saccoglossus bromphenolosus, 512
   sp., 478
Sacculinidae, 319
Sacoglossa, **234**, 235, 242-3
Saduria entomon, 338, 340
Sagitella kowalevskii, 126
Sagitta bierii, 479, 512
   decipiens, 478, 479, 512
   elegans, 478, 479
   friderici, 479, 512
   maxima, 479, 512
   minima, 479, 512
   scrippsae, 478, 479, 512
Sagittidae, 479, 512
Sagittoidea, 479
Sahnia sp., 313
Salmacina tribranchiata, 164
Salpida, 476
Samytha californiensis, 158
Sarcodictyon sp., 69
Sarcoptiformes, 300
Sarsia apicula, 488
   apicula, 488
   eximia, **47**, 488
   japonica, 48, 58, 488
   princeps, 34, **37**, **40**, 41
   tubulosa, 34, **40**, 42, **46**
   viridis, 34, 42
   spp., 34, 41, 42, 48, 58
Sarsiella sp., 312, 488

Sarsonuphis parva, 141
Saturnia brunnea, 269
   cervola, 269
   kennerlyi, 269
Saxidomus giganteus, 282
Scalibregma inflatum, **118**, 153
Scalibregmidae, **118**, 121, 153
Scalpellidae, 316, 501
Scalpellum columbianum, 316
Scaphopoda, 289-90
Schistocomus hiltoni, 159
Schistomeringos annulata, 144, 494
   caeca, **117**, 144, 494
   japonica, 144
   moniloceras, 144, 494
   pseudorubrovittata, 144, 494
   rudolphi, **142**, 144, 494
Schisturella cocula, 365, 381
Schizasteridae, 461
Schizobranchia insignis, 161
Schizomavella auriculata, 437, 445
Schizoplax brandtii, 185, 192
Schizoporella linearis subsp. inarmata, 438, 445
   unicornis, **425**, 437, 445
Schizoporellidae, 445, 510
Scintillona bellerophon, 275
Scionella estevanica, 160, 495
   japonica, 160
Scissurellidae, 194, 199
Scleractinia, 72
Scleraxonia, 70-1
Scleroconcha trituberculatum, 312
Sclerocrangon boreas, 402
Sclerodactylidae, 464
Scleroplax granulata, 416
Scleroptilidae, 71
Scleroptilum sp., 71
Scolelepis foliosa, 148
   squamata, 148
Scoloplos acmeceps, 145
   armiger, 117, 145
Scopularia, 13
Scotoplanes clarki, 466
   globosa, 466
Scrobiculariidae, 264, 280
Scrupocellaria californica, 426, **434**, 444
Scrupocellariidae, 444
Scutellina, 460
Scypha compacta, 9
   mundula, 9
   protecta, 9
   spp., 9
Scyphozoa, 65-7
Scyra acutifrons, **412**, 415
Searlesia dira, 198, **222**, 223
Seguenzia cervola, 204
   megaloconcha, 204
   quinni, 204
   stephanica, 204
Seguenziidae, 204

Seison sp., 105
Seisonida, 105
Seisonidae, 105
Seisonidea, 105
Semaeostomeae, 65-7
Semele rubropicta, 264, 280, **281**
Semibalanus balanoides, **315**, 317, 318
  cariosus, 316, 318
Sepiolidae, 293
Sepiolioidea, 293
Septibranchida, 288
Sergestidae, 393, 507
Sergia tenuiremis, 393
Serpula vermicularis, 119, 164
Serpulida, 164
Serpulidae, 114, **119**, 496
Serratosagitta bierii, 512
Serripes groenlandicus, 264, **277**, 278, 498
Sertularella tenella, **54**
  tricuspidata, **54**
  spp., 50, 60
Sertularia mirabilis, **54**
  robusta, **54**
  spp., 50, 60
Sertulariidae, 60
Sessiliflorae, 70-1
Siboglinidae, 182
Siboglinum fedotovi, 182
  pusillum, 182
Sigalion mathildae, 494
Sigalionidae, 114, 140, **142**, 493
Sigambra tentaculata, 129
Sigmadocia edaphus, 19, 20, 31
  spp., 20, 22, 31
Siliqua lucida, 279
  patula, 278
  sloati, 279
Sinistrella abnormis, **165**, 166, 496
  media, 166, 496
  verruca, 166, 496
Siphonaria thersites, **249**, 258
Siphonariidae, 193, 258
Siphonodentalium quadrifissatum, 500
Siphonophora, 62-5
Siphonostomatoida, 308-16
Sipuncula, 181-2
Sipunculida, 182
Smittina cordata, 437, 446
  landsborovi, **436**, 437, 446
Smittinidae, 446
Solariella nuda, 204
  obscura, 201, **215**
  peramabilis, 201, **215**
  vancouverensis, 201
Solaster dawsoni, 452, 453
  endeca, 451, 453
  paxillatus, 451, 453
  stimpsoni, 451, 453
Solasteridae, 453

Solemya reidi, 265
Solemyidae, 260, 265
Solemyoida, 265-6
Solen sicarius, 262, 278
Solenidae, 262, 278
Solenogastres, 185
Solenosmilia variabilis, 72
Solidobalanus engbergi, 317, 318
  hesperius, 317, 318
Solitaria, 422
Solmissus incisa, 41, 44
  marshalli, 41, 44
Sosaniopsis hesslei, 159
Spatangoida, 461
Specaria fraseri, 172
Sperosoma biseriatum, 460
  giganteum, 460
Sphaerodoridae, **116**, 120, **127**, 128
Sphaerodoropsis biserialis, 128
  minuta, **116**, **127**, 128
  sphaerulifer, **127**, 128
Sphaerodorum papillifer, 128
Sphaeromatidae, 336
Sphaeronectes gracilis, 63, 65
Sphaeronectidae, 65
Sphaerosyllis brandhorsti, 131
  californiensis, 493
  hystrix, 131, 493
  pirifera, 131, 493
  sp., 493
Sphaerothuria bitentaculata, 465
Sphenia ovoidea, 283
Spinulosida, 453-4
Spio butleri, 149
  cirrifera, 148
  filicornis, **117**
  sp., 149
Spiochaetopterus costarum, 149
Spionida, 145-9
Spionidae, 115, **117**, 145, **146**-9, 494
Spiophanes berkeleyorum, 148
  bombyx, 148
  kroyeri, 148
Spiromoelleria quadrae, 201
Spirontocaris arcuata, 399, 508
  holmesi, 399
  lamellicornis, **395**, 399, 508
  ochotensis, 399
  prionota, 399
  sica, 399
  spina, 399, 508
  synderi, 399
  truncata, 399
Spirophorida, 27
Spirorbidae, **111**, 114, 164, **165**-6, 496
Spirorbis bifurcatus, **165**
Spirularina, 71
Spisula falcata, 278
Splanchnotrophidae, 307
Spongionella sp., 487

Staurocalyptus dowlingi, 12, 13
Stauromedusae, 67
Staurophora, 59
Staurophora mertensi, 36, 43
Stegocephalexia penelope, 506
Stegocephalidae, 506
Stegocephaloidea, 382, 506
Stegopoma spp., 56, 59
Stelletta clarella, 19, 20, 27
Stellettidae, 27
Stelodoryx alaskensis, 18, 30
Stelotrochota hartmani, **21**, 25, 30
Stenolaemata, 424-30, 440-1
Stenoplax fallax, 186, 191
  heathiana, 186, 191
Stenothoidae, **353**, 379
Stenula spp., 379
Stephanauge annularis, 78
Stephanosella vitrea, 439, 445
Stephensoniella trevori, 173
Sternaspida, 153
Sternaspidae, 110, **118**, 153
Sternaspis scutata, 110, **118**, 153
Sthenelais berkeleyi, 140
  tertiaglabra, 140
Stibarobdella loricata, 179
Stichopodidae, 465
Stiliger fuscovittatus, 242, 243
Stiligeridae, 242
Stilipedidae, 382
Stilipes sp., 382
Stilomysis grandis, 323, 324
Stoecharthrum burresoni, 491
  fosterae, 491
Stolidobranchia, 476, 511
Stolonifera, 68, 69-70
Stomachetosella cruenta, 438, 445
  limbata, 438, 445
  sinuosa, 438, 445
Stomachetosellidae, 445
Stomatopora granulata, 428, 440
Stomotoca atra, 33, **37**, 42
Stomphia coccinea, 76, 78
  didemon, 76, 78
  sp., 76, 78
Streblosoma bairdi, 160
Streblospio benedicti, 148
Streptosyllis latipalpa, 130, 133
Strongylocentrotidae, 460
Strongylocentrotus droebachiensis, 460, 511
  franciscanus, 460
  pallidus, 460, 510
  purpuratus, 460
Styela clava, 511
  clavata, 476
  coriacea, 472, 476
  gibbsii, 472, 476
  montereyensis, 469, **470**, 476
  truncata, 476
Styelidae, 476, 511

Stygocapitella subterranea, 120, 151
Stylasterias forreri, 451, 454
Stylasteridae, 61-2
Stylasterina, 61, **62**
Stylatula elongata, 70, 71
Stylinos sp., 18, 29
Styliola subula, 241
Stylissa stipitata, 16, 28
Stylochidae, 88
Stylochoplana chloranota, 87, 88
Stylochus atentaculatus, 88
   tripartitus, 88
Stylopus arndti, 22, 24, 30
Stylostomum album, 87, 88
   sanjuania, 88
Suberites montiniger, 19, 27
   suberea forma *latus*, 14, 27
   simplex, 19, 27
   spp., 19, 20, 24, 27
Suberitidae, 27
Subselliflorae, 71
Swiftia kofoidi, 69
   simplex, 69
   spauldingi, 69
   torreyi, 69
Sycandra utriculus, 8, 10
Sycettida, 9-10
Sycettidae, 9-10
Syllidae, **112**, **116**, 120, 129-33, **130**, 493
Syllides japonica, 133
   longocirrata, **116**, 133
Syllis adamantea subsp.
   adamantea, 132
   alternata, 132
   armillaris, 131
   elongata, 131
   fasciata, 132
   gracilis, 131
   heterochaeta, 131
   hyalina, **130**, 131
   pulchra, **130**, 132
   spongiphila, 131
   stewarti, 131, 132
   variegata, 132
   sp., 132
Sylon hippolytes, 319
Sylonidae, 319
Symplectoscyphus spp., 50, 60
Synallactes gilberti, 465
Synallactidae, 465
Synaptidae, 466
Synchaeta baltica, 105
   johanseni, 105
Synchaetidae, 105
Synchelidium, **350**
   rectipalmum, 362, 378
   shoemakeri, 362, 378
Syndesmis dendrastrorum, 90
   inconspicua, 491
   neglecta, 491
   spp., 90

Syndisyrinx franciscanus, 90
Synidotea angulata, 338, 340
   bicuspida, 338, 340
   consolidata, 502
   cornuta, 502
   minuta, 502
   nebulosa, 338, 340
   nodulosa, 338, 340
   pettiboneae, 338, 340
   ritteri, 338, 340
Synnotum aegyptiacum, 426, **434**, 444
Synoicum parfustis, 474, 475
   spp., 475
Synopiidae, 381
Synopioidea, 381
Syphacidae, 345
Syringella amphispicula, 18, 28
Syrrhoe longifrons, 361, 381
Systellaspis brauerii, 396
   cristata, 396

Tachyrhynchus erosus, 212
   lacteolus, 212, **215**
Taenioglossa, 208-19
Talitridae, 380
Talitroidea, 379-80
Tanaidacea, 329-30
Tanaidae, 329
Tanaidomorpha, 329
Tanystylidae, 300
Tanystylum anthomasti, 297, **298**, 300
   occidentalis, 297, 300
   sp., 299, 300
Taonius pavo, 295
Tapes philippinarum, 282
Taranis strongi, 229
Tardigrada, 296
Tarsaster alaskanus, 454
Tauberia gracilis, 150, 494
Tecticeps pugettensis, 333, 336
Tectidrilus diversus, 175
   verrucosus, 175
Tectura fenestrata, 207
   paleacea, 206
   persona, 208
   rosacea, 206
   scutum, 207, **211**
Tedania fragilis, 30
   gurjanovae, 24, 30
Tedaniidae, 30
Tedanione obscurata, 24, 30
Tegella aquilirostris, 433, 443
   armifera, **432**, 433, 443
   robertsonae, **432**, 433, 443
Tegula brunnea, 201
   funebralis, 201
   pulligo, 201
Tellina bodegensis, 279, **281**
   carpenteri, 279
   modesta, 279
   nuculoides, 279, 499

Tellinidae, 265, 279-80
Telmessus cheiragonus, 415
Tenellia adspersa, 498
Tenonia priops, 137
Tenthrenodes sp., 8, 10
Tenuisagitta friderici, 512
Terebella ehrenbergi, 161
Terebellida, 158-61
Terebellidae, 115, **119**, **156**, 159-61, 495-6
Terebellides stroemi, **156**, 159
Terebratalia transversa, 419, 421
Terebratellidina, 421
Terebratulida, 421
Terebratulidae, 421
Terebratulidina, 421
Terebratulina unguicula, 419, 421
Teredinidae, 259, 285
Teredo navalis, 285
Tergipedidae, 257-8, 498
Tethya aurantia, 18, **21**, 28, 487
   californiana, 487
Tethydidae, 253
Tethyidae, 28, 487
Tetillidae, 27
Tetractinomorpha, 27-8
Tetranchyroderma pugetensis, 104
Tetrastemma bicolor, 98, 01
   bilineatum, 98, 101
   candidum, 101
   nigrifrons, 101
   phyllospadicola, 97, 101
   spp., 98, 101
Tetrastemmatidae, 101
Teuthoidea, 294-5
Tevniida, 183
Thaisidae, 221-3
Thalassema steinbecki, 181
Thalassinidea, 402-4
Thalestris rhodymeniae, 305
Thaliacea, 476
Thalysias laevigata, 30
Tharyx multifilis, 151, 495
   parvus, 151, 495
   secundus, 151, 495
   serratisetis, 151, 495
Thaumastodermatidae, 104
Thaumatoscyphus hexaradiatus, 67
Thecanephria, 182-3
Thecata, 43-4, 59-61
Thecocarpus spp., 60
Thecosomata, 233, 240-1
Thelepus cincinnatus, **156**, 160
   crispus, **156**, 160
   hamatus, 160
   japonicus, 160
   setosus, **156**, 160
Themiste dyscrita, 181, 182
   pyroides, 181, 182
Thompsonia sp., 319
Thoracica, 314-8
Thorlaksonius borealis, 505
   brevirostris, 505

538 Index

carinatus, 505
depressus, 503, 505
grandirostris, 505
subcarinatus, 505
truncatus, 505
Thracia beringi, 283
  challisiana, 283
  curta, 283
  trapezoides, 283
Thraciidae, 262, 283
Thrissacanthias penicillatus, 452
Thuiaria, **46**, 50
  distans, **55**
  lonchitis, **55**
  robusta, **55**
  tenera, **55**
  thuja, **54**
  spp., 60
Thyasira barbarensis, 274
  cygnus, 274
  gouldii, 274
Thyasiridae, 260, 264, 274, 498
Thyonicola americana, 219
  dogieli, 219
Thysanoessa raschii, 392
Thysanura, 302
Tiarannidae, 59
Tiaropsidae, 488
Tiaropsidium, 59, 488
  kelseyi, 33, 36, 43
Tiaropsis, 59, 488
  multicirrata, 36, 43
Tindaria compressa, 269
  dicofania, 269
  panamensis, 269
Tindariidae, 266, 269
Tiron biocellatu, 361, **362**, 381
Tochuina tetraquetra, **249**, 251, 253
Todarodes pacificus, 294
Tomopteridae, 110, **125**, 126
Tomopteris pacifica, **125**, 126
  septentrionalis, 126
Tonicella insignis, 186, 192
  lineata, 186, 192
Tontocythere sp., 313
Topsentia disparilis, 22, 28
Toxidocia spp., 20, 31
Toxoglossa, 228-32
Trachelobdella oregonensis, 179
Trachyleberididae, 313
Trachylina, 44, 488
Trachymedusae, 44, 488
Trachypleustes trevori, 505
  vancouverensis, 505
  spp., 504
Transennella confusa, 282
  tantilla, 282
Traskorchestia, **350**, **353**
  georgiana, 355, 380
  traskiana, 355, 380
Traskorchestianoetus brevipes, 500
  spiceri, 500

Travisia brevis, 153
  forbesii, 153
  japonica, 153
  pupa, 153
Travisiopsis lobifera, **125**, 126
Tresus, 265
  capax, **263**, 278
  nuttallii, 278
Tricellaria erecta, 426, **434**, 444
  occidentalis, 426, **434**, 444
  praescuta, 426, **434**, 444
  ternata, 426, 444
Trichobranchidae, 115, **156**, 159
Trichobranchus glacialis, **156**, 159
Trichocerca marina, 105
Trichocercidae, 105
Trichotropididae, 194, 216
Trichotropis bicarinata, 216
  borealis, 216
  cancellata, **214**, 216
  insignis, 216
Trichydra pudica, 35, 42, 59
Trichydridae, 42, 59
Tricladida, 92-3
Tricolia pulloides, 195, 201, **202**
Trididemnum opacum, **471**, 473, 475
  strangulatum, 475
Tridonta alaskensis, 276, **277**
Trigonostomidae, 91
Trilobodrilus nipponicus, 168
Trimusculidae, 498
Trimusculus reticulatus, 498
Triopha catalinae, 246, 250
  maculata, 498
Tritella laevis, 387, **389**, 390
  pilimana, 387, 390
Triticella pedicellata, 439, 442
Triticellidae, 442
Tritonia diomedea, **249**, 252, 253
  exsulans, 251, 253, 498
  festiva, **244**, 251, 253, 498
Tritoniidae, 253
Trochidae, 194, **196**, 201
Trochina, 200-4
Trochochaeta multisetosa, 115, **142**, 145
Trochochaetidae, 115, **142**, 145
Trombidiformes, 300, 301
Trophonopsis clathratus, **215**, 221
  dalli, 221
  disparilis, 221
  kamtchatkanus, 221
  lasius, **220**, 221
  macouni, 221
  pacificus, **215**, 221
  scitulus, 221
  staphylinus, 221
  tripherus, 221
Truncatellidae, 195, 210
Trypanosyllis gemmipara, **130**, 131

  ingens, 131
Trypetesa lateralis, 319
Trypostega claviculata, 438, 445
Tubificidae, 171, 174-8
Tubificoides apectinatus, 175
  brevicoleus, 175, **176**
  coatesae, 175
  foliatus, 175
  kozloffi, 175, **176**
  nerthoides, 175
  pseudogaster, 175
Tubonemertes wheeleri, 101
Tubulanidae, 99
Tubulanus albocinctus, 95, 99
  capistratus, 96, 99
  pellucidus, 96, 99
  polymorphus, 96, 99
  sexlineatus, 95, 99
Tubularia crocea, 48, 57, 488
  harrimani, 48, 57
  indivisa, 48, 57
  marina, 48, 57, 488
Tubulariidae, 41, 57
Tubulipora flabellaris, **427**, 430, 441
  pacifica, **425**, 430, 441
  tuba, **427**, 430, 441
Tubuliporidae, 441
Tubuliporina, 440-1
Turbanella cornuta, 104
  mustela, 104
  sp., 104
Turbanellidae, 104
Turbellaria, 84-93, 491-2
Turbinidae, 194, 195, 200-1
Turbonilla, 235
  alaskana, 236
  aurantia, 236
  barkleyensis, 236
  engbergi, 236
  lordi, 236
  lyalli, 236
  pesa, 236, **237**
  pugetensis, 236
  rinella, 236, **237**
  stylina, 236
  taylori, 236
  torquata, 236
Turridae, **196**, 197, 228-32
Turritellidae, 195, 212
Turritellopsis acicula, 212
Turtonia minuta, 282
Turtoniidae, 264, 282
Typhloplanoina, 91
Typhloscolecidae, 110, **125**, 126
Typhloscolex mülleri, **125**, 126

Ulmaridae, 67
Umagillidae, 90, 492
Umbellula lindahli, 71
Umbellulidae, 71a
Umbonula arctica, 439, 445
Umbonulidae, 445

Ungulinidae, 264, 275
Uniramia, 301-2
Upogebia pugettensis, 404
Upogebiidae, 403, 404, 508
Urechidae, 181
Urechinidae, 461
Urechinus loveni, 461
Urechis caupo, 181
Urochordata, 467-77, 511
Urosalpinx cinerea, 221, **220**
Urothoe spp., 381
Urothoidae, 381
Urticina columbiana, 76, 77
 coriacea, 75, 77
 crassicornis, 76, 77
 lofotensis, 75, 77
 piscivora, 76, 77
 sp., 77

Vadicola aprostatus, 174, **177**
Valbyteuthis oligobessa, 295
Valvatida, 453
Valvifera, 330, 338-40, 502
Vampyromorpha, 295
Vampyroteuthidae, 295
Vampyroteuthis infernalis, 295
Vanadis longissima, **116**, 124
Vannuccia rotundouncinata, 92
 tripapillosa subsp. *americana*, 92
Vargula sp., 312
Vaunthompsonia, 326, 327
 pacifica, 328
Vejdovskyella hellei, 172
Velella velella, 32, 41, 45, **51**, 58
Velellidae, 41, 58
Velutina plicatilis, 217, 497
 prolongata, 217
 velutina, 497
Velutinidae, 194, 217, 497
Veneridae, 264, 265, 280-2
Veneroida, 274-85
Vermetidae, 193, 212
Verongiida, 31
Verongiidae, 31
Verticordiidae, 287
Vesicomya lepta, 280
 ovalis, 280
 stearnsii, 280
Vesicomyidae, 280
Vesiculariidae, 442, 510
Vestimentifera, 183
Virgularia spp., 70, 71
Virgulariidae, 71
Vitrinella columbiana, 194, 210
Vitrinellidae, 194, 210
Volutharpa ampullacea, **224**, 225, 497
Volutidae, 197, 227
Volutomitra alaskana, **215**, 227
Volutomitridae, 198, 227
Volvulella cylindrica, 238, 239

Wahlia pulchella, 90

Weberella verrucosa, 18, 27
Wecomedon similis, 365, 381
 wecomus, 365, 381
Westwoodilla caecula, 362, **364**, 378

Xanthidae, 412, 416, 508
Xenacanthomysis pseudomacropsis, 323, 324
Xenobalanus globicipitis, 318
Xenopneusta, 181
Xestoleberididae, 313
Xestoleberis depressa, 313
 dispar, 313
Xestospongia trindanea, 20, 31
 vanilla, 20, 31
Xylophaga washingtona, 260, 285
Xylophagaidae, 260, 285

Yaquinaia microrhynchus, 91
Yoldia amygdalea, **267**, 268
 beringiana, 268
 martyria, 268
 montereyensis, 268
 myalis, 268
 scissurata, **267**, 268
 thraciaeformis, **267**, 268
Yoldiella capsa, 268
 dicella, 268
 orcia, 268
 sanesia, 268
Yoldiidae, 266, 268
Ypsilothuridae, 465

Zephyrinidae, 255
Zeuxo normani, 329
Zirfaea pilsbryii, 285
Zoantharia, 68-78
Zoanthidea, 72, 78
Zoroaster evermanni, 455
 ophiurus, 455
Zoroasteridae, 455
Zorocallida, 455
Zygherpe hyaloderma, 25, 29
Zygonemertes virescens, 98, 100, 492
 sp., 100
Zygophylax spp., 57, 60
Zygothuria lactea, 465